Asymmetric Synthesis of Three-Membered Rings

Asymmetric Synthesis of Three-Membered Rings

Hélène Pellissier, Alessandra Lattanzi and Renato Dalpozzo

Authors

Dr. Hélène Pellissier
Aix Marseille Université
Institut des Sciences Moléculaires de Marseille
UMR CNRS 7313 Avenue Escadrille
Normandie-Niemen
13397 Marseille
France

Professor Alessandra Lattanzi
Università di Salerno
Dipartimento di Chimica e Biologia A. Zambelli
Via Giovanni Paolo II
84084 Fisciano
Italy

Professor Renato Dalpozzo
Università della Calabria
Dipartimento di Chimica e Tecnolgie Chimiche
Ponte Bucci, Cubo 12/C
87036 Arcavacata di Rende (CS)
Italy

All books published by **Wiley-VCH** are carefully produced. Nevertheless, authors, editors, and publisher do not warrant the information contained in these books, including this book, to be free of errors. Readers are advised to keep in mind that statements, data, illustrations, procedural details or other items may inadvertently be inaccurate.

Library of Congress Card No.: applied for

British Library Cataloguing-in-Publication Data
A catalogue record for this book is available from the British Library.

Bibliographic information published by the Deutsche Nationalbibliothek
The Deutsche Nationalbibliothek lists this publication in the Deutsche Nationalbibliografie; detailed bibliographic data are available on the Internet at <http://dnb.d-nb.de>.

© 2017 Wiley-VCH Verlag GmbH & Co. KGaA, Boschstr. 12, 69469 Weinheim, Germany

All rights reserved (including those of translation into other languages). No part of this book may be reproduced in any form – by photoprinting, microfilm, or any other means – nor transmitted or translated into a machine language without written permission from the publishers. Registered names, trademarks, etc. used in this book, even when not specifically marked as such, are not to be considered unprotected by law.

Print ISBN: 978-3-527-34114-6
ePDF ISBN: 978-3-527-80201-2
ePub ISBN: 978-3-527-80203-6
Mobi ISBN: 978-3-527-80204-3
oBook ISBN: 978-3-527-80202-9

Cover Design Formgeber, Mannheim, Germany
Typesetting SPi Global Private Limited, Chennai, India
Printing and Binding betz-druck GmbH, Darmstadt, Deutschland

Printed on acid-free paper

Contents

Preface *ix*
List of Abbreviations *xi*

1 Asymmetric Cyclopropanation *1*
1.1 Introduction *1*
1.2 Simmons–Smith Cyclopropanation *2*
1.2.1 Chiral Substrates *3*
1.2.1.1 Chiral Allylic Alcohols *3*
1.2.1.2 Chiral Allylic Amines *7*
1.2.1.3 Chiral Acetal-Directed Cyclopropanations *7*
1.2.1.4 Simple Chiral Alkenes *9*
1.2.2 Chiral Auxiliaries *11*
1.2.3 Chiral Catalysts *15*
1.2.3.1 Charette's Ligand *15*
1.2.3.2 Other Stoichiometric Ligands *20*
1.2.3.3 Walsh' Procedure *22*
1.2.3.4 True Catalytic Procedures *24*
1.3 Transition-Metal-Catalyzed Decomposition of Diazoalkanes *27*
1.3.1 Intermolecular Cyclopropanation *28*
1.3.1.1 Chiral Auxiliaries *28*
1.3.1.2 Chiral Catalysts: Cobalt *32*
1.3.1.3 Chiral Catalysts: Copper *38*
1.3.1.4 Chiral Catalysts: Rhodium *56*
1.3.1.5 Chiral Catalysts: Ruthenium *69*
1.3.1.6 Chiral Catalyst: Other Metals *77*
1.3.2 Intramolecular Cyclopropanation *80*
1.3.2.1 Chiral Auxiliaries and Chiral Compounds *80*
1.3.2.2 Chiral Catalysts *82*
1.3.3 Chiral Stoichiometric Carbenes *93*
1.4 Michael-Initiated and Other Ring Closures *94*
1.4.1 Chiral Substrates *95*
1.4.2 Chiral Auxiliaries *100*
1.4.2.1 Chiral Michael Acceptors *100*
1.4.2.2 Chiral Nucleophiles *106*

1.4.3	Organocatalysis *115*	
1.4.3.1	Ylides *116*	
1.4.3.2	Nitrocyclopropanation *118*	
1.4.3.3	Halocarbonyl Compounds *128*	
1.4.4	Metal Catalysis *139*	
1.4.5	Other Ring Closures *140*	
1.5	Miscellaneous Reactions *148*	
1.5.1	Rearrangement of Chiral Oxiranes *148*	
1.5.2	Cycloisomerization of 1,*n*-Enynes *152*	
1.5.3	Denitrogenation of Chiral Pyrazolines *160*	
1.5.4	C–H Insertion *162*	
1.5.5	Addition to Cyclopropenes *164*	
1.5.6	Other Methods *167*	
1.6	Conclusions *172*	
	References *172*	
2	**Asymmetric Aziridination** *205*	
2.1	Introduction *205*	
2.2	Aziridination Based on the Use of Chiral Substrates *206*	
2.2.1	Addition to Alkenes *206*	
2.2.1.1	Aziridination via Nitrene Transfer to Alkenes *206*	
2.2.1.2	Aziridination via Addition–Elimination Processes *219*	
2.2.1.3	Miscellaneous Reactions *224*	
2.2.2	Addition to Imines *225*	
2.2.2.1	Methylidation of Imines *226*	
2.2.2.2	Aza-Darzens and Analogous Reactions *248*	
2.2.2.3	Addition/Elimination Processes *255*	
2.2.2.4	Miscellaneous Reactions *266*	
2.2.3	Addition to Azirines *267*	
2.2.4	Aziridination via Intramolecular Substitution *270*	
2.2.4.1	From 1,2-Amino Alcohols *270*	
2.2.4.2	From 1,2-Amino Halides *278*	
2.2.4.3	From 1,2-Azido Alcohols *282*	
2.2.4.4	From 1,2-Amino Sulfides and 1,2-Amino Selenides *285*	
2.2.4.5	From Epoxides *286*	
2.2.5	Miscellaneous Reactions *287*	
2.3	Aziridination Based on the Use of Chiral Catalysts *296*	
2.3.1	Aziridination via Nitrene Transfer to Alkenes *296*	
2.3.1.1	Cu-Catalyzed Aziridination *296*	
2.3.1.2	Rh-Catalyzed Aziridination *310*	
2.3.1.3	Ru-Catalyzed Aziridination *312*	
2.3.1.4	Catalysis by Other Metals *314*	
2.3.1.5	Organocatalyzed Aziridination *318*	
2.3.2	Aziridination via Carbene Transfer to Imines *332*	
2.3.2.1	Carbene Methodology *332*	
2.3.2.2	Sulfur-Ylide-Mediated Aziridination *350*	

2.3.3	Miscellaneous Reactions	*353*
2.3.4	Kinetic Resolutions of Aziridines	*357*
2.4	Conclusions	*363*
	References	*364*

3 Asymmetric Epoxidation *379*

- 3.1 Introduction *379*
- 3.2 Asymmetric Epoxidations Based on the Use of Chiral Auxiliaries *380*
- 3.3 Asymmetric Metal-Catalyzed Epoxidations *381*
- 3.3.1 Ti-, Zr-, Hf-Catalyzed Epoxidations *381*
- 3.3.2 V-, Nb-, Ta-Catalyzed Epoxidations *391*
- 3.3.3 Cr-, Mo-, W-Catalyzed Epoxidations *397*
- 3.3.4 Mn-, Re-, Fe-, Ru-Catalyzed Epoxidations *400*
- 3.3.5 Pt-, Zn-, Lanthanoid-Catalyzed Epoxidations *412*
- 3.4 Asymmetric Organocatalyzed Epoxidations *419*
- 3.4.1 Phase-Transfer Catalyst *419*
- 3.4.2 Polyamino Acids and Aspartate-Derived Peracids *423*
- 3.4.3 Chiral Dioxiranes, Iminium Salts, and Alkyl Hydroperoxides *431*
- 3.4.4 Chiral Amines *447*
- 3.5 Kinetic Resolution of Racemic Epoxides *468*
- 3.6 Asymmetric Sulfur-Ylide-Mediated Epoxidations *480*
- 3.7 Asymmetric Darzens-Type Epoxidations *489*
- 3.7.1 Chiral Auxiliary- and Reagent-Mediated Darzens Reactions *489*
- 3.7.2 Catalytic Asymmetric Darzens Reactions *492*
- 3.8 Other Ylide-Mediated Epoxidations *503*
- 3.9 Asymmetric Biocatalyzed Synthesis of Epoxides *505*
- 3.10 Conclusions *512*
- References *514*

4 Asymmetric Oxaziridination *539*

- 4.1 Introduction *539*
- 4.2 Oxaziridination Using Chiral Substrates *540*
- 4.3 Oxaziridination Using Chiral Catalysts *544*
- 4.4 Kinetic Resolutions *551*
- 4.5 Conclusions *554*
- References *555*

5 Asymmetric Azirination and Thiirination *559*

- 5.1 Introduction *559*
- 5.2 Asymmetric Azirination *559*
- 5.2.1 Neber Approaches *560*
- 5.2.2 Elimination Approaches *565*
- 5.2.3 Other Approaches *569*
- 5.3 Asymmetric Thiirination *570*
- 5.3.1 Conversion of Epoxides *571*

5.3.2	Condensation of Sulfur-Stabilized Carbanions to Carbonyl Compounds *572*	
5.3.3	Intramolecular Nucleophilic Substitution *574*	
5.3.4	Miscellaneous Reactions *576*	
5.4	Conclusions *577*	
	References *579*	

Index *583*

Preface

The importance of chirality is well recognized related to the fact that nearly all natural products are chiral and their physiological or pharmacological properties depend upon their recognition by chiral receptors, which will interact only with molecules of the proper absolute configuration. Indeed, the use of chiral drugs in enantiopure form is now a standard requirement for virtually every new chemical entity, and the development of new synthetic methods to obtain enantiopure compounds has become a key goal for pharmaceutical companies. Asymmetric synthesis constitutes one of the main strategies to gain access to enantioenriched compounds, involving the use of either chiral auxiliaries or catalysts.

Even 134 years after the synthesis of the first cyclopropane derivative, the synthesis of chiral three-membered (hetero)cycles remains a considerable challenge. Their strained structures, interesting bonding characteristics, and value as an internal mechanistic probe have attracted the attention of the physical organic chemistry community. Moreover, organic chemists have always been fascinated by these subunits, which have been playing a prominent role in organic chemistry. In fact, while three-membered rings are highly strained entities, they are nonetheless found in a wide variety of naturally occurring compounds including terpenes, pheromones, fatty acid metabolites, and unusual amino acids, among others. Indeed, the prevalence of three-membered-containing (hetero)compounds with biological activity, whether isolated from natural sources or rationally designed pharmaceutical agents, has inspired chemists to find novel and diverse approaches to their synthesis. The main strategy to gain access to these enantioenriched compounds involve the use of either chiral auxiliaries or catalysts that can in turn be metal-centered, small organic asymmetric molecules or enzymes.

This book collects all the developments achieved in the last 12 years in the fields of asymmetric cyclopropanation, aziridination, epoxidation, oxaziridination, azirination, and thiirination reactions. In addition to describing the large number of highly efficient processes based on the use of various chiral auxiliaries or substrates, this book demonstrates that the most important achievements in asymmetric synthesis of three-membered rings are the spectacular expansion of novel chiral catalysts, including the especially attractive chiral organocatalysts, which have been recently applied to these reactions. Indeed, a collection of new chiral Lewis-acid catalysts and organocatalysts have provided new opportunities for these enantioselective reactions and widely expanded their scope.

Each chapter of the book covers issues related to the title reactions and includes selected applications of the multiple synthetic methodologies discussed to prepare pharmaceuticals, natural or biologically active compounds. All the chapters include synthetic procedures based on the use of chiral pools and auxiliaries, which were employed in the earlier times, but also more convenient catalytic approaches based on the use of chiral metal catalysts and more recently organocatalysts.

Chapter 1, by R. Dalpozzo, deals with the synthesis of chiral cyclopropanes through asymmetric cyclopropanation. The more efficient methodologies employed are the well-known Simmons–Smith reaction, the transition-metal-catalyzed decomposition of diazo compounds, and the irreversible Michael-initiated ring closure (MIRC), among others. For all these procedures, the use of chiral substrates or auxiliaries as well as that of chiral metal- and organocatalysts is covered.

Chapter 2, by H. Pellissier, collects the recent developments in asymmetric aziridination. The use of chiral substrates in addition reactions to alkenes, imines, and azirines as well as in intramolecular substitutions among other reactions is developed in a first section. The second section deals with enantioselective metal- and organocatalyzed carbene transfers to imines and nitrene transfers to alkenes along with catalytic kinetic resolutions of racemic aziridines among other reactions promoted by chiral catalysts of all types.

Chapter 3, by A. Lattanzi, demonstrates the important progress achieved in the past decade in the vast area of asymmetric synthesis of epoxides. Important enantioselective metal- and organocatalyzed epoxidations of alkenes are firstly covered, while other sections deal with kinetic resolution of racemic epoxides, asymmetric sulfur-ylide-mediated epoxidations of carbonyl compounds, asymmetric Darzens reactions, and biocatalyzed synthesis of epoxides, among other methodologies.

Chapter 4, by H. Pellissier, deals with asymmetric oxaziridination, which can be achieved by using chiral substrates or chiral catalysts and kinetic resolutions. It is the smallest chapter of the book, demonstrating that this field is still in its infancy because it has been overshadowed for a long time by the fact that electron-deficient oxaziridines can be employed as convenient and stable sources of electrophilic oxygen.

Chapter 5, by H. Pellissier, collects the advances in asymmetric azirination and thiirination using chiral reagents as well as chiral catalysts, focusing on those published in the last 12 years.

The authors hope that this book will provide an insight into the present stage of asymmetric synthesis of three-membered rings and stimulate chemists to future discoveries to fulfill the enormous potential in this area, opening the way to the synthesis of a number of important products.

List of Abbreviations

2,6-DCPNO	2,6-dichloropyridine *N*-oxide
acac	acetylacetonate
ACDC	asymmetric counteranion-directed catalysis
Ad	1-adamantyl
AKR	aminolytic kinetic resolution
Anth	anthryl
Ar	aryl
BARF	tetrakis(3,5-bis(trifluoromethyl)phenyl)borate
BHT	2,6-di-*t*-butyl-4-methylphenyl
BINAM	1,1′-binaphthyl-2,2′-diamine
BINAP	2,2′-bis(diphenylphosphino)-1,1′-binaphthyl
BINOL	1,1′-bi-2-naphthol
BMEH	Bacillus megaterium epoxide hydrolase
Bn	benzyl
Boc	*tert*-butoxycarbonyl
Box	bis(oxazoline)
Bs	benzenesulfonyl
Bt	benzotriazole
BUDAM	tetra-*tert*-butyldianisylmethyl
Bz	benzoyl (PhCO)
CAN	ceric ammonium nitrate
CBS	oxazaborolidine Corey–Bakshi–Shibata catalyst
Cbz	benzyloxycarbonyl
CHP	cumyl hydroperoxide
Cod	1,5-cyclooctadiene
Cp	cyclopentadienyl
CPO	chloroperoxidase
CSA	camphorsulfonic acid
Cy	cyclohexyl
DABCO	1,4-diazabicyclo[2.2.2]octane
DAM	dianisylmethyl
DAP	diaminopimelic acid
DBN	3,5-dinitrobenzoyl
DBU	1,5-diazabicyclo[5.4.0]undec-5-ene
DCC	*N*,*N*′-dicyclohexylcarbodiimide

DCE	1,2-dichloroethane
de	diastereomeric excess
DEAD	diethyl azodicarboxylate
Dec	decyl
DFT	density functional theory
DIAD	diisopropyl azodicarboxylate
DIC	diisopropylcarbodiimide
DIPEA	diisopropyl ethyl amine
DMAP	4-(*N,N*-dimethylamino)pyridine
DMD	dimethyl dioxirane
DME	dimethoxyethane
DMF	*N,N*-dimethylformamide
DMM	dimethoxymethane
DMPU	1,3-dimethyl-3,4,5,6-tetrahydro-2(1*H*)-pyrimidinone
DMSO	dimethylsulfoxide
DNA	deoxyribonucleic acid
DPPA	diphenylphosphoryl azide
dr	diastereomeric ratio
E	electrophile
EDA	ethyl diazoacetate
ee	enantiomeric excess
EH	epoxide hydrolase
EPR	electron paramagnetic resonance
ESI-MS	electrospray ionization–mass spectrometry
Esp	α,α,α′,α′-tetramethyl-1,3-benzenedipropionate
EWG	electron-withdrawing group
FAD	flavin adenine dinucleotide
FG	functional group
Fu	furyl
GDH	glucose dehydrogenase
HDHH	halohydrin dehalogenase
Hept	heptyl
Hex	hexyl
HKR	hydrolytic kinetic resolution
HMDS	hexamethyldisilazide
HMPA	hexamethylphosphoramine
JHC	Jørgensen–Hayashi catalyst
KSAE	Katsuki–Sharpless asymmetric epoxidation
L	ligand
LA	Lewis acid
LDA	lithium diisopropylamide
LDHs	layered double hydroxides
L-DET	L-diethyl tartrate
L-DIPT	L-diisopropyl tartrate
LEH	limonene 1,2-epoxide hydrolase
LG	leaving group
LIDAKOR	lithium diisopropylamide-potassium *t*-butoxide

LTMP	lithium 2,2,6,6-tetramethylpiperidine
M	metal
MCPBA	3-chloroperoxybenzoic acid
MEDAM	tetramethyldianisylmethyl
MEDPM	tetramethyldiphenylmethyl
MEM	methoxyethoxymethyl
MEOX	methyl 1-oxo-(2-oxazolidine)-4-carboxylate
MEPY	methyl 2-oxopyrrolidine-5-carboxylate
Mes	mesyl
MIB	morpholino isoborneol
MIRC	Michael-initiated ring-closure
MOM	methoxymethyl
Ms	mesyl ($MeSO_2$)
MS	molecular sieves
MSH	*O*-mesitylenesulfonylhydroxylamine
MTBD	7-methyl-1,5,7-triazabicyclo[4.4.0]dec-5-ene
Mts	2,4,6-trimethylphenylsulfonyl
NADH	dihydronicotinamide adenine dinucleotide
NADPH	nicotinamide adenine dinucleotide phosphate
Naph	naphthyl
NBS	*N*-bromosuccinimide
NCS	*N*-chlorosuccinimide
NMP	*N*-methyl pyrrolidinone
NOBIN	1-amino-1'-hydroxylbinaphthyl
Non	nonyl
Npt	naphthyl
Ns	nosyl
NsNIPh	[(nosylimino)iodo]benzene
Nttl	1,8-naphthanoyl-*tert*-leucine
Nu	nucleophile
Oct	octyl
PEG	polyethylene glycol
Pf	phenylfluorenyl
Pfm	perfluorobutyramide
PG	protecting group
Phen	phenanthryl
Phth	phthaloyl
Piv	pivaloyl (*t*-BuCO)
PLA	poly-L-alanine
PLL	poly-L-leucine
PMB	4-methoxybenzyl
PMP	4-methoxyphenyl
PNB	*para*-nitrobenzyl
PNNP	*N*,*N*'-bis[*o*-(diphenylphosphino)-benzylidene]cyclohexane-1,2-diamine
PPTS	pyridinium *p*-toluenesulfonate
PTAB	phenyltrimethylammonium tribromide

PTC	phase-transfer catalyst
PTFE	polytetrafluoroethylene
Py	pyridyl
Pybox	pyridylbis(oxazoline)
r.t.	room temperature
Salen	salicylidenethanediamine
Segphos	5,5′-bis(diphenylphosphino)-4,4′-bi-1,3-benzodioxole
SEM	2-(trimethylsilyl)ethoxymethyl
Ses	trimethylsilylethanesulfonyl
SIPr	N,N'-bis(2,6-diisopropylphenyl)-4,5-dihydroimidazol-2-ylidene
SMO	styrene monooxygenase
SPC	sodium percarbonate
Su	succinimidyl
TADDOL	$\alpha,\alpha,\alpha',\alpha'$-tetraaryl-1,3-dioxolan-4,5-dimethanol
TASF	tris(dimethylamino)sulfonium difluorotrimethylsilicate
TBAC	tetrabutylammonium chloride
TBAF	tetrabutylammonium fluoride
TBAI	tetrabutyl ammonium iodide
TBDPS	*tert*-butyldiphenylsilyl
TBHP	*tert*-butyl hydroperoxide
TBME	*tert*-butyl methyl ether
TBS	*tert*-butyldimethylsilyl
TCPTTL	N-tetrachlorophthaloyl-(S)-*tert*-leucinate
TEA	triethylamine
TEBAC	benzyl triethyl ammonium chloride
TEEDA	tetraethylethylene diamine
TEMPO	2,2,6,6-tetramethylpiperidinyloxy
TES	triethylsilyl
Tf	trifluoromethanesulfonyl
TFA	trifluoroacetic acid
THF	tetrahydrofuran
Thio	thiophene
THP	tetrahydropyranyl
TIPS	triisopropylsilyl
TMEDA	tetramethylethylenediamine
TMG	1,1,3,3-tetramethylguanidine
TMOF	trimethylorthoformate
TMS	trimethylsilyl
TMSOTf	trimethylsilyl trifluoromethanesulfonate
Tol	4-methylphenyl
TON	turnover number
TPPP	tetraphenylphosphonium monoperoxybisulfate
TPS	triphenylsilyl
Tr	trityl (Ph$_3$C)
TRIP	3,3′-bis-(2,4,6-triisopropyl-phenyl)-1,1′-binaphthyl-2,2′-diyl hydrogen phosphate
Tris	2,4,6-triisopropylbenzenesulfonyl

Troc	2,2,2-trichloroethoxycarbonyl
Ts	tosyl (4-MePhSO$_2$)
UHP	urea hydrogen peroxide complex
XMO	xylene monooxygenase
Xyl	dimethylphenyl

1

Asymmetric Cyclopropanation

1.1 Introduction

Organic chemists have always been fascinated by the cyclopropane subunit [1]. Its strained structure[1] and interesting bonding characteristics have attracted the attention of the physical organic community [2]. Due to the limited degrees of freedom, these conformationally constrained molecules have very pronounced steric, stereoelectronic, and directing effects, which make them versatile probes for the study of regio-, diastereo-, and enantioselectivity [3].

On the other hand, the cyclopropane subunit is present in many biologically important compounds including terpenes, pheromones, fatty acid metabolites, and unusual amino acids [1j, 4], and it shows a large spectrum of biological properties, including enzyme inhibition and insecticidal, antifungal, herbicidal, antimicrobial, antibiotic, antibacterial, antitumor, and antiviral activities [5]. This fact has inspired chemists to find novel and diverse approaches to their synthesis, and thousands of cyclopropane compounds have been prepared [6]. In particular, naturally occurring cyclopropanes bearing simple or complex functionalities are chiral compounds; thus, the cyclopropane motif has long been established as a valuable platform for the development of new asymmetric technologies [7]. The enantioselective synthesis of cyclopropanes has remained a challenge, since it was demonstrated that members of the pyrethroid class of compounds were effective insecticides [8]. Asymmetric synthesis constitutes the main strategy to gain access to enantioenriched compounds, involving the use of either chiral auxiliaries or catalysts that in turn can be metal-centered, small organic asymmetric molecules or enzymes. New and more efficient methods employing all these methodologies to gain enantiomerically enriched cyclopropanes are still evolving, covering all the main cyclopropanation reactions: those are the well-known Simmons–Smith reaction [9], the transition-metal-catalyzed

---

[1] The strain energy is the difference between the observed heat of formation of a strained molecule and that expected for a strain-free molecule with the same number of atoms.

Asymmetric Synthesis of Three-Membered Rings, First Edition.
Hélène Pellissier, Alessandra Lattanzi and Renato Dalpozzo.
© 2017 Wiley-VCH Verlag GmbH & Co. KGaA. Published 2017 by Wiley-VCH Verlag GmbH & Co. KGaA

decomposition of diazo compounds [10],[2] and the irreversible Michael-initiated ring-closure (MIRC) [11].

1.2 Simmons–Smith Cyclopropanation

In the late 1950s, Simmons and Smith discovered that the reaction of alkenes with diiodomethane in the presence of activated zinc afforded cyclopropanes in high yields [12]. The reactive intermediate is an organozinc species, and the preparation of such species, including $RZnCH_2I$ or $IZnCH_2I$ compounds and samarium derivatives, was developed in the following years [13]. The popularity of the Simmons–Smith reaction arose from the broad substrate generality, the tolerance of a variety of functional groups, the stereospecificity with respect to the alkene geometry, and the *syn*-directing and rate-enhancing effect observed with proximal oxygen atoms [14].

In spite of the practical importance of the asymmetric Simmons–Smith cyclopropanation, the reaction pathway is not completely clear yet [15]. Theoretically, the Simmons–Smith cyclopropanation can proceed via a concerted [2+1] methylene transfer (Scheme 1.1, path A), in which the pseudo-trigonal methylene group of a halomethylzinc halide adds to an alkene π-bond and forms two new carbon—carbon bonds simultaneously, accompanying a 1,2-migration of the halide anion from the carbon to the zinc atom. Alternatively, a [2+2] carbometallation mechanism, in which the halomethyl group and the zinc halide add to both termini of the alkene π-bond followed by intramolecular nucleophilic substitution of the pseudo-carbanion, can be supposed (Scheme 1.1, path B). Experimental studies show that, using a zinc carbenoid, the cyclopropanation very likely proceeds by the [2+1] pathway, primarily because the carbon—zinc bond is covalent and unpolarized. In 2003, Nakamura *et al.* studied the reaction pathways of cyclopropanation using the Simmons–Smith reagent by means of the B3LYP

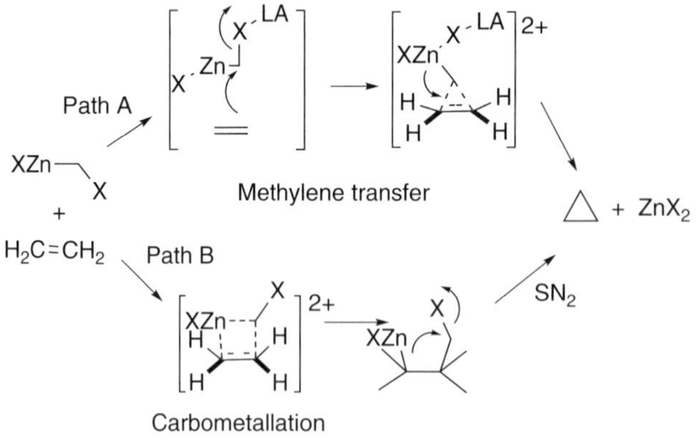

Scheme 1.1 Possible mechanisms for the Simmons–Smith reaction.

---

[2] The high reactivity of diazo compounds counterbalances the ring strain generated in the newly formed cyclopropane unit.

hybrid density functional method, confirming that the methylene-transfer pathway was the favored reaction course [15]. It took place through two stages, an S_N2-like displacement of the leaving group by the olefin, followed by a cleavage of the C—Zn bond to give the cyclopropane ring. However, the alternative carbometallation and cyclization pathway was found to be preferred when the carbon—metal bond is more polarized, such as in lithium carbenoids, and this hypothesis has received experimental support [16].

Kinetic studies on the cyclopropanation of dihydropyrroles show an induction period that is consistent with a change in the structure of the carbenoid reagent during the course of the reaction. This mechanistic transition is associated with an underlying Schlenk equilibrium that favors the formation of monoalkylzinc carbenoid $IZnCH_2I$ relative to dialkylzinc carbenoid $Zn(CH_2I)_2$, which is responsible for the initiation of the cyclopropanation. Density functional theory (DFT) computational studies were also conducted to study the factors influencing reaction rates and diastereoselectivities [17].

1.2.1 Chiral Substrates

The simplest method to obtain chiral compounds is to start from enantiopure substrates, and the built-in chirality is then preserved in the remainder of the reaction sequence. However, this requires the availability of enantiopure substances with the right configuration, and the cheapest ones are amino acids and sugars, which are available in nature as single enantiomers. In the present case, only the cyclopropanation of various asymmetric acyclic allylic alcohols has been widely developed instead, using the heteroatom as the directing group, by chelation with the zinc reagent. Most of them are prepared by enantioenriched reduction of unsaturated carbonyl compounds or by cleavage of chiral epoxides. This Simmons–Smith reaction has distinct advantages over the reaction with a simple olefin in relation to the reaction rate and stereocontrol [18]. Moreover, these reactions have been shown to be much faster than those with simple olefins, and the reaction with a cyclic allylic alcohol took place, forming the cyclopropane ring on the same side as the hydroxyl group [13, 19].

1.2.1.1 Chiral Allylic Alcohols

The cyclopropanations of 1-cycloalken-3-ols with five-, six-, and seven-membered rings generally produced very good *syn:anti* ratios, while a reversal of selectivity was observed with larger eight- or nine-membered ring [7a]. This can be explained on the basis of simple conformational analysis of the ground state [20]. For instance, in their approach to enantiomerically pure cyclopropyl ketones, Johnson and Barbachyn showed that β-hydroxysulfoximines derived from cyclic enones could produce the cyclopropane *syn* to the hydroxy group [21]. In addition, the synthesis of cyclopropanated sugars is diastereoselective. In particular, the *syn*-isomer was obtained as the major product with halomethylzinc reagents, whereas the *anti*-isomer could be prepared by a multistep sequence [22].

The stereoselective cyclopropanation of a chiral, acyclic allylic alcohol using the Simmons–Smith reagent (Zn–Cu, CH_2I_2) was first reported by Pereyre and coworkers in 1978 [23]. They observed that very high *syn*-selectivity (>200:1) was achieved with (Z)-disubstituted olefins, but much lower with (E)-disubstituted olefins (<2:1). Charette showed that the nature of the Zn carbenoid used in these reactions is very important for achieving high diastereoselectivities, especially with (E)-disubstituted olefins [24]. The stereochemical outcome of these reactions can be qualitatively predicted by assuming an oxygen-group-assisted delivery of the reagent from a conformation in which the minimization of the $A^{1,3}$-strain is the predominant controlling element, but other elements have to be taken into account.

Most of asymmetric cyclopropanations are key steps in the synthesis of natural products of biological interest. For instance, the elegant synthesis of (R)-muscone reported by Oppolzer features a diastereoselective cyclopropanation of a chiral macrocyclic (S,E)-allylic alcohol (Scheme 1.2) [25].

Takemoto and coworkers afforded an asymmetric total synthesis of halicholactone, in which the regio- and stereoselective cyclopropanation is the key step (Scheme 1.3) [26]. It should be noted that the right choice of protecting groups was crucial for the regioselectivity and the occurrence of the reaction.

Smith and coworkers afforded the total synthesis of the marine diolide (−)-clavosolide A by direct Simmons–Smith cyclopropanation of an N-methoxyamide (Scheme 1.4) [27].

White et al. developed a total synthesis of solandelactones E and F (two biologically active oxylipins), having another similar directed Simmons–Smith cyclopropanation as the key step, leading to a single diastereomer, as shown in Scheme 1.5. From this synthesis, authors confirmed that the structures of the two solandelactones were epimeric at C11 [28].

Brevipolides are extracted from the invasive tropical plant of *Hyptis brevipes* and showed interesting drug properties. A diastereoselective synthesis of C1–C12

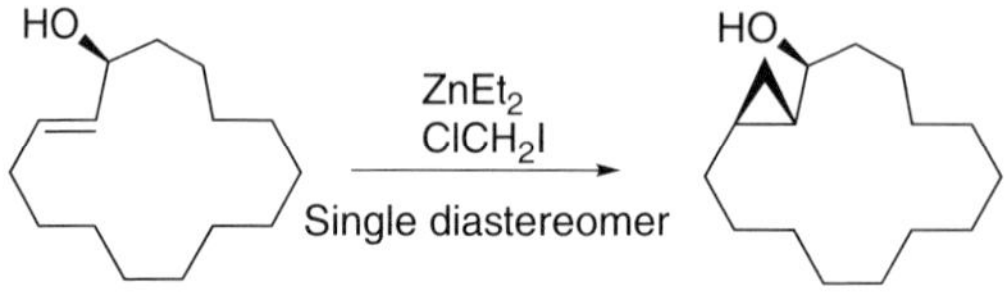

Scheme 1.2 Synthesis of (R)-muscone.

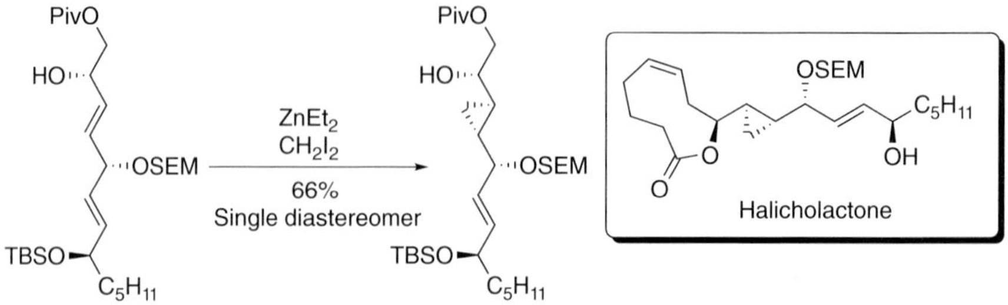

Scheme 1.3 Key step in the asymmetric total synthesis of halicholactone.

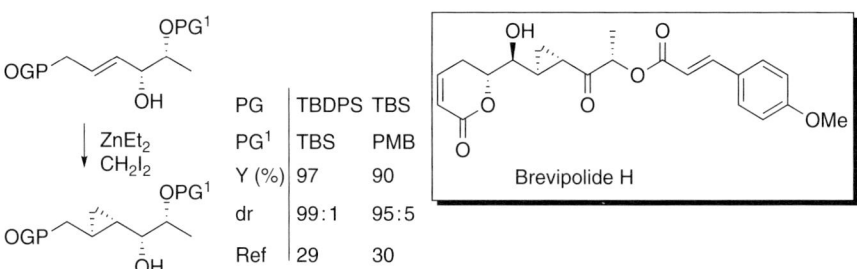

Scheme 1.4 Key step in the total synthesis of (−)-clavosolide A.

Scheme 1.5 Key step in the total synthesis of solandelactones E and F.

Scheme 1.6 Diastereoselective synthesis of C1–C12 fragment of brevipolide H.

fragment of brevipolide H was synthesized by Mohapatra's group (Scheme 1.6) [29]. More recently, a similar reaction was proposed by Kumaraswamy and coworkers, but with inferior results for the synthesis of 11′-*epi*-brevipolide H [30].

Schmalz's group developed a fully enantioselective synthesis of a C_2-symmetric bicyclo[4.4.1]undecanedione based on a diastereoselective cyclopropanation [31]. It should be noted that the usual Simmons–Smith conditions failed, due to complete decomposition; thus, the desired cyclopropanation was successfully

6 Asymmetric Synthesis of Three-Membered Rings

achieved using a $ZnEt_2/ClCH_2I$ reagent, providing the corresponding tricyclic diol as a single diastereomer (Scheme 1.7).

Charette et al. reported that the directed cyclopropanation of chiral acyclic allylic alcohols using *gem*-dizinc carbenoids was highly stereoselective, yielding either the *syn* or the *anti*-cyclopropane, depending upon the substitution pattern of the alkenes [32]. Thus, the zinc cyclopropanation of several *cis*-disubstituted allylic alcohols occurred with excellent facial selectivity for the attack of the *gem*-zinc carbenoid, leading to the corresponding *syn, cis*-cyclopropyl derivatives in high diastereomeric ratios for a wide range of sterically demanding substituents at the allylic position, even with protected allylic alcohol. The zinc cyclopropanation of the corresponding *trans*-isomer was less stereoselective. However, the introduction of a TMS substituent at either the R^1 or the R^2 position led to the exclusive formation of the *anti, cis* or of the *syn, trans*-isomer, as shown in Scheme 1.8.

Occhiato's group prepared substituted cyclopropane pipecolic acids as conformationally restricted templates for linear and cyclic peptidomimetics [33]. The synthesis started from commercially available enantiopure γ-hydroxymethyl-γ-butyrolactones, leading to product with complete stereoselectivity even with remote directing group (Scheme 1.9). It should be noted that, sometimes, the reaction conditions deprotected the nitrogen atom, thus avoiding cyclopropanation.

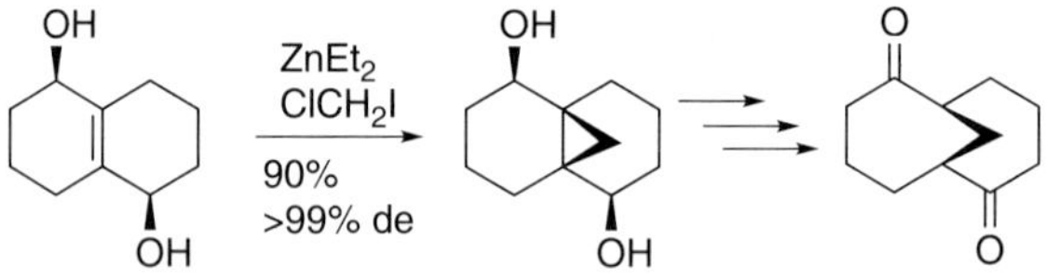

Scheme 1.7 Enantioselective synthesis of a C_2-symmetric bicyclo[4.4.1]undecanedione.

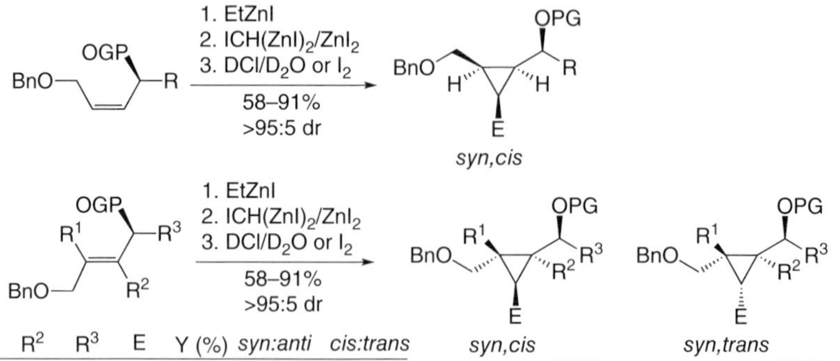

PG	R^1	R^2	R^3	E	Y (%)	syn:anti	cis:trans
H	H	H	Me	I	85	60:40	28:72
H	H	H	t-Bu	D	73	94:6	28:72
TIPS	H	H	t-Bu	D	64	>95:5	75:25
H	H	TMS	Me	D	68	<5:95	>95:5
H	H	TMS	t-Bu	D	77	<5:95	>95:5
H	TMS	H	t-Bu	D	84	>95:5	<5:95
H	TMS	H	t-Bu	I	81	>95:5	<5:95

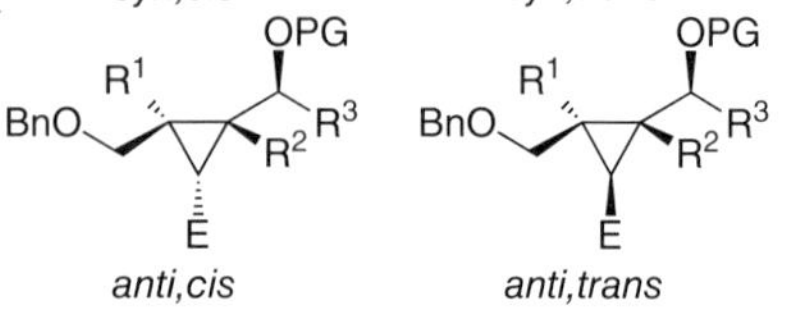

Scheme 1.8 Cyclopropanation of chiral allylic alcohols using *gem*-dizinc carbenoids.

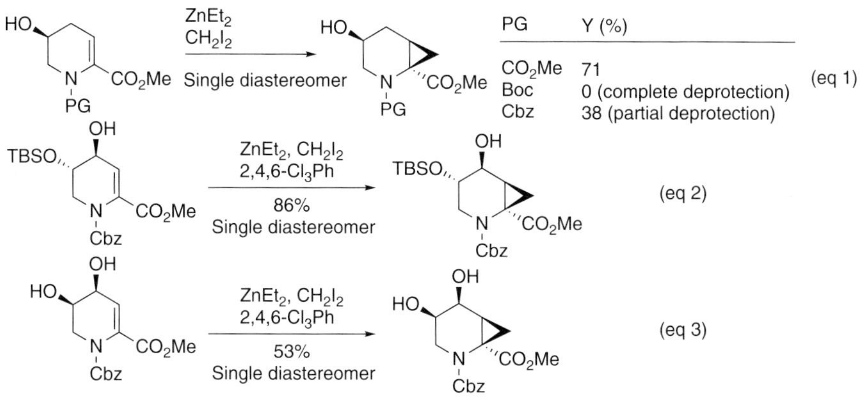

Scheme 1.9 Synthesis of substituted cyclopropane pipecolic acids.

1.2.1.2 Chiral Allylic Amines

Even though amines have the same potential for binding with the zinc reagent as oxygen functional groups, allylic amines have been much less explored compared to their corresponding alcohols.

Aggarwal and coworkers reported the first highly diastereoselective cyclopropanation of allylic tertiary amines using the Simmons–Smith reagent [34]. They found a divergent behavior of simple allylic amines and those bearing additional chelating groups. In both cases, the reaction was initiated by complexation of the amine with the zinc reagent. However, in the case of a simple allyl-substituted amine (R = BnCH$_2$, Scheme 1.10, eq 1), this species underwent a 1,2-shift to furnish a zinc-complexed ammonium ylide. In the case of an amino alcohol (R = (Ph)CHCH$_2$OH, Scheme 1.10, eq 1), a more stable chelate zinc complex was considered to be formed that did not readily undergo the 1,2-shift. Because of the proximity of the olefin to the tightly held zinc carbenoid, however, cyclopropanation occurred instead. On these bases, they used a range of chiral amino alcohols such as phenylglycinol (Scheme 1.10, eq 2), pseudoephedrine (Scheme 1.10, eq 3), and ephedrine (Scheme 1.10, eq 4), to achieve cyclopropanation with very high diastereoselectivity.

1.2.1.3 Chiral Acetal-Directed Cyclopropanations

Diastereoselective acetal-directed cyclopropanations constitute the key step of some important natural products or drugs containing cyclopropane moieties. The double asymmetric Simmons–Smith cyclopropanation of the (E)- and (Z)-bis(olefins) could be successfully used to prepare enantioenriched 1,2-bis(2-methylcyclopropyl) ethenes with excellent stereocontrol (Scheme 1.11) [35].

Diastereoselective acetal-directed cyclopropanations also constituted the key step of a total synthesis of solandelactone E (Scheme 1.12, eq 1) [36] and of a total synthesis of a marine fatty acid metabolite having lipoxygenase-inhibiting activity (Scheme 1.12, eq 2) [37], both providing the corresponding cyclopropyl derivative in excellent yield and with high stereoselectivity.

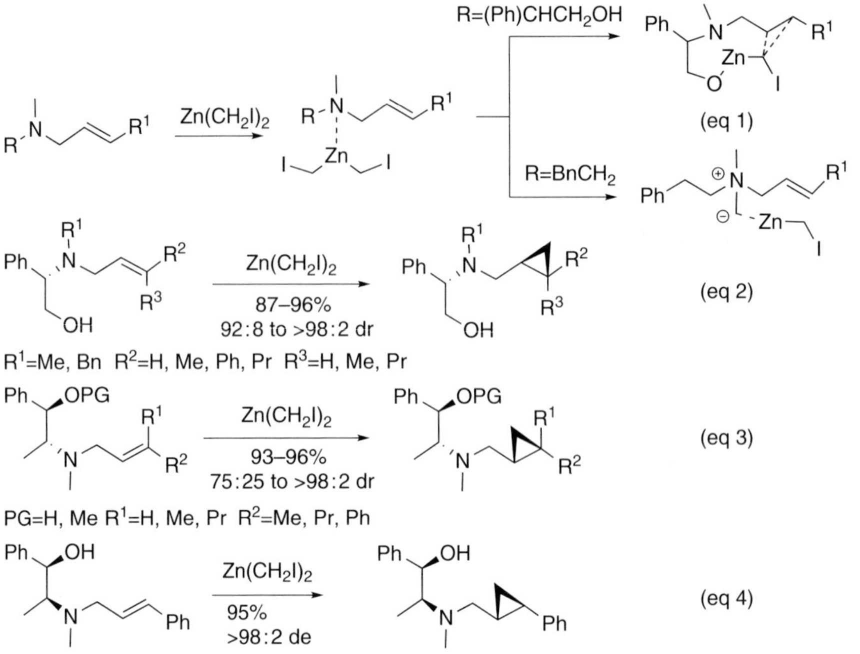

Asymmetric Synthesis of Three-Membered Rings

Scheme 1.10 Cyclopropanation of allylic tertiary amines.

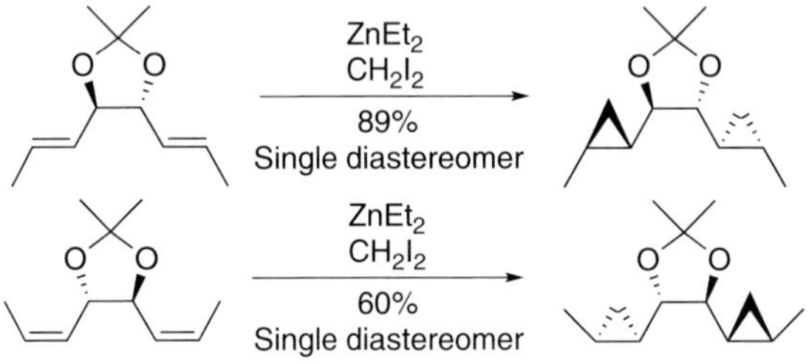

Scheme 1.11 Double asymmetric Simmons–Smith cyclopropanation of bis(olefins).

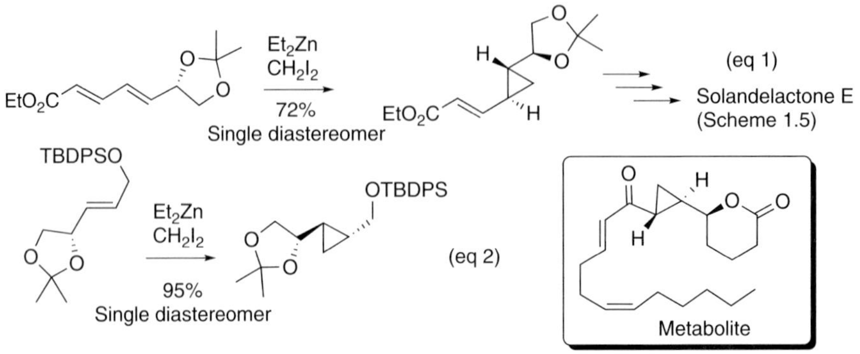

Scheme 1.12 Diastereoselective acetal-directed cyclopropanations.

Finally, fluorocyclopropanation of *trans*-styryldioxolane derived from D-glyceraldehyde acetonide afforded the desired cyclopropane in 73% yield, in 94:6 dr, and with 99% ee, with the fluorine substituent being oriented *trans* to the dioxolane. The *cis*-isomer led to a 75:25 dr, and the major isomer, isolated in 62% yield and with 99% ee, was found to be the all-*cis*-fluorocyclopropane [38].

1.2.1.4 Simple Chiral Alkenes

In the absence of a directing group, the cyclopropanation of cyclic olefins is generally subjected to steric effects. The level of stereochemical induction is usually very high, and the sense can be predicted on the basis of the prevailing ground-state conformation of the starting olefin. For instance, a stereoselective cyclopropanation from the more accessible β-face produced a key intermediate in the synthesis of (+)-acetoxycrenulide, as a single isomer (Scheme 1.13, eq 1) [39]. Another stereoselective cyclopropanation was used by Corey and Lee in their β-amyrin total synthesis (Scheme 1.13, eq 2) [40]. The regioselective methylenation of the 17–18 double bond should be also outlined, since the analogous reaction using dibromocarbene added exclusively to the 12–13 double bond.

The stereocontrol in the cyclopropanation of acyclic alkenes, in which the basic group that directed the reagent is not on a stereogenic center, usually was not very high, except when the allylic position bore a bulky dimethylphenylsilyl group. In fact, the cyclopropanation of functionalized (*E*)-crotylsilanes bearing a bis-homoallylic hydroxyl group gave reasonably good diastereoselectivities depending on the nature of the groups on the homoallylic position (the best results are 81% yield and 95:5 *anti:syn* ratio). It is worth noting that AlMe$_3$ was the organometallic species generating the carbenoid, because both the zinc- and samarium-derived reagents failed to produce the desired product [41].

Standard Simmons–Smith conditions were applied by Abad *et al.* to the cyclopropanation of a tetracyclic diterpene [42]. The cyclopropanation took place

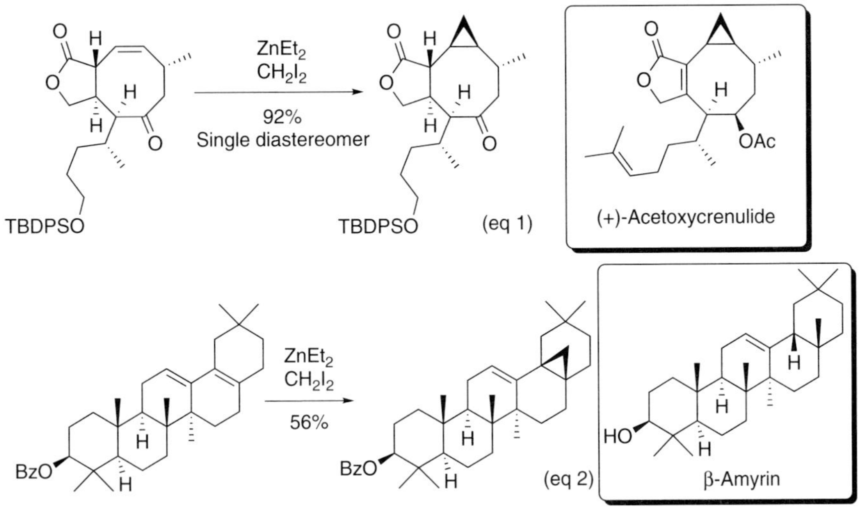

Scheme 1.13 Cyclopropanation of simple chiral cyclic olefins.

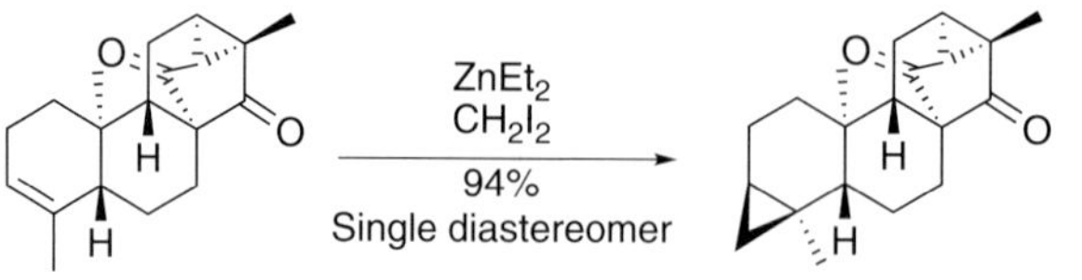

Scheme 1.14 Cyclopropanation of a tetracyclic diterpene.

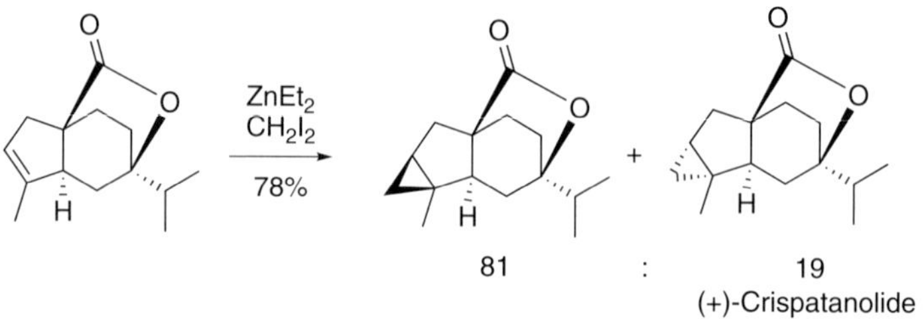

Scheme 1.15 Last step of a total synthesis of (+)-crispatanolide.

stereoselectively from the less hindered β-side of the double bond, affording the expected cyclopropane in excellent yield and diastereoselectivity (Scheme 1.14). This tricyclo[3.2.1.0]octane moiety was a key intermediate in the synthesis of trachylobane-, beyerane-, atisane-, and kaurane-type diterpenes.

Based on the same considerations on the steric effects of bulky polycyclic systems, Tori and coworkers applied standard Simmons–Smith conditions in the last step of a total synthesis of (+)-crispatanolide (Scheme 1.15) [43]. Surprisingly, the major product was not the expected (+)-crispatanolide, but a diastereomer, very likely because of the directing effect of the lactone carbonyl group. However, this synthesis allowed similarly assignment of the absolute configuration to the natural (+)-crispatanolide.

Moreover, 2-azabicyclo[3.1.0]hexane-3-carboxylic acids were obtained from chiral 2,3-dihydropyrroles derived from (R)-glutamic acid. The asymmetric Simmons–Smith reaction and hydrolysis reaction mainly led to the all-(R)-product. In this Simmons-Smith reaction the reaction time was found to influence the E/Z ratio and the best ratio was reached after 19.5 h (Scheme 1.16) [44].[3]

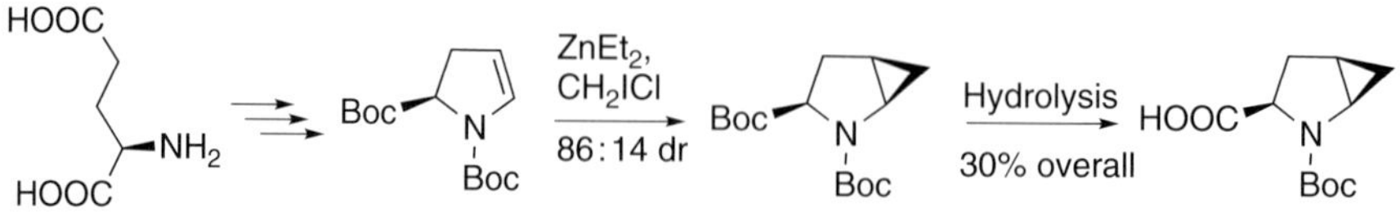

Scheme 1.16 Synthesis of 2-azabicyclo[3.1.0]hexane-3-carboxylic acid.

---

[3] The paper has only the abstract in English. There the formation of all-(S)-product is reported, but schemes report unnatural glutamic acid and it is always numbered R#. Sometimes among Chinese characters some products named S# are reported. Perhaps the reaction was performed from both enantiomers of glutamic acid.

1.2.2 Chiral Auxiliaries

The strategy that uses chiral auxiliaries is based on the transformation into "chiral product equivalents" by binding an enantiomerically pure derivative to the starting material. These compounds are then stereoselectively transformed into new chiral intermediates that contain new stereogenic centers in high diastereomeric excess, with diastereoselectivity being controlled by the presence of the chiral auxiliary fragment. Subsequent cleavage of the chiral auxiliary moiety affords a chiral compound containing a stereogenic center in high enantiomeric excess.[4] Thus, a number of auxiliary-based approaches, which can be encompassed in four general classes, have been reported for the Simmons–Smith cyclopropanation (Table 1.1). Most of these reactions led to cyclopropylmethanols (Scheme 1.17).

Table 1.1 Chiral auxiliaries for Simmons–Smith reaction using $ZnEt_2$, CH_2I_2.

Starting material	Yield (%)	de (%)	Product	References
Allylic ethers				
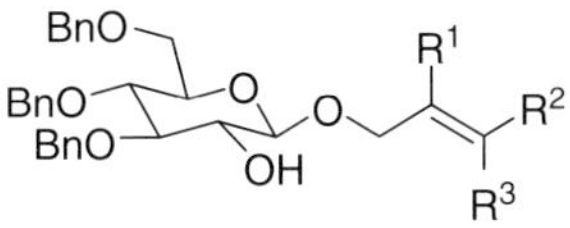 (BnO-sugar allyl ether with R^1, R^2, R^3)	≥95	≥98	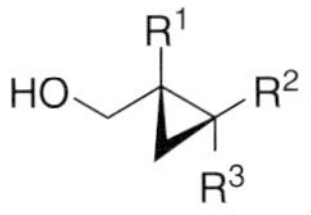 (cyclopropylmethanol with R^1, R^2, R^3)	[45]
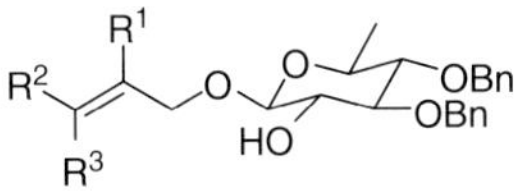 (allyl ether with OBn sugar, R^1, R^2, R^3)	≥95	≥98	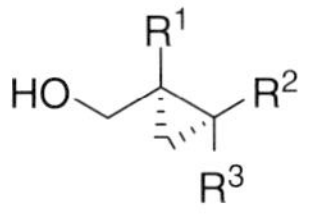 (cyclopropylmethanol with R^1, R^2, R^3)	[45]
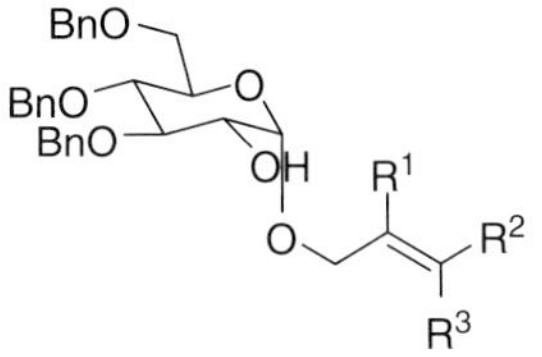 (BnO-sugar allyl ether with R^1, R^2, R^3)	83–93	92–94	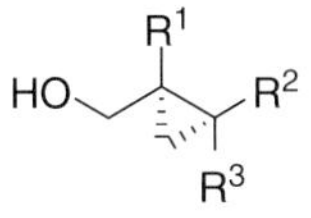 (cyclopropylmethanol with R^1, R^2, R^3)	[46]
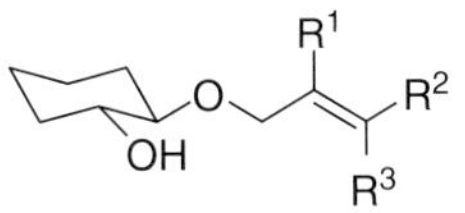 (cyclohexyl allyl ether with R^1, R^2, R^3)	90–98[a]	≥93	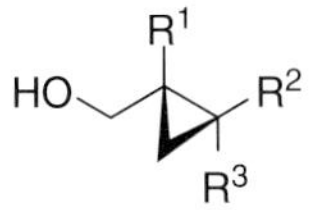 (cyclopropylmethanol with R^1, R^2, R^3)	[47]
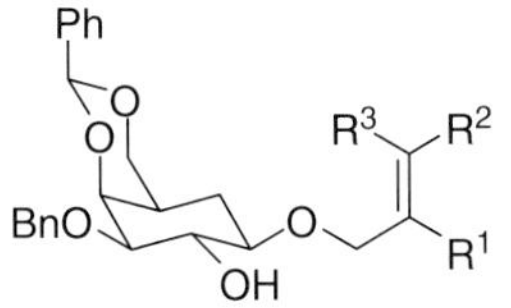 (Ph, BnO substituted allyl ether with R^1, R^2, R^3)	67–95[b]	Up to 100	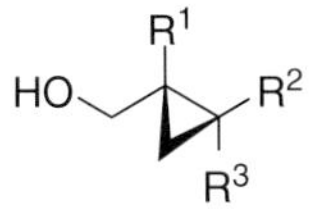 (cyclopropylmethanol with R^1, R^2, R^3)	[48]

(Continued)

---

[4] The need for additional steps to add and remove the chiral auxiliary reduces the overall yields and leads to wastage of material. However, this strategy was the first used by chemists to obtain enantioenriched products, and only later, the chiral catalysis emerged.

Table 1.1 (Continued)

Starting material	Yield (%)	de (%)	Product	References
Acetals				
(R^1, R^2 vinyl acetal with CO$_2$-i-Pr(Et) dioxolane)	50–95	93–97	R^1, R^2 cyclopropane-CHO	[49]
(R^1, R^2 vinyl acetal with Ar, dimethyl dioxane)	34–67	66–92	R^1, R^2 cyclopropane-CHO	[50]
(acetonide-OAr, R^1, R^2, R^3 vinyl)	45–90	21–81	R^1, R^2, R^3 cyclopropane-CHO	[51]
(sugar-derived bis-acetal with R^2, R^3 vinyl)	69–87[b]	50–100	HO-R^1, R^2, R^3	[52]
BnO, OBn dioxolane with R^1, R^2 cyclohexenyl	54–99[c]	88–95	bicyclic ketone with R^1, R^2 cyclopropane	[53]
α,β-Unsaturated carbonyl derivatives				
Cp(OC)(PPh$_3$)Fe-C(O)-CR^2=CHR^1	49–86[d]	94–97	R^1, R^2 cyclopropane-CO$_2$H	[54]
(camphor-derived N-acyl with TIPSO, cinnamoyl)	62[e]	99	Ph cyclopropane-CO$_2$H	[55]
(camphor-derived NH-acyl with TIPSO, cinnamoyl)	56[e]	99	Ph cyclopropane-CO$_2$H	[55]

Table 1.1 (Continued)

Starting material	Yield (%)	de (%)	Product	References
Other groups				
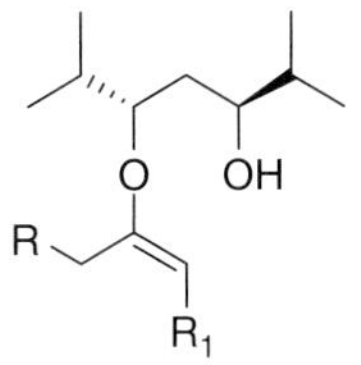 (alkene with OH, OR substituents)	57–80[f]	99	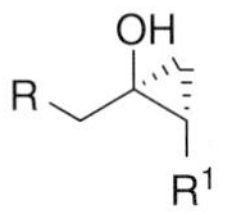 (cyclopropane with OH, R, R¹)	[56]
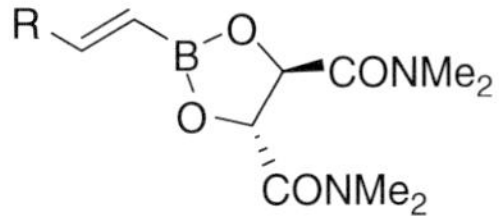 R-alkene with boronate CONMe₂ auxiliary	49–67[c]	94–97	 R-cyclopropane-OH	[57]
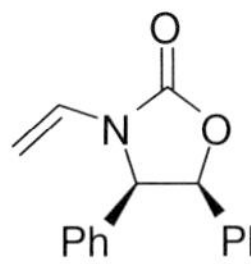 Vinyl oxazolidinone with Ph, Ph	76[g]	30	 R-cyclopropane-OH	[58]
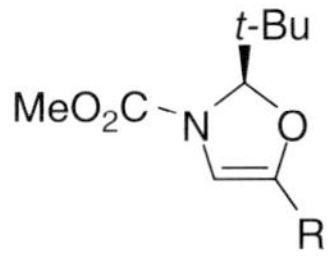 MeO₂C-N oxazoline with t-Bu, R	52–83[c]	>98	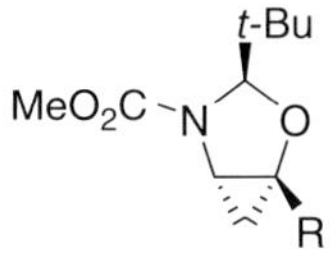 MeO₂C-N oxazoline cyclopropane with t-Bu, R	[59]
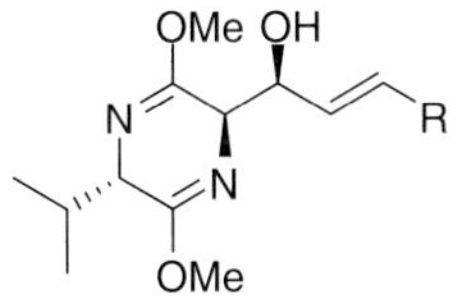 Piperazine with OMe, OH, alkene R	52–80[h]	36–96	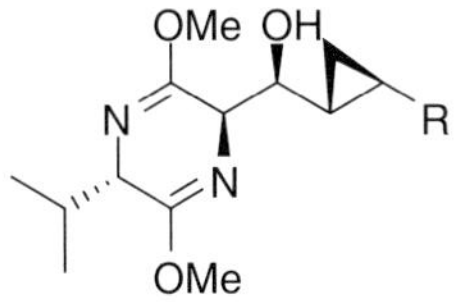 Piperazine with OMe, OH, cyclopropane R	[60]
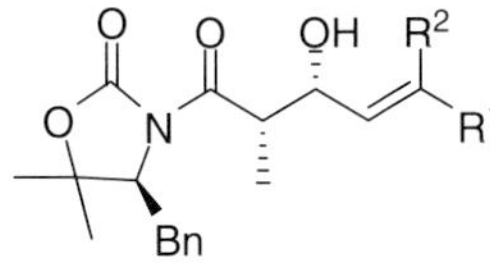 Oxazolidinone-Bn with OH, R¹, R² alkene	89–99[i]	>95	 R¹, R² cyclopropane-CHO	[61]
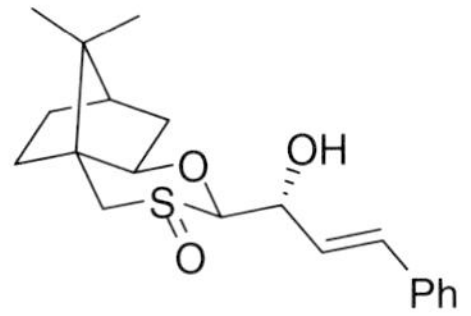 Camphor sulfonyl with OH, Ph alkene	95	>98	 Ph-cyclopropane-CHO	[62]

a) Carbohydrate-derived chiral auxiliaries acted as bidentate ligands in the complexation of the zinc reagent (ZnEt₂/ClCH₂I) with the oxygen atoms at C1 and C2. Therefore, simple chiral 1,2-cyclohexanediol is enough to induce chirality.
b) It should be noted that the reactive face is always the same in both reactions. The obtained opposite enantiomers arose from how the chiral auxiliaries were formed in the two cases. Among acetal-derived auxiliaries, the most efficient auxiliaries were derived from tartaric acid. The configuration of the cyclopropane can be rationalized by coordination of zinc carbenoid to tartrate oxygen atoms.
c) Zn/Cu, CH₂I₂.
d) Et₃Al, CH₂I₂.

(Continued)

Table 1.1 (Continued)

e) (+)-Diethyltartrate added.
f) This chiral auxiliary can produce one major diastereomer (87:13) from more complex diiodoalkane as precursors, in which the R group of the diiodoalkane accommodated in the product *trans* to the hydroxyl moiety.
g) $ZnEt_2$, $CHFI_2$. Chiral 2-fluorocyclopropylamine was also prepared by resolution [63]. It is a key intermediate in the synthesis of DU-6859 (Scheme 1.17), a quinolonecarboxylic acid exhibiting antibacterial activity and little side effects.
h) $ZnEt_2$, R^1CHI_2. Carbenoids of type CHR, from more complex diiodoalkane as precursors, afforded products, in which the R group was on the same side of the OH framework. However, the stereochemistry cannot be completely controlled.
i) The stereo-directing effect was exerted by the hydroxy group, and not by the oxazolidin-2-one group. Enantiomers can be obtained starting from the N-acyloxazolidin-2-one of opposite configuration. This methodology was applied to efficient total syntheses of the natural products grenadamide [64], cascarillic acid (Scheme 1.17) [65], and (−)-clavosolide B (Scheme 1.4) [66].

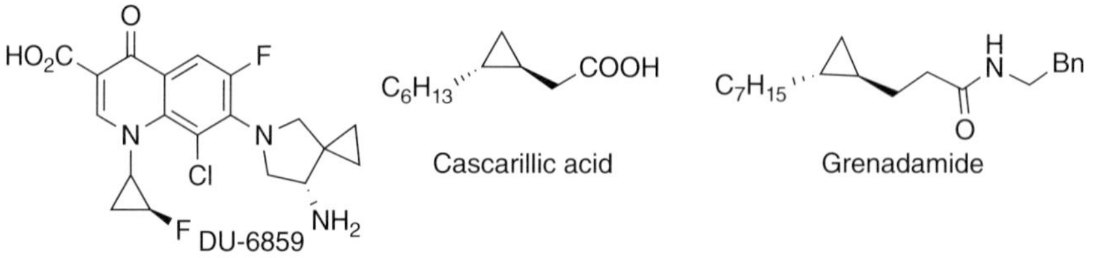

Scheme 1.17 Structures of DU-6859, grenadamide, and cascarillic acid.

The Simmons–Smith cyclopropanation of allenamides was explored in order to develop a direct construction of amidospiro[2.2]pentanes that are interesting potentially biologically active compounds [67]. Hsung and coworkers found that Simmons–Smith cyclopropanation of chiral enamides could be efficiently achieved in very high diastereomeric excesses [68]; thus, they envisaged extending their results to allenamides (Scheme 1.18). Unfortunately, the observed diastereoselectivity was very low, owing to a very little difference in energy between conformations **A** and **B** (from PM3 calculations with the Spartan Model™) much smaller than the corresponding difference between enamide conformations. The enantioselection increased with an increase of the bulkiness of the R substituent, because of enhanced steric crowding in conformation **B** (Scheme 1.18). The first cyclopropanation proceeds by an approach of the zinc carbenoid from the bottom π-face of the more favored conformer **A**. Since both π-faces of amido-methylene cyclopropane are sterically encumbered, the second cyclopropanation reaction occurs much more slowly, and mixtures of mono- and spiro-cyclopropyl derivatives were always obtained with mono/spiro ratio directly correlated with the degree of steric crowding of the π-face of the methylenecyclopropane intermediate.

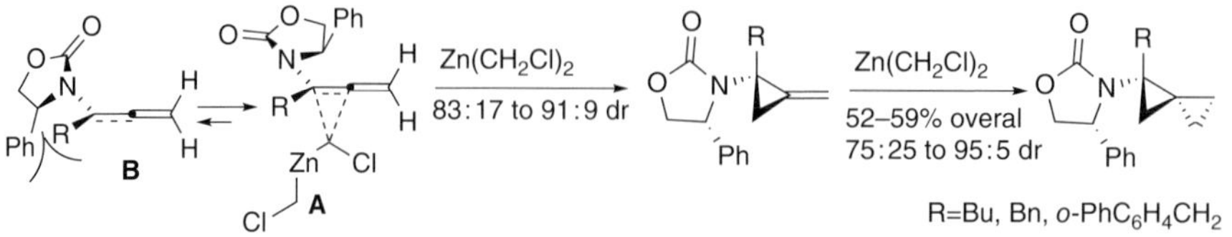

Scheme 1.18 Simmons–Smith cyclopropanation of chiral allenamides.

1.2.3 Chiral Catalysts

Catalytic methods offer many advantages over "chiral pool" or "chiral auxiliary" methods, namely, achiral starting materials or reduced reaction steps, respectively. Generally, they yield good enantioselectivities, but metal catalysis often requires inert reaction conditions and the use of heavy metals, traces of which could be retained in the products, making them not suitable for pharmaceutical preparation. On the other hand, organocatalysts offer some attractive benefits such as their ready availability and high stability, simple handling and storage, but they require the formation of labile intermediates or tight ion pairs. However, the most ancient papers used stoichiometric amounts of chiral ligands, and only in the past few years, true catalytic methodologies have been published.

1.2.3.1 Charette's Ligand

The first attempts to control the absolute stereochemistry in the cyclopropanation of substrates by adding external chiral ligands were reported in 1968, when a mixture of (−)-menthol and $IZnCH_2I$ was added to α,β-unsaturated esters or L-leucine was used as a co-additive in the cyclopropanation of vinyl ethers [69]. Also, (1R,2S)-N-methylephedrine-modified halomethylzinc afforded modest enantioselectivities [70]. The addition of diethyl tartrate to a mixture of the allylic alcohol, diethylzinc, and diiodomethane was the first practical stoichiometric system for the enantioselective cyclopropanation of allylic alcohol [71]. Moderate levels of enantioselection (34–92% ee) were observed in the cyclopropanation of silicon-substituted allyl alcohols.

A major breakthrough in this area occurred when a simple bifunctional, nonracemic ligand containing both acidic (different from zinc) and basic sites was found to allow simultaneous chelation of the acidic halomethylzinc reagent and of a basic zinc allylic alkoxide. These aggregates can be of three different types (Scheme 1.19):

i) Capture of an allyloxy(iodomethyl)zinc species by a Lewis acid site of the catalyst
ii) Capture of allyloxyzinc and iodomethylzinc species by a bifunctional chiral catalyst with two identical groups
iii) Capture of the two species by a bifunctional Lewis acid–Lewis base catalyst.

In 1994, Charette and Juteau found that the addition of a stoichiometric amount of a chiral dioxaborolane ligand prepared from the commercially available N,N,N',N'-tetramethyltartaric acid diamide and butylboronic acid (Scheme 1.20a) to the classical Simmons–Smith reagents provided asymmetric synthesis of cyclopropanes [72]. By DFT calculations, the Charette chiral dioxaborolane ligand was found to form an

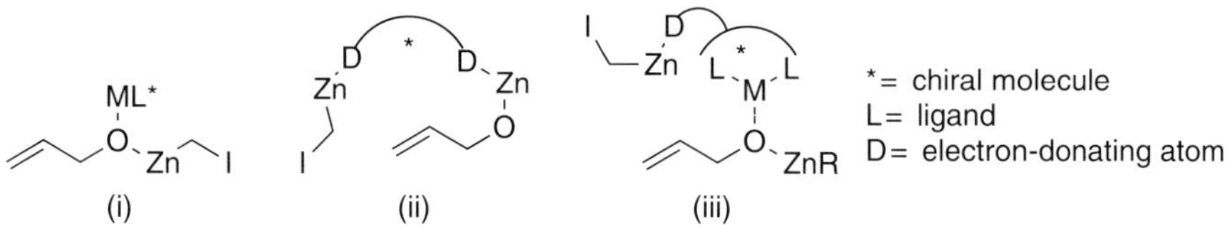

Scheme 1.19 Different types of aggregates among ligands, allylic alcohols, and zinc species.

* = chiral molecule
L = ligand
D = electron-donating atom

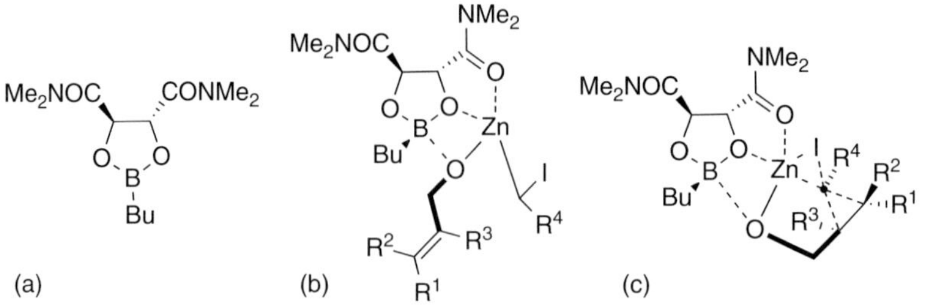

Scheme 1.20 Charette's chiral dioxaborolane ligand (a) and its alcoholate complex (b) and transition state (c).

energetically stable four coordinated chiral zinc–ligand complex, with the zinc bonded to the CH_2I group and coordinated by three oxygen atoms (Scheme 1.20b) [73]. The study of the relative energies of the different transition states allowed the identification of three main factors influencing the enantioselectivity:

i) the torsional strain along the carbon—carbon bond being formed,
ii) the $A^{1,3}$ strain caused by the chain conformation, and
iii) the ring strain generated in the transition states.

Generally, the effects of these three factors on the enantioselectivity are synergetic, resulting in generation of cyclopropylmethanols with enantiomeric excesses of up to 98% ee with *cis*- and *trans*-disubstituted allylic alcohols as well as with tetra-disubstituted ones. Charette also showed that the reaction could be used in the case of polyenes. Excellent regioselectivity favoring the allylic alcohol was observed, when the substrate contained more than one double bond [74]. This efficient methodology was used in the enantioselective cyclopropanation of important chiral building blocks for natural product synthesis, such as 3-tributylstannyl-2-propen-1-ol [75], 3-iodo-2-propen-1-ol [76], 2-chloro-2-propen-1-ol [77], the chiral cyclopropanes subunits of curacin A [78], FR-900848 [79], U-106305 [80], epothilone analogs [81], and (−)-doliculide (Scheme 1.21) [82]. The same reaction conditions were also applied to the enantio- and diastereoselective cyclopropanation of allenic alcohols, which led to enantiomerically enriched spiropentane derivatives (Scheme 1.22) [83].

The stereoselective synthesis of highly functionalized trisubstituted cyclopropanes is problematic. Even if high diastereoselectivities and enantioselectivities were often recorded with 1,2,3-substituted cyclopropanes, when treated with the reagent formed by mixing 1,1-diiodoalkanes and diethylzinc [84], some other trisubstituted allylic alcohols, such as 1-cyclohexenylmethanol, are converted to the corresponding cyclopropanes in good yield but with low enantiomeric excess. Thus, Charette and coworkers partially modified their *gem*-dizinc carbenoids (EtZnCHIZnEt) into $(IZn)_2CHI\cdot4Et_2O$ [85]. The initial deprotonation of the alcohol with diethylzinc was mandatory to prevent *in situ* formation of the classical Simmons–Smith reagent [86]. Under these experimental conditions, *trans*-allyl alcohols were converted into cyclopropylborinates, which, in turn, were subjected to Suzuki–Miyaura reaction conditions with iodoarenes to give products in 26–85% yield, with high diastereomeric ratio (>20 : 1) and 80–97% ee. The

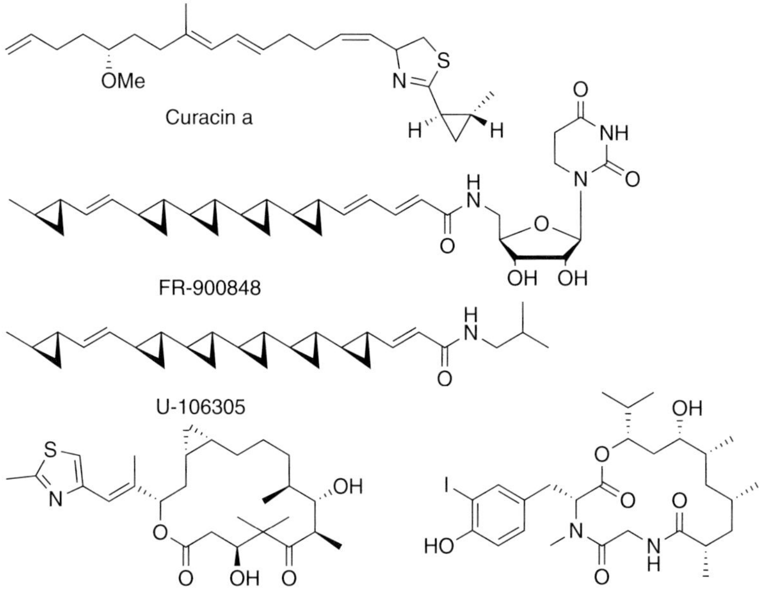

Scheme 1.21 Natural products obtained by enantioselective cyclopropanation with Charette's method.

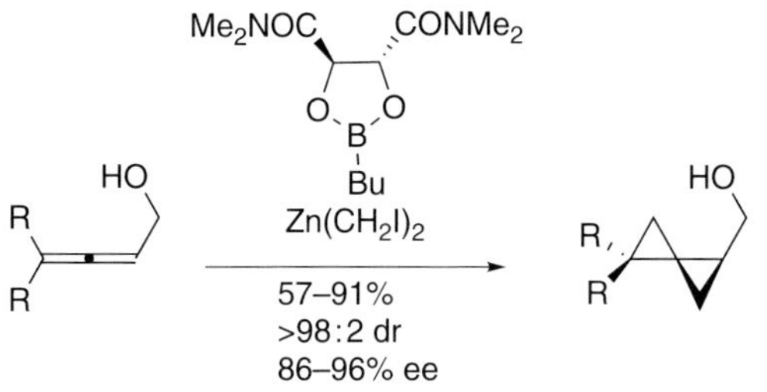

Scheme 1.22 Enantio- and diastereoselective cyclopropanation of allenic alcohols by Charette's method.

reaction of *cis*-allylic alcohols led to a mixture of diastereomeric cyclopropylzinc halides. The all-*cis* gave the expected reaction, while a *trans*-relationship between ZnI and the dioxaborolane moiety disfavored the zinc–boron exchange, which could only occur intramolecularly. The *cis*-cinnamyl alcohol represented an exception, because the favorable π-interaction between the phenyl ring and the zinc atom produced almost exclusively the all-*cis*-isomer, which was then incorporated in the final cyclopropane (Scheme 1.23).

The same research group was also able to adjust the stoichiometric ratio of R_2Zn relative to iodoform in order to increase the proportion of the RZnCHI$_2$ species with respect to the *gem*-dizinc carbenoid, thus allowing iodomethyl-cyclopropanation reactions. In particular, using 1:2 stoichiometric ratio of these reagents, complete conversion into iodocyclopropane was observed (Scheme 1.24) [87].

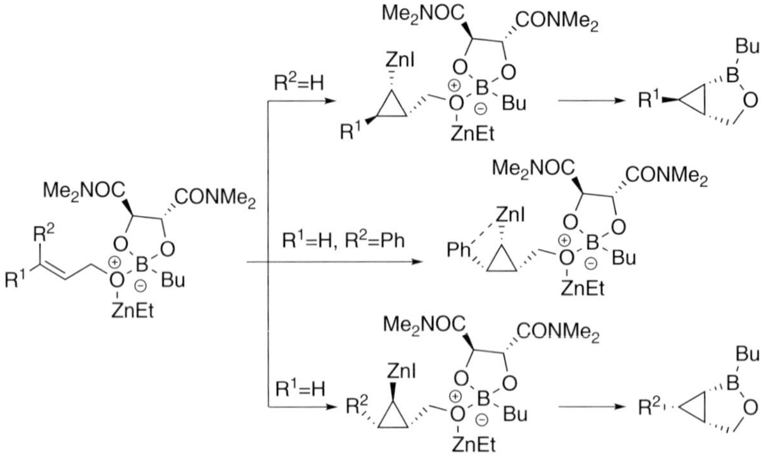
Me2NOC
CONMe2
R2
R1
O
B
O
Bu
ZnEt
R2=H
R1=H, R2=Ph
R1=H
ZnI
Ph
Bu
R2

Scheme 1.23 Stereoselective synthesis of cyclopropanes by partial modification of Charette's method.

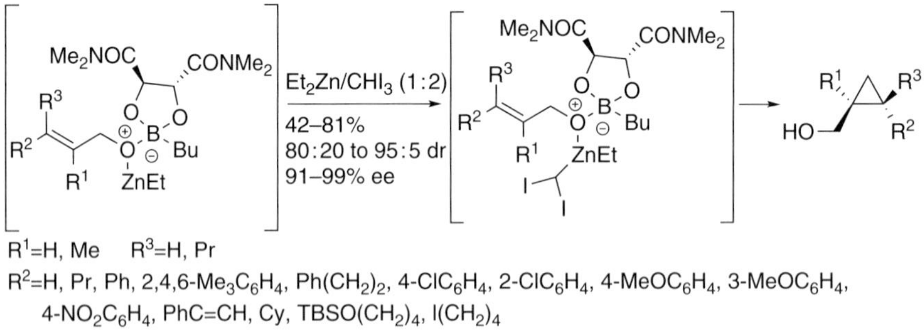
Et2Zn/CHI3 (1:2)
42–81%
80:20 to 95:5 dr
91–99% ee
Me2NOC
CONMe2
R3
R2
R1
ZnEt
Bu
HO
R1=H, Me R3=H, Pr
R2=H, Pr, Ph, 2,4,6-Me3C6H4, Ph(CH2)2, 4-ClC6H4, 2-ClC6H4, 4-MeOC6H4, 3-MeOC6H4,
4-NO2C6H4, PhC=CH, Cy, TBSO(CH2)4, I(CH2)4

Scheme 1.24 Stereoselective synthesis of cyclopropanes by adjust the stoichiometric ratio of R_2Zn and iodoform.

The authors proposed a transition-state model for this enantioselective iodocyclopropanation as depicted in Scheme 1.25. The spatial arrangement of substituents was based on the envisaged less sterically congested conformer of the two possible reactive ones. This hypothesis was also corroborated by DFT studies of analogous cyclopropanation reactions [73]. The reaction was scaled up to 10 mmol with comparable yields and selectivity, and, furthermore, an electrophilic trapping of the corresponding cyclopropyllithium occurred with retention of configuration to give access to a variety of enantioenriched 1,2,3-trisubstituted cyclopropanes. In particular, the synthesis of an HIV-1 protease inhibitor (Scheme 1.25) was performed [88].

The study of the reaction of chloroiodomethylzinc, diiodomethylzinc, and bromoiodomethylzinc carbenoids demonstrated that bromocyclopropanes were always recovered along with the iodocyclopropanes. In order to explain the different behavior between $BrCHI_2$ and $ClCHI_2$ as the carbenoid precursor, quenching experiments were performed, demonstrating the presence of a halogen scrambling that took place on the zinc carbenoid. Moreover, chloroiodomethylzinc carbenoid was more reactive

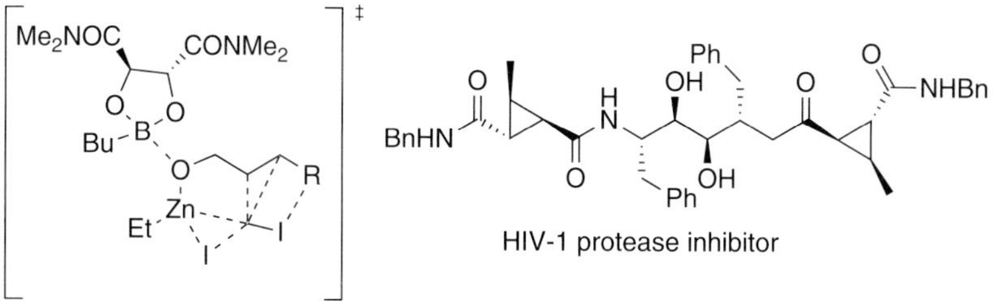

Scheme 1.25 Transition-state model for iodocyclopropanation and protease inhibitor prepared by this method.

than the diiodomethylzinc species, and the latter had reactivity similar to that of bromoiodomethylzinc. The same halogen scrambling was found in the monofluorocyclopropanation reaction [89]. The carbenoid precursor $FCHI_2$ was difficult to be prepared [58], but $ICHF_2$ was readily accessible from inexpensive reagents. The reaction allowed the preparation of monofluorocyclopropanes, where the fluorine atom and the hydroxy group had a *trans*-relationship (Scheme 1.26) [38]. It should be noted that *cis*-relationship could be obtained by an MIRC reaction on α,β-unsaturated amides (see Section 1.4.3). The reaction with secondary allylic alcohols was only performed in diastereoselective manner, that is, without the chiral promoter.

The direct synthesis of cyclopropylmethanols from arylzinc carbenoids was complicated by their instability and by the instability of the parent diiodoarenes. However, Charette's research group overcame this drawback by using diazo reagents as carbenoid precursors, which was a rare application of diazo compounds in a Simmons–Smith cyclopropanation reaction, despite the fact that diazo compounds occupy a vast area of research in cyclopropanation (see Section 1.3) [72f]. The stereo-arrangement of the three substituents on the product (Scheme 1.27),

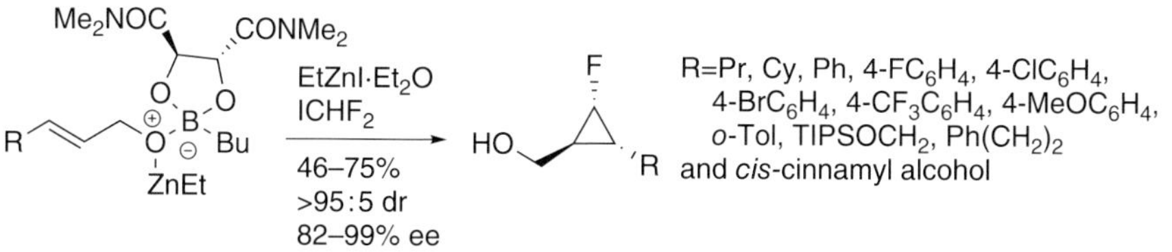

Scheme 1.26 Preparation of monofluorocyclopropanes.

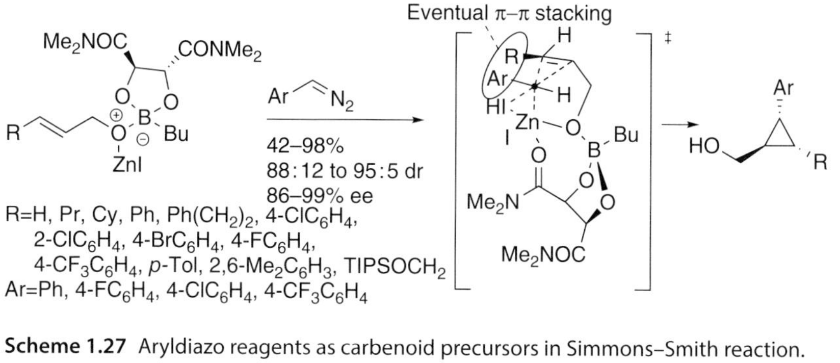

Scheme 1.27 Aryldiazo reagents as carbenoid precursors in Simmons–Smith reaction.

different from that reported in the analogous reaction depicted in Scheme 1.23, is worth of note. The authors proposed a mechanism for this reaction, in which aryldiazomethane reacted with the adventitious zinc iodide to generate the aryl-substituted carbenoid. The carbenoid species was then delivered selectively to one of the two faces of the dioxaborolane–zinc alcoholate complex, in order to minimize the steric repulsion of the directing group. Moreover, in such an assembly, a favorable π-stacking increased the diastereomeric ratios when R was an aryl rather than an alkyl group. In such a reaction pathway, zinc(II) iodide was regenerated, thus, starting from an enantiomerically pure allyl alcohol, the chiral promoter is unnecessary. Actually, 5 mol% of zinc iodide in dichloromethane and sodium hydride (1.0 equiv., to form the alcoholate) allowed the cyclopropanation reaction of in 95% yield, with 80:20 dr and 99% ee [72e].

Charette's chiral dioxaborolane ligand was also employed for the gram-scale synthesis of 3-[(1S,2S)-2-dimethylaminomethylcyclopropyl]-1H-indole-5-carbonitrile hydrochloride (38% overall yield after eight steps, 88% ee), a selective serotonin reuptake inhibitor, demonstrating that the reaction was easily scalable without detriment to the stereoselectivity [72d].

Finally, both Charette's (R,R)- and (S,S)-dioxaborolanes were employed by Kiyota and coworkers in the synthesis of both enantiomers of cyclopropanated analogs of geraniol (62% and 55% yield, >95% ee always, from (R,R)- and (S,S)-ligand, respectively), nerol (74% and 91% yield, >95% ee always), and nor-leaf alcohols (90% and 81% yield and 93% and 95% ee) [90]. It should be noted that cyclopropanation occurred only at the double bond nearest to the alkoxide.

1.2.3.2 Other Stoichiometric Ligands

Charette's group successfully proposed some other chiral promoters. One of these was a chiral 3,3'-disubstituted BINOL-derived phosphoric acid, thus designing a novel chiral zinc phosphate reagent (Scheme 1.28) [91]. The reaction worked well with protected allyl alcohols, while free alcohol gave only 39% ee, albeit in 95% yield. Even when used in catalytic amounts (10 mol%), this chiral phosphoric acid allowed the corresponding chiral cyclopropanes to be obtained in high yield and with up to 95% ee.

Katsuki and coworkers used another binaphthol derivative for the enantioselective cyclopropanation of allylic alcohols (Scheme 1.29) [92]. The scope of the

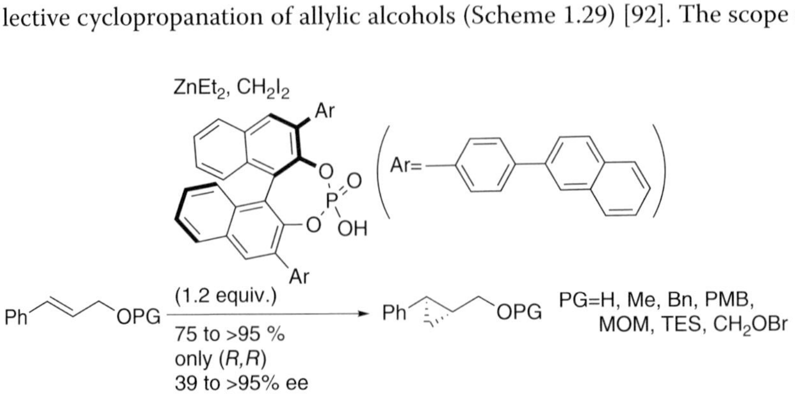

Scheme 1.28 Chiral zinc phosphate reagent for the enantioselective cyclopropanation.

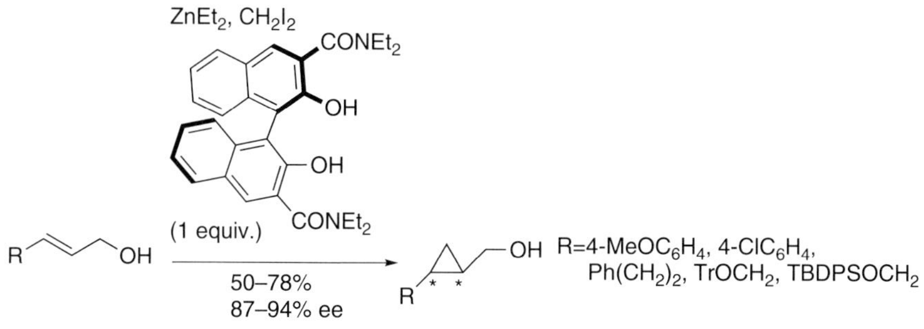

Scheme 1.29 Chiral zinc BINOL reagent for the enantioselective cyclopropanation.

reaction seemed somewhat limited, since only (E)-substituted allylic alcohols could be converted into the corresponding cyclopropanes with reasonably good yields and enantioselectivities, while the only example of (Z)-substituted allylic alcohols gave cyclopropane in only 34% yield and with 65% ee.

Charette's group also developed a family of chiral phosphates derived from TADDOL [93]. One of the members worked even with unfunctionalized olefins, representing the main advantage over the BINOL phosphate (Scheme 1.30).

Moreover, simple chiral dipeptide, N-Boc-L-Val-L-Pro-OMe, combined with $ZnEt_2$ and CH_2I_2, led to an active cyclopropanation system for unfunctionalized olefins with an encouragingly high enantioselectivity (Scheme 1.31) [94]. Then, the same authors found that catalytic amounts were enough for obtaining enantioselectivity (see Section 1.2.3.4).

Since zinc reagents modified by a covalent ligand ($RXZnCH_2I$) are effective for cyclopropanation, a number of chiral alcohols were tested using trans-β-methylstyrene as a substrate, in order to induce enantioselectivity in cyclopropanation

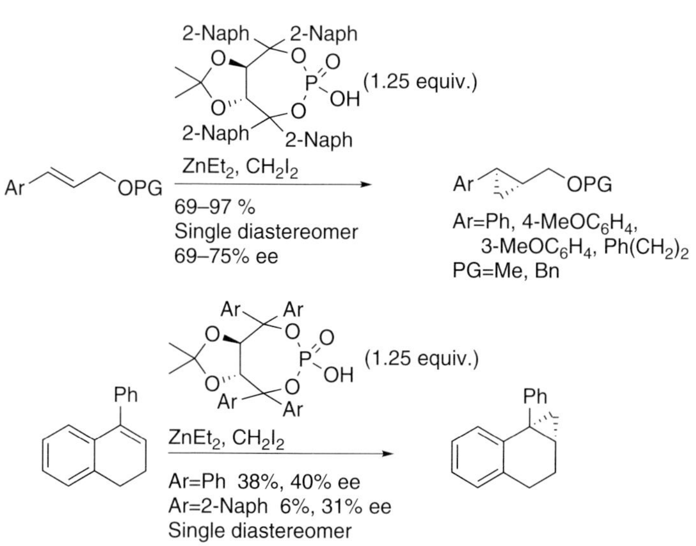

Scheme 1.30 Chiral zinc TADDOL reagents for the enantioselective cyclopropanation.

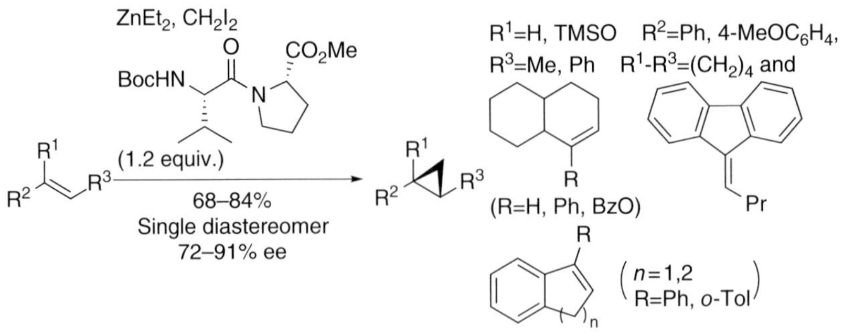

Scheme 1.31 Chiral zinc dipeptide reagent for the enantioselective cyclopropanation.

reactions [95]. Generally, the cyclopropanations were very sluggish, but they were accelerated by the addition of a catalytic amount of Et$_2$AlCl. The best enantioselectivity (51% ee) was achieved for the cyclopropane product using fructose-derived alcohol (Scheme 1.32).

1.2.3.3 Walsh' Procedure

In 2005, Walsh and coworkers proposed a highly enantio- and diastereoselective generation of cyclopropyl alcohols [96]. This methodology consisted of an initial enantioselective C—C bond formation generating a key allylic zinc alkoxide intermediate by asymmetric alkyl addition to α,β-unsaturated aldehydes in the presence of a catalytic amount of (−)-MIB. Then a diastereoselective cyclopropanation affords the corresponding cyclopropyl alcohols having up to four stereogenic centers with very high ee and de (Scheme 1.33). Actually, this methodology is not a strictly asymmetric Simmons–Smith cyclopropanation reaction. In fact, it adopts an asymmetric organocatalyzed synthesis of allyl alcohol, and the Simmons–Smith reaction works on an enantioenriched substrate. This addition/cyclopropanation sequence gives similar stereoselectivities to the cyclopropanation of isolated chiral allylic alcohols, but it is more efficient. As well as Charette's dioxaborolane, the scope of Walsh' methodology was extended to the synthesis of chiral halocyclopropyl alcohols by using, for instance, iodoform in place of diiodomethane as the carbenoid precursor [96]. Then the asymmetric bromo- and chlorocyclopropanation furnished halocyclopropyl alcohols with 89–99% ee and >95:5 dr but only in 56–80% yield [97].

A second tandem addition/cyclopropanation sequence based on hydroboration/transmetallation method was developed by the same group, generating a

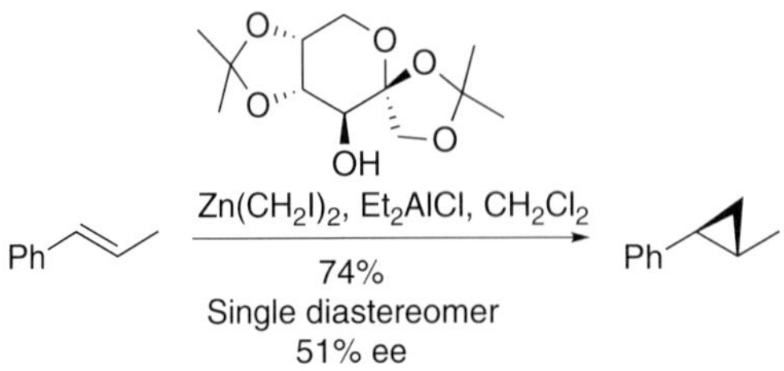

Scheme 1.32 Sugar zinc reagent for the enantioselective cyclopropanation.

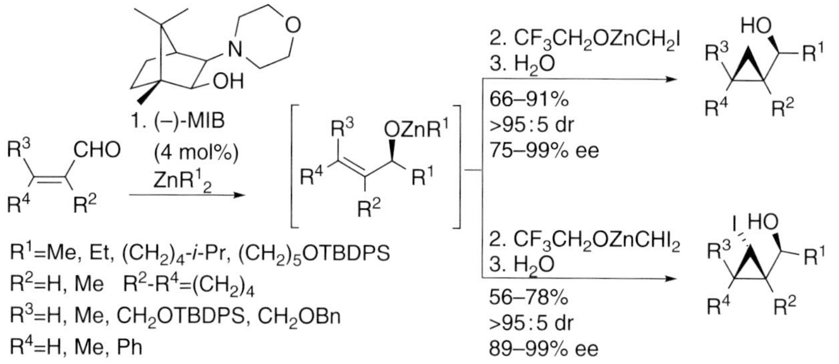

1 Asymmetric Cyclopropanation

Scheme 1.33 Walsh' procedure.

vinyl zinc reagent. In the presence of (−)-MIB, vinylation of an aldehyde proceeded to furnish the allylic alkoxide intermediate necessary for the cyclopropanation (Scheme 1.34) [96]. However, these substrates gave generally low conversions and variable diastereoselectivities, but later, Kim and Walsh were able to overcome this drawback and accelerate the cyclopropanation reaction rate by adding trifluoroacetic acid, minimizing the potential zinc(II)–π interactions [98]. This alternative procedure produced iodocyclopropyl alcohols in 50–87% yields with 86–99% ee and >95:5 dr [97]. The different configuration between iodocyclopropanes obtained from the two alternative ways is worth of note, although the *syn*-relationship between OH group and cyclopropyl ring is maintained.

Walsh' procedures were then applied to the synthesis of the following:

i) *syn*-Vinylcyclopropylalkanols, in which the carbenoid only reacted at the double bond nearest to the alkoxide (Scheme 1.35, eq 1) [99]
ii) *syn-cis*-Disubstituted cyclopropylalkanols in an eight-step *in situ* procedure from chloroalkynes (Scheme 1.35, eq 2) [99]
iii) *anti*-Cyclopropylmethanols from silyl ethers, in order to prevent *syn*-cyclopropanation typical of free alcohols (Scheme 1.35, eq 3) [99]

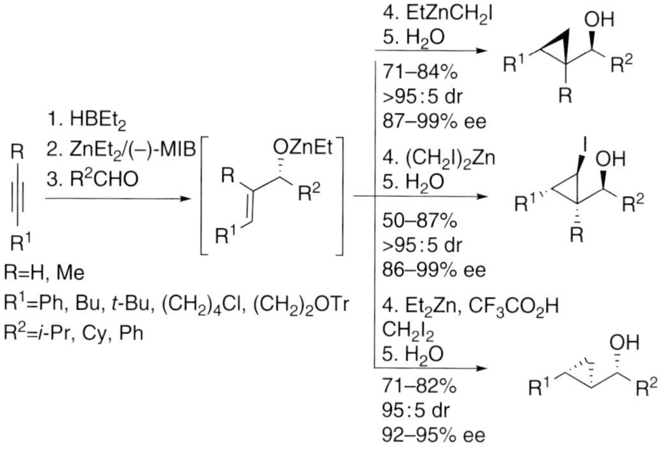

Scheme 1.34 Tandem addition/cyclopropanation sequence by Walsh' procedure.

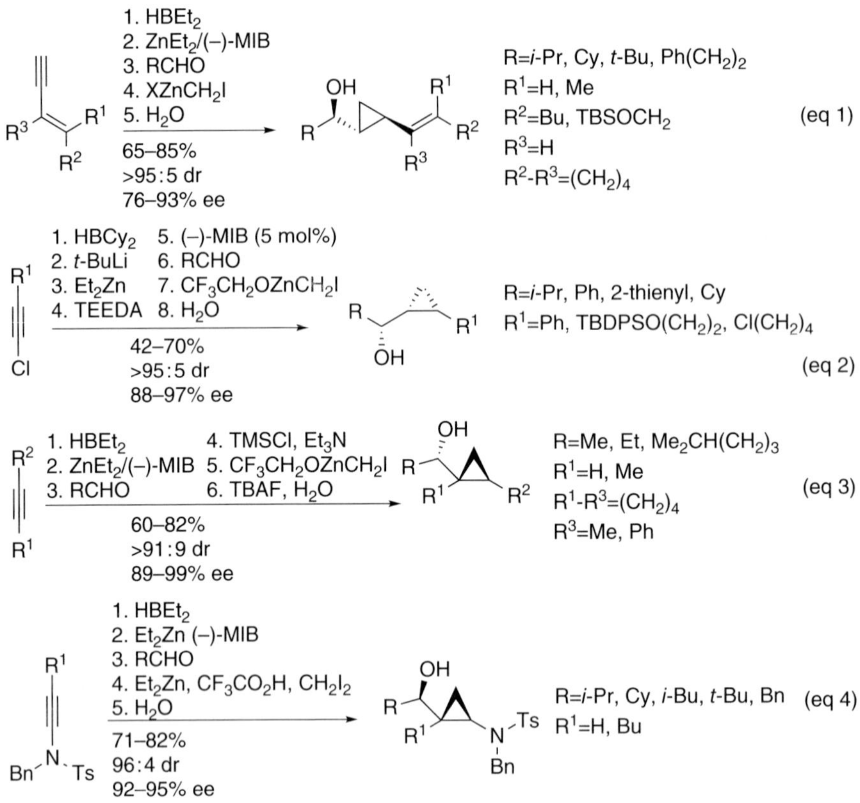

Scheme 1.35 Other applications of Walsh' procedure.

  iv) Aminocyclopropylmethanols with three contiguous stereocenters in a five-step one-pot procedure from ynamides (Scheme 1.35, eq 4) [100]
  v) Fluorocyclopropane by Charette's group [38]. They found that with increasing bulkiness of the zinc reagent, the yield declined, but diastereoselectivity increased and enantioselectivity did not show a clear trend.

In 2013, a chiral perhydrobenzoxazine was proposed for the enantio- and diastereoselective one-pot ethylation/or arylation/cyclopropanation reaction of α,β-unsaturated aldehydes instead of (−)-MIB (Scheme 1.36) [101]. The catalytic system tolerated a wide range of di- and trisubstituted enals, affording the corresponding *syn*-hydroxycyclopropanes. The presence of substituents at the α-position of the enal bulkier than methyl constituted a limitation to the substrate scope. The use of 1,1-diiodoethane afforded the corresponding cyclopropanes, with good enantiocontrol, although moderate diastereoselectivity.

1.2.3.4 True Catalytic Procedures

Only in the past decade, organocatalysis has been developed in Simmons–Smith reaction. For instance, as shown in Scheme 1.31, the cyclopropanation promoted by chiral dipeptide, *N*-Boc-L-Val-L-Pro-OMe, combined with ZnEt$_2$ and CH$_2$I$_2$,

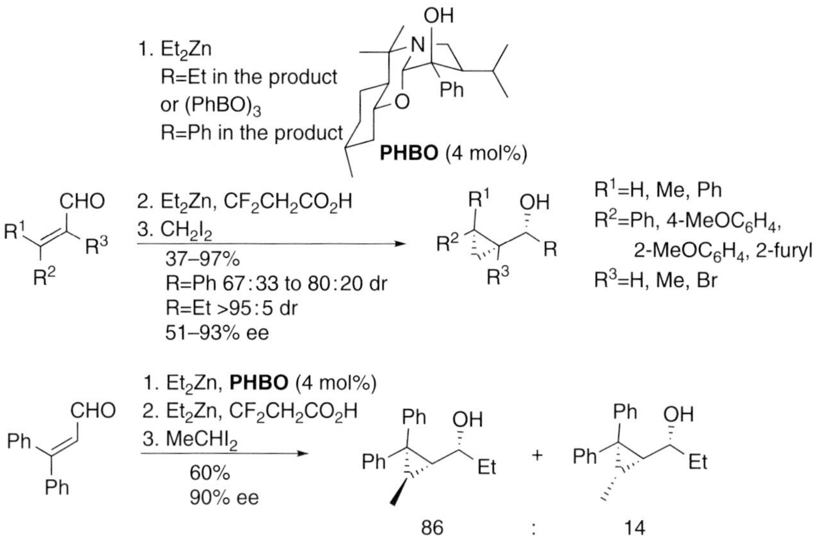

Scheme 1.36 Chiral perhydrobenzoxazine for the enantioselective cyclopropanation.

is described. In the presence of catalytic amounts (25 mol%), the enantioselectivity was decreased, due to an enhanced background reaction from Zn(CH$_2$I)$_2$. However, an achiral additive (ethyl methoxyacetate) that is able to coordinate Zn(CH$_2$I)$_2$ could reduce the background reaction, and the corresponding chiral cyclopropanes were obtained with comparable ee values to those obtained by the original stoichiometric procedure [102]. In 2006, the same group reported the use of another chiral dipeptide as ligand to promote the asymmetric Simmons–

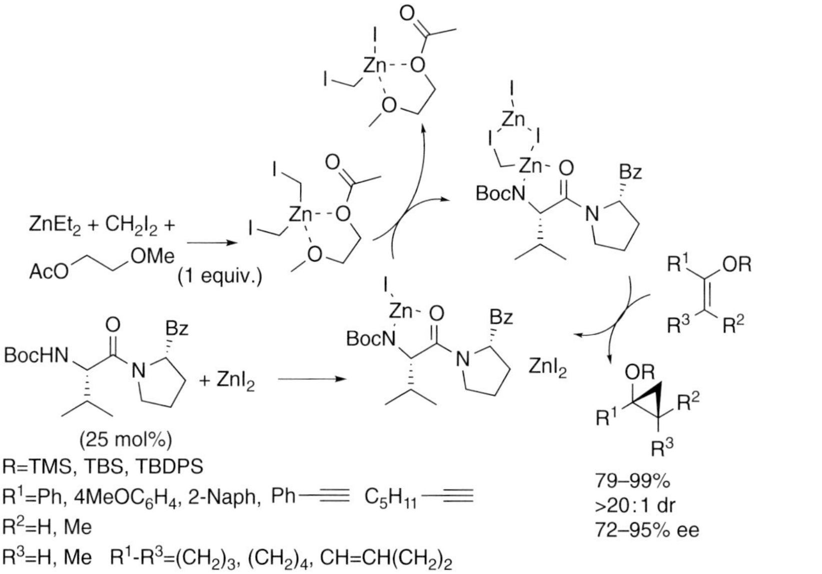

Scheme 1.37 Chiral zinc dipeptide catalyst for the enantioselective cyclopropanation.

Smith cyclopropanation of silyl enol ethers in the presence of ethyl methoxyacetate as additive (Scheme 1.37) [103].

In 2009, the same authors investigated the mechanism of the reaction in order to interpret the factors influencing the experimentally observed enantioselectivity [104]. They found that asymmetric cyclopropanation was achieved only if the organozinc compound deprotonated the N–H of the dipeptide ligand and that zinc(II) iodide played an important role in promoting the cyclopropanation. From these features, they envisaged the catalytic cycle as depicted in Scheme 1.37. Then, Zheng and Zhang, using B3LYP hybrid density functional methods, further supported the experimental evidence reported by Shi [105]. In fact, theoretical calculations allowed an interpretation of the enantioselectivity according to the "three-point contact model" involving ring formation and two steric repulsions.

A catalytic amount of (S)-phenylalanine-derived disulfonamides was successfully employed in the cyclopropanation reaction of a range of 3,3-diaryl-2-propen-1-ols in the presence of Et_2Zn and CH_2I_2, providing the corresponding cyclopropylmethanols with moderate-to-good enantioselectivity (Scheme 1.38) [106]. The chiral cyclopropane derived from the appropriate allyl alcohol was further converted into (+)-cibenzoline, an antiarrhythmic agent [107], (+)-tranylcypromine, a strong monoamineoxidase inhibitor, and (−)-milnacipran, a serotonin–noradrenaline reuptake inhibitor [108].

Shitama and Katsuki reported a Simmons–Smith cyclopropanation by asymmetric catalysis with an aluminum complex (10 mol%, Scheme 1.39a), in which the aluminum atom and a nitrogen atom from the ligand acted as the Lewis acid and the Lewis base site, respectively (Scheme 1.39b) [109]. Unfortunately, good substrates were limited to *trans*-disubstituted allylic alcohols [92–99% isolated yield with 63–94% ee of the (S,S)-cyclopropylmethanol], while the enantioselectivity of the reaction of (Z)-cinnamyl alcohol was moderate (58% ee of the R,S-isomer). Another enantioselective cyclopropanation of cinnamyl alcohol (93% yield, 53% ee of the R,R-isomer) has been accomplished in the presence of chiral sulfonamides derived from L-proline and L-proline methyl ester (Scheme 1.39c) [110].

Finally, the asymmetric Simmons–Smith cyclopropanation of allylic alcohols was also performed in the presence of an enantiomerically pure tridentate catalysts,

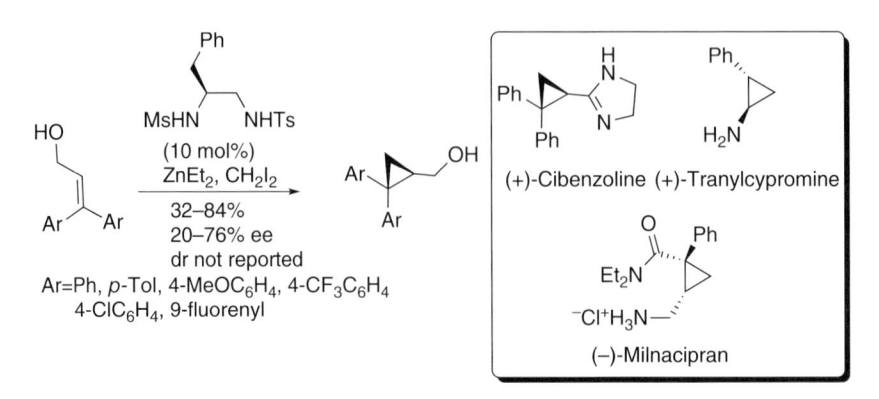

Scheme 1.38 (S)-Phenylalanine-derived disulfonamides as the organocatalysts in cyclopropanation reactions.

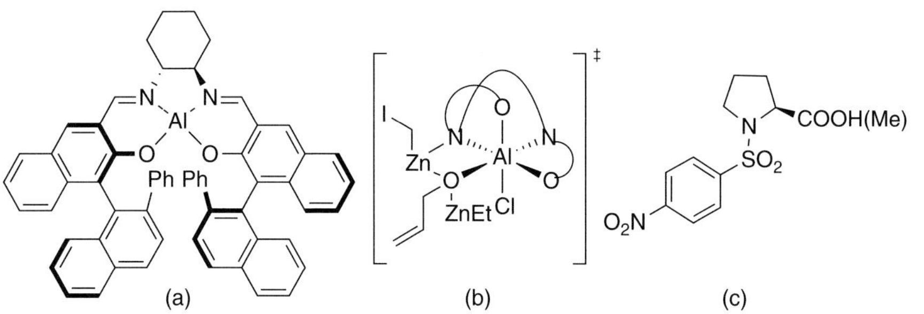

Scheme 1.39 Aluminum complexes and chiral sulfonamides derived from L-proline for catalytic Simmons–Smith cyclopropanation.

bearing hydroxy, stereogenic sulfinyl, and chiral aziridine moieties. It was found that both stereo-defined groups synergically exerted asymmetric induction on the product, but that only inversion of the aziridine configuration afforded the enantiomeric cyclopropane (Scheme 1.40) [111].

1.3 Transition-Metal-Catalyzed Decomposition of Diazoalkanes

Since the pioneering work of Nozaki *et al.* in 1966 [112], the transition-metal-catalyzed cyclopropanation of alkenes with diazo compounds has emerged as one of the most highly effective and stereocontrolled routes to functionalized cyclopropanes. The diasterocontrol in the cyclopropanation is often governed by the particular substituents on both the alkene and the diazo compounds, and thus, the catalyst must be cleverly designed in order to enhance selective formation of *cis* versus *trans* or *syn* versus *anti*-cyclopropanes. As already seen in the previous section, the most ancient attempts to achieve enantioenriched cyclopropanes used chiral auxiliaries. Since the 1990s, many chiral ligands surrounding the metal center of the catalyst have been introduced for obtaining the enantiocontrol. The accepted catalytic cycle of the carbenoid cyclopropanation reaction involves interaction of the catalyst with the diazo precursor to afford a metallo-carbene complex followed by transfer of the carbene species to the alkene (Scheme 1.41).

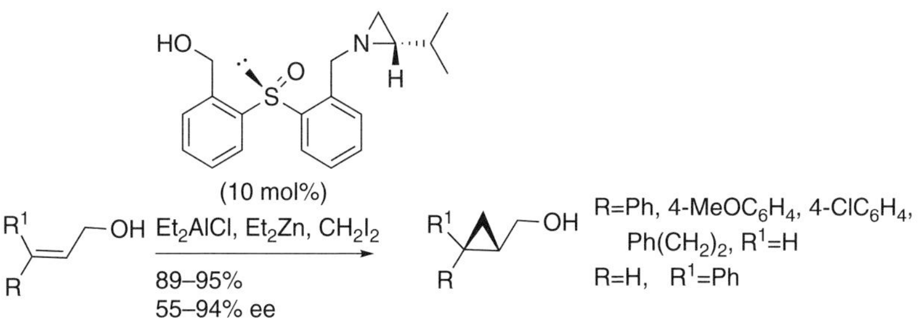

Scheme 1.40 Tridentate catalyst bearing hydroxy, sulfinyl, and aziridine moieties.

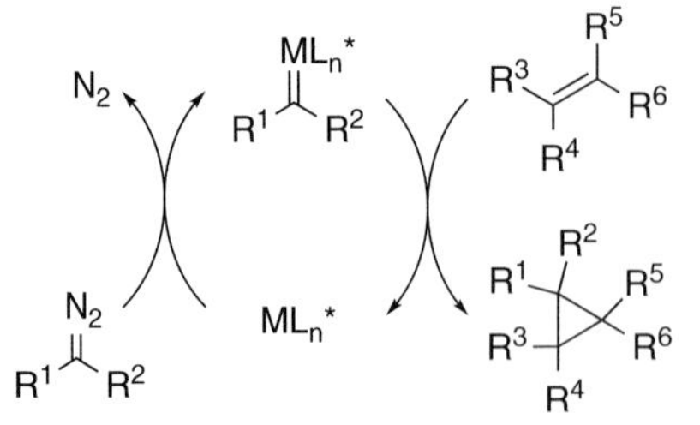

Scheme 1.41 Accepted catalytic cycle for the carbenoid cyclopropanation reaction.

The type of the reaction to be carried out (inter- vs intramolecular) plays a key role in the appropriate selection of the most efficient catalyst for a given transformation. In light of this, this section is divided into inter- and intramolecular cyclopropanation reactions, and in each subsection, chiral auxiliaries are described before and then chiral ligands are listed according to the involved metal ion.

1.3.1 Intermolecular Cyclopropanation

1.3.1.1 Chiral Auxiliaries

Chiral cyclic alkenes with sterically demanding groups near the olefin undergo cyclopropanation with good diasterocontrol under Pd catalysis or by 1,3-dipolar cycloaddition with diazomethane and nitrogen extrusion (Scheme 1.42) [113].

Moreover, in the asymmetric cyclopropanation of α'-acetoxy-α,β-unsaturated cyclopentanone and cyclohexanone, the five-membered-ring enone afforded only the *anti* diastereomer in 98% yield, whereas the six-membered enone afforded both the *syn*- and *anti*-diastereomers (*syn:anti* = 63:34) [114].

In the course of developing total asymmetric syntheses of pleocarpenene and pleocarpenone, Snapper's group observed a considerable stereochemical control in the cyclopropanation/deacetylation reaction of a chiral cyclobutene with ethyl diazoacetate (EDA), using $Cu(acac)_2$ as the catalyst (Scheme 1.43) [115].

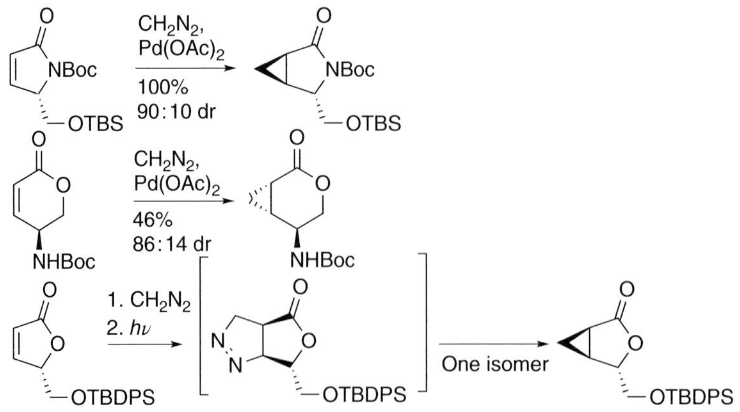

Scheme 1.42 Cyclopropanation under Pd-catalysis and 1,3-dipolar cycloaddition with diazomethane.

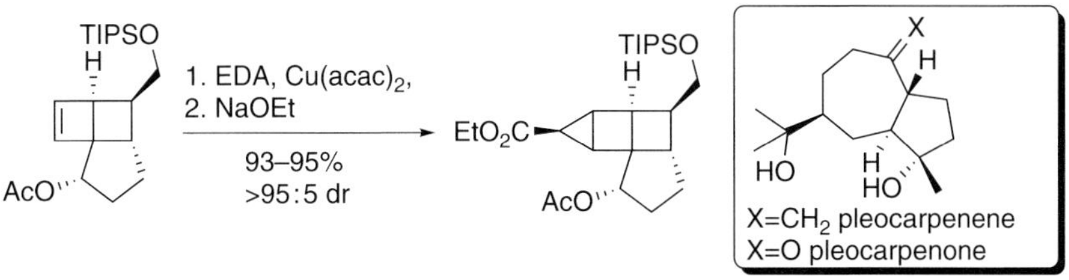

Scheme 1.43 EDA cyclopropanation/deacetylation reaction of chiral cyclobutene.

Finally, trifluoromethylcyclopropanes substituted with carbohydrate analogs were prepared by the addition of diazo compounds to 5-(R)-hydroxypyridazin-3-one derivative of D-(+)-galactose in high *syn*-selectivity. The observed π-facial selectivity was explained by taking into account the steric interaction and electronic factors of C5-OH atom with the CF_3 group of the diazo compound (Scheme 1.44) [116].

On the other hand, if the cyclopropanations of chiral acyclic alkenes with scarcely demanding chiral auxiliaries usually did not proceed with a high level of stereocontrol [117], the 1,3-dipolar cycloaddition of diazomethane to chiral acyclic alkenes led to cyclopropanes often as a single diastereomer (Scheme 1.45) [118].

A number of chiral auxiliaries were developed for the cyclopropanation of acyclic alkenes. For example, cinnamaldehyde, upon treatment with ephedrine, produced single oxazolidine diastereomer, then cyclopropanation also gave a single diastereomer; finally, the auxiliary was cleaved by treatment with silica gel [119]. Unfortunately, this auxiliary was tested only with cinnamaldehyde. Oppolzer's chiral sultam was also employed instead of ephedrine (Scheme 1.46) [120]. Although the diastereomeric excesses were quite modest, the diastereomeric products could be recrystallized to remove the minor isomer (from 98:2 dr to almost exclusively

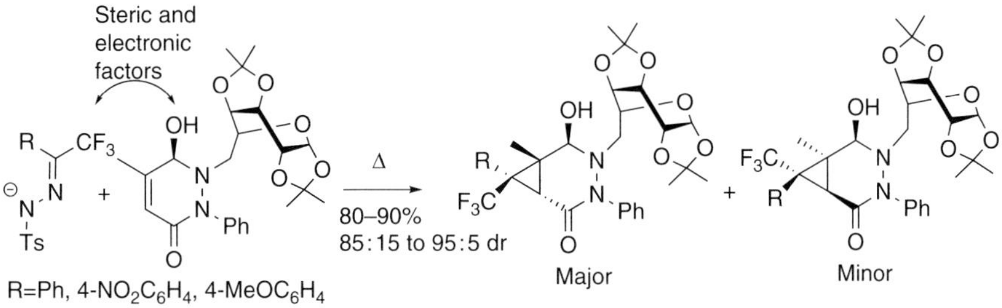

Scheme 1.44 Addition of diazo compounds with high *syn*-selectivity.

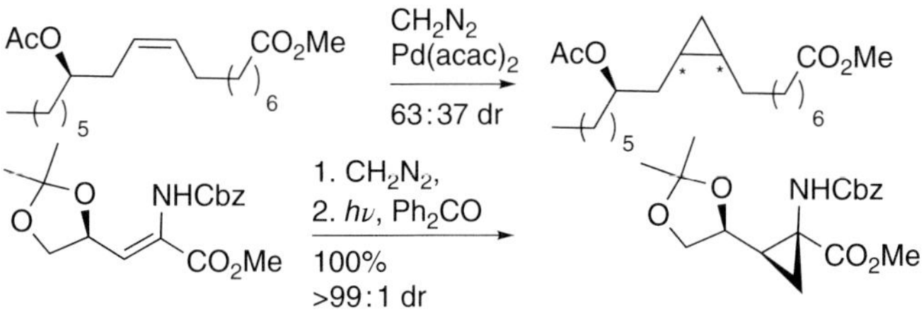

Scheme 1.45 1,3-Dipolar cycloaddition of diazomethane to chiral acyclic alkenes.

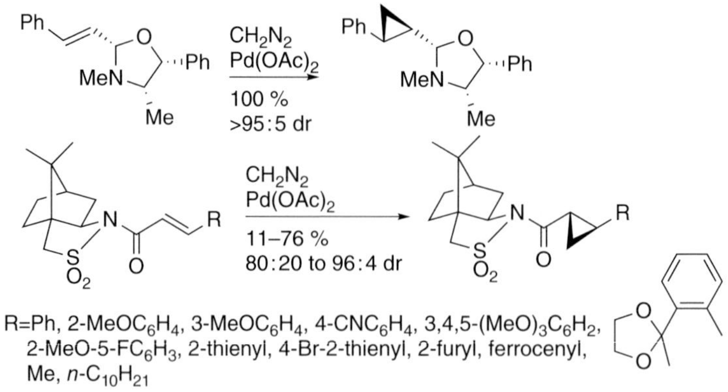

Scheme 1.46 Oppolzer's chiral sultam in cyclopropanation reaction.

one diastereomer). This synthesis was then applied to an important intermediate of novel melatonergic agents [121].

Pietruszka and coworkers prepared chiral cyclopropylboronic acid by a lithium–aluminum–hydride reduction, followed by hydrolysis of the resulting borohydride. This step limits the tolerance of substituents on the substrate. However, several derivatives were cyclopropanated with good diastereoselectivities, and diastereomeric products were easily separated by chromatography (Scheme 1.47) [122]. The cyclopropanations of the corresponding *cis*-isomer were tested, but the diastereoselectivities were significantly lower (65:35 to 87:13) [122g]. More recently, Florent and coworkers revisited the reaction of a (Z)-alkenylboron compound and, although the diastereomeric ratio between the two cyclopropane derivatives was not reported, more than 96% optical purity was achieved after removal of the chiral auxiliary [123].

Other chiral auxiliaries that allow control of the facial selectivity of the 1,3-dipolar cycloaddition of diazomethane have also been developed for specific substrates with dr always >95:5 (Scheme 1.48) [124]. However, these procedures often suffered from the stereoselective preparation of the alkene precursor or the photoinduced nitrogen extrusion.

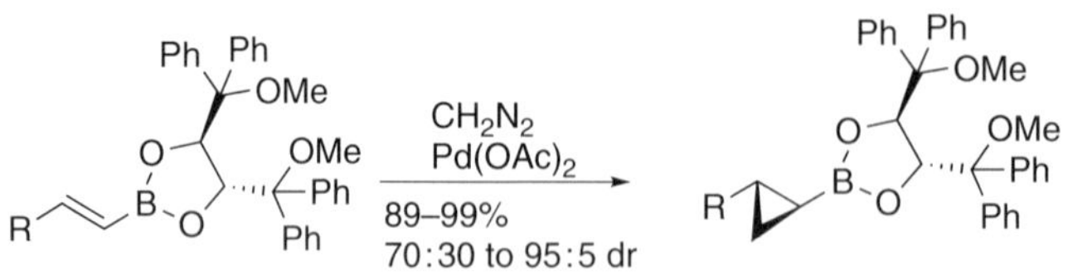

Scheme 1.47 Synthesis of cyclopropanes through a chiral boronic acid derivative.

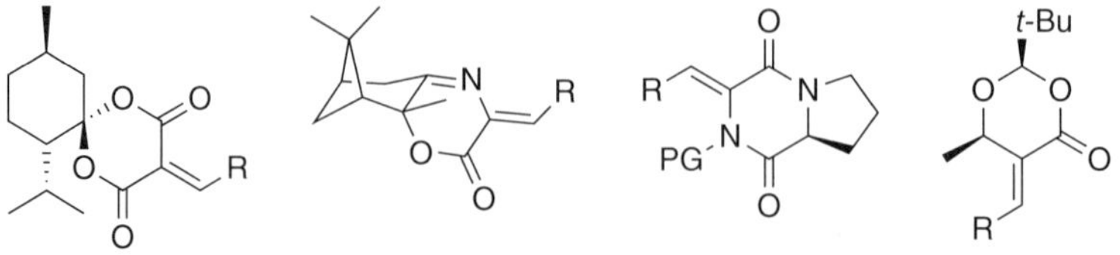

Scheme 1.48 Different chiral auxiliaries for the 1,3-dipolar cycloaddition of diazomethane.

In addition, carbohydrate derivatives have been employed as chiral auxiliaries in asymmetric cyclopropanation. However, the relatively few examples as chiral ligands in the Cu-catalyzed reactions of olefins with diazoacetates showed generally low E/Z ratios and enantioselectivities (see Section 1.3.1.3) [125]. Interestingly, Ferreira and coworkers reported the simultaneous use of an α-diazoacetate with a carbohydrate-derived chiral auxiliary and a chiral Cu(I) catalyst to induce chirality in the cyclopropanation reaction (Scheme 1.49) [126]. These authors studied the role of both the chiral auxiliary and the ligand, showing the remarkable importance of the carbohydrate-based chiral auxiliary on the enantioselectivities and the unexpected effect of the ligand on the E/Z ratios.

Finally, stable optically pure phosphino(silyl)carbenes, derived from photolysis of the corresponding diazo compounds, were cyclopropanated with methyl acrylate with high stereoselectivity [127]. A total *syn*-diastereoselectivity (with respect to the phosphino group) was observed (Scheme 1.50).

The cyclopropanation of TBS-protected D-glucal as well as other protected glycals in the presence of $Rh_2(OAc)_4$ as the catalyst was among the most ancient methods using chiral auxiliaries and rhodium catalysts (Scheme 1.51) [128].

Rhodium complexes were also found to be the best catalysts for the decomposition of aryl and vinyldiazoesters in the presence of alkenes, leading to the

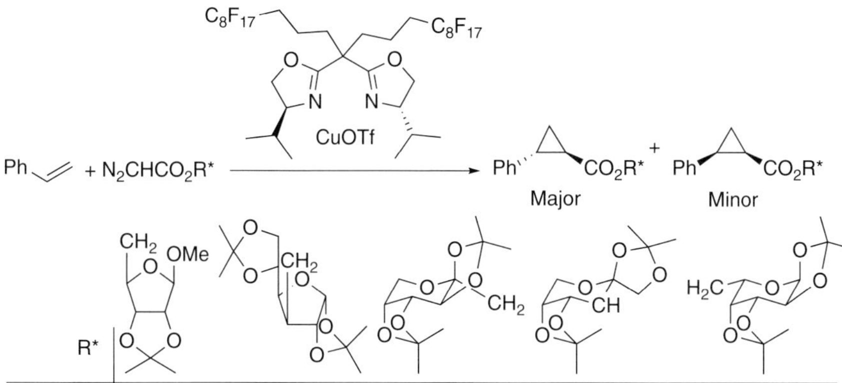

Yield (%)	30	25	62	35	15
de (%)	90	60	50	92	79
ee$_{trans}$ (%)	60	19	<10	34	2
ee$_{cis}$ (%)	53	26	<10	2	92

Scheme 1.49 Cyclopropanation reaction by the use of α-diazoacetate with a carbohydrate-derived chiral auxiliary and a chiral Cu(I) catalyst.

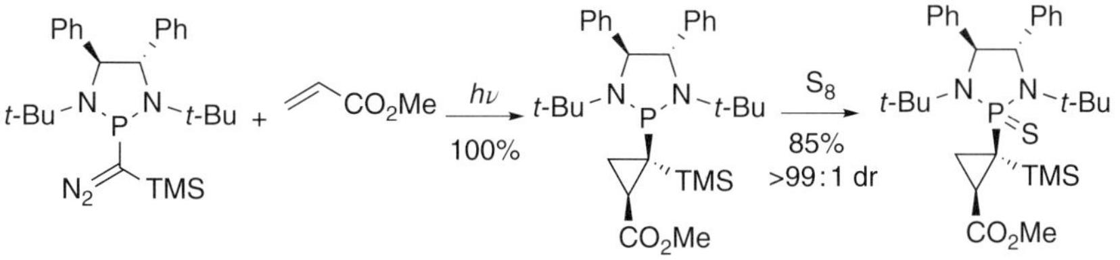

Scheme 1.50 Cyclopropanation with phosphino(silyl)carbenes from photolysis of the corresponding diazo compounds.

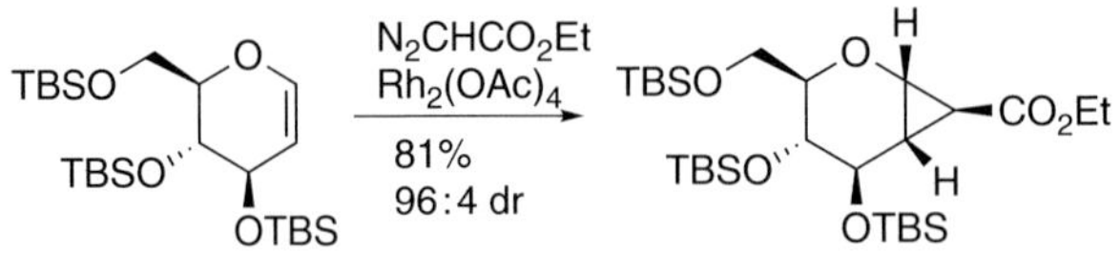

Scheme 1.51 Glucal as chiral auxiliary for Rh-catalyzed cyclopropanation.

corresponding cyclopropanes with a high level of diasterocontrol, and the introduction of a simple chiral auxiliary on the ester moiety generates an enantiomerically enriched product (Scheme 1.52) [129], except for the intermolecular cyclopropanation of styrene [130].

1.3.1.2 Chiral Catalysts: Cobalt

Cobalt complexes have been shown to be reactive catalysts for the α-diazoester decomposition, leading to a metal carbene that could convert alkenes to cyclopropanes. In 2009, Doyle published a highlight article collecting what was known on this topic [131]. The mechanism of this reaction was examined by EPR and electrospray ionization–mass spectrometry (ESI-MS) techniques, especially when cobalt–porphyrin catalysts were used, and evidence for a two-step mechanism was uncovered (Scheme 1.53) [132].

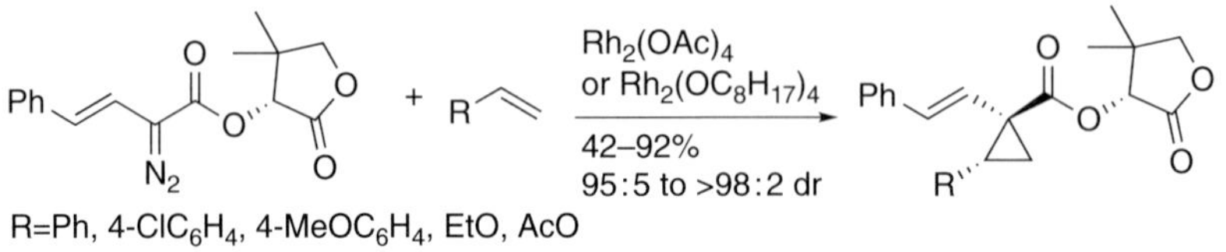

Scheme 1.52 Chiral auxiliary in the cyclopropanation of rhodium-catalyzed vinyldiazoester decomposition.

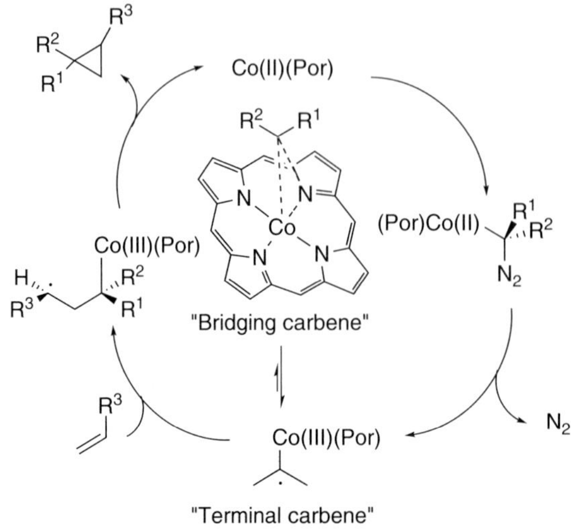

Scheme 1.53 Mechanism of cobalt–porphyrin catalysis.

The first step is an adduct formation that could exist as two isomers: the "terminal carbene" and the "bridging carbene." In the former, the "carbene" behaves as a redox noninnocent ligand having a d^6 cobalt center and the unpaired electron resides on the "carbene" carbon atom. In the latter, the "carbene" is bound to the metal and one of the pyrrolic nitrogen atoms of the porphyrin. DFT calculations suggested that the formation of the carbene is the rate-limiting step and that the cyclopropane ring formation proceeds by way of a stepwise radical process. Conclusive evidence for the existence of cobalt(III) carbene radicals has been obtained [133]. In fact, in the absence of the alkene substrate, the "terminal carbene" arising from diazoacetate dimerizes to afford binuclear cobalt(III)–porphyrin complex, characterized by X-ray structural analysis. DFT calculations prove the inability of the "terminal carbene" to abstract hydrogen atoms from the solvent, and the radical arising from allylic resonance can be trapped by TEMPO [133, 134]. Other calculations confirmed that the "terminal carbene" complex has a single bond from the metal to the carbon atom and radical character with localized spin density on the carbon. In addition, the carbon is nucleophilic in character and "tunable" through the introduction of different R substituents in order to achieve the desired reactivity. Based on these findings, rational design strategies to enhance catalytic activity can be proposed [135], highly increasing the low level of diastereo- and enantiocontrol of the early work in this area [136]. Applications of cobalt porphyrins in diastereo-and enantioselective cyclopropanation reactions are listed in Table 1.2.

Since 2003 [145], Zhang's research group used a series of novel *meso*-chiral cobalt–porphyrin complexes to catalyze the cyclopropanation of styrene with EDA, affording the desired cyclopropane ester as a *trans*-dominant form in excellent yields [146]. However, only low enantioselectivities were observed (≤12% ee), because of the orientation and flexibility of the chiral appendages. In addition, these authors found that similar reactions could be efficiently catalyzed by vitamin B_{12} derivatives such as aquocobalamin [147].

Sometimes, the use of DMAP as an additive allowed the *trans*- and enantioselectivities to be increased, suggesting a significant *trans* influence of potential coordinating ligands on the metal center [148]. Moreover, the high stereochemical outcome was explained through the potential hydrogen bonding interactions between the chiral amide N–H donor and acceptor units on the "terminal carbene" intermediate. The enantioselectivity was also improved by the use of bulkier ligands, for instance, bearing a *meso*-2,6-dimethoxyphenyl group and by further tightening the structure by intramolecular O···H—N hydrogen-bonding interactions. All of these interactions rigidified the intermediate toward its subsequent reaction with the alkene and thus led to a more selective catalytic process.

It should be noted that addition of α-nitrodiazoacetates to alkenes was reported to afford cyclopropanes as atypical dominant *Z*-isomers [140]. Conversely, rhodium-based chiral catalysts provided predominantly the *E*-isomers (see Section 1.3.1.4) [149].

Cobalt-salens and analogs are other potential efficient catalysts for asymmetric cyclopropanations (Table 1.3), since Yamada showed that 3-oxobutylideneaminatocobalt(II) complexes were quite effective in a *trans*-selective reaction [150]. The addition of a catalytic amount of *N*-methylimidazole (NMI) often

Table 1.2 Cobalt porphyrins in enantioselective cyclopropanation reactions.

$$R^1\text{-}C(N_2)\text{-}XO_2R + R^2\text{-}CH=CH\text{-}R^3 \longrightarrow \text{cyclopropane}(R^3, XO_2R, R^2, R^1)$$

CoP1: X=*t*-Bu; Y=H Z= 3,5-(*t*-Bu)$_2$C$_6$H$_3$
CoP2: X=*t*-Bu; Y=H Z= (cyclopropyl group)
CoP3: X=H; Y=OMe Z= (cyclopropyl group)
CoP4: X=H; Y=OMe Z= (tetrahydrofuranyl group)
CoP5: X=*t*-Bu; Y=H ent-Z= (cyclopropyl-Ph group)

Cobalt porphyrin	Reactants	Yield (%)	dr	ee (%)	References
CoP1 (1 mol%) DMAP (100 mol%)	X=C; R=Et, *t*-Bu; R^1=R^3=H; R^2=Ph	86–88	97:3 to >99:1 (E/Z)	78–95 (E)	[137]
CoP2 DMAP	X=C; R=Et, *t*-Bu; R^1=H; R^2=Ph, C$_6$F$_5$, 4-MeOC$_6$H$_4$, *p*-Tol, *m*-Tol, *o*-Tol, 4-*t*-BuC$_6$H$_4$, 4-BrC$_6$H$_4$, 4-ClC$_6$H$_4$, 4-FC$_6$H$_4$, 4-CF$_3$C$_6$H$_4$, 4-AcOC$_6$H$_4$, 2-Naph; R^3=H, Me, Ph	60–95	93:7 to >99:1 (E/Z)	68–98 (E)	[138]
CoP2 DMAP	X=C; R=Et, *t*-Bu; R^1=H; R^2=Me, Et, *n*-C$_5$H$_{11}$, CN, CO$_2$Et, CO$_2$*t*-Bu, CONH$_2$, CONMe$_2$, CONH*i*-Pr; R^3=H, CO$_2$Me, COMe, CN	40–94	62–38 to 99:1 (E/Z)	61–97 (E)	[139]
CoP2 (1 mol%)	X=C; R=Et, *t*-Bu; R^1=NO$_2$; R^2=Ph, *p*-Tol, *m*-Tol, *o*-Tol, 4-*t*-BuC$_6$H$_4$, 4-BrC$_6$H$_4$, 4-ClC$_6$H$_4$, 4-FC$_6$H$_4$, 4-CF$_3$C$_6$H$_4$, 3-NO$_2$C$_6$H$_4$, C$_6$F$_5$, Bu, Ph(CH$_2$)$_2$, CO$_2$Me, CO$_2$Et; R^3=H	42–97	10:1 to 99:1 (Z/E)	75–95 (Z)	[140]
CoP2 (5 mol%) DMAP (50 mol%)	X=C; R=Su; R^1=H; R^2=Ph, *p*-Tol, 4-*t*-BuC$_6$H$_4$, 4-MeOC$_6$H$_4$, 4-ClC$_6$H$_4$, 4-AcOC$_6$H$_4$, 4-CF$_3$C$_6$H$_4$, 3-NO$_2$C$_6$H$_4$, 2-Naph, CO$_2$Et, CONMe$_2$, Ac; R^3=H	33–90[a]	49:1 to >99:1 (E/Z)	89–98 (E)	[141]
CoP4 (1 mol%)	X=S; R=Me, *p*-Tol, 4-NO$_2$C$_6$H$_4$; R^1=H; R^2=Ph, 4-*t*-BuC$_6$H$_4$, 4-MeOC$_6$H$_4$, 4-CF$_3$C$_6$H$_4$, 3-NO$_2$C$_6$H$_4$, 2-Naph, CO$_2$Me, CO$_2$Et, CN, Ac; R^3=H	64–99	99:1 (E/Z)[b]	89–97 (E)[b]	[142]

Table 1.2 (Continued)

Cobalt porphyrin	Reactants	Yield (%)	dr	ee (%)	References
CoP2 (1 mol%)	X=C; R=t-Bu; R^1=CN; R^2=Ph, 4-MeOC$_6$H$_4$, 4-CF$_3$C$_6$H$_4$, 3-NO$_2$C$_6$H$_4$, C$_6$F$_5$, CO$_2$Me, CO$_2$Et, CONMe$_2$, CONH$_2$, Ac, CN, Bu, n-Hex, Ph(CH$_2$)$_2$, AcO, t-BuCO$_2$; R^3=H	72–99	>99:1 (E/Z)^c	71–99 (E)	[143]
CoP2 (5 mol%)	X=C; R=t-Bu; R^1=Ac; R^2=Ph, p-Tol, 4-MeOC$_6$H$_4$, 4-CF$_3$C$_6$H$_4$, 4-BrC$_6$H$_4$, 3-NO$_2$C$_6$H$_4$, C$_6$F$_5$, 3-CHOC$_6$H$_4$, N-Boc-Indol-3-yl, PhCH=CH, EtO$_2$CCH=CH, CO$_2$Me, CN, C$_6$H$_{13}$ R^3=H, Me	46–94	95:5 to >99:1 (E/Z)	60–99 (E)	[144]

CoP6: n=1
CoP7: n=2

Cobalt porphyrin	Reactants	Yield (%)	dr	ee (%)	References
CoP6 (0.5 mol%)	X=C; R=Et; R^1=H; R^2=Ph, p-Tol, 4-ClC$_6$H$_4$, CH$_2$=C(Me); R^3=H, Me, Ph^d	85–99	2:1 to 5.25:1 (E/Z)	8–90 (Z) 12–71 (E)	[134]
CoP7 (0.5 mol%)	X=C; R=Et; R^1=H; R^2=Ph, p-Tol, 4-ClC$_6$H$_4$, CH$_2$=C(Me); R^3=H, Me, Ph^d	85–98	2:1 to 8:1 (E/Z)	5–80 (Z) 8–75 (E)	[134]

a) The best enantiomeric excess value was attained with **CoP2**, but the use of **CoP1** afforded the best yield.
b) Except for acrylonitrile, where dr = 79:21 and ee = 61%. **CoP4** led to the 1R,2S-isomer as the main product (where C-1 is the carbon carrying the sulfonic moiety). With catalysts **CoP1** and **CoP2**, the 1S,2R-enantiomer was recovered, albeit in low enantiomeric excess.
c) Except for acrylonitrile, where dr = 74:26. Cyclopropanes could be chemoselectively reduced with full retention of configuration to the corresponding primary alcohols or primary amine without affecting the nitrile or the ester functionality, respectively.
d) Stereoselection was not predictable. In fact, (1R,2S)-cis and (1R,2R)-trans-isomers predominated with styrenes, but (1S,2R)-cis and (1S,2S)-trans-isomers predominated with α-methylstyrenes.

increased the rate of the reaction as well as the enantioselectivity, by occupying one additional coordination site on the catalyst. In 2010, mechanistic studies conducted in gas phase, and supported by DFT calculations, showed that the "bridging carbene" is energetically favored conversely from porphyrins, where it

Table 1.3 Cobalt-salen complexes in enantioselective cyclopropanation reactions.

$$R^1\text{-C(N}_2\text{)-CO}_2R + R^2\text{-CH=CH-}R^3 \longrightarrow R^3,R^2\text{-cyclopropane-}CO_2R, R^1$$

Cobalt-salen	Reactants	Yield (%)	dr	ee (%)	References
(5 mol%) NMI [salen with mesityl, cyclopentyloxy substituents]	R=t-Bu; R^1=R^3=H; R^2=Ph, 4-ClC$_6$H$_4$, 4-MeOC$_6$H$_4$, 2-Naph	85–99	82:18 to 91:9 (E/Z)^a	92–96 (E)	[150]
R=(R)-Ph, X=4-MeO (1 mol%)	R=t-Bu; R^1=R^3=H; R^2=Ph	91	96:4 (E/Z)	93 (E)	[151]
R=(S,S)-(CH$_2$)$_4$, X=3,4-Cl$_2$-6-i-BuO^b (10 mol%)	R^2=H; R^3=Ph, 4-ClC$_6$H$_4$, p-Tol, 4-CF$_3$C$_6$H$_4$, 4-MeOC$_6$H$_4$, 4-BrC$_6$H$_4$, m-Tol, 2,4-Me$_2$C$_6$H$_3$, 3-NO$_2$C$_6$H$_4$, 3-ClC$_6$H$_4$, o-Tol^c	49–95	2:1 to 180:1 (E/Z)	84–97 (R,R)	[152]
R=R^1=(R,R)-(CH$_2$)$_4$ (5 mol%) NMI (10 mol%)	R=t-Bu; R^1=H; R^2=Ph, 4-ClC$_6$H$_4$, 4-MeOC$_6$H$_4$, 2-Naph; R^3=H, Me	39–94	83:17 to 98:2 (Z/E)	95–99 (Z)	[153]

Table 1.3 (Continued)

Cobalt-salen	Reactants	Yield (%)	dr	ee (%)	References
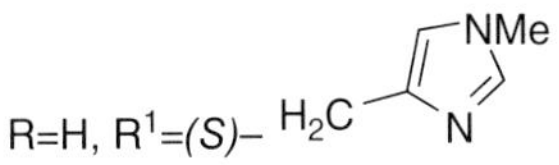 R=H, R¹=(S)- [structure with -NMe, H₂C, N]	R=t-Bu; R¹=H; R²=Ph, p-Tol, 4-ClC₆H₄, 4-MeOC₆H₄, 4-NO₂C₆H₄, 2-ClC₆H₄; R³=H	93–100	89:11 to 99:1 (Z/E)[d]	93–96 (Z)	[154]
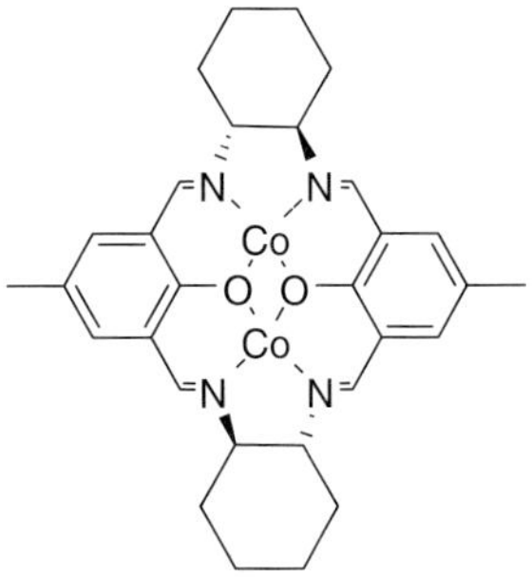 [Co-salen dimer structure with cyclohexane]	R=Et; R¹=R³=H; R²=Ph	92	74:26 (E/Z)	88 (E) 94 (Z)	[155]
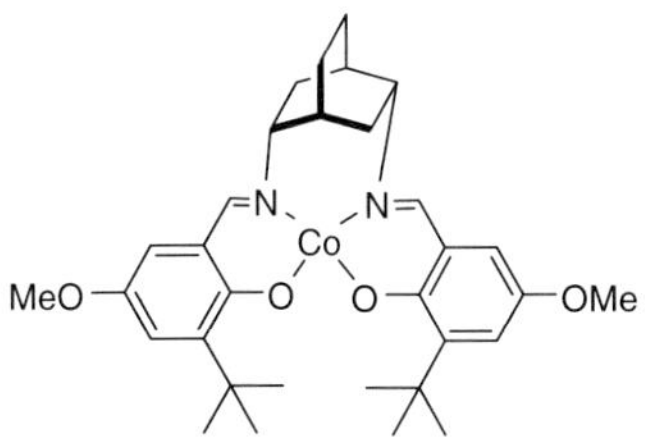 [MeO-substituted Co-salen structure with bicyclic groups] (5 mol%)	R=Et; R¹=H; R²=Ph, 2-MeOC₆H₄, 2-MeO₂CC₆H₄, p-Tol, 4CF₃C₆H₄, 4-MeOC₆H₄, SPh, 3,4-(MeO)₂C₆H₃, 2-furyl, 3-CO₂Et-5-Ph-2-furyl, 2-thienyl. 1-Naph, CH₂-1-Naph R³=Me, Et, Pr, Bu, CH₂CO₂Et, (CH₂)₂CO₂Me, cyclopropyl	87–97	16:1 to 50:1 (E/Z)	83–98 (E)[e]	[156]

a) The diastereoselectivity in the cyclopropanation of styrene decreases to 83:17 if methyl diazoacetate is used. The reaction with α-methylstyrene led only to 47% yield and 47:53 (E/Z) ratio.
b) Triphenylarsine in sulfuric acid–sodium acetate buffer was used as axial auxiliary ligand.
c) The diazo derivative is obtained *in situ* from 2,2,2-trifluoroethanamine.
d) The reaction with α-methylstyrene led only to 45% yield and 55:45 (Z/E) ratio.
e) KSAc was used as axial auxiliary ligand. The isolated product is enantiomeric with respect to that depicted in the figure above the table.

is excluded (Scheme 1.53). Moreover, the same studies established that the rate-determining step was the addition [157].

Other cobalt ligands have been introduced for the asymmetric cyclopropanation of diazo compounds such as a complex based on bis(2-pyridylimino)isoindoles (Scheme 1.54) [158].

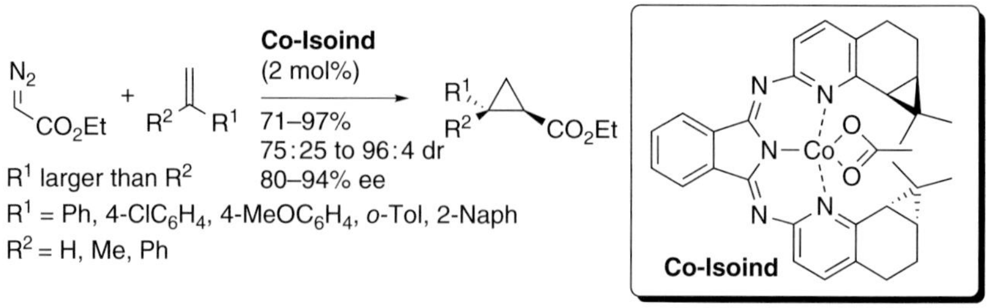

38 | Asymmetric Synthesis of Three-Membered Rings

Scheme 1.54 Bis(2-pyridylimino)isoindoles complex for asymmetric cyclopropanations.

1.3.1.3 Chiral Catalysts: Copper

Chiral copper-based catalysts are the most effective catalysts for the preparation of the *trans*-isomer of cyclopropanes with the widest reaction scope. Among them, nonracemic C_2-symmetric bidentate bisoxazoline (box) ligands have been used in cyclopropanation reactions with copper for more than 30 years [159]. Many investigations have shown that the ligand structure has a strong influence on the stereoselectivity of the cyclopropanation. Even very small structural changes often have drastic and sometimes unpredictable effects on the enantioselectivity, and the phenomenon comprehension is complicated by very low enthalpic barrier for the transition states leading to the *R*- and *S*-products. However, since 2001, using DFT calculations, Salvatella and coworkers rationalized the stereochemical prediction of the cyclopropanation. The calculated relative energies are in good agreement with the experimental enantiomeric excesses as well as with the *Z/E* ratio [160]. In 2004, Mend *et al.* studied again this reaction by means of DFT, showing that it was exothermic and that the turnover-limiting step was the formation of metal catalyst–cyclopropyl carboxylate complexes [161]. Then, Maseras and coworkers found a barrier, which arises from the entropic term, in the Gibbs free-energy surface compatible with the experimentally observed enantioselectivity [162]. The enantioselectivity of asymmetric catalysis was predicted based on quantitative quadrant-diagram representations of the catalysts and quantitative structure–selectivity relationship (QSSR) modeling [163]. The data set included 30 chiral ligands belonging to four different oxazoline-based ligand families. In a simpler approach, the derived stereochemical model indicated that an enantioselective catalyst could be obtained by placing very large groups at two diagonal quadrants and leaving free the two other quadrants. A higher-order approach revealed that bulky substituents in diagonal quadrants operate synergistically.

Chiral ligands for the copper-catalyzed cyclopropanation are listed in Table 1.4. Only the diastereo- and enantioselectivities achieved for the cyclopropanation of styrene are shown in order to compare efficiency, but most of them presented a wide spectrum of activity with many other alkenes (see Table 1.4 footnotes). In the table, bisoxazoline ligands derived from camphoric acid are not reported owing to the poor ee values obtained; somewhat better result was obtained for 1,1-diphenylethylene (81% ee) [201].

The use of ionic liquids offered the possibility of combining the positive aspects of both homogeneous and heterogeneous catalysis. The reaction took place in a

Table 1.4 Copper(I)-box catalysts employed in enantioselective cyclopropanation reactions.

Ph⟵ + N₂CHCO₂R →[Cu(I)L*] Ph⟶CO₂R (E=1R,2R) + Ph⟶CO₂R (Z=1R,2S)

Ligand	R	E/Z	ee (%)	References
box1	D-Menthyl	82:18[a]	97 (E) / 95 (Z)	[159]
box2	D-Menthyl	84:16	98 (E) / 95 (Z)	[164]
box3	D-Menthyl	86:14	98 (E) / 96 (Z)	[165, 166]
box4	D-Menthyl	83:17	90 (E) / 90 (Z)	[166]
box5	BHT	94:6[b,c]	99 (E)	[167]
box5	Et[d]	77:23	99 (E)	[168]
box5	Et[d]	72:28	98 (E)	[169]
box6	CHCy₂	94:6	36 (E) / 20 (Z)[e]	[170]
box7	L-Menthyl	85:15	89 (E) / 89 (Z)	[171]

(Continued)

Table 1.4 (Continued)

Ligand	R	E/Z	ee (%)	References
box8	L-Menthyl	68:32	95 (E) 97 (Z)	[172]
box9	L-Menthyl	81:19	84 (E) 92 (Z)	[173]
box10	L-Menthyl	77:23	94 (E) 79 (Z)	[174]
box11	Et	73:27	92 (E) 84 (Z)^f	[175]
R^1 = i-Pr (box12)	Et	62:38	63 (E) 63 (Z)	[176]
R^1 = t-Bu (box13)	Et	62:38	84 (E) 81 (Z)	[176]
R^1 = Ph (box14)	Et	69:31	64 (E) 56 (Z)	[176]
box15	Et	84:16	100 (E) 100 (Z)	[177]

Table 1.4 (Continued)

Ligand	R	E/Z	ee (%)	References
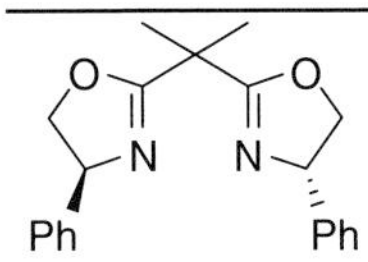 **box16** (Ph, Ph)	Et	70:30	66 (*E*) 54 (*Z*)[g]	[178]
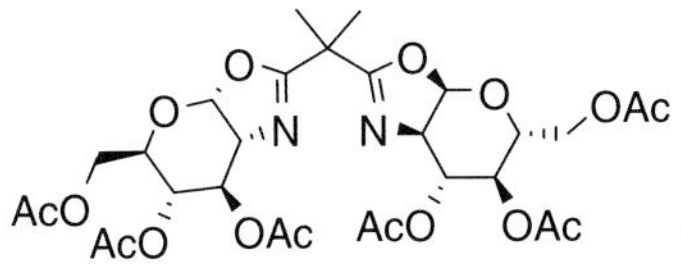 **box17**	Et	80:20 (*S,S*)	82(*E*) 82 (*Z*)[h]	[179]
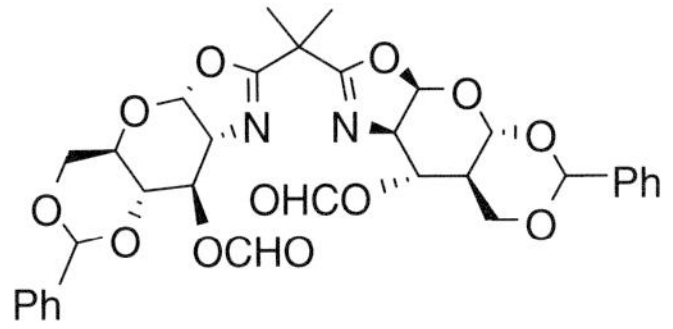 **box18**	Et	73:27 (*S,S*)	95 (*E*) 94 (*Z*)[i]	[180]
(box19) R¹=H, n=2	Et	93:7	65 (*E*) 59 (*Z*)	[181]
R¹=Me, n=3 (**box20**)	Et	94:6	75 (*E*) 31 (*Z*)	[181]
R¹=Ph, X=H (**box21**)[j]	Et	65:35	57 (*E*) 45 (*Z*)	[182]
R¹=*t*-Bu, X=H (**box22**)	Et	62:38	84 (*E*) 77 (*Z*)	[182]
R¹=*t*-Bu, X=Cl (**box23**)	Et	65:35	87 (*E*) 78 (*Z*)	[182]
R¹=*t*-Bu, X=OMe (**box24**)	Et	62:38	83 (*E*) 75 (*Z*)	[182]

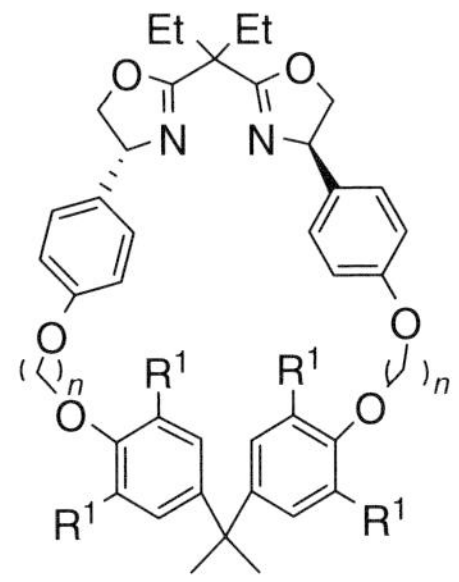

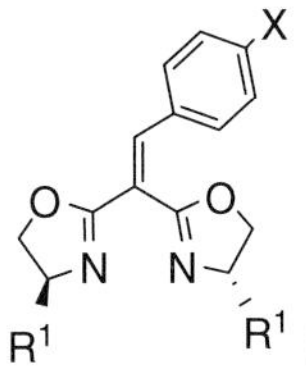

(Continued)

Table 1.4 (Continued)

Ligand	R	E/Z	ee (%)	References
box25	BHT	99:1	98(E)[k]	[183]
box26	Et	67:33	67 (E)	[184]
box27	Et	65:35	78 (E)	[185]
box28	Et	72:28	36 (E) 26 (Z)	[186]
box29	Et	59:41	82 (E) 81 (Z)	[187]
box30	Et	76:24	68 (E) 61 (Z)	[188]

Table 1.4 (Continued)

Ligand	R	E/Z	ee (%)	References
box31	Et	61:39	63 (E) 54 (Z)	[189]
box32	Et	89:11	88 (E) 84 (Z)[l]	[190]
box33	D-Menthyl	89:11 (S,S)	70 (E) 62 (Z)[m]	[191]
box34	D-Menthyl	83:17 (S,S)	89 (E) 96 (Z)	[192]
box35	Et	40:60 (S,S)	69 (E) 70 (Z)	[193]
box36	Et	37:63 (S,S)	61 (E) 64 (Z)	[193]

a) The relative low Lewis acidity causes low yields with inactivated alkenes. The main discriminating stereo-chemical element is the steric interaction between the ester group and the semicorrin substituent upon pyramidalization of the carbene center.
b) *ent*-**box5** is used, this catalyst is effective also in the cyclopropanation of some mono- (93:7 to 94:6 dr, ee >99% (E)) and 1,1-disubstituted alkenes (ee >99%)[167], 1,1-disubstituted-acyclic (1:1 dr, 95% ee (Z), 74–90% ee (E)) and cyclic silyl enol ethers (1:3 E/Z ratio, 92% ee (Z), 87% ee (E)) [194].
c) **box5** is effective also in the cyclopropanation of furans (91% ee) [195], protected allylic alcohols (8:2 to 91:9 E/Z ratio, 92–95% ee (E)) [196], with EDA, and α-fluorostyrene with *t*-butyl diazoacetate (81:19 E/Z ratio, 93% ee (E), 89% ee (Z)) [197].

Table 1.4 (Continued)

d) In ionic liquid [BMIM][BF$_4$].
e) **box6** is much more efficient in the cyclopropanation of trisubstituted and asymmetrically 1,2-disubstituted alkenes (86:14 to 99:1 dr, 82–95% ee).
f) In ionic liquid [EMIM] [OTf], the stability of the copper complex is increased and the reusability of the catalyst solution improved, with 83–98% ee [198].
g) In ionic liquid [BMIM] [BF$_4$], enantioselectivities rose up to 92% ee [199].
h) **box17** is also effective in the cyclopropanation of 4-MeO-styrene (65:35 dr, ee 77% (*E*), 80% (*Z*) and 1,1-phenylethene (ee 75% (*E*))). A similar pyridylbisthiazoline ligand affords only ee ≤ 28% [200].
i) **box18** is also effective in the cyclopropanation of non-1-ene or oct-1-ene: 73:27 dr, ee 77% (*E*), 80% (*Z*) in both cases.
j) α-Methylstyrene gives 56:44 *E/Z* ratio, and 86% ee.
k) **box25** is also effective in the cyclopropanation of 4-MeO-styrene (96:4 dr, 97% ee), 3-phenylpropene (92:8 dr, 97% ee) and 1-octene (99:1 dr, 95% ee).
l) **box32** is also effective in the cyclopropanation of 4-MeO-styrene (91:9 dr, ee 90% (*E*), 87% (*Z*)) and 4-Cl-styrene (88:12 dr, ee 85% (*E*), 82% (*Z*)). The substitution of the *t*-Bu with Et group in this catalyst allowed the cyclopropanation of 1,1,-diphenylethene in 91% ee.
m) **box33** is also effective in the cyclopropanation of substituted styrenes (82:18 to 86:14 dr, 52–82% ee (*E*)).

homogeneous phase with high activity and selectivity, and it is possible to easily separate the products after the reaction and reuse the catalyst, as in the case of heterogeneous catalysis [202].

Alkenes that were more complex were also involved in copper–bisoxazoline-catalyzed cyclopropanation with diazoalkanes. The cyclopropanation of 2,5-dimethyl-2,4-hexadiene with *t*-butyl diazoacetate (Scheme 1.55, eq 1) [203] and that of furans in the presence of protected α-amino-acid-containing bisoxazoline ligands (Scheme 1.55, eq 2) [204] are examples. In addition, cyclopentadienylsilane was also desymmetrized by Cu-catalyzed cyclopropanation (Scheme 1.55, eq 3) [205], and a carbohydrate-based bisoxazoline ligand allowed for the copper-catalyzed cyclopropanation of *N*-acyl indoles (Scheme 1.55, eq 4) [206].

Some of these copper(I)-box catalyzed reaction were then employed in multistep synthesis of natural products [207]. For instance, cyclopropanation of furans was applied to the total syntheses of some key intermediates of natural products and drugs [208]. More recently, the ligand **box18** was used in the reaction of non-1-ene and EDA for the total synthesis of unnatural (+)-grenadamide (Scheme 1.17, for the structure of the natural product) [180]. Moreover, the cyclopropanation of *N*-Boc-3-methylindole yielded a key building block for the synthesis of the indole alkaloid (−)-desoxyeseroline in 59% overall yield with 96% ee (Scheme 1.56) [206]. Moreover, ligand **box37** (Scheme 1.57) performed the stereoselective preparation of the tetracyclic core, and key intermediate, of cryptotrione (Scheme 1.56) in 93% yield with >91:9 dr. The enantiomeric excess was not reported, but a negative $α_D$ was given [209]. The use of a copper(II) salt instead of the classical copper(I) salt should be noted in this reaction.

The synthesis of new bicyclo[4.1.0] compounds was attempted with little success under copper-**box6** catalysis. In fact, good enantioselectivities were achieved only starting from enantiopure cyclic enamides [210]. Finally, the copper complexes of polytopic chiral ligands based on bis(oxazoline) units (Scheme 1.57) were tested in

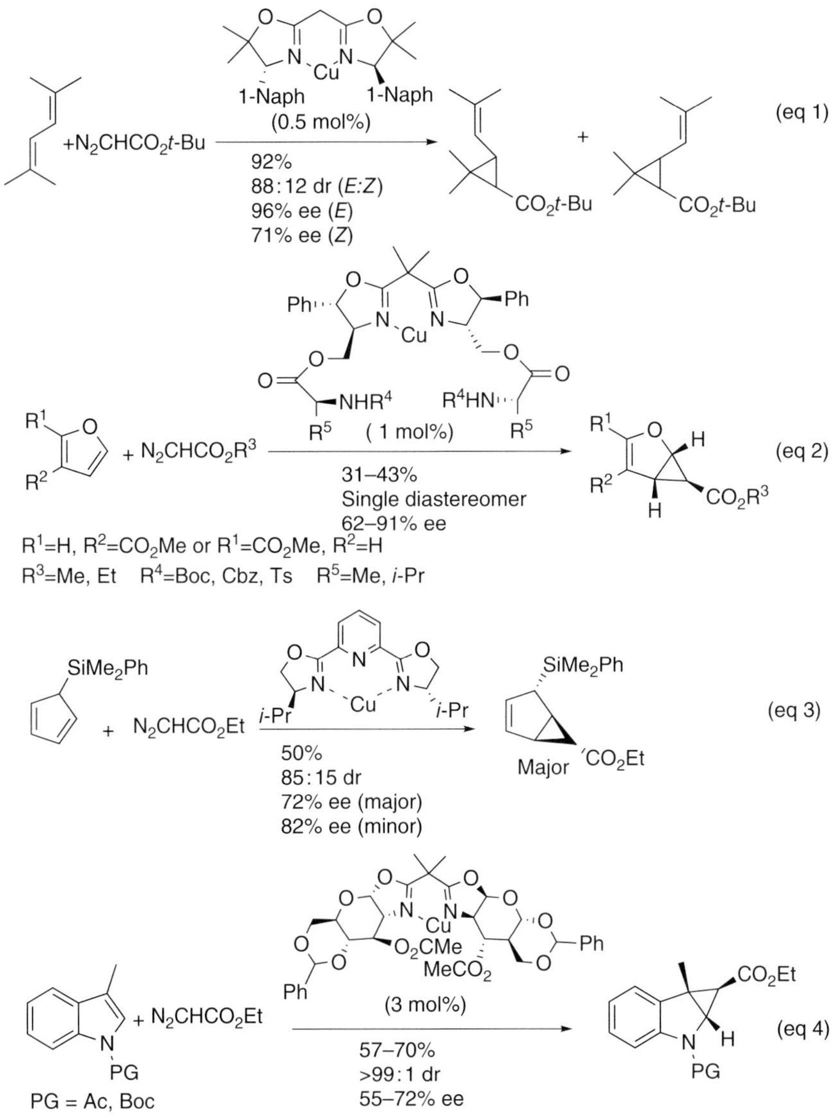

Scheme 1.55 Copper–bisoxazoline-catalyzed cyclopropanation of complex alkenes.

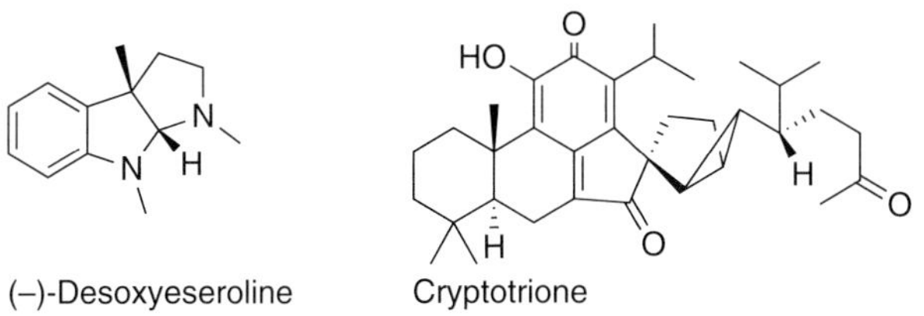

(−)-Desoxyeseroline

Cryptotrione

Scheme 1.56 Some natural products prepared by copper-box-catalyzed cyclopropanation.

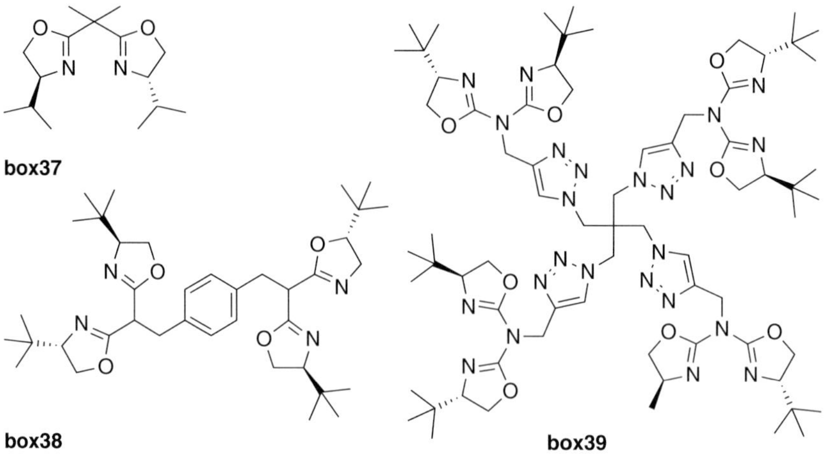

box37

box38

box39

Scheme 1.57 Other box ligands proposed for copper-box-catalyzed cyclopropanation.

the cyclopropanation reaction of styrene with EDA, using a release–capture strategy based on the formation of coordination polymers at the end of the reaction to recover and recycle catalysts [211]. Two points should be outlined: using the Cu-**box38**, the *cis*-cyclopropane constituted the predominant product, conversely from the corresponding Cu-**box5** (see Table 1.4) [212]; the tetratopic **box39** could be recovered and reused in up to 20 reaction cycles always with very good yields and enantioselectivities.

Diazoalkanes other than alkyl diazoacetates have also been employed in copper–bisoxazoline-catalyzed cyclopropanations. For instance, α-diazophosphonate diazomethane was used to obtain cyclopropylphosphonate derivatives under ***ent*-box5** catalysis (Scheme 1.58, eq 1). However, Nishiyama's ruthenium catalyst (see Section 1.3.1.5) gave better results and can be used with a wider range of substrates (88:12 to >98:2 *E*/*Z* ratio, 90–96% ee also with substituted styrenes, α-methylstyrene, and 1-phenylbuta-1,3-diene) [213]. Other examples are the reaction of diazomethane with *trans*-cinnamate esters (Scheme 1.58, eq 2) [149, 214], the reaction of (TMS) diazomethane with olefins (Scheme 1.58, eq 3) [215], the cyclopropanation of styrene with diazosulfonate esters (Scheme 1.58, eq 4) [216], and the cyclopropanation of alkenes with ethyl phenyldiazoacetate (Scheme 1.58, eq 5) [217].

Asymmetric carbene transfer involving diazo decomposition is almost exclusively restricted to the research laboratory owing to the often unjustified prejudice regarding diazo compounds and the reagents used to prepare them, which are believed to be toxic, carcinogenic, and potentially explosive. Therefore, the development of reactions avoiding diazo precursors is of considerable interest. For example, phenyliodonium ylides are potential substitutes of diazo compounds. In fact, they are generally superior carbene precursors, but they are often unstable. Thus, their use is advantageous only if the carbene transfer can be carried out in a one-pot procedure, in which the phenyliodonium ylide is generated and decomposed *in situ*, as in the copper(I)-catalyzed asymmetric cyclopropanations depicted in Scheme 1.59 [218].

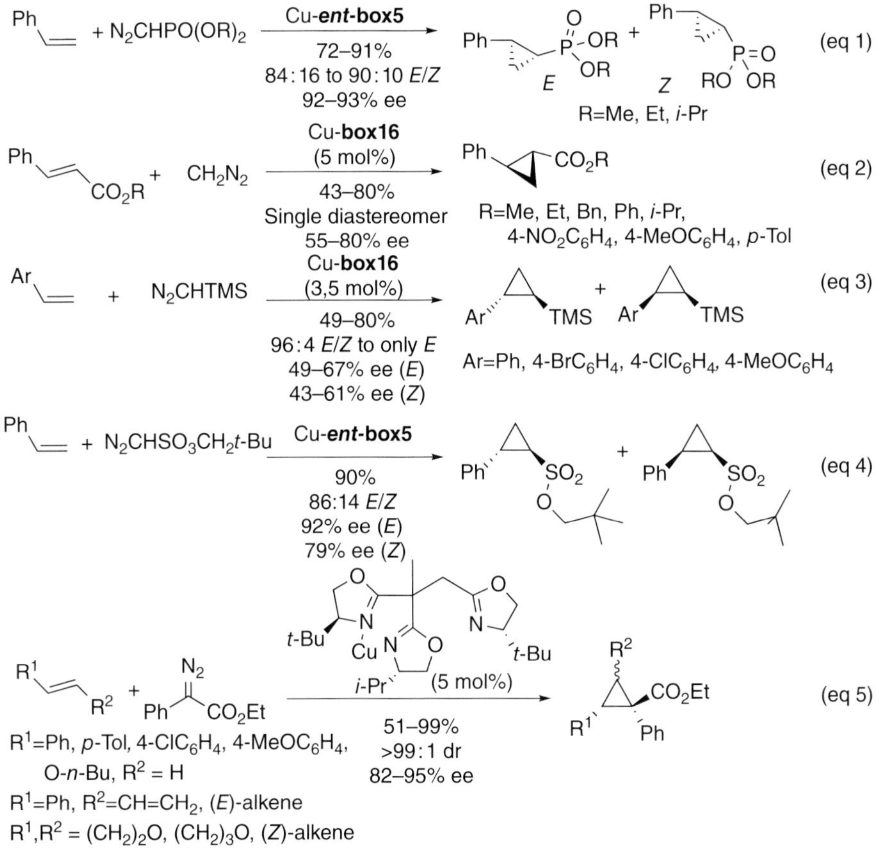

Scheme 1.58 Copper–bisoxazoline-catalyzed cyclopropanation of some diazoalkanes.

Many copper-box ligands were synthesized on polymer supports, but the most ancient attempts afforded enantioselectivities usually lower than those observed in solution [219]. Later, polystyrene or commercial Merrifield resin supported copper(II)–bisoxazoline catalysts were developed by Salvadori and coworkers, affording up to 71:29 dr and >90% ee or 71–93% ee, respectively, in the heterogeneous cyclopropanation of alkenes with EDA [220].

Mayoral group studied bisoxazoline ligands immobilized on laponite by electrostatic interactions, demonstrating a previously unknown role of the catalyst surface, although the enantioselectivity was low (≤55% ee) [221]. The same group also studied the use of homopolymers of bisoxazoline ligands [222]. The use of suitable dendrimers as cross-linkers in the polymerization process allowed high productivity of chiral cyclopropanes per molecule of chiral ligand and enantioselectivities of up to 78% ee were obtained. Box ligands onto silicaceous mesocellular foams (MCFs) afforded high enantioselectivity (up to 87% ee) and reactivity, but only 64:36 dr [223]. A partial modification of the surface of these MCFs with TMS groups prior to the use of the bisoxazolines enhanced enantioselectivities (up to 95% ee), but slowly increased dr (67:33) [224]. Due to their higher binding affinity toward copper,

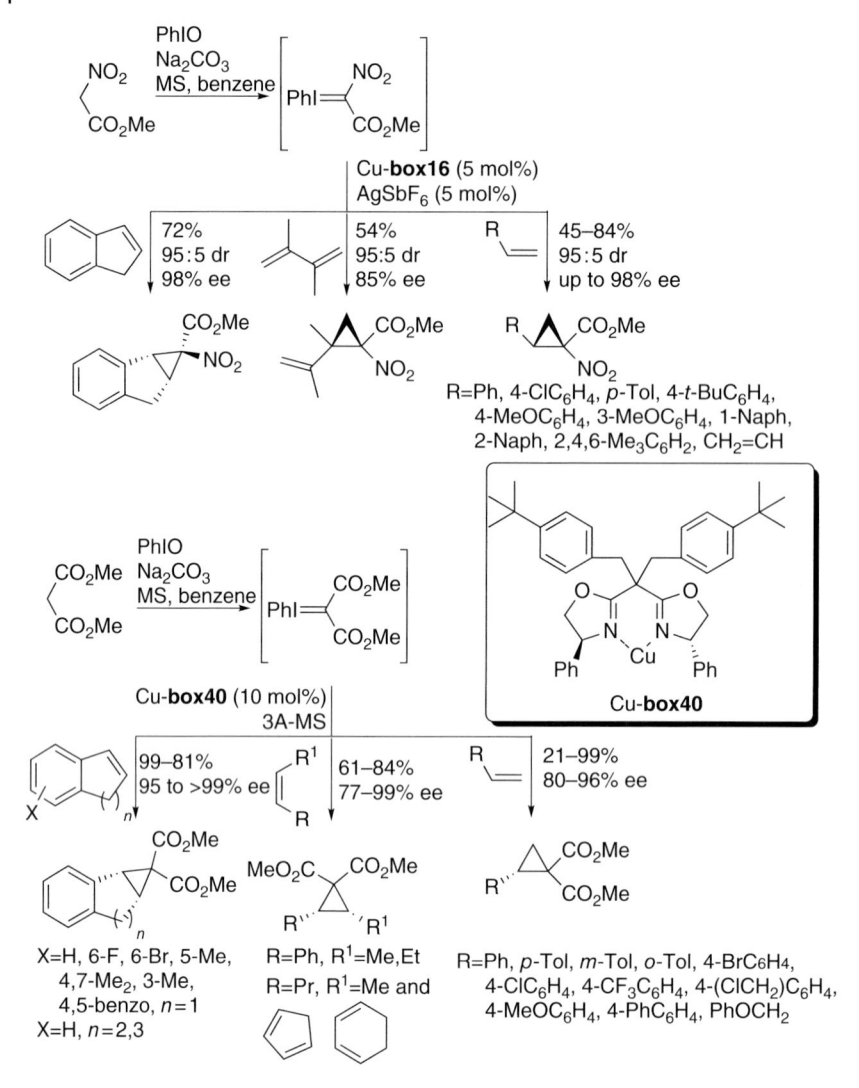

Scheme 1.59 Copper–bisoxazoline-catalyzed cyclopropanation with phenyliodonium ylides.

immobilized azabisoxazolines onto various polymeric supports are far superior, compared to their corresponding bisoxazolines. For instance, up to 99% ee was obtained for the reaction of styrene with EDA [225]. The use of polyisobutylene oligomers (in particular, PIB_{2300}) as soluble supports for the immobilization of bisoxazoline prepared from phenylglycine provides cyclopropane in 48% yield, 81:19 E/Z ratio, 92% (E), and 68% (Z) ee [226]. The catalyst could be reused five to six times. Moreover, a chiral ferrocenyl–bisoxazoline derivative with a biphenyl unit can be used as a ligand of the Cu-catalyzed cyclopropanation of styrene with EDA, providing a 55% yield of a mixture of *trans-* and *cis-*cycloadducts in a 65:35 ratio and 20% and 23% ee, respectively [227].

Burke's research group grafted a box ligand on a Wang resin. The copper(I) catalyst gave superior results than copper(II) and its selectivity remained constant over four cycles (5–61% yield, *E/Z* ratios of about 67:33, and 13–71% ee) [228]. The same catalyst was also immobilized in a noncovalent manner on an ion-exchange resin with comparable results, but leaching of the catalyst was an important factor after a few cycles [228]. Chiral copper(II)-box catalysts were also anchored onto mesoporous silica or encapsulated onto copper(II)-exchanged zeolites. Enantioselectivities of the cyclopropanation of styrene with diazoacetate were consistently lower than those obtained in homogeneous phase in all cases, although in zeolites, they appeared to increase with a decreasing pore size [229].

A polymersome nanoreactor was found to be able to catalyze asymmetric cyclopropanation reactions of some 4-substituted styrenes in water with 68:32 to 75:25 *E/Z* ratios and 53–84% ee (*E*-isomer). The hydrophobic environment around the catalyst was demonstrated to be substrate selective, because only hydrophobic substrates were readily converted into the corresponding cyclopropane products, while hydrophilic substrates (i.e., 4-COOH and 4-NH$_2$-styrenes) did not undergo any reaction [230].

Thus, in summary, the covalently immobilized catalysts could be superior in terms of activity, selectivity, and recycling compared to their noncovalently immobilized counterparts, but much work has yet to be carried out in terms of increasing the usability and efficiency of heterogeneous over homogeneous catalysts.

Regulating the electron density of the oxazoline ring has remained an unsolved problem. On the other hand, the two *N*-substituents of an imidazoline may serve as handles to tune the electronic and conformational properties of the ligand. For instance, the Cu-catalyzed cyclopropanation of styrene with EDA, performed in the presence of bis-imidazoline ligands, provided the corresponding cyclopropanes in high yield with good ee (Scheme 1.60) [231]. In addition, chiral dihydrodinaphthazepinyloxazolines were demonstrated to be effective ligands in the Cu-catalyzed cyclopropanation of styrene and its derivatives (Scheme 1.60) [232]. Finally, Kwong *et al.* have designed a series of chiral C_1-symmetric bidentate ligands, possessing two different nitrogen heterocycles (Scheme 1.60) [233].

Chiral thienyl pyridines as *N,S*-ligands [234] or 2-(2-phenylthiophenyl)-5,6,7,8-tetrahydroquinolines [235] provided effective copper catalysts in the cyclopropanation of styrene with EDA, but offering a low enantioselectivity (≤10% and ≤5% ee, respectively). Bis-azaferrocene ligand was also tested in the reaction of N$_2$CHCO$_2$BHT with monosubstituted alkenes, leading to good diastereo- and enantioselectivities (for instance, in the reaction with styrene: 96:4 *E/Z* ratio and 94% (*E*) and 79% (*Z*) ee) [236].

Several types of nitrogen-containing ligands, other than bisoxazolines and their derivatives, have been investigated for the asymmetric cyclopropanation of alkenes (Table 1.5). In addition, Baldwin and coworkers allowed L-menthyl diazoacetate to react with isotopically labeled styrenes, providing the corresponding four isotopically labeled L-menthyl-(1*S*,2*S*)-2-phenylcyclopropanecarboxylates with >99% ee, in the presence of a very simple diamine ligand derived from (1*S*,2*S*)-1,2-diphenylethylenediamine [253].

An iminodiazaphospholidine (Scheme 1.61) was used as the copper ligand in the asymmetric cyclopropanation of styrene with EDA leading to 99:1 (*E/Z* ratio) and 94% ee (*E*) [254]. Asymmetric pyridine derivatives, in particular pyridine-based

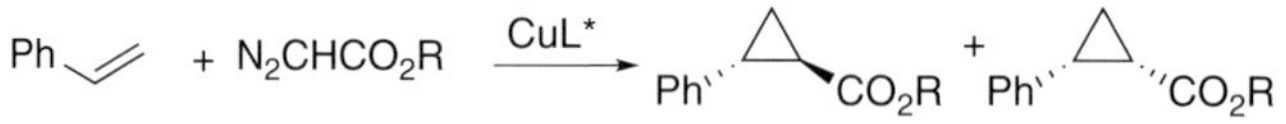

Scheme 1.60 Other bidentate ligands for copper catalyzed cyclopropanation.

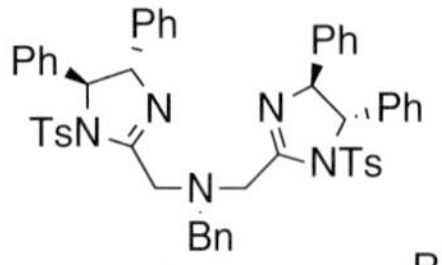

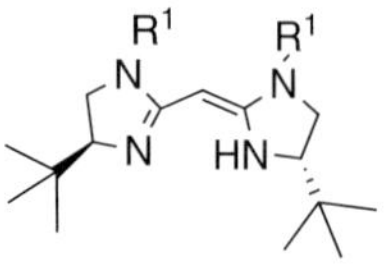

Cu(II) / L*	R	y (%)	E/Z	ee % (E)	ee % (Z)	
Ph₂-TsN/N-Bn-N/N-Ph (Ph, Ph, NTs)	Et	51	80:20	83	36	ref 231
R¹ = Bn; R¹ = 4-MeOC₆H₄ (bis-imidazoline, t-Bu)	Et; BHT	80; 88	69:31; 99:1	70; 84	50; —	
Cu(I) binaphthyl-oxazoline (Ph) (3 mol%)	Et; CHCy₂; L-menthyl; D-menthyl	60; 82; 96; 81	60:40; 79:21; 75:25; 81:19	66; 74; 83; 78	62; 82; 87; 70	ref 232
Cu(I) Ar–N= (2 mol%)	Ar = Thiazol-2-yl; 4-Methylthiazol-2-yl; 1-Methylthiazol-2-yl; Pyrazin-2-yl	Et	46–90	(R,R):(1R,2S) 68:32; 68:32; 66:34; 65:35	19; 38; 10; 24	11; 30; 20; 6
Cu(I) Ar–N= (2 mol%)	Ar = Thiazol-2-yl; 4-Methylthiazol-2-yl; 1-Methylthiazol-2-yl; Pyrazin-2-yl	Et	46–90	(S,S):(1S,2R) 70:30; 70:30; 69:31; 67:33	16; 14; 14; 8	16; 14; 12; 4	ref 233

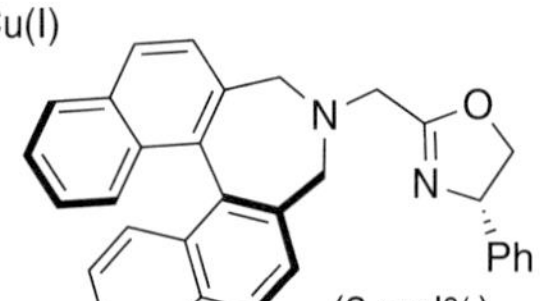

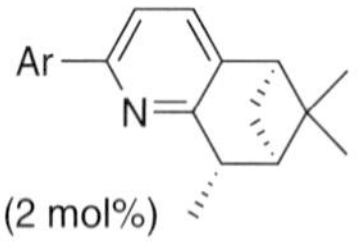

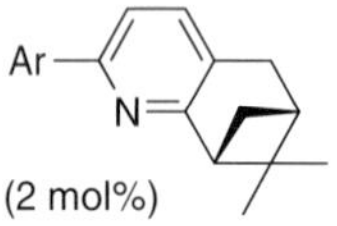

12-membered tetraaza-macrocycles (Scheme 1.61), are other used ligands. Catalytic reactions were run, and cyclopropanes were obtained in 45–83% yield with poor diastereomeric ratio (50:50 to 67:33) and enantiomeric excess (33–50% (Z) and 15–65% (E)). The absence of α-substituents on the styrene as well as the bulkier *tert*-butyl diazoacetate led to worse results in terms of both yield and stereoselectivity. The reaction was also applied to 2,5-dimethylhexa-2,4-diene, but the results were modest. The (R)-catalyst gave *cis*-(1R,2S)-2-methyl-2-phenylcyclopropanecarboxylate and *cis*-(1S,2R)-2-phenylcyclopropanecarboxylate, whereas the *trans*-(1R,2R)-isomers were obtained with the (S)-isomer. Similar results were obtained for other similar catalysts with different substituents and stereoselectivities at the 4, 8, and 13 positions, all leading to *trans*-(1R,2R)-isomers as the major products [255].

Planar chiral terpyridine–copper(II) complexes (Scheme 1.61) exhibited moderate diastereomeric excesses with good enantioselectivities of *cis*-cyclopropanes from reactions of various styrene derivatives with ethyl diazoacetate (4–68% yields, 23:77 to 68:32 E/Z ratio, 18–82% ee (1S,2S), 65–91% ee (1S,2R)). Only 4-(trifluoromethyl) styrene gave racemic mixture of *trans*-isomer and only 5% ee for the *cis*-isomer [256].

Table 1.5 Other nitrogen-containing ligands for copper catalysts employed in enantioselective cyclopropanation reactions.

Ph⟶ + N$_2$CHCO$_2$R $\xrightarrow{\text{Cu(I)L*}}$ Ph⋯△CO$_2$R + Ph△CO$_2$R
(E=1R,2R) (Z=1R,2S)

Ligand	R	E/Z	ee (%)	References
Bipyridine				
(bipyridine with TMS groups)	t-Bu	86:14	92 (E) 98 (Z)	[237]
(bipyridine with OMe/MeO)	Et	80:20	91 (E) 82 (Z)	[238]
	Et	80:20^a	90 (E) 82 (Z)	[239]
(bipyridine with Bu groups)	Et	67:33	74 (E) 78 (Z)	[240]
	BHT	99:1	70 (E)	[240]
(bipyridine with dioxolane Et groups)	Et^b	80:20	82 (E)	[241]
(binaphthyl bipyridine with OEt)	Et	64:36 (S,S)	75 (E)	[242]
(pyridine oxazoline with Ph)	Et^c	60:40	57 (E) 46 (Z)	[243]
(macrocyclic bipyridine)	Et^d	76:24	56 (E) <20 (Z)	[244]

(Continued)

Table 1.5 (Continued)

Ligand	R	E/Z	ee (%)	References
Chiral diamines and diimines				
(Ph/Ph diamine with mesityl groups)	L-Menthyl	93:7	96 (E) 66 (Z)	[245]
(Bn/Bn/Bn/Bn tetraazine)	Et	75:25	90 (E)	[246]
(cyclohexyl diimine with phenyl)	L-Menthyl	70:30	2 (E) 8 (Z)	[247]
(binaphthyl bis-Cl-salicylidene diimine)	L-Menthyl	90:10^e	68 (E) 81 (Z)	[248]
(tropane-naphthyl derivative)	Me	75:25	79 (E) 98 (Z)	[249]
(cyclohexane diamine with C$_8$F$_{17}$(CH$_2$)$_2$ chains)	Et^f	67:33	47 (E) 47 (Z)	[250]
(bis-imidazole cyclohexane ligand, X = O, nothing)	Et	67:33 to 81:19	1–20 (E) up to 94 (Z)	[251]

**=(S,S),(R,R) R^1=Me, Ph R^2=Me, Ph

Table 1.5 (Continued)

Ligand	R	E/Z	ee (%)	References
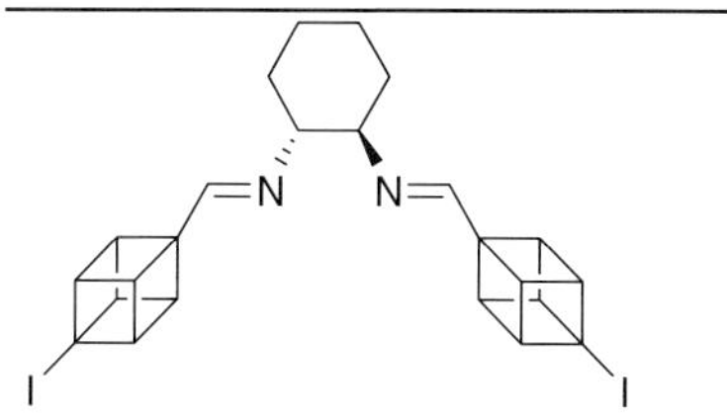 [structure]	Et	58:42	6 (E) 8 (Z)[g]	[252]

a) The enantiomer of this catalyst gives the enantiomeric cyclopropane in comparable selectivity [239].
b) This catalyst was also employed in the reaction of *t*-butyl diazoacetate with 4-F-styrene (92:8 E/Z ratio, 99% ee) and 4-Br-styrene (95:5 E/Z ratio, 83% ee). The high stereoselectivities were attributed to a structural rigidification provided by the chiral acetal moieties.
c) *t*-Butyl diazoacetate gave better E/Z ratios but lower enantioselectivities. Other styrenes were tested, but enantiomers of the corresponding *trans*-cyclopropanes could not be determined. A pyridylmonooxazoline ligand was also tested but gave somewhat lower ee values (up to 53%).
d) Phenanthroline-containing macrocycles gave somewhat better results (78:22 E/Z ratio, ee: 67% (E), <20% (Z)) and with phenanthroline a simple structural modification of the chiral cavity allowed the successful control of the *trans*- or *cis*-diastereoselectivity of the reaction (up to 9:91 E/Z ratio) [244].
e) Effective also in the cyclopropanation of 2-vinylnaphthalene (90:10 E/Z ratio, ee 76% (E), 62% (Z)), α-methylstyrene (72:28 E/Z ratio, ee 61% (E), 78% (Z)), 1,1-diphenylethene (98% ee).
f) These ligands showed similar activities, but lower enantioselectivities than those achieved using more synthetically demanding fluorous ligands such as fluorous bisoxazolines (see Table 1.4). Moreover, C_2-symmetric diimines derived provided better enantioselectivities (up to 67% ee) but low diastereoselectivities (≤14% de).
g) The poor stereoselectivity was explained by the lack of steric hindrance on either face of the copper complex in the pseudo six-membered ring conformation calculated as the most stable by both M06L and B3LYP calculations.

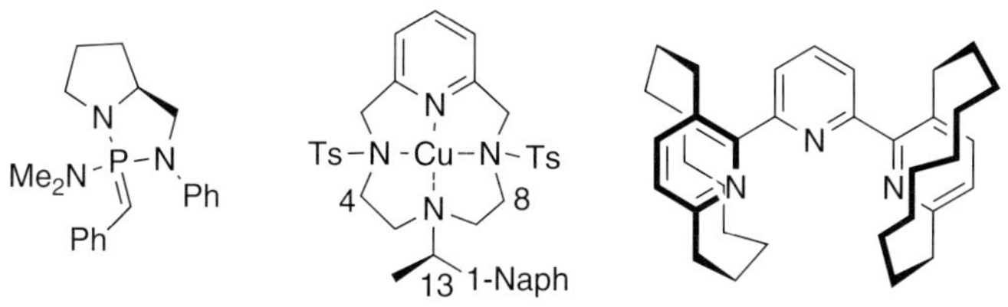

Scheme 1.61 Other nitrogen ligands for copper catalyzed cyclopropanation.

The copper–salicylaldimino complexes were also tested, but the decomposition of EDA in the presence of styrene gave scarce results (Scheme 1.62) [257]. Better results were obtained with a copper(II) planar (*R*)-paracyclophanylimine, and the enantiomeric induction was attributed to the restricted rotation around the cyclophane nitrogen [258]. Other Cu-salen catalysts are efficient especially in the synthesis of chrysantemate esters (Scheme 1.62, eq 1 [259], and eq 2 [260]). In eq 2, the best result was obtained by combining the copper Schiff base complex with a Lewis acid, such as Al(OEt)$_3$, which enhanced the catalytic efficiency. The reaction pathway of the asymmetric induction in the

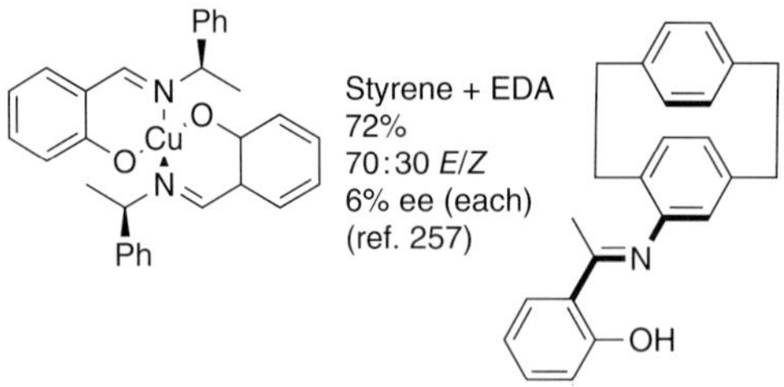

Asymmetric Synthesis of Three-Membered Rings

Styrene + EDA
72%
70:30 E/Z
6% ee (each)
(ref. 257)

Olefin	DA	E/Z	ee_Z (%)	ee_E (%)
Styrene	Et	67:34	61	76
	t-Bu	82:18	78	88

α-Methylstyrene up to 95% ee
(ref. 258)

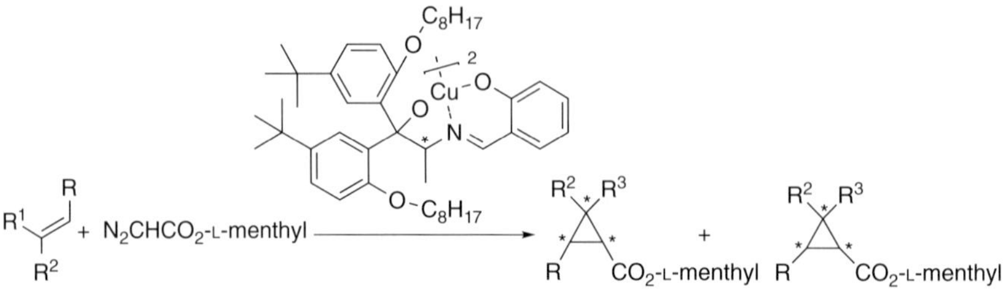

(eq 1)

Olefin	Catalyst Config	Z/E	ee_Z (%)	ee_E (%)
Styrene	R	14/86	54 (1S,2R)	69 (1S,2S)
	S	18/82	78 (1R,2S)	81 (1R,2R)
Oct-1-ene	R	17/83	46 (1S,2R)	76 (1S,2S)
	S	22/78	64 (1R,2S)	84 (1R,2R)
(E)-Oct-4-ene	R	—	—	82 (+)
	S	—	—	84 (−)
trans-Anethole	R	9/91	44 (+)	81 (+)
	S	12/88	60 (−)	89 (−)
1,1-Diphenylethene	R	—	—	75 (+)
	S	—	—	64 (−)
α-Methylstyrene	R	40/60	86 (nd)	68 (nd)
	S	36/64	68 (nd)	58 (nd)
2-Methyl-2-butene	R	79/21	—	—
	S	77/23	—	—
2-Methyl-5,5,5-trichloropent-2-ene	S	85/15	91 (1R,2S)	(with EDA)
2,5-dimethylhexa-2,4-diene	R	7/93	46 (1R,2S)	94 (1R,2R)

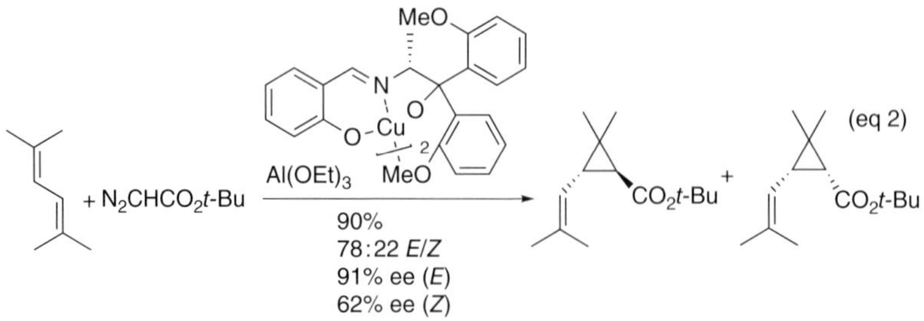

(eq 2)

Al(OEt)₃ MeO
90%
78:22 E/Z
91% ee (E)
62% ee (Z)

Scheme 1.62 Cu-salen catalysts for asymmetric cyclopropanation.

synthesis of (1R,3R)-chrysanthemic acid ester in the presence of Cu(I)-salen complexes was elucidated with the aid of hybrid DFT calculations [261]. The key features were that the alkoxycarbonyl carbene complex intermediate was intrinsically chiral and that the intramolecular hydrogen bonding in the carbene complex transmitted the chirality information from the side chain to the carbene complex.

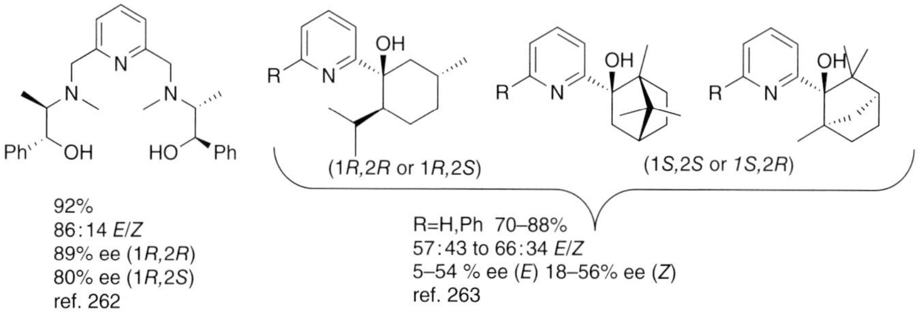
OH
R
N
OH
R
N
OH
R
N
N
N
N
Ph
OH
HO
Ph
(1R,2R or 1R,2S)
(1S,2S or 1S,2R)
92%
86:14 E/Z
89% ee (1R,2R)
80% ee (1R,2S)
ref. 262
R=H,Ph 70–88%
57:43 to 66:34 E/Z
5–54 % ee (E) 18–56% ee (Z)
ref. 263

Scheme 1.63 Chiral aminoalcohols as ligands for asymmetric cyclopropanation.

Chiral amino alcohols, for example, ligands derived from (1*R*,2*S*)-ephedrine [262] or various chiral *N*,*O*-pyridine alcohols [263] provided the expected cyclopropane from the reaction of styrene with EDA (Scheme 1.63).

Monodentate chiral ligands were only rarely employed in catalytic asymmetric synthesis for the main reason that they fail to form the compact catalytic manifolds that are pivotal for high asymmetric induction. On the other hand, monodentate ligands are in general less structurally complex, and thus, the synthesis, in principle, is less demanding. Among these examples, the Cu-catalyzed cyclopropanation of 1,1-diphenylethylene with menthyl diazoacetate was performed in the presence of a chiral imidazolidine ligand (Scheme 1.64) [264].

Chiral monodentate oxazoline ligands (Scheme 1.65) were proposed for the cyclopropanation reaction of styrene and α-methylstyrene and EDA with only

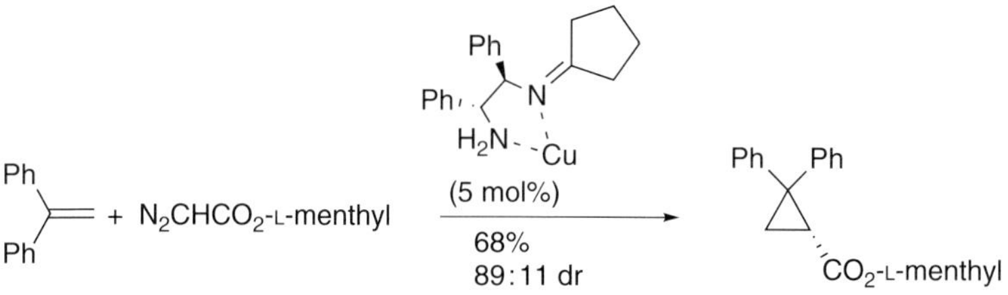
Ph
Ph
N
H2N
Cu
(5 mol%)
Ph
Ph
+ N2CHCO2-L-menthyl
68%
89:11 dr
Ph
Ph
CO2-L-menthyl

Scheme 1.64 Chiral imidazolidine ligand for asymmetric copper-catalyzed cyclopropanation.

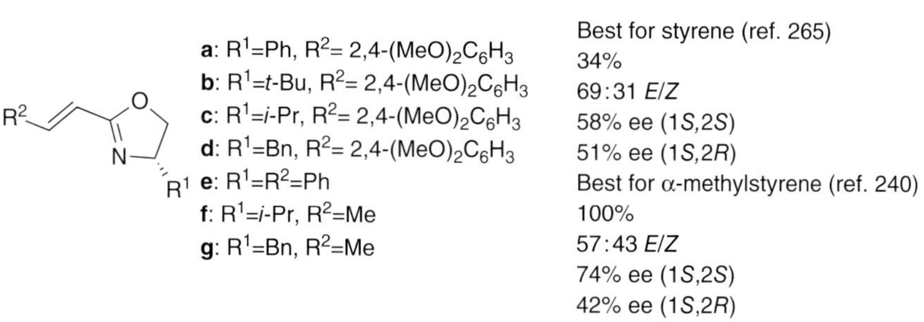
R2
O
N
R1
a: R1=Ph, R2= 2,4-(MeO)2C6H3
b: R1=t-Bu, R2= 2,4-(MeO)2C6H3
c: R1=i-Pr, R2= 2,4-(MeO)2C6H3
d: R1=Bn, R2= 2,4-(MeO)2C6H3
e: R1=R2=Ph
f: R1=i-Pr, R2=Me
g: R1=Bn, R2=Me
Best for styrene (ref. 265)
34%
69:31 E/Z
58% ee (1S,2S)
51% ee (1S,2R)
Best for α-methylstyrene (ref. 240)
100%
57:43 E/Z
74% ee (1S,2S)
42% ee (1S,2R)

Scheme 1.65 Chiral oxazoline ligands for asymmetric copper-catalyzed cyclopropanation.

moderate success [239, 265]. Computational calculations indicated that the active catalyst most likely incorporates two ligands per copper atom in an assembly similar to box ligands [239]. As a consequence, monodentate ligands were tested in both homo- and hetero-complexes with Cu (MeCN)$_4$PF$_6$, and homo-combinations were found to be superior to hetero-combinations.

Arndtsen and coworkers reported the preparation of a library of α-amino-acid-bound borate anions. Ion pairing of these anions to a copper cation could be used to induce enantioselectivity in the cyclopropanation of styrene with EDA (Scheme 1.66) [266].

Furusho's group showed that complementary double-helical molecules showing optical activity owing to their helicity could be enantioselectively synthesized and could catalyze the asymmetric cyclopropanation of styrene with EDA in the presence of a copper catalyst (Scheme 1.66) [267]. The results showed an almost linear relationship between the helix-sense excesses and the ee values of *trans*-cyclopropane. Moreover, the results suggested that the chiral space generated by the rigid double-helical structure was effective and indispensable for the high enantioselectivity.

1.3.1.4 Chiral Catalysts: Rhodium

Although many transition metal chiral complexes have been developed, dirhodium(II) complexes are among the most attractive catalysts, because of their activity and efficiency. Dirhodium(II) catalysts with very high turnover numbers have been reported, and consequently, the cost and toxicity of rhodium can be greatly overshadowed by the ability to use tiny amounts of the catalyst to generate large quantities of value-added products. The use of dirhodium(II) catalysts in inter- and intramolecular asymmetric cyclopropanation has been recently reviewed [268], and readers are invited to read this review for other information.

Rhodium-based chiral complexes were synthesized and tested in both inter- and intramolecular cyclopropanations. In particular, the development of dirhodium(II) carboxylate and carboxamidate catalysts (Scheme 1.67) has resulted in many highly chemo-, regio-, and stereoselective reactions of α-diazocarbonyl compounds [269].

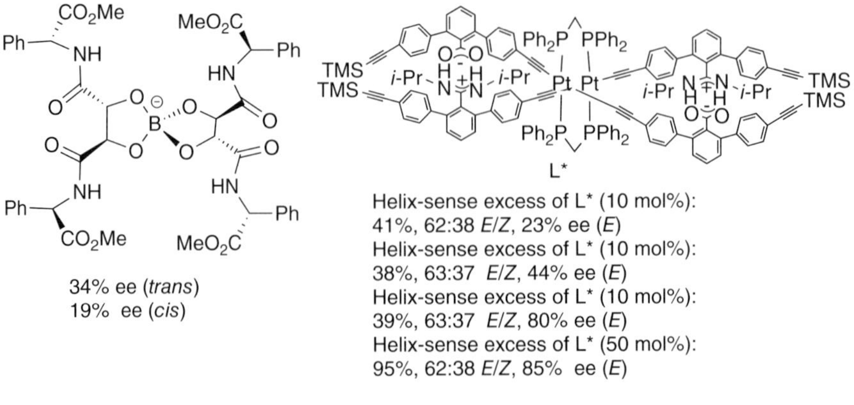

Scheme 1.66 α-Amino-acid-bound borate anions and double-helical molecules for asymmetric cyclopropanation.

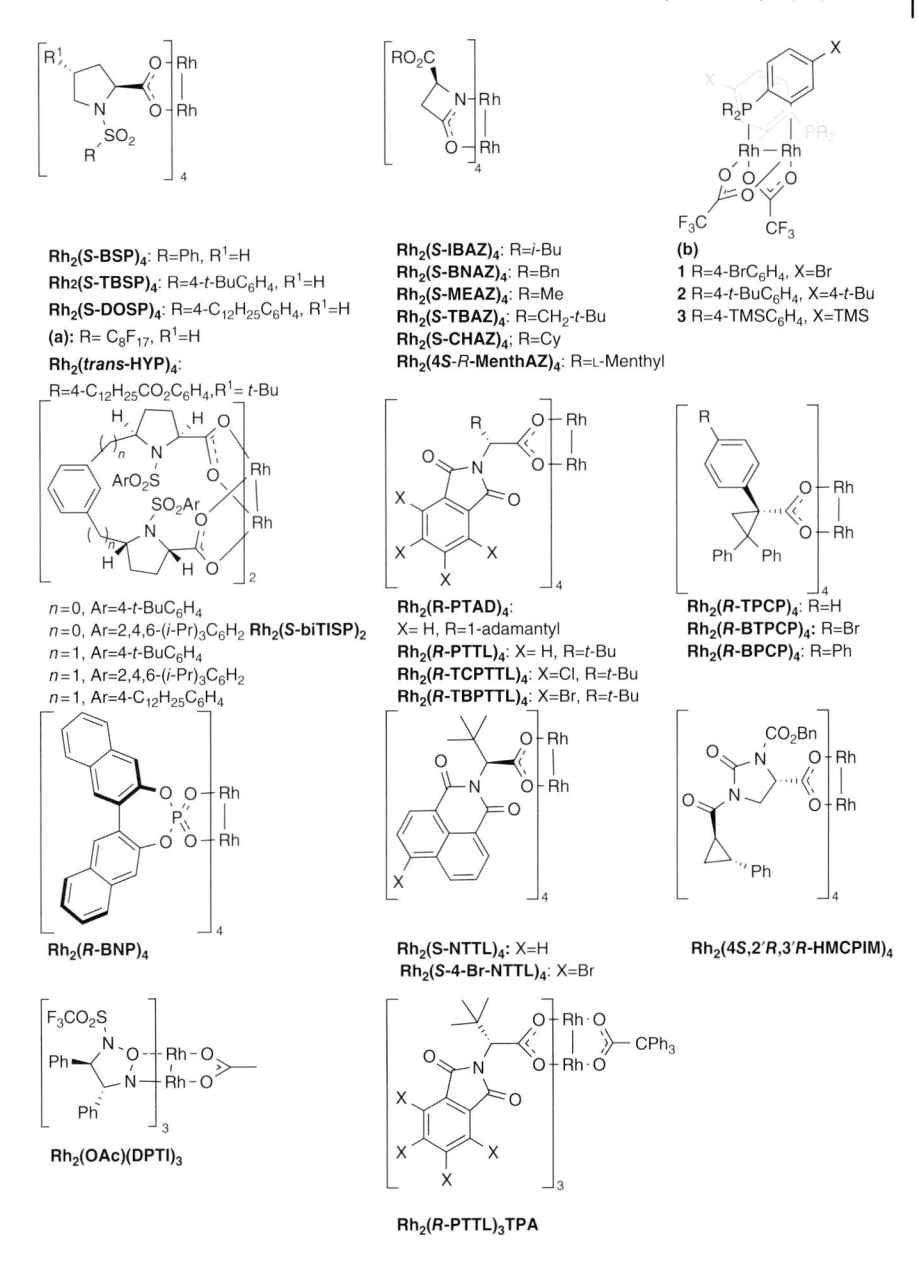

Rh₂(S-BSP)₄: R=Ph, R¹=H
Rh₂(S-TBSP)₄: R=4-t-BuC₆H₄, R¹=H
Rh₂(S-DOSP)₄: R=4-C₁₂H₂₅C₆H₄, R¹=H
(a): R= C₈F₁₇, R¹=H
Rh₂(trans-HYP)₄:
R=4-C₁₂H₂₅CO₂C₆H₄, R¹= t-Bu

Rh₂(S-IBAZ)₄: R=i-Bu
Rh₂(S-BNAZ)₄: R=Bn
Rh₂(S-MEAZ)₄: R=Me
Rh₂(S-TBAZ)₄: R=CH₂-t-Bu
Rh₂(S-CHAZ)₄: R=Cy
Rh₂(4S-R-MenthAZ)₄: R=L-Menthyl

(b)
1 R=4-BrC₆H₄, X=Br
2 R=4-t-BuC₆H₄, X=4-t-Bu
3 R=4-TMSC₆H₄, X=TMS

n=0, Ar=4-t-BuC₆H₄
n=0, Ar=2,4,6-(i-Pr)₃C₆H₂ **Rh₂(S-biTISP)₂**
n=1, Ar=4-t-BuC₆H₄
n=1, Ar=2,4,6-(i-Pr)₃C₆H₂
n=1, Ar=4-C₁₂H₂₅C₆H₄

Rh₂(R-PTAD)₄:
X= H, R=1-adamantyl
Rh₂(R-PTTL)₄: X= H, R=t-Bu
Rh₂(R-TCPTTL)₄: X=Cl, R=t-Bu
Rh₂(R-TBPTTL)₄: X=Br, R=t-Bu

Rh₂(R-TPCP)₄: R=H
Rh₂(R-BTPCP)₄: R=Br
Rh₂(R-BPCP)₄: R=Ph

Rh₂(R-BNP)₄

Rh₂(S-NTTL)₄: X=H
Rh₂(S-4-Br-NTTL)₄: X=Br

Rh₂(4S,2′R,3′R-HMCPIM)₄

Rh₂(OAc)(DPTI)₃

Rh₂(R-PTTL)₃TPA

Scheme 1.67 Chiral dirhodium catalysts for asymmetric cyclopropanations.

As a general guideline, the level of diasterocontrol with rhodium carbenes does not match that observed with copper, ruthenium, or cobalt carbenes, even when sterically hindered α-diazoesters are used. This drawback has minimized the use of rhodium catalysts in intermolecular processes involving simple α-diazoesters in the most ancient period of asymmetric synthesis. Then, better results were obtained when *ab initio* studies clarified the reaction mechanism.

Scheme 1.68 First model proposed for diastereoselective rhodium-catalyzed cyclopropanation.

The first model proposed to explain the diastereo- and enantiofacial selectivity using these rhodium catalysts is reported in Scheme 1.68 [270]. The presence of an electron-withdrawing substituent and an electron-donating substituent seemed crucial for high diastereoselectivity, because, when this combination of donor/acceptor functionality was lacking, much lower diastereoselectivities were observed.

Then, the approach of the alkene occurred on the side of the electron-withdrawing group, because the developing positive charge on the most substituted carbon could be stabilized by the oxygen lone pairs of the carboxyl group. This mechanism suggested that the reactive catalyst conformer had to possess D_2 symmetry, because the approach of the alkene to this conformer from the most accessible trajectory correctly predicted the sense of induction in these reactions. Thus, new rhodium(II) dicarboxylates with the prerequisite D_2 symmetry were prepared and tested ($Rh_2(S\text{-biTISP})_4$ and analogs), but three of them led to lower enantiomeric excesses than those observed with $Rh_2(DOSP)_4$ and only the triisopropylphenyl substituted catalyst with $n = 1$ was as effective as $Rh_2(DOSP)_4$. In 2003, Che et al. reported the use of a rhodium D_4-porphyrin to catalyze the cyclopropanation of various alkenes with EDA, providing TON>1000, but moderate enantioselectivities (up to 68% ee) and low E/Z ratios [271].

Then theoretical and kinetic isotope studies pointed out that the cyclopropanation occurs in a concerted nonsynchronous manner. The greater bulk of the electron-withdrawing group of the diazo compound results in high diastereoselectivity, because it avoids destabilization of the electrophilic carbenoid by aligning out of the plane of the rhodium–carbon π-bond. Based on X-ray structure and DFT calculations, Fox's research group proposed a structure ("all-up" conformation), in which a C_2-symmetric chiral cavity explained the enantioselectivity of typical carboxylate dirhodium complexes (Scheme 1.69) [272].

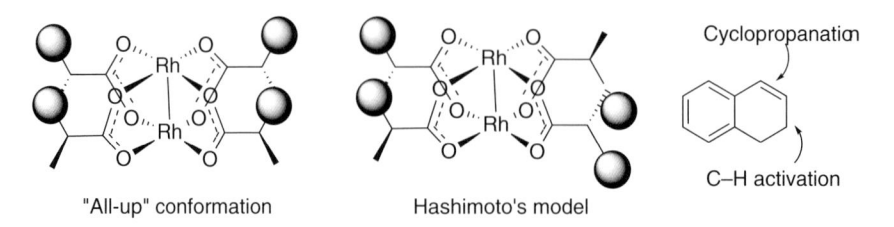

Scheme 1.69 Representation of the C_2-symmetric "all-up" chiral cavity and Hashimoto's model and cyclopropanation versus C–H activation reaction sites.

DFT calculations also demonstrated that the presence of chlorine or bromine atoms on the phthaloyl ring, such as in catalysts $Rh_2(S\text{-TCPTTL})_4$ and $Rh_2(S\text{-TBPTTL})_4$, significantly rigidifies the conformation in solution, through intramolecular halogen bonds between adjacent ligands and that the carbene leaves its *Si*-face accessible for reaction with the alkene. When the alkyl substituent is small, there is competing reactivity on the *Re*-face through a conformation, in which the carbene is aligned with the narrow dimension of the chiral cavity. In other cases, the stereochemistry was well predicted by Hashimoto's model [273], in which the chiral dirhodium complex adopts a C_2-symmetrical arrangement, with two adjacent groups positioned on the top face and the next two on the bottom face of the complex.

The TONs of these reactions were studied by tracking the disappearance of the $C=N_2$ stretch frequencies in the infrared spectra. In this way, in the reactions of styrene with diazoacetate, phenyl diazoacetate, and diazomalonate, the TONs were determined to be higher for the more stabilized donor/acceptor carbenoids [274]. Moreover, the highest TONs were achieved under solvent-free conditions with very active trapping agents, such as styrene and cyclopentadiene, in which the catalyst:substrate ratios were as low as 0.6 ppm, resulting in TONs of up to 1.8 million. However, since the reaction is exothermic, solvent-free reactions are difficult to perform on large scale.

Davies provided guidelines for choosing the optimal chiral dirhodium(II) catalyst for cyclopropanation of aryldiazoacetates. Depending on the aryl substituent, $Rh_2(R\text{-DOSP})_4$, $Rh_2(S\text{-PTAD})_4$, or $Rh_2(R\text{-BNP})_4$ can be utilized to obtain the desired cyclopropane with a high level of enantioinduction. The substitution on the alkene has instead only moderate influence on the level of asymmetric induction in cyclopropanation reactions [275].

Substrates with an allylic moiety have a potentially competing pathway to cyclopropanation: allylic C–H insertion (Scheme 1.69). The latter reaction will not be discussed here, but the factors influencing the two pathways should be mentioned. Davies and coworkers used *p*-bromophenyldiazoacetate as the carbenoid species, different electron-rich alkenes, and various dirhodium catalysts including $Rh_2(S\text{-biTISP})_4$, $Rh_2(S\text{-DOSP})_4$, and $Rh_2(S\text{-PTAD})_4$ [276]. C–H insertion more easily occurred in the reactions of *cis*-1,2-disubstituted alkenes or more highly substituted alkenes. In general, methyl allylic sites were much harder to functionalize than methylene allylic sites. Catalyst $Rh_2(S\text{-DOSP})_4$ gave predominantly the cyclopropanes in 70–88% yield, with >97:3 de and 89–95% ee. The other two catalysts were less selective toward the cyclopropanation, and $Rh_2(S\text{-biTISP})_4$ mainly favored C–H insertion. Finally, the $Rh_2(S\text{-DOSP})_4$ catalyst afforded opposite enantioinduction with respect to $Rh_2(S\text{-PTAD})_4$. Moreover, dihydronaphthalene, another good candidate for studies on cyclopropanation versus C–H activation, showed that the combination of methyl 2-diazopent-3-enoate and $Rh_2(S\text{-PTAD})_4$ strongly favored the cyclopropanation. On the other hand, the catalyst $Rh_2(S\text{-DOSP})_4$ and encumbered diazo compounds, such as methyl 3-(*tert*-butyldimethylsilyloxy)-2-diazopent-3-enoate, favored the C–H activation [277]. Interestingly in this reaction, C–H activation product was well predicted by Hashimoto's model [273].

Historically, the most ancient reports showed a low diasterocontrol, although the level of enantioselection was sometimes excellent [278]. Only $Rh_2(4S\text{-}R\text{-}$

MenthAZ)$_4$ led to good *cis*-selectivity (82:18 and 92:8 dr) and excellent enantiocontrol (97% ee, *cis*) in the cyclopropanation of substituted styrenes with *tert*-butyl diazoacetate, but with low yield relative to the alkene (44% and 21%) [279]. This catalyst also allowed the cyclopropanation of 3-diazo-3,6-dihydro-2*H*-pyran-2-one with various alkenes [74–86 yields, 83:17 to >95:5 dr, 73–86% ee (*S,S*)] [280]. Then the efficiency of a series of chiral azetidinone–dirhodium(II) catalysts was tested in the cyclopropanation of various olefins with diazomalonates. These reactions gave access to the corresponding cyclopropanes with enantioselectivities of up to 50% ee [278, 281]. Moreover, Rh$_2$(*S*-MEAZ)$_4$ afforded cyclopropanes from styrene and α-nitroesters in up to 76% yield, but with only 33% ee [149]. Charette's research group found Rh$_2$(*S*-IBAZ)$_4$ as an efficient catalyst for cyclopropanation of α-cyanodiazophosphonate and α-cyanodiazoacetate (Scheme 1.70) [282]. The particular electrophilicity of cyanocarbene intermediates permitted the use of allenes as substrates, affording the first catalytic asymmetric alkylidene cyclopropanation reaction using diazo compounds. In fact, α-cyanocarbenes are forced to stay in-plane, conversely from other electron-withdrawing groups, which adopt an out-of-plane conformation (see below). The in-plane conformation is highly energetic, thus leading to a more electron-deficient reactive carbene, allowing less nucleophilic π-systems such as allenes to react.

Charette *et al.* screened a variety of structurally diverse chiral dirhodium catalysts for enantioselectivity of styrene with α-nitro-α-diazo carbonyl compounds and observed modest-to-high yields in a wide range of solvents, but with only modest enantioselectivities (≤41% ee) [149]. A family of bisoxazoline complexes of coordinatively unsaturated monomeric rhodium(II) have been described by Tilley *et al.* and subsequently employed as catalysts for the cyclopropanation of olefins with EDA, giving excellent yields (66–94%) and enantioselectivities of up to 84% ee [283].

However, Rh$_2$(DOSP)$_4$ and Rh$_2$(TBSP)$_4$ were revealed as the most active catalysts [284]. For instance, Rh$_2$(DOSP)$_4$ provided the highest enantiocontrol for many alkenes, but *trans*-disubstituted alkenes did not react under these conditions. This Rh-catalyzed cyclopropanation was also applied to the synthesis of cyclopropane α-amino acids [285], from reaction of alkynyldiazoacetates (61–91% yields, 92:8 to 99:1 dr, 56–95% ee) [286], aryldiazoesters (82–90% yields, 60:40 to 98:2 dr, 80–97%

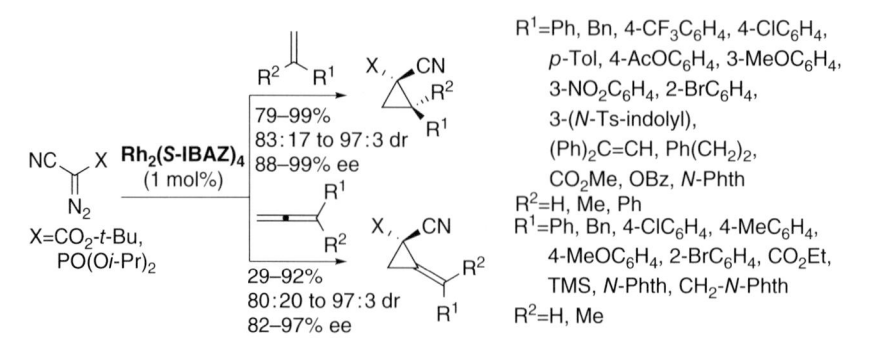

Scheme 1.70 Asymmetric cyclopropanation of α-cyanodiazophosphonate and α-cyanodiazoacetate.

ee) [287], and heteroaryldiazoacetates (92:8 to 99:1 dr, 23–89% ee) with olefins [288]. The reaction of ethyl 3,3,3-trifluoro-2-diazopropionate with various olefins catalyzed by $Rh_2(R\text{-}DOSP)_4$ was also explored [289]. The reaction between this diazo compound and 1,1-diphenylethylene gave 72% yield and 40% ee, whereas the cyclopropanation of monosubstituted olefins led to Z/E mixtures of the corresponding cyclopropanes with a maximum of 75% ee for 4-methoxystyrene.

On the other hand, $Rh_2(R\text{-}DOSP)_4$ was also used to induce the decomposition of aryldiazoacetates in the presence of pyrroles or furans, resulting in the formation of mono- or bis-cyclopropanes of the heterocycle (Scheme 1.71) [290]. It should be noted that the enantioinduction was markedly influenced by the structure of the heterocyclic substrate and depended upon which bond of the heterocycle initially interacted with the carbenoid, thus either face of the heterocycle can be attacked under the influence of the same chiral catalyst. The control was governed by a delicate interplay of steric and electronic influences. This methodology was applied to the total synthesis of a natural product, (+)-erogorgiaene, by the cyclopropanation of a dihydronaphthalene [291].

Moreover, this reaction was extended to a solid-phase cyclopropanation between phenyldiazoacetate and a resin-bound alkene by pyridine linker. The stereoselectivities were almost identical to those observed in solution [292]. However, this immobilization strategy showed high level of rhodium leaching upon recycling after the first reaction.

Therefore, a single ligand of $Rh_2(S\text{-}DOSP)_4$ was exchanged with a ligand that could undergo a grafting reaction to a solid support. As this substitution could break the high symmetry that is characteristic of these complexes, the attachment was carefully designed to limit negative influence on the enantioselectivity. In fact, the chiral pocket would be still maintained in the catalyst structures when one or more of the chiral carboxylate ligands of the rhodium complex were replaced (see the following discussion). The cyclopropanation catalyzed by this

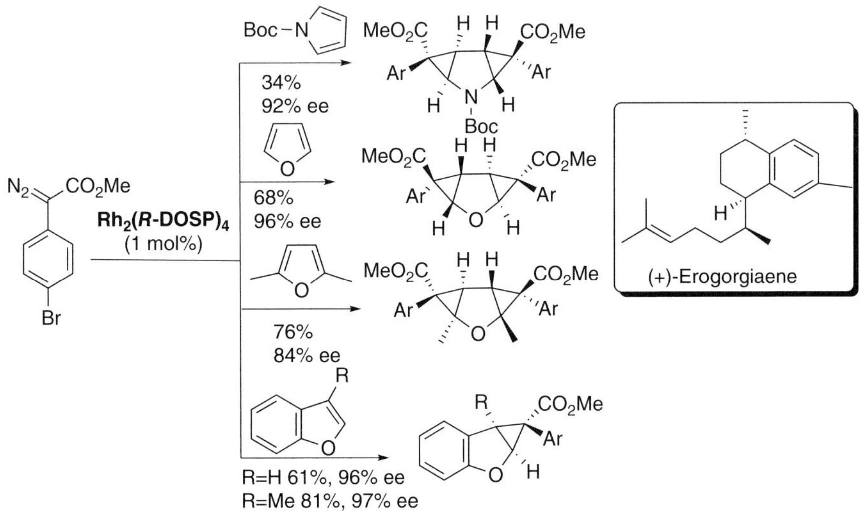

Scheme 1.71 Asymmetric cyclopropanation of pyrroles or furans catalyzed by $Rh_2[(R)\text{-}DOSP]_4$.

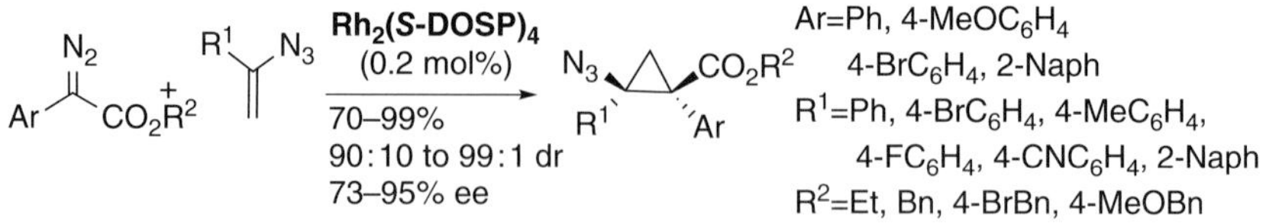

Scheme 1.72 Enantioselective preparation of cis-β-azidocyclopropane esters.

grafted catalyst was repeated over five consecutive reactions and always provided similar levels of both yield and enantioinduction [293]. Other chiral dirhodium catalysts could be immobilized on a polymer, providing yields and selectivities for cyclopropanations comparable to those with the homogeneous catalyst (up to 84% ee and 87% yield) [294]. In particular, Davies and Walji developed a very effective strategy for the heterogenization of chiral dirhodium catalysts, exhibiting all the reactivity features of the homogeneous catalysts, with the advantage of excellent recyclability [295].

Moreover, $Rh_2(S\text{-DOSP})_4$ provided mixtures of cyclopropanes and C–H insertion products in a 2:1 ratio in the reaction of diazoacetate and N-Boc ethyl alkenylamines. The (1S,2R)-cyclopropanes were recovered in 55–64% yield with 95:5 dr and 92–96% ee. Other rhodium catalysts were tested, but product distribution was not significantly affected [296].

β-Aminocyclopropane carboxylic acids are widely used in peptide syntheses, but they cannot be efficiently prepared from asymmetric cyclopropanation of N-protected enamines [297]. However, azidoalkenes could be regarded as alternative precursors of cis-β-aminocyclopropane carboxylic acids, and $Rh_2(S\text{-DOSP})_4$ was found to be an efficient catalyst for the preparation of (1R,2S)-isomers (Scheme 1.72) [298]. The reaction can be carried out on gram scale with the same yield but with slightly lower enantioselectivity and longer reaction time.

Finally, intermolecular asymmetric cyclopropanations of aryldiazoacetates with labile protecting groups on the ester and styrene derivatives can be catalyzed by chiral dirhodium(II) complexes $Rh_2(S\text{-DOSP})_4$ and $Rh_2(R\text{-BPCP})_4$. In particular, the trimethylsilylethyl aryldiazoacetates gave the best results with $Rh_2(S\text{-DOSP})_4$, while trichloroethyl aryldiazoacetates with $Rh_2(R\text{-BPCP})_4$ (Scheme 1.73) [299].

Similar dirhodium complex $Rh_2(R\text{-BTPCP})_4$ was found to be an effective chiral catalyst for the enantioselective cyclopropanation of styryldiazoacetates (Scheme 1.74) [300]. DFT computational studies at the B3LYP and UFF levels suggested that when the carbenoid binds to the catalyst, two of the 4-bromophenyl groups rotate outward to make room for the carbenoid. Then, the ester group aligns perpendicular to the carbene plane and blocks attack on its side. Thus, the substrate approaches over the donor group, but it finds the Re-face blocked by the aryl ring of the ligand and only the Si-face open for the attack, in agreement with the observed absolute configuration of the product.

Hansen's research group introduced catalysts $Rh_2(trans\text{-HYP})_4$ and $Rh_2(cis\text{-HYP})_4$ and compared them with $Rh_2(S\text{-DOSP})_4$ in the reaction of styrene and phenyldiazoacetate, finding very similar yields and enantiomeric excess values [301]. As expected, $Rh_2(trans\text{-HYP})_4$ and $Rh_2(cis\text{-HYP})_4$ displayed opposite enantioselectivity to each other, because the enantioselectivity of proline-derived

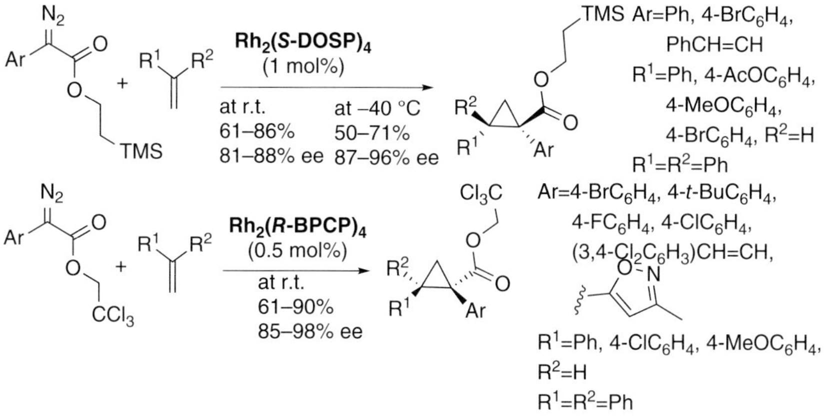

Scheme 1.73 Enantioselective synthesis of cyclopropanecarboxylates with labile protecting groups on the ester.

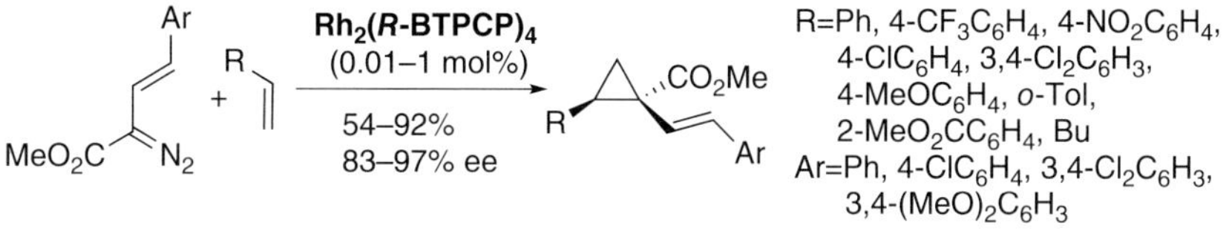

Scheme 1.74 Cyclopropanation of styryldiazoacetates.

catalysts is governed by the stereochemistry at the C2 of proline. These results are in agreement with an active "all-up" C_4-symmetric conformation.

Dirhodium complex, $Rh_2(R\text{-}PTAD)_4$ was found efficient in the reaction of 1-aryl-2,2,2-trifluorodiazoethanes with alkenes, generating the corresponding trifluoromethyl-substituted cyclopropanes, an example of the important class of fluorinated cyclopropanes [302], with very high diastereo- and enantioselectivities (Scheme 1.75) [303].

The enantiomer $Rh_2(S\text{-}PTAD)_4$ catalyzed the stereocontrolled synthesis of nitrile-substituted cyclopropanes from 2-diazo-2-phenylacetonitrile and aryl alkenes (Scheme 1.76, eq 1) [304]. It is worth noting that the small cyano acceptor group would be unable to influence the selectivity. In fact, as reported earlier, studies on the mechanism asserted that the bulkiness of the electron-withdrawing group drives the selectivity. This is true for alkylethenes (formed as 61:39 to

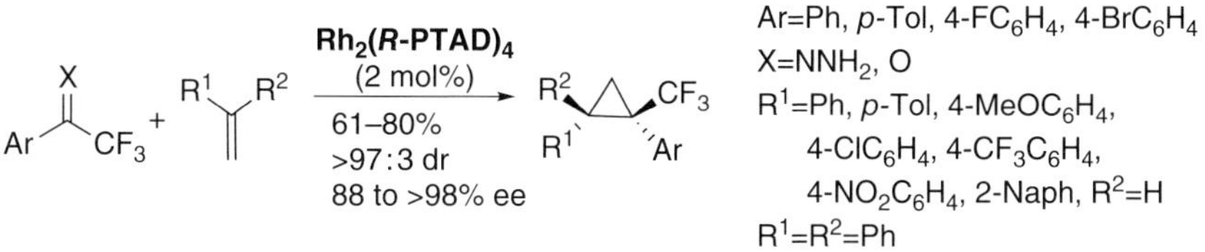

Scheme 1.75 Asymmetric synthesis of fluorinated cyclopropanes.

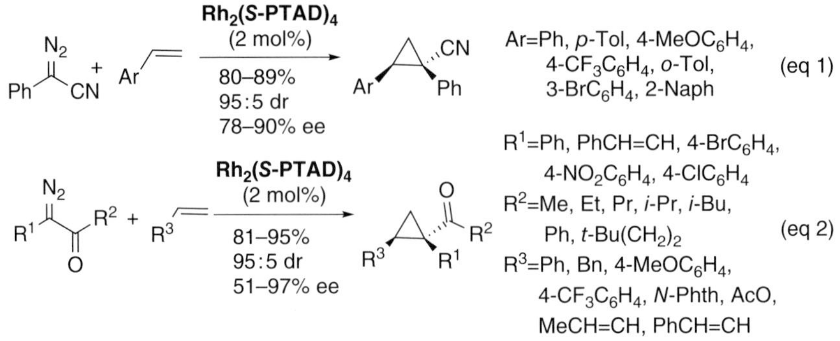

Scheme 1.76 Asymmetric synthesis of some cyclopropanes catalyzed by $Rh_2(S\text{-}PTAD)_4$.

46:54 mixtures of diastereomers), but high diastereoselectivity was observed with styrenes. Therefore, the authors claimed an attractive π-stacking interaction between the aryl rings of styrene and phthalimide during the cyclopropanation that is absent in alkyl-substituted alkenes. Catalyst $Rh_2(S\text{-}PTAD)_4$ also catalyzed the reaction of aryl-α-diazo ketones with activated alkenes (Scheme 1.76, eq 2) [305]. The enantioselectivity dropped when either the aryl group in R^1 was substituted with a styryl group or bulky groups were close to the carbonyl and increased when the alkyl chain of the ketone group was lengthened, but C–H insertion became a competing reaction for lengths beyond propyl. Vinyl acetate, dihydrofuran, and dienes were less enantioselective, while vinyl ether and inactivated alkenes were not effective substrates.

Similar results were obtained in the cyclopropanation of alkenes with diazoacetophenones under $Rh_2(S\text{-}TCPTTL)_4$ catalysis (Scheme 1.77) [306]. The reaction could be carried out on a multigram scale. Different substituted diazoacetophenones showed similar efficiency, except for the 4-dimethylamino derivative that achieved only 38% yield. Unfortunately, alkyl-substituted alkenes did not provide the corresponding cyclopropanes in useful yields, and dienes afforded only Cope-rearranged achiral products. The enantioselectivity outcome is in good agreement with an "all-up" conformation of the catalyst and once more with π-stacking interactions between the aryl ketone moiety and phthalimide, while the electron-withdrawing group (X) did not play a crucial role in the stereocontrol of the reaction. Thus, nitro, nitrile, or a methyl ester in the starting diazo derivative leads to high stereocontrol.

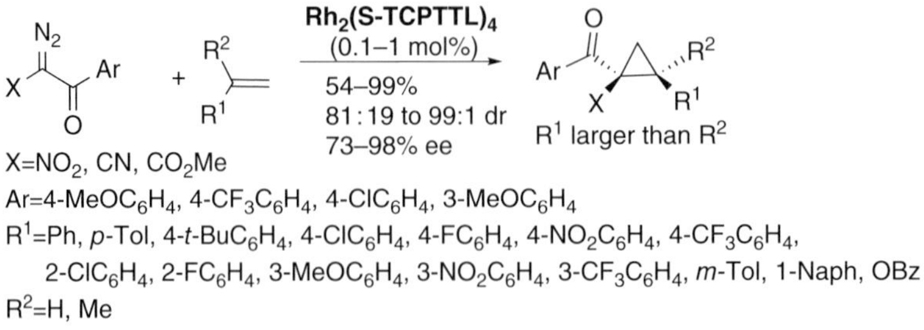

Scheme 1.77 Enantioselective synthesis of cis-cyclopropane α-amino acid precursors.

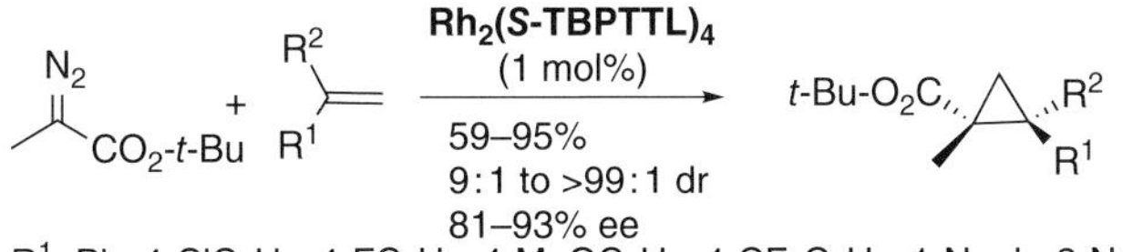

Scheme 1.78 Enantioselective cyclopropanation with α-diazopropionate.

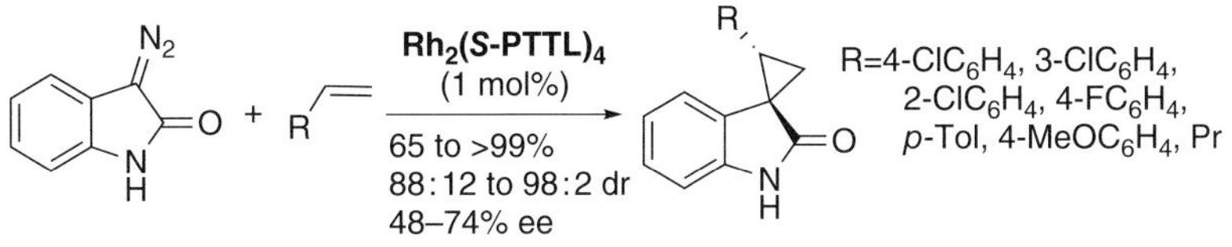

Reaction conditions: $Rh_2(S\text{-TBPTTL})_4$ (1 mol%), 59–95%, 9:1 to >99:1 dr, 81–93% ee.

R^1 = Ph, 4-ClC$_6$H$_4$, 4-FC$_6$H$_4$, 4-MeOC$_6$H$_4$, 4-CF$_3$C$_6$H$_4$, 1-Naph, 2-Naph, PhCH=CH, Bu
R^2 = H, Me, Ph

Scheme 1.79 Enantioselective synthesis of spirocyclopropyloxindoles.

Reaction conditions: $Rh_2(S\text{-PTTL})_4$ (1 mol%), 65 to >99%, 88:12 to 98:2 dr, 48–74% ee.

R = 4-ClC$_6$H$_4$, 3-ClC$_6$H$_4$, 2-ClC$_6$H$_4$, 4-FC$_6$H$_4$, p-Tol, 4-MeOC$_6$H$_4$, Pr

Hashimoto described that the reaction of 1-aryl-substituted and related conjugated alkenes with *tert*-butyl α-diazopropionate by catalysis with $Rh_2(S\text{-TBPTTL})_4$ led to the corresponding (1R,2S)-cyclopropanes containing a quaternary stereogenic center (Scheme 1.78) [307].

Awata and Arai achieved the asymmetric cyclopropanation of diazooxindoles with $Rh_2(S\text{-PTTL})_4$ as the catalyst. Spirocyclopropyloxindoles, which constitute biologically important compounds, were obtained in good yield and diastereoselectivity (Scheme 1.79) [308]. Then the mechanism of this reaction was detailed by DFT calculations, which demonstrated that the origin of the *trans*-diastereoselectivity lies in the π–π interactions between the *syn*-indole ring in carbenoid ligand and the phenyl group in styrene. The enantioselectivity could be ascribed both to steric interaction between the phenyl ring in styrene and the phthalimide ligand and to stabilization of π–π and CH–π interactions in the transition states [309].

$Rh_2(S\text{-biTISP})_2$ was able to catalyze the cyclopropanation of styrene with methyl phenyldiazoacetate with high turnover number (92 000) and turnover frequency (4000 h^{-1}) [310]. This is one of the rare examples of a large-scale reaction with rhodium catalysts; in fact, with a substrate/catalyst ratio of 100 000, 92% yield and 85% ee were obtained on a crude of 46 g. In addition, this catalyst, immobilized on highly cross-linked polystyrene resins with a pyridine attachment, provided up to 88% ee for this reaction [311]. Finally, the same catalyst was applied to the stereoselective synthesis of cyclopropylphosphonates containing quaternary stereocenters by the reaction of dimethyl aryldiazomethylphosphonates (Scheme 1.80) [312].

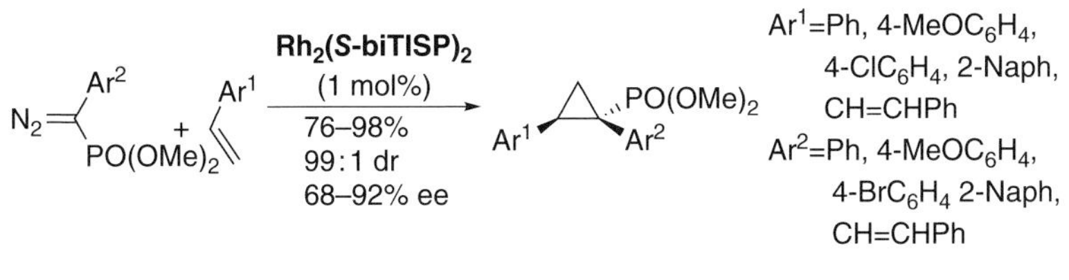

Reaction conditions: $Rh_2(S\text{-biTISP})_2$ (1 mol%), 76–98%, 99:1 dr, 68–92% ee.

Ar^1 = Ph, 4-MeOC$_6$H$_4$, 4-ClC$_6$H$_4$, 2-Naph, CH=CHPh
Ar^2 = Ph, 4-MeOC$_6$H$_4$, 4-BrC$_6$H$_4$, 2-Naph, CH=CHPh

Scheme 1.80 Asymmetric synthesis of cyclopropylphosphonates catalyzed by $Rh_2(S\text{-biTISP})_2$.

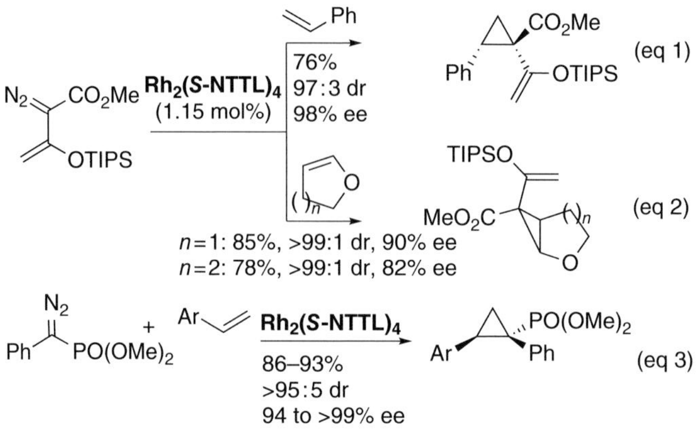

Scheme 1.81 Asymmetric cyclopropanations with $Rh_2(S\text{-}NTTL)_4$.

The $Rh_2(NTTL)_4$ catalyst allowed an exceptional diastereo- and enantioselective cyclopropanation of styrene (Scheme 1.81, eq 1), dihydrofuran, and dihydropyran (Scheme 1.81, eq 2) with (silanyloxyvinyl)diazoacetates [313]. This methodology, applied to alkyl diazo(trialkylsilyl)acetates, furnished the corresponding cyclopropanes in good yield, but with low enantioselectivity (<54% ee) [314]. Hence, the use of ethyl diazo(triethylsilyl)acetate led to 69% yield, with 82:18 dr and 54% ee. The same catalyst was also the best-performing catalyst in the synthesis of cyclopropylphosphonate derivatives (compare Schemes 1.80 and 1.81, eq 3) [315]. The selectivity is independent of the size of the phosphonate group and can be predicted by Hashimoto's model. The use of Meldrum's acid, methyl diazostyrylacetate, and methyl diazophenylacetate instead of the phosphonate ester group dramatically deteriorated the asymmetric induction.

Then, studies on the shape of the complex $Rh_2(S\text{-}NTTL)_4$ demonstrated that the 4-substituent lies at the cavity rim, thus exerting a strong influence on the enantiofacial discrimination of the incoming alkene during carbene transfer [316]. Therefore, larger 4-substituents should improve the outcome, and bromine was found to be the best choice. Actually, the complex $Rh_2(S\text{-}4\text{-}Br\text{-}NTTL)_4$ catalyzed the one-pot asymmetric cyclopropanation of alkenes with ylides from dimethyl malonate or Meldrum's acid producing the S- and R-isomers for aromatic and aliphatic alkenes, respectively (Scheme 1.82).

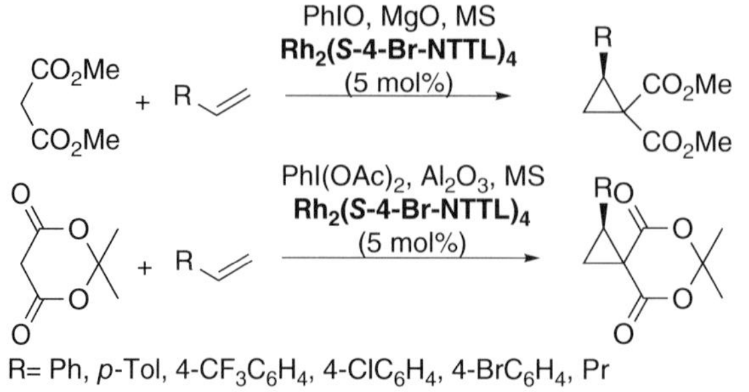

R = Ph, p-Tol, 4-CF$_3$C$_6$H$_4$, 4-ClC$_6$H$_4$, 4-BrC$_6$H$_4$, Pr

Scheme 1.82 Asymmetric cyclopropanation of alkenes by using *in situ* generated ylides.

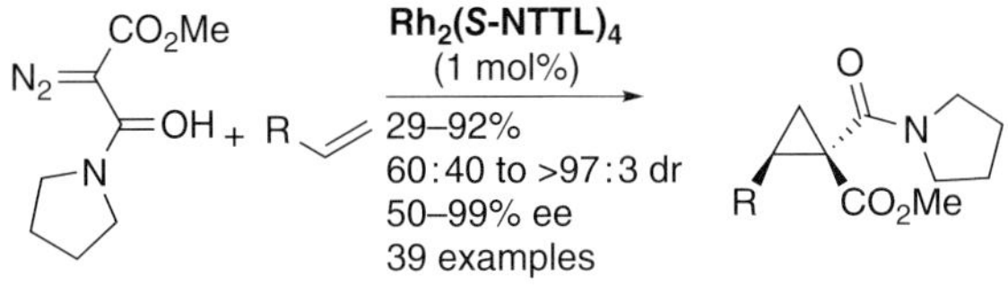

Scheme 1.83 Asymmetric cyclopropanation of alkenes with diazo reagents having "trans-directing ability" of the amide.

As suggested by Marcoux and Charette, the so-called *trans*-directing ability of the amide group present in the diazo derivative could increase $Rh_2(S\text{-}NTTL)_4$ selectivity (Scheme 1.83) [317]. The in–out conformation was preferred in the transition state, and the larger amide group was the out-of-plane group. Then the alkene attacked the electrophilic carbene by orienting its largest substituent on the side of the in-plane ester group and the chiral ligands discriminate between the two transition-state faces. Aliphatic alkenes afforded only trace amounts of the corresponding cyclopropane, but the problem was overcome by selective cyclopropanation of the less-hindered double bond of a diene, followed by hydrogenation. The synthetic versatility of these cyclopropanes was demonstrated by their transformation into different derivatives with preservation of the enantiomeric purity. The addition of a sulfonic amide for decomposing the diazo derivative did not affect enantioselectivity and yield; thus, various achiral additives were checked. If the diazo derivative possessed two acceptor groups (such as NO_2, CN, RCO, CO_2R), the presence of trifluoromethanesulfonamide and DMAP enhanced the stereoselectivity, thus suggesting a correlation between the additive and the corresponding symmetry of the catalyst, but how additives increase the enantiomeric excess and diastereomeric ratio remains unclear. Donor–acceptor rhodium(II) carbenes instead were negatively influenced by the additives.

N-Sulfonyl-1,2,3-triazoles are stable synthetic equivalents of unstable azavinyl carbenes and allow the synthesis of cyclopropanecarboxaldehydes from either aromatic or aliphatic alkenes (Scheme 1.84) [318]. It is surprising that *trans*-1-phenylpropene produced the corresponding cyclopropanecarboxaldehyde with excellent 98% enantioselectivity, while the *cis* analog delivered almost racemic product. Starting triazoles can be easily prepared either by cycloaddition of sulfonylazides and nitriles or more simply by *in situ* generation of N-triflyl azavinyl carbenes from NH-1,2,3-triazoles treated with triflic anhydride in the presence of $Rh_2(S\text{-}NTTL)_4$.

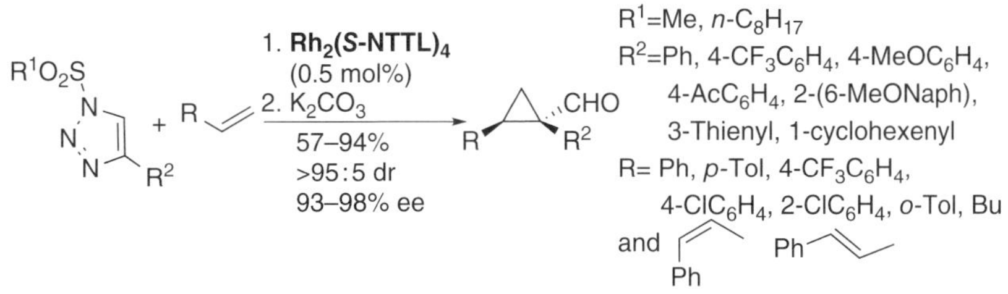

Scheme 1.84 Asymmetric cyclopropanation with N-sulfonyl-1,2,3-triazoles.

Complex Rh$_2$(4S,2′R,3′R-HMCPIM)$_4$ allowed the cyclopropanation of styrene with EDA in 68% ee and 59% yield, but with almost no diastereoselectivity [319]. Tetrakis-dirhodium(II)-(S)-N-(n-perfluorooctylsulfonyl)prolinate (named (**a**) in Scheme 1.67) has been used in a homogeneous or fluorous biphasic manner, displaying the best chemo- and enantioselectivities in the cyclopropanation of styrene with ethyl phenyldiazoacetate, (81% yield, 74% ee), when using perfluoro(methylcyclohexane) as the solvent [320].

Sambasivan and Ball described the preparation of rhodium–peptide catalysts for use in cyclopropanation reactions [321]. Over 200 sequences were prepared, and the so-called normal provided *Re*-face addition products and were characterized by bulk amino acids (i.e., leucine, isoleucine, phenylalanine) at the second and sixth positions. In contrast, polar carboxamide side chains (glutamine and asparagine) at the same positions characterized the "enantiomeric" sequences, providing *Si*-face addition. Both sequences carried two aspartate moieties in the third and seventh positions, which favored the formation of cyclopropanes. The reaction of aryldiazoacetates and monosubstituted alkenes with these catalysts (0.25 mol%) afforded (1R,2S)- or (1S,2R)-cyclopropanecarboxylates in 42–99% yields and with 38–97% ee.

The chiral pocket of dirhodium complexes is still maintained in the catalyst structures when one or more of the chiral carboxylate ligands of the rhodium complex are replaced by an achiral ligand (heteroleptic complexes). This chance has already been mentioned for the immobilization of Rh$_2$(DOSP)$_4$, but it was also applied in homogeneous catalysis. For instance, dirhodium complexes bearing bulky *ortho*-metallated arylphosphines (Scheme 1.67, group **b**) produced high *cis*-diastereo- and (1R,2S)-enantioselectivities in the cyclopropanation of styrene with EDA [322]. The substituents largely influenced the diastereoselectivity of the reaction, since increasing the size of the substituents, Br < *t*-Bu < TMS, Z/E ratios went from 53 : 47 to 90 : 10, but yields decreased from 80% to 39%. Enantiomeric excesses of the *cis*-isomer ranged from 81% to 91% ee without correspondence with substituent bulkiness. Interestingly, similar reaction could be performed in water as the solvent. Moreover, the immobilization of these catalysts on a cross-linked polystyrene resin gave higher yields, compared to those obtained with the standard homogeneous trifluoroacetate derivatives, whereas the diastereo- and enantioselectivities were generally lower in the cyclopropanation of styrene with EDA [323]. Corey and coworkers reported the reaction of styrene with EDA in the presence of another dirhodium catalyst, Rh$_2$(OAc)(DPTI)$_3$, affording the corresponding product in 84% yield with 67 : 33 Z/E ratio, 99% ee (*cis*), and 94% ee (*trans*) [324]. It is noteworthy that the major diastereomer was the *cis*-cyclopropane. Later, Charette's and Fox's research groups independently hypothesized that a large aromatic surface area in the achiral ligand was necessary for maintaining the selectivity [325]. In particular, Fox used Rh$_2$(S-PTTL)$_3$-TPA in the cyclopropanation reactions of α-alkyl-α-diazo esters with aliphatic and aromatic alkenes, obtaining 53–99% yields, with 71 : 29 to 99 : 1 dr and 65–97% ee, values comparable or superior to those achieved with Rh$_2$(S-PTTL)$_4$ as the catalyst [325b].[5] Charette's research group prepared various

---

[5] It should be noted that α-alkyl-α-diazo esters are known to undergo β-hydride elimination to form alkenes, but sterically demanding carboxylate ligands drastically decrease this side reaction.

heteroleptic complexes and tested them in the cyclopropanation reaction of styrene with α-nitrodiazoacetophenones [325a]. Thus, the replacement of one tetrachlorophthalimide ligand from $Rh_2(S\text{-TCPTTL})_4$ with phthalimide, succinimide, or 1,8-naphthalimide ligands did not significantly affect the asymmetric induction, whereas 2-naphthylacetate as the fourth ligand furnished a racemic product. The absence of enantioinduction was ascribed to a lack of rigidifying halogen bonds in the 2-naphthylacetate complex and to the absence of the N-imido moiety evidently necessary in all ligands to achieve a high asymmetric induction, independently of whether or not the fourth carboxylate is chiral. Charette also found that the asymmetric induction increased, replacing one of the four chiral ligands with a ligand that has a *gem*-dimethyl group instead of the chiral center, because of a conformational change in the catalyst owing to the presence of the two methyl groups in the fourth ligand.

Finally, just one rhodium(I) chiral catalyst was reported for the cyclopropanation of alkenes with dimethyl diazomalonate (Scheme 1.85) [326]. By using the (*R*,*R*)-configured tetrafluorobenzobarrelene complex, the *S*-configured cyclopropanes have been recovered. The reaction of α-methylstyrene gave only 57% ee, and in the reaction of 4-phenylbut-1-ene, as the representative of aliphatic alkenes, the enantioselectivity and yield were both low. Experimental evidence supported a transition state wherein the carbonyl oxygen on the ligand was coordinated to the rhodium(I) center. An active single coordination site on the rhodium cation was essential for the catalytic activity. In fact, the more bonded chloride ion, instead of the tetraborate, was not catalytically active.

1.3.1.5 Chiral Catalysts: Ruthenium

The success of the rhodium complexes in catalyzing carbene-transfer reactions is tempered by the high price of this metal. Therefore, ruthenium, a direct neighbor of rhodium in the periodic table, has been more recently introduced in the field of catalytic cyclopropanation, because it costs roughly one-tenth the price of rhodium. Another reason for focusing attention on ruthenium catalysts is the

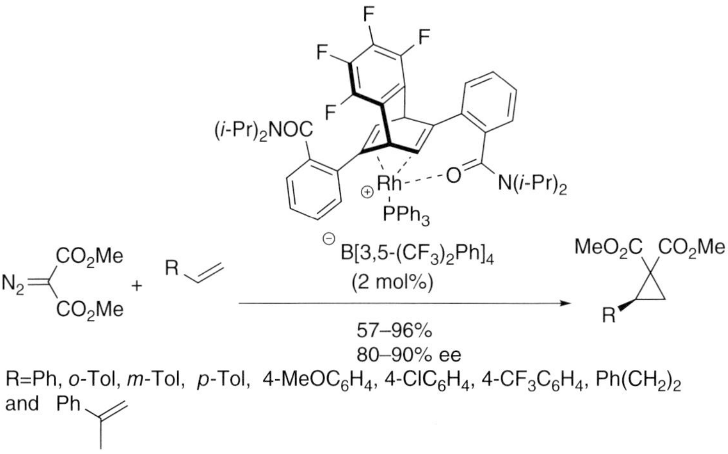

Scheme 1.85 Asymmetric cyclopropanation catalyzed by a rhodium(I) complex.

greater diversity of complexes to be evaluated, due to the richer coordination chemistry, as compared to rhodium [327].

Thus, many highly active and selective homogeneous catalysts have been introduced for the asymmetric cyclopropanation of alkenes (Scheme 1.86) [328]. Indeed, in a short time, ruthenium has emerged as the third important catalyst metal for the carbenoid chemistry of diazo compounds, besides copper and rhodium. However, a significant drawback of Ru catalysts is the rather low electrophilic character of the presumed ruthenium–carbene intermediates, which often restricts the application to terminal activated alkenes and double bonds with a higher degree of alkyl substitution. Another limitation of some ruthenium complexes is the ability to catalyze other alkene reactions as well as cyclopropanation leading to many by-products. However, if ruthenium catalysts work successfully, they often rival rhodium catalysts in terms of effectiveness and relative, as well as absolute, stereochemistry.

Some methods of heterogenization of ruthenium catalysts, for instance, supporting them on polymer or porous silica supports, have been investigated. Their activity, selectivity, and recyclability have all been compared to those of the analogous homogeneous catalysts.

By commenting on the Scheme 1.58, eq 1, Nishiyama's catalyst [329] has already been mentioned as better catalyst compared to Cu-*ent*-**box5** catalyst in the synthesis of cyclopropylphosphonate derivatives [213]. In the reactions with pybox ligands, the geometry of the recovered products is consistent with a model, in which the phenyl group of styrene approaches the carbenoid species away from the ester and isopropyl groups. Two interesting observations have been made.

i) A remote stereoelectronic effect exerted by the substituent in the 4-position of the pyridine ring has been reported [330]. Electron-donating substituents decrease (84% ee for the cyclopropanation of styrene if X = NMe$_2$), and electron-withdrawing groups increase the enantiomeric excess significantly (X = CO$_2$Me, 97% ee). The *E/Z* ratios were instead not affected by the substituents.

ii) Non-C_2-symmetric ligands are also quite effective in this reaction. For example, Ru-**Pybox4** afforded the cyclopropane not only with high enantioselectivity but also with an improved diastereoselectivity, very likely because the removal of one of the oxazoline substituents created more space for the ester group in the chiral pocket [331].

Garcia and coworkers reported an extensive comparison of the two enantioselective catalytic systems Ru-**Pybox** and Cu-**box** complexes by *ab initio* calculations in the cyclopropanation of alkenes with methyl diazoacetate [332]. The geometries of the key reaction intermediates and transition structures calculated at the QM/MM level were generally in satisfactory agreement with the full-QM calculated geometries. Furthermore, the QM/MM energies were often in better agreement with the stereoselectivity experimentally observed compared to full-QM calculations.

Later, Deshpande *et al.* used Nishiyama's catalyst to catalyze the cyclopropanation of styrene with EDA, providing the corresponding *trans*-cyclopropane in 98% yield, with 96:4 dr, and 86% ee (*trans*) [333].

Moreover, 1-tosyl-3-vinylindoles were excellently cyclopropanated by Nishiyama's catalyst with ethyl and *t*-butyl diazoacetate (Scheme 1.87) [334]. It

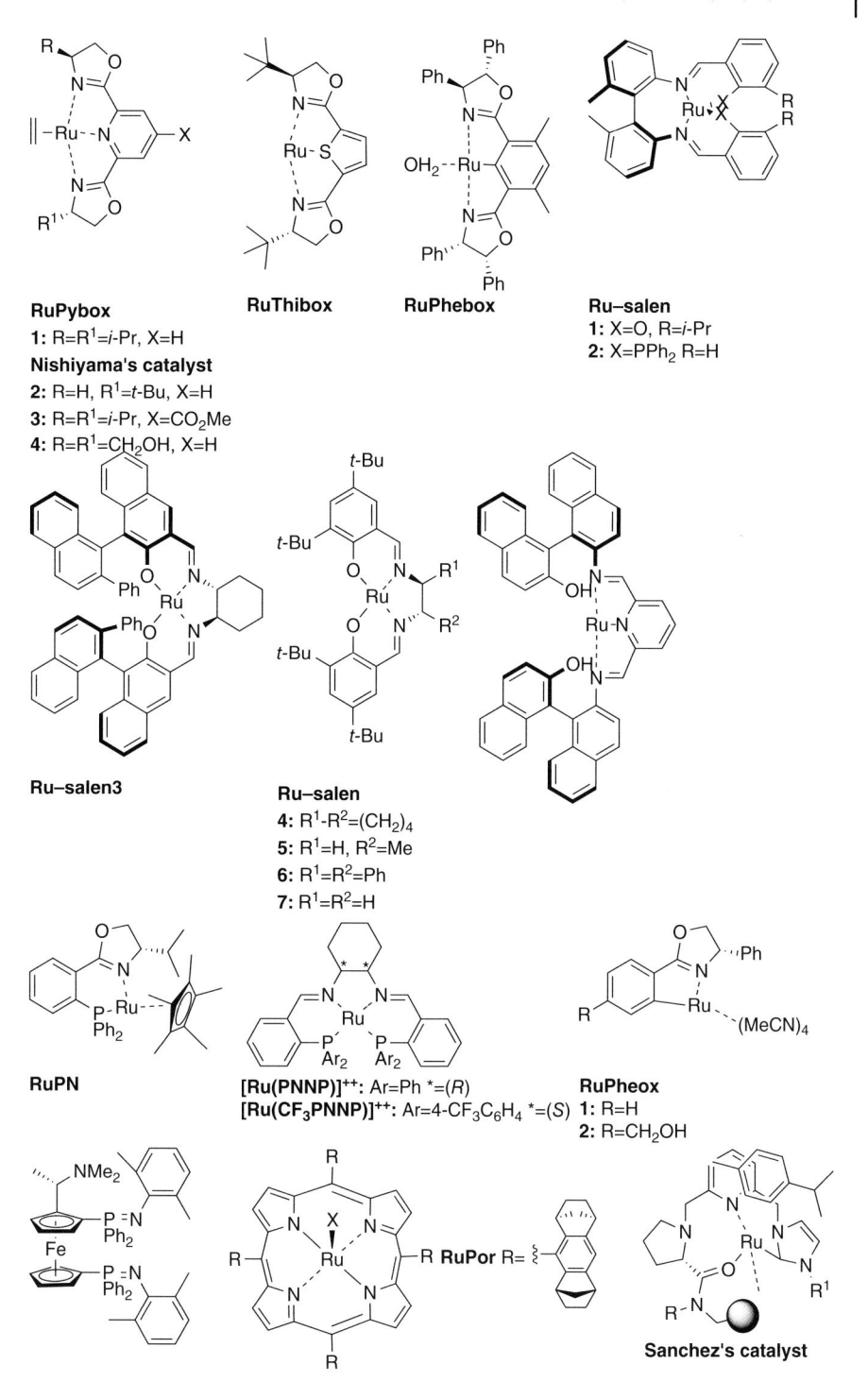

RuPybox
1: R=R¹=i-Pr, X=H
Nishiyama's catalyst
2: R=H, R¹=t-Bu, X=H
3: R=R¹=i-Pr, X=CO₂Me
4: R=R¹=CH₂OH, X=H

RuThibox

RuPhebox

Ru–salen
1: X=O, R=i-Pr
2: X=PPh₂ R=H

Ru–salen3

Ru–salen
4: R¹-R²=(CH₂)₄
5: R¹=H, R²=Me
6: R¹=R²=Ph
7: R¹=R²=H

RuPN

[Ru(PNNP)]⁺⁺: Ar=Ph *=(R)
[Ru(CF₃PNNP)]⁺⁺: Ar=4-CF₃C₆H₄ *=(S)

RuPheox
1: R=H
2: R=CH₂OH

RuPor R=

Sanchez's catalyst

Scheme 1.86 Chiral ruthenium catalysts for asymmetric cyclopropanations.

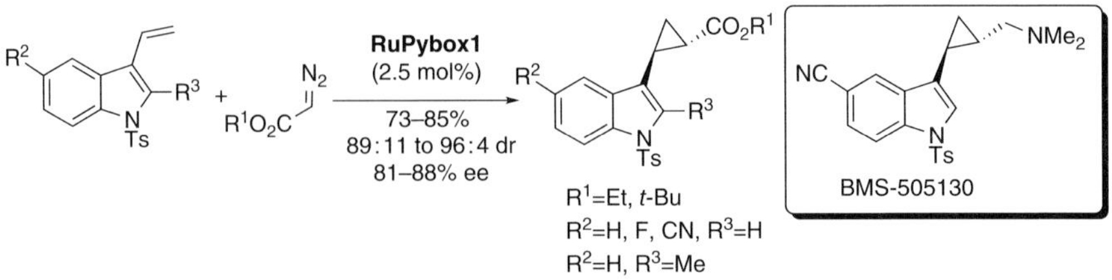

Scheme 1.87 Asymmetric cyclopropanation of 1-tosyl-3-vinylindoles.

should be noted that the *E/Z* diastereoselectivity was notably improved when using *t*-butyl diazoacetate. Moreover, the utility of this method was demonstrated by the conversion of one of the resulting chiral cycloadducts into BMS-505130, a selective serotonin reuptake inhibitor.

Nishiyama also developed the water-soluble hydroxymethyl derivative (Ru-**Pybox4**). The reaction of styrene with different diazoacetates in aqueous media provided the corresponding cyclopropanes in 24–75% yields, with 92:8 to 97:3 *E/Z* ratio, 57–94% ee (1*S*,2*S*), and 26–76% ee (1*R*,2*S*) [335].

Simmoneaux *et al.* examined chiral 2,6-bis(thiazolinyl)pyridines as ligands for the Ru-catalyzed cyclopropanation of olefins. The comparison of the enantiocontrol for the cyclopropanation of styrene with chiral ruthenium bisoxazoline and bisthiazoline allowed the evaluation of the different situation with regard to the diastereoselectivity and enantioselectivity when an oxygen atom was substituted by sulfur. They found many similarities with, in some cases, good enantiomeric excesses (up to 84% ee for *trans*-cyclopropylphosphonate were observed) [336].

Nishiyama's catalyst was immobilized in different manners:

i) By grafting on a Merrifield-type resin and on supports prepared by the polymerization of 4-vinyl-substituted ligands, providing yields of over 60% with up to 91% ee in four successive reactions. The enantioselectivity and the recyclability were strongly dependent upon the catalyst preparation method and the total exclusion of oxygen and moisture in the filtration process [337].
ii) By preparing Ru-**Pybox** monolithic miniflow reactors on styrene–divinylbenzene polymeric backbones having different compositions and pybox chiral moieties. Under conventional conditions and in supercritical carbon dioxide, the continuous flow cyclopropanation reaction between styrene and EDA led to good enantioselectivity (up to 83% ee) demonstrating these highly efficient and robust heterogeneous chiral catalyst and allowing the development of environmentally friendly reaction conditions [338].
iii) By microencapsulation into linear polystyrene (60–68% yields, 75–85% ee in up to four successive reactions were achieved in the benchmark cyclopropanation reaction between styrene and EDA) [339].
iv) On modified starch, providing the cyclopropanation of styrene with EDA in 67% yield, 89:11 dr and 77% ee for the *trans*-isomer [340].

Zingaro and coworkers tested a modified Nishiyama's catalyst (Ru-**Thibox**) and obtained 70–82% yields with 79:21 to 82:18 *E/Z* ratio and 87% to >99% ee

(1*R*,2*R*), 82% to >99%ee (1*S*,2*R*) for the cyclopropanation of styrenes and 1,1-diphenylethene with EDA [341].

Bis(oxazolinyl)phenyl ruthenium complex (Ru-**Phebox**) was efficient for the cyclopropanation reactions of various styrene derivatives with *tert*-butyl diazoacetate (85–92% yields with 82:18 to 96:4 *E/Z* ratio and 98–99% ee (1*R*,2*R*)) [342]. Only α-methylstyrene afforded the *cis*-isomer (80% overall yield, 67:33 dr, 98% ee (*cis*), and 93% ee (*trans*)). The cyclopropanation of aliphatic alkenes proceeded in lower yield but with good diastereo- and enantioselectivities, whereas cyclopropanation of 1,2-disubstituted alkenes, such as 1-phenylpropene or indene, did not occur. The ruthenium carbene intermediate should be obtained by replacement of the equatorial H_2O ligand with the diazoacetate group, and then the alkene approached the *Re*-face to minimize the steric repulsion between the *tert*-butyl group of the diazo compounds and the R group of the alkene.

Ru-**salen1–3** systems displayed *cis*-selectivity in the cyclopropanation reaction (83:17 to 93:7 *Z/E* ratios, >97% ee) [343]. In particular, catalyst Ru-**salen2** was effective for the cyclopropanation of 2,5-dimethyl-2,4-hexadiene, producing the *cis*-isomer in 75% ee (94:6 dr) but only in 18% recovered yield [343a–c]. Ru-**salen4–6**, with the two free coordinating sites occupied by pyridine ligands, gave excellent enantiomeric excesses in the cyclopropanation of mono or 1,1-disubstituted alkenes (30–97% yields, 66:34 to >99:1 *E/Z* ratios, 69–99% ee (*trans*)) [344]. The cyclopropanation of styrene with EDA in the presence of Ru-***ent*-salen4** catalysts (99% yield, 92:8 *E/Z* ratio, 99% ee (1*R*,2*R*) and 96% ee (1*S*,2*R*)) constituted the key step of the synthesis of chiral *trans*-cyclopropyl β-amino acid derivatives [345]. Jones and coworkers modified Ru-**salen4-(Py)$_2$** complex by substituting one of the *para-tert*-butyl groups with a linker for grafting onto polymer or mesoporous silica [346]. The heterogeneous catalysts generated the desired cyclopropanes, by reaction with EDA and styrenes, in good *trans*-selectivity but with moderate enantioselectivities. Aliphatic alkenes also reacted but in very low yields. The highest selectivities and yield were obtained with longer linker between the Ru-**salen** active site and the polymer support, probably for a lessened steric hindrance around the catalytic site. The silica support resulted in important background reactions on the silica surface, thus lowering enantioselectivity. The addition of pyridine during the washing steps between catalytic cycles stabilized the complex, preventing both leaching of this ligand and losses in activity and selectivities upon recycling. Better results were obtained with a pair of interpenetrated and noninterpenetrated chiral metal organic frameworks constructed by incorporation of Ru-**salen4** into the zinc–carboxylate cubic unit [347]. Both catalysts, differing by the open channel sizes, gave products with good diastereoselectivities (up to 91:9) and very high enantioselectivities (up to >99%), albeit in low yields (11–55%), in the reaction of terminal alkenes and diazoacetate. The noninterpenetrated catalyst gave higher isolated yields of cyclopropanation products, presumably because of the larger open channels.

A different approach in the Ru-salen catalyzed cyclopropanations was proposed by Nguyen and coworkers, who achieved successful induction of asymmetry by introducing (*R*)-methyl-*p*-tolylsulfoxide as axial ligand on achiral Ru-**salen7** catalyst (Scheme 1.88) [348]. The proposed mechanism to explain the asymmetric induction involved the axial coordination of the chiral sulfoxide to the ruthenium

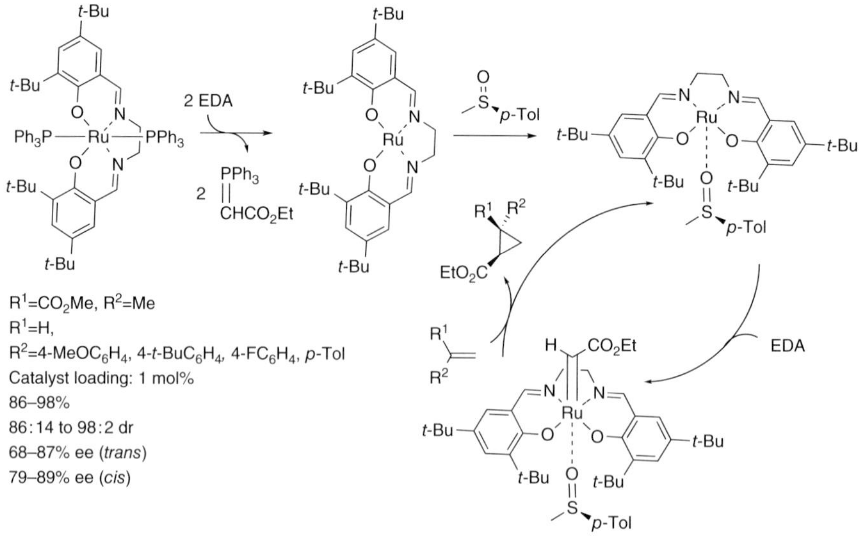

Scheme 1.88 Asymmetric cyclopropanation in the presence of a chiral sulfoxide additive.

center as the key induction step. The chiral additive bound preferentially to one of the two chiral conformers of the achiral (salen)ruthenium complex, thus effectively forcing the larger achiral salen ligand to adopt a preferred chiral conformation. The asymmetry of the additive was transmitted and amplified to the opposite axial position, where the other axial triphenylphosphine ligand was replaced by EDA, thus forming the ruthenium carbene. Finally, the carbene stereoselectively interacted with the olefin to complete the cyclopropanation cycle.

Several chiral diphosphines salen-like complexes, such as the chiral [RuCl(PNNP)]$^+$SbF$_6^-$, were used as ruthenium catalysts [349]. This catalyst was able to cyclopropanate various alkenes with EDA with cis-diastereoselectivity [33–69% yields, 85:15 to 99:1 Z/E ratio, 96–99% ee (1R,2S), 58–98% ee (1S,2S)]. Later, the same research group compared this catalyst with chiral [Ru(OH$_2$)$_2$(PNNP)]$^{2+}$ and the chloride-free catalysts formed from [RuCl$_2$(PNNP)] and AgSbF$_6$ and found both to be less effective than [RuCl(PNNP)]$^+$SbF$_6^-$ [350].

Mezzetti's group also investigated the possibility of controlling the absolute configuration at the metal center by means of both monodentate chiral phosphoramidite [351] and chiral phosphine [352] ligands (Scheme 1.86 for the structures). After activation with TlPF$_6$, AgSbF$_6$, or (Et$_3$O)PF$_6$ as halide scavenger, these complexes were used to catalyze the cyclopropanation of styrene and α-methylstyrene with EDA, but the results were always unsatisfactory.

Another chiral phosphine bidentate ligand (Ru**PN**) was tried out for the cis-selective cyclopropanation of styrene with EDA with 36% yield 75:25 Z/E ratio, and 74% ee [353]. Better results were obtained by ruthenium complexes of chiral (iminophosphoranyl)ferrocenes (Scheme 1.86 for the structure); in fact, cis-cyclopropanecarboxylates from a range of olefins with EDA were recovered in up to 79% yield, with up to 95:5 dr and up to 99% ee [354]. Indeed, this new type of ligands were shown to be comparable to, or better than, the well-known ligands, such as bisoxazolines or semicorrin, in terms of asymmetric induction.

Recently, Ru-**pheox1** was found to be an efficient catalyst for the cyclopropanation reaction of monosubstituted alkenes with succinimidyldiazoacetate [355]. The desired cyclopropanes were obtained in 94–98% yield with >99:1 E/Z ratio. The enantiomeric excess was calculated after reduction without epimerization of succinimidyl cyclopropanecarboxylates to cyclopropylmethanols and found in the range of 91–99% (1R,2R). The preferred face for the attack of the ruthenium carbene by the terminal alkene was determined by the steric crowding around the seven-membered ring resulting after the coordination between the succinimidyl carbonyl group and the ruthenium metal center. Then, the same research group applied the Ru-**Pheox1** catalyzed cyclopropanation:

1) To vinylcarbamates with diazoesters (77–99% yields, up to 96:4, with N,N-disubstituted vinylcarbamates and up to 99% ee). However, the reaction of carbobenzyloxyvinylamine and *tert*-butyldiazoacetate or succinimidyl diazoacetate led to about equimolecular amounts of *cis*- and *trans*-isomers, the latter with low enantiomeric excess [356].
2) To diazomethylphosphonate with alkenes (72–93% yields, 62:38 to 99:1 E/Z ratio, 94–99% ee (R,R)) and α,β-unsaturated carbonyl compounds (33–87% yields, 98:2 to 99:1 E/Z ratio, 78–98% ee (R,R)). These compounds were used as key intermediates in the synthesis of the analogs of nucleotide and L-Glu (Scheme 1.89) [357].
3) To allenes with succinimidyl diazoacetate (60–92% yields, 90:10 to 99:1 E/Z ratio, 84–99% ee (R,R)). It should be noted that only the less sterically hindered double bond was attacked and that the reduction of the exocyclic double bond was performed with high stereoselectivity, affording enantioenriched *cis*-cyclopropanes [358].

On the other hand, the cyclopropanation of styrene with diazoacetate catalyzed by Ru-**Pheox2** was attempted, but despite the high *trans*-enantioselectivity (97% ee), the cyclopropanation product was isolated in only 30% yield [359]. Better results are obtained with Ru-**Pheox2** supported on a macroporous polymer; in fact, the corresponding (R,R)-cyclopropanecarboxylates were obtained in 80–99% yield and with 91–99% ee [360]. The most relevant features of this catalyst were the best results achieved among the heterogeneous catalysts and its reusability, because it was recycled more than 10 times, even after 3 months of storage of the used catalyst, without any loss in its catalytic activity or selectivity.

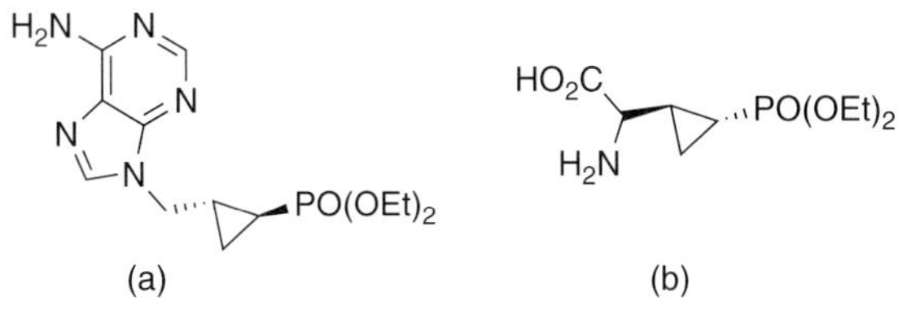

Scheme 1.89 Analogues of nucleotide (a) and L-Glu (b) from Ru-**Pheox1**-catalyzed cyclopropanation of diazomethylphosphonate with α,β-unsaturated esters.

Scheme 1.90

Ph–C(O)–R + EDA →(RuPor) cyclopropane: R, Ph, CO₂Et on ring

X (in **RuPor**)	R	Y (%)	dr	ee % (trans)	ee % (cis)
CO	Me	79	66:34	90	43
CO	H	80	96:4	87	34
PF$_3$	OTMS	98	70:30	83	38
PF$_3$	Ph	45	83:17	96%	14

(i-PrO)$_2$OP–CH=N$_2$ + Ar–CH=CH$_2$ →**RuPor** (X=CO)
90–97%
95:5 to 99:1 dr
87–92% ee (*trans*)
5–34% ee (*cis*)

Product: Ar, PO(Oi-Pr)$_2$ cyclopropane

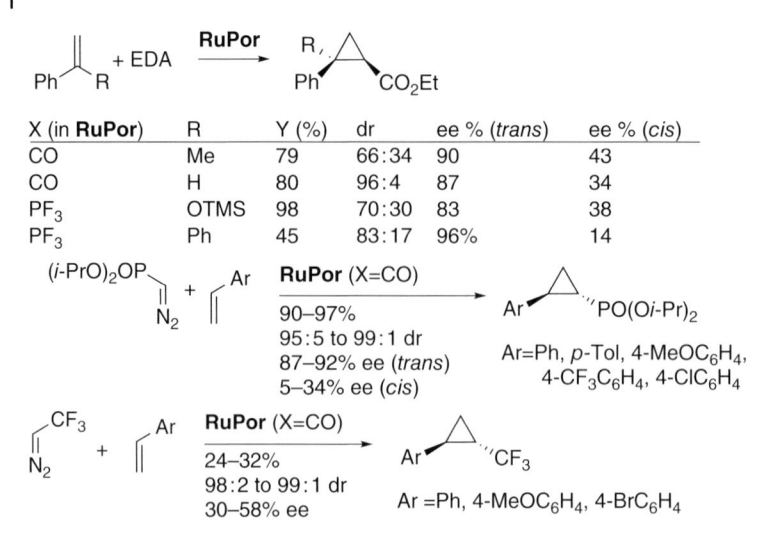

Ar = Ph, p-Tol, 4-MeOC$_6$H$_4$, 4-CF$_3$C$_6$H$_4$, 4-ClC$_6$H$_4$

CF$_3$–CH=N$_2$ + Ar–CH=CH$_2$ →**RuPor** (X=CO)
24–32%
98:2 to 99:1 dr
30–58% ee

Product: Ar, CF$_3$ cyclopropane

Ar = Ph, 4-MeOC$_6$H$_4$, 4-BrC$_6$H$_4$

Scheme 1.90 Asymmetric cyclopropanation catalyzed by ruthenium porphyrin.

Cobalt porphyrins have provided robust catalysts for asymmetric cyclopropanations as described in Table 1.2, but ruthenium–porphyrin catalysts were also often employed in this reaction. In 2003, a comparison between rhodium and ruthenium–porphyrin complexes for similar reactions showed that better *E/Z* ratios and higher ee values for the *trans*-isomer were obtained with ruthenium complexes than with the corresponding rhodium complexes [271]. Moreover, Ru-Por (Scheme 1.86 for the structure) afforded cyclopropanation of alkenes with EDA [361] and of substituted styrenes with diisopropyl diazomethylphosphonate [362] or with 2,2,2-trifluorodiazoethane [363] (Scheme 1.90). The reactions of EDA [364] and 2,2,2-trifluorodiazoethane [363] were also investigated, giving similar results, under heterogeneous conditions with the corresponding metalloporphyrin polymers, obtained from a chiral ruthenium–porphyrin complex, functionalized with four vinyl groups and copolymerized with divinylbenzene.

Another partially unsuccessful attempt to apply heterogeneous ruthenium catalysts in the cyclopropanation reaction was reported by Sanchez and coworkers [365]. They prepared neutral unsymmetrical pyridine pincer-type ligands with a lateral (*S*)-prolinamide donor functionality and an *N*-heterocyclic carbene moiety incorporated onto mesoporous silica gel. The complex obtained from the ruthenium chloride salt was inactive in the reaction of diazoacetates with styrene or α-methylstyrene, while the hexafluorophosphate salt afforded cyclopropanes (as *Z/E* mixtures) from various aryldiazoacetates and diazoacetates, except for *tert*-butyldiazoacetate, in 75–100% yield but with only 5% ee. Moreover, this supported catalysts required longer reaction times than the corresponding homogeneous catalyst to reach high yields, and the activity of such immobilized catalysts decreased somewhat after recycling.

Finally, to complete these last three sections, the copper(II), rhodium(III), and ruthenium(II) complexes, of formula [Cu(tpy)Cl$_2$], [Rh(tpy)Cl$_3$], and [Ru(tpy)(CO)Cl$_2$], respectively, should be mentioned (Scheme 1.91) [366]. It should be

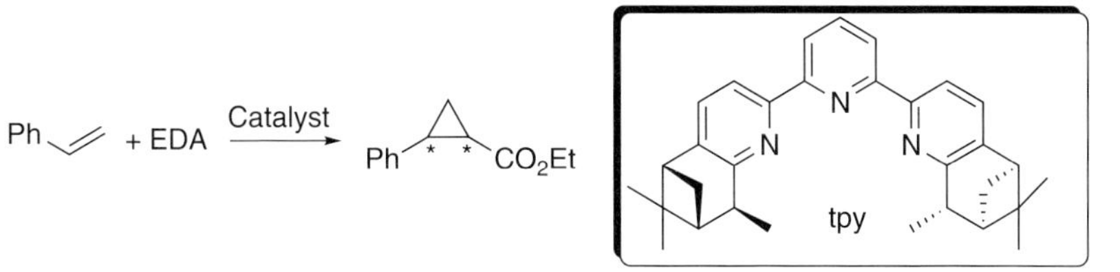

Ph⟋ + EDA →[Catalyst] Ph—△—CO₂Et   (tpy structure)

Catalyst	Y (%)	trans/cis Ratio	ee % (trans)	ee % (cis)
Cu(tpy)Cl$_2$ (10 mol%)	84	66:34	72 (1R,2R)	82 (1R,2S)
Rh(tpy)Cl$_3$ (2 mol%)	54	30:70	65 (1S,2S)	71 (1S,2R)
Ru(tpy)(CO)Cl$_2$ (2 mol%)	98	84:16	60 (1S,2S)	67 (1S,2R)

Scheme 1.91 Copper(II), rhodium(III), and ruthenium(II) salts complexed with chiral terpyridine ligands.

noted that the copper complex afforded the enantiomers opposite to those obtained with rhodium and ruthenium complexes, while the *cis*-isomer, unusually, prevailed with the rhodium complex.

1.3.1.6 Chiral Catalyst: Other Metals

Cobalt, copper, rhodium, and ruthenium are the most frequently metal ions used to prepare asymmetric complexes for cyclopropanation reactions, but some examples of other metal ion complexes have been reported as well.

Denmark and coworkers firstly introduced chiral ligands on the palladium complexes for the diazomethane-mediated cyclopropanation of alkenes. However, although a large number of chiral complexes have been tested, no enantioselection was observed, very likely because either partial or complete decomplexation of the ligand occurred during the course of the reaction [367].

On the other hand, the cheap, nontoxic, and environmentally benign iron was employed to catalyze cyclopropanations in the presence of chiral porphyrins. As an example, the same porphyrin depicted in Scheme 1.86, but complexed with iron chloride, induced the cyclopropanation of alkenes with EDA [368] and with 2,2,2-trifluorodiazoethane [363], providing the corresponding *trans*-cycloadducts in 56–70% yields with 92:8 to 96:4 and 74–86% ee and in 60–96% yields with 90:10 to 95:5 and 71–75% ee, respectively. Comparing these data with those reported in Scheme 1.90, it should be noted that iron porphyrin is a more efficient catalyst in the synthesis of chiral trifluoromethylphenylcyclopropanes. Thus, the scope of this methodology was extended to the use of the corresponding macroporous iron–porphyrin polymer, but yields and ee values were unsatisfactory [363].

Chiral iridium-salen complexes were developed for the *cis*-cyclopropanations of a wide range of alkenes with *tert*-butyl diazoacetate (Scheme 1.92) [369].

A large excess of alkene (up to 10 equiv.) was necessary to obtain high yields, and in contrast to most of the reactions reported earlier, (1R,2S)-cyclopropanecarboxylates were recovered as the major products. It should be noted that the reactions with *trans*-1-phenylpropene were sluggish, *cis*-1-phenylpropene gave only a modest yield, and reactions of both conjugated and nonconjugated enynes and dienes proceeded exclusively at the terminal alkene. The authors suggested a mechanism

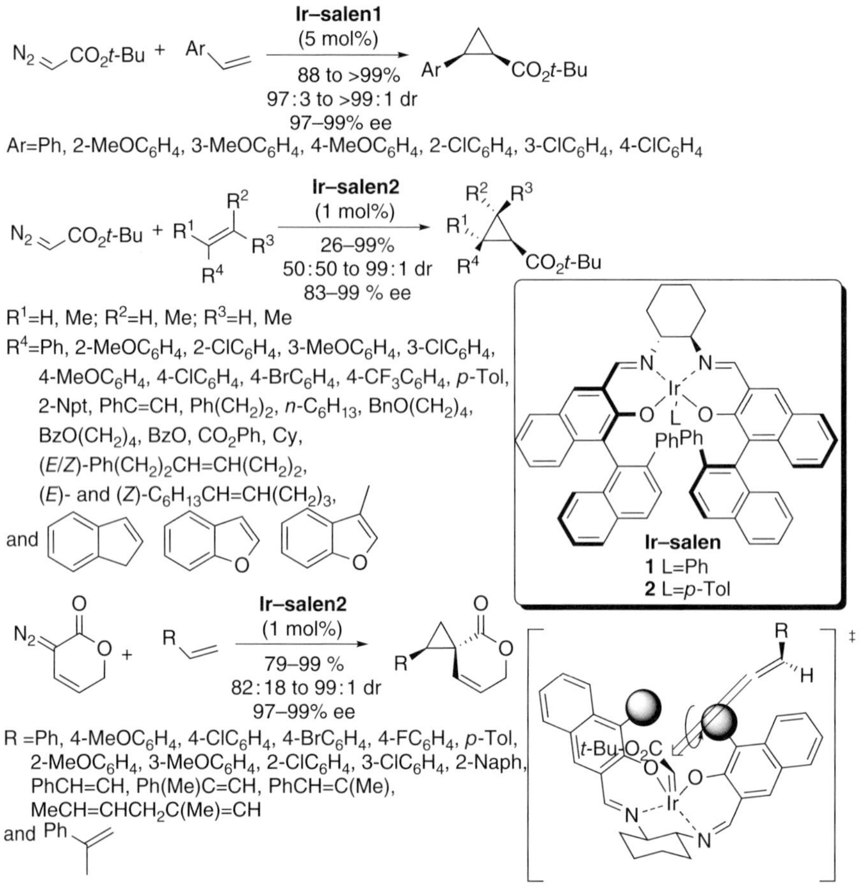

Scheme 1.92 Asymmetric cyclopropanations catalyzed by iridium–salen catalysts.

supported with *ab initio* calculations, in which the classical carbene species occupies the apical position in the iridium complex. The incoming alkene directed its larger substituent away from the carbenoid ester group, in a perpendicular approach. Then, a counterclockwise rotation led to the major product, while the clockwise rotation was less favorable due to steric repulsion between the substituent and the basal salen ligand.

The iridium–salen complex also allowed the synthesis of 5-oxaspiro[2.5]oct-7-en-4-ones, by reaction of a six-membered diazolactone with monosubstituted alkenes and dienes [370]. The most interesting features were the high *trans*-selectivity, in contrast with diazoacetate and the essential addition of molecular sieves to reduce the amount of alkene from ten- to twofold, by suppressing the competitive insertion of adventitious water on the lactone. Finally, it is worth noting that, in all Ir-salen catalyzed reactions depicted in Scheme 1.92, the less-hindered double bond was always attacked.

The cyclopropanation of substituted styrenes with EDA was also performed in the presence of chiral osmium complexes bearing sterically bulky Schiff-base ligands such as bis(3,5-di-*tert*-butylsalicylidene)-1,2-(*R*,*R*)-cyclohexane-diamine [up to 83:17 dr and up to 79% ee (*S*,*S*)] [371]. In the presence of chiral Mo_3CuS_4 clusters, both diastereo- and enantioselectivities were only moderate for both inter- and intramolecular processes [372].

Moreover, (*R*)-Difluorophos/$Hg(OTf)_2$-catalyzed cyclopropanation of diazooxindoles and alkenes has been reported (Scheme 1.93, eq 1) [373]. Unfortunately, the stereoselectivity for highly substituted cyclopropanes was not given. It should be noted that the obtained spiroxindoles were diastereomeric with respect to those obtained with rhodium catalysis (see Scheme 1.79). The stereochemistry was in agreement with an alkene approach to the carbenoid intermediate over the lactam group, with the substituent of the alkene away from the bulky metal, thus leading to the formation of the *trans*-isomer as the major product. A spiroketal bisphosphine derived chiral digold complex also allowed the cyclopropanation of diazooxindoles with *cis*- and *trans*-1,2-disubstituted alkenes (Scheme 1.93, eq 2) [374].

Finally, a particular metal-catalyzed decomposition of diazoalkanes is represented by the cyclopropanation of a variety of acrylamides and acrylates with ethyl diazoacetate catalyzed by a variant of cytochrome P450 from *Bacillus megaterium* five mutations away from wild type. (1*S*,2*R*)-Diethyl 1-(4-methylphenyl)cyclopropane-1,2-dicarboxylate and 13 different (1*R*,2*S*)-alkyl 2-(*N*,*N*-dialkylcarbamoyl)-2-arylcyclopropanecarboxylates were obtained in 50–99% yields with 89:11 to 99:1 dr and 57–98% ee [375].

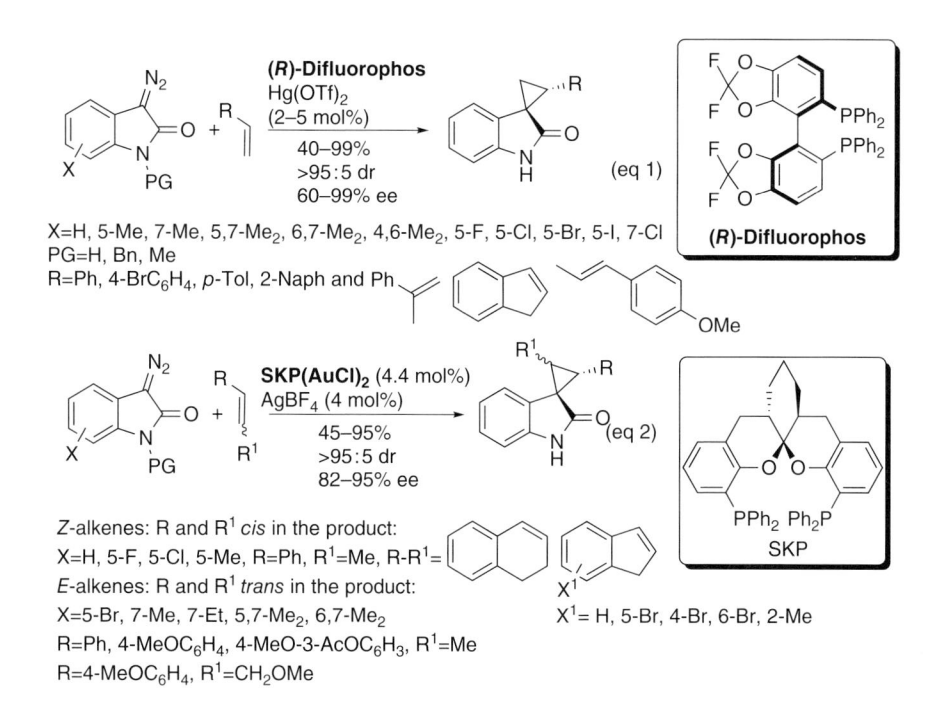

Scheme 1.93 Enantioselective mercury(II)-catalyzed cyclopropanation of diazooxindoles.

1.3.2 Intramolecular Cyclopropanation

When both functionalities, the diazo unit and the alkenes are in the same molecule, an intramolecular cyclopropanation is possible in the presence of the appropriate catalyst, thus generating synthetically versatile [n.1.0]bicycloalkanes. In contrast to the intermolecular reaction, only one diastereomer is obtained when forming five- or six-membered rings, thus the most successful systems involve cyclization of either γ,δ- or δ,ε-unsaturated diazocarbonyl systems. It is also possible to form macrocycles using the intramolecular cyclopropanation reaction, but the diastereoselectivity is no longer sure [376]. Finally, it is important to consider the chemoselectivity, as, in some cases, the C–H insertion may become the major pathway.

1.3.2.1 Chiral Auxiliaries and Chiral Compounds

Initially, intramolecular cyclopropanation reactions were carried out with an appropriate chiral substrate, and generally proceeded with complete stereocontrol, leading to the exclusive formation of one stereoisomeric bicyclic product. Homogeneous and heterogeneous copper and rhodium catalysts were more popular for this reaction. Systems that are less rigid, such as prostaglandin derivatives (Scheme 1.94), afforded a mixture of stereoisomers (91% yield, 69:31 dr) under rhodium catalysis [377].

Both copper and rhodium catalysts were used to synthesize terpenes such as dihydromayurone (36% and 57% yield of only one stereoisomer) [378], while

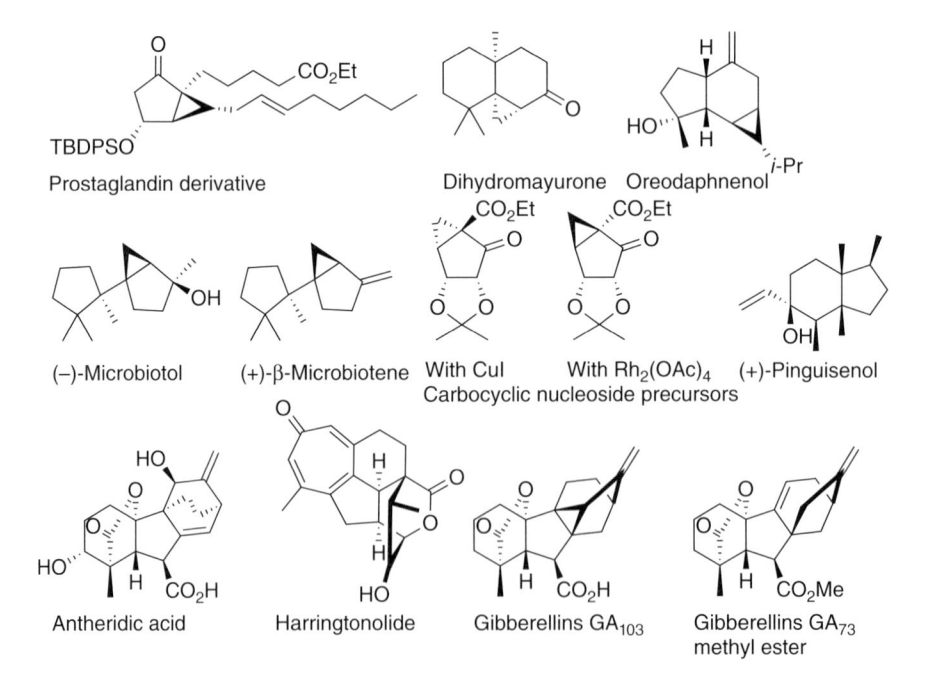

Scheme 1.94 Some molecules prepared by intramolecular cyclopropanation of chiral diazo compounds.

rhodium-catalyzed intramolecular cyclopropanation was employed in the synthesis of the tricarbocyclic framework (95% yield and with 97:3 dr) of oreodaphnenol (Scheme 1.94) [379]. A Cu-catalyzed intramolecular cyclopropanation (75% yield) allowed the synthesis of (−)-microbiotol and (+)-β-microbiotene from a chiral diazo ketone derived from the readily available cyclogeraniol [380].

Carbocyclic nucleoside precursors were prepared from ribose derivative (81% and 74% yield from copper and rhodium catalysis, respectively). It is worth noting that copper and rhodium catalysts gave rise to different stereoisomers (82:18 dr and 25:75 dr, respectively) [381].

The key step in the construction of the bicyclic system of (+)-pinguisenol was based on the intramolecular cyclopropanation of a chiral diazoketone (52% yield with $CuSO_4$), followed by regioselective cyclopropane cleavage [382]. The combination of the intramolecular cyclopropanation of dienes with the vinylcyclopropane-cyclopentene rearrangement was employed as a key step in the total synthesis of antheridic acid [383].

The Buchner cyclization was investigated under rhodium catalysis by Maguire with chiral α-diazoketone derivatives (Scheme 1.95) [384] and exploited in the synthesis of the natural product harringtonolide [385]. On the other hand, gibberellin derivatives were obtained in the presence of a copper catalyst that minimized the formation of the significant amounts of a C–H insertion product observed with rhodium catalysts [386].

The substituents were found to play an important role in the double diastereotopic differentiation strategy of α-diazophosphonate templates with the (R)-pantolactone auxiliary using $Rh_2(OAc)_4$-catalyzed intramolecular cyclopropanation. Furthermore, the double diastereoselective intramolecular cyclopropanation of a pseudo-C_2-symmetric phosphonate was performed with excellent diastereoselectivity (Scheme 1.96) [387].

The intramolecular cyclopropanation of diazoacetates prepared from butane-2,3-diacetals of L-threitol in the presence of $Rh_2(OAc)_4$ afforded two cyclopropane diastereomers in a ratio of 70:30. Up to 91:9 dr was instead obtained in the presence of a chiral catalyst such as $Rh_2(S\text{-MEPY})_4$, thus demonstrating that the trajectory of the double bond onto the metal carbene was dependent upon both the configurations of the catalyst and the reacting substrate (Scheme 1.97) [388].

Finally, a Mn(III)-mediated cyclopropanation of a chiral allyl acetoacetate led to the corresponding 3-oxabicyclo[3.1.0]hexan-2-one (Scheme 1.98) [389], then applied to the syntheses of several furofuranones and furofuran lignans. It should be noted that, in this case, the carbene was not obtained from diazoacetate decomposition.

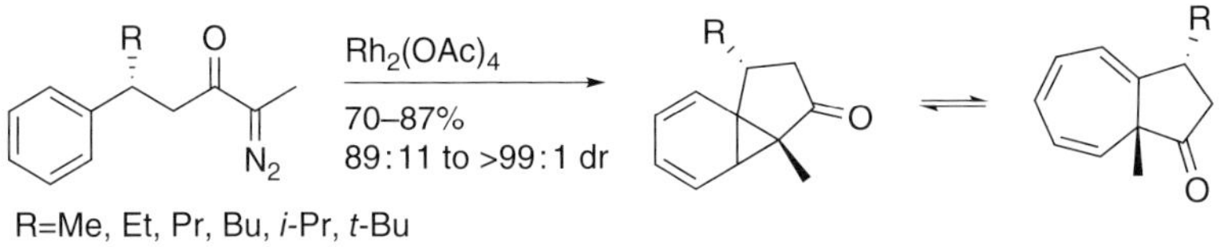

R=Me, Et, Pr, Bu, *i*-Pr, *t*-Bu

Scheme 1.95 Buchner cyclization.

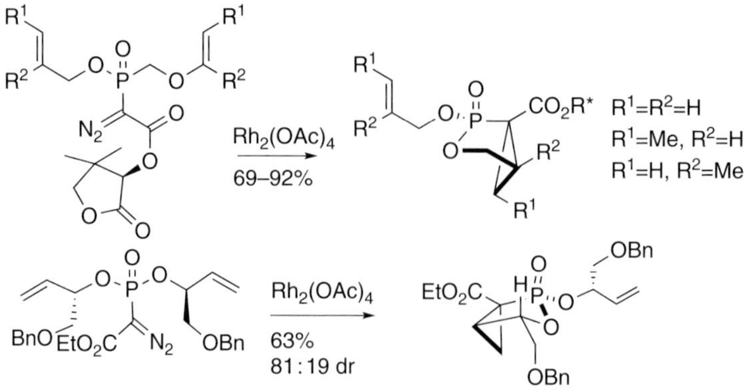

Scheme 1.96 Asymmetric cyclopropanation of chiral diazophosphonates.

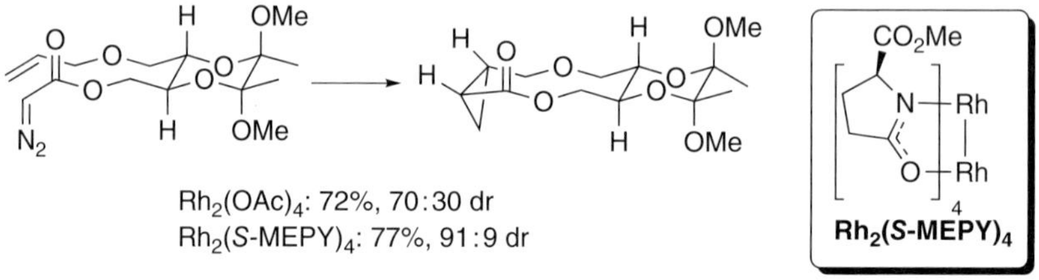

Scheme 1.97 Asymmetric cyclopropanation of butane-2,3-diacetal of (L)-threitol.

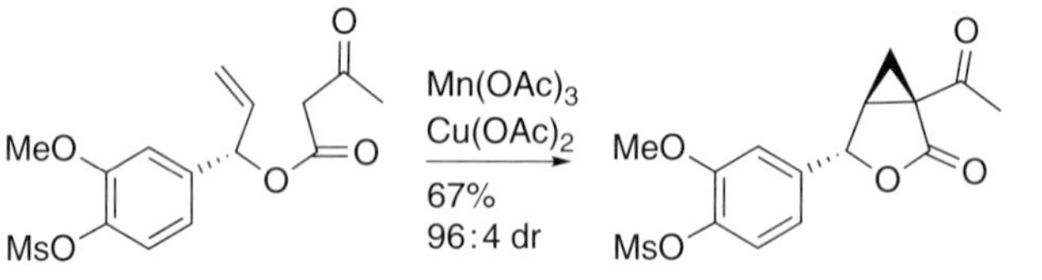

Scheme 1.98 Asymmetric cyclopropanation of chiral allyl acetoacetate.

1.3.2.2 Chiral Catalysts

Larger diffusion in the chemical community had found chiral catalysts for the enantioselective intramolecular cyclopropanation of unsaturated diazoketones. Cobalt, copper, rhodium, and ruthenium catalysts described in Sections 1.3.1.2–1.3.1.6 as well as new specific metal catalysts have been applied to the intramolecular cyclopropanation. The effectiveness of each catalyst changes with the substitution and the ring size of the fused cyclopropane [390].

In general, dirhodium(II) carboxamidate catalysts are superior for small-ring fused cyclopropane compounds, whereas the copper(I) bis(oxazoline) catalysts for medium/large-ring fused cyclopropane compounds. Ruthenium and cobalt catalysts are useful in the synthesis of trisubstituted cyclopropanes, especially when involving the formation of five- and six-membered rings. On the other hand, diazocarbonyl compounds in intramolecular processes can be divided into three major categories:

i) diazoketones
ii) diazoesters
iii) diazoacetamide derivatives.

Semicorrin copper complex **box1** was the most efficient and frequently used catalyst to ensure enantiocontrol for a long period of time [391]. Up to 95% ee was obtained, depending on the substrate, but the yields were always modest (<60%) In comparison, other catalysts led to a low level of induction [392]. Only *ortho*-metallated aryl phosphine dirhodium(II) complexes (Scheme 1.67, group **b**) provided the desired cyclopropane products in very high yield (>90%) and with enantioselectivities comparable to those achieved with Cu-**box1** [393].

The most efficient enantioselective intramolecular reaction is summarized here following the same order used earlier for intermolecular reactions.

Among chiral cobalt catalysts, salen complexes were found to be very efficient (Scheme 1.99). These catalysts were found to be superior to the corresponding Ru(II) salen complexes (see the following discussion) [394].

Cobalt(II) porphyrin (**CoP5**, Table 1.2) [395] and bis(2-pyridylimino)isoindoles complexes (Co-**Isoind**, Scheme 1.54) [158] were also employed to synthesize [*n*.1.0]bicyclic ring systems directly from linear unsaturated diazo precursors (Scheme 1.100). It should be noted that the two catalysts afforded enantiomeric products 1*R*,5*S*,6*S* and 1*S*,5*R*,6*R*, respectively. In particular, with the former complex, the absolute configuration was in agreement with the absolute (*R*,*R*)-configuration of the Z-substituent in **CoP5**.

Moreover, the corresponding 3-oxabicyclo[3.1.0]hexan-2-one was generated as a single diastereomer, except for compounds with R^2=PhCH=CH, R^1=H and R^2=H, R^1=Et. These results were consistent with the stepwise radical mechanism described in Scheme 1.53, also demonstrating that the last ring-closure step is a low-barrier or a barrierless process. The γ-butyrolactone unit was selectively opened to produce examples of cyclopropane derivatives difficult to be prepared directly by asymmetric intermolecular cyclopropanation, and the cyclopropane unit was successfully

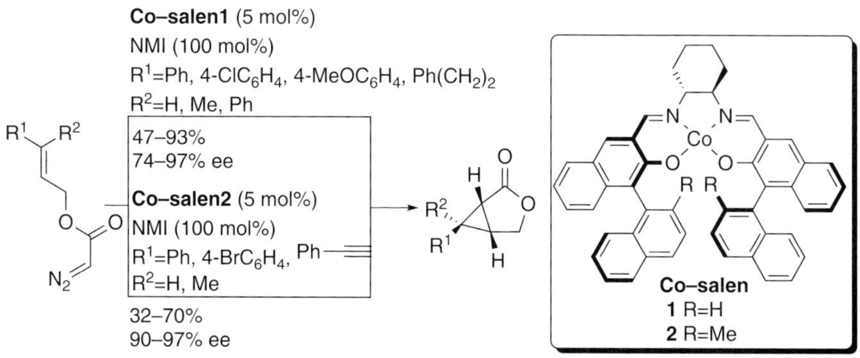

Scheme 1.99 Asymmetric intramolecular cyclopropanation catalyzed by Co-salen complexes.

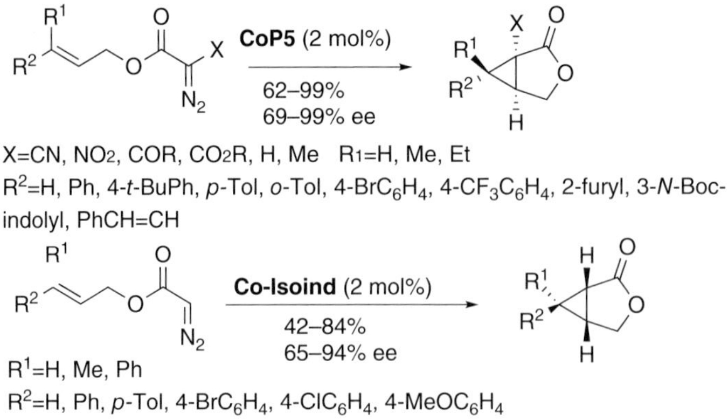
CoP5 (2 mol%)
62–99%
69–99% ee
X=CN, NO2, COR, CO2R, H, Me R1=H, Me, Et
R2=H, Ph, 4-t-BuPh, p-Tol, o-Tol, 4-BrC6H4, 4-CF3C6H4, 2-furyl, 3-N-Boc-indolyl, PhCH=CH
Co-Isoind (2 mol%)
42–84%
65–94% ee
R1=H, Me, Ph
R2=H, Ph, p-Tol, 4-BrC6H4, 4-ClC6H4, 4-MeOC6H4

Scheme 1.100 Other cobalt(II) complexes in intramolecular cyclopropanation.

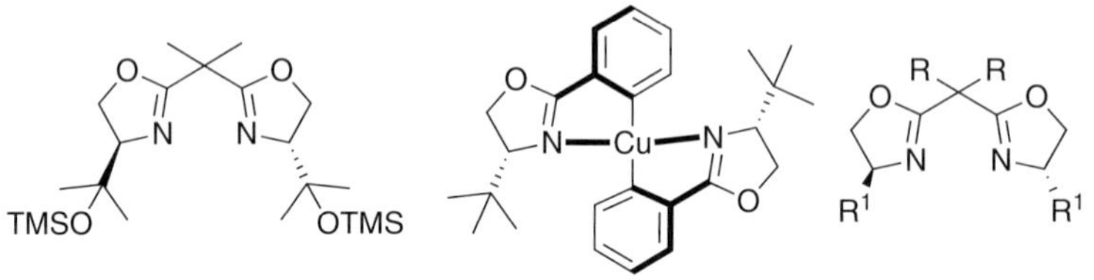
TMSO
OTMS
Cu

box41

Cu-box42

box
43: R=Bn, R^1=*i*-Pr
44: R=R^1=Me
45: R=H, R^1=Bn

Scheme 1.101 Other box ligands proposed for intramolecular copper-box-catalyzed cyclopropanation.

enlarged through 1,3-dipolar cycloaddition with dipolarophiles. Both reactions proceeded to form a single diastereomer without loss of optical purity.

Regarding copper catalysts, in addition to the aforementioned copper–semicorrin Cu-**box1**, other catalysts have been employed in asymmetric intramolecular cyclopropanations (Scheme 1.101). For instance, the synthesis of the phorbol CD-ring skeleton was achieved through the asymmetric intramolecular cyclopropanation of a silyl enol ether with the catalyst obtained mixing 15 mol% of the **box41** ligand and 5 mol% of $Cu(OTf)_2$ (Scheme 1.102, eq 1) [396]. Corey reported the synthesis of a synthetic precursor of sirenin by cyclization of a γ-diazocarbonyl derivative (Scheme 1.102, eq 2) [397]. The classical Cu-box catalysts were scarcely efficient; thus, Corey designed a novel Cu-box catalyst (Cu-**box42**).

Cu-**box16** (Table 1.4) was an efficient catalyst for the preparation of an enantiopure fluorobicycloketone (Scheme 1.102, eq 3) [398].

For a decade, Nakada's group was interested in the enantioselective Cu-**box**-catalyzed intramolecular cyclopropanations of a range of α-diazo-β-keto sulfones (Scheme 1.103) [399]. The cyclopropanes present a (1*R*,5*R*) stereochemistry, which is interestingly diastereomeric with respect to most of the other described

Scheme 1.102 Some copper-catalyzed asymmetric intramolecular cyclopropanations.

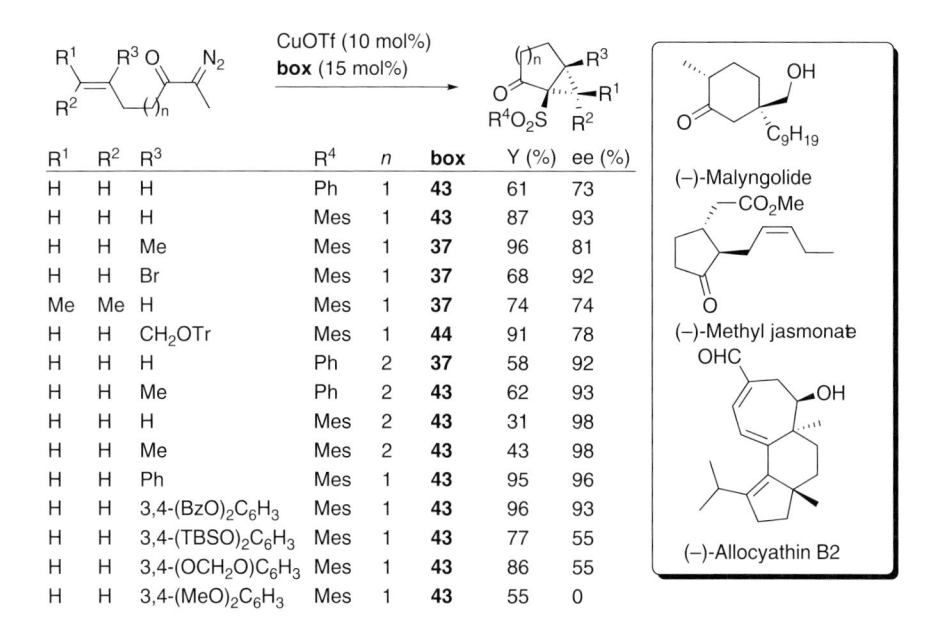

R^1	R^2	R^3	R^4	n	box	Y (%)	ee (%)
H	H	H	Ph	1	43	61	73
H	H	H	Mes	1	43	87	93
H	H	Me	Mes	1	37	96	81
H	H	Br	Mes	1	37	68	92
Me	Me	H	Mes	1	37	74	74
H	H	CH_2OTr	Mes	1	44	91	78
H	H	H	Ph	2	37	58	92
H	H	Me	Ph	2	43	62	93
H	H	H	Mes	2	43	31	98
H	H	Me	Mes	2	43	43	98
H	H	Ph	Mes	1	43	95	96
H	H	3,4-(BzO)$_2$C$_6$H$_3$	Mes	1	43	96	93
H	H	3,4-(TBSO)$_2$C$_6$H$_3$	Mes	1	43	77	55
H	H	3,4-(OCH$_2$O)C$_6$H$_3$	Mes	1	43	86	55
H	H	3,4-(MeO)$_2$C$_6$H$_3$	Mes	1	43	55	0

Scheme 1.103 Asymmetric intramolecular cyclopropanation of α-diazo-β-keto sulfones.

reactions. Studies on structure–enantioselectivity relationships were performed with the following results:

i) In compounds with R^3 = 3,4-R$_2$Ph, no selectivity was observed when R = MeO, moderate enantioselectivity when R–R=OCH$_2$O or R=TBSO, and high enantioselectivity without substituents or when R=BnO [400].

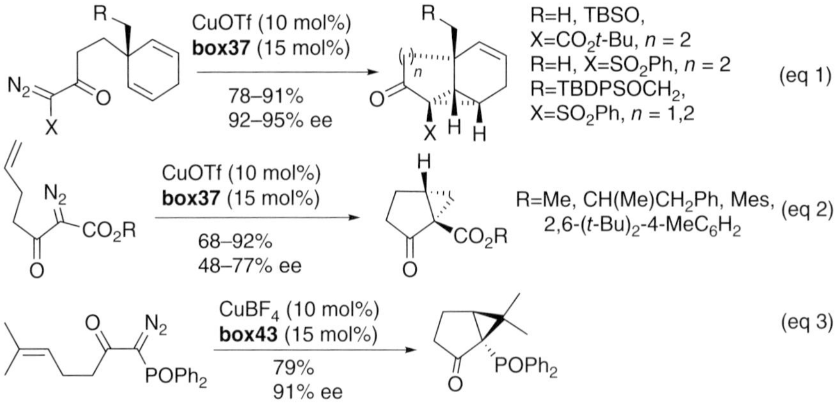

Scheme 1.104 Preparation of tricyclo[4.4.0.0^7]dec-2-ene and cyclopropanation of α-diazo-β-keto esters.

ii) High enantioselectivity was attained in particular when the methyl substitution on the phenyl ring of the sulfone moiety is in the 2-position [401].
iii) Regarding the R^1 group on the catalyst, the benzyl group was found to give the best enantiomeric excess, but the yield was only 67%, thus R^2 = i-Pr was preferred. The bulkiness of R group on the catalyst increased enantiomeric excess, but reaction yields decreased [402].

The success of this methodology prompted its application to the total syntheses of several biologically active products, such as (−)-allocyathin B$_2$ [403], (−)-malyngolide [404], and (−)-methyl jasmonate [405].

Nakada's group also developed the enantioselective preparation of tricyclo[4.4.0.0]dec-2-ene derivatives (Scheme 1.104, eq 1) [406] and tricyclo[4.3.0.0] nonenone in either catalytic (10 mol%) or stoichiometric amounts in 48% (66% conversion and 79% ee) and 69% yield (84% conversion and 80% ee), respectively [407]. The resulting chiral cyclopropanes were utilized for enantioselective syntheses of natural products (+)-busidarasin C, acetoxytubipofuran [406], (+)-digitoxigenin [408], (−)-platensimycin, (−)-platencin, nemorosone [407], garsubellin A, clusianone, and hyperforin (Scheme 1.105) [409].

The same methodology was also applied to the intramolecular cyclopropanation of various 2-diazo-3-oxo-6-heptenoic acid esters [the best enantioselectivities were achieved in the case of substrates bearing a bulky ester group (Scheme 1.104, eq 2)] [410] and to the reaction of α-diazo-β-oxo-5-hexenyl phosphonate in the presence of Cu-**box43** ligand (Scheme 1.104, eq 3) [411]. The obtained (1R,5S)-bicyclo[3.1.0]hexane was the key intermediate in the enantioselective total synthesis of (+)-colletoic acid (Scheme 1.105) [411].

A better enantiocontrol was observed with Cu-**box5** complex, increasing the ring size. Thus, allylic diazoacetate gave only 20% ee, whereas the formation of 10-membered rings proceeded with 87–90% ee. Moreover, Cu-**box5** complex exclusively favored the macrocyclic product over the allylic cyclopropanation product [412]. Comparable enantiomeric excesses (85% ee) were obtained for the

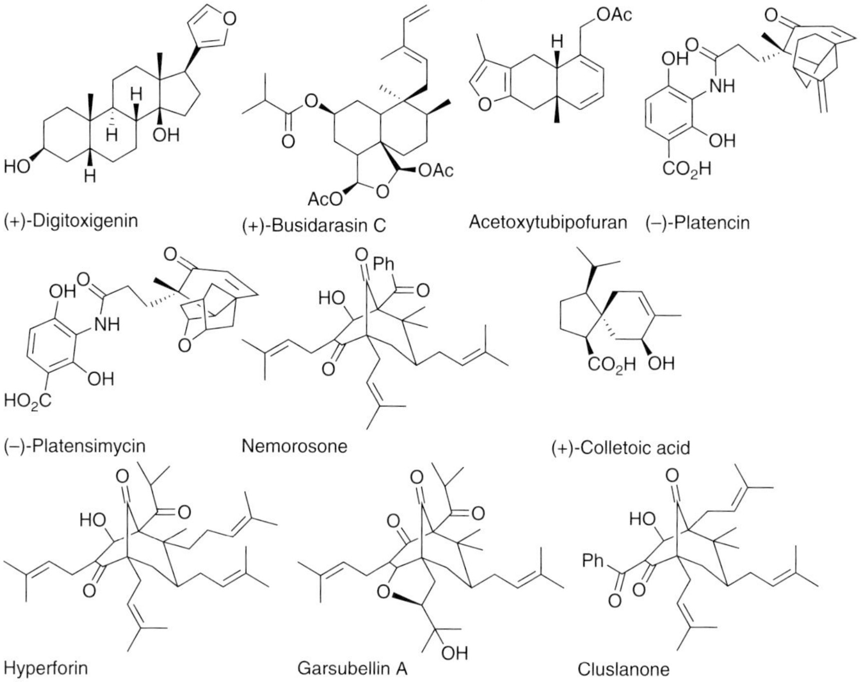

Scheme 1.105 Some natural products prepared from α-diazo-β-keto sulfones, esters, and phosphonates.

synthesis of 15- and 20-membered ring-fused cyclopropane products, although as a mixture of *cis*- and *trans*-isomers (69:31 dr).

The very unreactive diazomalonates were reported to react under Cu-**box45** catalysis in 72–73% yield with 11–35% ee [413].

Chiral carboxylate and carboxamidate ligands were mainly developed in the preparation of dirhodium catalysts. In most cases, C_1-symmetric ligands from simple amino acids are coordinated around the dirhodium core to give the chiral rhodium catalysts. Catalysts already used in intermolecular reactions and new specific rhodium catalyst were employed. Generally, these catalysts, such as $Rh_2(MEPY)_4$, are more electron-rich compared to the tetracarboxylates and have a different reactivity profile [414]. For instance, in Scheme 1.96, $Rh_2(S\text{-}MEPY)_4$ has already been mentioned in the reaction of a chiral substrate. Moreover, this catalyst and $Rh_2(S\text{-}MEOX)_4$ (Scheme 1.106) provided 3-oxabicyclo[3.1.0]hexan-2-one from intramolecular cyclopropanation of allyl diazoacetate in 95% ee (75% and 95% yield, respectively) [415], whereas lower stereochemical induction was observed with ruthenium, copper, or other dirhodium catalysts. The enantioselectivity depends on the double-bond substitution. Indeed, (*E*)- and (*Z*)-(phenyl)allyl diazoacetate afforded the corresponding product in 78% and 70% yield with 68% and >94% ee, respectively. On the other hand, the catalyst $Rh_2(S\text{-}MPPIM)_4$ produced the desired bicyclic system from the (*E*)-isomer with >95% ee (61% yield) [416]. In addition, the cyclopropanation of farnesyl diazoacetate with $Rh_2(S\text{-}MEPY)_4$ exclusively formed the γ-lactone

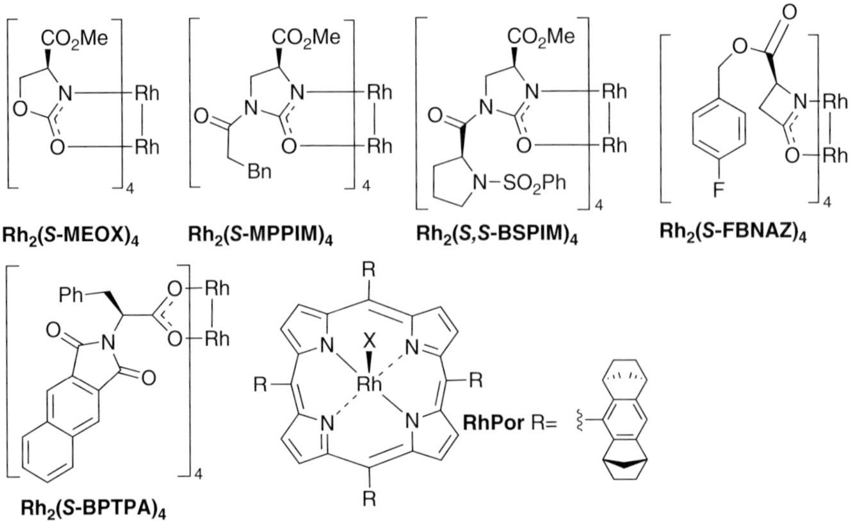

Scheme 1.106 Some dirhodium catalysts for intramolecular cyclopropanation.

[417]. It should be noted the different chemoselectivity with respect to Cu-**box5** complex, which always preferred the farthest double bond (see the previous discussion) [412].

Rh$_2$(S-MEPY)$_4$-catalyzed cyclization of allylic diazoacetates also led to 1,2,3-trisubstituted cyclopropanes, as rigid replacements of dipeptide arrays in several biological systems, with high enantiomeric excesses [418]. Moreover, Martin reported that secondary divinyldiazoacetates underwent cyclopropanation with exceptional enantiocontrol (Scheme 1.107) [419]. The presence of a methyl group in the diene system drastically decreased diastereoselectivity [420].

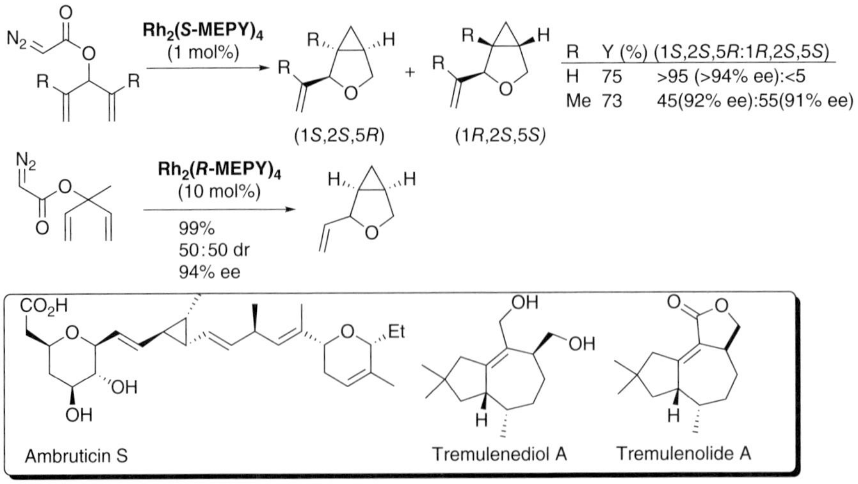

Scheme 1.107 Cyclopropanation of divinyldiazoacetates.

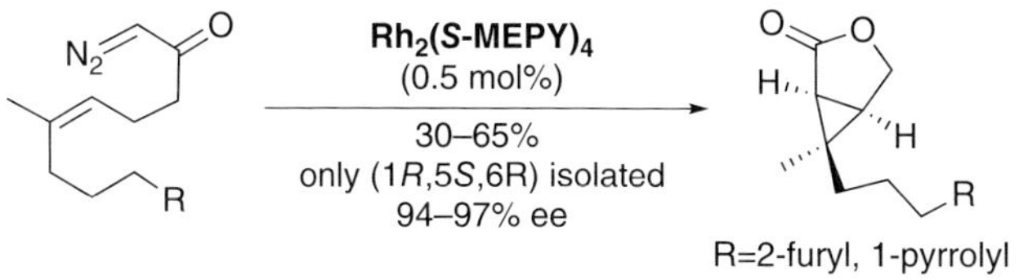

Scheme 1.108 Asymmetric intramolecular cyclopropanation of allyldiazoacetates.

This reaction was applied to the synthesis of ambruticin S [421], tremulenediol A, and tremulenolide A [422] and to the synthesis of various cyclopropane-derived peptidomimetics [423].

Homoallylic diazoacetates afforded the bicyclic products under Rh$_2$(S-MEPY)$_4$ catalysis in 55–80% yields with 71–90% ee [424], and it was found that the enantioselectivity was not highly dependent on the substitution pattern of the double bond. Rh$_2$(S-MEOX)$_4$ provided similar levels of enantiocontrol, whereas Rh$_2$(S-MPPIM)$_4$ was less enantioselective. It is also worth noting, but not too surprising, that the enantioselectivity decreased, increasing the ring size conversely from copper complexes (see the previous discussion) [412]. In fact, the formation of a macrocycle is more similar to intermolecular cyclopropanation, for which copper catalysts usually provided higher enantiomeric excesses. Finally, another application of Rh$_2$(R-MEPY)$_4$ is depicted in Scheme 1.108 [425].

Both Rh$_2$(S-MEPY)$_4$ and Rh$_2$(S-MEAZ)$_4$ were immobilized on an ArgoPore resin and tested in the enantioselective intramolecular cyclopropanation of allyl diazoacetates [295]. Up to 95% ee along with about 80% product yield was obtained.

Diazoacetamides (20–95% yields, 75–93% ee) underwent intramolecular cyclopropanation under Rh$_2$(S-MPPIM)$_4$ catalysis with an enantiocontrol similar to that found with diazoacetates (65–75% yields, 89–95% ee), especially with a methyl nitrogen substituent, which favors the desired *S-trans*-conformer, minimizing the amount of the undesired dipolar addition reaction [426]. The homoallylic diazoacetamide provided the bicyclic product, with slightly lower enantiomeric excesses. Moreover, one of the resulting cyclopropanated products from diazoacetate was converted into a highly potent group 2 and group 3 glutamate receptor agonist [319].

Doyle showed that Rh$_2$(S-MEOX)$_4$ provided the best enantiocontrol in the intramolecular cyclopropanation of substituted allyl diazoacetates, leading to the desired cyclopropanes with 71–85% ee and 43–52% ee for the *trans* and the *cis* starting materials, respectively, with yields ranging from 46% to 81% [427].

The diazo decomposition of vinyl- and aryl-substituted diazoacetates required a more reactive catalyst [428]. For instance, the cyclization of allyl α-styryl-α-diazoacetate proceeded in 81%, 59%, and 56% yields with 28%, 58%, and 59% ee with Rh$_2$(R-DOSP)$_4$, Rh$_2$(S-IBAZ)$_4$, or Rh$_2$(S-MEAZ)$_4$, respectively. Rh$_2$(R-DOSP)$_4$ catalyst was also examined with a wide range of substituted allyl α-styryl-α-diazoacetates, affording the expected products in 46–72% yields with 25–87% ee. The reactions of *cis*-alkenes resulted in much higher asymmetric induction compared to *trans*-alkenes, while the highest enantioselectivity was achieved with a disubstituted terminal alkene. On the other hand, Rh$_2$(S-MEAZ)$_4$ proved to be equally effective or more effective than Rh$_2$(R-DOSP)$_4$ for the enantioselective cyclization of allyl α-phenyl-α-diazoacetates. With an unsubstituted allyl group, the desired bicyclic system was isolated with 68% ee, instead of 36% ee, but the yield was 80% instead of 93%, whereas with a 1,1-disubstituted alkene, the enantiomeric excess was 45%, and the product was recovered in 85% yield. It should be noted that the bicyclic lactones have opposite configuration in the reaction of allyl α-styryl-α-diazoacetates and allyl α-phenyl-α-diazoacetates, although the catalysts have the same configuration. The reaction catalyzed with azetidinone-carboxylate (AZ) ligands was then improved and a series of substituted allyl α-phenyl-α-diazoacetates were cyclized in 74–93% yields with 45–84% ee [429].

The intramolecular cyclopropanation of vinyldiazoacetates and dienes, followed by a Cope rearrangement of the resulting divinylcyclopropane intermediate, afforded fused cycloheptadienes, which are intermediates in the asymmetric synthesis of the *epi*-tremulane skeleton (Scheme 1.109) [430].

Charette et al. reported an intramolecular cyclopropanation involving α-nitro- [149] and α-cyano- [431] α-diazo carbonyl compounds. The former reaction, catalyzed by $Rh_2(R-DOSP)_4$, led to the formation of nine-membered nitrocyclopropyllactones (Scheme 1.110), while the latter used $Rh_2(S-FBNAZ)_4$ as the chiral catalyst.

Furthermore, the asymmetric intramolecular cyclopropanation of allyl diazoacetates was performed in the presence of RhPor in 31–65% yields with 31–49% ee [271]. In the presence of *ortho*-metallated arylphosphines dirhodium complex **b1** (Scheme 1.67 for the structure), allyl diazoacetate cyclized in 93% yield and with 56% ee [322c, 432]. Other chiral dirhodium(II) catalysts (**b2** and **b3**, Scheme 1.67) were employed in the enantioselective intramolecular cyclopropanation of 1-diazo-

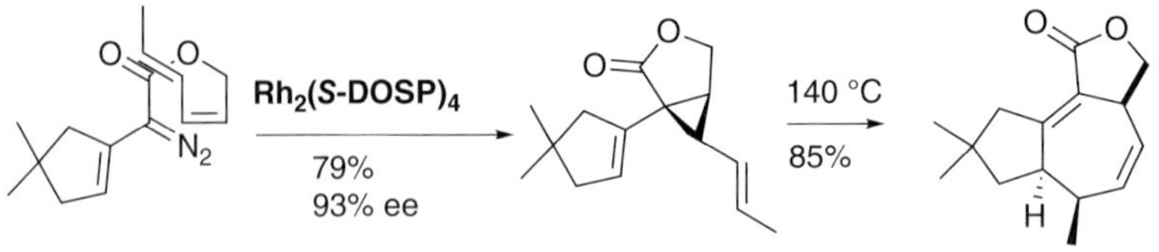

Scheme 1.109 Synthesis of the *epi*-tremulane skeleton.

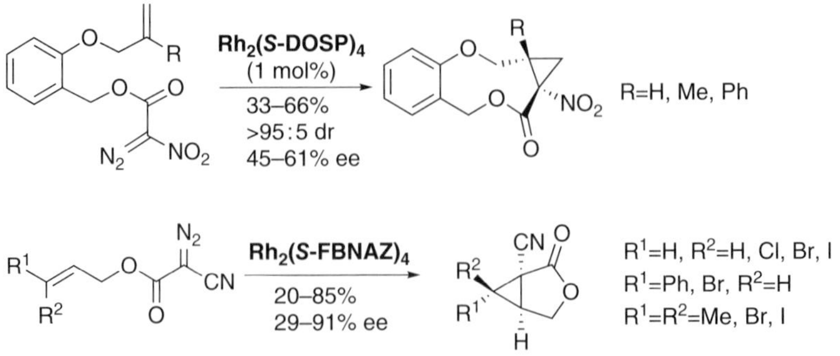

Scheme 1.110 Asymmetric intramolecular cyclopropanation of α-nitro- and α-cyano-α-diazo carbonyl compounds.

6-methyl-3-(2-propenyl)-5-hepten-2-one (up to 90% ee) [433] and in a total synthesis of sabina lactone [434].

An interesting and original sequence of two successive intramolecular cyclopropanations involved a bis-diazoacetate under Rh$_2$(S,S-BSPIM)$_4$ catalysis (Scheme 1.111) [435].

Moreover, (triethylsilyl)-substituted allyl diazoacetate was decomposed with Rh$_2$(S-NTTL)$_4$ or Rh$_2$(S-BPTPA)$_4$ and the latter was more efficient (82% vs 76% yield and 56% vs 38% ee) [314]. The scope of this methodology was extended to allyl 2-diazo-3-silanyloxybut-3-enoates under Rh$_2$(S-PTTL)$_4$ catalysis and provided the corresponding lactones in 57–93% yields with 78–67% ee (1R,5S-isomer) [313c, 436].

Finally, several chiral ruthenium complexes have been used, in recent years, to catalyze the intramolecular cyclopropanation of allyl diazoacetates. As an example, RuPor led to the corresponding (1S,5S)-lactones in 65–85% yields with 30–85% ee [271]. Ru-**salen3** (Scheme 1.86) and Ru-**salen8** (Scheme 1.112) were efficient catalysts to induce chirality for the intramolecular cyclopropanation of various alkenyl α-diazoacetates (33–82% yields with 33–89% ee) [343g, 394]. The cyclization was strongly affected by the substitution pattern of the alkenyl group of the substrate and the length of the carbon chain connecting the alkenyl and diazomethyl moieties. Ru-**salen4** was found to be a general catalyst for the asymmetric intramolecular cyclopropanation of *trans*-allylic diazoacetates leading to (1S,5R,6R)-3-oxabicyclo[3.1.0]hexan-2-ones (75–91% yields and 58–98% ee), while a *cis*-allylic diazoacetate gave lower enantiomeric excess value (51%) and

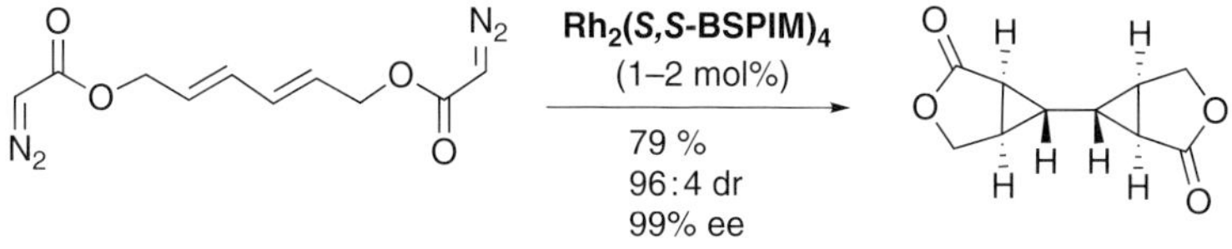

Scheme 1.111 Double intramolecular cyclopropanation of bis-diazoacetate.

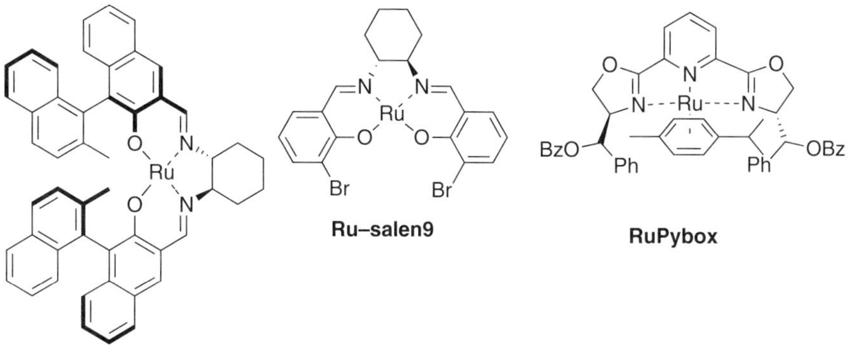

Scheme 1.112 Chiral ruthenium catalysts for asymmetric intramolecular cyclopropanations.

yield (48%), and the disubstitution gave similar yield (77%) but low enantiomeric excess value (58%) [437].

On the other hand, Ru-**salen9**, containing PPh$_3$ ligands, was effective in the intramolecular cyclopropanation of *cis*-substituted allylic diazoacetate (46–71% yields, 44–90% ee (1*S*,5*R*,6*S*-isomer)) [438]. The yields and stereochemical outcomes were consistent with those reported in Scheme 1.99 for the same reaction catalyzed by **CoP5**. DFT calculations at the 6-31G level suggested a reaction pathway involving a carbene, which replaced the apical ligand, thus the irradiation with light became essential, because the apical CO ligand is difficult to remove. In contrast, Ru**Phebox** allowed the intramolecular reaction of *trans*-cinnamyl diazoacetate in 96% yield and with 99% ee with a opposite (1*R*,5*S*,6*S*)-stereochemistry, with respect to **CoP5** or Rh-**salen** [342]. Ru**Pybox** catalyst allowed the same reaction in up to 86% yield and with up to 77% ee depending on the solvent used [335].

Ru**Pheox2** is completely water soluble and insoluble in diethyl ether; thus, Iwasa's research group reported an interesting intramolecular cyclopropanation in water as reaction medium. The easy separation of the ether phase, containing the cyclopropane product, from the catalyst in the water phase allowed for a very simple reuse of the catalyst at least five times without significant decrease in reactivity or enantioselectivity [359]. The reaction of *trans*-allylic diazoacetates afforded (1*S*,5*R*,6*R*)-3-oxabicyclo[3.1.0]hexan-2-ones in 89–99% yield with 83–98% ee, disubstituted allylic diazoacetates gave lower enantioselection (76–95% yield, 36–97% ee), while *cis*-allylic diazoacetates were not tested. It should be noted that Ru**Pheox2** supported on a macroporous polymer was more effective than the unsupported catalyst. In fact, *trans*-allylic diazoacetates reacted in less than a minute to give products in comparable yields and ee values.

Ru**Pheox1** successfully carried out the intramolecular cyclopropanation of various symmetric and racemic allenic diazoacetates to afford the corresponding bicyclic alkylidenecyclopropane fused γ-lactones (Scheme 1.113) [358]. In particular, racemic allenic diazoacetates with increasing R^1 substituent size improved diastereoselectivity (up to *E*/*Z* = 81 : 19), but the enantioselectivity of the *E*-isomer fell up to 48% ee. Moreover, lower yield (up to 40%) and enantioselectivity (up to 8%) were obtained increasing R^3 substituent size.

Finally, the enantioselective intramolecular cyclopropanation of electron-deficient allylic diazoacetates, in the presence of **RuPheox1**, was reported (Scheme 1.114). One of the corresponding cyclopropane-fused γ-lactones was the key intermediate in the synthesis of DCG-IV and dysibetaine CPa [439].

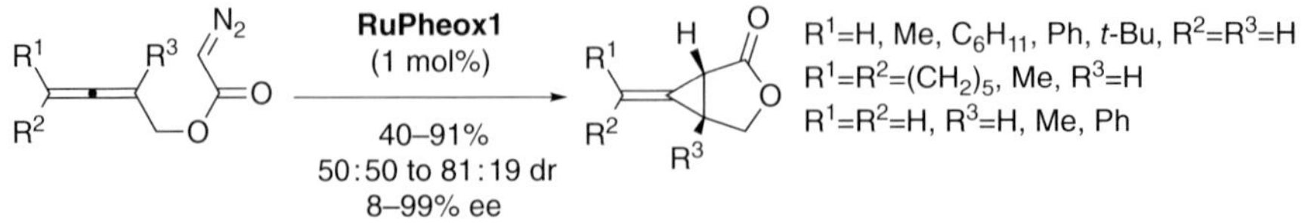

Scheme 1.113 Asymmetric intramolecular cyclopropanation of various allenic diazoacetates.

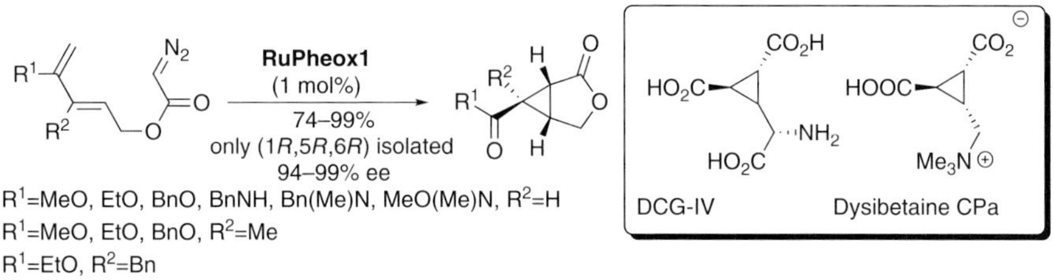

Scheme 1.114 Asymmetric intramolecular cyclopropanation of various electron-deficient allylic diazoacetates.

1.3.3 Chiral Stoichiometric Carbenes

In the previous sections, transition-metal-catalyzed cyclopropanation of olefins with diazo reagents have been presented. In addition, some chiral stoichiometric metal carbene systems as cyclopropanating reagents have been reported in the literature, although they have not been used extensively.

Since 1974, chiral iron complexes have been used in the stereoselective synthesis of cyclopropanes, but only 40% ee was observed for the methylene transfer [440]. Much higher enantioselectivities could be achieved when the carbene carbon was prochiral (Scheme 1.115) [441]. Typically, *trans*-methyl *cis*-phenyl-cyclopropanes were preferentially obtained.

Hossain reported the cyclopropanation reactions of optically pure (+) or (−) iron carbene complexes, prepared from iron and chromium complexes and aldehydes (Scheme 1.116) [442]. The reaction was also applied to the synthesis of a precursor for cilastatin. The origin of diastereoselectivity resided in the stability of the carbene and in a late transition state, which accounted for the *trans*-selectivity observed with nonaromatic alkenes, while the *cis*-isomer with aromatic alkenes was explained by the formation of a strong π-stacking effect.

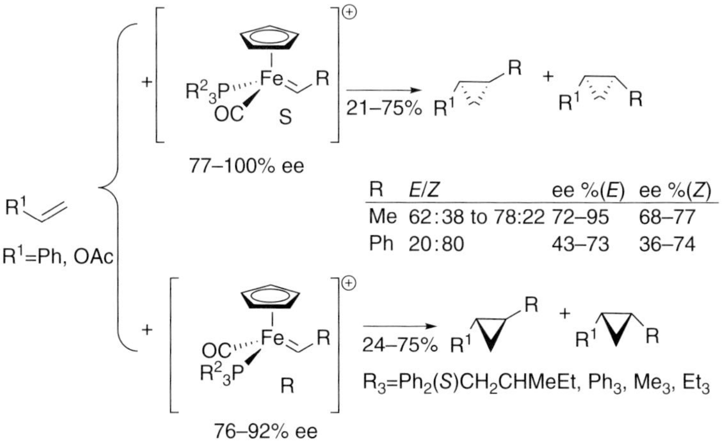

Scheme 1.115 Alkylidene asymmetric transfer from chiral cationic iron–carbene complexes.

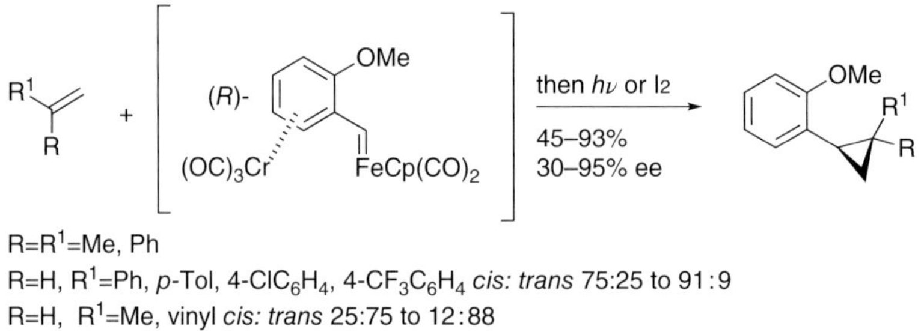

94 | Asymmetric Synthesis of Three-Membered Rings

Scheme 1.116 Carbene transfer from chiral cationic iron–chromium complexes.

R=R¹=Me, Ph
R=H, R¹=Ph, p-Tol, 4-ClC₆H₄, 4-CF₃C₆H₄ cis: trans 75:25 to 91:9
R=H, R¹=Me, vinyl cis: trans 25:75 to 12:88

Conditions shown: (R)-(OC)₃Cr·FeCp(CO)₂ carbene complex with OMe substituent; then hν or I₂; 45–93%; 30–95% ee.

Chiral oxazolines were used as chiral auxiliaries for the cyclopropanation with chromium complex with 55:45 to 94:6 dr and 70–99% ee, but, unfortunately, the chiral auxiliary could not be removed without destroying the cyclopropyl moiety [443].

Chiral Fischer carbene complexes derived from tungsten allowed the cyclopropanation of 2-methoxyfuran. The reaction involved the conjugate nucleophilic addition of 2-methoxyfuran to the carbene complex, followed by ring closure of the resulting zwitterionic intermediate species. Finally, oxidation of the resulting carbene gave rise to the corresponding enantiopure 1,2,3-trisubstituted cyclopropane (Scheme 1.117) [444].

1.4 Michael-Initiated and Other Ring Closures

Michael-initiated ring-closure (MIRC) reactions are a very successful method for obtaining cyclopropanes. These reactions involve a conjugate addition to an electrophilic alkene generally to produce an enolate, which then undergoes an

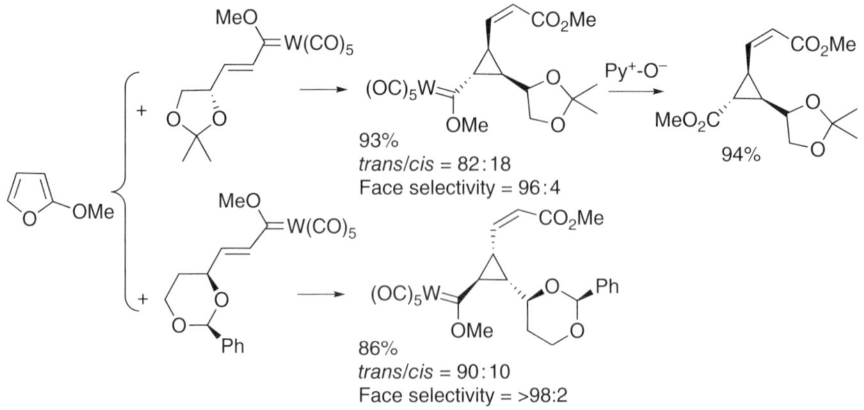

Scheme 1.117 Asymmetric cyclopropanations with Fischer tungsten-derived carbene complexes.

Top pathway: 93%, trans/cis = 82:18, Face selectivity = 96:4; then Py⁺-O⁻, 94%.
Bottom pathway: 86%, trans/cis = 90:10, Face selectivity = >98:2.

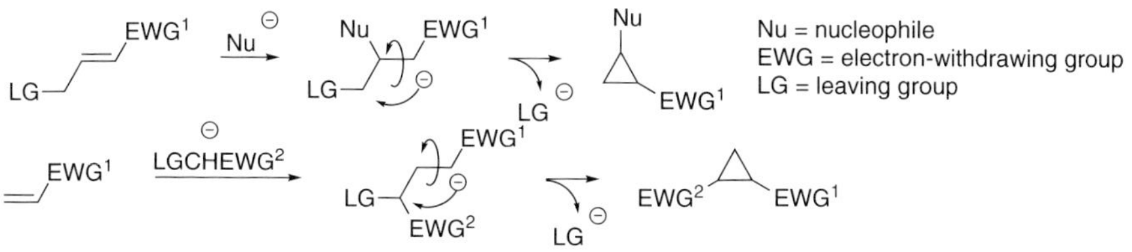

Scheme 1.118 Michael-initiated ring-closure cyclopropanation reaction.

intramolecular ring closure. Stereospecific cyclopropanation reactions using the MIRC reaction are observed only when the ring-closure process is faster than the rotation around the single bond in the first intermediate formed (Scheme 1.118), or when the first intermediate is configurationally stable, for example, by intramolecular hydrogen or coordination bonds. Two types of substrates/reactants can give rise to MIRC reactions:

i) Nucleophiles, such as alkoxides, thiolates, cyanides, enolates, Grignard reagents, hydrides, phosphites, and phosphonites, which add to electrophilic substrates containing a leaving group
ii) Nucleophiles containing the leaving group, such as α-halo carbanions or sulfur, phosphorus, arsenium, and telluronium ylides, which add to electron-deficient olefins.

Although chiral auxiliaries linked to the electrophilic moiety have been the choice for ensuring asymmetric induction in the past, more recently, the growing popularity of organocatalysis has prompted organic chemists to meet the challenge and to carry out MIRC reaction using this approach.

1.4.1 Chiral Substrates

The reactions involving sulfur ylides and enantiomerically pure cyclohexanone derivatives are well known to proceed with very high diastereoselectivities under steric control.

For example, the cyclopropanation of (R)-carvone with methylenedimethylsulfoxonium provided the desired cyclopropylcarvone, as the single diastereoisomer resulting from the attack of the ylide on the less-hindered face (Scheme 1.119, eq 1) [445]. Enantioenriched 2,3-methanopipecolic acid was prepared under similar conditions (Scheme 1.119, eq 2) [446]. Only one diastereoisomer of 7,8-cyclopropyltaxol (a potent taxol analogue) was successfully obtained when methylenedimethylsulfoxonium was used. The selectivity could be easily explained by steric arguments, as the bottom face of the enone is blocked by the A ring (Scheme 1.119, eq 3) [447]. A diastereoselective intermolecular cyclopropanation involving the addition of methylenedimethylsulfoxonium ylide to a chiral cyclopentanone derivative allowed the formation of bicyclo[3.1.0]hexane (Scheme 1.119, eq 4) [448]. Another example of sulfur ylide chemistry was the synthesis of an enantiomerically pure cyclopropatryptophan derivative (Scheme 1.119, eq 5) [449]. In the last two reactions, the observed stereochemical outcome is originated by the sulfur ylide attack from the more accessible convex face of the starting compound.

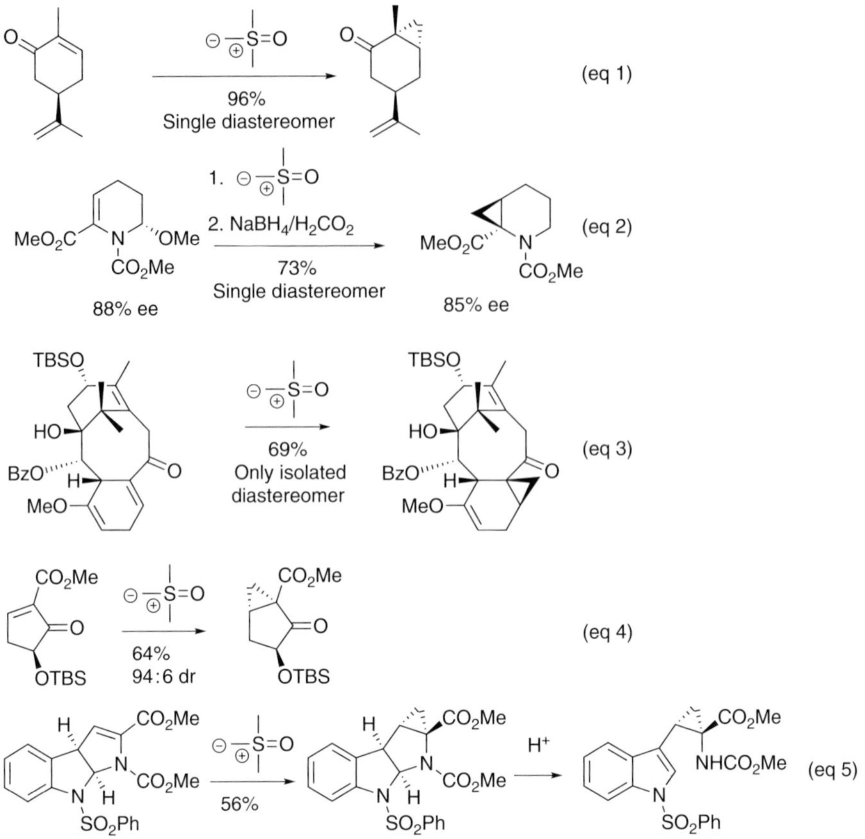

Scheme 1.119 Reactions involving methylenedimethylsulfoxonium ylide.

Isopropylidenediphenylsulfonium ylide reacted with suitably 4-functionalized 5,5-dimethyl-2-cyclopentenones to give only one stereoisomer of bicyclo[3.1.0]hexane derivatives, which was converted into (+)-cis-chrysanthemic acid (Scheme 1.120, eq 1) [450]. The cyclopropanation of (+)-dicyclopentadienone, readily available by enzymatic resolution, exclusively occurred from the convex face of the bicyclic system and afforded a single diastereoisomer (Scheme 1.120, eq 2) [451].

The asymmetric synthesis of LY354740, a potent and selective agonist for a glutamate receptor involved in the mammalian central nervous system, was achieved by the cyclopropanation of a protected dihydroxycyclopentenone, which exclusively afforded the exo-product (Scheme 1.120, eq 3) [452]. Analogously, the total syntheses of 4-acylamino analogs of LY354740 were achieved by the cyclopropanation of (S)-tert-butyl 4-oxocyclopent-2-enylcarbamate with $Me_2S=CHCO_2Et$, which worked in 73% yield producing a single diastereomer [453]. 3-Aza-bicyclo[3.1.0]hexane ring systems (a proline template amino acid) were obtained from O,N-acetal and N-Boc-pyrrolinone with sulfur ylides by the cyclopropanation of the less-hindered exo-face [454]. Only the unsubstituted cyclopropane derived from Boc-pyrrolinone was recovered in trace from $Me_2S(O)=CH_2$ and in 19% yield from

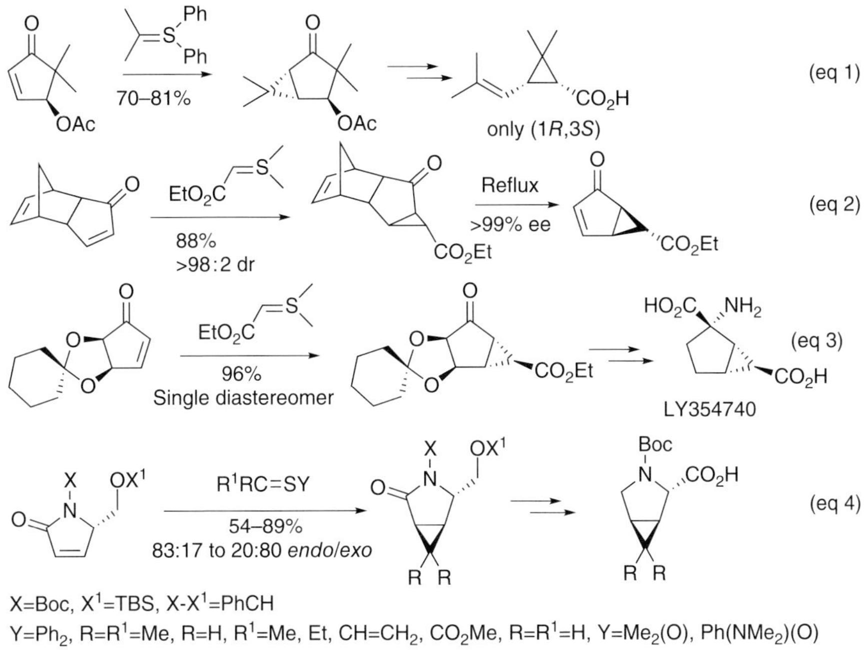

Scheme 1.120 Reactions involving other sulfur ylides.

Ph(NMe)$_2$S(O)=CH$_2$ (Scheme 1.120, eq 4). The *endo:exo* ratio varied with the substituent on the ylide and the substrate. It is worth noting that, in all the reactions described in eqs 2–4, the acidity of the methine proton was not a problem.

Lithium dienolate anion from ethyl 2-bromocrotonate was another reagent for cyclopropanation reactions as depicted in Scheme 1.121 [455]. The resulting cyclopropane ring was built at the more accessible face, but the diastereomeric ratio at the ester center could not be controlled. This compound is a key intermediate in the synthesis of pentalenene, involved in the biogenesis of the antibiotic pentalenolactone.

Chattopadhyaya used a 5′-protected α,β-ene-3′-phenylselenone as a synthetic equivalent of a (CH$_2$)$_2^{2+}$ ion in the reaction with carbon dinucleophiles such as sodium malonate, and conjugate bases of nitromethane, acetophenone, isobutyl-cyanoacetate, and both methylene bis-diethylphosphate and bis-phenylsulfone, to give bicyclo[3.1.0]cyclopropane analogues of 2′,3′-dideoxyuridine (Scheme 1.122) [456]. In all cases, the isomer resulting from the delivery of the methylene group on the more accessible face of the dideoxynucleoside was exclusively formed. The same procedure was then applied to the preparation of a conformationally constrained dimeric building block containing cyclopropylamide functionality [457].

In acyclic systems, the highest selectivities were generally observed with conformationally constrained alkenes. For example, (*S*,*E*)-methyl 3-(2,2-dimethyl-1,3-dioxolan-4-yl)acrylate reacted with 2-lithio-2-propyl-*N*-tosylisopropylsulfoximide and isopropylidenetriphenylphosphorane mainly leading to the corresponding (1*S*,3*S*)-2,2-dimethylcyclopropanecarboxylate from attack at the *Si*-face (93 : 7 and

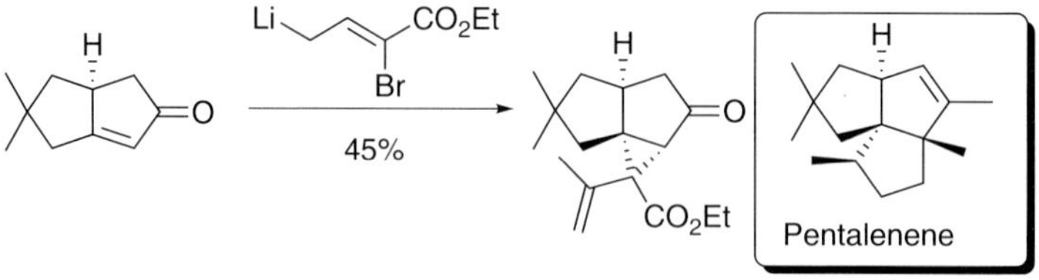

Scheme 1.121 MIRC Reaction of an enone.

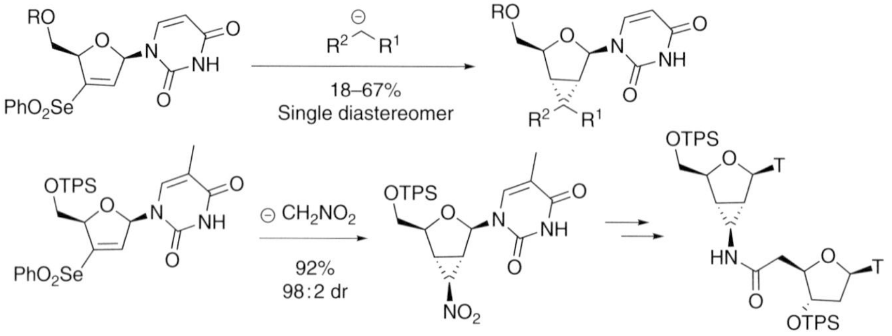

Scheme 1.122 Synthesis of cyclopropyl nucleoside derivatives.

90:10 dr, respectively) [458]. On the other hand, isopropylidenediphenylsulfurane attacked the *Re*-face of the alkene (98:2 dr in favor of the (1*R*,3*R*)-isomer).

With the (*S*,*Z*)-acrylate, the attack of all three reagents occurred on the *Re* face. However, the reaction of the sulfur ylide led to the formation of the *cis*-cyclopropane (98:2 dr in favor of the (1*S*,3*R*)-isomer), while both the phosphorus and lithium reagents led to the isomerized *trans*-cyclopropanes (94:6 and 61:39 dr in favor of the (1*R*,3*R*)-isomer, respectively). The authors proposed models to rationalize these results only with phosphorus and lithium reagents, but not with sulfur ylide. Other acrylates from D-glyceraldehyde gave similar results [459], as well as the cyclopropanation of dimethyl alkylidene malonate with either sulfur or phosphorus ylides occurred in >98:2 dr (2*S*) and 76% and 80% yields, respectively [460]. The methodology was also applied to the synthesis of (2*S*,1′*S*,2′*S*)-(carboxycyclopropyl) glycine, and best results were obtained with methylenedimethylsulfoxonium ylide at low temperature (73% yield, 95:5 dr) [461]. A series of 1,2,3-trisubstituted cyclopropanes were also prepared from enones derived from (*S*)-glyceraldehyde acetonide and Garner's aldehyde [(*S*)-3-Boc-2,2-dimethyloxazolidine-4-carboxaldehyde] and sulfur ylides in 24–96% yields [462]. Good diastereoselectivities were achieved with ester-substituted ylides, leading mainly to *all-R* products and (1*R*,2*S*,3*R*,ω*R*)-isomer[6] from *E* and *Z* isomers, respectively, whereas with amide-substituted ylides, the yields and the selectivities were unsatisfactory. This approach was extended to C_2-symmetric bis-(unsaturated esters) from tartaric acid, which afforded products easily transformable into *trans*-chrysanthemic acid or its analogs (Scheme 1.123)

---

[6] Pay attention that the new substituents also make (*R*) the stereocenter of the glyceraldehyde moiety, although it is not changed.

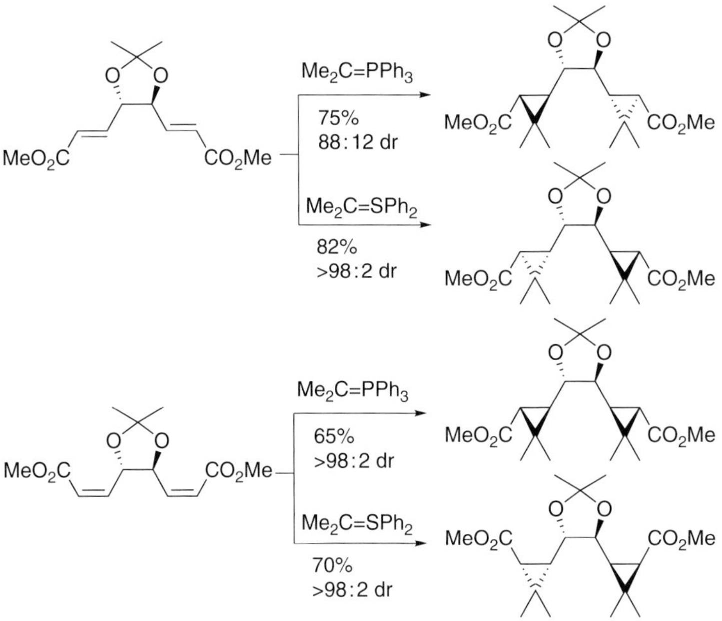

Scheme 1.123 Diastereoselective cyclopropanation of bis(unsaturated esters).

[463]. In this case, the Z-to-E isomerization of the Z,Z isomer with phosphorus ylide is worth noting. A further evidence was the reaction with only 1 equiv. of phosphorus ylide, in which a 1:1:1 mixture of starting material, (E)-monoadduct, and product was recovered. Conversely, the (Z)-monoadduct resubmitted to the same reaction conditions affords the (1S,1′S,3S,3′S)-isomer and not the expected (1R,1′S,3R,3′S)-isomer.

The cyclopropanation of other derivatives of chiral 1,2-O-isopropylideneglyceraldehyde, namely (Z)-oxazolone and nitroalkenes derivatives, was performed with methylene(diethylamino)phenylsulfoxonium ylide (Scheme 1.124, eq 1) [464] and with isopropylidenediphenylsulfonium ylide, respectively (Scheme 1.124, eq 2) [465]. The sense of induction was predicted by the conformation depicted in Scheme 1.124.

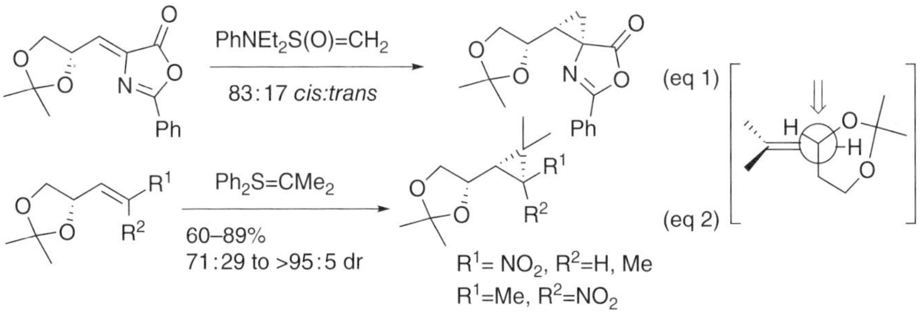

Scheme 1.124 Cyclopropanation of chiral (Z)-oxazolone and nitroalkenes.

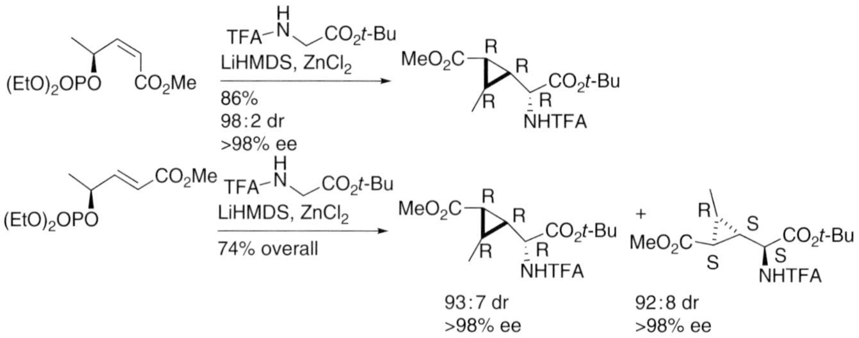

Scheme 1.125 Syntheses of chiral cyclopropyl amino acids.

The Zn-chelate-glycine ester enolates were highly efficient nucleophiles in MIRC reaction on methyl 4-(diethoxyphosphoryloxy)pent-2-enoate (Scheme 1.125) [466]. In particular, the reaction of the Z-isomer gave only one stereoisomer with up to four stereogenic centers created in one step.

The asymmetric cyclopropanation of a chiral olefin derived from D-glucose with trimethylsulfoxonium ylide afforded the corresponding cyclopropane as a single product, which was the key intermediate in a carbohydrate-based approach for the enantioselective synthesis of the polyketide acid unit present in nagahamide A (Scheme 1.126) [467].

1.4.2 Chiral Auxiliaries

1.4.2.1 Chiral Michael Acceptors

The first class of chiral auxiliaries, used as Michael acceptors, were α,β-unsaturated chiral esters, but low diastereoselectivities were generally observed in the most ancient papers [468]. Later, menthyl esters as chiral auxiliaries were employed (Scheme 1.127, eq 1) [469]. The surmised reaction course for the cyclopropanation with 1-seleno-2-silylethene (Scheme 1.127, eq 2) contemplated the nucleophilic addition of the vinylselenide to the unsaturated ester, activated by the Lewis acid. Steric repulsion between the selenium and the isopropyl moiety of the menthyl group favored the approach from the less-hindered *Re*-face.

(−)-8-Phenylmenthyl α,β-unsaturated esters were able to react with silylated allylic telluronium ylide. The aryl group very likely blocked the *Si*-face of the alkene by a π-stacking effect between the phenyl and the dienyl groups of the α,β-unsaturated ester, thus the telluronium ylide attacked at C-3 on the *Re*-face (Scheme 1.128) [470].

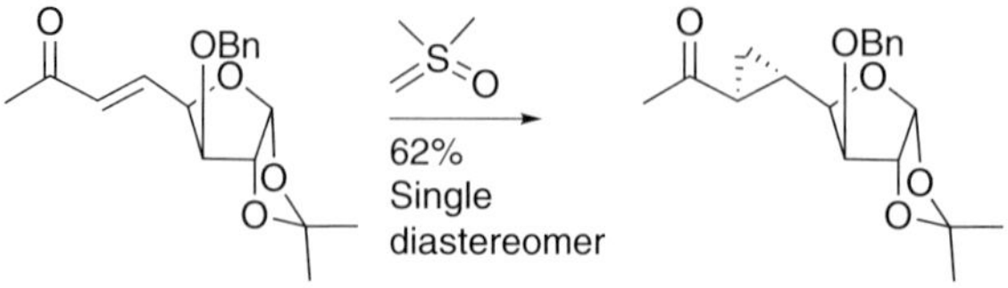

Scheme 1.126 Asymmetric cyclopropanation of D-glucose-derived enone.

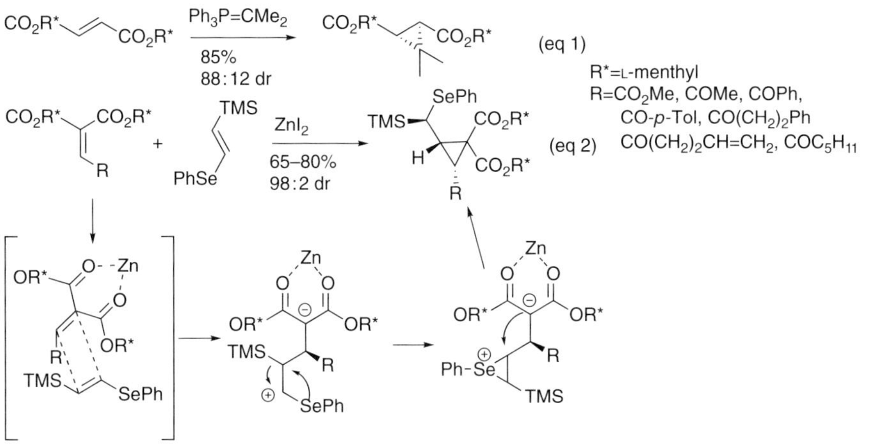

Scheme 1.127 Menthyl esters as chiral auxiliaries.

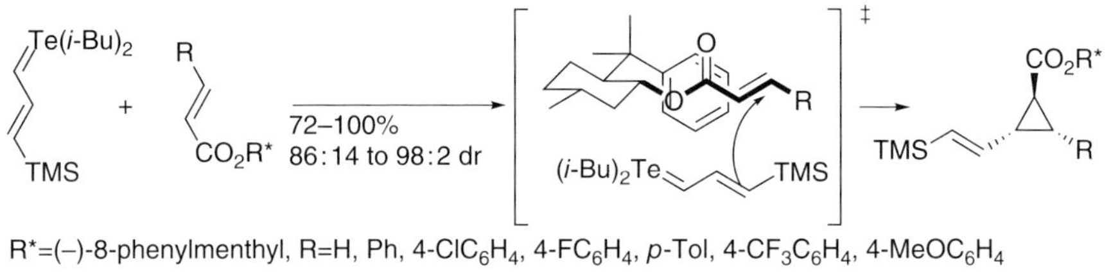

Scheme 1.128 Telluronium ylide addition to chiral α,β-unsaturated esters.

Scolastico described one example involving the reaction of a chiral γ-oxazolidine-substituted conjugated ester with isopropylidenetriphenylphosphorane with excellent π-face selectivity. The proposed transition state involved the conformer with minimum $A^{1,3}$-strain and with the most electronegative substituent (oxygen) perpendicular to the π orbital of the alkene (Scheme 1.129, eq 1) [471]. The cyclopropanation of chiral N-enoyloxazolidinones (Scheme 1.129, eq 2) with isopropylidenediphenylsulfonium ylide in the presence of Lewis acids proceeded with a good level of diasterocontrol, but cinnamoyloxazolidinone gave the lowest selectivity [472]. The Lewis acid, chelating both carbonyl groups, afforded a dominant reactive rotamer, in which one face of the β-carbon was shielded from attack. On the other hand, in the absence of the Lewis acid, free rotation around the C—N bond afforded low diastereoselectivity, and the rotamer with the isopropyl group far from the incoming ylide was favored, leading to slight predominance of the diastereomer with the cyclopropane and isopropyl moieties in *syn*-relationship.

Chiral bicyclic lactams prepared from L-valinol or L-*tert*-leucinol reacted with methylenedimethylsulfoxonium ylide with a high degree of *exo:endo*-diastereoselectivity, which, however, depends on the angular substituent of the unsaturated lactam (Scheme 1.130) [473]. Substituted sulfonium ylides were also employed:

i) Isopropylidenediphenylsulfonium ylide afforded the *gem*-dimethyl cyclopropyl adduct in 94% yield and >99:1 dr.

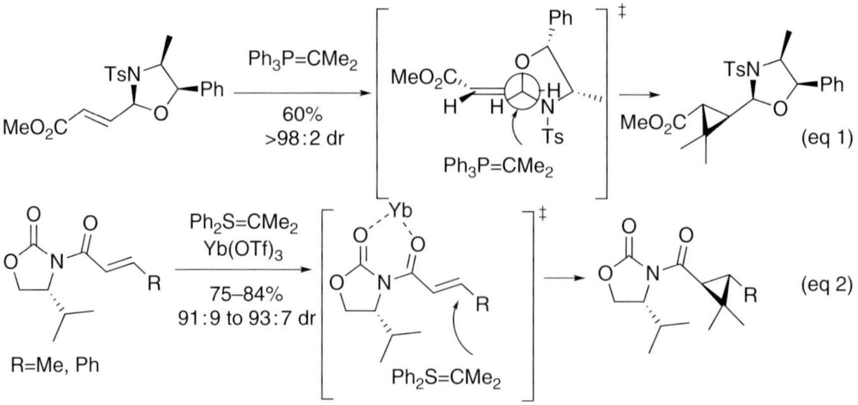

Scheme 1.129 Reaction of unsaturated chiral γ-oxazolidine derivatives.

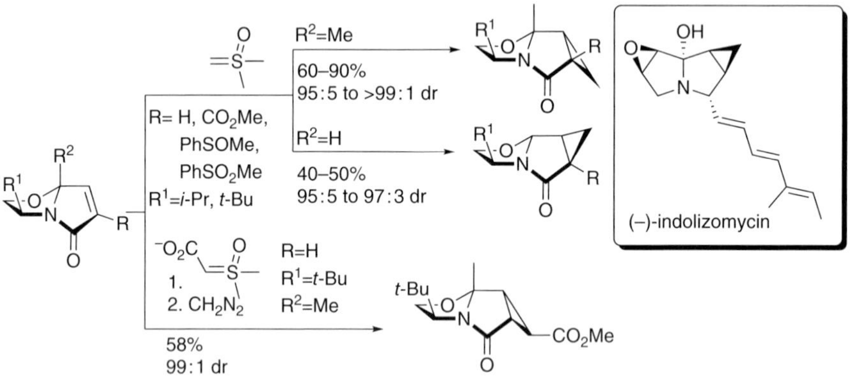

Scheme 1.130 Cyclopropanation of chiral unsaturated lactams.

ii) Unsimmetrically substituted sulfonium ylides produced cyclopropanes in 41–95% yields with almost exclusive formation of the *endo*-isomer and 83:17 to >99:1 *syn:anti* ratios under kinetic control.

iii) Carboxymethylenesulfonium ylide led to the exclusive formation of the *anti*-isomer probably as a result of thermodynamic equilibration.

This method was also used for the preparation of a key precursor to (−)-indolizomycin [474].

Chiral auxiliaries were elaborated in order to enable a stereoselective cyclopropanation of an exocyclic double bond for the preparation of enantiomerically pure cyclopropyl α-amino acids [475].

An optically active vinyl sulfoxide was used as Michael acceptor with an allyl Grignard reagent leading to one diastereoisomer (Scheme 1.131, eq 1) [476].

The coordination of the Grignard reagent to the oxygen atom of the sulfinyl group was envisaged to explain the selectivity, the energetically more favorable conformer then allowed the preferential delivery from the bottom face. Acyclic chiral vinyl sulfoxides have also been converted into cyclopropanes (Scheme 1.131, eq 2) [477].

The proposed transition state involved a nonchelate model in which 1,3-elimination occurred preferentially from the lone pair side of the *p*-toluenesulfinyl group. This strategy was extended to the addition of the bromomalonate carbanion to a chiral α-ketovinylsulfoxide (Scheme 1.131, eq 3) [478]. The nature of the solvent, the base, and the counterion influenced both the yield and the selectivity. It should be noted that, with respect to the previous reaction, the formation of a chelate between the carbonyl and sulfoxide oxygen, which led to an opposite diasterofacial selectivity, by the addition of bromomalonate enolate from the sterically less crowded lone-pair side of the chiral sulfinyl substituent. In addition, chiral sulfinylcyclopentenones efficiently reacted with the α-bromoacetate carbanion (Scheme 1.131, eq 4) [479].

The cyclopropanation of chiral (*E*,*S*)-(1-dimethoxyphosphoryl-2-phenyl)vinyl *p*-tolyl sulfoxide with various sulfur ylides allowed the synthesis of 2-amino-3-phenyl-1-cyclopropanephosphonic esters, constrained analogs of phaclofen (Scheme 1.131, eq 5) [480]. Moreover, the reaction between lithiated phenyl phenylthiomethylsulfone and 1-(phenylthio)vinylsulfoximide produced a mixture of two stereoisomeric cyclopropanes in 75% yield and 75:25 ratio [481]. Chiral sulfinylfuranones reacted with sulfonium ylides via the addition of the sulfur nucleophile to the double bond from the opposite face to that occupied by the OEt group, with nonstabilized ylides [482]. The formation of the *endo*- or *exo*-adducts was completely dependent upon the steric hindrance of the substituents at sulfur. Consequently, dimethylsulfonium derivatives yielded mainly the *exo*-products, while, with the diphenylsulfonium ylides, the *endo*-products predominated (Scheme 1.131, eq 6).

In 2008, Marek and coworkers exploited an MIRC reaction on chiral alkylidene bis(*p*-tolylsulfoxides) with the trimethylsulfoxonium ylide (Scheme 1.132) [483]. The corresponding chiral bis(*p*-tolylsulfinyl) cyclopropanes were then used to prepare enantiomerically enriched polyalkylated cyclopropane derivatives. They found, in fact, that a *syn*-selective sulfoxide–lithium exchange with retention of configuration at the carbon center occurred at the more hindered position, because strain was released. Then, a second sulfoxide–lithium exchange can be carried with the same features, leading to enantiomerically pure products. Permutation of the electrophiles introduced after each of the two sulfoxide–lithium exchange reactions allowed for the formation of two products, having opposite configurations at the quaternary center. This methodology was used for the preparation of (9*R*,10*S*)-dihydrosterculic acid [484].

The reaction of (*S*)-(1-diethoxyphosphoryl)vinyl *p*-tolylsulfoxide and (*E*,*S*)-(1-dimethoxyphosphoryl-2-phenyl)vinyl *p*-tolyl sulfoxide with (dimethylsulfuranylidene)acetate mainly allowed for the preparation of (1*S*,2*S*)-(1-diethoxyphosphoryl-2-ethoxycarbonyl)cyclopropyl *p*-tolylsulfoxides (Scheme 1.133). However, the two diastereomers were easily separable, and they could be converted into enantiomerically enriched 2-(2′-phosphonocyclopropyl)glycines, which are among the most important sources of active analogs for the glutamic acid receptors [485]. The same substrate was allowed to react with a sulfonium

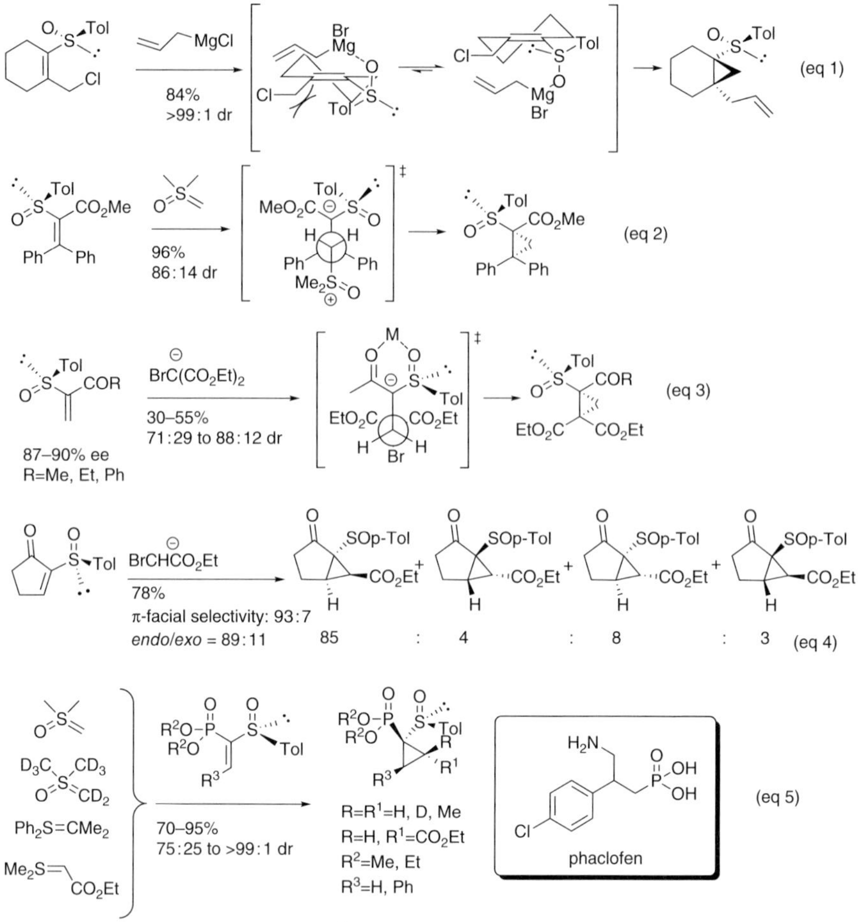

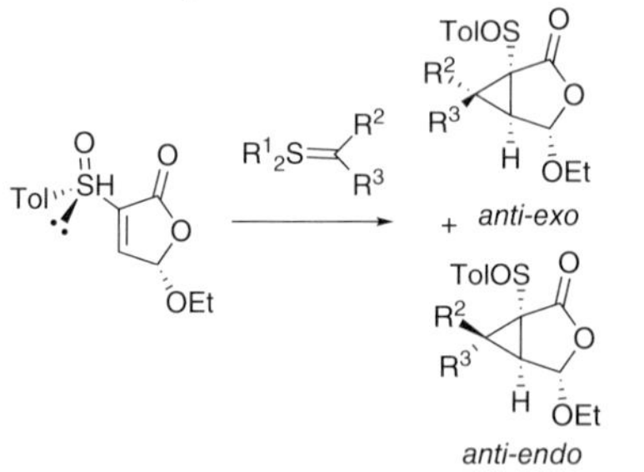

R^1	R^2	R^3	Y (%)	*anti:syn*	*endo:exo*
Ph	H	H	59	100:0	—
Ph	Me	Me	91	100:0	—
Ph	Me	H	77	100:0	78:22
Ph	CO_2Et	H	68	95.5	96:4
Me	CO_2Et	H	50	66:34	4:96
Me	Ph	H	25	90:10	23:77

(eq 6)

Scheme 1.131 Asymmetric cyclopropanation of sulfinyl compounds.

ylide containing a ketone moiety bonded to the ylide carbon atom, instead of CO_2Et group. The cyclopropane was recovered in 86% yield, the major diastereomer (5:1 dr) having the ketone in *cis*-relationship with phosphonate, differently from that depicted in Scheme 1.133. Unfortunately, the chiral auxiliary sulfinyl group could not be removed from the product [486].

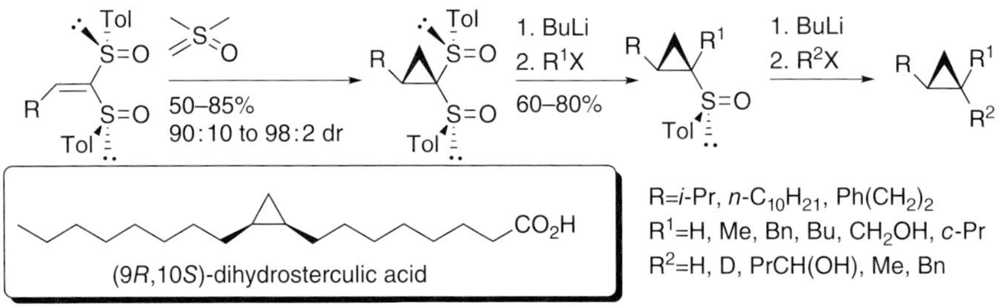

Scheme 1.132 Preparation of enantiomerically pure polyalkylated cyclopropane derivatives.

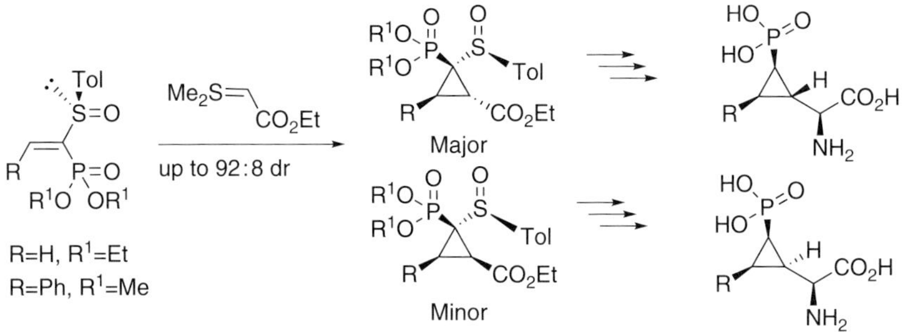

Scheme 1.133 Stereocontrolled synthesis of 2-(2′-phosphonocyclopropyl)glycines.

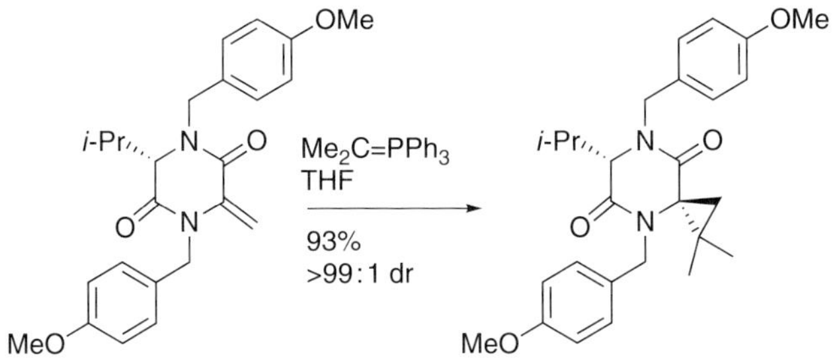

Scheme 1.134 Asymmetric cyclopropanation of dehydroalanine acceptor.

The conjugate addition of an *in situ* generated phosphonium-derived ylide to a chiral dehydroalanine acceptor allowed the preparation of the corresponding cyclopropane-substituted diketopiperazine (Scheme 1.134) [487].

After cross metathesis with ethyl vinyl ketone, the readily available (3R,4R)-1,5-hexadiene-3,4-diol provided the chiral substrate for cyclopropanation with dimethylsulfoxonium ylide, leading to the disubstituted *trans*-cyclopropane in 79% yield and >20 : 1 dr, the key intermediate in the total synthesis of brevipolides (see Scheme 1.6 for the structure) [488].

Finally, the asymmetric cyclopropanation using two sulfinyl auxiliaries, one on the Michael acceptor and the other on the nucleophile, can usefully close this

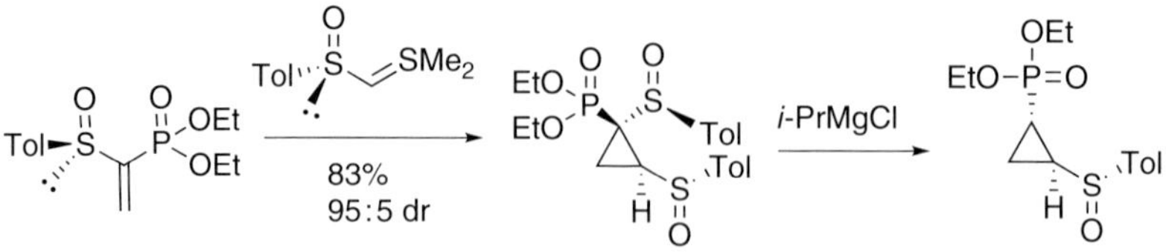

Scheme 1.135 Asymmetric cyclopropanation using two sulfinyl auxiliaries.

chapter and introduce the next one. The product was obtained with full diastereoselectivity and very high facial selectivity. The crowding of the phosphonate moiety is usually considered as a *trans*-directing driving force, but also dipole–dipole interaction between both sulfinyl substituents has to be considered, and leads to the same isomer. The reaction of with isopropylmagnesium chloride allowed the removal of only one sulfinyl group selectively (Scheme 1.135) [489]. A possible explanation of the observed stereochemistry of the final product could be the attack of the Grignard reagent on sulfinyl sulfur atom α to phosphoryl group with subsequent epimerization of carbanionic center before quenching.

In addition, the key step of the synthesis of (−)-*trans*-2-aminomethylcyclopropanecarboxylic acid, a partial agonist for GABAc receptor, was the asymmetric cyclopropanation of L-menthyl acrylate with ammonium ylides derived from *tert*-butyl bromoacetate and quinidine or quinine [490]. *Trans*-substituted cyclopropanes were exclusively obtained. Unfortunately, the reaction did not go to completion; however, 89% and 63% yields (based on recovered starting material) were obtained for 75:25 and 7:93 (*R*,*R*):(*S*,*S*) mixtures with quinidine or quinine ylides, respectively. From these observations, L-menthyl acrylate and quinidine or quinine derivative could be considered stereochemically mismatched or matched, respectively.

1.4.2.2 Chiral Nucleophiles

A variety of stoichiometric chiral nucleophiles can perform enantioselective cyclopropanation of alkenes. Chiral sulfur, selenium, nitrogen, phosphorus, and arsenic ylides as well as chiral enolates have been added to α,β-unsaturated carbonyl derivatives. The asymmetric cyclopropanation using chiral sulfur ylides has been carried out since the 1960s, albeit with low enantioselectivities [491].

It is worth of the first mention that, owing to the analogy with the reaction sequence reported in Scheme 1.135, the reaction of deuterated achiral 2-phosphonateacrylates using (S)-dimethylsulfonium-(*p*-tolylsulfinyl)methylide, allowed the preparation of (1R)-ethyl-1-diethylphosphono-2,2-dideuteriocyclopropanecarboxylate, in 67% overall yield and 80% ee [492]. It should be noted that, when the phosphoryl and sulfinyl substituents were on the same side of the cyclopropane ring, during sulfoxide–metal exchange, 1,2-migration of a phosphoryl group could take place, leading to desulfinylated cyclopropane of an entirely different structure. Then, the asymmetric cyclopropanation of vinylphosphonates using other optically active sulfonium and selenonium ylides derived from terpenes led to ethyl 1-(diethoxyphosphoryl)-2-phenylcyclopropanecarboxylate, diethyl 1-(diethoxyphosphoryl)cyclopropane-1,2-dicarboxylate, diethyl 1-[(phenylsulfonyl)-2-phenylcyclopropane]-phosphonate,

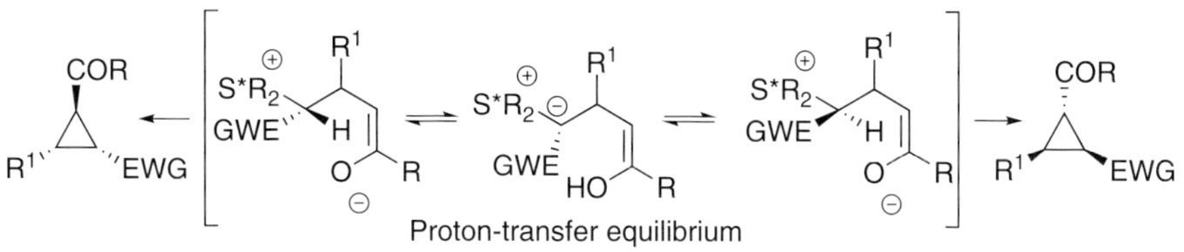

Scheme 1.136 Mechanism accounting for enantioselectivity in the asymmetric ylide-promoted MIRC reaction.

and diethyl-[(phenylsulfonyl)-2-carboethoxycyclopropane]-phosphonate in 53–94% yields with 56:44 to 81:19 *trans:cis* ratio and up to 98% ee. With regard to the *trans*-isomer, menthol-derived selenonium ylide mainly gave (*S*,*R*)-cyclopropane, whereas an isothiocineole-derived ylide led to an (*R*,*S*)-isomer. The stereochemical outcome of these reactions is easily rationalized by the mechanism of the reaction described in Scheme 1.136 [493]. Finally, the asymmetric synthesis of (1*S*,2*R*)-1-amino-2-methylcyclopropanephosphonic acid, a phosphonic analog of (−)-norcoronamic acid, was set up, starting from 1-cyanovinylphosphonic acid and (*S*)-dimethylsulfonium-(*p*-tolylsulfinyl)methylide. It is worth noting that the nitrile group changed the stereoselectivity of the reaction with respect to the CO_2R framework, thus avoiding 1,2-migration of the phosphoryl group on the cyclopropane ring [494].

Aggarwal's research group studied the mechanism of the cyclopropanation of acyclic enones with chiral sulfur ylides [495]. However, this model could also be applied to nitrogen and phosphorus ylides and was able to rationalize the outcome of a variety of substituents on the ylides in terms of chemo- and enantioselectivity. In particular, a clear correlation between ylide stability (ketone > ester > amide) and enantiomeric excess (25%, 46%, and 89% ee, respectively) was observed. The enantioselectivity was governed by an unusual proton-transfer step prior to ring closure: the more stable the ylide is, the slower the ring closure and higher the enantioselectivity (Scheme 1.136). The proposed mechanism is strongly supported by experiments with deuterium-labeled ylides, in which higher enantioselectivities were observed in agreement with the slower deuteron transfer.

Chiral lithiated *N*-tosylsulfoximines were efficiently used, because they overcome the additional step needed to remove the chiral auxiliary, being spontaneously removable in the elimination step (Scheme 1.137).

In fact, the lithiated sulfoximines afforded diastereoselective Michael reactions at low temperature, arranging a transition state in which the large sulfonimidoyl moiety and the substituent (R) in the Michael acceptor were *anti* in order to minimize steric interactions. Then ring closure occurred upon warming to room temperature by an intramolecular displacement of the sulfonimidoyl group with inversion of the configuration. They were found to produce enantioenriched cyclopropanes when reacted with enones (Scheme 1.137, eq 1) [496] or cinnamyl hydroxamic acid derivatives (Scheme 1.137, eq 2) [497]. In addition to the examples reported in Scheme 1.137, eq 2, alkyl-substituted hydroxamic acid derivatives were tested and led to products in 62% yield with 94% ee for the major diastereoisomer, but three diastereoisomers were detected, in a ratio of 75:19:6.

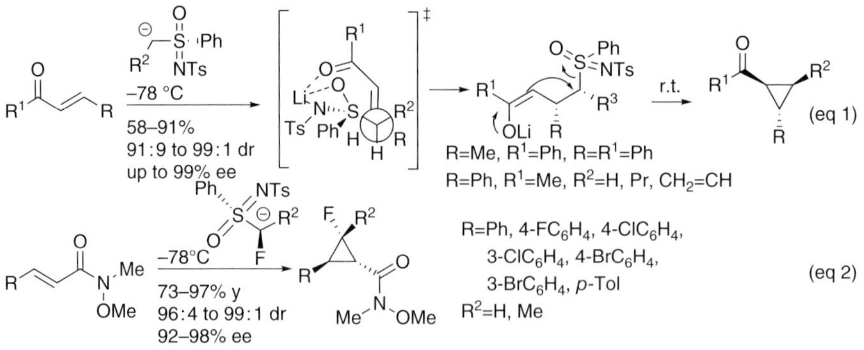

Scheme 1.137 Cyclopropanation of enones with chiral lithiated sulfoximines.

Moreover, two examples of cyclopropanes with a quaternary fluorinated center (R^2=Me) were reported in 70–75% yield, 7:1 dr, and 76–87% ee.

A single diastereoisomer (about 80% yield) was observed in the synthesis of β-(trimethylsilyl)ethyl cyclopropanecarboxylate derivatives (chrysantemate precursors) from chiral p-tolyl β-(trimethylsilyl)ethyl sulfoxide [498].

The reaction of chiral oxathiane with arylmethyl alcohols in the presence of triflic anhydride produced a series of (arylmethyl)sulfonium salts, which can be converted in situ into chiral sulfonium ylides, before reacting with ethyl acrylate or methyl vinyl ketone (Scheme 1.138) [499]. It should be noted that acrolein led mainly to the corresponding epoxides.

A large literature produced by Tang and coworkers explained the reaction of the chiral sulfonium salt prepared from D-camphor. The first example reported the preparation of (1S,2S,3R)-1,2-disubstituted 3-(trimethylsilylvinyl)cyclopropanes with high enantioselectivities (Scheme 1.139, eq 1) [500]. Low yields were observed with aliphatic or Z-configured alkenes.

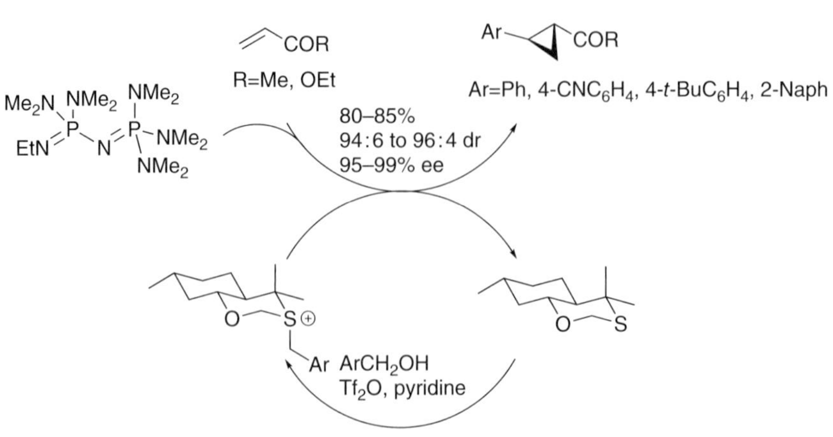

Scheme 1.138 MIRC reaction with chiral oxathiane.

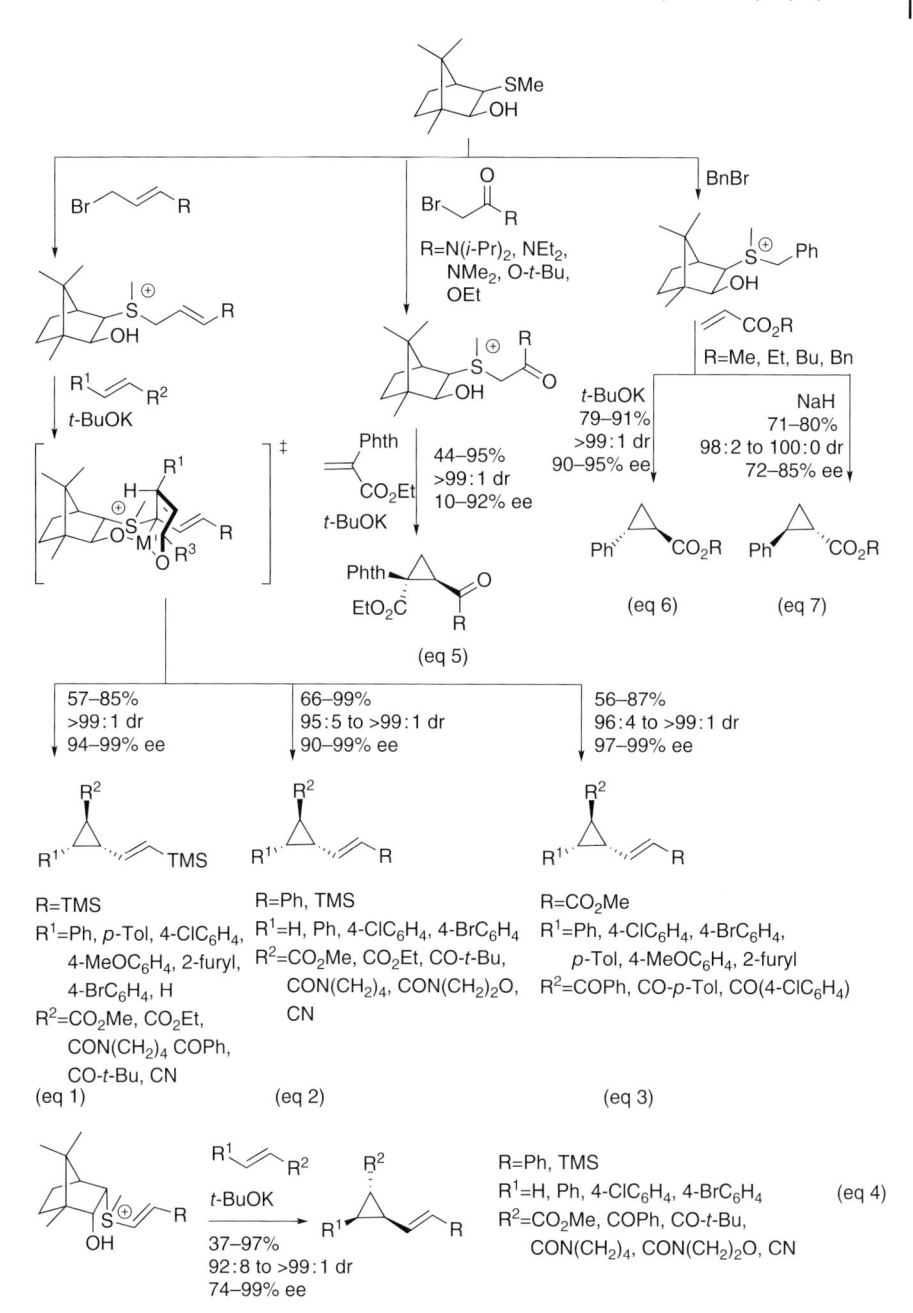

Scheme 1.139 Asymmetric cyclopropanations via camphor-derived sulfonium ylides.

The transition-state model, accounting for the observed stereochemical outcome, contemplated the carbonyl group of the substrate coordinated to the metal of the ylide via a six-membered ring. Then, the substrate reacts at the *Re*-face of the ylide to avoid the steric interaction between the R group and the methyl substituent of the sulfur. Later, the same group prepared other 1,2-disubstituted-3-vinylcyclopropanes (Scheme 1.139, eq 2 [501], and 3 [502]). It should be noted that the use of *endo*-type sulfonium ylides allowed the isolation of the products with opposite absolute configuration (Scheme 1.139, eq 4) [501]. Moreover, dehydroaminoacid derivatives allowed the preparation of 1-aminocyclopropanecarboxylic acids (Scheme 1.139, eq 5) [503]. The steric hindrance of the amide sulfonium salts significantly influenced the cyclopropanation process. In fact, with increasing crowding yield, enantiomeric excess decreased. The phthaloyl group could be easily removed at room temperature to provide the amino acid in moderate overall yield.

In a similar manner, Huang and Huang developed an enantioselective synthesis of 2-phenyl-1-cyclopropane-carboxylates [504]. Two features are worth noting: the starting chiral sulfide could be recovered almost quantitatively and reused conveniently, and the enantioselectivity could be tuned by changing the base. Indeed, when potassium *tert*-butoxide or sodium hydride was used, (1*R*,2*R*) or (1*S*,2*S*)-2-phenyl-1-cyclopropane carboxylates were obtained, respectively (Scheme 1.139, eqs 6 and 7, respectively). Acrylonitrile as electrophile can also be used, providing the corresponding cyclopropane in 75% yield, as a single diastereomer with 91% ee. In the light of Aggarwal's mechanism, which is more recent, the different bases have to influence the proton transfer step.

Interestingly, the reaction of camphor selenonium ylide afforded 1,2,3-trisubstituted cyclopropanes, with a different stereochemistry, notwithstanding the close similarity of sulfonium and selenonium ylides. In fact, the *exo*-isomer led to (1*R*,2*R*,3*R*)-1,2,3-trisubstituted cyclopropanes and the *endo*-isomer to (1*S*,2*S*,3*S*)-cyclopropanes (Scheme 1.140) [505].

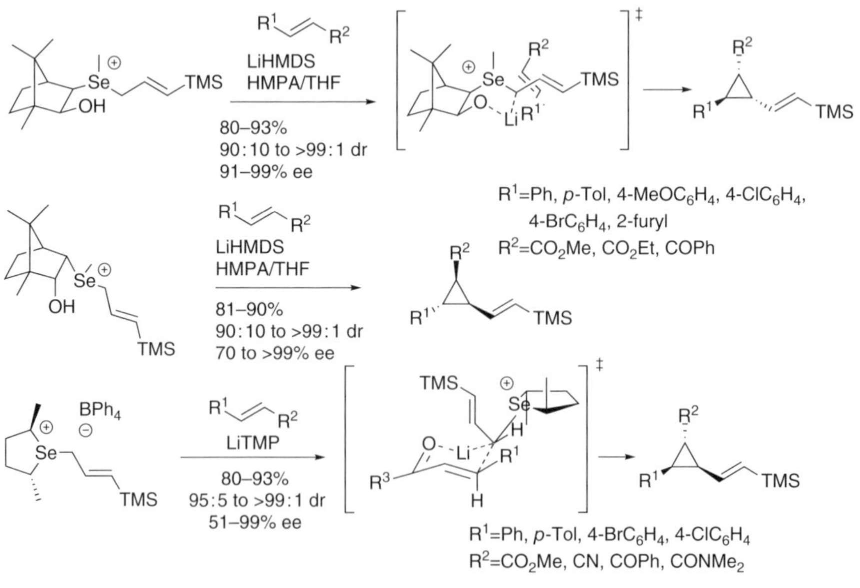

Scheme 1.140 Asymmetric synthesis of cyclopropanes via chiral selenonium ylides.

The enantioselectivity was higher with free OH group with respect to the methoxy derivative, suggesting that a negatively charged oxygen atom might play an important role in assembling the transition state. Although not mentioned in the paper, a role of the base in directing the selectivity cannot be excluded, because the base strength of LiHMDS/HMPA is quite different from *tert*-butoxide, as Huang and Huang demonstrated [504]. Moreover, in the same paper, ylides derived from (2R,5R)-dimethyltetrahydroselenophene reacted with various α,β-unsaturated compounds, affording the corresponding (1R,2R,3S)-cyclopropanes. In all reactions, selenium by-products could be easily recovered and reused to prepare the starting ylides.

In the literature, also, chiral allylic ylides derived from (2R,5R)-dimethyltetrahydrotellurophene were reported to be able to react with α,β-unsaturated esters, amides, imines, and ketones, providing the corresponding 1,3-disubstituted 2-vinylcyclopropanes (Scheme 1.141) [506].

An instance of phosphorus ylide, that is, a chiral phosphonamide derived auxiliary, was reported to be added to cyclic enones, leading to the formation of the corresponding *endo,endo*-cyclopropane after ring closure (Scheme 1.142, eq 1) [507].

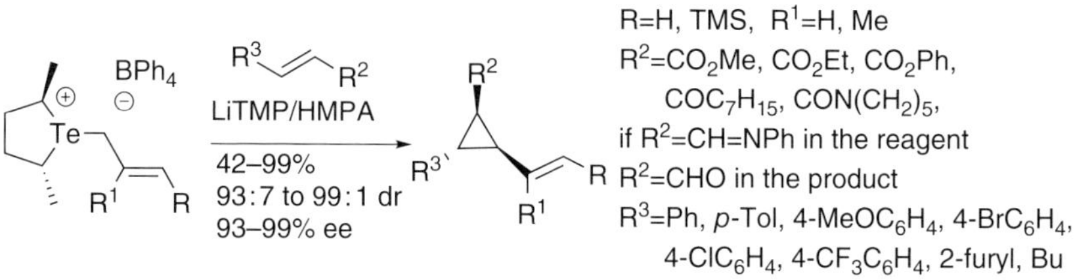

Scheme 1.141 Asymmetric allylic telluronium-ylide-mediated cyclopropanations.

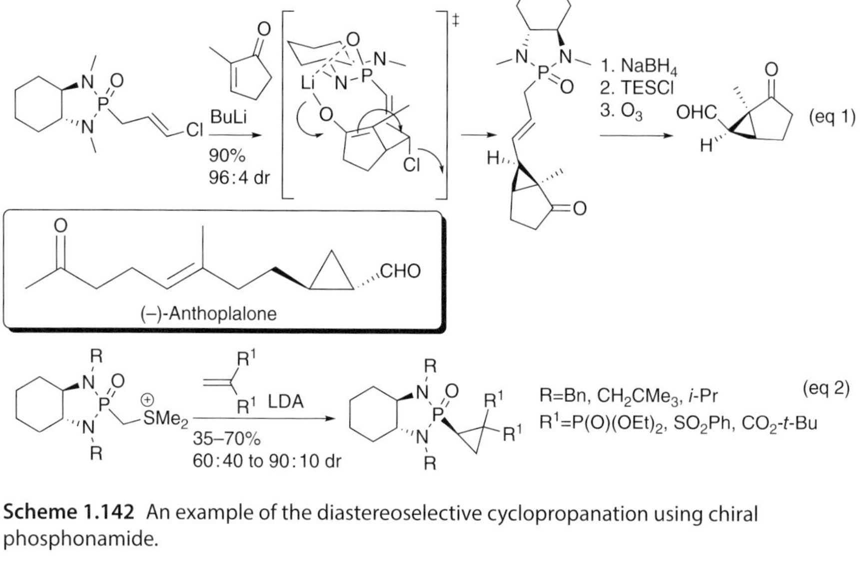

Scheme 1.142 An example of the diastereoselective cyclopropanation using chiral phosphonamide.

The chiral auxiliary was removed by ozonolysis to generate the cyclopropanecarboxaldehyde. The *exo,endo*-isomer could be obtained both by epimerization of the *endo,endo*-cyclopropanecarboxaldehyde or from reaction with the *cis*-chloroallylphosphonamide. This method was extended to α,β-unsaturated γ-lactones, δ-lactones, γ-lactams, and acyclic esters. In all reactions, the *endo,endo*-isomers were recovered in 51–88% yields with 95:8 to >98:2 dr. The phosphonamide derivatives could be manipulated to generate functionally diverse cyclopropanes without affecting the stereochemistry, for example, in the total synthesis of the cytotoxic agent (−)-anthoplalone [508]. The suggested mechanism claimed an attack of the γ-chloroallylic anion on the *Re*-face of the cyclic enone, leading to a Li-chelated enolate intermediate, with the enolate and chloride in *anti* relationship, thus favoring the leaving atom expulsion.

In addition, a sulfur–phosphorus ylide reacted with olefins to give a mixture of two cyclopropane diastereomers (Scheme 1.142, eq 2) [509]. It should be noted that the presence of two electron-withdrawing groups in ethylene was essential for the reaction occurrence. In fact, only one activating group did not lead to the corresponding cyclopropanes.

Chiral pyridinium ylides, generally prepared *in situ* from the corresponding pyridinium salts and triethylamine, were also applicable for cyclopropanation, especially with β-substituted methylidenemalonitriles, giving rise to the corresponding *trans*-cyclopropanes. In particular, α-pyridinium acetate or acetamide bearing an 8-phenylmenthyl group as the chiral auxiliary (Scheme 1.143, eq 1) [510] and *para*-pyridinophane ylides (Scheme 1.143, eq 2) [511] led to the (1*R*,3*S*)-isomers, while chiral pyridinium ylides, the conformation of which was fixed through a cation–π interaction, afforded the (1*S*,3*R*)-isomers (Scheme 1.143, eq 3) [512].

The stereochemical outcome of the first reaction was explained by surmising that the enolate, in which the large pyridyl and 8-phenylmenthyloxy groups were in a *trans*-relationship, was expected to be predominant upon deprotonation. The reaction with the Michael acceptor occurred at the less sterically encumbered face of the enolate.

However, the zwitterionic intermediate had severe steric hindrance; thus, under the basic reaction conditions, it epimerized before cyclization. In the second reaction, the steric bulk of the R^1 substituent in the ylide drove the enantioselectivity, but also a moderate "remote stereocontrol" effect was exerted by the R^2 substituents. DFT (B2LYP/6-31G**) calculations were in good agreement with the experimental results. Finally, in the third reaction, the electron-deficient olefin would approach the ylide from the less-hindered side, with the R group far from COR^1 group in order to avoid steric repulsion.

In addition, ephedrine-derived azetidinium ylides showed a remarkable ability to perform the cyclopropanation of Michael acceptors, allowing the formation of tri- or tetrasubstituted products, bearing one or two quaternary carbon centers along with one or two tertiary centers (Scheme 1.144) [513].

Enolates bearing chiral auxiliaries were also used in MIRC reactions. For example, the enolate derived from Evans's chiral auxiliary and (*E*)-ethyl-3-(trifluoromethyl) acrylate afforded cyclopropanes as a mixture of two out of eight possible stereoisomers (Scheme 1.145, eq 1) [514]. The stereochemical outcome was explained on the basis of the formation of the *anti*-isomer after 1,4-addition

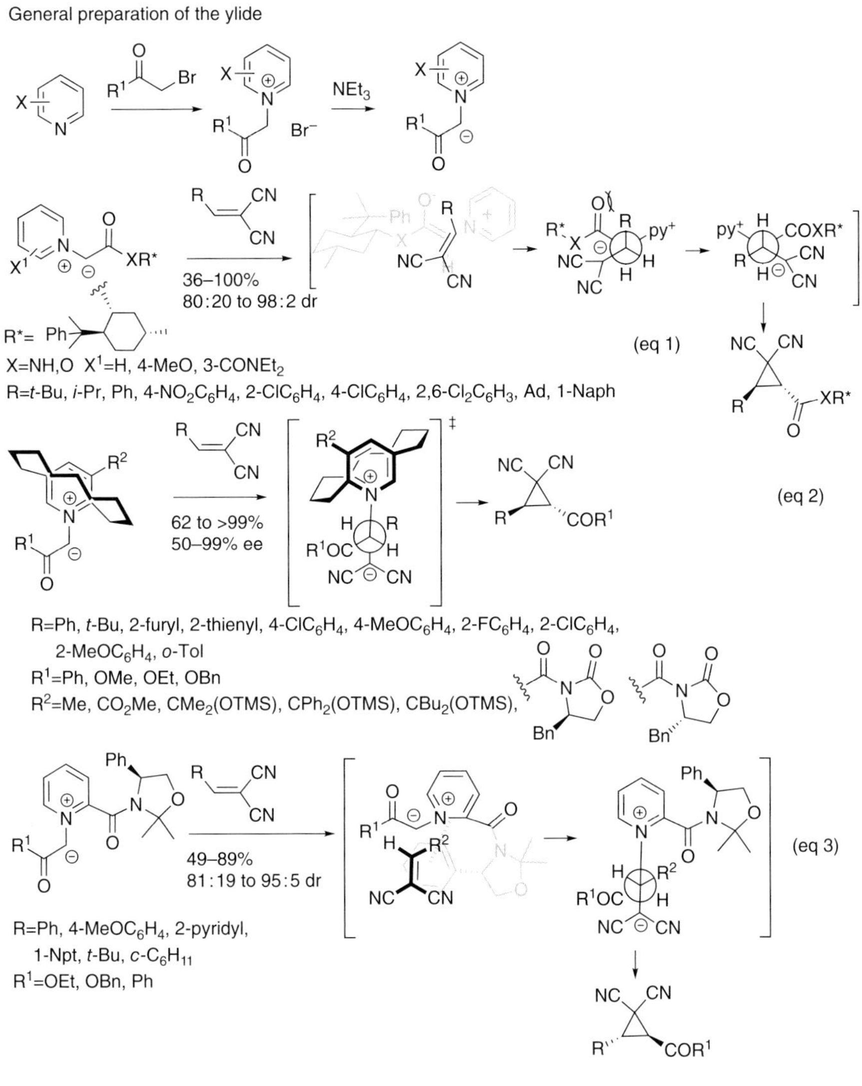

Scheme 1.143 Asymmetric pyridinium-ylide-mediated cyclopropanation.

in agreement with other oxazolidinones. Elimination provided two possible cyclopropyl derivatives, and a subsequent epimerization at the ester moiety accounted for the erosion of the diastereoselectivities. Moreover, the addition of lithium enolate of chiral N-acylimidazolidinone to 2,4,6-trimethylphenyl 4-bromo-4,4-difluorocrotonate allowed the synthesis of difluorocyclopropanecarboxylates (Scheme 1.145, eq 2) [515]. Finally, a borderline reaction between an MIRC and a Simmons–Smith reaction should be mentioned [516]. Since the authors proposed the formation of a chelated zinc enolate in the Z-configuration favored by the steric effect of the voluminous bromine atom instead of a carbenoid species, it is included here. The enolate underwent a highly selective 1,4-conjugate addition,

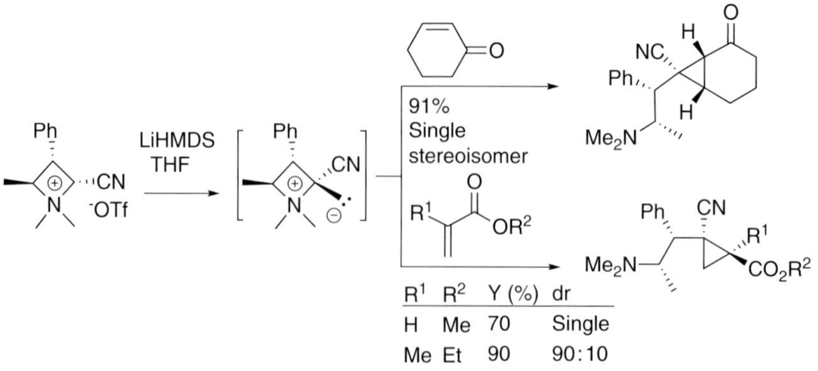

Scheme 1.144 Asymmetric azetidinium-ylide-mediated cyclopropanation.

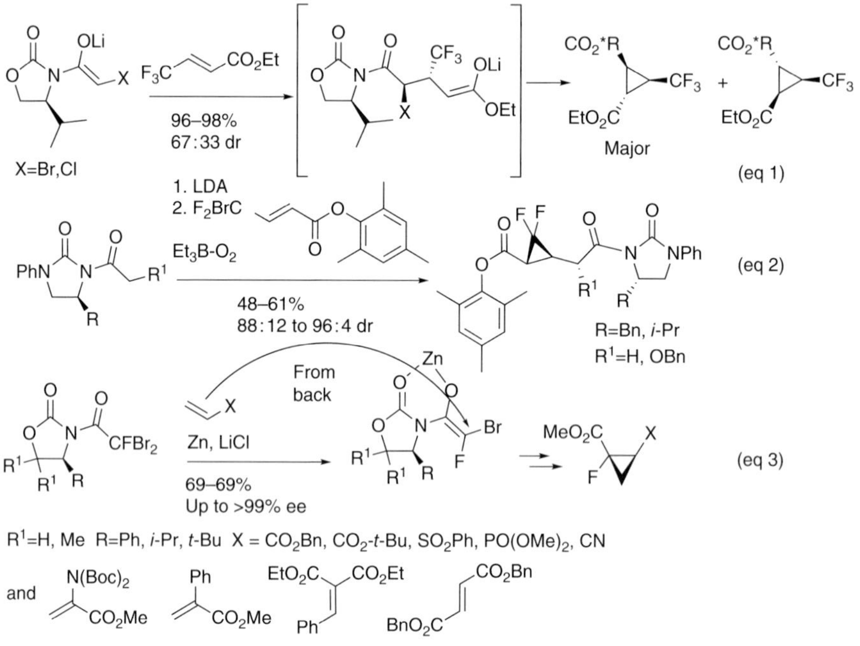

Scheme 1.145 Asymmetric cyclopropanation with Evans's chiral auxiliary.

followed by a less selective cyclization, where the steric effect and the chelating power of the X group played a fundamental role in the formation of the *trans*- or the *cis*-configured cyclopropane (Scheme 1.145, eq 3). Cleavage of the chiral auxiliary proceeded smoothly with no epimerization of the cyclopropanes and with very good recovery of the chiral auxiliary.

The last instance is the alkylation of a lithiated chiral bislactim ether with 4-alkyl-4-bromobut-2-enoates, which led to only 50:50 mixture of the corresponding diastereoisomers, but fortunately, they could be separated by flash

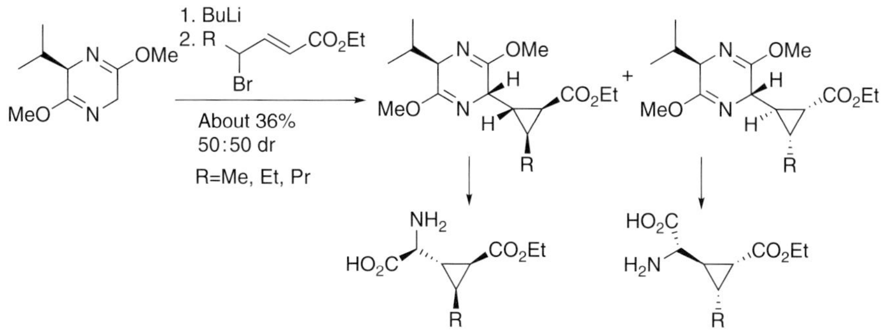

Scheme 1.146 Enantiomerically pure α-cyclopropyl-α-amino acids from chiral bislactim ether.

chromatography and thus hydrolyzed to enantiomerically pure α-cyclopropyl-α-amino acids (Scheme 1.146) [517].

The preparation of stereochemically defined cyclopropanes could also be achieved from vinyl sulfones derived from sugars equipped with a leaving group at the 4-position [518]. The advantage of the large available variety of both nucleophiles and sugars, which allow for the preparation of a myriad of cyclopropanes with all three cyclic carbons stereodefined, was counteracted by the tedious preparation of starting materials.

1.4.3 Organocatalysis

More recently, a strong emphasis has been placed on the development of catalytic methods to generate chiral cyclopropanes via MIRC reactions. Classical iminium–enamine, carbene, and noncovalent organocatalysis approaches have been proposed as elegant asymmetric methods for the cyclopropanation of electron-poor alkenes, such as enals, enones, α,β-unsaturated esters, amides, and nitriles. The aim of the catalyst is the shielding of one of the double bond faces and then the prevention of the rotation around the newly formed single bond before ring closure. One of the first examples was the synthesis of 1,2,3-trisubstituted cyclopropanes from halocycloalkenones and stabilized carbanions by phase-transfer organocatalysts, but only 49–83% ee values were obtained [519].

More recently, Russo and Lattanzi studied the addition of 1,3-dicarbonyl compounds on enone γ-diphenylphosphinates (leaving group = $OPOPh_2$), under basic conditions. Among the reported examples, they employed some cinchona alkaloids derivatives in stoichiometric amounts as the bases and observed promising levels of enantiocontrol, but the reaction was restricted to only one example with different promoters [520].

The addition of an α-substituted cyanoacetate to a vinylselenone, catalyzed by a bifunctional urea catalyst provided adducts, which cyclized smoothly by elimination of benzeneseleninate in a one-pot sequence (Scheme 1.147) [521]. It should be noted that, ethylate-promoted cyclization always gave lower yields but higher enantiomeric excesses compared to chloride-promoted ones, with the same substrate. It is also noteworthy that this procedure complemented the rho-

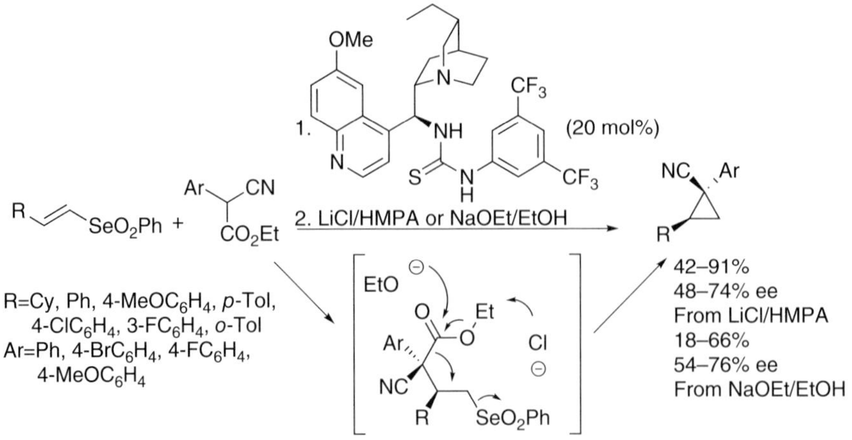

Scheme 1.147 Cyclopropanes from organocatalytic MIRC of vinyl selenones.

dium-catalyzed cyclopropanation of alkenes by 2-diazo-2-phenylacetonitrile, in which an *E* relationship between the nitrile and the R group was obtained (Scheme 1.76, eq 1) [304].

However, ylides, bromonitromethane, and halocarbonyl compounds have been the most used substrates in MIRC reactions, and the following sections are dedicated to them.

1.4.3.1 Ylides

In the previous sections, ylides were often the reagent of choice for MIRC reactions. They carried asymmetric moieties or reacted on chiral substrates. However, under organocatalysis, both partners can be asymmetric and the catalyst provides the chiral environment. The reaction set up by Aggarwal's group, who mixed carbene and sulfur ylide chemistry in the synthesis of enantioenriched cyclopropanes, represents a borderline between MIRC reactions carried out under stoichiometric or catalytic conditions and under organo or metal catalysis (Scheme 1.148) [522]. In fact, the chiral sulfur ylide was generated both stoichiometrically and catalytically from a metal carbene intermediate, which derived from the reaction of rhodium acetate and a diazo reagent (see Section 1.3). Under the best conditions [phase-transfer catalyst and S*R$_2$(**c**)], the yield were anyway low (10–73%) and the diastereoselectivity about 80:20 and 80–92% ee. Later, a further application, by using Cu(acac)$_2$ instead of rhodium acetate and S*R$_2$(**b**), allowed access to a precursor to the pharmacologically important compound (+)-LY354740 (see Scheme 1.120, eq 3 for the structure) [523]. The reaction conditions employed had influence on both the diastereo- and enantioselectivities: under catalytic conditions, a good enantioselectivity with a low diastereoselectivity was observed, while under stoichiometric conditions, a low enantioselectivity with a high diastereoselectivity was observed.

However, diazo compounds could themselves react as ylides under the appropriate reactions conditions. Actually, diazoacetates and acroleins underwent cyclo-

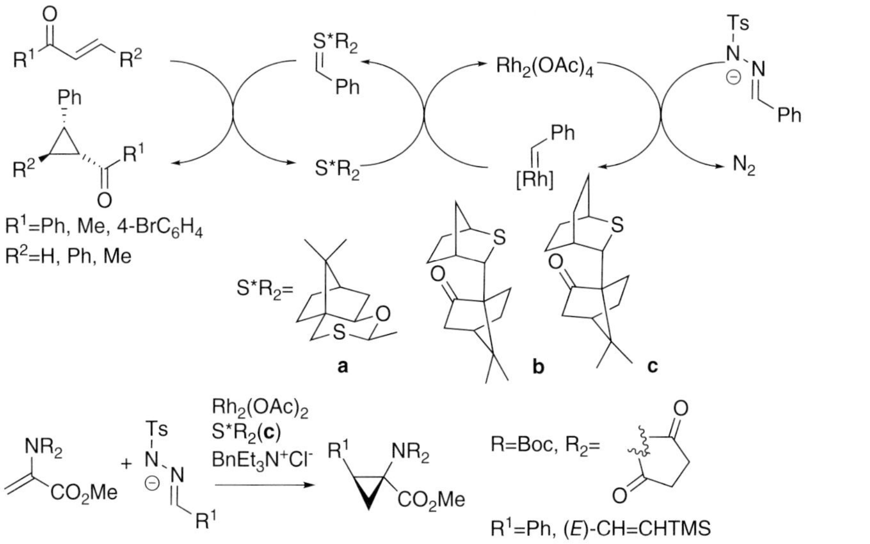

Scheme 1.148 Rhodium-catalyzed asymmetric cyclopropanation with chiral sulfides.

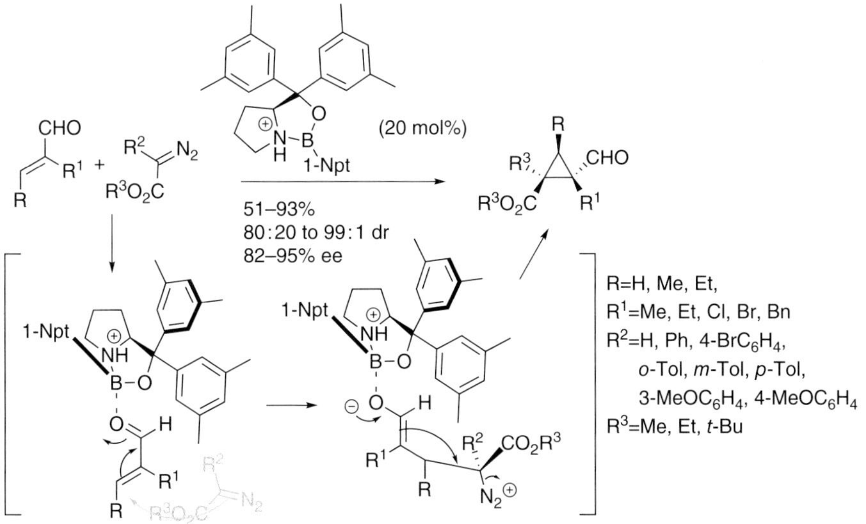

Scheme 1.149 Oxazaborolidinium ion-catalyzed cyclopropanation of acroleins with diazoacetates.

propanation with chiral oxazaborolidinium ions as the catalyst (Scheme 1.149) [524]. Unfortunately, 4-methoxyphenyldiazoacetate yielded an unsatisfactory *trans*/*cis* ratio (55:45), although enantiocontrol was still high (95% ee for *trans*, 94% ee for *cis*). Moreover, every reaction needed different temperatures and times, which had to be optimized whenever required. The mechanism envisaged by the authors and depicted in Scheme 1.149 successfully predicted the experimentally determined absolute configuration.

In Table 1.6, the most important reactions involving asymmetric cyclopropanation by the use of sulfur ylides are summarized.

Studer envisaged the mechanism depicted in Scheme 1.150 to explain the stereochemical outcome of the reaction with N-heterocyclic carbene [528]. The oxidant allowed for the formation of an acyl derivative from the alcohol obtained by the addition of the catalyst to the enal. Then, ylide addition should occur at the less-hindered face, and substituents should accommodate in order to minimize steric interactions. Finally, since all three stereocenters were set before the final esterification, the different selectivities observed with different alcohols were attributed to a competitive alcoholysis reaction of the acyl derivative, before cyclopropanation reaction. In fact, these α,β-unsaturated esters obviously led to a racemic product.

Chen, Xiao, and coworkers supposed two intermediates to be involved in the mechanism of their reaction [529] in agreement with NMR studies and DFT calculations (Scheme 1.151). The hydrogen bonds directed the reactants into the proper positions and induced the observed facial selectivities.

Liu, Feng, and coworkers surmised an iminium ion intermediate, in which its *Re*-face of the ketone was less hindered, and the sulfonium ylide was addressed to this face by hydrogen bonding with the secondary amine moiety (Scheme 1.152) [530].

1.4.3.2 Nitrocyclopropanation

Nitrocyclopropanes are versatile intermediates, since they are substructures in a variety of biologically active natural and unnatural compounds, and the nitro group is the way to other synthetically useful groups. Thus, the organocatalyzed MIRC addition of bromonitromethane to electron-deficient alkenes under basic conditions has attracted great attention from organic chemists. Many examples have appeared in the literature, and they are listed in Table 1.7.

Interestingly, under their reaction conditions, Wang and coworkers kinetically resolved racemic 4-substituted cyclohex-2-en-1-ones. In fact, only the 4*S*-isomer reacted with 0.6 equiv. of bromonitromethane, leading to the corresponding cyclopropane and, thus, enriching the mixture with the unreactive 4*R*-isomer [535].

The asymmetric synthesis of (1*S*,2*R*,5*R*,6*S*)-3-thiabicyclo[3.1.0]hexanes has also been achieved by a one-pot, domino sulfa-Michael/aldol condensation/MIRC sequence of α,β-unsaturated aldehydes with 1,4-dithiane-2,5-diol catalyzed by **JHC** (Scheme 1.153) [539]. Very likely, organocatalyst generated an iminium ion, which was attacked by *in situ* generated mercaptoacetaldehyde from the less-hindered *Re*-face. Then, an intramolecular aldol condensation reaction occurred to form (*R*)-2-aryl-2,5-dihydrothiophene-3-carbaldehyde iminium ion. Finally, MIRC reaction proceeded on this iminium ion, and the initial carbon–carbon bond formation had to occur at its less-hindered *Si*-face.

The reaction of alkylidene oxindoles [540], tetralones, indanones, chroman-4-ones [541], and 1,3-indandiones [542] was also explored (Scheme 1.154). In particular, the reaction with oxindoles also worked with 1-bromonitroethane in 65–91% yield with 60:40 to 83:17 dr and 92% to >99% ee (1*R*,2*R*,3*S*-isomer C1 = spiro,

Table 1.6 Organocatalytic asymmetric MIRC reaction with sulfur ylides.

Reactants	Catalyst	Yield (%)	dr	ee (%)	References
R^1-CH=CH-CHO + S=R R^1 = Me, n-Pr, i-Pr, Ph, $CH_2OCH_2CH=CH_2$, $CH_2=CH(CH_2)_4$ R = Bz, 4-BrC$_6$H$_4$, p-Tol, CO-t-Bu	Indoline-2-carboxylic acid (CO$_2$H) (20 mol%)	85–63	86:14 to 99:1	89–96 (1R,2S,3R)	[525]
R^1-CH=CH-CHO + S=R R^1 = Me, n-Pr, Ph, n-PrOCH$_2$ R = Ph, 4-BrC$_6$H$_4$	Indoline-2-carboxamide N-sulfonyl (R^2 = NO$_2$, MeO) (20 mol%)	21–60	>97:3	88–99 (1R,2S,3R)	[526]
R^1-CH=CH-CHO + S=R R^1 = Me, n-Pr, Ph, CH_2=CHOCH$_2$, c-C$_5$H$_9$ R = Ph, 4-BrC$_6$H$_4$	Indoline-2-(tetrazolyl) (20 mol%)	74–91	98:2 to 99:1	99 (1R,2S,3R)	[527]
Ar^1-CO-CH=CH-Ar^2 + Ph-CH=CH-CH$_2$Br Ar^1 = Ph, 4-ClC$_6$H$_4$, 4-BrC$_6$H$_4$, p-Tol, 4-MeOC$_6$H$_4$ Ar^2 = Ph, p-Tol	Camphor-derived (SMe, OH) (20 mol%)/Cs$_2$CO$_3$	66–92	75:25 to 87:13	78–82 (1S,2S,3R)	[502]

(Continued)

Table 1.6 (Continued)

Reactants	Catalyst	Yield (%)	dr	ee (%)	References
R¹⟶CHO + ⟶S⟶(=O)R R¹=Ph, 4-MeOC₆H₄, p-Tol, 4-NO₂C₆H₄, 4-MeO₂CC₆H₄, 4-AcC₆H₄, 4-ClC₆H₄, 2-NO₂C₆H₄, 2-Naph, 2-furyl R=Ph, 4-MeOC₆H₄, 4-NO₂C₆H₄, 4-ClC₆H₄, 4-BrC₆H₄	(indane-fused imidazoline with N-Mes) (5 mol%) 3,3′,5,5′-tetra-*tert*-butyldiphenoquinone	24–74	67:34 to 95:5	62–99 (1R,2S,3S)	[528]
R¹⟶CHO + Ph⟶(=O)⟶S (thiolactone) R¹=p-Tol, 4-NO₂C₆H₄, 4-MeOC₆H₄	(morpholine-fused imidazoline with N-Mes, CH₂Ph substituent) (5 mol%) 3,3′,5,5′-tetra-*tert*-butyldiphenoquinone DABCO	30–43	91:9 to 95:5	76–87 (1S,5R,6R,Z)	[528]

Substrates	Catalyst	Yield (%)	dr	ee (%) (config)	Ref.
R^1=Ph, 4-MeOC$_6$H$_4$, p-Tol, 4-FC$_6$H$_4$, 4-ClC$_6$H$_4$, 4-BrC$_6$H$_4$, 3-BrC$_6$H$_4$, 2-MeOC$_6$H$_4$, 2-thienyl, PhCH=CH, Ph(CH$_2$)$_2$	3,5-(CF$_3$)$_2$C$_6$H$_3$-NHC(O)-thiourea-type bis-urea (10 mol%)	35–86	67:33 to 94:6	42–80 (1S,2R,3R)	[529]
R=Ph, 4-MeOC$_6$H$_4$, p-Tol, 4-FC$_6$H$_4$, 4-ClC$_6$H$_4$, 4-BrC$_6$H$_4$, 3-BrC$_6$H$_4$, 2-FC$_6$H$_4$, 2-ClC$_6$H$_4$, PhCH=CH	H$_2$N-CHPh-CHPh-NHCy (20 mol%) PhCOOH (20 mol%)	50–68	>95:5	67–93 (1S,2R,3S)	[530]
R^1=Ph, 4-FC$_6$H$_4$, 4-ClC$_6$H$_4$, 4-BrC$_6$H$_4$, 3-CF$_3$C$_6$H$_4$, 4-PhC$_6$H$_4$, 3-MeOC$_6$H$_4$, 2-Naph, 2-furyl, Pr, i-Pr, R=Ph, 4-FC$_6$H$_4$, 4-ClC$_6$H$_4$, 4-BrC$_6$H$_4$, p-Tol, 2-Naph	Supported peptide (20 mol%)	82–88	93:4:3 to 97:1:2	98–99 (1R,2S,3R)	[531]
R^1=Ph, 4-ClC$_6$H$_4$, 4-BrC$_6$H$_4$, 3-CF$_3$C$_6$H$_4$, 4-NO$_2$C$_6$H$_4$, 4-MeOC$_6$H$_4$, 2-NO$_2$C$_6$H$_4$, 2-Naph, 2-thienyl					

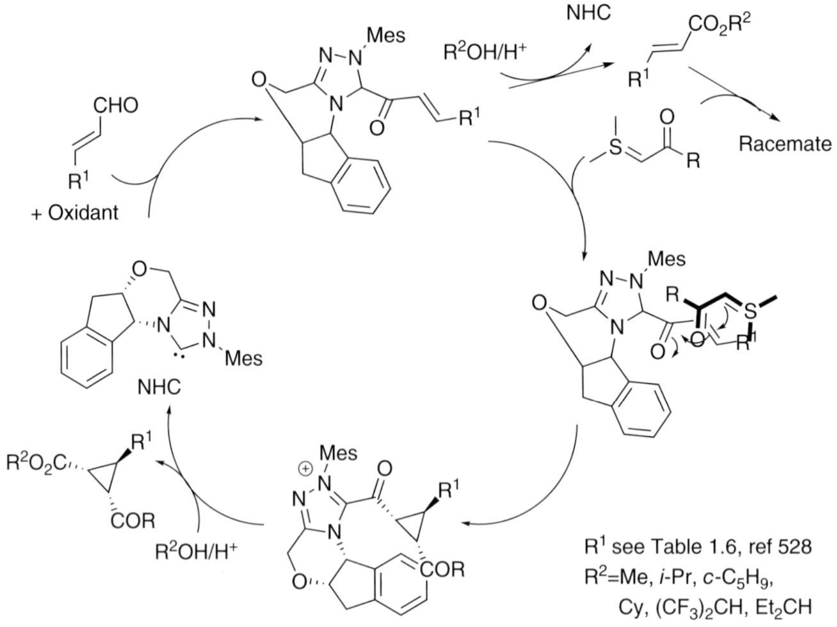

Scheme 1.150 Mechanism of the enantioselective cyclopropanation of unsaturated aldehydes by oxidative NHC catalysis.

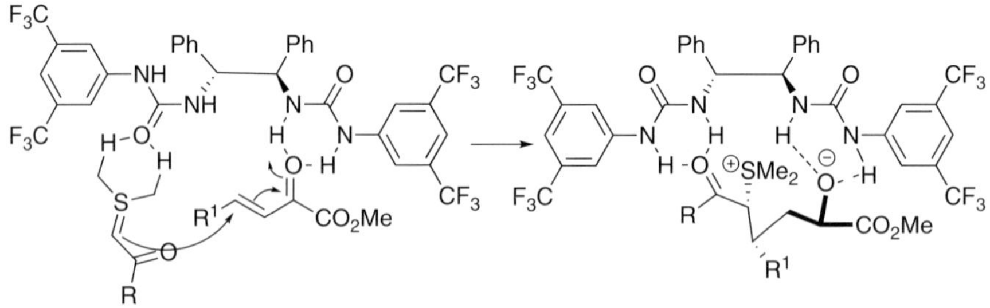

Scheme 1.151 Envisaged intermediates in the asymmetric cyclopropanation of sulfur ylides with β,γ-unsaturated-α-keto esters.

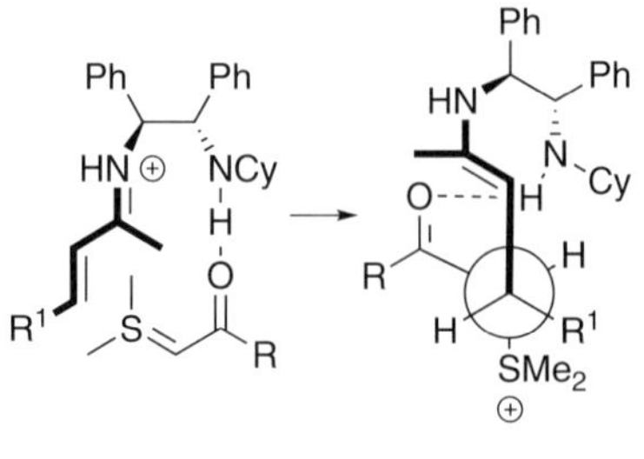

Scheme 1.152 Envisaged intermediates in the asymmetric organocatalytic cyclopropanation with a simple diamine.

Table 1.7 Asymmetric cyclopropanation of electron-deficient alkenes with bromonitromethane.

Alkene	Catalyst, base	Yield (%)	dr	ee (%)	References
R⌒CHO R=Ph, Pr, 4-BrC$_6$H$_4$, 4-ClC$_6$H$_4$, 2-Naph	Ph, Ph, OTMS pyrrolidine **JHC** (20 mol%), Et$_3$N (1 equiv)	42–95[a]	50:50 to 75:25[b]	91–99	[532]
R⌒CHO R=Ph, 4-ClC$_6$H$_4$, p-Tol, 3-ClC$_6$H$_4$, 2-ClC$_6$H$_4$, 2-MeOC$_6$H$_4$, 4-MeOC$_6$H$_4$, 4-NO$_2$C$_6$H$_4$, 2-furyl, 3-pyridyl[c]	Ph, Ph, OTES pyrrolidine (10 mol%), NaOAc (1 equiv)	46–68	50:50 to 70:30[b]	86–96	[533]
CN, R⌒C(O)NH-(2-F-C$_6$H$_4$) R=Ph, p-Tol, 4-ClC$_6$H$_4$, 3-ClC$_6$H$_4$, 2-ClC$_6$H$_4$, 4-BrC$_6$H$_4$, 1-Naph	3,5-(CF$_3$)$_2$C$_6$H$_3$-NHC(=S)NH-cyclohexyl-NMe$_2$ (10 mol%), Et$_3$N (1.5 equiv)	75–81	50:50 to 73:17[d]	98–99	[534]

(Continued)

Table 1.7 (Continued)

Alkene	Catalyst, base	Yield (%)	dr	ee (%)	References
R⟶R¹, ‖O R=Ph, 4-ClC₆H₄, 3-PhC₆H₄, p-Tol, 4-NO₂C₆H₄, 3-furyl, R¹=Ph, 4-ClC₆H₄, 4-CF₃C₆H₄, 4-MeOC₆H₄, 2-Naph	[cinchona-derived catalyst with OMe, NH₃⁺ and O₂C-CH(OH)-C₆H₄-Me counterion] (20 mol%), then trans-dimethylpiperazine (1.2 equiv)	60–85	69:31 to 99:1[e]	71–99	[535]
[cyclohexenone with R¹ substituents], n = 1–3, R=R¹=H; n = 2, R=R¹=Me, R=H, R¹=CMe₂Et[f]	[cinchona-derived catalyst with OMe, NH₃⁺ and O₂C counterion] (20 mol%), 4-methylmorpholine (1 equiv)	83–99	—[g]	83–99	[535]
[cyclohexenone with R¹ substituents], n = 1–3, R=H; n = 2, R=Me	[thiourea catalyst with CF₃ groups, Ph, NH₃⁺ PhCO₂⁻] (20 mol%), 4-methylmorpholine (1 equiv)	31–98	—[g]	88–99	[536]

| | | 80 | —[g] | 77 | [537] |

$n = 1$–3, R=R^1=R^2=H; $n = 1$, R^1=Me, R=R^2=H; $n = 1$–2, R^2=Me, R=R^1=H; $n = 2$, R=Ph, R^1=R^2=H[h]

| | | 8–100 | —[g] | 14–90 | [538] |

(30 mol%), morpholine (3 equiv)

| | | 42–84 | 61:39 to 75:14:11[j] | 29–76 | [538] |

R=Me, Pr, *i*-Pr, *n*-C$_5$H$_{11}$ R^1 =Me, Et, Ph[i]

a) In the case of R=Pr, large amounts of unreacted starting material were recovered.
b) Major isomer: R and CHO have a *cis*-relationship and the nitro group is *trans* to these two groups. Minor isomer: R and CHO have a *trans*-relationship and the nitro group is *cis* to CHO.
c) The reaction was also applied to 1-bromonitroethane, 1-bromonitropropane, and 1-bromo-1-phenylnitromethane (28–68% yield, 95–99% ee), with prevalence of the isomer with R and CHO in a *trans*-relationship and the nitro group *cis* to CHO. Under similar reaction conditions, acyclic α,β-unsaturated ketones were unreactive [536].
d) Major isomer: R and CN have a *cis*-relationship and the nitro group is *trans* to these two groups. Minor isomer: R, CN, and the nitro group have a *cis*-relationship.
e) Major isomer: R and COR^1 have a *trans*-relationship and the nitro group is *cis* to COR^1. Minor isomer: R and COR^1 have a *cis*-relationship and the nitro group is *trans* to these two groups. Note the obtained different diastereomers with respect to those obtained in entries 1 and 2.
f) 1-Bromonitroethane reacted in 54% yield and 97% ee, but phenylbromonitromethane was unreactive.
g) Major isomer: the nitro group is *trans* to the cycle.
h) 3-Methoxycyclohexanone was unreactive, owing to the electron-donating power of this substituent.
i) A twofold excess of bromonitropropane was used. β-Substituted alkenes and vinyl ethers gave no reaction. Only 1*S*,2*S*,3*R* and 1*S*,2*S*,3*S* isomers were recovered. The third diastereomer was not formed very likely for unfavorable interaction with the nitro group.
j) (1*S*,2*S*,3*R*):(1*S*,2*S*,3*S*):(1*R*,2*S*,3*R*) ratio, with C1 is the carbon carrying COR^1 and C2 that carrying NO$_2$.

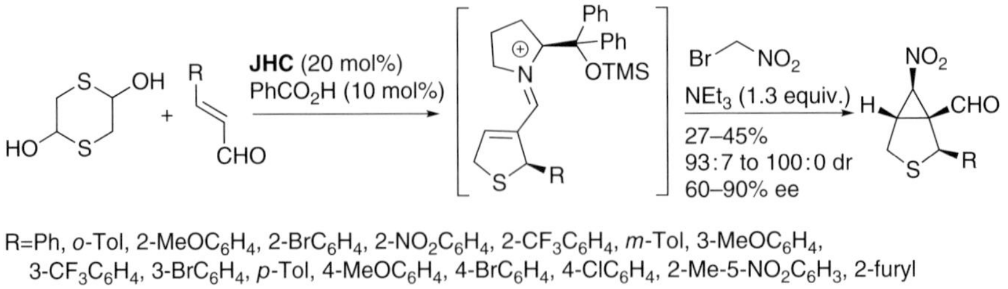

R=Ph, o-Tol, 2-MeOC$_6$H$_4$, 2-BrC$_6$H$_4$, 2-NO$_2$C$_6$H$_4$, 2-CF$_3$C$_6$H$_4$, m-Tol, 3-MeOC$_6$H$_4$, 3-CF$_3$C$_6$H$_4$, 3-BrC$_6$H$_4$, p-Tol, 4-MeOC$_6$H$_4$, 4-BrC$_6$H$_4$, 4-ClC$_6$H$_4$, 2-Me-5-NO$_2$C$_6$H$_3$, 2-furyl

Scheme 1.153 Asymmetric synthesis of 3-thiabicyclo[3.1.0]hexanes.

C2 = C-NO$_2$). Moreover, the pseudo-enantiomeric thiourea catalyst led to the enantiomer in comparable yields and ee values [540]. In addition to the examples reported in Scheme 1.154, eq 2, squaramide was also tested with tert-butyl (E)-3-(2-ethoxy-2-oxoethylidene)-2-oxoindoline-1-carboxylate leading to 3-spirocyclopropane-2-oxindole skeleton in 98% yield with 94:6 dr and 96% ee [541]. Finally, an alternative MIRC reaction employs β-bromo-β-nitroalkenes as the 2C synthon. For example, the enantiomer of the spiro derivative depicted in Scheme 1.154 eq 3 was obtained from bromonitrostyrene and 1,3-indandione, although in moderate yield and enantioselectivity [542].

However, the most important instances of this reaction were reported by Dou and Lu [543] and Alèman's group [544], who described the synthesis of spiroxindoles and simple cyclopropanes, respectively (Scheme 1.155). In the former reaction, it should be noted that, in the absence of DABCO, spirocyclopropyloxindoles were obtained without diastereoselectivity. In the presence of DABCO, the cyclopropane opened the ring and the following ring closure led mainly to the most stable isomer, which had the spiro stereocenter inverted with respect to the products obtained by Bencivenni's group (compare eq 1 in both Schemes 1.154 and 1.155) [543].

O-Alkylation to produce furoindoles was never observed, and the reaction was scalable to a 0.5 mmol scale. Alèman proposed a mechanism based on DFT calculations, in which, in the first step, aldehydes and bromo nitroalkenes reacted following the accepted reaction course for Jørgensen–Hayashi catalyst, and in the second step, the intramolecular bromo substitution was catalyzed by DABCO.

The stereochemistry was based on the inversion of the anion formed with DABCO. With alkyl substituents, π-stacking could occur only with C=O moiety, favoring the eclipsed conformation, which formed **A**. With R^1=Ph, the favorable π-stacking occurred between the two aryl groups, leading to the eclipsed conformation, which formed **B** [544]. Actually, the reaction of (E)-1-bromo-1-nitrohex-1-ene, in which π-stacking cannot be formed, did not give cyclopropanation.

Finally, the synthesis of nitrocyclopropanes was achieved from reaction of nitroalkenes and malonate derivatives under oxidative conditions (Scheme 1.156) [545]. It is interesting to compare this reaction and that depicted in Scheme 1.112 [316h]. In fact, both reactions used a malonate and a phenyliodoacetate to form cyclopropanes, but here phenyliodoacetate was used as an oxidant of a Michael adduct, while in Ghanem's reaction, phenyliodoacetate was used to give a phe-

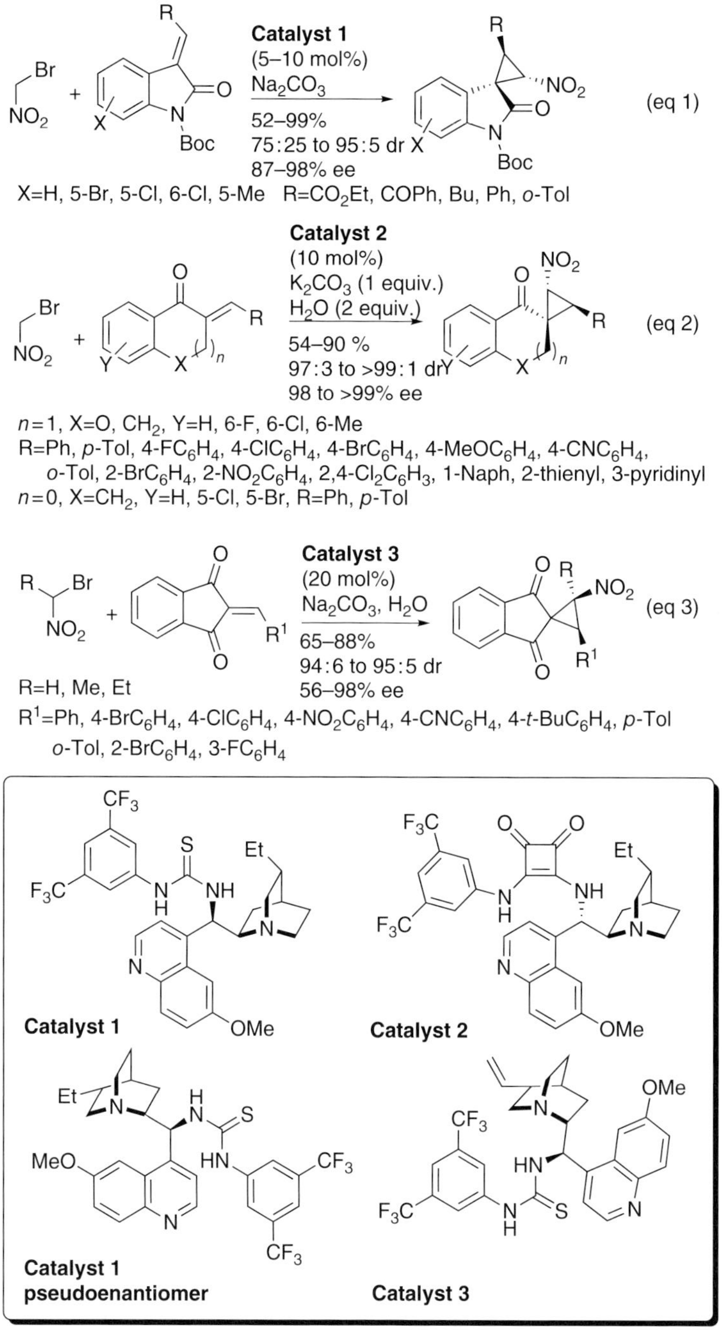

1 Asymmetric Cyclopropanation | 127

Catalyst 1 (5–10 mol%), Na_2CO_3, 52–99%, 75:25 to 95:5 dr, 87–98% ee (eq 1)

X=H, 5-Br, 5-Cl, 6-Cl, 5-Me   R=CO_2Et, COPh, Bu, Ph, o-Tol

Catalyst 2 (10 mol%), K_2CO_3 (1 equiv.), H_2O (2 equiv.), 54–90 %, 97:3 to >99:1 dr, 98 to >99% ee (eq 2)

n=1, X=O, CH_2, Y=H, 6-F, 6-Cl, 6-Me
R=Ph, p-Tol, 4-FC_6H_4, 4-ClC_6H_4, 4-BrC_6H_4, 4-$MeOC_6H_4$, 4-CNC_6H_4,
    o-Tol, 2-BrC_6H_4, 2-$NO_2C_6H_4$, 2,4-$Cl_2C_6H_3$, 1-Naph, 2-thienyl, 3-pyridinyl
n=0, X=CH_2, Y=H, 5-Cl, 5-Br, R=Ph, p-Tol

Catalyst 3 (20 mol%), Na_2CO_3, H_2O, 65–88%, 94:6 to 95:5 dr, 56–98% ee (eq 3)

R=H, Me, Et
R^1=Ph, 4-BrC_6H_4, 4-ClC_6H_4, 4-$NO_2C_6H_4$, 4-CNC_6H_4, 4-t-BuC_6H_4, p-Tol
    o-Tol, 2-BrC_6H_4, 3-FC_6H_4

Catalyst 1

Catalyst 2

Catalyst 1 pseudoenantiomer

Catalyst 3

Scheme 1.154 Enantioselective nitrocyclopropanation of alkylidene oxindoles, tetralones, indanones, and chroman-4-ones.

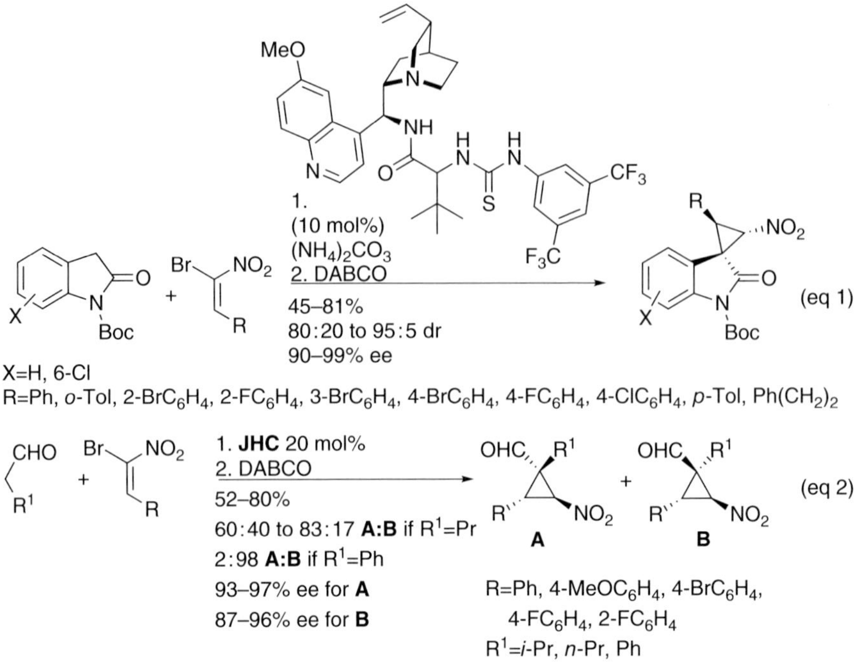

128 | Asymmetric Synthesis of Three-Membered Rings

Scheme 1.155 Enantioselective cyclopropanation with β-bromo-β-nitroalkenes.

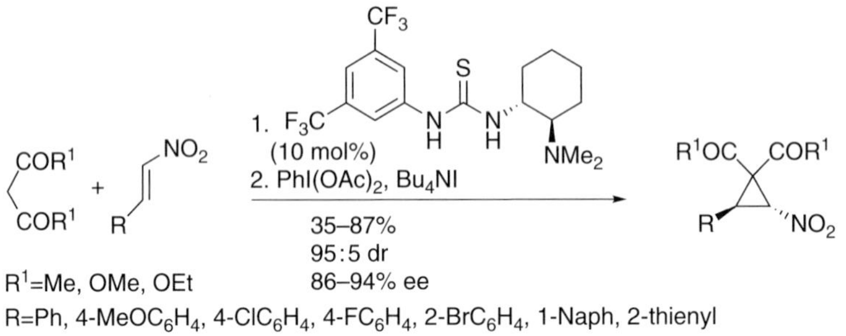

Scheme 1.156 Synthesis of nitrocyclopropanes by oxidative cyclization.

nyliodonium ylide with malonate. However, the introduction of bromomalonates in this reaction avoided the oxidation step (see next section and Table 1.8).

1.4.3.3 Halocarbonyl Compounds

As mentioned at the end of the previous section, bromomalonate and analogs provide a straightforward tool for MIRC reactions. Many examples are reported in the literature with different electron-poor alkene and organocatalysts, and they are listed in Table 1.8.

Generally, Jørgensen–Hayashi catalyst and analogs worked via classical iminium or enamine catalysis. On the other hand, with thiourea catalysts from cinchona

Table 1.8 Asymmetric cyclopropanation of electron-deficient alkenes with bromodicarbonyl compounds.

Alkene	Dicarbonyl compound	Catalyst, base	Yield (%)	dr	ee (%)	References
R⌒CHO R=Ph(CH$_2$)$_2$, Pr, Me, MeCH=CH, CO$_2$Et, Ph, 4-NO$_2$C$_6$H$_4$, 4-ClC$_6$H$_4$, 4-BrC$_6$H$_4$, 4-MeOC$_6$H$_4$, 3-ClC$_6$H$_4$, 2-Naph	X⌒COR^1 COR^1 R^1=OEt, OBn, Me X=Cl, Br	JHC (10–20 mol%), NEt$_3$ (1 equiv.)	50–88[a]	90:10 to 96:4	90–99 (2R,3S)	[546]
R⌒CHO R=Ph, 4-NO$_2$C$_6$H$_4$, 4-FC$_6$H$_4$, 4-CF$_3$C$_6$H$_4$, 4-MeOC$_6$H$_4$, 2-NO$_2$C$_6$H$_4$, 2-ClC$_6$H$_4$, 2-MeOC$_6$H$_4$, 2-furyl, Et, n-C$_5$H$_{11}$	Br⌒CO$_2$R^1 C R^1=Me, Et, Bn, i-Pr	JHC (10 mol%), 2,6-lutidine (1 equiv.)	42–94	>97:3	90–98 (2R,3S)	[547]
R⌒CHO R=Ph, 4-NO$_2$C$_6$H$_4$, 4-CNC$_6$H$_4$, 4-BrC$_6$H$_4$, Et, Bu	Br⌒COR^1 CO$_2$R^2 R^1=Me, t-Bu R^2=Et, Me, t-Bu	JHC (20 mol%), NEt$_3$ (1.2 equiv.)	68–95	70:30 to 96.4:1[b]	63–99 (1S,2R,3S)	[548]
R⌒CHO R=Ph, 4-NO$_2$C$_6$H$_4$, 2-NO$_2$C$_6$H$_4$, 2-BrC$_6$H$_4$, 3-BrC$_6$H$_4$, 4-BrC$_6$H$_4$, 4-FC$_6$H$_4$, 4-ClC$_6$H$_4$, p-Tol, 4-CF$_3$C$_6$H$_4$	Br⌒CO$_2$Et CO$_2$Et	JHC (20 mol%), NEt$_3$ (0.7 equiv.)	66–88	—[b]	93–98 (2R,3S)	[549]
R⌒CHO R=H, Bn, Me, Et, Bu, MeO$_2$C(CH$_2$)$_2$, i-Pr	Br⌒CO$_2$Et CO$_2$Et	JHC (20 mol%), lutidine (3 equiv.) or NMI (5 equiv.)	35–81	—	79–97 (R)	[550]

(Continued)

Table 1.8 (Continued)

Alkene	Dicarbonyl compound	Catalyst, base	Yield (%)	dr	ee (%)	References
R⌒CHO R=Ph, 4-NO₂C₆H₄, 4-MeOC₆H₄, 2-MeOC₆H₄, 2-furyl	Br⌒CO₂Et / CO₂Et	(3,5-(CF₃)₂C₆H₃)₂C(OTMS)-pyrrolidine (20 mol%), H₂O^d	56–84	91:9	92–99 (2R,3S)	[551]
R⌒NO₂ R=Ph, p-Tol, 4-CF₃C₆H₄, 4-BrC₆H₄, 4-MeOC₆H₄, 4-ClC₆H₄, 2-ClC₆H₄, 1-Naph, 2-thienyl, 2-furyl^e	Br⌒CO₂Me / CO₂Me	(2-Naph)(2-Naph)C(OTMS)-pyrrolidine (30 mol%), DABCO (1 equiv)	45–69	—	17–49 (2S,3R)	[552]
R⌒NO₂ R=Ph, 4-BrC₆H₄, 2-Naph, 2-thienyl	Cl⌒CO₂Me / CO₂Me	quinine-derived thiourea (OMe, 3,5-(CF₃)₂C₆H₃) (2 mol%) DBU	65–73	>99:1	25–47 (2S,3R)	[553]

| R⌒NO₂ | Br—CO₂Me / CO₂Me | [cinchona alkaloid catalyst with OH, quinoline-OH, allyl] | 47–78 | — | 75–99 (2S,3R) | [554] |

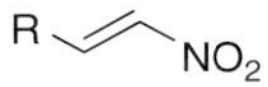

R=Ph, 4-ClC₆H₄, 2-ClC₆H₄, 3-ClC₆H₄, 4-BrC₆H₄, p-Tol, 4-MeOC₆H₄, 4-NO₂C₆H₄, 2-NO₂C₆H₄, 1-Naph, 2-thienyl, 2-furyl[e,f]

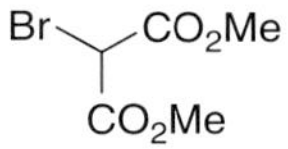

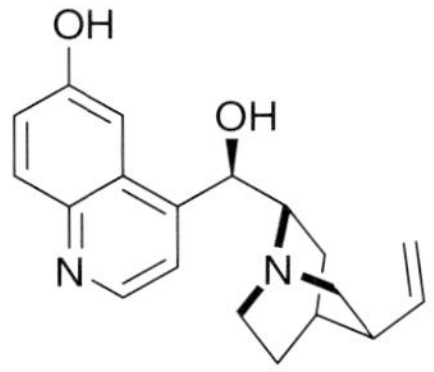

(5 mol%), DABCO (1 equiv)

| R⌒Ar (C=O) | Br—CO₂Me / CO₂Me | [cinchona ammonium catalyst with anthracenylmethyl, OMe, OH] | 58–98 | — | 68–82[g] | [555] |

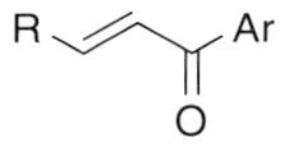

Ar=Ph, 4-ClC₆H₄, 4-FC₆H₄, p-Tol,
R=Ph, 4-ClC₆H₄, 3-NO₂C₆H₄, 4-NO₂C₆H₄

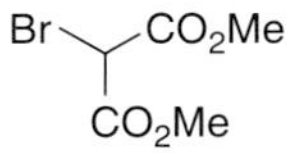

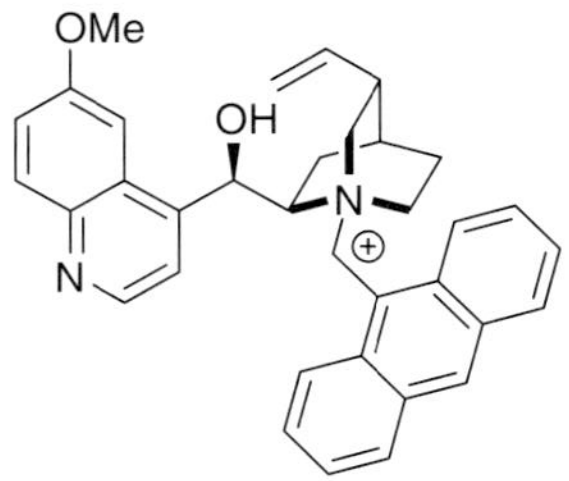

(10 mol%), K₂CO₃ (0.5 equiv)

(Continued)

Table 1.8 (Continued)

Alkene	Dicarbonyl compound	Catalyst, base	Yield (%)	dr	ee (%)	References
R⌒COPh (R=Ph, 4-ClC₆H₄, 4-NO₂C₆H₄)	Br-CH(CO₂Et)₂	Ph-sugar-crown ether catalyst (15 mol%), Na₂CO₃ (2 equiv.)	28–52	—	42–88[h]	[556]
R-C(CN)=CH-CN (R=Ph, 4-ClC₆H₄, 4-NO₂C₆H₄, p-Tol, 3-ClC₆H₄, 3-NO₂C₆H₄, m-Tol, 2-ClC₆H₄, 2-NO₂C₆H₄, o-Tol)	Br-CH(CO₂Et)₂		41–84	—	17–92[g]	[556]
R⌒COPh=CH-COPh	Br-CH(CO₂Et)₂		31–52	—	5–60[h]	[556]
R⌒CO₂Me (R=Ph, 4-NO₂C₆H₄, 4-MeOC₆H₄, 3-MeOC₆H₄)	Br-CH(CO₂Me)₂	cinchona alkaloid catalyst (5 mol%), K₃PO₄/H₂O (5 equiv, 1:1)	91–99	—	84–96 (2S,3S)[i]	[557]

R=Ph, 4-MeOC₆H₄, 2,3-Cl₂C₆H₃, 2-ClC₆H₄, 4-FC₆H₄, 2-MeOC₆H₄, 2-BrC₆H₃, 2,4-(MeO)₂C₆H₃, 2-furyl, 2-thienyl, n-C₇H₁₅

| | | 73–99 | — | 58–82 (2S,3R) | [558] |

(10 mol%), K$_2$CO$_3$ (1 equiv)

a) In the reaction of hexa-2,4-dienal, only the double bond in the 2-position reacted. Increasing steric bulk of the ester fragment, the enantioselectivity rose, but yields decreased. The reaction was also applied to ethyl 2-bromo-3-oxobutanoate (84% yield, 94% ee and 75:25 dr) in favor of diastereomer with CO$_2$Et and CHO in a cis-relationship.
b) Aldehydes were prepared in situ by oxidation of the corresponding alcohol with MnO$_2$, thus yields were calculated based on the allylic alcohol.
c) NMI as the base greatly accelerated the reaction rate (3 h vs 144 h), but the yields were generally lowered.
d) An O-(trimethylsilyldiaryl)prolinol catalyst incorporating a hydrophobic alkyl side chain in the 4-position was also tested with poor results.
e) Unsuccessful with β-alkylnitroalkenes.
f) The reaction suffered from steric hindrance of the aryl substituent on the nitroalkene.
g) Major isomer: (−); absolute configuration not given.
h) Major isomer: (+); absolute configuration not given.
i) A multigram scale-up of the reaction was possible, without loss of yield or enantioselectivity. The use of a pseudo-enantiomeric catalyst gave the opposite enantiomers in similar yield but lower enantiomeric excess.

alkaloids, hydrogen bonds linked the thiourea motif to the carbonyl derivative and the quinuclidine moiety to the β-dicarbonyl nucleophile. The final cyclization geometry was forced by this first stereoselective carbon–carbon bond formation.

The Wang reaction conditions [547] were applied in the key step of a total synthesis of podophyllic aldehydes [559]. Both (+)- and (−)-podophyllic aldehydes are obtained by switching the organocatalyst in the asymmetric cyclopropanation, which was achieved with **JHC** and ***ent*-JHC** in 91% and 88% yield, respectively, and with always 95% ee.

Jørgensen–Hayashi catalyst (20 mol%) was also applied in the synthesis of chiral cyclopropane-fused tetrahydroquinolines. (1*S*,3*R*)-3-Formylcyclopropanes were obtained from MIRC reaction of dimethyl bromomalonate with *ortho*-N-protected aminophenyl α,β-unsaturated aldehydes in 61–94% yields and 92–97% ee, using $(ClCH_2)_2$ as the solvent with NEt_3 (1.5 equiv.) as the base. These products were then submitted to aza-cyclization under basic conditions for the formation of the tetrahydroquinolines. The two steps could be carried out in a one-pot procedure without affecting stereoselectivity [560].

In addition to alkylidenes 1,3-indandiones and oxindoles being useful substrates for MIRC reaction with bromonitroalkanes, they were used for 2-bromo-1,3-dicarbonyl compounds. In fact, (*S*)-cyclopropanes-1,3-indandiones were recovered from the reaction of 1,3-indandiones under modified Jørgensen–Hayashi catalyst (Scheme 1.157, eq 1) [561]. The reaction was also performed with chlorodicarbonyls and oxindoles (Scheme 1.157, eq 2) [562].

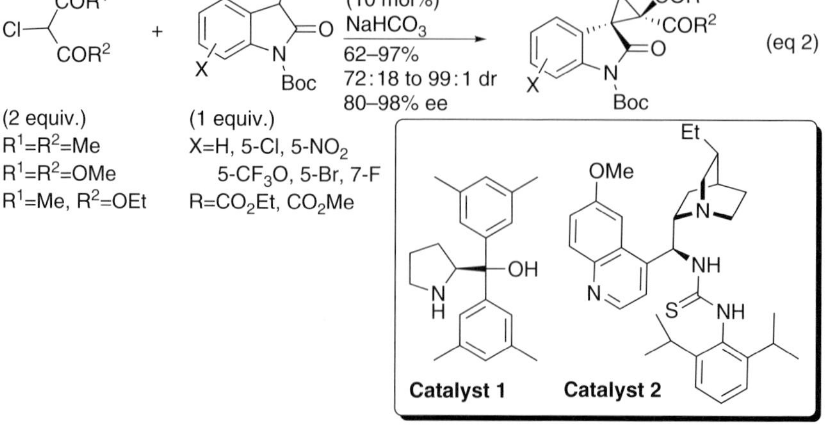

Scheme 1.157 Spirocyclopropanation with halodicarbonyl compounds.

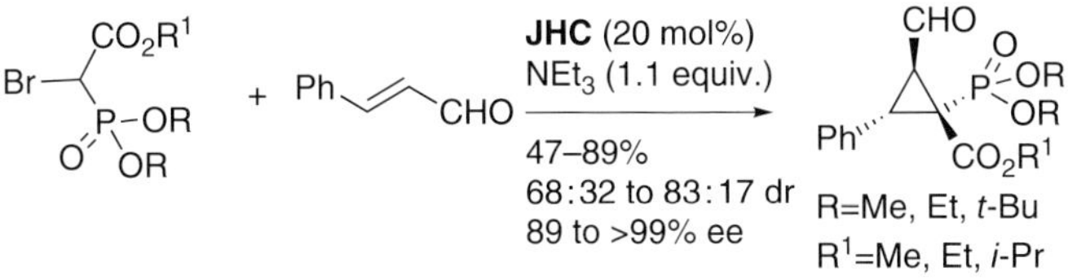

Scheme 1.158 Organocatalytic synthesis of α-cyclopropylphosphonates.

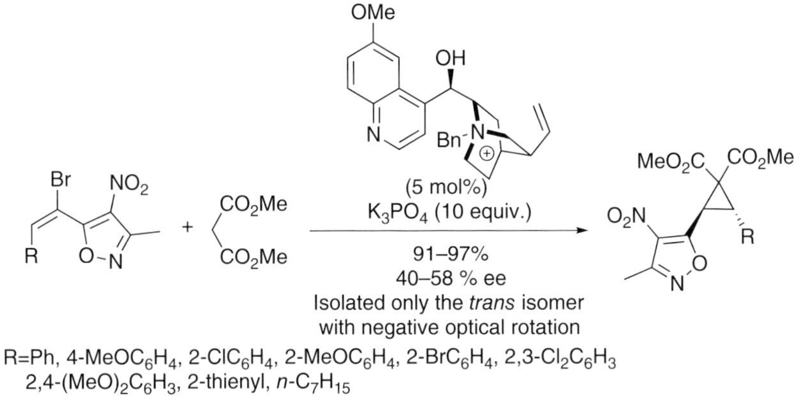

Scheme 1.159 MIRC reaction of 4-nitro-5-bromostyrylisoxazoles with dimethyl malonate.

Although α-bromophosphonoacetate synthesis is not simple, Phillips and Barros were able to find a valuable route to these compounds and then to study their reactivity in MIRC reactions (Scheme 1.158) [563]. The (1R,2S,3S)-product stereochemistry was easily predicted by a classical iminium ion mechanism followed by an S_N2 ring-closure step.

As well as in nitrocyclopropanation reaction, β-bromo-β-nitroalkenes as the 2C synthon can be employed in MIRC reactions. An instance was the reaction of 4-nitro-5-bromostyrylisoxazoles with malonate esters under the catalysis of a cinchona-derived phase-transfer catalyst (Scheme 1.159) [564], providing an alternative route to the reaction proposed by the same research group [557]. It should be noted that malonates with bulkier substituents compared to methyl provided products with lower enantioselectivity, with diphenyl malonate giving only a racemic mixture.

The authors compared these results with those previously obtained and justified the lower enantioselectivity observed with 4-nitro-5-bromostyrylisoxazoles (compare the results reported in Scheme 1.159 with those reported in Table 1.8 [557]) because of a shield of bromine on the NO_2, thus limiting its fundamental interaction with the catalyst.

Finally, the cyclopropanation reaction also proceeds with simple halocarbonyl derivatives (Table 1.9).

Gaunt and coworkers extended their reaction conditions [565] to the intramolecular version, allowing (1S,6R,7S)-[4.1.0]bicycloheptanes to be formed in

Table 1.9 Asymmetric cyclopropanation of electron-deficient alkenes with α-halocarbonyls.

Alkene	Carbonyl compound	Catalyst, base	Yield (%)	dr	ee (%)	References
CH$_2$=C(R^1)COR^2; R^1=H, Me, NHBoc, R^2=Ph, 4-(NEt$_2$)C$_6$H$_4$, 4-BrC$_6$H$_4$, OBn, OMe	RCOCH$_2$Br; R=O-t-Bu, NEt$_2$	Methoxyquinine (20 mol%) Cs$_2$CO$_3$ (1.2 equiv.)	63–96	—	80–97 (1S,2S)[a]	[565]
R^1(CN)C=C(CN); R^1=Ph, 2-ClC$_6$H$_4$, t-Bu	RCOCH$_2$X; X=Cl, Br R^1=Ph, 2-ClC$_6$H$_4$	Cinchonidine (1 mol%) Na$_2$CO$_3$ (2 equiv.)	27–81	—	20–82 (2S,3R)	[566]
R^{1}CH=CHCHO; R^1=Ph, m-Tol, 2-FC$_6$H$_4$, p-Tol, 4-BrC$_6$H$_4$, 4-ClC$_6$H$_4$, 4-FC$_6$H$_4$	RCOCH$_2$Cl; R=Ph, p-Tol, 4-MeOC$_6$H$_4$, 4-NO$_2$C$_6$H$_4$, 2-FC$_6$H$_4$, 2-ClC$_6$H$_4$, 3-BrC$_6$H$_4$, 2-furyl, N-Boc-3-indolyl	ent-JHC (20 mol%) NEt$_3$ (1 equiv.)	62–85	—	90–99 (1S,2S,3S)[b]	[567]
R^{1}CH=CHCHO; R^1=Ph, 4-MeOC$_6$H$_4$, 4-BrC$_6$H$_4$, 4-NO$_2$C$_6$H$_4$, 2-thienyl, 2-furyl	3-chloro-2-oxindole; X=H, 4-Br, 5-Br	[diaryl prolinol OTMS catalyst, Ar=3,5-(CF$_3$)$_2$C$_6$H$_3$] (20 mol%) NaHCO$_3$	44–76	83:17 to 95:5[c]	96 to >99 (1S,2R,3S)	[562b, 568]

Substrate	Catalyst / Conditions	Yield (%)	dr / ee	Config.	Ref.
(3-alkylidene oxindole with Boc) Y=H, Cl, Br, Me, R=CO₂Me, CO₂Et, CN, 4-NO₂C₆H₄	Squaramide catalyst (cinchona-derived, OMe-quinoline, 3,5-(CF₃)₂C₆H₃NH) (20 mol%) NaHCO₃	71–99	91:9 to 95:5	83–99 (1S,2S,3S)	[562b, 568]
3-Cl oxindole (NH) X=H, 5-NO₂, 5-Br, 7-F					
OHC–CH=CH–C(R)=CH– (dienal) R=Ph, 4-MeOC₆H₄, 4-BrC₆H₄, 4-CF₃C₆H₄, 3-ClC₆H₄, o-Tol, 2-thienyl	JHC (5 mol%) Na₂CO₃ (1 equiv.)	56–91	86:14 to 94:6	92–99 (E, 1R,2S,3S)ᵈ	[569]
3-Cl oxindole (NH), t-Bu X=H, 5-Br, 5-Cl, 5-Me, 6-F					
R–CO–CH=CH–CO–R¹ R=R¹=Ph, 4-BrC₆H₄, p-Tol, 4-ClC₆H₄, 4-NO₂C₆H₄ R=Me, R¹=Ph, 4-NO₂C₆H₄	Thiourea catalyst (cinchona-derived, OMe-quinoline, 3,5-(CF₃)₂C₆H₃NH) (10 mol%) NaHCO₃ (2 equiv.)	53–81ᵉ	90:10 to 95:5	68–87 (2S,3S)	[570]
3-Cl oxindole (N-Boc) X=H, Br					

(Continued)

Table 1.9 (Continued)

Alkene	Carbonyl compound	Catalyst, base	Yield (%)	dr	ee (%)	References
![R group alkene with R¹ and N-N structure] R=Ph, 4-ClC₆H₄, 3-NO₂C₆H₄, 2-BrC₆H₄, R¹=Ph, 4-ClC₆H₄	![chloro-oxindole] X=H, 4-Cl, 4-Br, 6-Cl	![methoxyquinidine-squaramide catalyst with OMe, CF₃] (5 mol%) K₂CO₃ (1 equiv.)	91–99	70:30 to 87:13	40–74[f]	[571]

a) The other enantiomer could be obtained by using methoxyquinidine.
b) The absolute configuration of the catalyst affected the stereochemistry of C-2, by directing the approach of the chloroketone to the iminium intermediate.
c) Crotonaldehyde and 3-methylbut-2-enal gave moderate diastereoselectivity: 67 (75% ee):33 (87% ee) and 80 (36% ee):20 (89% ee), respectively.
d) The exclusive δ-site selectivity should be noted.
e) Unsaturated keto esters gave noncyclized compound as the main product.
f) Major isomer: (+); absolute configuration not given.

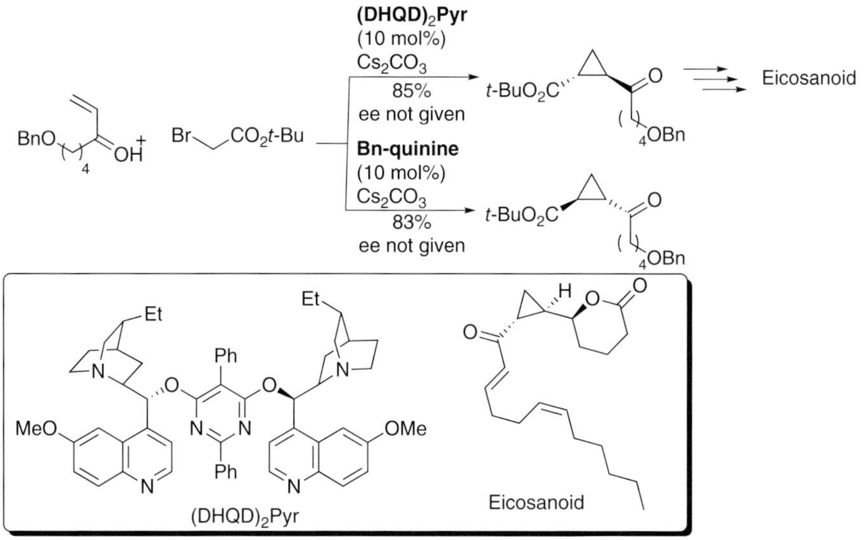

Scheme 1.160 Organocatalytic synthesis of the cyclopropyl moiety of eicosanoid.

27–84% yields, as single diastereomers, and with 95–99% ee [572]. Kumaraswamy, instead, applied this reaction [565] to prepare the cyclopropyl moiety of eicosanoid, one example of the oxylipin class of natural products (Scheme 1.160) [573]. Both enantiomers could be obtained with (DHQD)$_2$Pyr and the benzyl ether of quinine, but the two enantiomeric excesses were not reported. However, at the end of the multistep synthesis, the stereochemistry of eicosanoid from the product obtained with (DHQD)$_2$Pyr was found to be superimposable with the natural compound, thus confirming the stereochemistry. The same research group then reported some changes in the reaction procedure: *tert*-butyldiphenylsilyl protection of the starting hydroxy ketone led to the cyclopropane in low yields; the use of 2-bromo-*N*,*O*-dimethylacetamide in place of *tert*-butylbromoacetate yielded the two enantiomeric cyclopropanes in 88% and 85% isolated yield, respectively, but the ee values were not reported again [574].

1.4.4 Metal Catalysis

Only a few instances of such methods are reported. The first example of chiral bis(oxazoline) complexed to Lewis acid that mediated asymmetric cyclopropanations of a Michael acceptor with a sulfur ylide was the cyclopropanation of *N*-enoyloxazolidinones with isopropylidenediphenylsulfonium [472]. Stoichiometric amounts of zinc salts complexed with (4*R*,4′*R*)-2,2′-(propane-2,2-diyl)bis(4-phenyl-4,5-dihydrooxazole) gave the best results (53–63% yield and 92–95% ee) with (*E*)-3-but-2-enoyloxazolidin-2-one, but the asymmetric reaction of the cinnamate derivatives proceeded in much lower enantiomeric excess, in analogy with that observed with the chiral auxiliary (see Scheme 1.129, eq 2). Moreover, a loss of stereoselectivity was observed with catalytic amounts of Lewis acid.

Chiral Lewis acids complexed with BINOL derivatives were tested in the cycloaddition reactions of 1-seleno-2-silylethene, but low yields (33–41%) and enantioselectivities (43–47%) were observed [575].

Then, bicyclo[3.1.0]hexanes were obtained in 54–99% yields >99:1 dr, 74–96% ee by an enantioselective intramolecular cyclopropanation of sulfur ylide allylic esters through a synergistic effect of a bimetallic gold catalyst from dimeric TADDOL–phosphoramidite ligand and AgNTf$_2$ [576]. It should be noted that both the *E* and *Z* isomers and even regioisomer of the allylic moiety (i.e., (*E*)- and (*Z*)-2-buten-1-yl and 1-methyl-2-propen-1-yl frameworks) gave the cyclopropane with the same configuration. Thus, it is possible to use starting materials with low regio and geometric purity to produce diastereomerically pure, enantioenriched (1*S*,4*S*,5*R*)-bicyclo[3.1.0]hexanes. Very likely, after the formation of the allylic–gold complex, the substrate isomers could interconvert before ring-closure reaction. This reaction was also employed for the asymmetric synthesis of butenolide natural products, such as *trans*-(−)-cognac lactone.

The reaction of dibenzylideneacetone with allyl iodide in the presence of (*S*)-methyl mandelate (1 equiv.) and indium led to the cleavage of the C=O bond and addition of two allyl fragments from the reagent, providing the cyclopropane in 71% yield and 60% ee (*S*-isomer) [577]. This cyclopropane was further submitted to a Ru-catalyzed ring-closing metathesis to generate a norcarene unit ((1*S*,6*S*)-1-(*E*)-2′-(phenylethenyl)-bicyclo[4.1.0]hept-2-ene) without loss of enantiopurity.

The conjugate addition of Grignard reagents to 4-chloro-α,β-unsaturated esters, thioesters, and ketones, catalyzed by CuI-(*R*)-TolBINAP (1.5 mol%), allowed the formation of (*R*,*R*)-1-alkyl-2-substituted cyclopropanes in 50–92% yield and 26–98% ee [578]. The precise control of the amount of Grignard reagent (the best amount was 1.2 equiv.) was essential for obtaining good yields of the cyclopropanes, because a deficiency of the Grignard reagent led to a significant amount of acyclic product, whereas with large excesses, a subsequent addition of the Grignard reagent to the carbonyl moiety of the corresponding cyclopropane occurred. Moreover, the lowest yield and enantioselectivity were observed with phenylmagnesium bromide. The reaction was applied to the syntheses of cascarillic acid and grenadamide (Scheme 1.17 for the structures).

Finally, the enantioselective synthesis of (2*S*,3*R*)-configured nitrocyclopropanes (70–99% yield, >99 de, and 85–99% ee) was achieved by reaction of bromomalonates and nitroalkenes promoted by a chiral nickel complex (5 mol%) in the presence of DBU [579]. The authors proposed that the 1,2-cyclohexanediamine nickel catalyst activated the bromomalonate anion in a bidentate fashion, forcing the attack at the *Si*-face of the double bond of the nitroalkene.

1.4.5 Other Ring Closures

Various other ring-closure reactions forming chiral cyclopropanes have been reported, generally involving the intramolecular displacement of a leaving group, such as halides or sulfonates, with an enolate. One of the fist examples was reported by Sharpless, who allowed reaction of cyclic sulfates of vicinal diols with malonate anions, leading to cyclopropanes with complete inversion of the

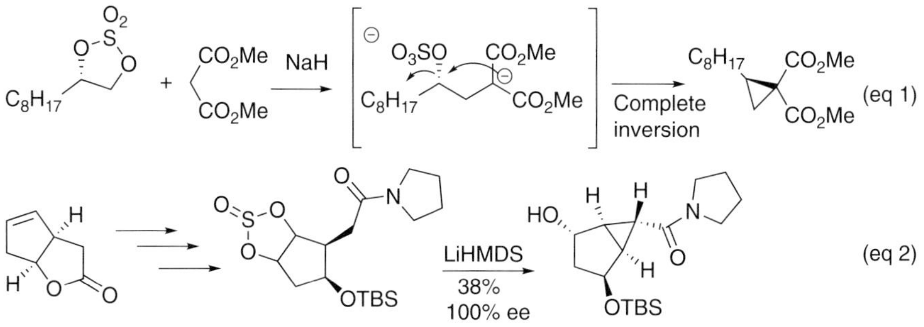

Scheme 1.161 Cyclopropanes from vicinal diol cyclic sulfur derivatives.

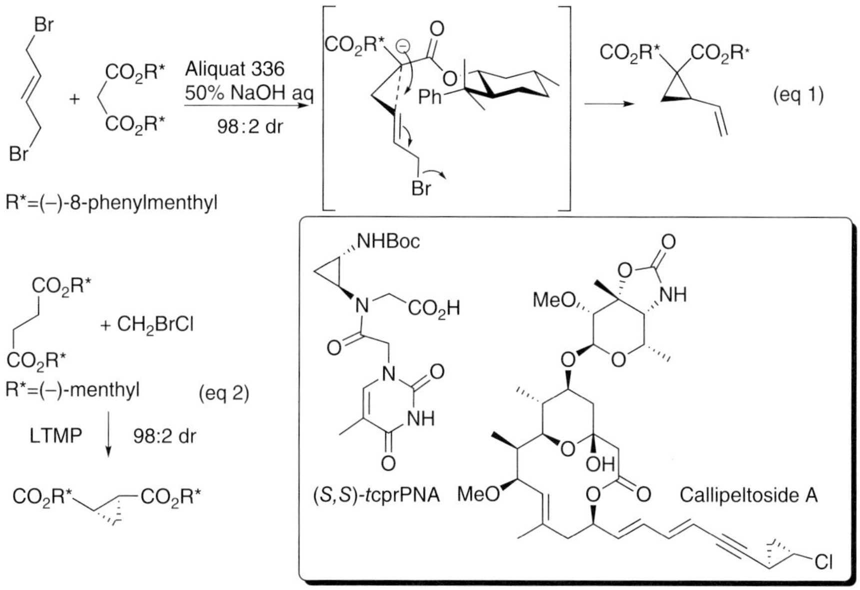

Scheme 1.162 Cyclopropanes from chiral dicarboxylates.

configuration (Scheme 1.161, eq 1) [580]. Then, these compounds were used for the synthesis of other enantioenriched methylenecyclopropanes [581].

On the other hand, the synthesis of bicyclo[3.1.0]hexane carboxylic acid derivatives was performed by intramolecular cyclopropanation of chiral cyclic sulfites as the key step (Scheme 1.161, eq 2) [582]. It should be noted with respect to the previous reaction that the chiral centers did not involve the sulfite moiety.

Enantioenriched vinylcyclopropane derivative was also obtained by using bis-(−)-8-phenylmenthyl malonates in reaction with 1,4-dibromo-2-butene, because the chiral auxiliary allowed a stable conformation in which the allyl bromide moiety is opposite to the bulky phenylmenthyl substituent (Scheme 1.162, eq 1) [583].

Alternatively, the dianion of (−)-dimenthylsuccinate could be added to bromochloromethane (Scheme 1.162, eq 2) [584]. This method was used in the total synthesis of ambruticin (Scheme 1.107 for the structure) [585], callipeltoside A [586], and peptide nucleic acids (PNA) with a cyclopropane in the backbone, such as (S,S)-tcprPNA [587].

The diastereoselective iodocarbocyclization of bis-(−)-8-phenylmenthyl allylmalonate produced cyclopropylmethyl iodide in 89% yield and with 93% ee when titanium tetra(tert-butoxide) and pyridine were used to trap the HI generated during the reaction [588]. It should be noted that attempts to induce the enantioselectivity using a chiral titanium complex instead of the chiral auxiliary failed [589].

The intramolecular rearrangement of both isomers of (allyloxy)dimesitylsilyllithium reagent bearing a phenyl group on the olefin part exclusively led to dimesityl-[(1R,2S,3S)-2-methyl-3-phenylcyclopropyl]silanol (68–79% yield, 98% ee) [590]. The substituent effect was revealed by *ab initio* calculations in terms of the regioselectivity in the reaction of silyllithiums with an olefin. Tandem lithiumene cyclization, followed by thiophenoxide expulsion, produced only one vinylcyclopropane diastereoisomer, when starting from a chiral allylic alcohol [591].

The allylic sulfones are also acidic enough to easily generate anions, which can cyclize if a leaving group is present in an appropriate position. Thus, Diez et al. developed a highly diastereoselective synthesis of chiral cyclopropanols [592] and of chiral N-diphenylmethylene-2-vinyl-substituted cyclopropylamines (Scheme 1.163) [593]. It is worth noting the exclusive three-membered instead of the five-membered ring closure and, in the case of imines, the influence of the E,Z-isomerism on the diastereoselection of the reaction. This methodology was then applied to the synthesis of a large variety of cyclopropanol amino acids due to its simplicity, high yield, and high diastereocontrol [594].

Ring-closing reactions involving phosphorus transfer, to generate both nucleophile and leaving group, yielded cyclopropanes. In fact, under basic conditions, chiral silyloxytetrahydrofuranes led to a mixture of two *trans*-cyclopropanes with a poor diastereoselectivity via phosphine oxide-mediated cascade reactions (Scheme 1.164, eq 1) [595]. Similarly, γ-azido ketones yielded *trans*-cyclopropanes (Scheme 1.164, eqs 2 and 3) [596].

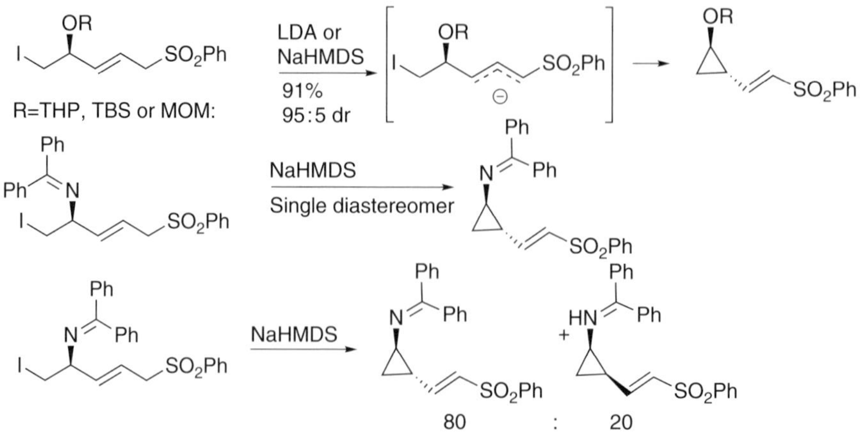

Scheme 1.163 Asymmetric cyclopropanations of allyl sulfones.

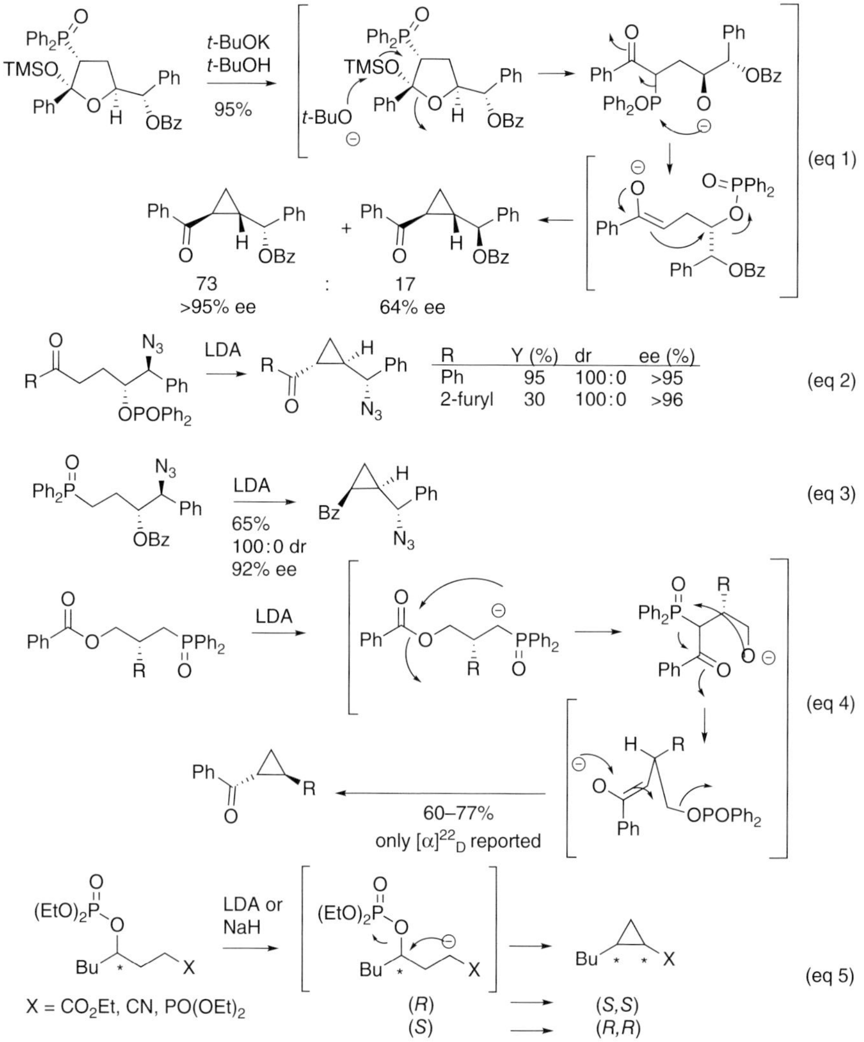

R	Y (%)	dr	ee (%)
Ph	95	100:0	>95
2-furyl	30	100:0	>96

Scheme 1.164 Phosphorus-mediated asymmetric cyclopropanations.

More recently, a three-step cascade reaction was reported to produce cyclopropanes from phosphine oxides (Scheme 1.164, eq 4) [597]. Unfortunately, starting materials had to be prepared via a long, moderate yielding and tedious reaction procedure, in which a chiral oxazolidinone was the chiral auxiliary. Krawczyk and coworkers were able to easily prepare enantioenriched phosphates, and the cyclization of (*S*)-phosphates was found to lead to (1*R*,2*R*)-cyclopropanes (Scheme 1.164, eq 5) [598]. Interestingly, the enantiomeric excess of the products remained very close to that of the starting materials in both reactions.

Cationic cyclizations that lead to cyclopropanes are scarce, because the behavior of the C_4H_9 cation cannot be efficiently controlled, except when two methyl groups (*tert*-cyclopropylmethyl) [599] or a silicon substituent gives more stable cations. In

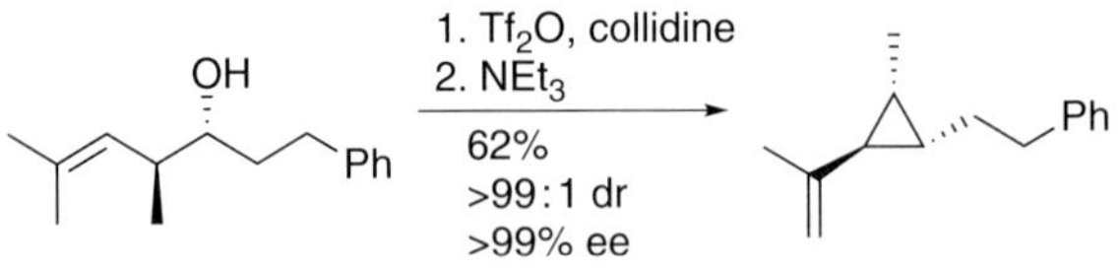

Scheme 1.165 Cyclopropane from cyclization of chiral homoallylic alcohols.

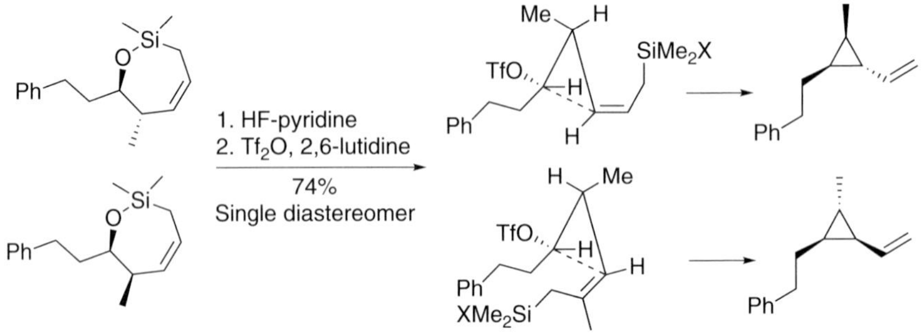

Scheme 1.166 Cyclopropane from chiral silyloxycycloheptene.

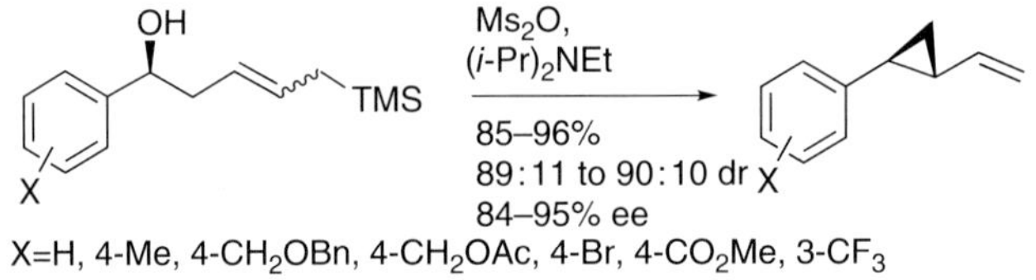

Scheme 1.167 Asymmetric cyclopropanation of allylsilane homoallylic alcohols.

the first case, cyclopropanes were obtained in 61–93% yields, and the chirality was efficiently transferred, as starting from the enantiopure homoallylic alcohol, the corresponding product was obtained with greater than 99% ee (Scheme 1.165 for an example).

In the second case, silyloxycycloheptene was opened by a nucleophile. The two diastereoisomers of the silyloxycycloheptene led to *trans-* or *cis-*cyclopropane, owing to two different transition states in which the $A^{1,3}$ strain was minimized for the -CH$_2$SiMe$_2$X substituent (Scheme 1.166) [600].

Moreover, allylsilane homoallylic alcohols were a source of vinylcyclopropanes through an intermediate β-silylcyclopropylcarbinyl cation (Scheme 1.167) [601]. It should be noted that electron-withdrawing substituted phenyl rings showed higher enantiospecificity for the cyclization, whereas an appreciable loss of enantiomeric purity was observed with electron-rich ones.

In addition, White and coworkers studied the corresponding reaction with a tributylstannyl substituent instead of TMS [602]. Both (*E*)- and (*Z*)-olefins were submitted to the same methodology, affording the same products with a higher diastereoselectivity from the (*Z*)-olefin (Scheme 1.168). Moreover, a nonconju-

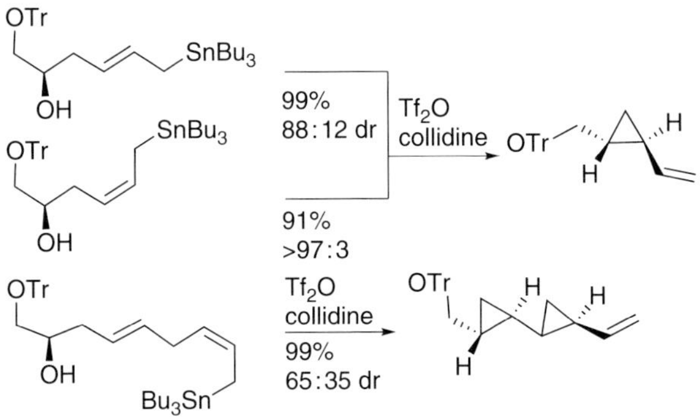

Scheme 1.168 Asymmetric cyclopropanations of tributylstannane homoallylic alcohols.

gated diene led to the quantitative formation of three enantiomerically pure, stereoisomeric bicyclopropanes in which the *trans,syn,trans*-isomer was predominant. This bicyclopropane could be converted into an intermediate employable in the synthesis of FR-900848 (Scheme 1.21 for the structure) [602] and into the key cyclopropane intermediate for the synthesis of solandelactones (Scheme 1.5 for the structure) [603].

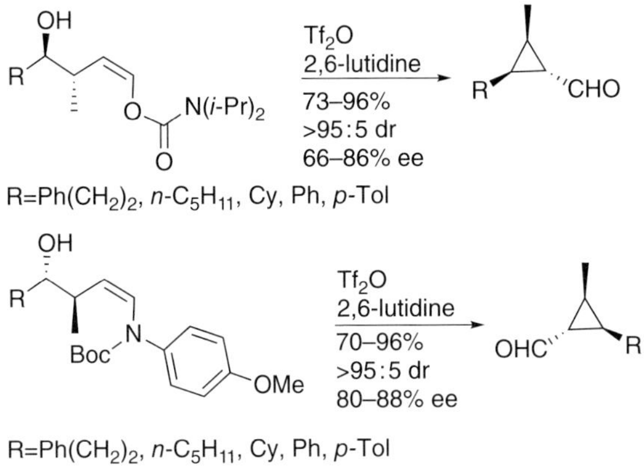

R=Ph(CH$_2$)$_2$, *n*-C$_5$H$_{11}$, Cy, Ph, *p*-Tol

R=Ph(CH$_2$)$_2$, *n*-C$_5$H$_{11}$, Cy, Ph, *p*-Tol

Scheme 1.169 Asymmetric cyclopropanations of *O*- and *N*-enecarbamates.

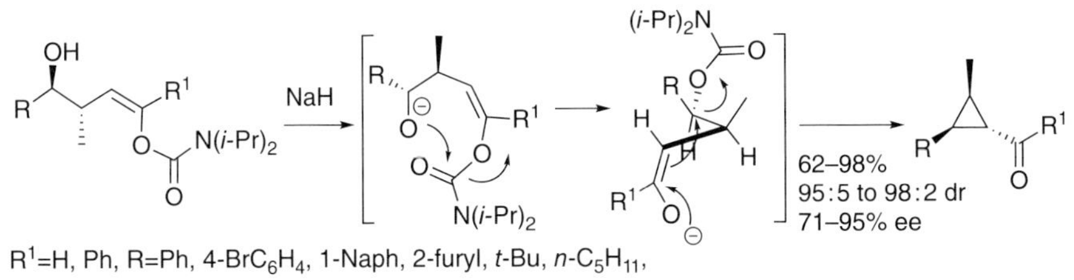

R^1=H, Ph, R=Ph, 4-BrC$_6$H$_4$, 1-Naph, 2-furyl, *t*-Bu, *n*-C$_5$H$_{11}$,

Scheme 1.170 Asymmetric cyclopropanation of homoallylic alcohols.

Asymmetric Synthesis of Three-Membered Rings

Then the reaction was extended to O- and N-enecarbamates, thus providing both enantiomers of the cyclopropanecarboxaldehydes (Scheme 1.169) [604]. The ring closure took place with inversion of configuration at the alcohol center in a stereospecific manner.

However, the simple treatment of chiral homoallylic alcohols with NaH furnished the corresponding cyclopropanes with excellent diastereoselectivity, complete chirality transfer, and high efficiency (Scheme 1.170) [605]. Very likely, the *N,N*-diisopropylcarbamoyl group migrated to the O-4 atom, forming a (*Z*)-enolate, which underwent cycloalkylation by nucleophilic substitution of the carbamate group with stereoinversion. In this step, the enolate moiety had to occupy an *anti*-position to avoid steric repulsion with the methyl and R groups.

The Ti-mediated reductive cyclopropanation reactions of chiral 3,4-dehydroprolinol derivatives in the presence of dibenzylformamide should also be mentioned (Scheme 1.171) [606]. Application of similar conditions to L-*N*-allyl-(*N,N*-dibenzyl) prolineamide allowed the synthesis of tricyclic cyclopropylamine [607].

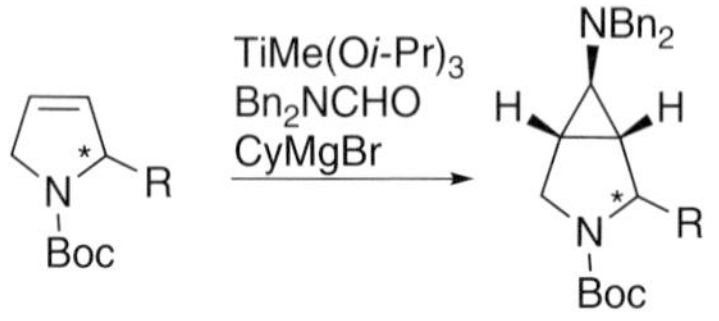

R	Y (%)	dr	ee (%)
(S)-CO$_2$-*t*-Bu	50	100:0	95
(R)-CH$_2$OH	69	100:0	100

Scheme 1.171 Ti-catalyzed reductive asymmetric cyclopropanation reactions.

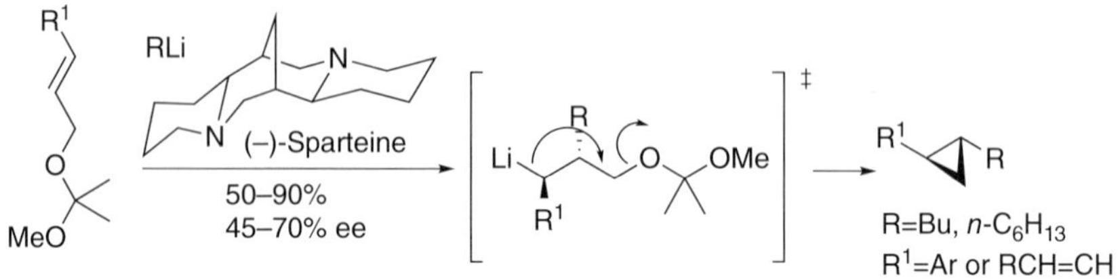

Scheme 1.172 Enantioselective carbolithiation of alkenes.

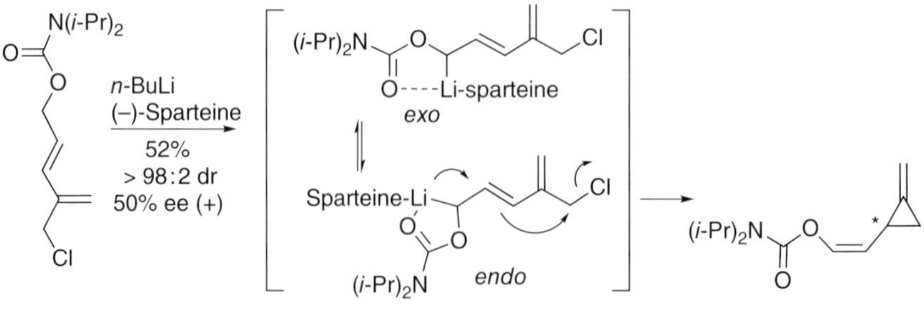

Scheme 1.173 Asymmetric synthesis of via (−)-sparteine-induced intramolecular S'$_E$-cycloalkylation reaction.

The methods reported until now start from chiral substrates, but some ring-closing reactions are also reported under chiral catalysis. As an example, chiral cyclopropane subunits were prepared from alkenes via (−)-sparteine-catalyzed carbolithiation reaction (Scheme 1.172) [608].

In other reactions, the alkyllithium reagent can act as a base, analogously with reaction depicted in Scheme 1.163. In such a manner, an asymmetric synthesis of (Z)-1-methylene-2-vinylcyclopropanes by intramolecular S'$_E$-cycloalkylation reaction was reported (Scheme 1.173) [609]. The stereoselectivity of the reaction was explained by the fact that the α-lithiated intermediate reacted from the (2E)-endo rather than (2E)-exo conformation. However, the stereochemistry of the chiral center was not attributed, and only a positive optical rotation was reported.

Later, further studies on the mechanism were conducted, starting from (E)-5-chloro(bromo)-4,4-dimethylpent-2-enyl diisopropylcarbamate [610]. Chloro derivative gave better yields (90% vs 79%) and selectivity (57% vs 38% ee), and the product was recognized as a (Z)-configured enantioenriched (S)-vinylcyclopropane. The deprotonation was demonstrated to give mainly (S)-lithium derivative. However, the product was demonstrated to arise through an *anti*-S'$_E$ pathway, thus, the reaction should proceed via a dynamic kinetic resolution in which the rates of epimerization and cycloalkylation were of similar magnitude.

The preparation of (1R,2S)-1-amino-2-vinylcyclopropane-carboxylic-acid-derived sulfonamide and ethyl ester was attempted by a cinchona phase-transfer organocatalysis, starting from (E)-N-alkylideneglycine ethyl ester and 1,4-dibromo(chloro)-2-butenes. Unfortunately, only the reaction of (E)-N-benzylideneglycine ethyl ester and *trans*-1,4-dibromo-2-butene led to satisfactory yields (88%), diastereo- (95:5), and enantioselectivity (80%) [611]. However, these compounds are common building blocks in the preparations of potent HCV NS3 protease inhibitors.

A quite different manner of using chiral catalysis was represented by the synthesis of an enantioenriched compound, which subsequently underwent stereoselective ring closure. For instance, 4-aryl-4-oxobutanonic esters, readily available by Friedel–Crafts acylation reaction, were asymmetrically and chemoselectively reduced to an alcohol, the alcohol activated, and the ring closed through the ester enolate. The cyclization step afforded the (S,S)-configured cyclopropanes without affecting the enantiomeric excess of the starting alcohol (Scheme 1.174) [612].

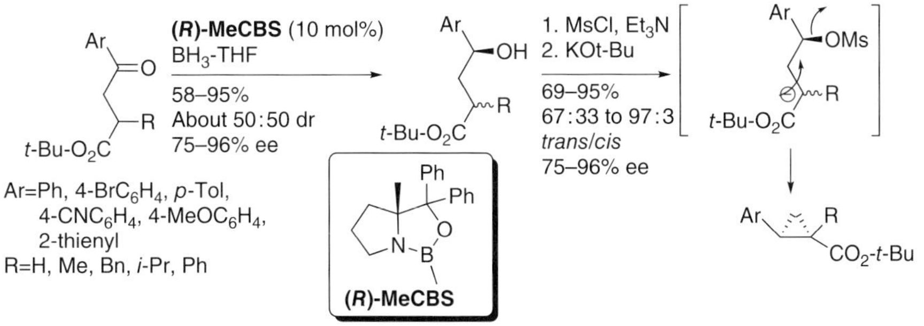

Scheme 1.174 Intramolecular ring closure of activated chiral benzyl alcohols.

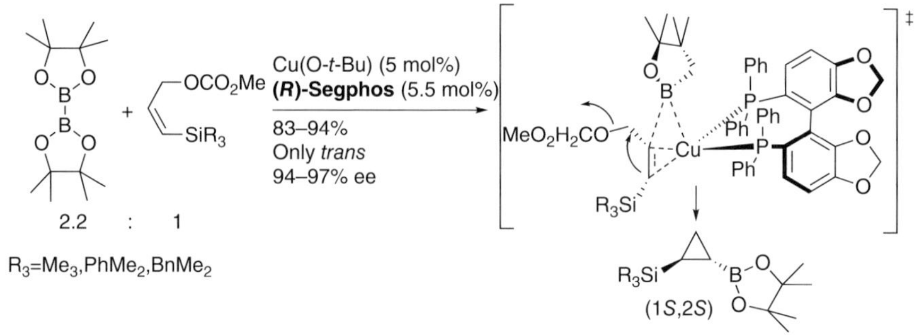

Scheme 1.175 Synthesis of optically active boron–silicon bifunctional cyclopropanes.

Finally, allylcarbonates with a γ-silicon substituent with bis(pinacolato)diboron afforded (S,S)-cyclopropylboronates using copper(I)-(R)-segphos catalyst [613]. The reaction rate greatly depended on the configuration of the substrates, and Z-isomers gave the best results. The stereochemical outcome was explained surmising a transition state free from steric repulsion between the substituents of the (Z)-allylcarbonate and the phenyl groups of the ligand (Scheme 1.175). Then, the boron functionality was stereoselectively modified by Suzuki–Miyaura coupling and converted to cyclopropylamines [613, 614].

1.5 Miscellaneous Reactions

Various other reactions forming chiral cyclopropanes with different mechanisms with respect to those reported in the previous sections have been reported, and they are listed in this section.

1.5.1 Rearrangement of Chiral Oxiranes

Epichloro- or epibromohydrins and glycidol derivatives have been commonly used to form cyclopropanes, as they are readily available. Two pathways are possible depending on the nature of the leaving group: the displacement of the leaving group after (path a) or before the ring opening (path b) of the epoxide (Scheme 1.176) [615]. It should be noted that, depending on what mechanism is

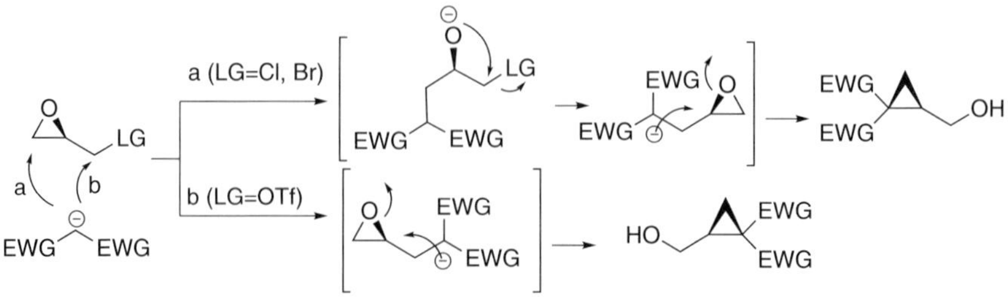

Scheme 1.176 Mechanisms for the formation of chiral cyclopropanes from chiral epoxides.

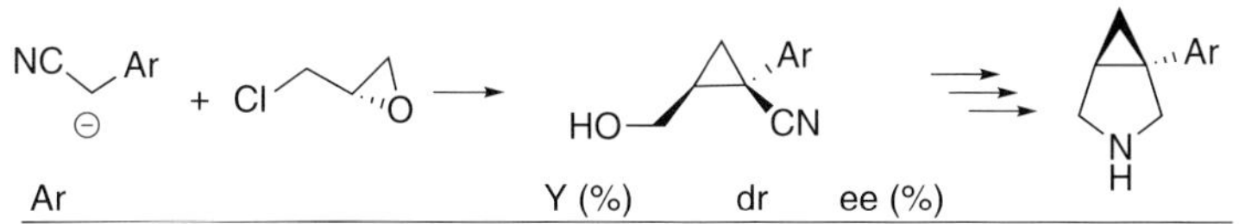

Scheme 1.177 Asymmetric epoxy nitrile coupling.

Ar	Y (%)	dr	ee (%)	
3,4-Cl$_2$C$_6$H$_3$	95	85:15	96	DOV21947
p-Tol	95	85:15	>97	Bicifadine

operative, the opposite enantiomers are formed. Nucleophiles include carbanions derived from malonates, β-phosphonates, ketones, sulfones, and nitriles.

For instance, the reaction between sodium dimethylmalonate and optically pure epichlorohydrin proceeded through "path a" and the corresponding product was isolated as bicyclic lactone by transesterification reaction between the carboxylate and hydroxy moieties in cis-relationship (36% yield and 93.4% ee) [616]. Phenyl- or phenylsulfonylacetonitrile and epichlorohydrin led to substituted chiral cyclopropane lactones in 67% and 82% yield and with 96% and 98% ee, respectively [617]. Other arylacetonitriles afforded the corresponding hydroxyl nitriles in high ee values, when the aggregation state of the nitrile anion was properly manipulated [618]. This methodology was successfully applied to the total syntheses of two neurotransmitters, bicifadine and DOV21947, in a single-stage process without the isolation of any intermediates (Scheme 1.177).

Conversely, glycidyl triflate (92% ee) and sodium di-t-butyl malonate led to the formation of the lactone with the opposite configuration (therefore by "path b") in 48% yield and 91% ee [619]. The use of glycidyl 3-nitrobenzenesulfonate and cesium fluoride improved the yields [620]. Robinson and Aggarwal applied this strategy in the total synthesis of solandelactone E (Scheme 1.5 for the structure) [621]. The leaving group was a sulfonate from (2R,3R)-2-hydroxymethyl-3-allyloxirane, the nucleophile lithiated acetonitrile, and the (1R,2R)-2-((R)-1-hydroxybut-3-enyl)cyclopropanecarbonitrile was recovered in 77% yield (trans/cis = 85:15).

In Scheme 1.164, phosphorus derivatives mediated asymmetric cyclopropanations. Optically pure epoxides were used to transfer the enantiomeric excess to

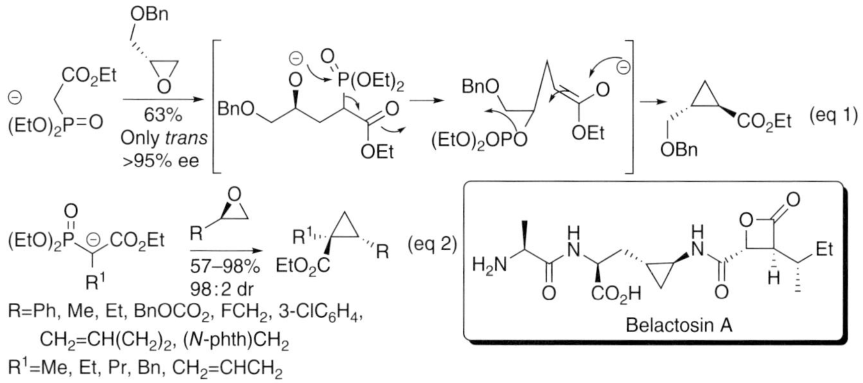

Scheme 1.178 Wadsworth–Emmons cyclopropanation of chiral epoxides.

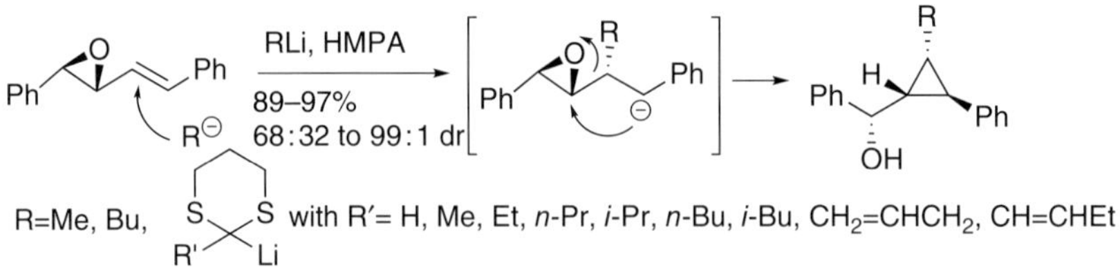

Scheme 1.179 Asymmetric cyclopropanation of phenyl vinyl epoxide with lithiated carbanions.

cyclopropane by the Wadsworth–Emmons cyclopropanation. For example, (S)-glycidol benzyl ether led to the corresponding enantiopure cyclopropane. The proposed mechanism involved an epoxide opening, followed by a migration of the phosphonate group from carbon to oxygen and a subsequent S_N2 ring closure (Scheme 1.178, eq 1) [622]. This reaction was also applied to a total synthesis of a potent antitumor agent, belactosin A. Later, Bray and Minicone extended the reaction scope of this reaction (Scheme 1.178, eq 2) [623]. They did not report enantiomeric excesses, but they affirmed that the chiral center of the epoxides retained the configuration in the final product. Wadsworth–Emmons cyclopropanations are known to be easily scalable, and therefore, the potential utility of these procedures could be greater if they are scalable, in turn.

Chiral phenyl vinyl epoxide reacted with lithiated 2-alkyl-1,3-dithiane or alkyllithium in the presence of HMPA (Scheme 1.179) [624]. The reaction was considered a tandem conjugated addition–epoxide opening sequence.

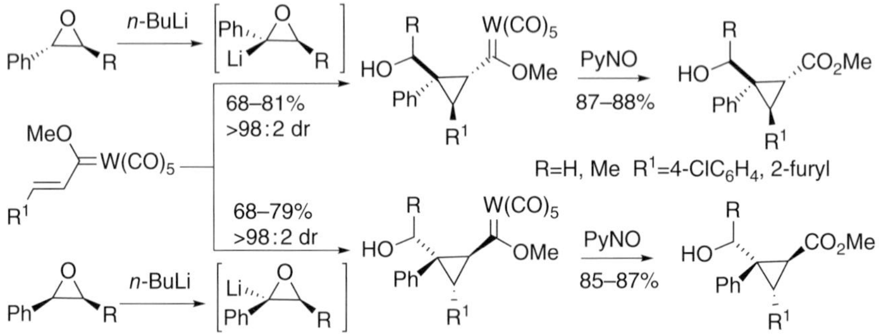

Scheme 1.180 Asymmetric cyclopropanations of α,β-unsaturated Fischer carbene complexes with lithiated oxiranes.

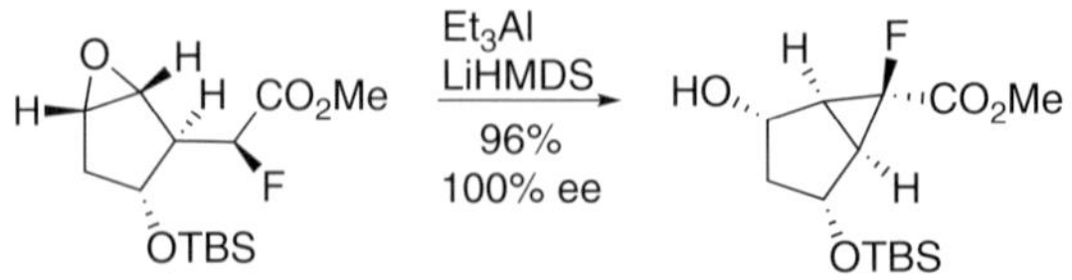

Scheme 1.181 Asymmetric Et$_3$Al-mediated intramolecular opening of epoxide.

Lithiated chiral oxiranes were also condensed onto α,β-unsaturated Fischer tungsten-derived carbene complexes in a diastereo- and enantiospecific manner, providing the corresponding tetrasubstituted cyclopropane carbenes, then converted into the corresponding cyclopropanecarboxylates (Scheme 1.180) [625].

In addition, several examples of intramolecular opening of chiral epoxides have been reported. The Et$_3$Al-mediated synthesis of bicyclo[3.1.0]hexane system was an example (Scheme 1.181) [626]. Interestingly, this transformation provided perfect F-*endo* selectivity. It was also applied to the total syntheses of metabotropic glutamate receptor agonists (mGluR2/3 agonists).

LTMP-induced intramolecular cyclopropanation of unsaturated terminal epoxides provided an efficient and completely stereoselective entry to bicyclo[3.1.0]hexan-2-ols and bicyclo[4.1.0]heptan-2-ols. This synthesis was applied in a total synthesis of (+)-cuparenone, starting from a chiral chlorohydrin converted into the corresponding bicyclohexanol (Scheme 1.182) [627].

Superbasic mixtures, such as LIDAKOR, allowed the 3-*exo*-cyclization of suitably substituted oxiranes lacking strong electron-withdrawing substituents to be achieved [628]. The reaction was highly stereoselective, leading to the corresponding *trans*-cyclopropanes. Moreover, the outcome of the rearrangement process did not depend on the configuration of the starting chiral oxirane ring or in the relative stereochemistry of the silyloxy substituent. The reaction mixture always contained two isomers, owing to a *tert*-butyldimethylsilyl

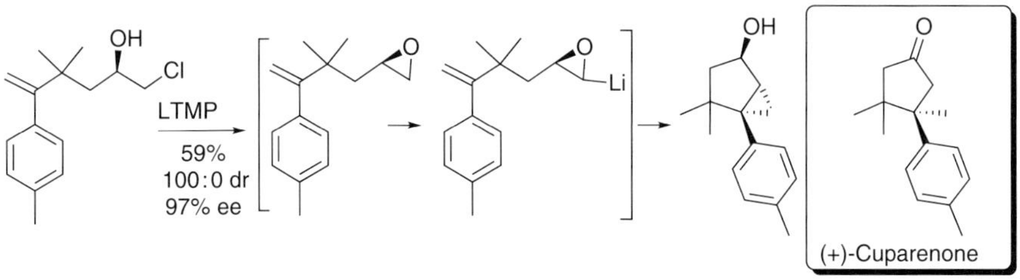

Scheme 1.182 Asymmetric LTMP-mediated intramolecular cyclopropanation of chlorohydrin.

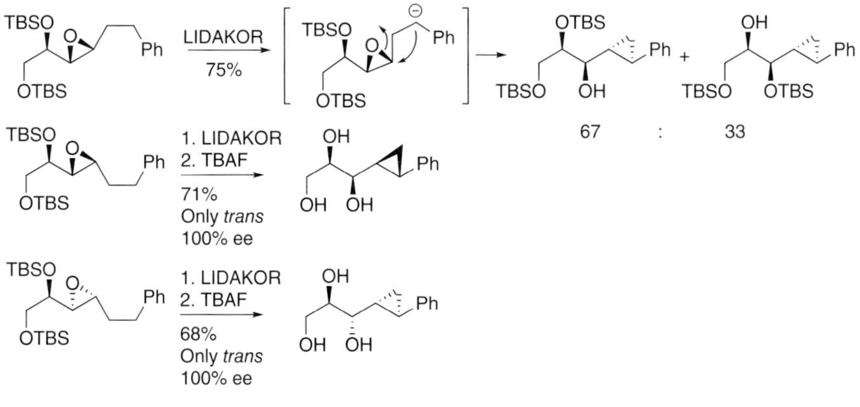

Scheme 1.183 LIDAKOR-promoted rearrangements of oxiranes to cyclopropanes.

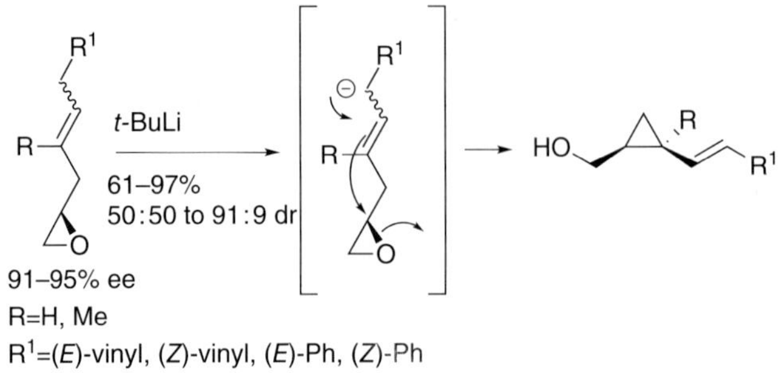

Scheme 1.184 Synthesis of cyclopropane from terminal enantioenriched epoxides.

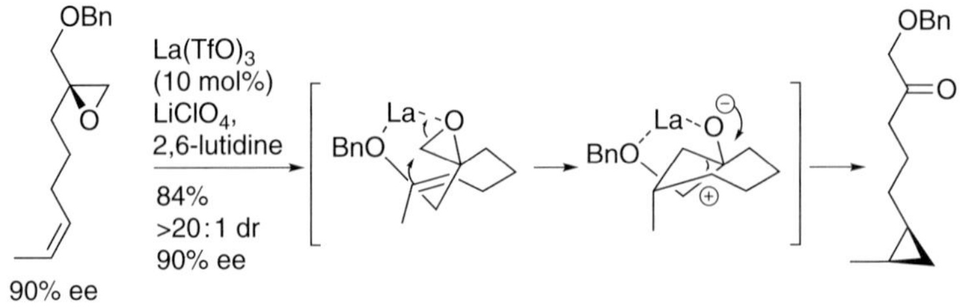

Scheme 1.185 Methylene-transfer from enantioenriched epoxides.

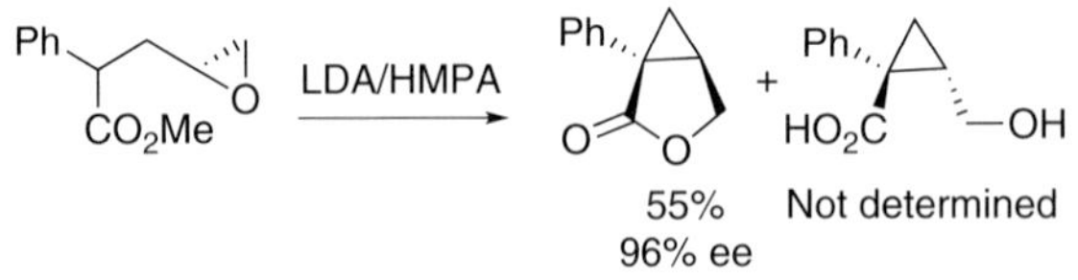

Scheme 1.186 An enantioselective synthesis of levomilnacipran.

group migration to the neighboring oxygen during the isomerization process. However, fluoride deprotection of silyl groups overcame this drawback (Scheme 1.183).

Terminal enantioenriched epoxides were prepared in 53–98%, with 91–95% ee via organocatalytic asymmetric enamine chlorination with the MacMillan catalyst. If a Z-double bond was present in the 2-position of the side chain of the epoxides, these compounds were stereoselectively rearranged into cyclopropanes (Scheme 1.184) [629].

Just one example of a methylene-transfer reaction from enantioenriched epoxides obtained from Sharpless asymmetric epoxidation to furnish the enantioenriched cyclopropane was reported (Scheme 1.185) [630].

Starting from a bicyclolactone prepared by a stereoselective rearrangement of the corresponding enantioenriched epoxide, an enantioselective synthesis of

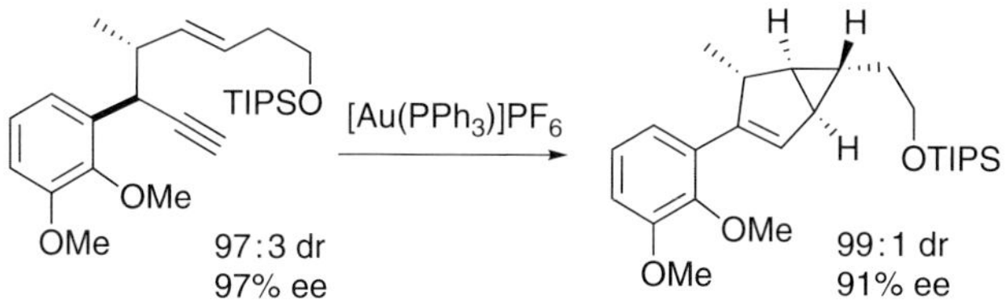

Scheme 1.187 Au-catalyzed asymmetric cyclopropanation of 1,5-enynes.

(−)-milnacipran was set up (Schemes 1.186 and 1.38 for the structure of (−)-milnacipran) [631].

1.5.2 Cycloisomerization of 1,n-Enynes

The transition-metal-catalyzed cycloisomerization reaction of 1,n-enynes has emerged as an atom-economical process for the preparation of diverse cyclic compounds over the last 90 years [632]. Obviously, in the recent years, some enantioselective versions of this reaction have been set up. For instance, in the presence of a gold catalyst, enantioenriched 1,5-enynes underwent cycloisomerization into bicyclo[3.1.0]hexane derivatives (Scheme 1.187 depicts an example) [633]. In another example, trichloro(pyridine)gold(III) stereospecifically rearranged the enantioenriched enyne derivatives into the corresponding bicyclo[3.1.0]ketones with over 19:1 dr [634].

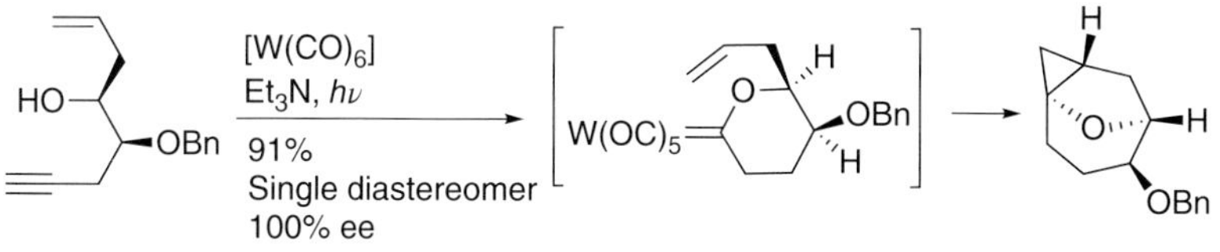

Scheme 1.188 W-catalyzed asymmetric cyclopropanation of hydroxylated enynes.

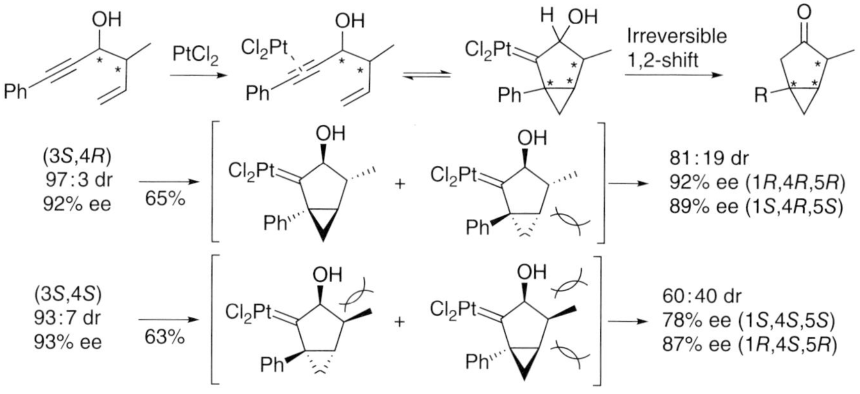

Scheme 1.189 Pt-catalyzed asymmetric cyclopropanations of hydroxylated enynes.

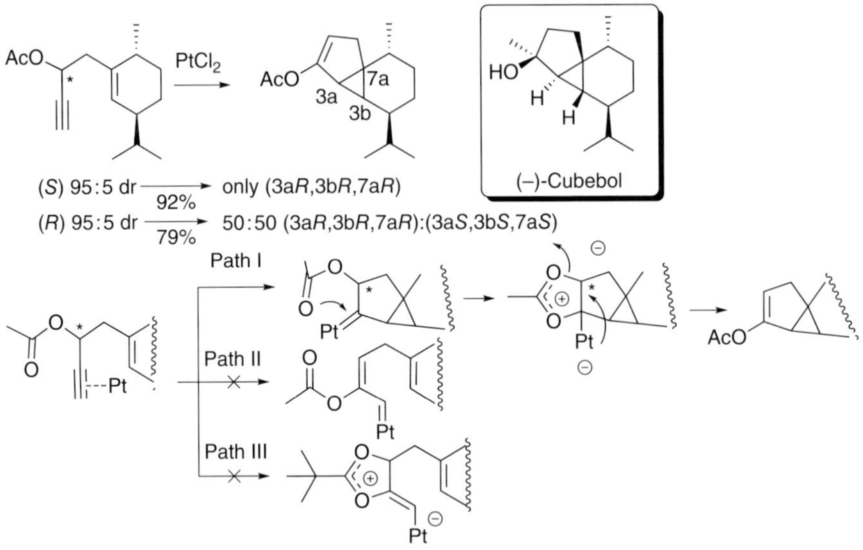

Scheme 1.190 Pt-catalyzed asymmetric cyclopropanations of propargyl esters.

Irradiation of a chiral hydroxylated enyne in the presence of W(CO)$_6$ provided the corresponding tricyclic product as a single stereoisomer (Scheme 1.188) [635]. In addition, a platinum catalyst can catalyze this reaction if a hydroxyl group at the propargylic position is present on the chiral enyne (Scheme 1.189) [636]. An irreversible 1,2-hydrogen shift from the bicyclo[3.1.0]hexane carbene skeleton moved the equilibrium platinum complex–platinum carbene toward products. Steric hindrance in the bicyclo[3.1.0]hexane skeleton was invoked to explain the significantly higher de in the *syn* series.

Platinum also catalyzed cycloisomerization of a chiral propargyl ester (Scheme 1.190) [637]. It should be noted that in the case of acetates, (S)-propargyl ester afforded the corresponding tricyclic product with high selectivity, paving an efficient route to (−)-cubebol, while with (R)-configured isomer, the reaction was not stereoselective. This feature persuaded the authors that the configuration of the stereogenic center carrying the acetate unit translated into the stereochemistry of the product (Scheme 1.190, path I). Consequently, a vinylcarbene intermediate cannot be formed prior to cyclization (Scheme 1.190, path II), because it implied that both isomers led to the same product distribution. Almost at the same time, another research group published the same strategy, starting from pivaloates instead of acetates and providing similar results (70–74% yields, (R)-pivaloate gave 80:20 dr,

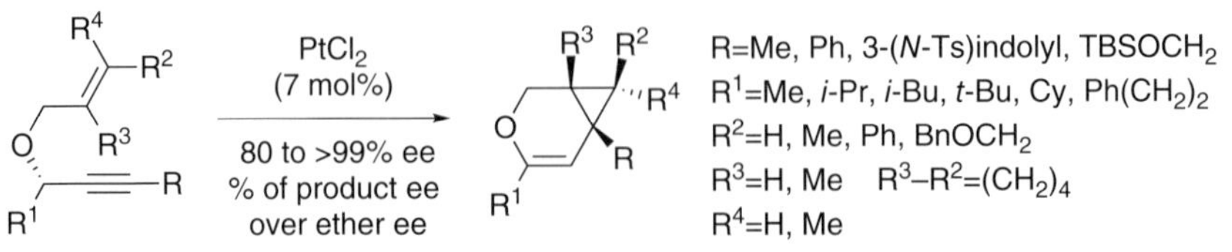

Scheme 1.191 Cycloisomerization of enantioenriched allyl propargyl ethers.

whereas (*S*)-pivaloate gave 60 : 40 dr) [638]. In order to justify their results, these authors postulated a concerted C—C bond-formation/C—O bond-breaking pathway, (Scheme 1.190, path III). Soriano and Marco-Contelles settled the issue by computational methods performing a detailed relaxed potential energy surface scan [639]. The result of these calculations allowed for retaining the mechanism depicted in Scheme 1.190, and path I is the most plausible to justify the stereochemical outcome of the reactions.

Finally, enantioenriched allyl propargyl ethers were reported to give an enantiospecific cycloisomerization to oxabicycloheptanes (Scheme 1.191) [640]. The stereochemistry depicted in Scheme 1.191 was confirmed by X-ray structural analysis.

However, the use of chiral catalysts represents an improvement also in this procedure. Actually, catalytic methodologies have gradually replaced those starting from chiral reagents more recently.

For instance, the enantioselective cycloisomerization of nitrogen-bridged 1,6-enynes, affording azabicyclo[4.1.0]hept-4-enes, was widely studied (Table 1.10). The reaction mechanism depicted in Scheme 1.192 is generally accepted for these reactions. The relative steric bulk of both the nitrogen and the catalyst substituents assembled the more stable conformation, which directed the subsequent 6-*endo*-*dig* cyclization, responsible for the final stereochemistry.

A quite different approach to the cyclization of 1,6-enynes was performed using a cyclopentadienylruthenium catalyst containing a tethered chiral sulfoxide (Scheme 1.193) [645]. In fact, 1,6-enynes containing a racemic propargyl alcohol moiety assembled [3.1.0] instead of [4.1.0] bicycles with (*R*,*R*)-stereochemistry. Besides sterically hindered sulfonamide, diphenylphosphoramidate and malonate derivatives worked well in this reaction and a [4.1.0]bicyclic piperidine was prepared from a 1,7-enyne (56% yield, 70% ee). Interestingly, when the reaction was carried out with enantioenriched propargyl alcohols, 84% and 36% ee were obtained with the (*R*)- and the (*S*)-propargyl alcohol and the 60% ee obtained from the racemic compound was almost the average value.

The preparation of enantioenriched cyclopropanes was also performed by combining the power of asymmetric synthesis of enynes with gold-catalyzed cycloisomerization chemistry. Some of these compounds are key intermediates for the synthesis of terpenoid compounds. For instance, chiral phosphine–gold(I) catalyzed the cycloisomerization of propargyl derivatives to cycloheptenes or cyclooctenes (Scheme 1.194) [646]. It is worth noting that yields and enantiomeric excesses are slightly better than those reported in Table 1.10 for azabicyclic compounds [643]. However, the best catalyst was found to vary for different examples, and the reaction was highly substrate-dependent, thus difficult to reproduce. About the mechanism, a gold-stabilized vinyl carbenoid was surmised after gold-mediated 1,2-shift of the OR^1 group. The two isomers of this carbenoid could equilibrate, and the less abundant but required (*E*)-TS had to be subtracted from the equilibrium. On the other hand, an external and more reactive alkene such as 1,1-diphenylethylene led to the products arising from the most stable (*Z*)-TS [647].

A P,N-bidentate ligand with a C_2-symmetric *N*-heterocycle enabled reactive α-oxo gold carbene intermediates generated *in situ* to undergo asymmetric intramolecular

Table 1.10 Organocatalytic cycloisomerization of 1,6-enynes.

Reactant	Catalyst	Yield (%)	ee (%)	References
ArO$_2$SN⟶≡, R Ar=Me, o-Tol, R=Ph, 4-ClC$_6$H$_4$, p-Tol, 2-Naph	[IrCl(Cod)]$_2$ (10 mol%), p-TolBINAP (20 mol%) AgOTf (24 mol%) CO (1 atm)	57–71	64–71 (—)[a]	[641]
R^1, R^2 structures with Ph groups	Rh complex with PPh$_3$, B[3,5-(CF$_3$)$_2$Ph]$_4$ (5 mol%)	71–94	68–99 (1S,6R,7R)	[642]
X=NTs, NTf, O R=H, Ph R^1=H, Me, R^2=H, Ph, 4-MeOC$_6$H$_4$, 2-MeOC$_6$H$_4$, 2,6-Me$_2$C$_6$H$_3$, 4-ClC$_6$H$_4$				
R, R^1, R^2 enyne structure	biaryl diphosphine ligand R=3,5-(t-Bu)$_2$-4-MeOC$_6$H$_2$ (3 mol%) AgOTf (6 mol%)	8–64	13–99[b]	[643]
X=NTs, O, R^1=H, Me, R^2=H, Ph, 4-MeOC$_6$H$_4$, R=H, Ph, 4-MeOC$_6$H$_4$, 4-NO$_2$C$_6$H$_4$, 3,5-Me$_2$C$_6$H$_3$, 4-MeO$_2$CC$_6$H$_4$, 3-BrC$_6$H$_4$				

TsN≡
R¹
R

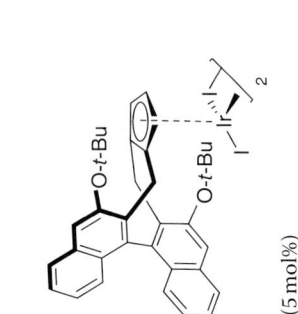

R=H, Ph, 4-MeOC₆H₄, 4-CF₃OC₆H₄, p-Tol, 4-BrC₆H₄, 4-NO₂C₆H₄, 3-ClC₆H₄, 3,5-Me₂C₆H₃, 2-Naph, 2-thienyl, 3-furyl, 2-benzothienyl, 2-benzofuryl, 5-benzo[d][1,3]dioxolyl, CH₂OBn, R¹ =H, Me, Ph, 5-benzo[d][1,3]dioxolyl

(5 mol%)

47–87^c 63–94   [644]
           (1S,6S,7S)

a) Only negative optical rotation was reported.
b) Only optical rotation (mainly negative) was reported. In schemes, the same configuration as under rhodium catalysis was depicted where the optical rotation values were positive. However, these are enantiomeric with respect to those reported in Scheme 1.191 [640]. Although, the two papers do not have compounds with the same substituents, the opposite optical rotations reported indirectly confirmed this analytically unsupported stereochemistry.
c) Olefin with cis-relationship between R and CH₂NTs groups was not reactive.

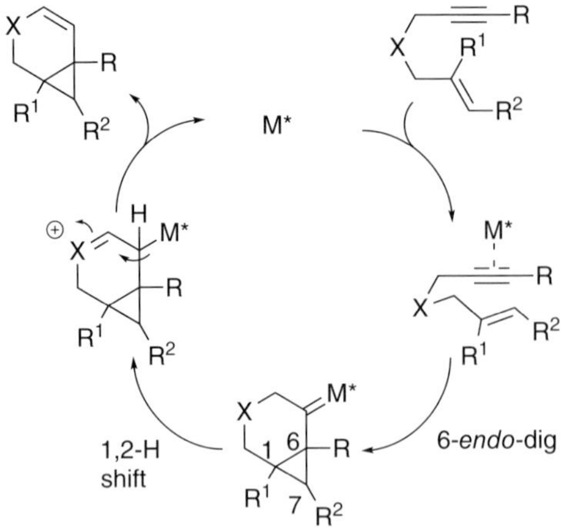

Scheme 1.192 Proposed catalytic cycle for the cyclization of 1,6-enynes.

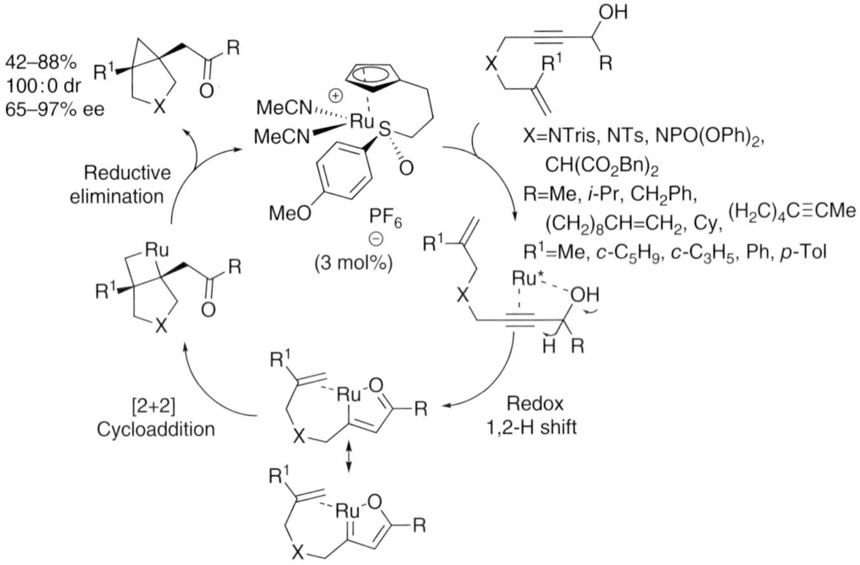

Scheme 1.193 Ru-catalyzed asymmetric redox bicycloisomerization reaction.

cyclopropanation (Scheme 1.195) [648]. This reaction was applied to the synthesis of LY354740 (Scheme 1.120, eq 3 for the structure). The reaction of (*E*)-ethyl 3-(2-ethynylphenyl)but-2-enoate proceeded in 78% yield but only with 11% ee. Moreover, ethyl 3-(2-ethynylphenyl)-2-methylacrylate (prepared as an inseparable 5:1 mixture of (*E*/*Z*)-isomers) was converted into the corresponding cyclopropane diastereomers in an almost identical ratio (91% combined yield), thus indicating that the cyclopropanation was concerted and stereospecific. The stereochemical outcome was

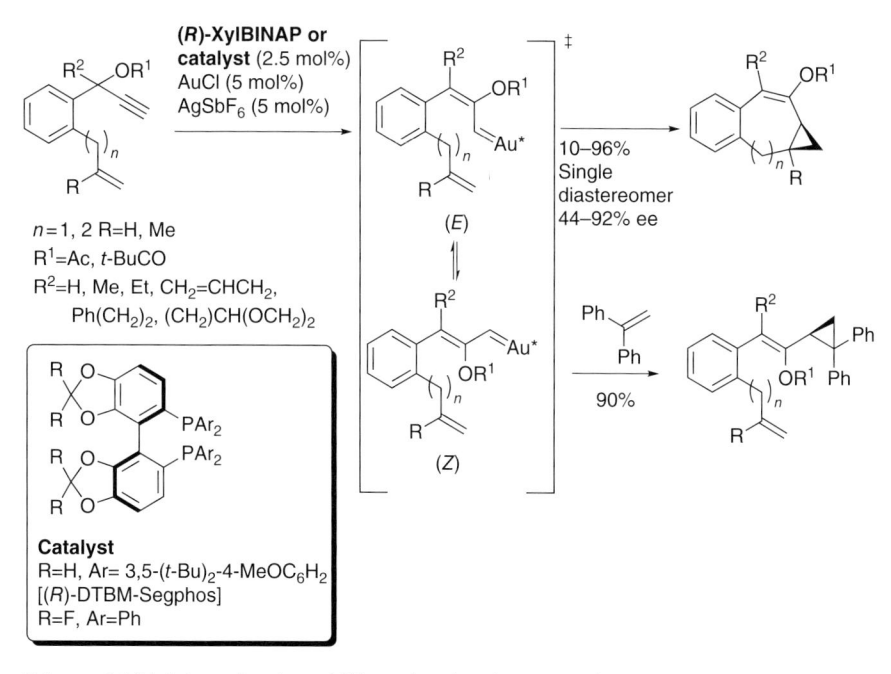

Scheme 1.194 Intramolecular gold(I)-catalyzed cyclopropanations.

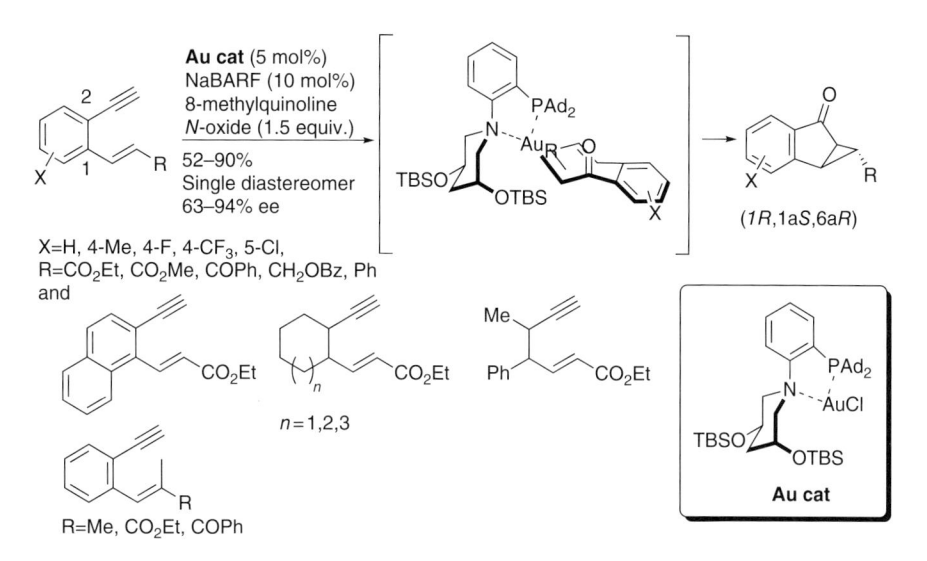

Scheme 1.195 Enantioselective oxidative intramolecular cyclopropanation.

envisaged arising from a transition state in which the axial TBSO group of the piperidine ring shielded the *Re*-face of the carbene center from alkene attack.

The intermolecular version of this reaction was also developed. For instance, propargyl esters were used for the preparation of vinylcyclopropanes. Chiral imidazolium- and triazolium-based gold(I)–NHC catalysts were tested with poor

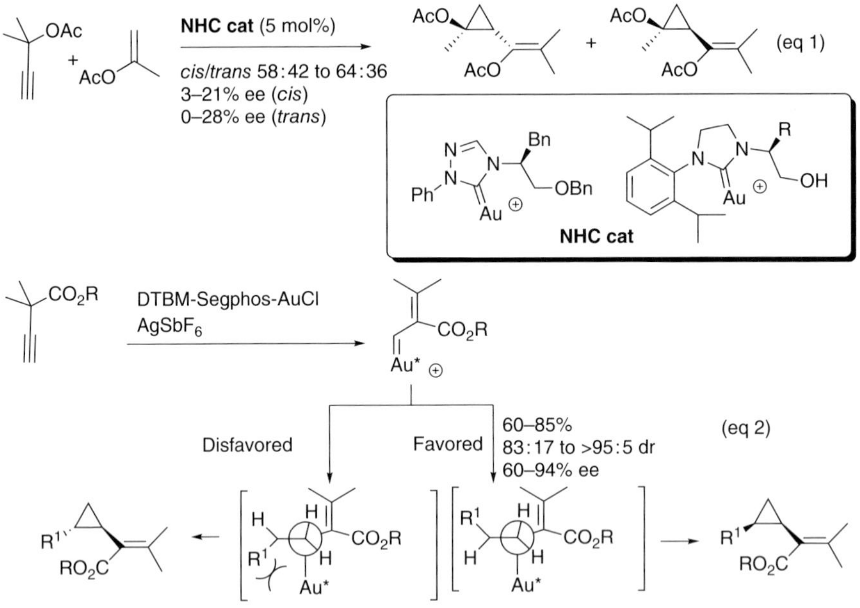

Scheme 1.196 Au-catalyzed cyclopropanation of olefins with propargyl esters.

results (Scheme 1.196, eq 1) [649], while DTBM-Segphos gave high *cis*-selectivity and enantioselectivity (Scheme 1.196, eq 2) [650].

Finally, these reactions were also attempted in heterogeneous catalysis. In particular, asymmetric catalyst was prepared by encapsulating metallic nanoclusters in chiral self-assembled monolayers immobilized on mesoporous SiO_2 support and activated through metal oxidation by $PhICl_2$. Under these conditions, Au–diproline/MCF-17 catalyzed intermolecular and intramolecular cyclopropanation reactions with up to 51% and 30% enantiomeric excess, respectively [651].

To close this section, a related reaction should be mentioned, since it anyway involves double and triple carbon–carbon bonds, that is, the cotrimerization of alkenes and diethyl acetylenedicarboxylate, leading to functionalized (*R*,*R*)-furylcyclopropanes in 34–62% yield with 97–99% ee and >19:1 dr [652].

1.5.3 Denitrogenation of Chiral Pyrazolines

The decomposition of pyrazolines has proved to be an excellent method for the preparation of cyclopropanes. The reaction works under photo, thermal, acid, and microwave decomposition.

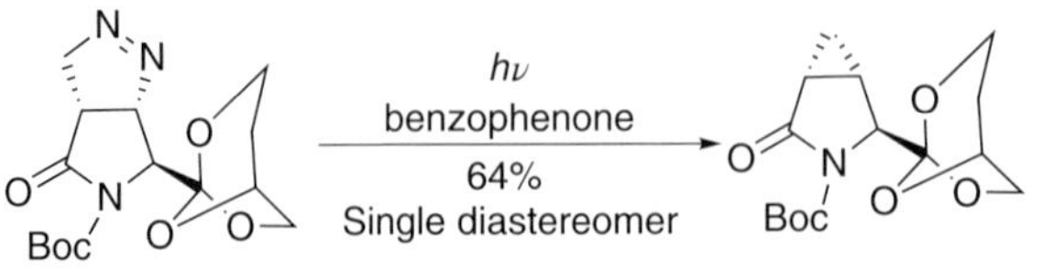

Scheme 1.197 Photolysis of pyrazoline.

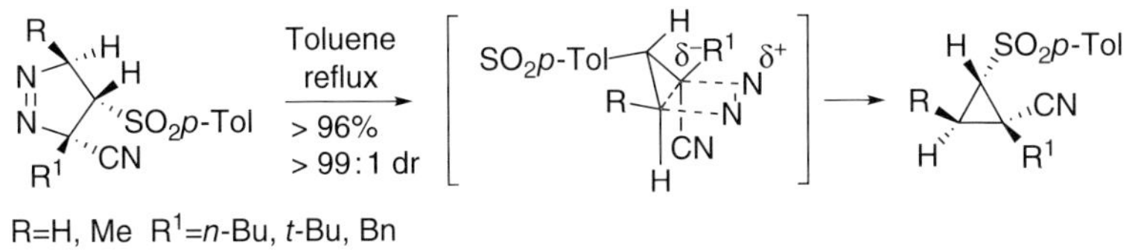

Scheme 1.198 Thermolysis of sulfonylpyrazolines.

As an example of photolysis, in the presence of benzophenone as a photosensitizer, chiral pyrazolines afforded the corresponding cyclopropanes as unique isomers (Scheme 1.197) [653]. This process very likely involved the formation of diradicals. This reaction was applied to the syntheses of pharmacologically important cyclopropane proline amino acids.

The thermal extrusion of nitrogen from pyrazolines mainly gave olefins, and, when yielding cyclopropanes as the main products, they usually gave low stereoselectivity. However, Garcia Ruano's research group developed a successful thermolysis of chiral sulfonylpyrazolines (Scheme 1.198) [654]. This extrusion took place with complete retention of configuration of all the chiral centers. A polar transition state was consistent with this concerted thermal decomposition [655].

Garcia Ruano also described the addition of diazomethane or diazoethane to enantiopure (S)-3-[(4-methylphenyl)sulfinyl]lactones with almost complete π-facial selectivity. The configurations of pyrazolines could be rationalized by the approach of the diazoalkane from the less-hindered bottom face (far from the tolyl group) of

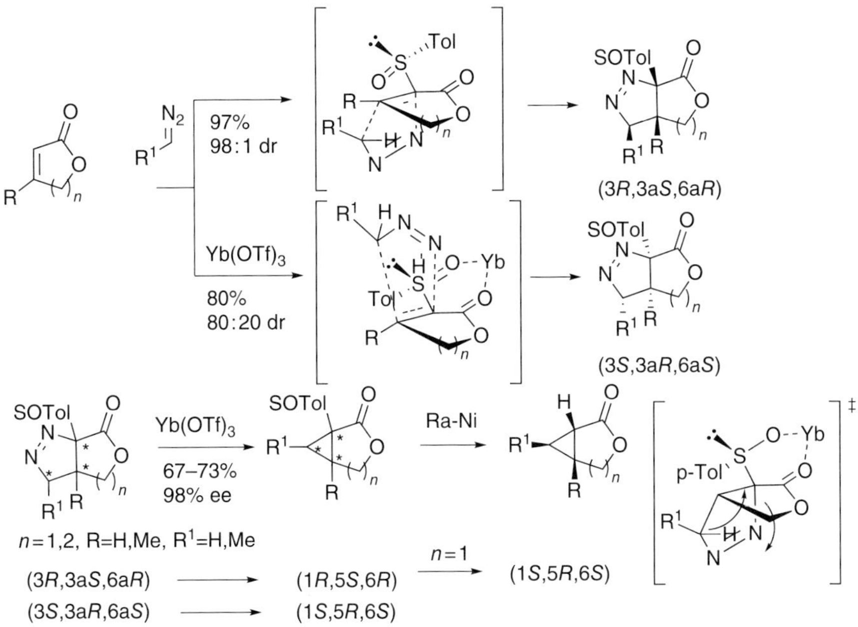

Scheme 1.199 Yb(OTf)$_3$-catalyzed synthesis and denitrogenation of sulfinylpyrazolines.

the conformation, in which the electrostatic repulsion between the sulfinyl and carbonyl oxygen atoms was minimized. In the presence of Yb(OTf)$_3$, the formation of a chelated species was preferred, and the change in the spatial arrangement of the tolyl group caused the inversion of the facial selectivity. Then, the isolated pseudoenantiomers were submitted to the denitrogenation in the presence of a substoichiometric amount of Yb(OTf)$_3$ (Scheme 1.199) [656].

The reaction evolved with complete retention of configuration at both carbons flanking the nitrogen atoms and the subsequent desulfinylation by treatment with Ni Raney resulted in the formation of enantiomerically pure cyclopropanecarboxylic acid derivatives. Regarding the mechanism, the metal could form a chelated species with the sulfinyl and carbonyl oxygens; thus, the enhanced electronic deficiency of this carbon provoked the concerted migration of C3 with extrusion of nitrogen. The reaction of γ-lactones was more efficient, and 87–97% and 68–98% yields were obtained from (3R,3aS,6aR) and (3S,3aR,6aS) isomers, respectively [656]. It should be noted that decomposition of the pyrazolines did not occur at −78 °C, thus allowing the synthesis and purification of the (3S,3aR,6aS)-lactones in the first step of this synthesis.

Finally, chiral spiropyrazolinepenicillanates, obtained from 1,3-dipolar cycloaddition reactions of 6-alkylidenepenicillanates with diphenyldiazomethane, phenyldiazomethane, and diazomethane, were submitted to stereospecific microwave-induced ring contraction to chiral spirocyclopropylpenicillanates (Scheme 1.200) [657]. However, the preparation of starting material was closely dependent on the diazoalkanes. Thus, phenyldiazomethane reacted regio- and stereoselectively with all tested 6-alkylidenepenicillanates, whereas diphenyldiazomethane only with 6-(Z)-(1-methoxycarbonylmethylene)- and 6-(Z)-(1-benzoylmethylene)penicillanates, albeit to give a different regioisomer. Both diphenyldiazomethane and diazomethane with 6-(Z)-(1-*tert*-butoxycarbonylmethylene)penicillanate and diazomethane with 6-(Z)-(1-methoxycarbonylmethylene)penicillanates led to two regioisomers. Finally, the 1,3-dipolar cycloaddition reactions of diazomethane with the acetyl and benzoyl derivatives were regiospecific.

1.5.4 C–H Insertion

Satoh and coworkers set up a series of syntheses of enantioenriched bicyclo[*n*.1.0] alkanes by employing a magnesium carbenoid C–H insertion reaction (Scheme 1.201) [658]. The starting materials were prepared from various carbonyl derivatives and (R)-chloromethyl-*p*-tolylsulfoxide, then a highly stereoselective reaction with the lithium enolate of a carboxylic acid *tert*-butyl ester occurred, exclusively leading to compounds in which the absolute configuration of the chlorine bearing carbon atom was S.

The reaction was also extended to the lithium enolate of *tert*-butyl 4-phenylbutyrate, and the absolute configuration of the carbon bearing the 2-phenylethyl group in the chlorosulfoxides was determined to be R. A detailed mechanistic investigation was conducted, and it was found that:

i) the sulfoxide–magnesium exchange took place remarkably with retention of configuration;

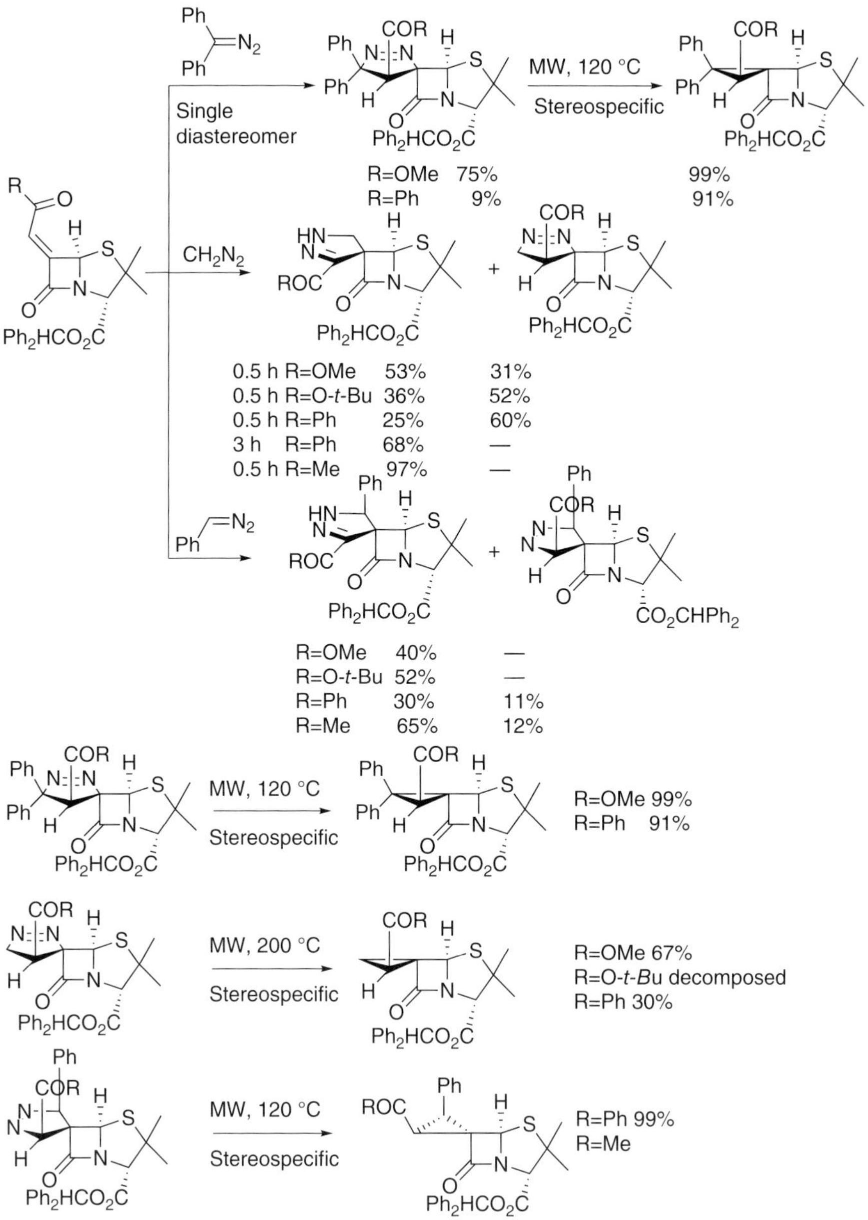

Scheme 1.200 Microwave decomposition of 6-alkylidenepenicillanates.

ii) the C—H bond on the methylene carbon of the cycloalkene attacked the chlorine atom from its backside to give the cyclopropane;
iii) the reaction occurred only at the methylene carbon on the cycloalkene, because the methylene carbon adjacent to the carbonyl has no hydrogen atoms on the backside of the chlorine atom;

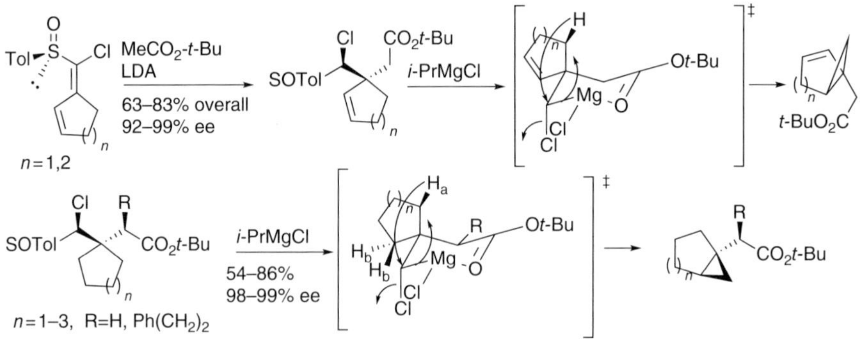

Scheme 1.201 Asymmetric cyclopropanation via magnesium carbenoid 1,3-CH insertion reaction in cyclic compounds.

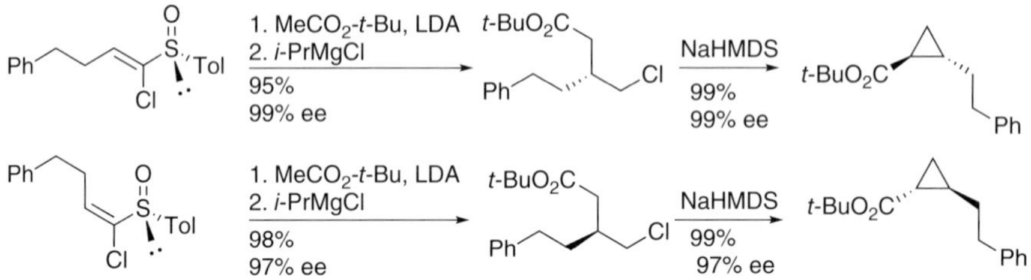

Scheme 1.202 Asymmetric cyclopropanation via magnesium carbenoid 1,3-CH insertion reaction in straight-chain compounds.

iv) (Z)-1-chlorovinyl-p-tolylsulfoxides, having no hydrogen atoms on the backside of the chlorine atom, because the carbon is sp^2 instead of sp^3, gave no reaction or intractable mixtures;

v) in saturated cyclic derivatives, only the C—H_b bond is placed almost vertically to the carbon—chlorine bond.

In straight-chain sulfoxides, the magnesium carbenoids were protonated before the 1,3-C–H insertion took place and the corresponding 4-chloro-substituted ester was obtained. Cyclopropanes were recovered after treatment with a base (Scheme 1.202) [659]. Both E- and Z-isomers reacted, because straight-chain sulfoxides are less conformationally demanding, leading to a 20:1 trans/cis mixture of the optically active tert-butyl cyclopropanecarboxylates.

At the end of this section, a reaction that is not properly an asymmetric synthesis of cyclopropanes should be mentioned. In fact, a TADDOL-based phosphoramidite palladium(0) complex enabled an enantioselective Friedel–Crafts reaction by C–H insertion on aryl cyclopropanes (Scheme 1.203) [660]. This method provided efficient access to cyclopropyl-dihydroquinolones and dihydroisoquinolones as well as the construction of the seven-membered ring of the cyclopropylindolobenzazepine core of BMS-791325. A substrate with another aryl bromide groups as the nitrogen substituent allowed for the synthesis of a pentacyclic dihydroquinolinone in 95% yield and 93% ee. Poor results were instead obtained with 3-ammi-

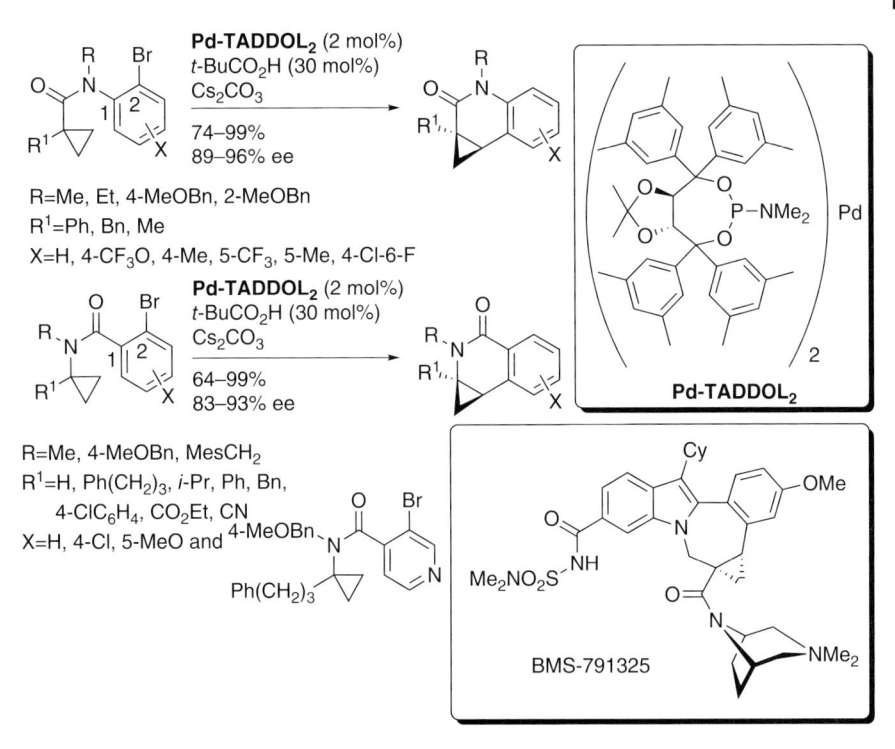

Scheme 1.203 Palladium(0)-catalyzed intramolecular cyclopropane functionalization.

nopyridine, trimethylsilylcyclopropane, and 3-thienylketone as the substrates and secondary amides are unreactive. Finally, it is worth noting that the destruction of the cyclopropane moiety in the synthesis of dihydroisoquinolones was never observed, conversely from similar Pd(0)-catalyzed C—H bond functionalization reactions [661].

1.5.5 Addition to Cyclopropenes

The diastereoselective addition reactions of cyclopropenes constitute an attractive alternative to the more mainstream routes to chiral cyclopropanes [3, 662]. For example, enantiopure 2,2-disubstituted cyclopropylboronates could be easily prepared by hydroboration of cyclopropenes under rhodium catalysis. In the presence of a chiral ligand such as (R)-BINAP, (1S,2R)-cyclopropanes were the main product (Scheme 1.204, eq 1), while (S)-TolBINAP afforded the enantiomeric (1R,2S)-product in comparable yield and selectivity [663]. More recently, a diastereo- and enantioselective Cu-catalyzed hydroboration of cyclopropanes, which did not need a directing group, was described (Scheme 1.204, eq 2) [664]. Additionally, the capture of the cyclopropylcopper intermediate with electrophilic amines was possible, but the enantioselective version of this reaction has never been reported.

Under chiral rhodium catalysis, hydrostannation of cyclopropenes provided optically active cyclopropylstannanes [665]. The hydrostannation had a more

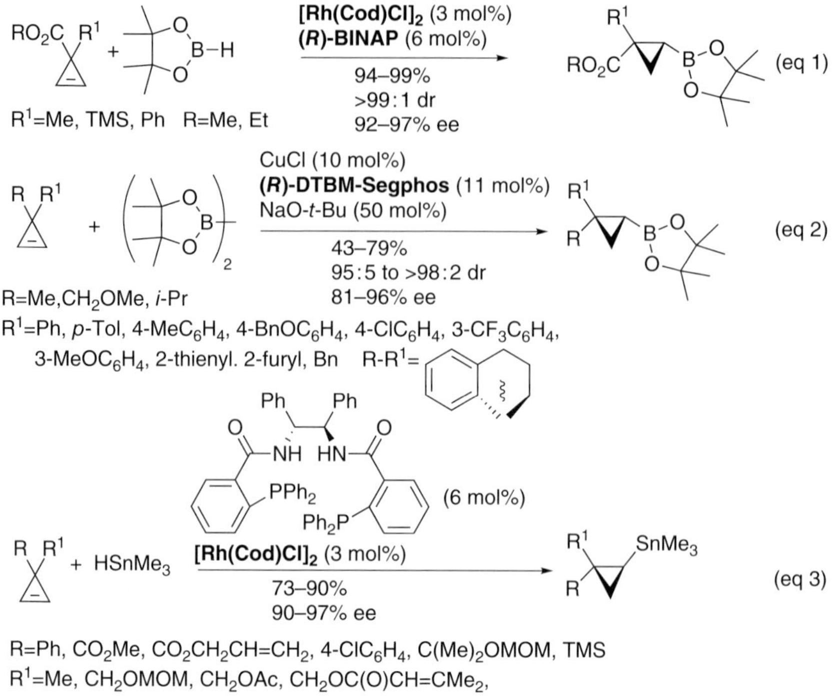

Scheme 1.204 Asymmetric hydroboration and hydrostannation of cyclopropenes.

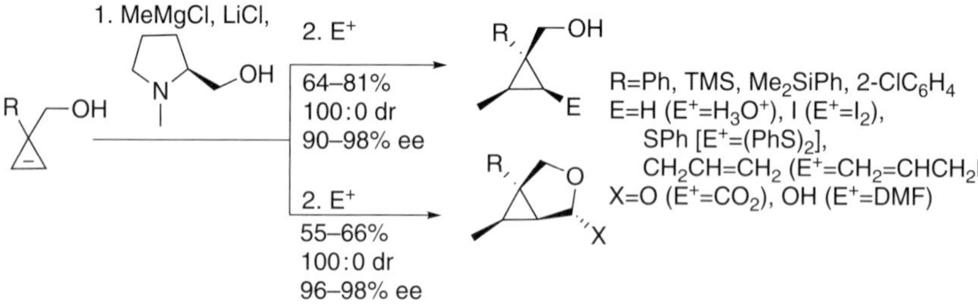

Scheme 1.205 Asymmetric carbomagnesations of cyclopropenes.

general scope than hydroboration under similar catalysis, because also in this case, no directing group was required (Scheme 1.204, eq 3). All these reactions allowed further functionalization, by selective removing of boron or tin groups.

In the presence of N-methylprolinol as a chiral ligand, the addition of MeMgCl allowed an enantio- and facially selective carbomagnezation of cyclopropanes (Scheme 1.205) [666]. The introduction of electrophiles created three stereocenters with high enantioselectivity.

Chiral cyclopropene derivatives, obtained by reaction of alkynes and diazoalkanes under $Rh_2(DOSP)_4$ catalysis (see Section 1.3.1.4), and Grignard reagents allowed a regio- and diastereoselective synthesis of chiral methylenecyclopro-

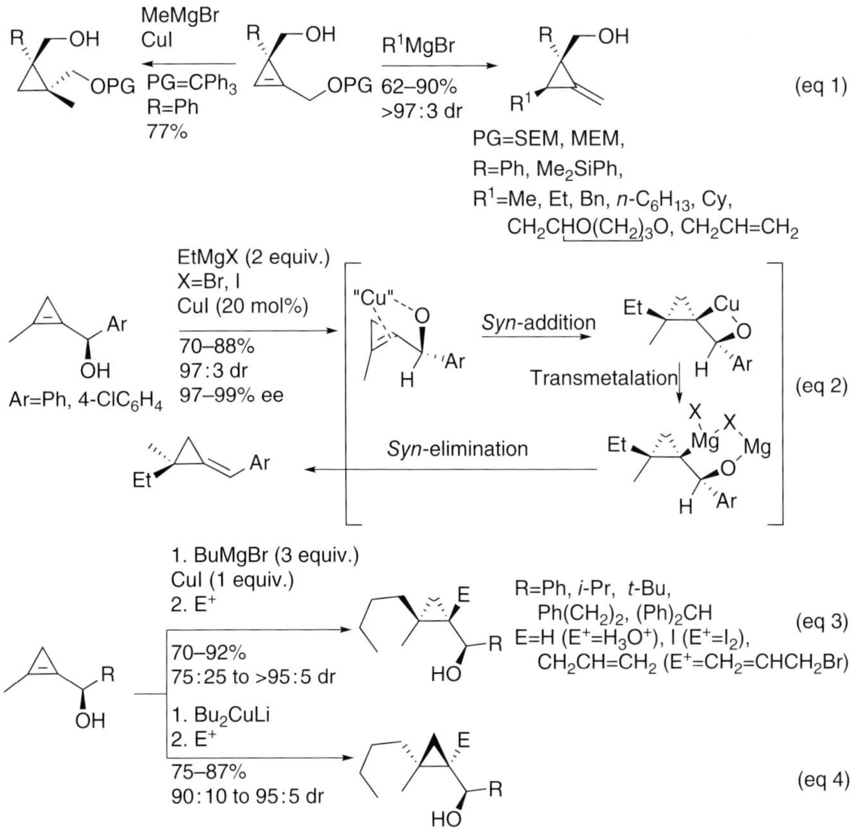

Scheme 1.206 Asymmetric carbometallation of cyclopropenylcarbinols.

panes (Scheme 1.206, eq 1) [667]. It should be noted that a single stereoisomer was obtained only when using bromide as the Grignard counterion and that trityl protection reversed the regioselectivity. In addition, unprotected chiral cyclopropenylcarbinol led to elimination of alkylidenecyclopropane, regardless of the nature of the alkylmagnesium halides in the presence of a catalytic amount of CuI (Scheme 1.206, eq 2) [668]. It is worth noting that, as discussed in the previous reaction, CuI was essential to reverse regioselectivity of the addition (Scheme 1.206, eq 1 arrow leftward). The stereoselectivity of this process was explained by envisaging that the most stable conformer of the cyclopropenylcarbinolate had the smallest hydrogen "inside" and the largest aryl group "outside" to minimize $A^{1,3}$-strain. Then, the reaction proceeded through a *syn*-addition/*syn*-elimination mechanism. Later, this research group demonstrated that the carbon–copper bond was less prone to β-elimination compared to the carbon–magnesium bond [669]. Thus, a stoichiometric addition of CuI allowed for quenching of the metallated species with different electrophiles to give the corresponding chiral functionalized cyclopropylcarbinol derivatives (Scheme 1.206, eq 3). Interestingly, the use of an organocuprate reversed the observed diastereomeric ratio in favor of the *syn*-isomer (Scheme 1.206, eq 4).

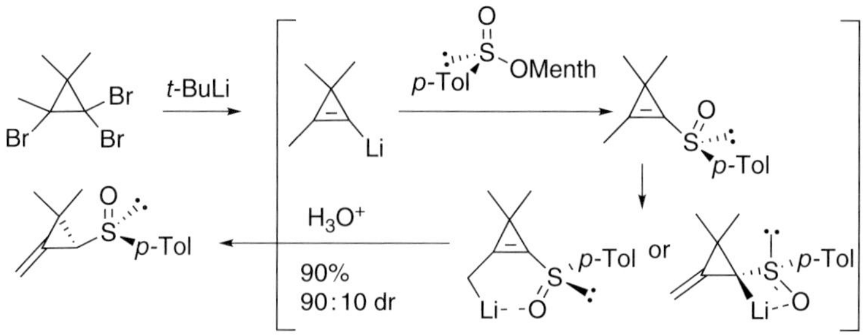

Scheme 1.207 Synthesis of chiral methylenecyclopropane via three-component reaction.

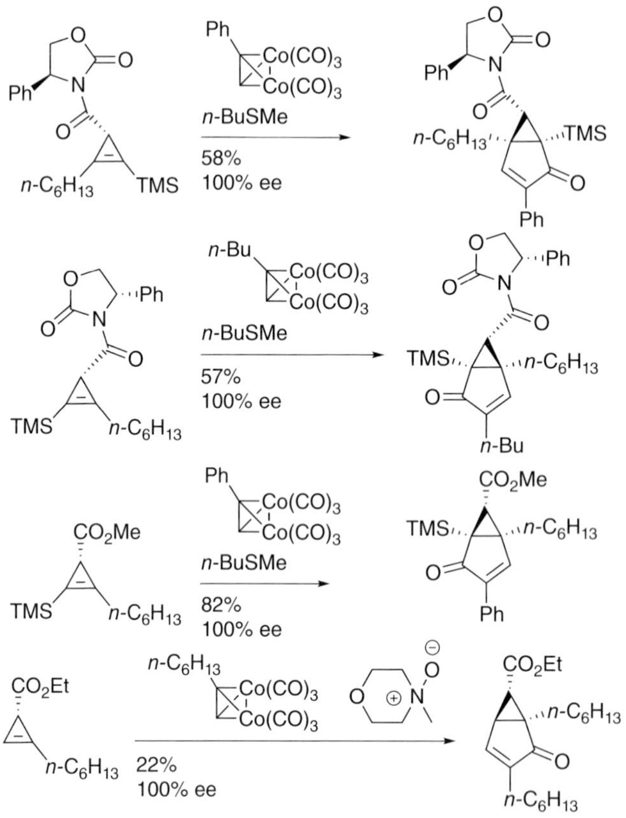

Scheme 1.208 Asymmetric Pauson–Khand reactions of cyclopropenes.

Chiral methylenecyclopropane derivatives were also prepared via a three-component reaction from 1,1,2-tribromocyclopropanes, (−)-menthyl (S)-p-toluenesulfinate, and electrophiles (Scheme 1.207) [670].

Chiral cyclopropanes also promoted regio- and stereoselective intermolecular Pauson–Khand reactions [671]. A single enantiomerically pure bicyclohexene was always isolated (Scheme 1.208).

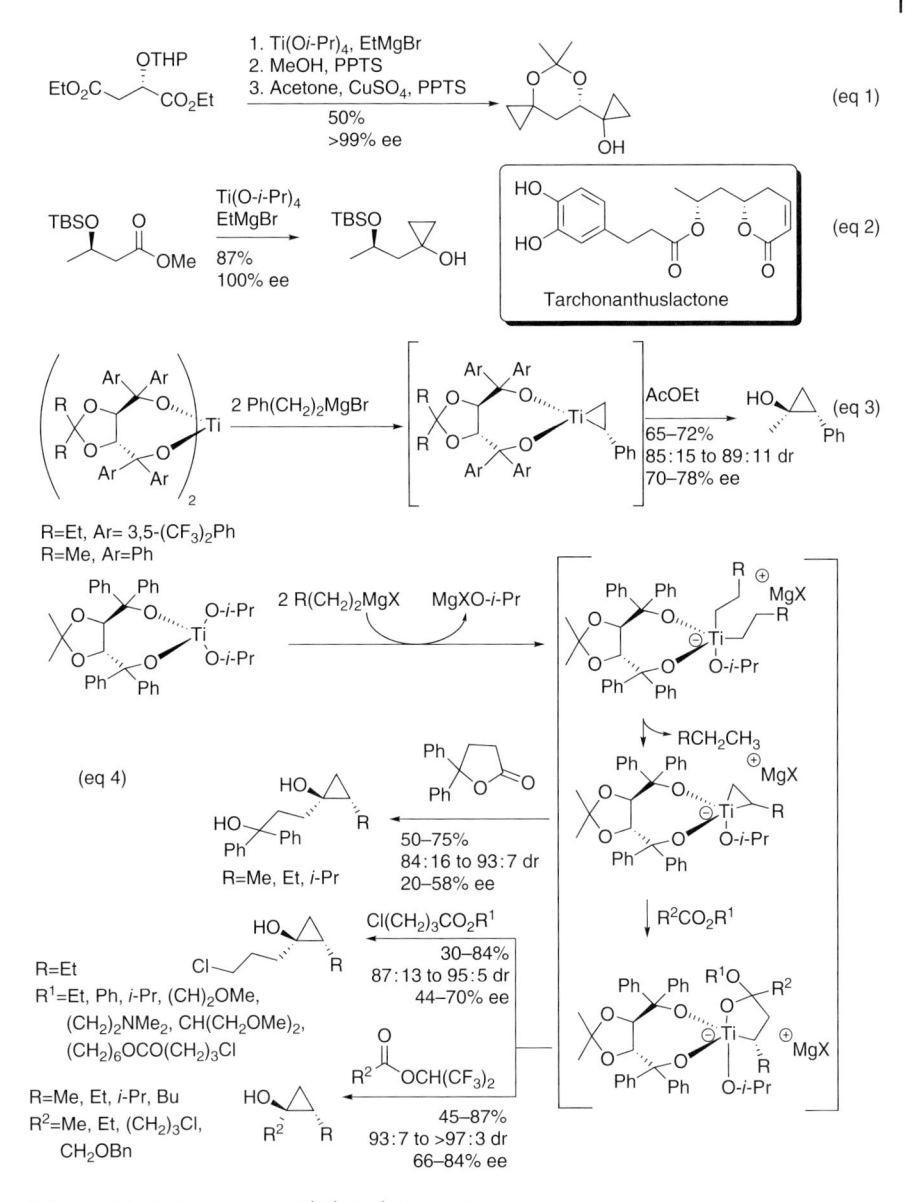

Scheme 1.209 Asymmetric Kulinkovich reaction.

1.5.6 Other Methods

Since 1989, Kulinkovich's group discovered that the reaction of esters with a mixture of Ti(O-*i*-Pr)$_4$ and an excess of a Grignard reagents led to the corresponding substituted cyclopropanols [672].

More recently, enantioselective synthesis of cyclopropanols by the Kulinkovich method has been investigated. The use of either chiral substrates or chiral titanium ligands was successfully tested. As an example, Kulinkovich's group itself reported the asymmetric reductive cyclopropanation of chiral THP-protected diethyl malate (Scheme 1.209, eq 1) [673]. Almost at the same time, Singh and coworkers reported that, under Kulinkovich reaction conditions, a chiral β-alkoxy ester afforded the corresponding enantiomerically pure cyclopropanol, which was the key intermediate for all the stereoisomers of tarchonanthuslactone (Scheme 1.209, eq 2) [674]. Moreover, titanium TADDOLates mediated the asymmetric Kulinkovich reaction (Scheme 1.209, eqs 3 and 4) [675]. In order to rationalize the stereochemical outcome of the reaction, a mechanistic study was undertaken by Kulinkovich and coworkers [675c].

They proposed a model for the cyclopropanation involving the participation of five-coordinated alkyltitanium species (Scheme 1.209, eq 4), unlike Corey's idea (Scheme 1.209, eq 3) [675a]. Under this hypothesis, the stereogenic carbon in titanacyclopropane intermediates could occupy either apical or equatorial positions, but, in both cases, the (S)-diastereomers were less sterically hindered compared to the corresponding (R)-diastereomers, in agreement with Corey's suggestion.

In fact, in the most stable diastereomers, the R group was far from the pseudo-axial phenyl groups of the catalyst. The formation of titanium alkoxides by-products should be avoided, because they were more reactive compared to the TADDOLate and, catalyzing the background reaction, strongly affected the reaction stereoselectivity. The influence of the structure of the leaving alkoxy group on the stereoselectivity of cyclopropanation of methyl 4-chlorobutanoate was tentatively rationalized by introducing an oxatitanacyclopentane species, which could in turn undergo transformation with retention of configuration at the stereogenic carbanion center. In fact, the mechanism of the three-carbon ring for-

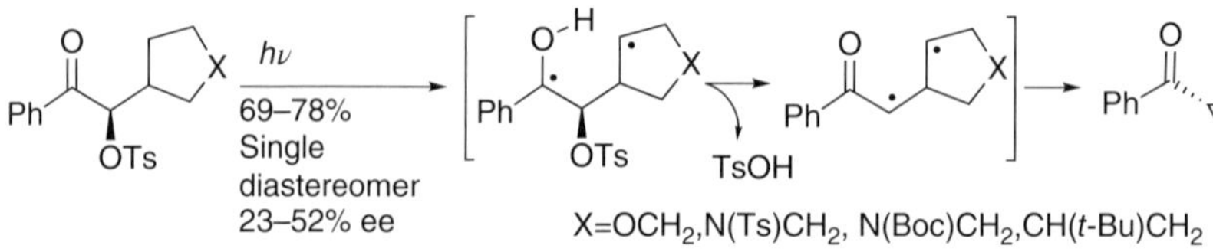

Scheme 1.210 Asymmetric photochemical cyclization of ketones.

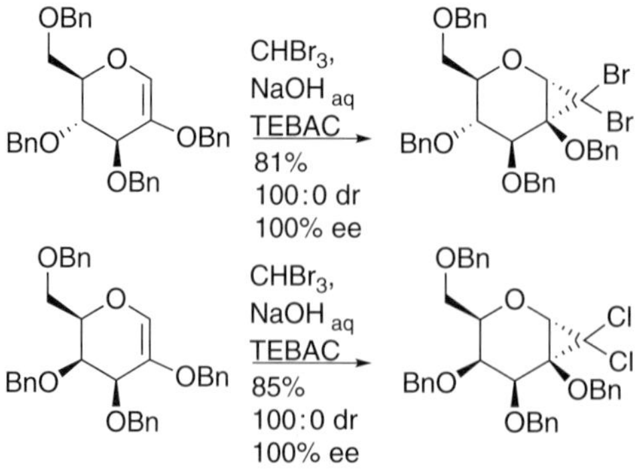

Scheme 1.211 Asymmetric cyclopropanations of glycals.

mation did not explain the leaving alkoxy group influence, since it departed before the cyclopropane ring-forming step.

The irradiation of a chiral ketone led to the formation of the corresponding *exo* cyclopropylated product with complete stereoselectivity, in agreement with the concept of spin-center shift [676]. The 1,4-diradical generated by irradiation shifted the radical center from the carbonyl to the adjacent carbon atom with *p*-toluenesulfonic acid elimination. The resulting 1,3-diradical cyclized with complete diastereoselectivity to give the final cyclopropane (Scheme 1.210).

Another diastereoselective methodology to form cyclopropane was performed with bromoform or chloroform on glycal under phase-transfer conditions (Scheme 1.211), in order to prepare unnatural septanoside derivatives [677].

Vinylcyclopropanecarboxylates could be prepared on the basis of an organoiron methodology (Scheme 1.212) [678]. In fact, a chiral (1-methoxycarbonylpentadienyl) iron cation treated with MeLi predominantly gave the corresponding (pentenediyl) iron complex, which afforded the corresponding enantiomerically pure vinylcyclopropanes by treatment with excess ceric ammonium nitrate. The synthesis was applied to the synthesis of the C9–C16 segment of ambruticin (Scheme 1.107 for the structure) [678b].

In the same period, enantiomerically pure [60]fullerene tris-adducts with an *e,e,e*-addition pattern have been prepared by cyclopropanation of C_{60} with chiral D_3-symmetrical cyclo-tris(malonate) tethers bearing chiral C_8-spacers connecting the reactive malonate [679].

A strained ketene hemithioacetal underwent the addition of methyl pyruvate in the presence of a chiral copper catalyst. The obtained chiral product was the key intermediate in a total synthesis of (−)-acylfulven and (−)-irofulven (Scheme 1.213) [680].

Enantioenriched α-*N*-homoallylaminonitriles could be stereoselectively cyclized to azabicyclo[3.1.0]hexanes in the presence of a base and zinc bromide.

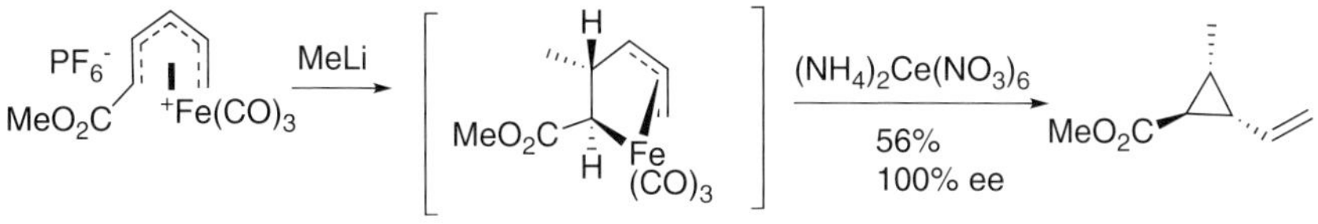

Scheme 1.212 Asymmetric synthesis of vinylcyclopropane via organoiron methodology.

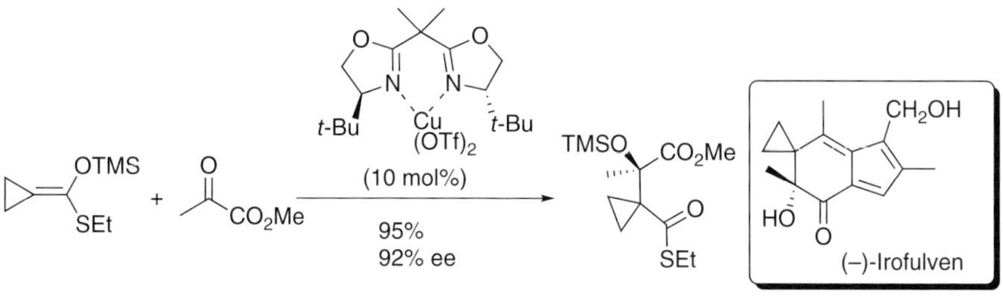

Scheme 1.213 Asymmetric aldol reaction of cyclopropanated ketene hemithioacetal.

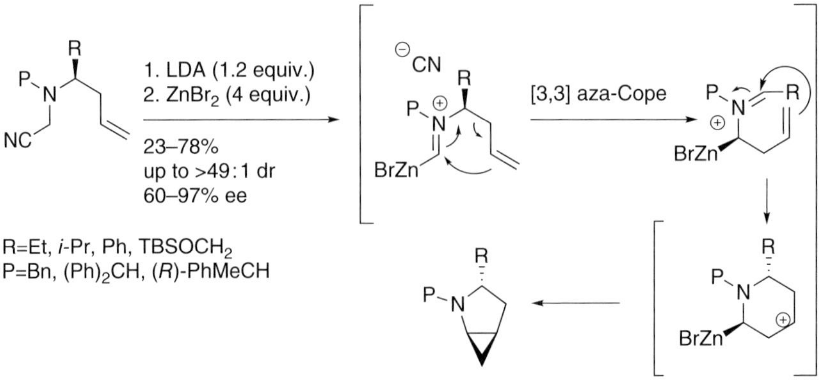

Scheme 1.214 Synthesis of enantiopure 3-substituted 2-azabicyclo[3.1.0]hexanes.

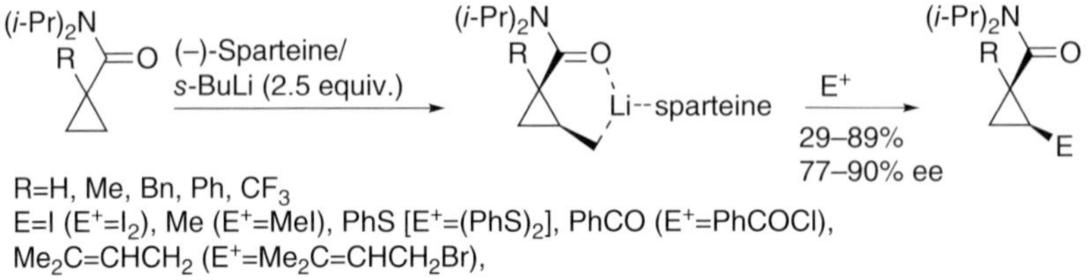

Scheme 1.215 Enantioselective synthesis of cyclopropylcarboxamides by metallation.

It is worth noting that the stereospecific inversion of the homoallylic stereogenic center with substrates derived from α-branched nitrogen-protecting groups (Scheme 1.214) [681]. However, benzyl-protected substrates showed a different stereochemical outcome for the cyclization, indicative of a not-explained mechanistic difference.

The last example of synthesis of chiral cyclopropanes reported here is the chiral-base-mediated desymmetrization reaction of β-metallated cyclopropylcarboxamides (Scheme 1.215) [682]. In the presence of (−)-sparteine (Scheme 1.172 for the structure), the *syn*-metallation of the cyclopropane with respect to the directing amide substituent asymmetrically occurred. The predominant β-metallation even in the presence of possible enolate formation (R=H) was noteworthy. The tendency of the lithiated cyclopropanes to undergo decomposition and self-condensation represented a serious drawback.

1.6 Conclusions

As cyclopropane rings are widespread in natural and biologically active compounds, their synthesis remains a considerable challenge, even over a century after the synthesis of the first cyclopropane derivative, and new syntheses appear in the literature. With the development of asymmetric synthesis, strategies involving chiral pools, chiral auxiliaries, chiral metal catalysts, and organocatalysts have led to other significant developments in the scope of the cyclopropanation reaction.

Enzymatic methods cannot be forgotten, but they are not considered here. The asymmetric cyclopropanation is therefore a very important tool for organic synthesis, and significant contributions are still needed.

References

1 (a) De Meijere, A. (ed.) (1986) *Topics in Current Chemistry*, vol. 133, Springer, Heidelberg; (b) De Meijere, A. (ed.) (1987) *Topics in Current Chemistry*, vol. 135, Springer, Heidelberg; (c) Rappoport, Z. and Patai, S. (eds) (1987) *The Chemistry of the Cyclopropyl Group*, vol. 1, John Wiley & Sons, Ltd, Chichester; (d) De Meijere, A. (ed.) (1988) *Topics in Current Chemistry*, vol. 144, Springer, Heidelberg; (e) De Meijere, A. (ed.) (1990) *Topics in Current Chemistry*, vol. 155, Springer, Heidelberg; (f) Rappoport, Z. and Patai, S. (eds) (1995) *The Chemistry of the Cyclopropyl Group*, vol. 2, John Wiley & Sons Ltd, Chichester; (g) De Meijere, A. (ed.) (1996) *Topics in Current Chemistry*, vol. 178, Springer, Heidelberg; (h) De Meijere, A. (ed.) (1997) *Houben-Weyl: Carbocyclic Three and Four-Membered Ring Compounds*, vol. E17a–c, Thieme, Stuttgart; (i) De Meijere, A. (ed.) (2000) *Topics in Current Chemistry*, vol. 207, Springer, Berlin; (j) Kulinkovich, O.G. (2015) *Cyclopropanes in Organic Synthesis*, John Wiley & Sons, Ltd, Chichester.

2 (a) de Meijere, A. (1979) *Angew. Chem. Int. Ed. Engl.*, 18, 809–826; (b) Wiberg, K.B. (1987) in *Structures, Energies and Spectra of Cyclopropanes in Cyclopropyl Group*, vols. 1 and 2, Chapter 1, 1–26 (ed. Z. Rappoport), John Wiley & Sons, Ltd, Chichester; (c) Goudreau, S.R. and Charette, A.B. (2010) *Angew. Chem. Int. Ed.*, 49, 486–488.

3 Marek, I., Simaan, S., and Masarwa, A. (2007) *Angew. Chem. Int. Ed.*, 46, 7364–7376.

4 (a) Salaun, J. and Baird, M.S. (1995) *Curr. Med. Chem.*, 2, 511–542; (b) Gnad, F. and Reiser, O. (2003) *Chem. Rev.*, 103, 1603–1624; (c) Wessjohann, L.A., Brandt, W., and Thiemann, T. (2003) *Chem. Rev.*, 103, 1625–1648; (d) Brackmann, F. and de Meijere, A. (2007) *Chem. Rev.*, 107, 4493–4537; (e) Brackmann, F. and de Meijere, A. (2007) *Chem. Rev.*, 107, 4538–4583.

5 (a) Liu, H.-W. and Walsh, C.T. (1987) in *Biochemistry of the Cyclopropyl Group in Cyclopropyl Group*, vols. 1 and 2, Chapter 16 (ed. Z. Rappoport), John Wiley & Sons, Ltd, Chichester; (b) Gagnon, A., Duplessis, M., and Fader, L. (2010) *Org. Prep. Proced. Int.*, 42, 1–69.

6 (a) Charette, A.B., Lebel, H., and Roy, M.-N. (2014) *Copper-Catalyzed Asymmetric Synthesis*, Chapter 8, Wiley-VCH Verlag GmbH; (b) Wang, H., Zhou, X., and Mao, Y. (2014) *Heterocycles*, 89, 1767–1800; (c) Qian, D. and Zhang, J. (2015) *Chem. Soc. Rev.*, 44, 677–698.

7 (a) Lebel, H., Marcoux, J.-F., Molinaro, C., and Charette, A.B. (2003) *Chem. Rev.*, 103, 977–1050; (b) Pellissier, H. (2008) *Tetrahedron*, 64, 7041–7095; (c) Bartoli, G., Bencivenni, G., and Dalpozzo, R. (2014) *Synthesis*, 46, 979–1029.

8 Arlt, D., Jautelat, M., and Lantzsch, R. (1981) *Angew. Chem. Int. Ed. Engl.*, 20, 703–722.

9. Charette, A.B. and Beauchemin, A. (2004) *Simmons–Smith cyclopropanation reaction*, in *Organic Reactions* (ed. L.E. Overman), John Wiley & Sons, Inc..
10. Davies, H.M.L. and Antoulinakis, E.G. (2004) Intermolecular metal-catalyzed carbenoid cyclopropanations, in *Organic Reactions* (ed. L.E. Overman), John Wiley & Sons, Inc..
11. (a) Little, R.D. and Dawson, J.R. (1980) *Tetrahedron Lett.*, 21, 2609–2612; (b) Sun, X.-L. and Tang, Y. (2008) *Acc. Chem. Res.*, 41, 937–948.
12. (a) Simmons, H.E. and Smith, R.D. (1958) *J. Am. Chem. Soc.*, 80, 5323–5324; (b) Simmons, H.E. and Smith, R.D. (1959) *J. Am. Chem. Soc.*, 81, 4256–4264.
13. Chan, J.H.H. and Rickborn, B. (1968) *J. Am. Chem. Soc.*, 90, 6406–6411.
14. (a) Boche, G. and Lohrenz, J.C.W. (2001) *Chem. Rev.*, 101, 697–756; (b) Cornwall, R.G., Wong, O.A., Du, H., Ramirez, T.A., and Shi, Y. (2012) *Org. Biomol. Chem.*, 10, 5498–5513.
15. Nakamura, M., Hirai, A., and Nakamura, E. (2003) *J. Am. Chem. Soc.*, 125, 2341–2350.
16. Stiasny, H.C. and Hoffmann, R.W. (1995) *Chem. Eur. J.*, 1, 619–624.
17. Ramirez, A., Truc, V.C., Lawler, M., Ye, Y.K., Wang, J., Wang, C., Chen, S., Laporte, T., Liu, N., Kolotuchin, S., Jones, S., Bordawekar, S., Tummala, S., Waltermire, R.E., and Kronenthal, D. (2014) *J. Org. Chem.*, 79, 6233–6243.
18. (a) Hoveyda, A.H., Evans, D.A., and Fu, G.C. (1993) *Chem. Rev.*, 93, 1307–1370; (b) Charette, A.B. and Marcoux, J.-F. (1995) *Synlett*, 1197–1207; (c) Hermann, H., Lohrenz, J.C.W., Kühn, A., and Boche, G. (2000) *Tetrahedron*, 56, 4109–4115; (d) Fang, W.-H., Phillips, D.L., Wang, D.-Q., and Li, Y.-L. (2002) *J. Org. Chem.*, 67, 154–160.
19. Staroscik, J.A. and Rickborn, B. (1972) *J. Org. Chem.*, 37, 738–740.
20. (a) Poulter, C.D., Friedrich, E.C., and Winstein, S. (1969) *J. Am. Chem. Soc.*, 91, 6892–6894; (b) Molander, G.A. and Harring, L.S. (1989) *J. Org. Chem.*, 54, 3525–3532; (c) Molander, G.A. and Etter, J.B. (1987) *J. Org. Chem.*, 52, 3942–3944.
21. Johnson, C.R. and Barbachyn, M.R. (1982) *J. Am. Chem. Soc.*, 104, 4290–4291.
22. (a) Boyer, F.-D. and Lallemand, J.-Y. (1992) *Synlett*, 969–971; (b) Murali, R., Ramana, C.V., and Nagarajan, M. (1995) *J. Chem. Soc., Chem. Commun.*, 217–218.
23. Ratier, M., Castaing, M., Godet, J.-Y., and Pereyre, M. (1978) *J. Chem. Res. Miniprint*, 2309–2318.
24. Charette, A.B. and Lebel, H. (1995) *J. Org. Chem.*, 60, 2966–2967.
25. Oppolzer, W. and Radinov, R.N. (1993) *J. Am. Chem. Soc.*, 115, 1593–1594.
26. (a) Takemoto, Y., Baba, Y., Saha, G., Nakao, S., Iwata, C., Tanaka, T., and Ibuka, T. (2000) *Tetrahedron Lett.*, 41, 3653–3656; (b) Baba, Y., Saha, G., Nakao, S., Iwata, C., Tanaka, T., Ibuka, T., Ohishi, H., and Takemoto, Y. (2001) *J. Org. Chem.*, 66, 81–88.
27. Smith, A.B. and Simov, V. (2006) *Org. Lett.*, 8, 3315–3318.
28. White, J.D., Martin, W.H.C., Lincoln, C., and Yang, J. (2007) *Org. Lett.*, 9, 3481–3483.
29. Mohapatra, D.K., Kanikarapu, S., Naidu, P.R., and Yadav, J.S. (2015) *Tetrahedron Lett.*, 56, 1041–1044.

30 Kumaraswamy, G., Jayaprakash, N., Rambabu, D., Gangulyb, A., and Banerjeeb, R. (2014) *Org. Biomol. Chem.*, 12, 1793–1803.
31 Sheikh, S.E., Kausch, N., Lex, J., Neudörfl, J.-M., and Schmalz, H.-G. (2006) *Synlett*, 10, 1527–1530.
32 Fournier, J.-F., Mathieu, S., and Charette, A.B. (2005) *J. Am. Chem. Soc.*, 127, 13140–13141.
33 Sernissi, L., Petrović, M., Scarpi, D., Guarna, A., Trabocchi, A., Bianchini, F., and Occhiato, E.G. (2014) *Chem. Eur. J.*, 20, 11187–11203.
34 Aggarwal, V.K., Fang, G.Y., and Meek, G. (2003) *Org. Lett.*, 5, 4417–4420.
35 (a) Barrett, A.G.M., Kasdorf, K., and Williams, D.J. (1994) *J. Chem. Soc., Chem. Commun.*, 1781–1782; (b) Barrett, A.G.M., Doubleday, W.W., Kasdorf, K., and Tustin, G.J. (1996) *J. Org. Chem.*, 61, 3280–3288; (c) Onoda, T., Shirai, R., Koiso, Y., and Iwasaki, S. (1996) *Tetrahedron Lett.*, 37, 4397–4400; (d) Onoda, T., Shirai, R., Kawai, N., and Iwasaki, S. (1996) *Tetrahedron*, 52, 13327–13338.
36 Davoren, J.E. and Martin, S.F. (2007) *J. Am. Chem. Soc.*, 129, 510–511.
37 Mohapatra, D.K. and Yellol, G.S. (2005) *Arkivoc*, 2005 (3), 144–155.
38 Navuluri, C. and Charette, A.B. (2015) *Org. Lett.*, 17, 4288–4291.
39 (a) Ezquerra, J., He, W., and Paquette, L.A. (1990) *Tetrahedron Lett.*, 31, 6979–6982; (b) Paquette, L.A., Wang, T.-Z., and Pinard, E. (1995) *J. Am. Chem. Soc.*, 117, 1455–1456; (c) Wang, T.-Z., Pinard, E., and Paquette, L.A. (1996) *J. Am. Chem. Soc.*, 118, 1309–1318.
40 Corey, E.J. and Lee, J. (1993) *J. Am. Chem. Soc.*, 115, 8873–8874.
41 Panek, J.S., Garbaccio, R.M., and Jain, N.F. (1994) *Tetrahedron Lett.*, 35, 6453–6456.
42 Abad, A., Agullo, C., Cunat, A.C., de Alfonso Marzal, I., Navarro, I., and Gris, A. (2006) *Tetrahedron*, 62, 3266–3283.
43 Nakashima, K., Kawano, H., Kumano, M., Kodama, H., Kameoka, M., Yamamoto, A., Mizutani, R., Sono, M., and Tori, M. (2015) *Tetrahedron Lett.*, 56, 4912–4915.
44 Fu, X., Fan, J., Zou, A., Yu, J., He, G., and Zhu, H. (2014) *Chin. J. Org. Chem.*, 34, 1616–1622.
45 (a) Charette, A.B., Côté, B., and Marcoux, J.-F. (1991) *J. Am. Chem. Soc.*, 113, 8166–8167; (b) Charette, A.B. and Côté, B. (1995) *J. Am. Chem. Soc.*, 117, 12721–12732.
46 Charette, A.B., Turcotte, N., and Marcoux, J.-F. (1994) *Tetrahedron Lett.*, 35, 513–516.
47 Charette, A.B. and Marcoux, J.-F. (1993) *Tetrahedron Lett.*, 34, 7157–7160.
48 Vega-Pérez, J.M., Perinàn, I., Palo-Nieto, C., Vega-Holm, M., and Iglesias-Guerra, F. (2010) *Tetrahedron: Asymmetry*, 21, 81–95.
49 (a) Arai, I., Mori, A., and Yamamoto, H. (1985) *J. Am. Chem. Soc.*, 107, 8254–8256; (b) Mori, A., Arai, I., Yamamoto, H., Nakai, H., and Arai, Y. (1986) *Tetrahedron*, 42, 6447–6458.
50 Ebens, R. and Kellogg, R.M. (1990) *Recl. Trav. Chim. Pays-Bas*, 109, 552–560.
51 Kang, J., Lim, G.J., Yoon, S.K., and Kim, M.Y. (1995) *J. Org. Chem.*, 60, 564–577.
52 (a) Vega-Pérez, J.M., Perinàn, I., Vega, M., and Iglesias-Guerra, F. (2008) *Tetrahedron: Asymmetry*, 19, 1720–1729; (b) Vega-Pèrez, J.M., Perinàn, I., and Iglesias-Guerra, F. (2009) *Tetrahedron: Asymmetry*, 20, 1065–1072.

53 (a) Mash, E.A. and Nelson, K.A. (1985) *J. Am. Chem. Soc.*, 107, 8256–8258; (b) Mash, E.A. and Nelson, K.A. (1987) *Tetrahedron*, 43, 679–692; (c) Yeh, S.-M., Huang, L.-H., and Luh, T.-Y. (1996) *J. Org. Chem.*, 61, 3906–3908.

54 (a) Ambler, P.W. and Davies, S.G. (1988) *Tetrahedron Lett.*, 29, 6979–6982; (b) Ambler, P.W. and Davies, S.G. (1988) *Tetrahedron Lett.*, 29, 6983–6984.

55 Tanaka, K., Uno, H., Osuga, H., and Suzuki, H. (1994) *Tetrahedron: Asymmetry*, 5, 1175–1178.

56 (a) Sugimura, T., Futagawa, T., Yoshikawa, M., and Tai, A. (1989) *Tetrahedron Lett.*, 30, 3807–3810; (b) Sugimura, T., Futagawa, T., and Tai, A. (1988) *Tetrahedron Lett.*, 29, 5775–5778; (c) Sugimura, T., Yoshikawa, M., Futagawa, T., and Tai, A. (1990) *Tetrahedron*, 46, 5955–5966; (d) Sugimura, T., Futagawa, T., and Tai, A. (1990) *Chem. Lett.*, 19, 2295–2298.

57 Imai, T., Mineta, H., and Nishida, S. (1990) *J. Org. Chem.*, 55, 4986–4988.

58 (a) Tamura, O., Hashimoto, M., Kobayashi, Y., Katoh, T., Nakatani, K., Kamada, M., Hayakawa, I., Akiba, T., and Terashima, S. (1992) *Tetrahedron Lett.*, 33, 3487–3490; (b) Akiba, T., Tamura, O., Hashimoto, M., Kobayashi, Y., Katoh, T., Nakatani, K., Kamada, M., Hayakawa, I., and Terashima, S. (1994) *Tetrahedron*, 50, 3905–3914.

59 (a) Seebach, D. and Stucky, G. (1988) *Angew. Chem. Int. Ed. Engl.*, 27, 1351–1353; (b) Seebach, D., Stucky, G., and Pfammatter, E. (1989) *Chem. Ber.*, 122, 2377–2389.

60 (a) Schöllkopf, U., Tiller, T., and Bardenhagen, J. (1988) *Tetrahedron*, 44, 5293–5305; (b) Groth, U., Schöllkopf, U., and Tiller, T. (1991) *Liebigs Ann. Chem.*, 857–860.

61 Cheeseman, M., Feuillet, F.J.P., Johnson, A.L., and Bull, S.D. (2005) *Chem. Commun.*, 2372–2374.

62 Ko, K.-Y. and Yun, H.-S. (2012) *Bull. Korean Chem. Soc.*, 33, 2415–2418.

63 (a) Tamura, O., Hashimoto, M., Kobayashi, Y., Katoh, T., Nakatani, K., Kamada, M., Hayakawa, I., Akiba, T., and Terashima, S. (1994) *Tetrahedron*, 50, 3889–3904; (b) Tamura, O., Hashimoto, M., Kobayashi, Y., Katoh, T., Nakatani, K., Kamada, M., Hayakawa, I., Akiba, T., and Terashima, S. (1992) *Tetrahedron Lett.*, 33, 3483–3486.

64 Green, R., Cheeseman, M., Duffill, S., Merritt, A., and Bull, S.D. (2005) *Tetrahedron Lett.*, 46, 7931–7934.

65 Cheeseman, M., Davies, I.R., Axe, P., Johnson, A.L., and Bull, S.D. (2009) *Org. Biomol. Chem.*, 7, 3537–3548.

66 Son, J.B., Hwang, M.-H., Lee, W., and Lee, D.-H. (2007) *Org. Lett.*, 9, 3897–3900.

67 Lu, T., Hayashi, R., Hsung, R.P., DeKorver, K.A., Lohse, A.G., Song, Z., and Tang, Y. (2009) *Org. Biomol. Chem.*, 7, 3331–3337.

68 Song, Z., Lu, T., Hsung, R.P., Al-Rashid, Z.F., Ko, C., and Tang, Y. (2007) *Angew. Chem. Int. Ed.*, 46, 4069–4072.

69 (a) Sawada, S., Takehana, K., and Inouye, Y. (1968) *J. Org. Chem.*, 33, 1767–1770; (b) Sawada, S., Oda, J., and Inouye, Y. (1968) *J. Org. Chem.*, 33, 2141–2143; (c) Furukawa, J., Kawabata, N., and Nishimura, J. (1968) *Tetrahedron Lett.*, 9, 3495–3498.

70 Denmark, S.E. and Edwards, J.P. (1992) *Synlett*, 229–230.

71 (a) Ukaji, Y., Nishimura, M., and Fujisawa, T. (1992) *Chem. Lett.*, 21, 61–64; (b) Ukaji, Y., Sada, K., and Inomata, K. (1993) *Chem. Lett.*, 22, 1227–1230.

72 (a) Charette, A.B. and Juteau, H. (1994) *J. Am. Chem. Soc.*, 116, 2651–2652; (b) Charette, A.B., Prescott, S., and Brochu, C. (1995) *J. Org. Chem.*, 60, 1081–1083; (c) Charette, A.B., Juteau, H., Lebel, H., and Molinaro, C. (1998) *J. Am. Chem. Soc.*, 120, 11943–11952; (d) Anthes, R., Benoit, S., Chen, C.-K., Corbett, E.A., Corbett, R.M., DelMonte, A.J., Gingras, S., Livingston, R.C., Pendri, Y., Sausker, J., and Soumeillan, M. (2008) *Org. Process Res. Dev.*, 12, 178–182; (e) Goudreau, S.R. and Charette, A.B. (2009) *J. Am. Chem. Soc.*, 131, 15633–15635.

73 Wang, T., Liang, Y., and Yu, Z.-X. (2011) *J. Am. Chem. Soc.*, 133, 9343–9353.

74 (a) Charette, A.B., Juteau, H., Lebel, H., and Deschênes, D. (1996) *Tetrahedron Lett.*, 37, 7925–7928; (b) Charette, A.B. and Juteau, H. (1997) *Tetrahedron*, 53, 16277–16286.

75 (a) Falck, J.R., Mekonnen, B., Yu, J., and Lai, J.-Y. (1996) *J. Am. Chem. Soc.*, 118, 6096–6097; (b) Itoh, T., Inoue, H., and Emoto, S. (2000) *Bull. Chem. Soc. Jpn.*, 73, 409–416; (c) Hoffmann, R.W. and Koberstein, R. (2000) *J. Chem. Soc., Perkin Trans. 2*, 595–602.

76 Charette, A.B. and Giroux, A. (1996) *J. Org. Chem.*, 61, 8718–8719.

77 Paterson, I., Davies, R.D.M., and Marquez, R. (2001) *Angew. Chem. Int. Ed.*, 40, 603–607.

78 (a) White, J.D., Kim, T.-S., and Nambu, M. (1995) *J. Am. Chem. Soc.*, 117, 5612–5613; (b) Nagle, D.G., Geralds, R.S., Yoo, H.-D., Gerwick, W.H., Kim, T.-S., Nambu, M., and White, J.D. (1995) *Tetrahedron Lett.*, 36, 1189–1192; (c) Wipf, P. and Xu, W. (1996) *J. Org. Chem.*, 61, 6556–6562; (d) Lai, J.-Y., Yu, J., Mekonnen, B., and Falck, J.R. (1996) *Tetrahedron Lett.*, 37, 7167–7170.

79 (a) Barrett, A.G.M., Doubleday, W.W., Kasdorf, K., Tustin, G.J., White, A.J.P., and Williams, D.J. (1995) *J. Chem. Soc., Chem. Commun.*, 407–408; (b) Barrett, A.G.M. and Kasdorf, K. (1996) *J. Chem. Soc., Chem. Commun.*, 325–327; (c) Barrett, A.G.M. and Kasdorf, K. (1996) *J. Am. Chem. Soc.*, 118, 11030–11037.

80 (a) Barrett, A.G.M., Hamprecht, D., White, A.J.P., and Williams, D.J. (1996) *J. Am. Chem. Soc.*, 118, 7863–7864; (b) Charette, A.B. and Lebel, H. (1996) *J. Am. Chem. Soc.*, 118, 10327–10328.

81 Nicolaou, K.C., Namoto, K., Ritzén, A., Ulven, T., Shoji, M., Li, J., D'Amico, G., Liotta, D., French, C.T., Wartmann, M., Altmann, K.-H., and Giannakakou, P. (2001) *J. Am. Chem. Soc.*, 123, 9313–9323.

82 Ghosh, A.K. and Liu, C. (2001) *Org. Lett.*, 3, 635–638.

83 Charette, A.B., Jolicoeur, E., and Bydlinski, G.A.S. (2001) *Org. Lett.*, 3, 3293–3295.

84 Charette, A.B. and Lemay, J. (1997) *Angew. Chem. Int. Ed. Engl.*, 36, 1090–1092.

85 Charette, A.B., Gagnon, A., and Fournier, J.-F. (2001) *J. Am. Chem. Soc.*, 124, 386–387.

86 Zimmer, L.E. and Charette, A.B. (2009) *J. Am. Chem. Soc.*, 131, 15624–15626.

87 Beaulieu, L.-P.B., Zimmer, L.E., and Charette, A.B. (2009) *Chem. Eur. J.*, 15, 11829–11832.

88 Beaulieu, L.-P.B., Zimmer, L.E., Gagnon, A., and Charette, A.B. (2012) *Chem. Eur. J.*, 18, 14784–14791.

89 Beaulieu, L.-P.B., Schneider, J.F., and Charette, A.B. (2013) *J. Am. Chem. Soc.*, 135, 7819–7822.

90 Asao, H., Sakauchi, H., Kuwahara, S., and Kiyota, H. (2007) *Tetrahedron: Asymmetry*, 18, 537–541.
91 Lacasse, M.-C., Poulard, C., and Charette, A.B. (2005) *J. Am. Chem. Soc.*, 127, 12440–12441.
92 Kitajima, H., Aoki, Y., Ito, K., and Katsuki, T. (1995) *Chem. Lett.*, 24, 1113–1114.
93 Voituriez, A. and Charette, A.B. (2006) *Adv. Synth. Catal.*, 348, 2363–2370.
94 Long, J., Yuan, Y., and Shi, Y. (2003) *J. Am. Chem. Soc.*, 125, 13632–13633.
95 Lorenz, J.C., Long, J., Yang, Z., Xue, S., Xie, Y., and Shi, Y. (2004) *J. Org. Chem.*, 69, 327–334.
96 Kim, H.Y., Lurain, A.E., Garcia-Garcia, P., Carroll, P.J., and Walsh, P.J. (2005) *J. Am. Chem. Soc.*, 127, 13138–13139.
97 Kim, H.Y., Salvi, L., Carroll, P.J., and Walsh, P.J. (2009) *J. Am. Chem. Soc.*, 131, 954–962.
98 Kim, H.Y. and Walsh, P.J. (2012) *J. Phys. Org. Chem.*, 25, 933–938.
99 Kim, H.Y., Salvi, L., Carroll, P.J., and Walsh, P.J. (2010) *J. Am. Chem. Soc.*, 132, 402–412.
100 Valenta, P., Carroll, P.J., and Walsh, P.J. (2010) *J. Am. Chem. Soc.*, 132, 14179–14190.
101 Infante, R., Nieto, J., and Andrés, C. (2014) *Org. Biomol. Chem.*, 12, 345–354.
102 Long, J., Du, H., Li, K., and Shi, Y. (2005) *Tetrahedron Lett.*, 46, 2737–2740.
103 Du, H., Long, J., and Shi, Y. (2006) *Org. Lett.*, 8, 2827–2829.
104 Long, J., Xu, L., Du, H., Li, K., and Shi, Y. (2009) *Org. Lett.*, 11, 5226–5229.
105 Zheng, Y. and Zhang, J. (2010) *Adv. Synth. Catal.*, 352, 1810–1817.
106 Miura, T., Murakami, Y., and Imai, N. (2006) *Tetrahedron: Asymmetry*, 17, 3067–3069.
107 Koyata, N., Miura, T., Akaiwa, Y., Sasaki, H., Sato, R., Nagai, T., Fujimori, H., Noguchi, T., Kirihara, M., and Imai, N. (2009) *Tetrahedron: Asymmetry*, 20, 2065–2071.
108 Ishizuka, Y., Fujimori, H., Noguchi, T., Kawasaki, M., Kishida, M., Nagai, T., Imai, N., and Kirihara, M. (2013) *Chem. Lett.*, 42, 1311–1313.
109 Shitama, H. and Katsuki, T. (2008) *Angew. Chem. Int. Ed.*, 47, 2450–2453.
110 Xiao-hua, S., Cun, L., Zhao, J., Cai-rong, Z., and Deng-gao, J. (2011) *J. Chem. Res.*, 35, 582–584.
111 Rachwalski, M., Kaczmarczyk, S., Leśniak, S., and Kiełbasiński, P. (2014) *ChemCatChem*, 6, 873–875.
112 Nozaki, H., Moriuti, S., Takaya, H., and Noyori, R. (1966) *Tetrahedron Lett.*, 7, 5239–5244.
113 (a) Hanessian, S. and Murray, P.J. (1987) *J. Org. Chem.*, 52, 1170–1172; (b) Shimamoto, K. and Ohfune, Y. (1989) *Tetrahedron Lett.*, 30, 3802–3804; (c) Shimamoto, K., Ishida, M., Shinozaki, H., and Ohfune, Y. (1991) *J. Org. Chem.*, 56, 4167–4176; (d) Hanafi, N. and Ortuño, R.M. (1994) *Tetrahedron: Asymmetry*, 5, 1657–1660; (e) Martìn-Vila, M., Hanafi, N., Jiménez, J.M., Alvarez-Larena, A., Piniella, J.F., Branchadell, V., Oliva, A., and Ortuño, R.M. (1998) *J. Org. Chem.*, 63, 3581–3589.
114 Özdemirhan, F.D., Celik, M., Atli, S., and Tanyeli, C. (2006) *Tetrahedron: Asymmetry*, 17, 287–291.

115 Williams, M.J., Deak, H.L., and Snapper, M.L. (2007) *J. Am. Chem. Soc.*, 129, 486–487.
116 Ben Hamadi, N. (2015) *J. Fluorine Chem.*, 170, 47–51.
117 Davletbakova, A.M., Maidanova, I.O., Baibulatova, N.Z., Dokichev, V.A., Tomilov, Y.V., Yunusov, M.S., and Nefedov, O.M. (2001) *Russ. J. Org. Chem.*, 37, 608–611.
118 Jiménez, J.M., Rifé, J., and Ortuño, R.M. (1996) *Tetrahedron: Asymmetry*, 7, 537–558.
119 (a) Carrié, R. (1980) *Heterocycles*, 14, 1529–1544; (b) Abadallah, H., Grée, R., and Carrié, R. (1982) *Tetrahedron Lett.*, 23, 503–506.
120 (a) Vallgarda, J. and Hacksell, U. (1991) *Tetrahedron Lett.*, 32, 5625–5628; (b) Vallgarda, J., Appelberg, U., Csoregh, I., and Hacksell, U. (1994) *J. Chem. Soc., Perkin Trans. 1*, 461–470.
121 Sun, L.-Q., Takaki, K., Chen, J., Bertenshaw, S., Iben, L., Mahla, C.D., Ryan, E., Wu, D., Gao, Q., and Xu, C. (2005) *Bioorg. Med. Chem. Lett.*, 15, 1345–1349.
122 (a) Pietruszka, J. and Widenmeyer, M. (1997) *Synlett*, 977–979; (b) Luithle, J.E.A. and Pietruszka, J. (1997) *Liebigs Ann.*, 1997, 2297–2302; (c) Luithle, J.E.A., Pietruszka, J., and Witt, A. (1998) *Chem. Commun.*, 2651–2652; (d) Luithle, J.E.A. and Pietruszka, J. (1999) *J. Org. Chem.*, 64, 8287–8297; (e) Pietruszka, J. and Witt, A. (2000) *J. Chem. Soc., Perkin Trans. 1*, 4293–4300; (f) Luithle, J.E.A. and Pietruszka, J. (2000) *J. Org. Chem.*, 65, 9194–9200; (g) Luithle, J.E.A. and Pietruszka, J. (2000) *Eur. J. Org. Chem.*, 2000, 2557–2562; (h) Pietruszka, J., Witt, A., and Frey, W. (2003) *Eur. J. Org. Chem.*, 2003, 3219–3229; (i) Pietruszka, J. and Witt, A. (2003) *Synlett*, 91–94; (j) Garcia Garcia, P., Hohn, E., and Pietruszka, J. (2003) *J. Organomet. Chem.*, 680, 281–285.
123 Ty, N., Pontikis, R., Chabot, G.G., Devillers, E., Quentin, L., Bourg, S., and Florent, J.-C. (2013) *Bioorg. Med. Chem.*, 21, 1357–1366.
124 (a) Sato, M., Hisamichi, H., Kaneko, C., Suzaki, N., Furuya, T., and Inukai, N. (1989) *Tetrahedron Lett.*, 30, 5281–5284; (b) Alami, A., Calmes, M., Daunis, J., Escale, F., Jacquier, R., Roumestant, M.-L., and Viallefont, P. (1991) *Tetrahedron: Asymmetry*, 2, 175–178; (c) Alcaraz, C., Herrero, A., Marco, J.L., Fernàndez-Alvarez, E., and Bernabé, M. (1992) *Tetrahedron Lett.*, 33, 5605–5608; (d) Alcaraz, C., Fernàndez, M.D., de Frutos, M.P., Marco, J.L., Bernabé, M., Foces-Foces, C., and Cano, F.H. (1994) *Tetrahedron*, 50, 12443–12456; (e) Bartels, A., Jones, P.G., and Liebscher, J. (1998) *Synthesis*, 1645–1654.
125 Diéguez, M., Pamies, O., Ruiz, A., Diaz, Y., Castillon, S., and Claver, C. (2004) *Coord. Chem. Rev.*, 248, 2165–2192.
126 Ferreira, V.F., Leao, R.A.C., da Silva, F. de C., Pinheiro, S., Lhoste, P., and Sinou, D. (2007) *Tetrahedron: Asymmetry*, 18, 1217–1223.
127 Krysiak, J., Lyon, C., Baceiredo, A., Gornitzka, H., Mikolajczyk, M., and Bertrand, G. (2004) *Chem. Eur. J.*, 10, 1982–1986.
128 (a) Henry, K.J. Jr. and Fraser-Reid, B. (1995) *Tetrahedron Lett.*, 36, 8901–8904; (b) Hoberg, J.O. and Claffey, D.J. (1996) *Tetrahedron Lett.*, 37, 2533–2536; (c) Timmers, C.M., Leeuwenburgh, M.A., Verheijen, J.C., Vandermarel, G.A., and Vanboom, J.H. (1996) *Tetrahedron: Asymmetry*, 7, 49–52.

129 (a) Davies, H.M.L., Clark, T.J., and Church, L.A. (1989) *Tetrahedron Lett.*, 30, 5057–5060; (b) Davies, H.M.L. and Cantrell, W.R. (1991) *Tetrahedron Lett.*, 32, 6509–6512; (c) Davies, H.M.L., Huby, N.J.S., Cantrell, W.R., and Olive, J.L. (1993) *J. Am. Chem. Soc.*, 115, 9468–9479.

130 (a) Doyle, M.P., Dorow, R.L., Terpstra, J.W., and Rodenhouse, R.A. (1985) *J. Org. Chem.*, 50, 1663–1666; (b) Doyle, M.P., Protopopova, M.N., Brandes, B.D., Davies, H.M.L., Huby, N.J.S., and Whitesell, J.K. (1993) *Synlett*, 151–153; (c) Haddad, N. and Galili, N. (1997) *Tetrahedron: Asymmetry*, 8, 3367–3370; (d) Gross, Z., Galili, N., and Simkhovich, L. (1999) *Tetrahedron Lett.*, 40, 1571–1574; (e) Gross, Z., Simkhovich, L., and Galili, N. (1999) *Chem. Commun.*, 599–600.

131 Doyle, M.P. (2009) *Angew. Chem. Int. Ed.*, 48, 850–852.

132 Dzik, W.I., Xu, X., Zhang, X.P., Reek, J.N.H., and de Bruin, B. (2010) *J. Am. Chem. Soc.*, 132, 10891–10902.

133 Lu, H., Dzik, W.I., Xu, X., Wojtas, L., de Bruin, B., and Zhang, X.P. (2011) *J. Am. Chem. Soc.*, 133, 8518–8521.

134 Fantauzzi, S., Gallo, E., Rose, E., Raoul, N., Caselli, A., Issa, S., Ragaini, F., and Cenini, S. (2008) *Organometallics*, 27, 6143–6151.

135 Belof, J.L., Cioce, C.R., Xu, X., Zhang, X.P., Space, B., and Woodcock, H.L. (2011) *Organometallics*, 30, 2739–2746.

136 (a) Tatsuno, Y., Konishi, A., Nakamura, A., and Otsuka, S. (1974) *J. Chem. Soc., Chem. Commun.*, 588–589; (b) Nakamura, A., Konishi, A., Tatsuno, Y., and Otsuka, S. (1978) *J. Am. Chem. Soc.*, 100, 3443–3448; (c) Nakamura, A., Konishi, A., Tsujitani, R., Kudo, M., and Otsuka, S. (1978) *J. Am. Chem. Soc.*, 100, 3449–3461; (d) Jommi, G., Pagliarin, R., Rizzi, G., and Sisti, M. (1993) *Synlett*, 833–834.

137 Chen, Y., Fields, K.B., and Zhang, X.P. (2004) *J. Am. Chem. Soc.*, 126, 14718–14719.

138 Chen, Y. and Zhang, X.P. (2007) *J. Org. Chem.*, 72, 5931–5934.

139 Chen, Y., Ruppel, J.V., and Zhang, X.P. (2007) *J. Am. Chem. Soc.*, 129, 12074–12075.

140 Zhu, S., Perman, J.A., and Zhang, X.P. (2008) *Angew. Chem. Int. Ed.*, 47, 8460–8463.

141 Ruppel, J.V., Gauthier, T.J., Snyder, N.L., Perman, J.A., and Zhang, X.P. (2009) *Org. Lett.*, 11, 2273–2276.

142 Zhu, S., Ruppel, J.V., Lu, H., Wojtas, L., and Zhang, X.P. (2008) *J. Am. Chem. Soc.*, 130, 5042–5043.

143 Zhu, S., Xu, X., Perman, J.A., and Zhang, X.P. (2010) *J. Am. Chem. Soc.*, 132, 12796–12799.

144 Xu, X., Zhu, S., Cui, X.X., Wojtas, L., and Zhang, X.P. (2013) *Angew. Chem. Int. Ed.*, 52, 11857–11861.

145 Huang, L., Chen, Y., Gao, G.-Y., and Zhang, X.P. (2003) *J. Org. Chem.*, 68, 8179–8184.

146 Chen, Y., Gao, G.-Y., and Zhang, X.P. (2005) *Tetrahedron Lett.*, 46, 4965–4969.

147 Chen, Y. and Zhang, X.P. (2004) *J. Org. Chem.*, 69, 2431–2435.

148 Chen, Y. and Zhang, X.P. (2006) *Synthesis*, 1697–1700.

149 Charette, A.B. and Wurz, R. (2003) *J. Mol. Catal. A: Chem.*, 196, 83–91.

150 (a) Ikeno, T., Sato, M., and Yamada, T. (1999) *Chem. Lett.*, 28, 1345–1346; (b) Yamada, T., Ikeno, T., Sekino, H., and Sato, M. (1999) *Chem. Lett.*, 28, 719–720; (c) Ikeno, T., Nishizuka, A., Sato, M., and Yamada, T. (2001) *Synlett*, 406–408; (d) Ikeno, T., Sato, M., Sekino, H., Nishizuka, A., and Yamada, T. (2001) *Bull. Chem. Soc. Jpn.*, 74, 2139–2150; (e) Ikeno, T., Iwakura, I., and Yamada, T. (2001) *Bull. Chem. Soc. Jpn.*, 74, 2151–2160; (f) Ikeno, T., Iwakura, I., Yabushita, S., and Yamada, T. (2002) *Org. Lett.*, 4, 517–520; (g) Ikeno, T., Sato, M., Iwakura, I., Kokura, A., Nagata, T., and Yamada, T. (2008) *J. Synth. Org. Chem Jpn.*, 66, 110–123.

151 (a) Fukuda, T. and Katsuki, T. (1995) *Synlett*, 825–826; (b) Fukuda, T. and Katsuki, T. (1997) *Tetrahedron*, 53, 7201–7208.

152 Morandi, B., Mariampillai, B., and Carreira, E.M. (2011) *Angew. Chem. Int. Ed.*, 50, 1101–1104.

153 (a) Ito, Y.N. and Katsuki, T. (1999) *Bull. Chem. Soc. Jpn.*, 72, 603–619; (b) Niimi, T., Uchida, T., Irie, R., and Katsuki, T. (2000) *Tetrahedron Lett.*, 41, 3647–3651; (c) Niimi, T., Uchida, T., Irie, R., and Katsuki, T. (2001) *Adv. Synth. Catal.*, 343, 79–88.

154 Shitama, H. and Katsuki, T. (2007) *Chem. Eur. J.*, 13, 4849–4858.

155 Gao, J., Woolley, F.R., and Zingaro, R.A. (2005) *Org. Biomol. Chem.*, 3, 2126–2128.

156 White, J.D. and Shaw, S. (2014) *Org. Lett.*, 16, 3880–3883.

157 Koehn, S.K., Gronert, S., and Aldajaei, J.T. (2010) *Org. Lett.*, 12, 676–679.

158 Langlotz, B.K., Wadepohl, H., and Gade, L.H. (2008) *Angew. Chem. Int. Ed.*, 47, 4670–4674.

159 (a) McManus, H.A. and Guiry, P.J. (2004) *Chem. Rev.*, 104, 4151–4202; (b) Desimoni, G., Faita, G., and Jorgensen, K.A. (2006) *Chem. Rev.*, 106, 3561–3651.

160 Fraile, J.M., Garcìa, J.I., Martìnez-Merino, V., Mayoral, J.A., and Salvatella, L. (2001) *J. Am. Chem. Soc.*, 123, 7616–7625.

161 Mend, Q., Li, M., Tang, D., Shen, W., and Zhang, J. (2004) *THEOCHEM*, 711, 193–199.

162 Drudis-Solé, G., Maseras, F., Lledós, A., Vallribera, A., and Moreno-Mañas, M. (2008) *Eur. J. Org. Chem.*, 2008, 5614–5621.

163 Aguado-Ullate, S., Urbano-Cuadrado, M., Villalba, I., Pires, E., García, J.I., Bo, C., and Carbó, J.J. (2012) *Chem. Eur. J.*, 18, 14026–14036.

164 Leutenegger, U., Umbricht, G., Fahrni, C., von Matt, P., and Pfaltz, A. (1992) *Tetrahedron*, 48, 2143–2156.

165 Lowenthal, R.E., Abiko, A., and Masamune, S. (1990) *Tetrahedron Lett.*, 31, 6005–6008.

166 Müller, D., Umbricht, G., Weber, B., and Pfaltz, A. (1991) *Helv. Chim. Acta*, 74, 232–240.

167 (a) Evans, D.A., Woerpel, K.A., Hinman, M.M., and Faul, M.M. (1991) *J. Am. Chem. Soc.*, 113, 726–728; (b) Evans, D.A., Woerpel, K.A., and Scott, M.J. (1992) *Angew. Chem. Int. Ed.*, 31, 430–432.

168 Knight, J.G. and Belcher, P.E. (2005) *Tetrahedron Lett.*, 16, 1415–1418.

169 Davies, D.L., Kandola, S.K., and Patel, R.K. (2004) *Tetrahedron: Asymmetry*, 15, 77–80.

170 Lowenthal, R.E. and Masamune, S. (1991) *Tetrahedron Lett.*, 32, 7373–7376.

171 (a) Bedekar, A.V. and Andersson, P.G. (1996) *Tetrahedron Lett.*, 37, 4073–4076; (b) Harm, A.M., Knight, J.G., and Stemp, G. (1996) *Tetrahedron Lett.*, 37, 6189–6192.
172 Uozumi, Y., Kyota, H., Kishi, E., Kitayama, K., and Hayashi, T. (1996) *Tetrahedron: Asymmetry*, 7, 1603–1606.
173 (a) Imai, Y., Zhang, W.B., Kida, T., Nakatsuji, Y., and Ikeda, I. (1997) *Tetrahedron Lett.*, 38, 2681–2684; (b) Imai, Y., Zhang, W.B., Kida, T., Nakatsuji, Y., and Ikeda, I. (2000) *J. Org. Chem.*, 65, 3326–3333.
174 Kim, S.-G., Cho, C.-W., and Ahn, K.H. (1997) *Tetrahedron: Asymmetry*, 8, 1023–1026.
175 Glos, M. and Reiser, O. (2000) *Org. Lett.*, 2, 2045–2048.
176 Bayardon, J., Holczknecht, O., Pozzi, G., and Sino, D. (2006) *Tetrahedron: Asymmetry*, 17, 1568–1572.
177 Itagaki, M. and Yamamoto, Y. (2006) *Tetrahedron Lett.*, 47, 523–525.
178 Portada, T., Roje, M., Hamersak, Z., and Zinic, M. (2005) *Tetrahedron Lett.*, 46, 5957–5959.
179 Irmak, M., Groschner, A., and Boysen, M.M.K. (2007) *Chem. Commun.*, 177–179.
180 Minuth, T. and Boysen, M.M.K. (2010) *Synthesis*, 2799–2803.
181 Portada, T., Roje, M., Raza, Z., Caplar, V., Zinic, M., and Sunjic, V. (2007) *Eur. J. Org. Chem.*, 2007, 838–856.
182 (a) da Palma Carreiro, E., Chercheja, S., Burke, A.J., Prates Ramalho, J.P., and Rodrigues, A.I. (2005) *J. Mol. Catal. A: Chem.*, 236, 38–45; (b) da Plama Carreiro, E., Chercheja, S., Moura, N.M.M., Gertrudes, C.S.C., and Burke, A.J. (2006) *Inorg. Chem. Commun.*, 9, 823–826; (c) Burke, A.J., da Palma Carreiro, E., Chercheja, S., Moura, N.M.M., Prates Ramalho, J.P., Rodrigues, A.I., and dos Santos, C.I.M. (2007) *J. Organomet. Chem.*, 692, 4863–4874.
183 Mazet, C., Köhler, V., and Pfaltz, A. (2005) *Angew. Chem. Int. Ed.*, 44, 4888–4891.
184 Benaglia, M., Benincori, T., Mussini, P., Pilati, T., Rizzo, S., and Sannicolo, F. (2005) *J. Org. Chem.*, 70, 7488–7495.
185 Annunziata, R., Benaglia, M., Cinquini, M., Cozzi, F., and Pozzi, G. (2003) *Eur. J. Org. Chem.*, 2003, 1191–1197.
186 Atodiresei, I., Schiffers, I., and Bolm, C. (2006) *Tetrahedron: Asymmetry*, 17, 620–633.
187 (a) Khanbabaee, K., Basceken, S., and Flörke, U. (2006) *Tetrahedron: Asymmetry*, 17, 2804–2812; (b) Khanbabaee, K., Basceken, S., and Flörke, U. (2007) *Eur. J. Org. Chem.*, 2007, 831–837.
188 Wang, Z.-Y., Du, D.-M., and Xu, J.-X. (2005) *Synth. Commun.*, 35, 299–313.
189 Du, D.-M., Fu, B., and Hua, W.-T. (2003) *Tetrahedron*, 59, 1933–1938.
190 Gao, M.Z., Wang, B., Kong, D., Zingaro, R.A., Clearfield, A., and Xu, Z.L. (2005) *Synth. Commun.*, 35, 2665–2673.
191 Liu, B., Zhu, S.-F., Wang, L.-X., and Zhou, Q.-L. (2006) *Tetrahedron: Asymmetry*, 17, 634–641.
192 Zhang, W., Xie, F., Matsuo, S., Imahori, Y., Kida, T., Nakatsuji, Y., and Ikeda, I. (2006) *Tetrahedron: Asymmetry*, 17, 767–777.

193 Kellehan, D., Kirby, F., Frain, D., Rodríguez-García, A.M., García, J.I., and O'Leary, P. (2013) *Tetrahedron: Asymmetry*, 24, 750–757.
194 (a) Schumacher, R., Dammast, F., and Reissig, H.-U. (1997) *Chem. Eur. J.*, 3, 614–619; (b) Ebinger, A., Heinz, T., Umbricht, G., and Pfaltz, A. (1998) *Tetrahedron*, 54, 10469–10480.
195 Böhm, C. and Reiser, O. (2001) *Org. Lett.*, 3, 1315–1318.
196 Østergaard, N., Jensen, J.F., and Tanner, D. (2001) *Tetrahedron*, 57, 6083–6088.
197 Haufe, G., Rosen, T.C., Meyer, O.G.J., Fröhlich, R., and Rissanen, K. (2002) *J. Fluorine Chem.*, 114, 189–198.
198 Fraile, J.M., Garcia, J.I., Herrerias, C.I., Mayoral, J.A., Reiser, O., and Vaultier, M. (2004) *Tetrahedron Lett.*, 45, 6765–6768.
199 (a) Fraile, J.M., Garcia, J.I., Herrerias, C.I., Mayoral, J.A., Gmough, S., and Vaultier, M. (2004) *Green Chem.*, 6, 93–98; (b) Castillo, M.R., Fousse, L., Fraile, J.M., Garcia, J.I., and Mayoral, J.A. (2007) *Chem. Eur. J.*, 13, 287–291.
200 Irmak, M., Lehnert, T., and Boysen, M.M.K. (2007) *Tetrahedron Lett.*, 48, 7890–7893.
201 Tan, Q., Wen, J., Li, D., Li, H., and You, T. (2005) *J. Mol. Catal. A: Chem.*, 242, 113–118.
202 Wasserscheid, P. (2000) in *Ionic Liquids in Synthesis* (eds P. Wasserscheid and T. Welton), Wiley-VCH Verlag GmbH, Weinheim, p. 213.
203 (a) Itagaki, M., Masumoto, K., and Yamamoto, Y. (2005) *J. Org. Chem.*, 70, 3292–3295; (b) Itagaki, M., Masumoto, K., Suenobu, K., and Yamamoto, Y. (2006) *Org. Process Res. Dev.*, 10, 245–250.
204 Schinnerl, M., Böhm, C., Seitz, M., and Reiser, O. (2003) *Tetrahedron: Asymmetry*, 14, 765–771.
205 Allais, F., Angelaud, R., Camuzat-Dedenis, B., Julienne, K., and Landais, Y. (2003) *Eur. J. Org. Chem.*, 2003, 1069–1073.
206 Ozoduru, G., Schubach, T., and Boysen, M.M.K. (2012) *Org. Lett.*, 14, 4990–4993.
207 Honma, M., Takeda, H., Takano, M., and Nakada, M. (2009) *Synlett*, 1695–1712.
208 (a) Chhor, R.B., Nosse, B., Sörgel, S., Böhm, C., Seitz, M., and Reiser, O. (2003) *Chem. Eur. J.*, 9, 260–270; (b) Nosse, B., Chhor, R.B., Jeong, W.B., Böhm, C., and Reiser, O. (2003) *Org. Lett.*, 5, 941–944; (c) Kalidindi, S., Jeong, W.B., Schall, A., Bandichhor, R., Nosse, B., and Reiser, O. (2007) *Angew. Chem. Int. Ed.*, 46, 6361–6363; (d) Jezek, E., Schall, A., Kreitmeier, P., and Reiser, O. (2005) *Synlett*, 6, 915–918.
209 Chen, S., Rong, C., Feng, P., Li, S., and Shi, Y. (2012) *Org. Biomol. Chem.*, 10, 5518–5520.
210 Suárez del Villar, I., Gradillas, A., and Pérez-Castells, J. (2010) *Eur. J. Org. Chem.*, 2010, 5850–5862.
211 García, J.I., Gracia, J., Herrerías, C.I., Mayoral, J.A., Miñana, A.C., and Sáenz, C. (2014) *Eur. J. Org. Chem.*, 2014, 1531–1540.
212 This inversion of diastereoselectivity has been previously observed substituting the two methyl groups in the catalyst Cu-**box5** with two benzyl groups.

Burguete, M.I., Fraile, J.M., García, J.I., García-Verdugo, E., Herrerías, C.I., Luis, S.V., and Mayora, J.A. (2001) *J. Org. Chem.*, 66, 8893–8901.

213 Charette, A.B. and Bouchard, J. (2005) *Can. J. Chem.*, 83, 533–542.

214 Charette, A.B., Janes, M.K., and Lebel, H. (2003) *Tetrahedron: Asymmetry*, 14, 867–872.

215 France, M.B., Milojevich, A.K., Stitt, T.A., and Kim, A.J. (2003) *Tetrahedron Lett.*, 44, 9287–9290.

216 Ye, T. and Zhou, C. (2005) *New J. Chem.*, 29, 1159–1163.

217 Xu, Z.-H., Zhu, S.-N., Sun, X.-L., Tang, Y., and Dai, L.-X. (2007) *Chem. Commun.*, 1960–1962.

218 (a) Wurz, R.P. and Charette, A.B. (2003) *Org. Lett.*, 5, 2327–2329; (b) Moreau, B. and Charette, A.B. (2005) *J. Am. Chem. Soc.*, 127, 18014–18015; (c) Moreau, B., Alberico, D., Lindsay, V.N.G., and Charette, A.B. (2012) *Tetrahedron*, 68, 3487–3496; (d) Deng, C., Wang, L.-J., Zhu, J., and Tang, Y. (2012) *Angew. Chem. Int. Ed.*, 51, 11620–11623.

219 (a) Fernàndez, M.J., Fraile, J.M., Garcìa, J.I., Mayoral, J.A., Burguete, M.I., Garcìa-Verdugo, E., Luis, S.V., and Harmer, M.A. (2000) *Top. Catal.*, 13, 303–309; (b) Burguete, M.I., Fraile, J.M., Garcìa, J.I., Garcìa-Verdugo, E., Luis, S.V., and Mayoral, J.A. (2000) *Org. Lett.*, 2, 3905–3908; (c) Fraile, J.M., Garcìa, J.I., Herrerìas, C.I., Mayoral, J.A., Carrié, D., and Vaultier, M. (2001) *Tetrahedron: Asymmetry*, 12, 1891–1894; (d) Clarke, R.J. and Shannon, I.J. (2001) *Chem. Commun.*, 1936–1937; (e) Annunziata, R., Benaglia, M., Cinquini, M., Cozzi, F., and Pitillo, M. (2001) *J. Org. Chem.*, 66, 3160–3166.

220 (a) Mandoli, A., Orlandi, S., Pini, D., and Salvadori, P. (2003) *Chem. Commun.*, 2466–2467; (b) Mandoli, A., Garzelli, R., Orlandi, S., Pini, D., Lessi, M., and Salvadori, P. (2009) *Catal. Today*, 140, 51–57.

221 Cornejo, A., Fraile, J.M., Garcia, J.I., Gil, M.J., Herrerias, C.I., Legarreta, G., Martinez-Merino, V., and Mayoral, J.A. (2003) *J. Mol. Catal. A: Chem.*, 196, 101–108.

222 Diez-Barra, E., Fraile, J.M., Garcia, J.I., Garcia-Verdugo, E., Herrerias, C.I., Luis, S.V., Mayoral, J.A., Sanchez-Verdu, P., and Tolo, J. (2003) *Tetrahedron: Asymmetry*, 14, 773–778.

223 Lancaster, T.M., Lee, S.S., and Ying, J.Y. (2005) *Chem. Commun.*, 3577–3579.

224 (a) Lee, S.S., Hadinoto, S., and Ying, J.Y. (2006) *Adv. Synth. Catal.*, 348, 1248–1254; (b) Lee, S.S. and Ying, J.Y. (2006) *J. Mol. Catal. A: Chem.*, 256, 219–224.

225 Werner, H., Herrerias, C.I., Glos, M., Gissibl, A., Fraile, J.M., Pérez, I., Mayoral, J.A., and Reiser, O. (2006) *Adv. Synth. Catal.*, 348, 125–132.

226 Bergbreiter, D.E. and Tian, J. (2007) *Tetrahedron Lett.*, 48, 4499–4503.

227 Patti, A. and Pedotti, S. (2005) *Chirality*, 17, 233–236.

228 Carreiro, E.P., Moura, N.M.M., and Burke, A.J. (2012) *Eur. J. Org. Chem.*, 2012, 518–528.

229 (a) Silva, A.R., Albuquerque, H., Borges, S., Siegel, R., Mafra, L., Carvalho, A.P., and Pires, J. (2012) *Microporous Mesoporous Mater.*, 158, 26–38; (b) Fakhfakh, F., Baraket, L., Ghorbel, A., Fraile, J.M., and Mayoral, J.A. (2015) *React. Kinet. Mech. Cat.*, 116, 119–130.

230 van Oers, M.C.M., Abdelmohsen, L.K.E.A., Rutjes, F.P.J.T., and van Hest, J.C.M. (2014) *Chem. Commun.*, 50, 4040–4043.

231 (a) Arai, T., Mizukami, T., Yokoyama, N., Nakazato, D., and Yanagisawa, A. (2005) *Synlett*, 2670–2672; (b) Ramalingam, B., Neuburger, M., and Pfaltz, A. (2007) *Synthesis*, 572–582.
232 Ma, J.-A., Wan, J.-H., Zhou, Y.-B., Wang, L.-X., Zhang, W., and Zhou, Q.-L. (2003) *J. Mol. Catal. A: Chem.*, 196, 109–115.
233 Teng, P.-F., Tsang, C.-S., Yeung, H.-L., Wong, W.-L., Wong, W.-T., and Kwong, H.-L. (2006) *J. Organomet. Chem.*, 691, 5664–5672.
234 Chelucci, G., Muroni, D., Saba, A., and Soccolini, F. (2003) *J. Mol. Catal. A: Chem.*, 197, 27–35.
235 Chelucci, G., Muroni, D., Pinna, G.A., Saba, A., and Vignola, D. (2003) *J. Mol. Catal. A: Chem.*, 191, 1–8.
236 (a) Lo, M.M.-C. and Fu, G.C. (1998) *J. Am. Chem. Soc.*, 120, 10270–10271; (b) Rios, R., Liang, J., Lo, M.M.C., and Fu, G.C. (2000) *Chem. Commun.*, 377–378.
237 (a) Ito, K. and Katsuki, T. (1993) *Tetrahedron Lett.*, 34, 2661–2664; (b) Ito, K. and Katsuki, T. (1993) *Synlett*, 638–640.
238 Kwong, H.-L., Lee, W.-S., Ng, H.-F., Chiu, W.-H., and Wong, W.-T. (1998) *J. Chem. Soc., Dalton Trans.*, 1043–1046.
239 Carreiro, E.P., Ramalho, J.P.P., and Burke, A.J. (2011) *Tetrahedron*, 67, 4640–4648.
240 Malkov, A.V., Pernazza, D., Bell, M., Bella, M., Massa, A., Teply, F., Meghani, P., and Kocovsky, P. (2003) *J. Org. Chem.*, 68, 4727–4742.
241 (a) Lyle, M.P.A. and Wilson, P.D. (2004) *Org. Lett.*, 6, 855–857; (b) Lyle, M.P.A., Draper, N.D., and Wilson, P.D. (2006) *Org. Biomol. Chem.*, 4, 877–885.
242 Bai, X.-L., Kang, C.-Q., Liu, X.-D., and Gao, L.-X. (2005) *Tetrahedron: Asymmetry*, 16, 727–731.
243 Bouet, A., Heller, B., Papamicaël, C., Dupas, G., Oudeyer, S., Marsais, F., and Levacher, V. (2007) *Org. Biomol. Chem.*, 5, 1397–1404.
244 Puglisi, A., Benaglia, M., Annunziata, R., and Bologna, A. (2003) *Tetrahedron Lett.*, 44, 2947–2951.
245 Kanemasa, S., Hamura, S., Harada, E., and Yamamoto, H. (1994) *Tetrahedron Lett.*, 35, 7985–7988.
246 Tanner, D., Johansson, F., Harden, A., and Andersson, P.G. (1998) *Tetrahedron*, 54, 15731–15738.
247 Hechavarria Fonseca, M., Eibler, E., Zabel, M., and König, B. (2003) *Inorg. Chim. Acta*, 352, 136–142.
248 Suga, H., Kakehi, A., Ito, S., Ibata, T., Fudo, T., Watanabe, Y., and Kinoshita, Y. (2003) *Bull. Chem. Soc. Jpn.*, 76, 189–199.
249 Lesma, G., Cattenati, C., Pilati, T., Sacchetti, A., and Silvani, A. (2007) *Tetrahedron: Asymmetry*, 18, 659–663.
250 (a) Bayardon, J., Sinou, D., Holczknecht, O., Mercs, L., and Pozzi, G. (2005) *Tetrahedron: Asymmetry*, 16, 2319–2327; (b) Shepperson, I., Quici, S., Pozzi, G., Nicoletti, M., and O'Hagan, D. (2004) *Eur. J. Org. Chem.*, 2004, 4545–4551.
251 Mloston, G., Mucha, P., and Heimgartner, H. (2012) *Lett. Org. Chem.*, 9, 89–96.
252 Biegasiewicz, K.F., Ingalsbe, M.L., Denis, J.D.S., Gleason, J.L., Ho, J., Coote, M.L., Savage, G.P., and Priefer, R. (2012) *Beilstein J. Org. Chem.*, 8, 1814–1818.

253 Keliher, E.J., Burrell, R.C., Chobanian, H.R., Conkrite, K.L., Shukla, R., and Baldwin, J.E. (2006) *Org. Biomol. Chem.*, 4, 2777–2784.
254 Brunel, J.M., Legrand, O., Reymond, S., and Buono, G. (1999) *J. Am. Chem. Soc.*, 121, 5807–5808.
255 (a) Caselli, A., Cesana, F., Gallo, E., Casati, N., Macchi, P., Sisti, M., Celentano, G., and Cenini, S. (2008) *Dalton Trans.*, 4202–4206; (b) Castano, B., Guidone, S., Gallo, E., Ragaini, F., Casati, N., Macchi, P., Sisti, M., and Caselli, A. (2013) *Dalton Trans.*, 42, 2451–2462.
256 Mugishima, N., Kanomata, N., Akutsu, N., and Kubota, H. (2015) *Tetrahedron Lett.*, 56, 1898–1903.
257 (a) Li, Z., Zheng, Z., and Chen, H. (2000) *Tetrahedron: Asymmetry*, 11, 1157–1163; (b) Li, Z., Liu, G.S., Zheng, Z., and Chen, H. (2000) *Tetrahedron*, 56, 7187–7191; (c) Li, Z., Zheng, Z., Wan, B., and Chen, H. (2001) *J. Mol. Catal. A: Chem.*, 165, 67–71.
258 Masterson, D.S., Shirley, C., and Glatzhofer, D.T. (2012) *J. Mol. Catal. A: Chem.*, 361–362, 111–115.
259 (a) Aratani, T., Yoneyoshi, Y., and Nagase, T. (1977) *Tetrahedron Lett.*, 18, 2599–2602; (b) Aratani, T. (1985) *Pure Appl. Chem.*, 57, 1839–1844.
260 (a) Itagaki, M. and Suenobu, K. (2007) *Org. Process Res. Dev.*, 11, 509–518; (b) Itagaki, M., Hagiya, K., Kamitamari, M., Masumoto, K., Suenobu, K., and Yamamoto, Y. (2004) *Tetrahedron.*, 60, 7835–7843.
261 Suenobu, K., Itagaki, M., and Nakamura, E. (2004) *J. Am. Chem. Soc.*, 126, 7271–7280.
262 Gao, J. and Zhong, S.H. (2003) *J. Mol. Catal. A: Chem.*, 191, 23–27.
263 Lee, W.-S., Leung, H.-K., Cheng, L.-S., Ng, L.-Y., Lee, C.-S., Huang, K.-H., Wong, W.-T., and Kwong, H.-L. (2004) *Inorg. Chim. Acta*, 357, 4389–4395.
264 Tepfenhart, D., Moisan, L., Dalko, P.I., and Cossy, J. (2004) *Tetrahedron Lett.*, 45, 1781–1783.
265 Carreiro, E.P., Burke, A.J., Ramalho, J.P.P., and Rodrigues, A.I. (2009) *Tetrahedron: Asymmetry*, 20, 1272–1278.
266 Llewellyn, D.B. and Arndtsen, B.A. (2005) *Tetrahedron: Asymmetry*, 16, 1789–1799.
267 Hasegawa, T., Furusho, Y., Katagiri, H., and Yashima, E. (2007) *Angew. Chem. Int. Ed.*, 46, 5885–5888.
268 Adly, F.G. and Ghanem, A. (2014) *Chirality*, 26, 692–711.
269 Merlic, C.A. and Zechman, A.L. (2003) *Synthesis*, 1137–1156.
270 Davies, H.M.L., Bruzinski, P.R., Lake, D.H., Kong, N., and Fall, M.J. (1996) *J. Am. Chem. Soc.*, 118, 6897–6907.
271 Teng, P.-F., Lai, T.-S., Kwong, H.-L., and Che, C.-M. (2003) *Tetrahedron: Asymmetry*, 14, 837–844.
272 DeAngelis, A., Dmitrenko, O., Yap, G.P.A., and Fox, J.M. (2009) *J. Am. Chem. Soc.*, 131, 7230–7231.
273 Watanabe, N., Ohtake, Y., Hashimoto, S.-i., Shiro, M., and Ikegami, S. (1995) *Tetrahedron Lett.*, 36, 1491–1494.
274 Pelphrey, P., Hansen, J., and Davies, H.M.L. (2010) *Chem. Sci.*, 1, 254–257.
275 Chepiga, K.M., Qin, C., Alford, J.S., Chennamadhavuni, S., Gregg, T.M., Olson, J.P., and Davies, H.M.L. (2013) *Tetrahedron*, 69, 5765–5771.

276 Ventura, D.L., Li, Z., Coleman, M.G., and Davies, H.M.L. (2009) *Tetrahedron*, 65, 3052–3061.
277 Nadeau, E., Ventura, D.L., Brekan, J.A., and Davies, H.M.L. (2010) *J. Org. Chem.*, 75, 1927–1939.
278 (a) Doyle, M.P., Zhou, Q.L., Simonsen, S.H., and Lynch, V. (1996) *Synlett*, 697–698; (b) Ishitani, H. and Achiwa, K. (1997) *Synlett*, 781–782; (c) Doyle, M.P., Davies, S.B., and Hu, W. (2000) *Org. Lett.*, 2, 1145–1147; (d) Doyle, M.P., Davies, S.B., and Hu, W. (2000) *Chem. Commun.*, 867–868; (e) Barberis, M., Lahuerta, P., Pérez-Prieto, J., and Sanau, M. (2001) *Chem. Commun.*, 439–440.
279 Hu, W., Timmons, D.J., and Doyle, M.P. (2002) *Org. Lett.*, 4, 901–904.
280 Bykowski, D., Wu, K.-H., and Doyle, M.P. (2006) *J. Am. Chem. Soc.*, 128, 16038–16039.
281 Doyle, M.P. and Hu, W. (2003) *Arkivoc*, 2003 (7), 15–22.
282 Lindsay, V.N.G., Fiset, D., Gritsch, P.J., Azzi, S., and Charette, A.B. (2013) *J. Am. Chem. Soc.*, 135, 1463–1470.
283 Krumper, J.R., Gerisch, M., Suh, J.M., Bergman, R.G., and Tilley, T.D. (2003) *J. Org. Chem.*, 68, 9705–9710.
284 (a) Kennedy, M., McKervey, M.A., Maguire, A.R., and Roos, G.H.P. (1990) *J. Chem. Soc., Chem. Commun.*, 361–362; (b) McKervey, M.A. and Ye, T. (1992) *J. Chem. Soc., Chem. Commun.*, 823–824; (c) Roos, G.H.P. and McKervey, M.A. (1992) *Synth. Commun.*, 22, 1751–1756; (d) Davies, H.M.L. and Hutcheson, D.K. (1993) *Tetrahedron Lett.*, 34, 7243–7246; (e) Yoshikawa, K. and Achiwa, K. (1995) *Chem. Pharm. Bull.*, 43, 2048–2053.
285 (a) Moye-Sherman, D., Welch, M.B., Reibenspies, J., and Burgess, K. (1998) *Chem. Commun.*, 2377–2378; (b) Moye-Sherman, D., Jin, S., Ham, I., Lim, D., Scholtz, J.M., and Burgess, K. (1998) *J. Am. Chem. Soc.*, 120, 9435–9443.
286 Davies, H.M.L. and Boebel, T.A. (2000) *Tetrahedron Lett.*, 41, 8189–8192.
287 (a) Doyle, M.P., Zhou, Q.L., Charnsangavej, C., Longoria, M.A., McKervey, M.A., and Garcìa, C.F. (1996) *Tetrahedron Lett.*, 37, 4129–4132; (b) Davies, H.M.L., Bruzinski, P.R., and Fall, M.J. (1996) *Tetrahedron Lett.*, 37, 4133–4136; (c) Davies, H.M.L. and Rusiniak, L. (1998) *Tetrahedron Lett.*, 39, 8811–8812; (d) Davies, H.M.L., Nagashima, T., and Klino, J.L. III (2000) *Org. Lett.*, 2, 823–826.
288 Davies, H.M.L. and Townsend, R.J. (2001) *J. Org. Chem.*, 66, 6595–6603.
289 Müller, P., Grass, S., Shahi, S.P., and Bernardinelli, G. (2004) *Tetrahedron*, 60, 4755–4763.
290 Hedley, S.J., Ventura, D.L., Dominiak, P.M., Nygren, C.L., and Davies, H.M.L. (2006) *J. Org. Chem.*, 71, 5349–5356.
291 Davies, H.M.L., Dai, X., and Long, M.S. (2006) *J. Am. Chem. Soc.*, 128, 2485–2490.
292 Nagashima, T. and Davies, H.M.L. (2001) *J. Am. Chem. Soc.*, 123, 2695–2696.
293 Chepiga, K.M., Feng, Y., Brunelli, N.A., Jones, C.W., and Davies, H.M.L. (2013) *Org. Lett.*, 15, 6136–6139.
294 Doyle, M.P. and Yan, M. (2003) *Org. Lett.*, 5, 561–563.
295 Davies, H.M.L. and Walji, A.M. (2005) *Org. Lett.*, 7, 2941–2944.
296 McMills, M.C., Humes, R.J., and Pavlyuk, O.M. (2012) *Tetrahedron Lett.*, 53, 849–851.

297 Melby, T., Hughes, R.A., and Hansen, T. (2007) *Synlett*, 2277–2279.
298 Gu, P., Su, Y., Wu, X.-P., Sun, J., Liu, W., Xue, P., and Li, R. (2012) *Org. Lett.*, 14, 2246–2249.
299 Negretti, S., Cohen, C.M., Chang, J.J., Guptill, D.M., and Davies, H.M.L. (2015) *Tetrahedron*, 71, 7415–7420.
300 Qin, C., Boyarskikh, V., Hansen, J.H., Hardcastle, K.I., Musaev, D.G., and Davies, H.M.L. (2011) *J. Am. Chem. Soc.*, 133, 19198–19204.
301 Bonge, H.T., Kaboli, M., and Hansen, T. (2010) *Tetrahedron Lett.*, 51, 5375–5376.
302 Dolbier, W.R. and Battiste, M.A. (2003) *Chem. Rev.*, 103, 1071–1098.
303 Denton, J.R., Sukumaran, D., and Davies, H.M.L. (2007) *Org. Lett.*, 9, 2625–2628.
304 Denton, J.R., Cheng, K., and Davies, H.M.L. (2008) *Chem. Commun.*, 1238–1240.
305 Denton, J.R. and Davies, H.M.L. (2009) *Org. Lett.*, 11, 787–790.
306 (a) Lindsay, V.N.G., Lin, W., and Charette, A.B. (2009) *J. Am. Chem. Soc.*, 131, 16383–16385; (b) Lindsay, V.N.G., Nicolas, C., and Charette, A.B. (2011) *J. Am. Chem. Soc.*, 133, 8972–8981.
307 Goto, T., Takeda, K., Anada, M., Ando, K., and Hashimoto, S. (2011) *Tetrahedron Lett.*, 52, 4200–4203.
308 Awata, A. and Arai, T. (2013) *Synlett*, 24, 29–32.
309 Xue, Y.-S., Cai, Y.-P., and Chen, Z.-X. (2015) *RSC Adv.*, 5, 57781–57791.
310 Davies, H.M.L. and Venkataramani, C. (2003) *Org. Lett.*, 5, 1403–1406.
311 Davies, H.M.L., Walji, A.M., and Nagashima, T. (2004) *J. Am. Chem. Soc.*, 126, 4271–4280.
312 Davies, H.M.L. and Lee, G.H. (2004) *Org. Lett.*, 6, 2117–2120.
313 (a) Müller, P., Bernardinelli, G., Allenbach, Y.F., Ferri, M., and Flack, H.D. (2004) *Org. Lett.*, 6, 1725–1728; (b) Müller, P., Bernardinelli, G., Allenbach, Y.F., Ferri, M., and Grass, S. (2005) *Synlett*, 1397–1400; (c) Müller, P., Allenbach, Y.F., Chappellet, S., and Ghanem, A. (2006) *Synthesis*, 1689–1696.
314 Müller, P. and Lacrampe, F. (2004) *Helv. Chim. Acta*, 87, 2848–2859.
315 Adly, F.G., Maddalena, J., and Ghanem, A. (2014) *Chirality*, 26, 764–774.
316 (a) Müller, P. and Ghanem, A. (2003) *Synlett*, 12, 1830–1833; (b) Müller, P., Allenbach, Y.F., and Robert, E. (2003) *Tetrahedron: Asymmetry*, 14, 779–785; (c) Müller, P., Allenbach, Y.F., and Bernardinelli, G. (2003) *Helv. Chim. Acta*, 86, 3164–3178; (d) Müller, P. and Ghanem, A. (2004) *Org. Lett.*, 6, 4347–4350; (e) Ghanem, A., Aboul-Enein, H.Y., and Müller, P. (2005) *Chirality*, 17, 44–50; (f) Ghanem, A., Lacrampe, F., and Schurig, V. (2005) *Helv. Chim. Acta*, 88, 216–239; (g) Ghanem, A., Lacrampe, F., Aboul-Enein, H.Y., and Schurig, V. (2005) *Monatsh. Chem.*, 136, 1205–1219; (h) Ghanem, A., Gardiner, M.G., Williamson, R.M., and Müller, P. (2010) *Chem. Eur. J.*, 16, 3291–3295.
317 (a) Marcoux, D. and Charette, A.B. (2008) *Angew. Chem. Int. Ed.*, 47, 10155–10158; (b) Marcoux, D., Goudreau, S.R., and Charette, A.B. (2009) *J. Org. Chem.*, 74, 8939–8955; (c) Marcoux, D., Azzi, S., and Charette, A.B. (2009) *J. Am. Chem. Soc.*, 131, 6970–6972; (d) Marcoux, D., Lindsay, V.N.G., and Charette, A.B. (2010) *Chem. Commun.*, 46, 910–912.

318 (a) Chuprakov, S., Kwok, S.W., Zhang, L., Lercher, L., and Fokin, V.V. (2009) *J. Am. Chem. Soc.*, 131, 18034–18043; (b) Grimster, N., Zhang, L., and Fokin, V.V. (2010) *J. Am. Chem. Soc.*, 132, 2510–2511.

319 Doyle, M.P., Morgan, J.P., Fettinger, J.C., Zavalij, P.Y., Colyer, J.T., Timmons, D.J., and Carducci, M.D. (2005) *J. Org. Chem.*, 70, 5291–5301.

320 Biffis, A., Braga, M., Cadamuro, S., Tubaro, C., and Basato, M. (2005) *Org. Lett.*, 7, 1841–1844.

321 (a) Sambasivan, R. and Ball, Z.T. (2012) *Angew. Chem. Int. Ed.*, 51, 8568–8572; (b) Sambasivan, R. and Ball, Z.T. (2013) *Chirality*, 25, 493–497.

322 (a) Estevan, F., Lahuerta, P., Lloret, J., Pérez-Prieto, J., and Werner, H. (2004) *Organometallics*, 23, 1369–1372; (b) Estevan, F., Lahuerta, P., Lloret, J., Sanau, M., Ubeda, M.A., and Vila, J. (2004) *Chem. Commun.*, 2408–2409; (c) Estevan, F., Lahuerta, P., Lloret, J., Penno, D., Sanau, M., and Ubeda, M.A. (2005) *J. Organomet. Chem.*, 690, 4424–4432; (d) Estevan, F., Lloret, J., Sanau, M., and Ubeda, M.A. (2006) *Organometallics*, 25, 4977–4984.

323 Lloret, J., Estevan, F., Bieger, K., Villanueva, C., and Ubeda, M.A. (2007) *Organometallics*, 26, 4145–4151.

324 Lou, Y., Horikawa, M., Kloster, R.A., Hawryluk, N.A., and Corey, E.J. (2004) *J. Am. Chem. Soc.*, 126, 8916–8918.

325 (a) Lindsay, V.N.G. and Charette, A.B. (2012) *ACS Catal.*, 2, 1221–1225; (b) Boruta, D.T., Dmitrenko, O., Yap, G.P.A., and Fox, J.M. (2012) *Chem. Sci.*, 3, 1589–1593.

326 Nishimura, T., Maeda, Y., and Hayashi, T. (2010) *Angew. Chem. Int. Ed.*, 49, 7324–7327.

327 Maas, G. (2004) *Chem. Soc. Rev.*, 33, 183–190.

328 Nishiyama, H. (2004) *Top. Organomet. Chem.*, 11, 81–92.

329 (a) Nishiyama, H., Park, S.-B., Haga, M., Aoki, K., and Itoh, K. (1994) *Chem. Lett.*, 24, 1111–1114; (b) Nishiyama, H., Itoh, Y., Matsumoto, H., Park, S.-B., and Itoh, K. (1994) *J. Am. Chem. Soc.*, 116, 2223–2224; (c) Nishiyama, H., Itoh, Y., Sugawara, Y., Matsumoto, H., Aoki, K., and Itoh, K. (1995) *Bull. Chem. Soc. Jpn.*, 68, 1247–1262; (d) Nishiyama, H. (1999) *Enantiomer*, 4, 569–574.

330 Park, S.-B., Murata, K., Matsumoto, H., and Nishiyama, H. (1995) *Tetrahedron: Asymmetry*, 6, 2487–2494.

331 Nishiyama, H., Soeda, N., Naito, T., and Motoyama, Y. (1998) *Tetrahedron: Asymmetry*, 9, 2865–2869.

332 Garcia, J.I., Jiménez-Osés, G., Martinez-Merino, V., Mayoral, J.A., Pires, E., and Villalba, I. (2007) *Chem. Eur. J.*, 13, 4064–4073.

333 Simpson, J.H., Godfrey, J., Fox, R., Kotnis, A., Kacsur, D., Hamm, J., Totelben, M., Rosso, V., Mueller, R., Delaney, E., and Deshpande, R.P. (2003) *Tetrahedron: Asymmetry*, 14, 3569–3574.

334 Marcin, L.R., Denhart, D.J., and Mattson, R.J. (2005) *Org. Lett.*, 7, 2651–2654.

335 Iwasa, S., Tsushima, S., Nishiyama, K., Tsuchiya, Y., Takazawa, F., and Nishiyama, H. (2003) *Tetrahedron: Asymmetry*, 14, 855–865.

336 Le Maux, P., Abrunhosa, I., Berchel, M., Simmoneaux, G., Gulea, M., and Masson, S. (2004) *Tetrahedron: Asymmetry*, 15, 2569–2573.

337 Cornejo, A., Fraile, J.M., Garcia, J.I., Gil, M.J., Luis, S.V., Martinez-Merino, V., and Mayoral, J.A. (2005) *J. Org. Chem.*, 70, 5536–5544.

338 Burguete, M.I., Cornejo, A., Garcia-Verdugo, E., Gil, M.J., Luis, S.V., Mayoral, J.A., Martinez-Merino, V., and Sokolova, M. (2007) *J. Org. Chem.*, 72, 4344–4350.

339 Cornejo, A., Fraile, J.M., Garcia, J.I., Gil, M.J., Luis, S.V., Martinez-Merino, V., and Mayoral, J.A. (2005) *Tetrahedron*, 61, 12107–12110.

340 Cornejo, A., Martinez-Merino, V., Gil, M.J., Valerio, C., and Pinel, C. (2006) *Chem. Lett.*, 35, 44–45.

341 Gao, M.Z., Kong, D., Clearfield, A., and Zingaro, R.A. (2004) *Tetrahedron Lett.*, 45, 5649–5652.

342 Ito, J.-I., Ujiie, S., and Nishiyama, H. (2010) *Chem. Eur. J.*, 16, 4986–4990.

343 (a) Stoop, R.M., Bauer, C., Setz, P., Wörle, M., Wong, T.Y.H., and Mezzetti, A. (1999) *Organometallics*, 18, 5691–5700; (b) Bachmann, S., Furler, M., and Mezzetti, A. (2001) *Organometallics*, 20, 2102–2108; (c) Bachmann, S. and Mezzetti, A. (2001) *Helv. Chim. Acta*, 84, 3063–3074; (d) Uchida, T., Irie, R., and Katsuki, T. (1999) *Synlett*, 1793–1795; (e) Uchida, T., Irie, R., and Katsuki, T. (1999) *Synlett*, 1163–1165; (f) Uchida, T., Irie, R., and Katsuki, T. (2000) *Tetrahedron*, 56, 3501–3509; (g) Uchida, T. and Katsuki, T. (2006) *Synthesis*, 1715–1723.

344 Miller, J.A., Jin, W.C., and Nguyen, S.T. (2002) *Angew. Chem. Int. Ed.*, 41, 2953–2956.

345 Miller, J.A., Hennessy, E.J., Marshall, W.J., Scialdone, M.A., and Nguyen, S.T. (2003) *J. Org. Chem.*, 68, 7884–7886.

346 Gill, C.S., Venkatasubbaiah, K., and Jones, C.W. (2009) *Adv. Synth. Catal.*, 351, 1344–1354.

347 Falkowski, J.M., Liu, S., Wang, C., and Lin, W. (2012) *Chem. Commun.*, 48, 6508–6510.

348 Miller, J.A., Gross, B.A., Zhuravel, M.A., Jin, W., and Nguyen, S.T. (2005) *Angew. Chem. Int. Ed.*, 44, 3885–3889.

349 (a) Bonaccorsi, C., Bachmann, S., and Mezzetti, A. (2003) *Tetrahedron: Asymmetry*, 14, 84–854; (b) Bonaccorsi, C. and Mezzetti, A. (2005) *Organometallics*, 24, 4953–4960.

350 Bonaccorsi, C., Santoro, F., Gischig, S., and Mezzetti, A. (2006) *Organometallics*, 25, 2002–2010.

351 (a) Huber, D. and Mezzetti, A. (2004) *Tetrahedron: Asymmetry*, 15, 2193–2197; (b) Huber, D., Kumar, P.G.A., Pregosin, P.S., Mikhel, I.S., and Mezzetti, A. (2006) *Helv. Chim. Acta*, 89, 1696–1715.

352 Grabulosa, A., Mannu, A., Mezzetti, A., and Muller, G. (2012) *J. Organomet. Chem.*, 696, 4221–4228.

353 Lasa, M., Lopez, P., Cativiela, C., Carmona, D., and Oro, L.A. (2005) *J. Mol. Catal. A: Chem.*, 234, 129–135.

354 Hoang, V.D.M., Reddy, P.A.N., and Kim, T.-J. (2007) *Tetrahedron Lett.*, 48, 8014–8017.

355 Chanthamath, S., Phomkeona, K., Shibatomi, K., and Iwasa, S. (2012) *Chem. Commun.*, 48, 7750–7752.

356 Chanthamath, S., Nguyen, D.T., Shibatomi, K., and Iwasa, S. (2013) *Org. Lett.*, 15, 772–775.
357 Chanthamath, S., Ozaki, S., Shibatomi, K., and Iwasa, S. (2014) *Org. Lett.*, 16, 3012–3015.
358 Chanthamath, S., Chua, H.W., Kimura, S., Shibatomi, K., and Iwasa, S. (2014) *Org. Lett.*, 16, 3408–3411.
359 Abu-Elfotoh, A.-M., Nguyen, D.P.T., Chanthamath, S., Phomkeona, K., Shibatomi, K., and Iwasa, S. (2012) *Adv. Synth. Catal.*, 354, 3435–3439.
360 Abu-Elfotoh, A.-M., Phomkeona, K., Shibatomi, K., and Iwasa, S. (2010) *Angew. Chem. Int. Ed.*, 49, 8439–8443.
361 Berkessel, A., Kaiser, P., and Lex, J. (2003) *Chem. Eur. J.*, 9, 4746–4756.
362 (a) Paul-Roth, C., De Montigny, F., Rethoré, G., Simmoneaux, G., Gulea, M., and Masson, S. (2003) *J. Mol. Catal. A: Chem.*, 201, 79–91; (b) Ferrand, Y., Le Maux, P., and Simmoneaux, G. (2004) *Org. Lett.*, 6, 3211–3214.
363 Le Maux, P., Juillard, S., and Simmoneaux, G. (2006) *Synthesis*, 1701–1704.
364 (a) Ferrand, Y., Le Maux, P., and Simmoneaux, G. (2005) *Tetrahedron: Asymmetry*, 16, 3829–3836; (b) Ferrand, Y., Poriel, C., Le Maux, P., Rault-Berthelot, J., and Simmoneaux, G. (2005) *Tetrahedron: Asymmetry*, 16, 1463–1472.
365 del Pozo, C., Corma, A., Iglesias, M., and Sanchez, F. (2011) *Green Chem.*, 13, 2471–2481.
366 Yeung, C.-T., Lee, W.-S., Tsang, C.-S., Yiu, S.-M., Wong, W.-T., Wong, W.-Y., and Kwong, H.-L. (2010) *Polyhedron*, 29, 1497–1507.
367 Denmark, S.E., Stavenger, R.A., Faucher, A.-M., and Edwards, J.P. (1997) *J. Org. Chem.*, 62, 3375–3389.
368 Lai, T.-S., Chan, F.-Y., So, P.-K., Ma, D.-L., Wong, K.-Y., and Che, C.-M. (2006) *Dalton Trans.*, 4845–4851.
369 (a) Kanchiku, S., Suematsu, H., Matsumoto, K., Uchida, T., and Katsuki, T. (2007) *Angew. Chem. Int. Ed.*, 46, 3889–3891; (b) Suematsu, H., Kanchiku, S., Uchida, T., and Katsuki, T. (2008) *J. Am. Chem. Soc.*, 130, 10327–10337.
370 Ichinose, M., Suematsu, H., and Katsuki, T. (2009) *Angew. Chem. Int. Ed.*, 48, 3121–3123.
371 Zhang, J., Liang, J.-L., Sun, X.-R., Zhou, H.-B., Zhu, N.-Y., Zhou, Z.-Y., Chan, P.W.H., and Che, C.-M. (2005) *Inorg. Chem.*, 44, 3942–3952.
372 Feliz, M., Guillamon, E., Llusar, R., Vicent, C., Stiriba, S.-E., Pérez-Prieto, J., and Barberis, M. (2006) *Chem. Eur. J.*, 12, 1486–1492.
373 Cao, Z.-Y., Zhou, F., Yu, Y.-H., and Zhou, J. (2013) *Org. Lett.*, 15, 42–45.
374 Cao, Z.-Y., Wang, X., Tan, C., Zhao, X.-L., Zhou, J., and Ding, K. (2013) *J. Am. Chem. Soc.*, 135, 8197–8200.
375 Renata, H., Wang, Z.J., Kitto, R.Z., and Arnold, F.H. (2014) *Catal. Sci. Technol.*, 4, 3640–3643.
376 Doyle, M.P. and Hu, W. (2001) *Synlett*, 1364–1370.
377 (a) Taber, D.F. and Hoerrner, R.S. (1992) *J. Org. Chem.*, 57, 441–447; (b) Taber, D.F. and Kanai, K. (1999) *J. Org. Chem.*, 64, 7983–7987.
378 (a) Lee, E., Shin, I.J., and Kim, T.S. (1990) *J. Am. Chem. Soc.*, 112, 260–264; (b) Srikrishna, A. and Anebouselvy, K. (2001) *J. Org. Chem.*, 66, 7102–7106.
379 Sarkar, T.K. and Nandy, S.K. (1998) *Tetrahedron Lett.*, 39, 2411–2412.

380 Srikrishna, A., Nagamani, S.A., and Jagadeesh, S.G. (2005) *Tetrahedron: Asymmetry*, 16, 1569–1571.
381 (a) Gallos, J.K., Koftis, T.V., and Koumbis, A.E. (1994) *J. Chem. Soc., Perkin Trans. 1*, 611–612; (b) Gallos, J.K., Massen, Z.S., Koftis, T.V., and Dellios, C.C. (2001) *Tetrahedron Lett.*, 42, 7489–7491.
382 Srikrishna, A. and Vijaykumar, D. (2000) *J. Chem. Soc., Perkin Trans. 1*, 2583–2589.
383 Corey, E.J. and Kigoshi, H. (1991) *Tetrahedron Lett.*, 32, 5025–5028.
384 (a) Maguire, A.R., Buckley, N.R., O'Leary, P., and Ferguson, G. (1996) *J. Chem. Soc., Chem. Commun.*, 2595–2596; (b) Maguire, A.R., Buckley, N.R., O'Leary, P., and Ferguson, G. (1998) *J. Chem. Soc., Perkin Trans. 1*, 4077–4092.
385 (a) Zhang, H., Appels, D.C., Hockless, D.C.R., and Mander, L.N. (1998) *Tetrahedron Lett.*, 39, 6577–6580; (b) Frey, B., Wells, A.P., Rogers, D.H., and Mander, L.N. (1998) *J. Am. Chem. Soc.*, 120, 1914–1915.
386 (a) King, G.R., Mander, L.N., Monck, N.J.T., Morris, J.C., and Zhang, H. (1997) *J. Am. Chem. Soc.*, 119, 3828–3829; (b) Morris, J.C. (1998) *Synthesis*, 455–467.
387 Moore, J.D. and Hanson, P.R. (2003) *Tetrahedron: Asymmetry*, 14, 873–880.
388 Weathers, T.M., Doyle, M.P., and Carducci, M.D. (2006) *Adv. Synth. Catal.*, 348, 449–455.
389 Swain, N.A., Brown, R.C.D., and Bruton, G. (2004) *J. Org. Chem.*, 69, 122–129.
390 Doyle, M.P., Peterson, C.S., Zhou, Q.L., and Nishiyama, H. (1997) *Chem. Commun.*, 211–212.
391 Piqué, C., Fähndrich, B., and Pfaltz, A. (1995) *Synlett*, 491–492.
392 (a) Doyle, M.P., Eismont, M.Y., and Zhou, Q.L. (1997) *Russ. Chem. Bull.*, 46, 955–958; (b) Kim, S.-G., Cho, C.-W., and Ahn, K.H. (1999) *Tetrahedron*, 55, 10079–10086; (c) Park, S.-W., Son, J.-H., Kim, S.-G., and Ahn, K.H. (1999) *Tetrahedron: Asymmetry*, 10, 1903–1911.
393 (a) Barberis, M., Pérez-Prieto, J., Stiriba, S.-E., and Lahuerta, P. (2001) *Org. Lett.*, 3, 3317–3319; (b) Barberis, M., Pérez-Prieto, J., Herbst, K., and Lahuerta, P. (2002) *Organometallics*, 21, 1667–1673.
394 Saha, B., Uchida, T., and Katsuki, T. (2003) *Tetrahedron: Asymmetry*, 14, 823–836.
395 Xu, X., Lu, H., Ruppel, J.V., Cui, X., Lopez de Mesa, S., Wojtas, L., and Zhang, X.P. (2011) *J. Am. Chem. Soc.*, 133, 15292–15295.
396 Tokunoh, R., Tomiyama, H., Sodeoka, M., and Shibasaki, M. (1996) *Tetrahedron Lett.*, 37, 2449–2452.
397 Gant, T.G., Noe, M.C., and Corey, E.J. (1995) *Tetrahedron Lett.*, 36, 8745–8748.
398 Wong, A., Welch, C.J., Kuethe, J.T., Vazquez, E., Shaimi, M., Henderson, D., Davies, I.W., and Hughes, D.L. (2004) *Org. Biomol. Chem.*, 2, 168–174.
399 (a) Honma, M., Sawada, T., Fujisawa, Y., Utsugi, M., Watanabe, H., Umino, A., Matsumura, T., Hagihara, T., Takano, M., and Nakada, M. (2003) *J. Am. Chem. Soc.*, 125, 2860–2861; (b) Honma, M. and Nakada, M. (2003) *Tetrahedron Lett.*, 44, 9007–9011.
400 Sawada, T. and Nakada, M. (2005) *Adv. Synth. Catal.*, 347, 1527–1532.
401 Takeda, H. and Nakada, M. (2006) *Tetrahedron: Asymmetry*, 17, 2896–2906.
402 Sawada, T. and Nakada, M. (2012) *Tetrahedron: Asymmetry*, 23, 350–356.
403 Takano, M., Umino, A., and Nakada, M. (2004) *Org. Lett.*, 6, 4897–4900.
404 Miyamoto, H., Iwamoto, M., and Nakada, M. (2005) *Heterocycles*, 66, 61–67.
405 Takeda, H., Watanabe, H., and Nakada, M. (2006) *Tetrahedron*, 62, 8054–8063.

406 Ida, R. and Nakada, M. (2007) *Tetrahedron Lett.*, 48, 4855–4859.
407 (a) Hirai, S. and Nakada, M. (2010) *Tetrahedron Lett.*, 51, 5076–5079; (b) Hirai, S. and Nakada, M. (2011) *Tetrahedron*, 67, 518–530.
408 Honma, M. and Nakada, M. (2007) *Tetrahedron Lett.*, 48, 1541–1544.
409 Uetake, Y., Uwamori, M., and Nakada, M. (2015) *J. Org. Chem.*, 80, 1735–1745.
410 Takeda, H., Honma, M., Ida, R., Sawada, T., and Nakada, M. (2007) *Synlett*, 4, 579–582.
411 Sawada, T. and Nakada, M. (2013) *Org. Lett.*, 15, 1004–1007.
412 (a) Doyle, M.P. and Hu, W. (2000) *J. Org. Chem.*, 65, 8839–8847; (b) Doyle, M.P. and Hu, W. (2000) *Tetrahedron Lett.*, 41, 6265–6269.
413 Koskinen, A.M.P. and Hassila, H. (1993) *J. Org. Chem.*, 58, 4479–4480.
414 Doyle, M.P. and Forbes, D.C. (1998) *Chem. Rev.*, 98, 911–936.
415 (a) Doyle, M.P., Pieters, R.J., Martin, S.F., Austin, R.E., Oalmann, C.J., and Mueller, P. (1991) *J. Am. Chem. Soc.*, 113, 1423–1424; (b) Doyle, M.P., Austin, R.E., Bailey, A.S., Dwyer, M.P., Dyatkin, A.B., Kalinin, A.V., Kwan, M.M.Y., Liras, S., Oalmann, C.J., Pieters, R.J., Protopopova, M.N., Raab, C.E., Roos, G.H.P., Zhou, Q.L., and Martin, S.F. (1995) *J. Am. Chem. Soc.*, 117, 5763–5775.
416 Doyle, M.P., Zhou, Q.-L., Dyatkin, A.B., and Ruppar, D.A. (1995) *Tetrahedron Lett.*, 36, 7579–7582.
417 Rogers, D.H., Yi, E.C., and Poulter, C.D. (1995) *J. Org. Chem.*, 60, 941–945.
418 (a) Martin, S.F., Austin, R.E., Oalmann, C.J., Baker, W.R., Condon, S.L., DeLara, E., Rosenberg, S.H., Spina, K.P., Stein, H.H., Cohen, J., and Kleinert, H.D. (1992) *J. Med. Chem.*, 35, 1710–1721; (b) Hillier, M.C., Davidson, J.P., and Martin, S.F. (2001) *J. Org. Chem.*, 66, 1657–1671.
419 Martin, S.F., Spaller, M.R., Liras, S., and Hartmann, B. (1994) *J. Am. Chem. Soc.*, 116, 4493–4494.
420 Ashfeld, B.L. and Martin, S.F. (2006) *Tetrahedron*, 62, 10497–10506.
421 Berberich, S.M., Cherney, R.J., Colucci, J., Courillon, C., Geraci, L.S., Kirkland, T.A., Marx, M.A., Schneider, M.F., and Martin, S.F. (2003) *Tetrahedron*, 59, 6819–6832.
422 Ashfeld, B.L. and Martin, S.F. (2005) *Org. Lett.*, 7, 4535–4537.
423 Reichelt, A. and Martin, S.F. (2006) *Acc. Chem. Res.*, 39, 433–442.
424 (a) Martin, S.F., Oalmann, C.J., and Liras, S. (1992) *Tetrahedron Lett.*, 33, 6727–6730; (b) Doyle, M.P., Dyatkin, A.B., Kalinin, A.V., Ruppar, D.A., Martin, S.F., Spaller, M.R., and Liras, S. (1995) *J. Am. Chem. Soc.*, 117, 11021–11022.
425 Fillion, E. and Beingessner, R.L. (2003) *J. Org. Chem.*, 68, 9485–9488.
426 (a) Doyle, M.P., Eismont, M.Y., Protopopova, M.N., and Kwan, M.M.Y. (1994) *Tetrahedron*, 50, 4519–4528; (b) Doyle, M.P. and Kalinin, A.V. (1996) *J. Org. Chem.*, 61, 2179–2184; (c) Collado, I., Pedregal, C., Bueno, A.B., Marcos, A., Gonzalez, R., Blanco-Urgoiti, J., Pérez-Castells, J., Schoepp, D.D., Wright, R.A., Johnson, B.G., Kingston, A.E., Moher, E.D., Hoard, D.W., Griffey, K.I., and Tizzano, J.P. (2004) *J. Med. Chem.*, 47, 456–466.
427 Doyle, M.P. and Zhou, Q.-L. (1995) *Tetrahedron: Asymmetry*, 6, 2157–2160.
428 (a) Davies, H.M.L. and Doan, B.D. (1999) *J. Org. Chem.*, 64, 8501–8508; (b) Doyle, M.P., Hu, W., Phillips, I.M., Moody, C.J., Pepper, A.G., and Slawin, A.M.Z. (2001) *Adv. Synth.Catal.*, 343, 112–117; (c) Doyle, M.P. and Hu, W. (2001) *Adv. Synth. Catal.*, 343, 299–302.

429 Doyle, M.P., Hu, W., and Weathers, T.M. (2003) *Chirality*, 15, 369–373.
430 Davies, H.M.L. and Doan, B.D. (1996) *Tetrahedron Lett.*, 37, 3967–3970.
431 Lin, W. and Charette, A.B. (2005) *Adv. Synth. Catal.*, 347, 1547–1552.
432 Pérez-Prieto, J., Stiriba, S.-E., Moreno, E., and Lahuerta, P. (2003) *Tetrahedron: Asymmetry*, 14, 787–790.
433 Escudero, C., Pérez-Prieto, J., and Stiriba, S.-E. (2006) *Inorg. Chim. Acta*, 359, 1974–1978.
434 Barberis, M. and Pérez-Prieto, J. (2003) *Tetrahedron Lett.*, 44, 6683–6685.
435 Doyle, M.P., Wang, Y., Ghorbani, P., and Bappert, E. (2005) *Org. Lett.*, 7, 5035–5038.
436 Müller, P., Allenbach, Y.F., and Grass, S. (2005) *Tetrahedron: Asymmetry*, 16, 2007–2013.
437 Xu, Z.-J., Fang, R., Zhao, C., Huang, J.-S., Li, G.-Y., Zhu, N., and Che, C.-M. (2009) *J. Am. Chem. Soc.*, 131, 4405–4417.
438 Li, G.-Y., Zhang, J., Wai Hong Chan, P., Xu, Z.-J., Zhu, N., and Che, C.-M. (2006) *Organometallics*, 25, 1676–1688.
439 Nakagawa, Y., Chanthamath, S., Shibatomi, K., and Iwasa, S. (2015) *Org. Lett.*, 17, 2792–2795.
440 (a) Davison, A., Krusell, W.C., and Michaelson, R.C. (1974) *J. Organomet. Chem.*, 72, C7–C10; (b) Flood, T.C., DiSanti, F.J., and Miles, D.L. (1976) *Inorg. Chem.*, 15, 1910–1918.
441 (a) Brookhart, M., Timmers, D., Tucker, J.R., Williams, G.D., Husk, R., Brunner, H., and Hammer, B. (1983) *J. Am. Chem. Soc.*, 105, 6721–6723; (b) Brookhart, M. and Buck, R.C. (1989) *J. Organomet. Chem.*, 370, 111–127; (c) Brookhart, M., Liu, Y., Goldman, E.W., Timmers, D.A., and Williams, G.D. (1991) *J. Am. Chem. Soc.*, 113, 927–939.
442 (a) Theys, R.D. and Hossain, M. (1995) *Tetrahedron Lett.*, 36, 5113–5116; (b) Wang, Q., Yang, F., Du, H., Hossain, M., Bennett, D., and Grubisha, D.S. (1998) *Tetrahedron: Asymmetry*, 9, 3971–3977; (c) Wang, Q., Mayer, M.F., Brennan, C., Yang, F., Hossain, M., Grubisha, D.S., and Bennett, D. (2000) *Tetrahedron*, 56, 4881–4891; (d) Wang, Q. and Hossain, M. (2001) *J. Organomet. Chem.*, 617, 751–754.
443 Barluenga, J., Suàrez-Sobrino, A.L., Tomàs, M., Garcià-Granda, S., and Santiago-Garcìa, R. (2001) *J. Am. Chem. Soc.*, 123, 10494–10501.
444 Barluenga, J., de Prado, A., Santamaria, J., and Tomas, M. (2007) *Chem. Eur. J.*, 13, 1326–1331.
445 (a) Rocquet, F. and Sevin, A. (1974) *Bull. Soc. Chim. Fr.*, 111, 888–894; (b) de Faria, M.L., Magalhães, R. de A., Silva, F.C., Matias, L.G. de O., Ceschi, M.A., Brocksom, U., and Brocksom, T.J. (2000) *Tetrahedron: Asymmetry*, 11, 4093–4238.
446 Matsumura, Y., Inoue, M., Nakamura, Y., Talib, I.L., Maki, T., and Onomura, O. (2000) *Tetrahedron Lett.*, 41, 4619–4622.
447 Wender, P.A., Glass, T.E., Krauss, N.E., Mühlebach, M., Peschke, B., and Rawlins, D.B. (1996) *J. Org. Chem.*, 61, 7662–7663.
448 Dauben, W.G. and Lewis, T.A. (1995) *Synlett*, 857–858.
449 (a) Bruncko, M. and Crich, D. (1992) *Tetrahedron Lett.*, 33, 6251–6254; (b) Bruncko, M. and Crich, D. (1994) *J. Org. Chem.*, 59, 4239–4249.

450 Krief, A. and Swinnen, D. (1996) *Tetrahedron Lett.*, 37, 7123–7126.
451 Moher, E.D. (1996) *Tetrahedron Lett.*, 37, 8637–8640.
452 Collado, I., Domìnguez, C., Ezquerra, J., Pedregal, C., and Monn, J.A. (1997) *Tetrahedron Lett.*, 38, 2133–2136.
453 Lee, W.L. and Miller, M.J. (2004) *J. Org. Chem.*, 69, 4516–4519.
454 Zhang, R., Mamai, A., and Madalengoitia, J.S. (1999) *J. Org. Chem.*, 64, 547–555.
455 (a) Hudlicky, T., Radesca, L., Luna, H., and Anderson, F.E. (1986) *J. Org. Chem.*, 51, 4746–4748; (b) Hudlicky, T., Natchus, M.G., and Sinai-Zingde, G. (1987) *J. Org. Chem.*, 52, 4641–4644.
456 (a) Wu, J.-C. and Chattopadhyaya, J. (1989) *Tetrahedron*, 45, 4507–4522; (b) Wu, J.-C. and Chattopadhyaya, J. (1990) *Tetrahedron*, 46, 2587–2592.
457 Haly, B., Bharadwaj, R., and Sanghvi, Y.S. (1996) *Synlett*, 687–689.
458 Krief, A., Dumont, W., Pasau, P., and Lecomte, P. (1989) *Tetrahedron*, 45, 3039–3052.
459 Krief, A. and Lecomte, P. (1993) *Tetrahedron Lett.*, 34, 2695–2698.
460 Krief, A., Provins, L., and Froidbise, A. (1998) *Tetrahedron Lett.*, 39, 1437–1440.
461 Ma, D. and Ma, Z. (1997) *Tetrahedron Lett.*, 38, 7599–7602.
462 (a) Ma, D., Cao, Y., Yang, Y., and Cheng, D. (1999) *Org. Lett.*, 1, 285–288; (b) Ma, D. and Jiang, Y. (2000) *Tetrahedron: Asymmetry*, 11, 3727–3736; (c) Ma, D., Cao, Y., Wu, W., and Jiang, Y. (2000) *Tetrahedron*, 56, 7447–7456.
463 (a) Krief, A., Dumont, W., and Pasau, P. (1988) *Tetrahedron Lett.*, 29, 1079–1082; (b) Krief, A. and Dumont, W. (1988) *Tetrahedron Lett.*, 29, 1083–1084.
464 Cativiela, C., Dìaz-de-Villegas, M.D., and Jiménez, A.I. (1995) *Tetrahedron*, 51, 3025–3032.
465 Galley, G., Hübner, J., Anklam, S., Jones, P.G., and Pätzel, M. (1996) *Tetrahedron Lett.*, 37, 6307–6310.
466 Pohlman, M. and Kazmaier, U. (2003) *Org. Lett.*, 5, 2631–2633.
467 Mohapatra, D.K., Ray Chaudhuri, S.R., Sahoo, G., and Gurjar, M.K. (2006) *Tetrahedron: Asymmetry*, 17, 2609–2616.
468 Morrison, J.A. and Mosher, H.S. (1976) *Asymmetric Organic Reactions*, American Chemical Society, Washington, DC, p. 241.
469 (a) De Vos, M.J. and Krief, A. (1983) *Tetrahedron Lett.*, 24, 103–106; (b) Yamazaki, S., Kataoka, H., and Yamabe, S. (1999) *J. Org. Chem.*, 64, 2367–2374.
470 Ye, S., Tang, Y., and Dai, L.-X. (2001) *J. Org. Chem.*, 66, 5717–5722.
471 (a) Bernardi, A., Scolastico, C., and Villa, R. (1989) *Tetrahedron Lett.*, 30, 3733–3734; (b) Bernardi, A., Cardani, S., Poli, G., and Scolastico, C. (1986) *J. Org. Chem.*, 51, 5041–5043.
472 Mamai, A. and Madalengoitia, J.S. (2000) *Tetrahedron Lett.*, 41, 9009–9014.
473 Romo, D. and Meyers, A.I. (1991) *Tetrahedron*, 47, 9503–9569.
474 Groaning, M.D. and Meyers, A.I. (1999) *Tetrahedron Lett.*, 40, 4639–4642.
475 (a) Alami, A., Calmes, M., Danis, J., and Jacquier, R. (1993) *Bull. Soc. Chim. Fr.*, 130, 5–24; (b) Burgess, K., Ho, K.-K., and Moye-Sherman, D. (1994) *Synlett*, 575–583.
476 (a) Imanishi, T., Ohra, T., Sugiyama, K., Ueda, Y., Takemoto, Y., and Iwata, C. (1992) *J. Chem. Soc., Chem. Commun.*, 269–270; (b) Imanishi, T., Ohra, T.,

Sugiyama, K., Takemoto, Y., and Iwata, C. (1995) *Chem. Pharm. Bull.*, 43, 571–577.
477 Hamdouchi, C. (1992) *Tetrahedron Lett.*, 33, 1701–1704.
478 (a) Hiroi, K. and Arinaga, Y. (1994) *Chem. Pharm. Bull.*, 42, 985–987; (b) Hiroi, K. and Arinaga, Y. (1994) *Tetrahedron Lett.*, 35, 153–156.
479 Mikolajczyk, M., Midura, W.H., Michedkina, E., Filipczak, A.D., and Wieczorek, M.W. (2005) *Helv. Chim. Acta*, 88, 1769–1775.
480 (a) Midura, W.H. and Mikolajczyk, M. (2002) *Tetrahedron Lett.*, 43, 3061–3065; (b) Midura, W.H., Krysiak, J.A., and Mikolajczyk, M. (2003) *Tetrahedron: Asymmetry*, 14, 1245–1249; (c) Midura, W.H. (2005) *Phosphorus, Sulfur Silicon Relat. Elem.*, 180, 1285–1290; (d) Mikolajczyk, M. (2005) *Pure Appl. Chem.*, 77, 2091–2098; (e) Mikolajczyk, M. (2005) *J. Organomet. Chem.*, 690, 2488–2496; (f) Midura, W.H., Krysiak, J.A., Cypryk, M., Mikolajczyk, M., Wieczorek, M.W., and Filipczak, A.D. (2005) *Eur. J. Org. Chem.*, 2005, 653–662.
481 Bayley, P.L., Hewkin, C.T., Clegg, W., and Jackson, R.F.W. (1993) *J. Chem. Soc., Perkin Trans. 1*, 577–584.
482 Ruano, J.L.G., Fajardo, C., Martin, M.R., Midura, W.H., and Mikolajczyk, M. (2004) *Tetrahedron: Asymmetry*, 15, 2475–2482.
483 Abramovitch, A., Fensterbank, L., Malacria, M., and Marek, I. (2008) *Angew. Chem. Int. Ed.*, 47, 6865–6868.
484 Palko, J.W., Buist, P.H., and Manthorpe, J.M. (2013) *Tetrahedron: Asymmetry*, 24, 165–168.
485 Midura, W.H., Krysiak, J., Rzewnicka, A., Supel, A., Lizwa, P., and Ewas, A.M. (2013) *Tetrahedron*, 69, 730–737.
486 Krysiak, J.A., Rzewnicka, A., and Midura, W.H. (2013) *Phosphorus, Sulfur Silicon Relat. Elem.*, 188, 483–486.
487 Bünuel, E., Bull, S.D., Davies, S.G., Garner, A.C., Savory, E.D., Smith, A.D., Vickers, R.J., and Watkin, D.J. (2003) *Org. Biomol. Chem.*, 1, 2531–2542.
488 Lin, J.-W., Kurniawan, Y.D., Chang, W.-J., Leu, W.-J., Chan, S.-H., and Hou, D.-R. (2014) *Org. Lett.*, 16, 5328–5331.
489 Midura, W.H. and Krysiak, J. (2015) *Chirality*, 27, 816–819.
490 Sugeno, Y., Ishikawa, Y., and Oikawa, M. (2014) *Synlett*, 25, 987–990.
491 (a) Johnson, C.R., Janiga, E.R., and Haake, M. (1968) *J. Am. Chem. Soc.*, 90, 3890–3891; (b) Johnson, C.R. (1973) *Acc. Chem. Res.*, 6, 341–347; (c) Barbachyn, M.R. and Johnson, C.R. (1984) in *Asymmetric Synthesis*, vol. 4 (eds J.D. Morrison and J.W. Scott), Academic Press, New York, p. 227.
492 Midura, W.H. and Rzewnicka, A. (2013) *Tetrahedron: Asymmetry*, 24, 937–941.
493 Midura, W.H., Ścianowski, J., Banach, A., and Zając, A. (2014) *Tetrahedron: Asymmetry*, 25, 1488–1493.
494 Midura, W.H. and Rzewnicka, A. (2014) *Synlett*, 25, 2213–2216.
495 Riches, S.L., Saha, C., Filgueira, N.F., Grange, E., McGarrigle, E.M., and Aggarwal, V.K. (2010) *J. Am. Chem. Soc.*, 132, 7626–7630.
496 Pyne, S.G., Dong, Z., Skelton, B.W., and White, A.H. (1997) *J. Org. Chem.*, 62, 2337–2343.
497 Shen, X., Zhang, W., Zhang, L., Luo, T., Wan, X., Gu, Y., and Hu, J. (2012) *Angew. Chem. Int. Ed.*, 51, 6966–6970.

498 Toru, T., Nakamura, S., Takemoto, H., and Ueno, Y. (1997) *Synlett*, 449–450.
499 Solladié-Cavallo, A., Diepvohuule, A., and Isarno, T. (1998) *Angew. Chem. Int. Ed. Engl.*, 37, 1689–1691.
500 Ye, S., Huang, Z.-Z., Xia, C.-A., Tang, Y., and Dai, L.-X. (2002) *J. Am. Chem. Soc.*, 124, 2432–2433.
501 Deng, X.-M., Cai, P., Ye, S., Sun, X.-L., Liao, W.-W., Li, K., Tang, Y., Wu, Y.-D., and Dai, L.-X. (2006) *J. Am. Chem. Soc.*, 128, 9730–9740.
502 Zhu, B.-H., Zhou, R., Zheng, J.-C., Deng, X.-M., Sun, X.-L., Sheng, Q., and Tang, Y. (2010) *J. Org. Chem.*, 75, 3454–3457.
503 Zhou, R., Deng, X., Zheng, J., Sheng, Q., Sun, X., and Tang, Y. (2011) *Chin. J. Chem.*, 29, 995–1000.
504 Huang, K. and Huang, Z.-Z. (2005) *Synlett*, 10, 1621–1623.
505 Wang, H.-Y., Yang, F., Li, X.-L., Yan, X.-M., and Huang, Z.-Z. (2009) *Chem. Eur. J.*, 15, 3784–3789.
506 (a) Liao, W.-W., Li, K., and Tang, Y. (2003) *J. Am. Chem. Soc.*, 125, 13030–13031; (b) Zheng, J.-C., Liao, W.-W., Tang, Y., Sun, X.-L., and Dai, L.-X. (2005) *J. Am. Chem. Soc.*, 127, 12222–12223.
507 (a) Hanessian, S., Andreotti, D., and Gomtsyan, A. (1995) *J. Am. Chem. Soc.*, 117, 10393–10394; (b) Hanessian, S., Cantin, L.-D., Roy, S., Andreotti, D., and Gomtsyan, A. (1997) *Tetrahedron Lett.*, 38, 1103–1106; (c) Hanessian, S., Griffin, A., and Devasthale, P.V. (1997) *Bioorg. Med. Chem. Lett.*, 7, 3119–3124.
508 (a) Hanessian, S., Cantin, L.-D., and Andreotti, D. (1999) *J. Org. Chem.*, 64, 4893–4900; (b) Marinozzi, M. and Pellicciari, R. (2000) *Tetrahedron Lett.*, 41, 9125–9128.
509 Owsianik, K., Wieczorek, W., Balińska, A., and Mikołajczyk, M. (2014) *Heteroat. Chem*, 25, 690–697.
510 (a) Kojima, S., Hiroike, K., and Ohkata, K. (2004) *Tetrahedron Lett.*, 45, 3565–3568; (b) Kojima, S., Fujitomo, K., Itoh, Y., Hiroike, K., Murakami, M., and Ohkata, K. (2006) *Heterocycles*, 67, 679–694.
511 Kanomata, N., Sakaguchi, R., Sekine, K., Yamashita, S., and Tanaka, H. (2010) *Adv. Synth. Catal.*, 352, 2966–2978.
512 Yamada, S., Yamamoto, J., and Ohta, E. (2007) *Tetrahedron Lett.*, 48, 855–858.
513 Couty, F., David, O., Larmanjat, B., and Marrot, J. (2007) *J. Org. Chem.*, 72, 1058–1061.
514 Shinohara, N., Haga, J., Yamazaki, T., Kitazume, T., and Nakamura, S. (1995) *J. Org. Chem.*, 60, 4363–4674.
515 Taguchi, T., Shibuya, A., Sasaki, H., Endo, J.-I., Morikawa, T., and Shiro, M. (1994) *Tetrahedron: Asymmetry*, 5, 1423–1426.
516 Ivashkin, P., Couve-Bonnaire, S., Jubault, P., and Pannecoucke, X. (2012) *Org. Lett.*, 14, 5130–5133.
517 Màzon, A., Pedregal, C., and Prowse, W. (1999) *Tetrahedron*, 55, 7057–7064.
518 Atta, A.K. and Pathak, T. (2009) *J. Org. Chem.*, 74, 2710–2717.
519 Arai, S., Nakayama, K., Ishida, T., and Shioiri, T. (1999) *Tetrahedron Lett.*, 40, 4215–4218.
520 Russo, A. and Lattanzi, A. (2011) *Org. Biomol. Chem.*, 9, 7993–7996.
521 Marini, F., Sternativo, S., Del Verme, F., Testaferri, L., and Tiecco, M. (2009) *Adv. Synth. Catal.*, 351, 1801–1806.

522 (a) Aggarwal, V.K., Abdelrahman, H., Thompson, A., Mattison, B., and Jones, R.V.H. (1994) *Phosphorus, Sulfur Silicon Relat. Elem.*, 95–96, 283–292; (b) Aggarwal, V.K., Smith, H.W., Jones, R.V.H., and Fieldhouse, R. (1997) *Chem. Commun.*, 1785–1786; (c) Aggarwal, V.K., Smith, H.W., Hynd, G., Jones, R.V.H., Fieldhouse, R., and Spey, S.E. (2000) *J. Chem. Soc., Perkin Trans. 1*, 3267–3276; (d) Aggarwal, V.K., Alonso, E., Fang, G.Y., Ferrara, M., Hynd, G., and Porcelloni, M. (2001) *Angew. Chem. Int. Ed.*, 40, 1433–1436.

523 Aggarwal, V.K. and Grange, E. (2006) *Chem. Eur. J.*, 12, 568–575.

524 Gao, L., Hwang, G.-S., and Ryu, D.H. (2011) *J. Am. Chem. Soc.*, 133, 20708–20711.

525 Kunz, R.K. and MacMillan, D.W.C. (2005) *J. Am. Chem. Soc.*, 127, 3240–3241.

526 Hartikka, A., Slosarczyk, A.T., and Arvidsson, P.I. (2007) *Tetrahedron: Asymmetry*, 18, 1403–1409.

527 Hartikka, A. and Arvidsson, P.I. (2007) *J. Org. Chem.*, 72, 5874–5877.

528 Biswas, A., De Sarkar, S., Tebben, L., and Studer, A. (2012) *Chem. Commun.*, 48, 5190–5192.

529 Cheng, Y., An, J., Lu, L.-Q., Luo, L., Wang, Z.-Y., Chen, J.-R., and Xiao, W.-J. (2011) *J. Org. Chem.*, 76, 281–284.

530 Wang, J., Liu, X., Dong, S., Lin, L., and Feng, X. (2013) *J. Org. Chem.*, 78, 6322–6327.

531 Akagawa, K., Takigawa, S., Nagamine, I.S., Umezawa, R., and Kudo, K. (2013) *Org. Lett.*, 15, 4964–4967.

532 Vesely, J., Zhao, G.-L., Bartoszewicz, A., and Còrdova, A. (2008) *Tetrahedron Lett.*, 49, 4209–4212.

533 Zhang, J.-M., Hu, Z.-P., Dong, L.-T., Xuan, Y.-N., Lou, C.-L., and Yan, M. (2009) *Tetrahedron: Asymmetry*, 20, 355–361.

534 Inokuma, T., Sakamoto, S., and Takemoto, Y. (2009) *Synlett*, 1627–1630.

535 Lv, J., Zhang, J., Lin, Z., and Wang, Y. (2009) *Chem. Eur. J.*, 15, 972–979.

536 Dong, L.-T., Du, Q.-S., Lou, C.-L., Zhang, J.-M., Lu, R.-J., and Yan, M. (2010) *Synlett*, 266–270.

537 Hansen, H.M., Longbottom, D.A., and Ley, S.V. (2006) *Chem. Commun.*, 4838–4840.

538 Wascholowski, V., Hansen, H.M., Longbottom, D.A., and Ley, S.V. (2008) *Synthesis*, 1269–1275.

539 Zaghi, A., Bernardi, T., Bertolasi, V., Bortolini, O., Massi, A., and De Risi, C. (2015) *J. Org. Chem.*, 80, 9176–9184.

540 Pesciaioli, F., Righi, P., Mazzanti, A., Bartoli, G., and Bencivenni, G. (2011) *Chem. Eur. J.*, 17, 2842–2845.

541 Zhao, B.-L. and Du, D.-M. (2015) *Eur. J. Org. Chem.*, 2015, 5350–5359.

542 Das, U., Tsai, Y.-L., and Lin, W. (2013) *Org. Biomol. Chem.*, 11, 44–47.

543 Dou, X. and Lu, Y. (2012) *Chem. Eur. J.*, 18, 8315–8319.

544 Luis-Barrera, J., Mas-Ballestè, R., and Alemàn, J. (2015) *ChemPlusChem*, 80, 1595–1600.

545 Fan, R., Ye, Y., Li, W., and Wang, L. (2008) *Adv. Synth. Catal.*, 350, 2488–2492.

546 (a) Rios, R., Sundén, H., Vesely, J., Zhao, G.-L., Dziedzic, P., and Cordova, A. (2007) *Adv. Synth. Catal.*, 349, 1028–1032; (b) Ibrahem, I., Zhao, G.-L., Rios, R., Vesely, J., Sundén, H., Dziedzic, P., and Córdova, A. (2008) *Chem. Eur. J.*, 14, 7867–7879.

547 Xie, H., Zu, L., Li, H., Wang, J., and Wang, W. (2007) *J. Am. Chem. Soc.*, 129, 10886–10894.

548 Companyó, X., Alba, A.-N., Cárdenas, F., Moyano, A., and Rios, R. (2009) *Eur. J. Org. Chem.*, 2009, 3075–3080.

549 Rueping, M., Sunden, H., Hubener, L., and Sugiono, E. (2012) *Chem. Commun.*, 48, 2201–2203.

550 Terrasson, V., van der Lee, A., Marcia de Figueiredo, R., and Campagne, J.M. (2010) *Chem. Eur. J.*, 16, 7875–7880.

551 Uria, U., Vicario, J.L., Badìa, D., Carrillo, L., Reyes, E., and Pesquera, A. (2010) *Synthesis*, 701–713.

552 Russo, A. and Lattanzi, A. (2010) *Tetrahedron: Asymmetry*, 21, 1155–1157.

553 McCooey, S.H., McCabe, T., and Connon, S.J. (2006) *J. Org. Chem.*, 71, 7494–7497.

554 Xuan, Y.-N., Nie, S.-Z., Dong, L.-T., Zhang, J.-M., and Yan, M. (2009) *Org. Lett.*, 11, 1583–1586.

555 Herchl, R. and Waser, M. (2013) *Tetrahedron Lett.*, 54, 2472–2475.

556 Bakó, P., Rapia, Z., Grüna, A., Nemcsoka, T., Hegedűsb, L., and Keglevicha, G. (2015) *Synlett*, 26, 1847–1851.

557 Del Fiandra, C., Piras, L., Fini, F., Disetti, P., Moccia, M., and Adamo, M.F.A. (2012) *Chem. Commun.*, 48, 3863–3865.

558 Aitken, L.S., Hammond, L.E., Sundaram, R., Shankland, K., Brown, G.D., and Cobb, A.J.A. (2015) *Chem. Commun.*, 51, 13558–13561.

559 Ito, J., Sakuma, D., and Nishii, Y. (2015) *Chem. Lett.*, 44, 297–299.

560 Kim, C. and Kim, S.-G. (2014) *Tetrahedron: Asymmetry*, 25, 1376–1382.

561 Russo, A., Meninno, S., Tedesco, C., and Lattanzi, A. (2011) *Eur. J. Org. Chem.*, 2011, 5096–5103.

562 (a) Noole, A., Sucman, N.S., Kabeshov, M.A., Kanger, T., Macaev, F.Z., and Malkov, A.V. (2012) *Chem. Eur. J.*, 18, 14929–14933; (b) Noole, A., Malkov, A.V., and Kanger, T. (2013) *Synthesis*, 45, 2520–2524.

563 Phillips, A.M.F. and Barros, M.T. (2014) *Eur. J. Org. Chem.*, 2014, 152–163.

564 Piras, L., Moccia, M., Cortigiani, M., and Adamo, M.F.A. (2015) *Catalysts*, 5, 595–605.

565 Papageorgiou, C.D., Cubillo de Dios, M.A., Ley, S.V., and Gaunt, M.J. (2004) *Angew. Chem. Int. Ed.*, 43, 4641–4644.

566 Kojima, S., Suzuki, M., Watanabe, A., and Ohkata, K. (2006) *Tetrahedron Lett.*, 47, 9061–9065.

567 Li, W., Li, X., Ye, T., Wu, W., Liang, X., and Ye, J. (2011) *Tetrahedron Lett.*, 52, 2715–2718.

568 Noole, A., Ošeka, M., Pehk, T., Oeren, M., Jarving, I., Elsegood, M.R.J., Malkov, A.V., Lopp, M., and Kanger, T. (2013) *Adv. Synth. Catal.*, 355, 829–835.

569 César da Silva, R., Chatterjee, I., Escudero-Adán, E., Weber Paixão, M., and Melchiorre, P. (2014) *Asian J. Org. Chem.*, 3, 466–469.

570 Ošeka, M., Noole, A., Žari, S., Oeren, M., Järving, I., Lopp, M., and Kanger, T. (2014) *Eur. J. Org. Chem.*, 2014, 3599–3606.
571 Li, J.-H., Feng, T.-F., and Du, D.-M. (2015) *J. Org. Chem.*, 80, 11369–11377.
572 Johansson, C.C.C., Bremeyer, N., Ley, S.V., Owen, D.R., Smith, S.C., and Gaunt, M.J. (2006) *Angew. Chem. Int. Ed.*, 45, 6024–6028.
573 Kumaraswamy, G. and Padmaja, M. (2008) *J. Org. Chem.*, 73, 5198–5201.
574 Kumaraswamy, G., Ramakrishna, G., and Sridhar, B. (2011) *Tetrahedron Lett.*, 52, 1778–1782.
575 Yamazaki, S., Tanaka, M., and Yamabe, S. (1996) *J. Org. Chem.*, 61, 4046–4050.
576 Klimczyk, S., Misale, A., Huang, X., and Maulide, N. (2015) *Angew. Chem. Int. Ed.*, 54, 10365–10369.
577 Lloyd-Jones, G.C., Wall, P.D., Slaughter, J.L., Parker, A.J., and Laffan, D.P. (2006) *Tetrahedron*, 62, 11402–11412.
578 den Hartog, T., Rudolph, A., Macià, B., Minnaard, A.J., and Feringa, B.L. (2010) *J. Am. Chem. Soc.*, 132, 14349–14351.
579 Lee, H.J., Kim, S.M., and Kim, D.Y. (2012) *Tetrahedron Lett.*, 53, 3437–3439.
580 Gao, Y. and Sharpless, K.B. (1988) *J. Am. Chem. Soc.*, 110, 7538–7539.
581 Ramaswamy, S., Prasad, K., and Repic, O. (1992) *J. Org. Chem.*, 57, 6344–6347.
582 Yoshikawa, N., Tan, L., Yasuda, N., Volante, R.P., and Tillyer, R.D. (2004) *Tetrahedron Lett.*, 45, 7261–7264.
583 Quinkert, G., Schwartz, U., Stark, H., Weber, W.-D., Adam, F., Baier, H., Frank, G., and Dürner, G. (1982) *Liebigs Ann. Chem.*, 1999–2040.
584 Misumi, A., Iwanaga, K., Furuta, K., and Yamamoto, H. (1985) *J. Am. Chem. Soc.*, 107, 3343–3345.
585 (a) Kende, A.S., Fujii, Y., and Mendoza, J.S. (1990) *J. Am. Chem. Soc.*, 112, 9645–9646; (b) Kende, A.S., Mendoza, J.S., and Fujii, Y. (1993) *Tetrahedron*, 49, 8015–8038.
586 Trost, B.M., Dirat, O., and Gunzner, J.L. (2002) *Angew. Chem. Int. Ed.*, 41, 841–843.
587 Pokorski, J.K., Myers, M.C., and Appella, D.H. (2005) *Tetrahedron Lett.*, 46, 915–917.
588 Inoue, T., Kitagawa, O., Ochiai, O., and Taguchi, T. (1995) *Tetrahedron: Asymmetry*, 6, 691–692.
589 Kitagawa, O. and Taguchi, T. (1999) *Synlett*, 1191–1199.
590 Kawachi, A., Maeda, H., and Tamao, K. (2000) *Chem. Lett.*, 29, 1216–1217.
591 Cheng, D., Knox, K.R., and Cohen, T. (2000) *J. Am. Chem. Soc.*, 122, 412–413.
592 (a) Diez, D., Garcia, P., Marcos, I.S., Garrido, N.M., Basabe, P., and Urones, J.G. (2003) *Synthesis*, 53–62; (b) Diez, D., Garcia, P., Marcos, I.S., Garrido, N.M., Basabe, P., Broughton, H.B., and Urones, J.G. (2003) *Org. Lett.*, 5, 3687–3690.
593 Diez, D., Garcia, P., Fernandez, P., Marcos, I.S., Garrido, N.M., Basabe, P., Broughton, H.B., and Urones, J.G. (2005) *Synlett*, 1, 158–160.
594 Diez, D., Garcia, P., Marcos, I.S., Garrido, N.M., Basabe, P., Broughton, H.B., and Urones, J.G. (2005) *Tetrahedron*, 61, 699–707.
595 Boesen, T., Fox, D.J., Galloway, W., Pedersen, D.S., Tyzack, C.R., and Warren, S. (2005) *Org. Biomol. Chem.*, 3, 630–637.

596 (a) Fox, D.J., Parris, S., Pedersen, D.S., Tyzack, C.R., and Warren, S. (2006) *Org. Biomol. Chem.*, 4, 3108–3112; (b) Fox, D.J., Pedersen, D.S., and Warren, S. (2006) *Org. Biomol. Chem.*, 4, 3113–3116.

597 Clarke, C., Foussat, S., Fox, D.J., Pedersen, D.S., and Warren, S. (2009) *Org. Biomol. Chem.*, 7, 1323–1328.

598 Krawczyk, H., Wasek, K., and Kedzia, J. (2009) *Synthesis*, 1473–1476.

599 (a) Nagasawa, T., Handa, Y., Onoguchi, Y., Ohba, S., and Suzuki, K. (1995) *Synlett*, 739–741; (b) Nagasawa, T., Onoguchi, Y., Matsumoto, T., and Suzuki, K. (1995) *Synlett*, 1023–1024; (c) Nagasawa, T., Handa, Y., Onoguchi, Y., and Suzuki, K. (1996) *Bull. Chem. Soc. Jpn.*, 69, 31–39.

600 (a) Taylor, R.E., Engelhardt, F.C., and Yuan, H. (1999) *Org. Lett.*, 1, 1257–1260; (b) Taylor, R.E., Schmitt, M.J., and Yuan, H. (2000) *Org. Lett.*, 2, 601–603; (c) Taylor, R.E., Engelhardt, F.C., Schmitt, M.J., and Yuan, H. (2001) *J. Am. Chem. Soc.*, 123, 2964–2969.

601 Melancon, B.J., Perl, N.R., and Taylor, R.E. (2007) *Org. Lett.*, 9, 1425–1428.

602 Lincoln, C.M., White, J.D., and Yokochi, F.T. (2004) *Chem. Commun.*, 2846–2847.

603 White, J.D., Lincoln, C.M., Yang, J., Martin, W.H.C., and Chan, D.B. (2008) *J. Org. Chem.*, 73, 4139–4150.

604 Risatti, C.A. and Taylor, R.E. (2004) *Angew. Chem. Int. Ed.*, 43, 6671–6672.

605 Kalkofen, R., Brandau, S., Wibbeling, B., and Hoppe, D. (2004) *Angew. Chem. Int. Ed.*, 43, 6667–6669.

606 (a) de Meijere, A., Kozhushkov, S.I., and Savchenko, A.I. (2004) *J. Organomet. Chem.*, 689, 2033–2055; (b) Brackmann, F., Schill, H., and de Meijere, A. (2005) *Chem. Eur. J.*, 11, 6593–6600; (c) Brackmann, F., Colombo, N., Cabrele, C., and de Meijere, A. (2006) *Eur. J. Org. Chem.*, 2006, 4440–4450.

607 Gensini, M. and de Meijere, A. (2004) *Chem. Eur. J.*, 10, 785–790.

608 (a) Norsikian, S., Marek, I., Klein, S., Poisson, J.F., and Normant, J.F. (1999) *Chem. Eur. J.*, 5, 2055–2068; (b) Majumdar, S., de Meijere, A., and Marek, I. (2002) *Synlett*, 423–426.

609 Brandau, S., Fröhlich, R., and Hoppe, D. (2005) *Tetrahedron Lett.*, 46, 6709–6711.

610 Brandau, S. and Hoppe, D. (2005) *Tetrahedron*, 61, 12244–12255.

611 Lou, S., Cuniere, N., Su, B.-N., and Hobson, L.A. (2013) *Org. Biomol. Chem.*, 11, 6796–6805.

612 Kallemeyn, J.M., Mulhern, M.M., and Ku, Y.-Y. (2011) *Synlett*, 535–538.

613 Ito, H., Kosaka, Y., Nonoyama, K., Sasaki, Y., and Sawamura, M. (2008) *Angew. Chem. Int. Ed.*, 47, 7424–7427.

614 Pietruszka, J. and Solduga, G. (2009) *Eur. J. Org. Chem.*, 2009, 5998–6008.

615 McClure, D.E., Arison, B.H., and Baldwin, J.J. (1979) *J. Am. Chem. Soc.*, 101, 3666–3668.

616 Pirrung, M.C., Dunlap, S.E., and Trinks, U.P. (1989) *Helv. Chim. Acta*, 72, 1301–1310.

617 (a) Shuto, S., Ono, S., Hase, Y., Kamiyama, N., Takada, H., Yamasihita, K., and Matsuda, A. (1996) *J. Org. Chem.*, 61, 915–923; (b) Shuto, S., Ono, S., Hase, Y., Kamiyama, N., and Matsuda, A. (1996) *Tetrahedron Lett.*, 37, 641–644; (c) Kazuta, Y., Matsuda, A., and Shuto, S. (2002) *J. Org. Chem.*, 67, 1669–1677.

618 Xu, F., Murry, J.A., Simmons, B., Corley, E., Fitch, K., Karady, S., and Tschaen, D. (2006) *Org. Lett.*, 8, 3885–3888.
619 (a) Burgess, K. and Ho, K.K. (1992) *J. Org. Chem.*, 57, 5931–5936; (b) Burgess, K. and Ho, K.K. (1992) *Tetrahedron Lett.*, 33, 5677–5680.
620 Kitaori, K., Mikami, M., Furukawa, Y., Yoshimoto, H., and Otera, J. (1998) *Synlett*, 499–500.
621 Robinson, A. and Aggarwal, V.K. (2010) *Angew. Chem. Int. Ed.*, 49, 6673–6675.
622 Armstrong, A. and Scutt, J.N. (2003) *Org. Lett.*, 5, 2331–2334.
623 Bray, C.D. and Minicone, F. (2010) *Chem. Commun.*, 46, 5867–5869.
624 Xie, X., Yue, G., Tang, S., Huo, X., Liang, Q., She, X., and Pan, X. (2005) *Org. Lett.*, 7, 4057–4059.
625 Capriati, V., Florio, S., Luisi, R., Perna, F.M., and Barluenga, J. (2005) *J. Org. Chem.*, 70, 5852–5858.
626 Tan, L., Yasuda, N., Yoshikawa, N., Hartner, F.W., Eng, K.K., Leonard, W.R., Tsay, F.-R., Volante, R.P., and Tillyer, R.D. (2005) *J. Org. Chem.*, 70, 8027–8034.
627 Hodgson, D.M., Chung, Y.K., Nuzzo, I., Freixas, G., Kulikiewicz, K.K., Cleator, E., and Paris, J.-M. (2007) *J. Am. Chem. Soc.*, 129, 4456–4462.
628 Mordini, A., Peruzzi, D., Russo, F., Valacchi, M., Reginato, G., and Brandi, A. (2005) *Tetrahedron*, 61, 3349–3360.
629 Winter, P., Swatschek, J., Willot, M., Radtke, L., Olbrisch, T., Schafer, A., and Christmann, M. (2011) *Chem. Commun.*, 47, 12200–12202.
630 Hardee, D.J. and Lambert, T.H. (2009) *J. Am. Chem. Soc.*, 131, 7536–7537.
631 Alliot, J., Gravel, E., Pillon, F., Buisson, D.-A., Nicolas, M., and Doris, E. (2012) *Chem. Commun.*, 48, 8111–8113.
632 Bruneau, C. (2005) *Angew. Chem. Int. Ed.*, 44, 2328–2334.
633 Luzung, M.R., Markham, J.P., and Toste, F.D. (2004) *J. Am. Chem. Soc.*, 126, 10858–10859.
634 Fürstner, A. and Schlecker, A. (2008) *Chem. Eur. J.*, 14, 9181–9191.
635 Barluenga, J., Diéguez, A., Rodriguez, F., and Fananas, F.J. (2005) *Angew. Chem. Int. Ed.*, 44, 126–128.
636 Mamane, V., Gress, T., Krause, H., and Fürstner, A. (2004) *J. Am. Chem. Soc.*, 126, 8654–8655.
637 Fürstner, A. and Hannen, P. (2006) *Chem. Eur. J.*, 12, 3006–3019.
638 Fehr, C. and Galindo, J. (2006) *Angew. Chem. Int. Ed.*, 45, 2901–2904.
639 Soriano, E. and Marco-Contelles, J. (2007) *J. Org. Chem.*, 72, 2651–2654.
640 Newcomb, E.T. and Ferreira, E.M. (2013) *Org. Lett.*, 15, 1772–1775.
641 Shibata, T., Kobayashi, Y., Maekawa, S., Toshida, N., and Takagi, K. (2005) *Tetrahedron*, 61, 9018–9024.
642 Nishimura, T., Kawamoto, T., Nagaosa, M., Kumamoto, H., and Hayashi, T. (2010) *Angew. Chem. Int. Ed.*, 49, 163–1641.
643 Pradal, A., Chao, C.-M., Toullec, P.Y., and Michelet, V. (2011) *Beilstein J. Org. Chem.*, 7, 1021–1029.
644 Dieckmann, M., Jang, Y.-S., and Cramer, N. (2015) *Angew. Chem. Int. Ed.*, 54, 12149–12152.

645 Trost, B.M., Ryan, M.C., Rao, M., and Markovic, T.Z. (2014) *J. Am. Chem. Soc.*, 136, 17422–17425.

646 Watson, I.D.G., Ritter, S., and Toste, F.D. (2009) *J. Am. Chem. Soc.*, 131, 2056–2057.

647 Soriano, E. and Marco-Contelles, J. (2008) *Chem. Eur. J.*, 14, 6771–6779.

648 Ji, K., Zheng, Z., Wang, Z., and Zhang, L. (2015) *Angew. Chem. Int. Ed.*, 54, 1245–1249.

649 Strand, R.B., Helgerud, T., Solvang, T., Dolva, A., Sperger, C.A., and Fiksdahl, A. (2012) *Tetrahedron: Asymmetry*, 23, 1350–1359.

650 Johansson, M.J., Gorin, D.J., Staben, S.T., and Dean Toste, F. (2005) *J. Am. Chem. Soc.*, 127, 18002–18003.

651 Gross, E., Liu, J.H., Alayoglu, S., Marcus, M.A., Fakra, S.C., Toste, F.D., and Somorjai, G.A. (2013) *J. Am. Chem. Soc.*, 135, 3881–3886.

652 Shibata, Y., Noguchi, K., Hirano, M., and Tanaka, K. (2008) *Org. Lett.*, 10, 2825–2828.

653 Oba, M., Nishiyama, N., and Nishiyama, K. (2005) *Tetrahedron*, 61, 8456–8464.

654 (a) Garcia Ruano, J.L., Alonso de Diego, S.A., Martin, M.R., Torrente, E., and Martin Castro, A.M. (2004) *Org. Lett.*, 6, 4945–4948; (b) Garcia Ruano, J.L., Martin Castro, A.M., and Torrente, E. (2005) *Phosphorus, Sulfur Silicon Relat. Elem.*, 180, 1445–1446.

655 McGreer, D.E., Masters, I.M.E., and Liu, M.T.H. (1975) *J. Chem. Soc., Perkin Trans. 2*, 1791–1794.

656 (a) Garcia Ruano, J.L., Peromingo, M.T., Martin, M.R., and Tito, A. (2006) *Org. Lett.*, 8, 3295–3298; (b) Cruz Cruz, D., Yuste, F., Martin, M.R., Tito, A., and Garcia Ruano, J.L. (2009) *J. Org. Chem.*, 74, 3820–3826.

657 Santos, B.S., Gomes, C.S.B., and Pinho e Melo, T.M.V.D. (2014) *Tetrahedron*, 70, 3812–3821.

658 (a) Satoh, T., Ogata, S., and Wakasugi, D. (2006) *Tetrahedron Lett.*, 47, 7249–7253; (b) Ogata, S., Saitoh, H., Wakasugi, D., and Satoh, T. (2008) *Tetrahedron*, 64, 5711–5720; (c) Satoh, T., Kuramoto, T., Ogata, S., Watanabe, H., Saitou, T., and Tadokoro, M. (2010) *Tetrahedron: Asymmetry*, 21, 1–5.

659 Kimura, T., Hattori, Y., Momochi, H., Nakaya, N., and Satoh, T. (2013) *Synlett*, 24, 483–486.

660 Pedroni, J., Saget, T., Donets, P.A., and Cramer, N. (2015) *Chem. Sci.*, 6, 5164–5171.

661 (a) Rousseaux, S., Liegault, B., and Fagnou, K. (2012) *Chem. Sci.*, 3, 244–248; (b) Ladd, C.L., Roman, D.S., and Charrette, A.B. (2013) *Tetrahedron*, 69, 4479–4487.

662 (a) Rubina, M. and Gevorgyan, V. (2004) *Tetrahedron*, 60, 3129–3159; (b) Rubin, M., Rubina, M., and Gevorgyan, V. (2006) *Synthesis*, 1221–1245.

663 Rubina, M., Rubin, M., and Gevorgyan, V. (2003) *J. Am. Chem. Soc.*, 125, 7198–7199.

664 Parra, A., Amenós, L., Guisán-Ceinos, M., López, A., García Ruano, J.L., and Tortosa, M. (2014) *J. Am. Chem. Soc.*, 136, 15833–15836.

665 Rubina, M., Rubin, M., and Gevorgyan, V. (2004) *J. Am. Chem. Soc.*, 126, 3688–3689.

666 Liu, X. and Fox, J.M. (2006) *J. Am. Chem. Soc.*, 128, 5600–5601.

667 Yang, Z., Xie, X., and Fox, J.M. (2006) *Angew. Chem. Int. Ed.*, 45, 3960–3962.

668 Simaan, S., Masarwa, A., Bertus, P., and Marek, I. (2006) *Angew. Chem. Int. Ed.*, 45, 3963–3965.

669 Simaan, S. and Marek, I. (2007) *Org. Lett.*, 9, 2569–2571.

670 Zohar, E., Stanger, A., and Marek, I. (2005) *Synlett*, 2239–2241.

671 Pallerla, M.K. and Fox, J.M. (2005) *Org. Lett.*, 7, 3593–3595.

672 (a) Kulinkovich, O.G., Sviridov, S.V., Vasilevskii, D.A., and Pritytskaya, T.S. (1989) *Zh. Org. Khim.*, 25, 2244–2245; (b) Kulinkovich, O.G. and de Meijere, A. (2000) *Chem. Rev.*, 100, 2789–2834; (c) Kulinkovich, O.G. (2003) *Chem. Rev.*, 103, 2597–2632; (d) Casey, C.P. and Strotman, N.A. (2004) *J. Am. Chem. Soc.*, 126, 1699–1704.

673 Bekish, A.V., Isakov, V.E., and Kulinkovich, O.G. (2005) *Tetrahedron Lett.*, 46, 6979–6981.

674 Baktharaman, S., Selvakumar, S., and Singh, V.K. (2005) *Tetrahedron Lett.*, 46, 7527–7529.

675 (a) Corey, E.J., Rao, S.A., and Noe, M.C. (1994) *J. Am. Chem. Soc.*, 116, 9345–9346; (b) Konik, Y.A., Kananovich, D.G., and Kulinkovich, O.G. (2013) *Tetrahedron*, 69, 6673–6678; (c) Kulinkovich, O.G., Kananovich, D.G., Lopp, M., and Snieckus, V. (2014) *Adv. Synth. Catal.*, 356, 3615–3626.

676 Wessig, P. and Mühling, O. (2005) *Angew. Chem. Int. Ed.*, 44, 6778–6781.

677 Ganesh, N.V. and Jayaraman, N. (2007) *J. Org. Chem.*, 72, 5500–5504.

678 (a) Yun, Y.K., Godula, K., Cao, Y., and Donaldson, W.A. (2003) *J. Org. Chem.*, 68, 901–910; (b) Lukesh, J.M. and Donaldson, W.A. (2005) *Chem. Commun.*, 110–112.

679 Chronakis, N. and Hirsch, A. (2005) *Chem. Commun.*, 3709–3711.

680 Movassaghi, M., Piizzi, G., Siegel, D.S., and Piersanti, G. (2006) *Angew. Chem. Int. Ed.*, 45, 5859–5863.

681 (a) Ouizem, S., Cheramy, S., Botuha, C., Chemla, F., Ferreira, F., and Pérez-Luna, A. (2010) *Chem. Eur. J.*, 16, 12668–12677; (b) Ouizem, S., Chemla, F., Ferreira, F., and Perez-Luna, A. (2012) *Synlett*, 23, 1374–1378.

682 Lauru, S., Simpkins, N.S., Gethin, D., and Wilson, C. (2008) *Chem. Commun.*, 5390–5392.

2
Asymmetric Aziridination

2.1 Introduction

Since the first synthesis of an aziridine reported by Gabriel in 1888 [1], the synthetic scope of aziridine chemistry has blossomed [2]. Aziridines are among the most fascinating heterocyclic intermediates in organic synthesis, acting as precursors of many complex molecules including biologically active compounds due to the strain incorporated in their skeletons. The past decade had witnessed tremendous activity in the area of discovering new routes for their synthesis as well as developing novel transformations of these heterocycles. This considerable attention owes to their striking chemical properties. The high strain energy (27 kcal mol^{-1}) associated with the aziridine ring enables easy cleavage of the C—N bond, allowing regio- and stereoselective installation of a wide range of functional groups in a 1,2-relationship to nitrogen. Aziridines as nitrogen equivalents of epoxides can undergo either ring-cleavage reactions with a range of nucleophiles through a stereocontrolled manner or cycloaddition reactions with dipolarophiles, providing access to a wide range of important nitrogen-containing products [3]. However, they are less widely used in the synthesis compared to their oxygen counterparts, partly because there are fewer efficient methods for aziridination relative to epoxidation. This is particularly true when enantioselective methods are considered. Chiral aziridines can be prepared either by asymmetric catalytic methods or from chiral substrates. The main approaches to the synthesis of chiral aziridines can be classified as transfer of carbon to imines, transfer of nitrogen to olefins, reactions of ylides, cyclization reactions through addition/elimination processes, and miscellaneous reactions such as intramolecular substitutions. Obtaining aziridines, especially optically active aziridines, has become of great importance in organic chemistry for many reasons including the antitumor and antibiotic properties, among other biological activities of a great number of aziridine-containing compounds [4]. Indeed, as powerful alkylating agents, aziridines have an inherent *in vivo* potency by their ability to act as DNA cross-linking agents via nucleophilic ring opening of the aziridine moiety. Structure–activity relationships have identified the aziridine ring as being essential for the

Asymmetric Synthesis of Three-Membered Rings, First Edition.
Hélène Pellissier, Alessandra Lattanzi and Renato Dalpozzo.
© 2017 Wiley-VCH Verlag GmbH & Co. KGaA. Published 2017 by Wiley-VCH Verlag GmbH & Co. KGaA

antitumor activity, and a vast amount of work has concentrated on synthesizing derivatives of these natural products with increased potency. Various antitumor agents related to mitosanes and mitomycins, for example, have been synthesized and demonstrated to possess activity against a variety of cancers. A number of other synthetic chiral aziridines have also been shown to exhibit other useful biological properties such as enzyme-inhibitory activities. In addition to their important biological activities, their easy transformation into pharmacologically and biologically active compounds, their appearance as structural subunits in naturally occurring substances, chiral aziridines are used as precursors for a number of chiral ligands [5] and applied as chiral building blocks for the construction of various chiral nitrogen compounds, such as chiral amines, amino acids, β-amino sulfonic acids, amino alcohols, alkaloids, and β-lactam antibiotics.

The goal of this chapter is to collect the advances in asymmetric aziridination using chiral substrates (auxiliaries) as well as chiral catalysts, focusing on those published in the last 12 years. Previously, this area has been the subject of several reports [6]. This chapter is subdivided into two parts, dealing successively with asymmetric aziridination based on the use of chiral substrates (auxiliaries) and asymmetric aziridination based on the use of chiral catalysts. The first part is subdivided into five sections, according to the different methods employed to prepare chiral aziridines, such as asymmetric aziridination of alkenes, asymmetric aziridination of imines, asymmetric addition to azirines, asymmetric aziridination achieved through intramolecular substitution, and miscellaneous reactions. The second part of the chapter dedicated to the use of chiral catalysts is subdivided into four sections, dealing successively with enantioselective aziridination of alkenes, enantioselective aziridination of imines, miscellaneous enantioselective reactions, and kinetic resolutions of aziridines.

2.2 Aziridination Based on the Use of Chiral Substrates

2.2.1 Addition to Alkenes

Nitrogen-atom transfer to alkenes is a particularly appealing strategy for the generation of aziridines because of the ready availability of olefinic starting materials and the direct nature of such a process. There are two general methods for the addition of nitrenes and nitrenoids to alkenes, involving a one- or a two-step mechanism. Nitrenes and metallonitrenes add to alkenes by a direct aziridination reaction, whereas nonmetallic nitrenoids usually react through an addition–elimination process.

2.2.1.1 Aziridination via Nitrene Transfer to Alkenes
The aziridination of olefins is typically accomplished by using a nitrene-transfer reagent. The nitrene source for this reaction, a nitrene or a nitrenoid, can be generated from various methodologies, such as the metal-catalyzed reaction of [N-(p-toluenesulfonyl)imino]aryliodinanes [7], the oxidation of primary amines,

the α-elimination of HX from an amine or amide with an electronegative atom X (X = halogen or oxygen) attached to the NH group, the thermolytic or photolytic decomposition of organyl azides, and the α-elimination of metal halides from metal *N*-arenesulfonyl-*N*-haloamines. In 1993, Evans and Jacobsen independently reported the copper-catalyzed asymmetric alkene aziridination using [*N*-(*p*-toluenesulfonyl)imino]phenyliodinane (PhI=NTs) as the nitrene source [8]. This process had been successfully applied to the total synthesis of several natural or biologically active products, such as (+)-agelastatin A, possessing nanomolar activity against several cancer cell lines. Furthermore, this natural product inhibits glycogen synthase kinase-3β, a behavior that might provide an approach for the treatment of Alzheimer's disease. In 2006, Trost and Dong reported a total synthesis of (+)-agelastatin A based on the aziridination of a chiral piperazinone depicted in Scheme 2.1 [9]. This reaction was performed in the presence of PhI=NTs as the nitrene source and a catalytic amount of a novel copper *N*-heterocyclic carbene complex, providing the corresponding enantiomerically pure aziridine in moderate yield. This chiral aziridine was further converted into the expected (+)-agelastatin A in four steps.

In 2007, Flock and Frauenrath applied this type of methodology to the aziridination of chiral dioxins [10]. Unexpectedly, the Cu-catalyzed reaction of chiral 5-methyl-4*H*-1,3-dioxins with the nitrene generated from PhI=NTs led to the corresponding 4-methyl-1,3-oxazolidine-4-carbaldehydes according to an aziridination–rearrangement process with diastereoselectivities of up to 73% de (Scheme 2.2). These authors assumed that the aziridination of chiral dioxins intermediately afforded the corresponding aziridines, which immediately rearranged via ring opening/ring contraction to give the corresponding diastereomeric oxazolidine-carbaldehydes. When the reaction was carried out in solvents other than TBME, such as acetonitrile, acetone, or dichloromethane, a mixture of similar diastereomers was obtained, but with respect to the diastereoselectivity of the intermediate nitrogen-transfer step, the reaction proved to be unselective (Scheme 2.2).

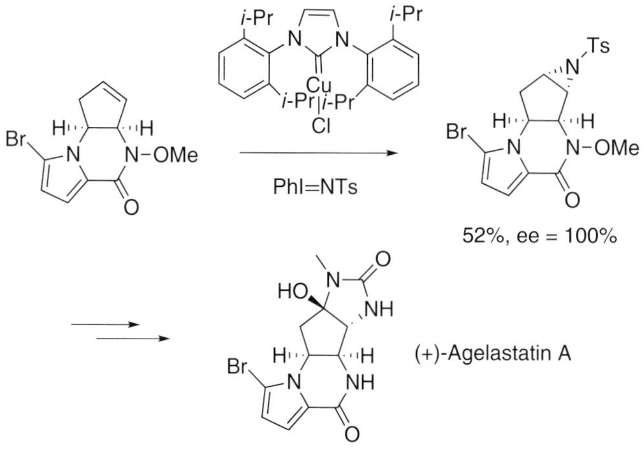

Scheme 2.1 Synthesis of (+)-agelastatin A through Cu-catalyzed aziridination of a chiral piperazinone using PhI=NTs.

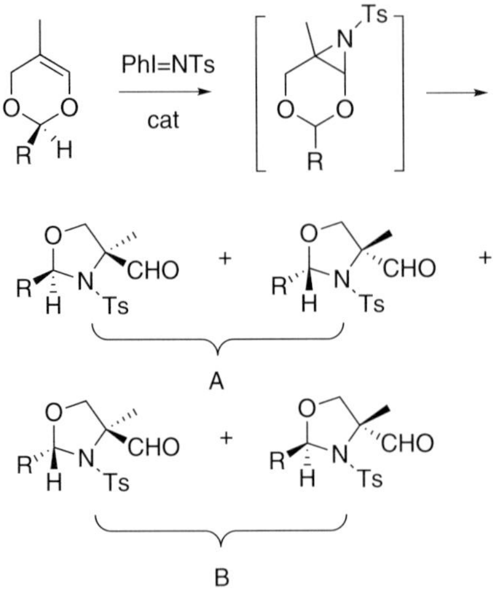

Scheme 2.2

R = *t*-Bu, cat = CuOTf, solvent = TBME:
41%, de = 73%, A/B = 40/60
R = *i*-Pr, cat = CuClO$_4$, solvent = TBME:
51%, de = 30%, A/B = 65/35
R = *i*-Pr, cat = CuClO$_4$, solvent = MeCN:
60%, de = 0%, A/B = 90/10
R = *i*-Pr, cat = CuClO$_4$, solvent = CH$_2$Cl$_2$:
51%, de = 0%, A/B = 71/29
R = *i*-Pr, cat = CuClO$_4$, solvent = acetone:
58%, de = 0%, A/B = 87/13

Scheme 2.2 Cu-catalyzed aziridination of chiral 5-methyl-4*H*-1,3-dioxins using PhI=NTs.

However, a disadvantage of this methodology is the often troublesome removal of the stable *N*-arylsulfonyl group, both in the intact aziridine and in the products of subsequent nucleophilic aziridine ring opening. It was for this reason that Dauban and Dodd developed, as an alternative, the trimethylsilylethanesulfonyl (PhI=NSes) version of these hypervalent iodine reagents, [*N*-(arenesulfonyl) imino]aryliodinanes, in order to allow facile removal of the alkylsulfonyl group using a source of fluoride anion [11]. In 2003, these authors applied the Ses iminoiodane to the Cu-catalyzed aziridination of 11-pregnane derivatives in order to prepare a chiral 11,12-aziridino analog of neuroactive steroids [12]. Using this method, the reaction of chiral 11-pregnene-3,20-dione or 3-α-acetoxy-11-pregnen-20-one with PhI=NSes in the presence of copper(I) triflate provided the corresponding α,α-11,12-aziridino steroids in moderate yields, as shown in Scheme 2.3. The 3α-acetoxy-11-pregnen-20-one derivative was further converted via TASF-mediated removal of the *N*-Ses blocking group into *N*-methyl-11,12-aziridino-3α-hydroxy-5β-pregnan-20-one, a conformationally constrained analog of the endogenous neurosteroid, pregnanolone, and a structural analog of the synthetic general anesthetic, minaxolone.

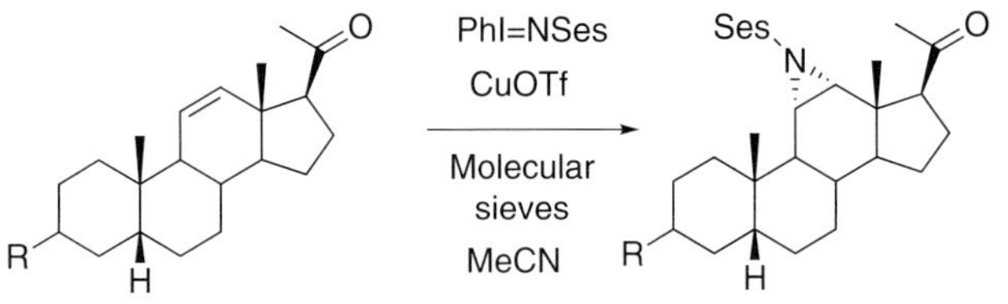

Scheme 2.3 Cu-catalyzed aziridination of chiral 11-pregnene derivatives using PhI=NSes.

R = O: 53%, ee = 100%
R = α-OAc: 45%, ee = 100%

One of the drawbacks of the described methods is the need to isolate the nitrene precursors (PhI=NR), some of which were reported to be unstable and explosive [13]. In order to simplify the procedure for the preparation of the nitrene precursor, Dauban et al. have reported a modified procedure for the copper-catalyzed asymmetric alkene aziridination [14]. In their procedures, PhI=NR was generated in situ by treating sulfonamides with iodosylbenzene (PhI=O) [14a]. These processes greatly simplified the troublesome handling of iminoiodanes but, more importantly, enhanced the variety of nitrogenous compounds that could be transformed into the hypervalent iodine(III) reagents. In this context, Dauban et al. reported, in 2004, that chiral N-(p-toluenesulfonyl)-p-toluenesulfonimidamide reacted with iodosylbenzene to afford an in situ chiral iminoiodane, which gave, in the presence of a copper(I) catalyst, a nitrene that was very efficiently transferred under stoichiometric conditions to a wide range of alkenes with diastereoselectivities of up to 60% de [15]. As shown in Scheme 2.4, the corresponding chiral aziridines were isolated in good-to-excellent yields, particularly in the case of poorly reactive α,β-unsaturated esters.

The combination of sulfonamide SesNH$_2$ with PhI=O was employed by these authors to achieve the copper-catalyzed aziridination of a chiral N-phenylfluorenyl(Pf)-protected α-allylglycine derivative [16]. This reaction con-

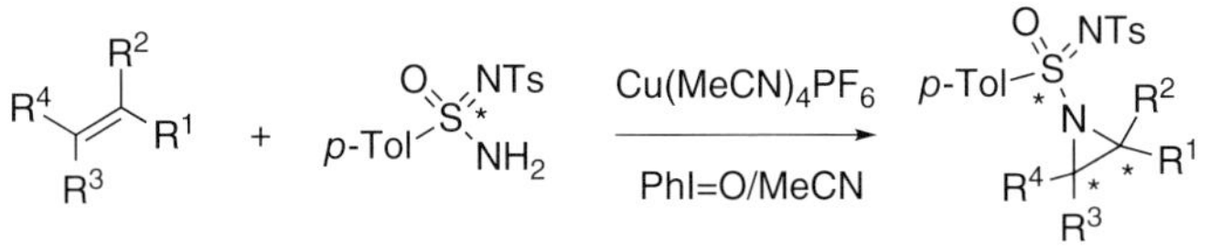

R^1 = CO$_2$Me, R^2 = R^3 = R^4 = H: 93%, de = 55%
R^1 = CO$_2$Me, R^2 = Me, R^3 = R^4 = H: 96%, de = 41%
R^1 = CO$_2$Me, R^2 = R^3 = H, R^4 = Me: 63%, de = 60%
R^1 = CO$_2$Me, R^2 = R^4 = Me, R^3 = H: 90%, de = 40%
R^1 = n-Pent, R^2 = R^3 = R^4 = H: 78%, de = 10%
R^1 = R^4 = H, R^2 = Me, R^3 = n-Pr: 73%, de <10%
R^1 = Me, R^2 = R^4 = H, R^3 = n-Pr: 79%, de <10%
R^1,R^3 = (CH$_2$)$_4$, R^2 = R^4 = H: 89%, de = 0%
R^1,R^3 = (CH$_2$)$_3$, R^2 = R^4 = H: 91%, de = 0%
R^1 = Ph, R^2 = R^3 = R^4 = H: 96%, de <10%

Scheme 2.4 Cu-catalyzed aziridination of alkenes using a chiral sulfonimidamide and PhI=O.

stituted the key step of a novel strategy to prepare new chiral rigid analogs of arginine. Several attempts were made to improve the yield and/or the diastereoselectivity of the aziridination, but these all proved to be somewhat unsatisfactory. Thus, while using benzene or dichloromethane instead of acetonitrile as the reaction solvent allowed moderately higher diastereoselectivities to be obtained; this was at the detriment of the yields, as shown in Scheme 2.5. Moreover, no substantial changes in both the yields and the diastereoselectivities were observed when the reaction was performed at lower (5 °C) or higher (45 °C) temperatures than at room temperature. Even doubling the quantity of the copper catalyst led to a slightly decreased aziridine yield. Finally, the best diastereoselectivity (50% de), albeit combined with a low yield (16%), was achieved when the reaction was carried out in the presence of a chiral bisoxazoline ligand, as shown in Scheme 2.5.

Earlier in 2000, Che et al. demonstrated that PhI=NR could also be simply generated by treating sulfonamides with a commercially available reagent, iodobenzene diacetate (PhI(OAc)$_2$) [17]. In this context, Du Bois et al. have developed the asymmetric intramolecular aziridination of chiral homoallyl sulfamates performed in the presence of a rhodium catalyst, MgO, and PhI(OAc)$_2$ [18]. This Rh-catalyzed alkene oxidation process allowed the corresponding bicyclic chiral aziridines to be obtained in high yields and diastereoselectivities of up to 90% de (Scheme 2.6). It must be noted that these stable, heteroatom-substituted aziridines constituted rather unusual structural motifs, access to which was not readily apparent in the absence of this novel method. A number of differently configured chiral unsaturated sulfamates have been tested to map the scope of this new aziridination process. While only moderate stereocontrol was noted for sulfamates derived from secondary alcohols, a sharp increase in diastereoselectivity was recorded for substrates containing a stereogenic element at the β-site. These results have led these authors to propose a stereochemical model to account for the observed sense of induction. It was consistent with an aziridination proceeding through a chair-like transition state (Scheme 2.6), which minimized gauche and A$_{1,3}$-type interactions. It is interesting to note that even

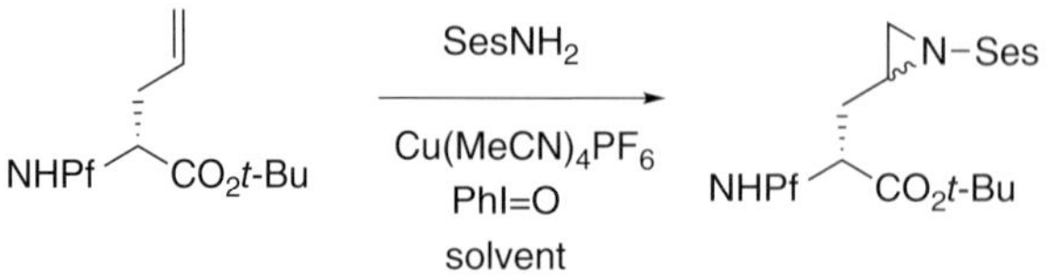

Solvent = MeCN: 32%, de = 30%
Solvent = C$_6$H$_6$: 26%, de = 34%
Solvent = CH$_2$Cl$_2$: 24%, de = 40%
Solvent = MeCN with 35 mol% of L*: 16%, de = 50%

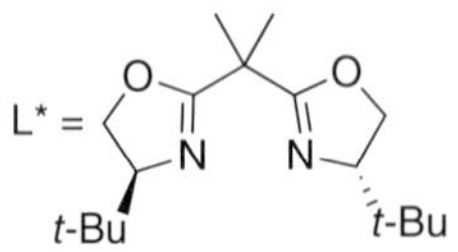

Scheme 2.5 Cu-catalyzed aziridination of a chiral α-allylglycine derivative using SesNH$_2$ and PhI=O.

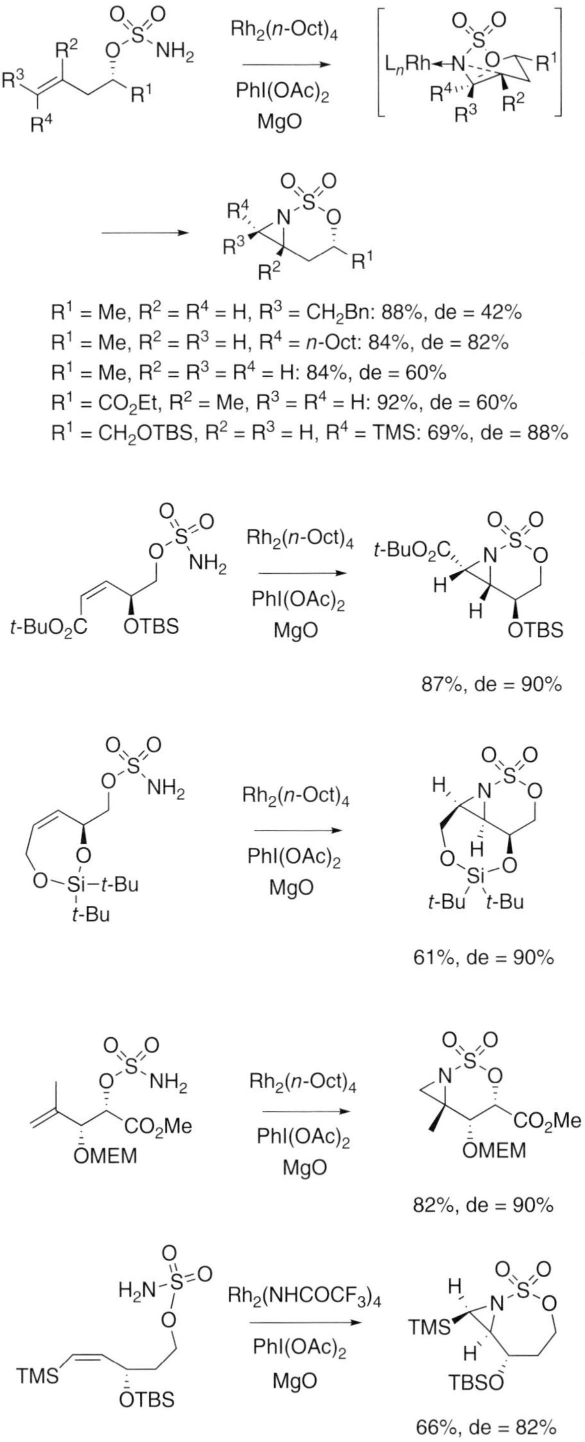

R^1 = Me, R^2 = R^4 = H, R^3 = CH_2Bn: 88%, de = 42%
R^1 = Me, R^2 = R^3 = H, R^4 = n-Oct: 84%, de = 82%
R^1 = Me, R^2 = R^3 = R^4 = H: 84%, de = 60%
R^1 = CO_2Et, R^2 = Me, R^3 = R^4 = H: 92%, de = 60%
R^1 = CH_2OTBS, R^2 = R^3 = H, R^4 = TMS: 69%, de = 88%

87%, de = 90%

61%, de = 90%

82%, de = 90%

66%, de = 82%

Scheme 2.6 Rh-catalyzed intramolecular aziridinations of chiral homoallyl sulfamates using $PhI(OAc)_2$.

seven-membered ring oxathiazepane-fused aziridines could also be generated with useful levels of diastereocontrol (Scheme 2.6).

As an alternative to the use of $Rh_2(NHCOCF_3)_4$ as the catalyst, which proved to be both tedious to prepare and difficult to isolate in high purity, Keaney and Wood have explored the use of rhodium perfluorobutyramide $(Rh_2(pfm)_4)$ for the aziridination of olefins [19]. Thus, the treatment of a chiral olefin depicted in Scheme 2.7 by trichloroethylsulfamate ester in the presence of a combination of $Rh_2(pfm)_4$ with $PhI(OAc)_2$ provided the expected trichloroethoxysulfonylaziridine in good yield and with moderate diastereoselectivity. This compound was a potent intermediate for the synthesis of (+)-kalihinol A.

With the aim of developing a synthesis of the orally active neuraminidase inhibitor, (−)-oseltamivir, Trost and Zhang have recently studied the asymmetric aziridination of a chiral diene depicted in Scheme 2.8 [20]. In this case, the best result for the aziridination was obtained when the reaction was performed in the presence of $SesNH_2$ as the nitrene source, $PhI(OPiv)_2$ as the oxidant, bis-[rhodium($\alpha,\alpha,\alpha',\alpha'$-tetramethyl-1,3-benzenedipropionate)] $[Rh_2(esp)_2]$ as the catalyst, and chlorobenzene as the solvent, as shown in Scheme 2.8. Indeed, the corresponding enantiopure γ,δ-aziridine was most satisfyingly isolated as the sole product of the reaction in 86% yield. It must be noted that the use of copper catalysts were unable to achieve the expected regioselective aziridination reaction of the starting functionalized diene since they showed no selectivity between the α,β- and the γ,δ-double bonds, forming an unseparable mixture of products. This enantiopure product was further converted into required (−)-oseltamivir through four supplementary steps with an overall yield of 30% from starting commercially available materials.

In 2009, Du Bois et al. reported the asymmetric synthesis of chiral N-allyl hydroxylamine-O-sulfamates through enantioselective palladium-catalyzed allylic aminations [21]. These products could be readily transformed into the corresponding enantiopure amino aziridines via diastereoselective oxidative cyclization upon treatment with $[Rh_2(esp)_2]$ as catalyst, $PhI(OAc)_2$, and MgO. As

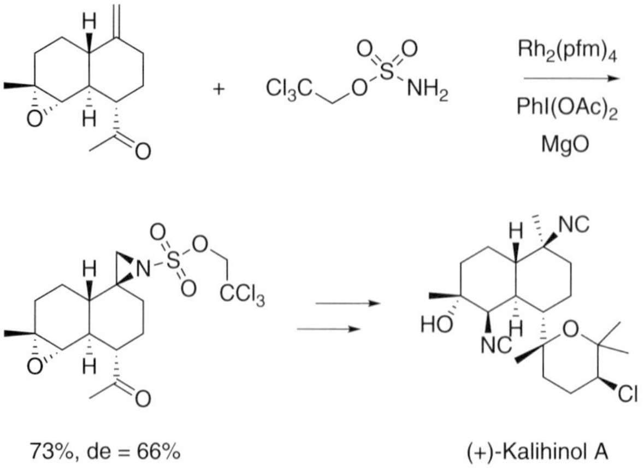

Scheme 2.7 Synthesis of (+)-kalihinol A though Rh-catalyzed aziridination.

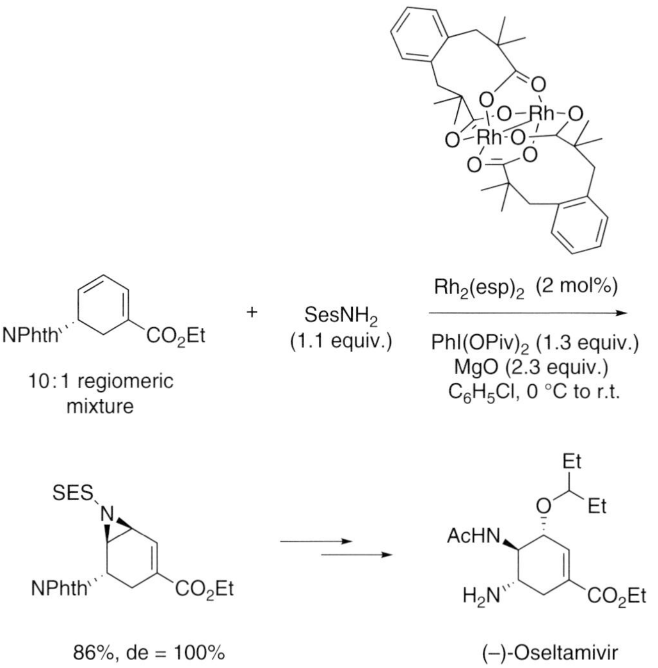

Scheme 2.8 Synthesis of (−)-oseltamivir through Rh-catalyzed aziridination.

shown in Scheme 2.9, both cyclic and acyclic olefins underwent the oxidative cyclization, providing the corresponding tricyclic or bicyclic aziridines, both in high yields and diastereoselectivities.

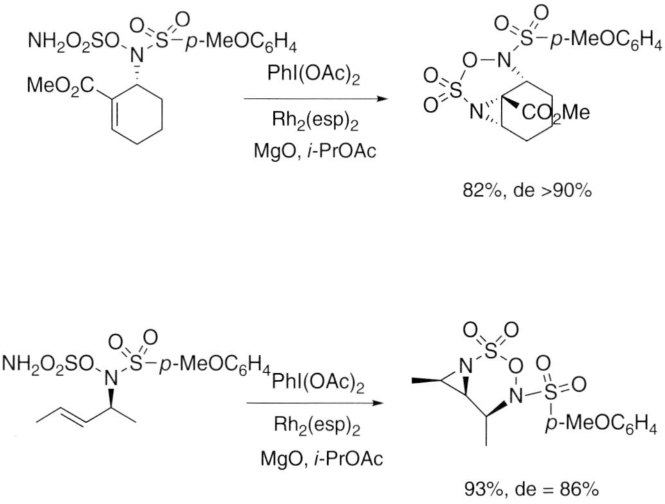

Scheme 2.9 Rh-catalyzed intramolecular aziridinations of chiral N-allyl hydroxylamine-O-sulfamates using PhI(OAc)$_2$.

Another methodology to generate nitrenes consists of the *in situ* oxidation of hydrazine derivatives in the presence of lead tetraacetate. In 2005, Vederas et al. successfully applied this methodology to the asymmetric aziridination of a camphor derivative in the presence of 3-amino-2-ethyl-3,4-dihydroquinazolin-4-one combined with Pb(OAc)$_4$ and hexamethyldisilazide (HMDS) [22]. Thus, the oxidative addition of this aminoquinazolinone mediated by Pb(OAc)$_4$ produced the corresponding aziridine in good yield and with a diastereoselectivity of 80% de, as shown in Scheme 2.10.

As an extension of this methodology, these authors have investigated the use of chiral 3-acetoxyaminoquinazolinones for the aziridination of an unsaturated α-aminopimelic ester in order to prepare aziridine analogs of diaminopimelic acid (DAP), which is an inhibitor of DAP epimerase [22]. Thus, the aziridination of this chiral alkene performed in the presence of the chiral aminoquinazolinone depicted in Scheme 2.11 and Pb(OAc)$_4$ provided the expected corresponding aziridine in moderate yield and with a diastereoselectivity of 72% de.

Moreover, an asymmetric intramolecular version of this methodology was developed by the same authors, providing the corresponding 10-membered polycyclic aziridines with almost complete diastereoselectivity, as shown in Scheme 2.12 [22].

Later, Chen et al. have studied the treatment of a range of N- and O-enones derived from various camphor-based chiral auxiliaries with N-aminophthalimide in the presence of Pb(OAc)$_4$ [23]. In general, both high diastereoselectivities of up to 98% de and high yields of up to 95% were obtained for the formed chiral N-phthalimidoaziridines, as shown in Scheme 2.13.

Another methodology to prepare aziridines is based on the thermolytic or photolytic decomposition of organyl azides [24]. In 2008, Tanaka et al. employed this method as the key step of a total synthesis of (−)-agelastatin A, a potent antineoplastic agent [25]. Thus, the requisite nitrogen functionality of the agelastatin

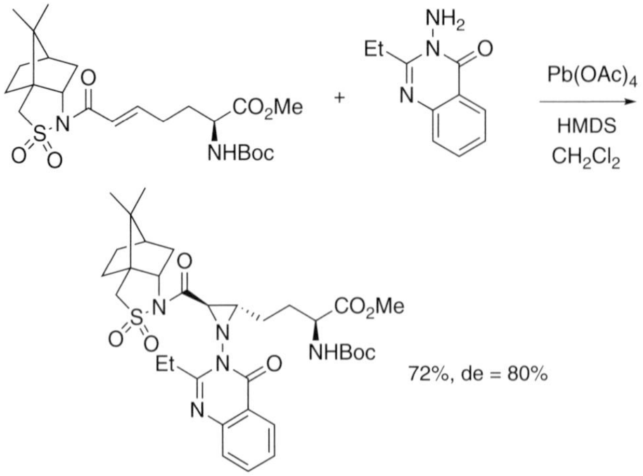

Scheme 2.10 Pb-catalyzed aziridination of a camphor derivative with 3-amino-2-ethyl-3,4-dihydroquinazolin-4-one.

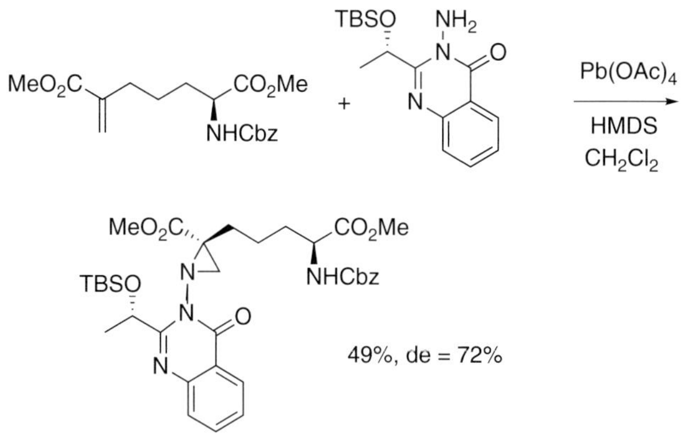

Scheme 2.11 Pb-catalyzed aziridination of a chiral unsaturated α-aminopimelic ester with a chiral aminoquinazolinone.

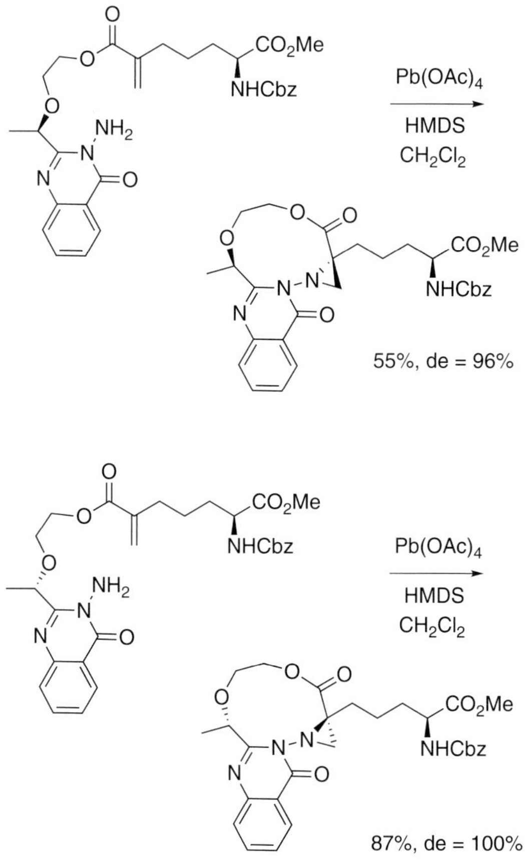

Scheme 2.12 Pb-catalyzed intramolecular aziridinations of chiral unsaturated aminoquinazolinones.

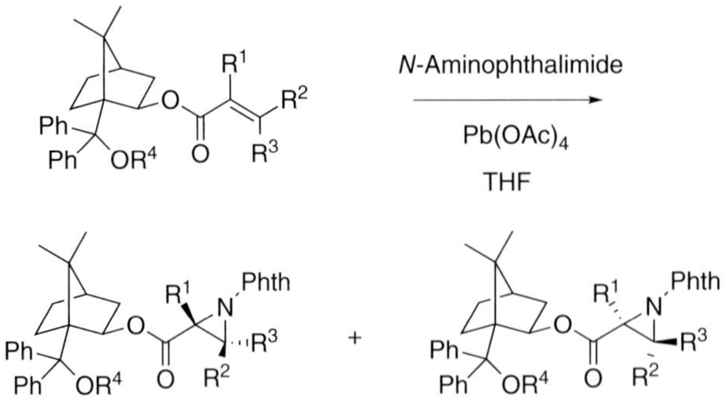

$R^1 = R^3 = R^4 = H$, $R^2 = Me$: 90%, 82:18
$R^1 = R^3 = R^4 = H$, $R^2 = Et$: 95%, 93:7
$R^1 = R^3 = R^4 = H$, $R^2 = n\text{-Pr}$: 90%, 93:7
$R^1 = R^3 = R^4 = H$, $R^2 = i\text{-Pr}$: 95%, 95:5
$R^1 = R^3 = R^4 = H$, $R^2 = Ph$: 75%, 90:10
$R^1 = Me$, $R^2 = Et$, $R^3 = R^4 = H$: 95%, 61:39
$R^1 = R^2 = Me$, $R^3 = R^4 = H$: 95%, 95:5
$R^1 = Me$, $R^2 = Et$, $R^3 = R^4 = H$: 95%, 95:5
$R^1 = R^4 = H$, $R^2 = R^3 = Me$: 70%, 5:95
$R^1 = R^2 = R^3 = H$, $R^4 = Me$: 90%, 87:13

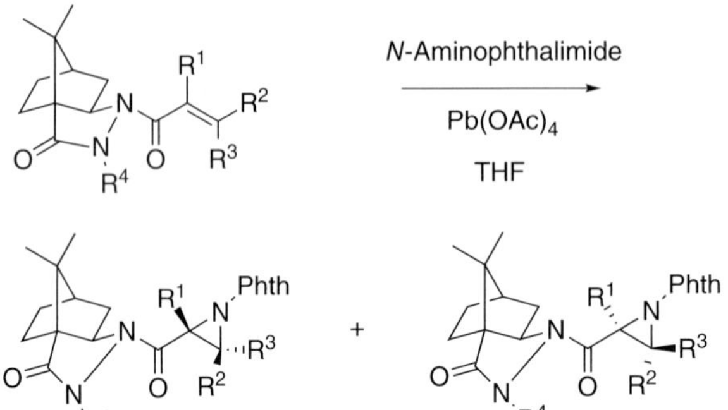

$R^1 = R^2 = R^3 = H$, $R^4 = Ph$: 94%, >95:5
$R^1 = R^3 = H$, $R^2 = Me$, $R^4 = Ph$: 90%, >95:5
$R^1 = R^2 = R^3 = H$, $R^4 = Ts$: 50%, >95:5
$R^1 = R^3 = H$, $R^2 = Me$, $R^4 = Ts$: 53%, >95:5

Scheme 2.13 Pb-catalyzed aziridinations of camphor-derived *N*- and *O*-enones with *N*-aminophthalimide.

core was installed by thermolytic intramolecular aziridination of a chiral azidoformate. The formed enantiopure tricyclic aziridine was further submitted to a regioselective azidation, leading to *trans*-diamination of the double bond. The obtained chiral azide was subsequently converted into the expected (−)-agelastatin A (Scheme 2.14).

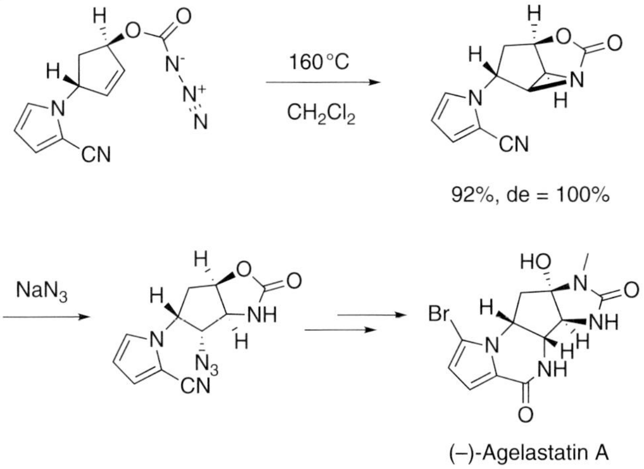

Scheme 2.14 Synthesis of (−)-agelastatin A through thermal intramolecular aziridination of a chiral azidoformate.

In 2006, Lowary et al. reported the synthesis of L-daunosamine and L-ristosamine glycosides on the basis of photoinduced intramolecular aziridinations of acylnitrenes derived from L-rhamnose [26]. As shown in Scheme 2.15, a chiral acyl azide allowed, upon exposure to UV light (254 nm), the corresponding aziridine to be obtained by generation of a presumed acylnitrene intermediate in good yield (79%) and with total diastereoselectivity. This aziridine was further converted into a glycoside L-daunosamine derivative. Similarly, the irradiation of a 2:1 α:β

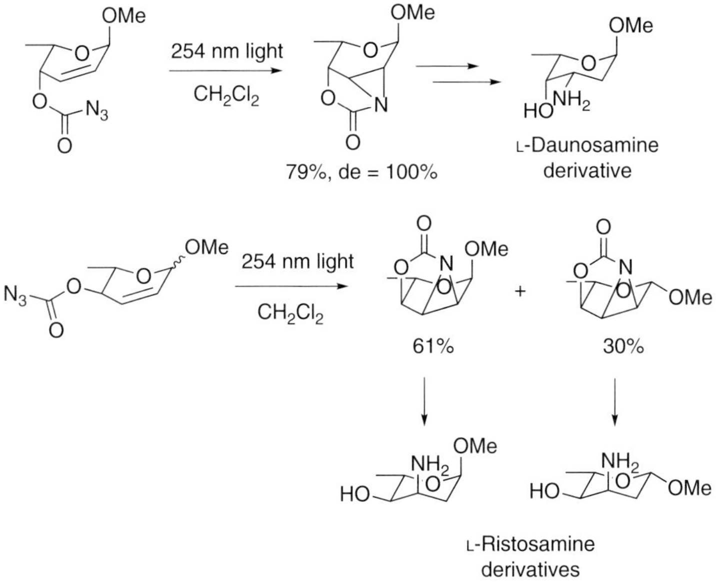

Scheme 2.15 Syntheses of L-daunosamine and L-ristosamine glycosides through intramolecular photoinduced aziridinations of chiral acyl azides.

anomeric mixture of an acyl azide of an L-*erythro*-hex-2-enopyranoside derivative provided a mixture of the corresponding aziridines in 91% yield (Scheme 2.15). These products were further separated by chromatography and then converted into glycoside L-ristosamine derivatives.

The aziridination of alkenes can also be achieved by using chloramine T (*N*-chloro-*N*-sodio-*p*-toluenesulfonamide) as the nitrene source. This methodology was recently applied by Chandrasekaran et al. for the synthesis of both enantiomers of *cis*-aziridino epoxides from (*R*)-(−)- and (*S*)-(+)-carvones [27]. Thus, treatment of enantiopure epoxycarvones by chloramine T in combination with phenyltrimethylammonium tribromide (PTAB) provided, for each enantiomer, a diasteromeric mixture of the corresponding aziridino epoxides in high yield, as shown in Scheme 2.16. Unfortunately, these authors did not mention the

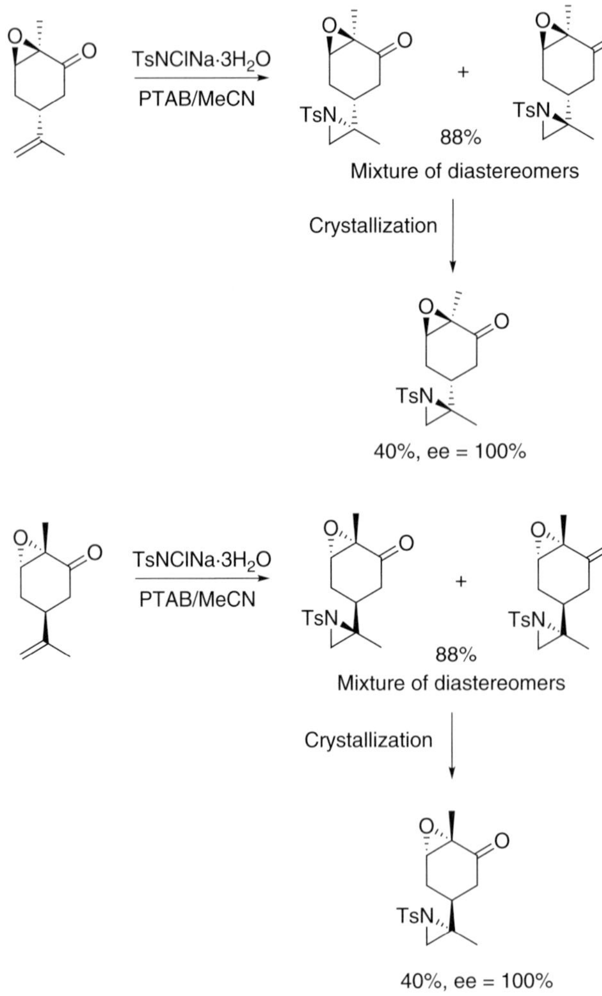

Scheme 2.16 Syntheses of *cis*-aziridino epoxides derived from (*R*)-(−)- and (*S*)-(+)-carvones by using chloramine T.

diastereoselectivity of the reactions, but only the yields of the required enantiopure *cis*-aziridino epoxides, which were further purified by crystallization.

2.2.1.2 Aziridination via Addition–Elimination Processes

The Gabriel–Cromwell aziridine synthesis involves a nucleophilic addition of a formal nitrene equivalent to a 2-haloacrylate or similar reagent. Thus, there is an initial Michael addition, followed by protonation and 3-*exo-tet* ring closure. In 2003, Maycock et al. reported the Gabriel–Cromwell reaction of a chiral α-iodocyclohexenone derived from (−)-quinic acid [28]. The aziridination performed in the presence of 4-methoxybenzylamine and Cs_2CO_3 as a base afforded the corresponding aziridine in good yield as an 80:20 mixture of diastereomers, as shown in Scheme 2.17. This reaction constituted the key step of a short route to (+)-bromoxone, the acetate of which showed potent antitumor activity.

In the same context, Dodd et al. have developed a total synthesis of the nonnatural enantiomer of polyoxamic acid on the basis of a tandem Michael-type addition/elimination involving a chiral triflate derived from D-ribonolactone [29]. As shown in Scheme 2.18, this triflate reacted with 3,4-dimethoxybenzylamine to provide the corresponding aziridine as a single diastereomer in good yield. The diastereoselectivity of the reaction was explained as a result of a Michael-type addition of 3,4-dimethoxybenzylamine to the face opposite to that

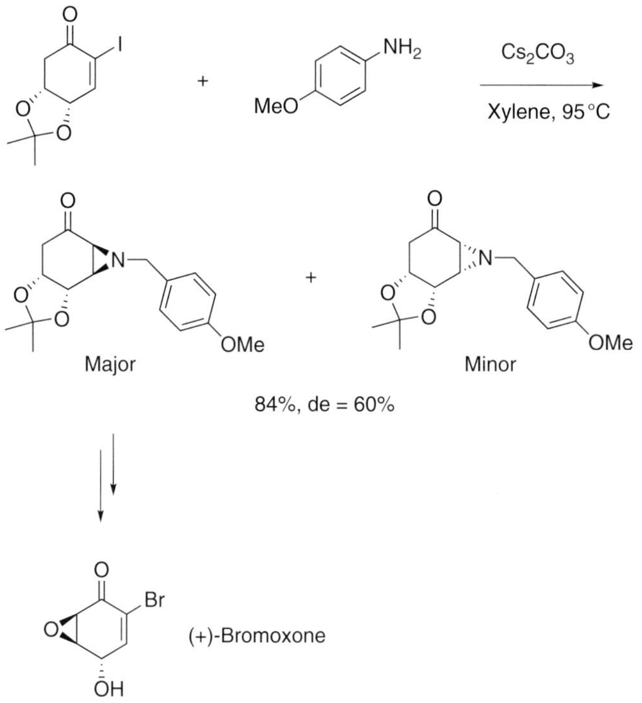

Scheme 2.17 Synthesis of (+)-bromoxone through Gabriel–Cromwell reaction.

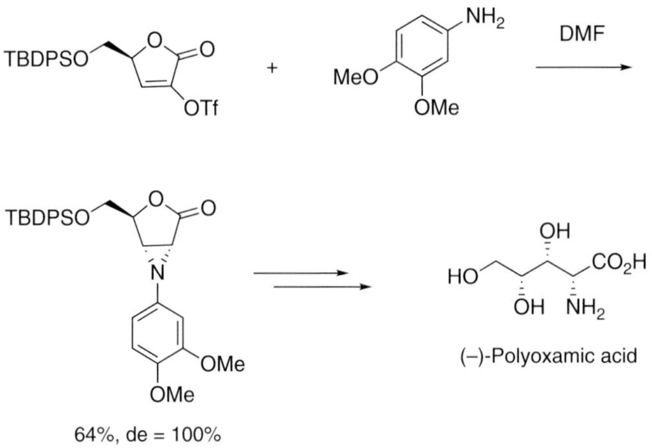

Scheme 2.18 Synthesis of (−)-polyoxamic acid through Gabriel–Cromwell reaction.

of the bulky silyl group at C-5. This aziridine was further converted through six steps into the expected (−)-polyoxamic acid with an overall yield of 10%.

In 2010, Lobo et al. described the Michael addition of the lithium salt of a chiral sulfinamide to cyclic α-bromo enones followed by intramolecular cyclization to provide the corresponding aziridines as a diastereomeric mixture in low-to-excellent yields (11–96%) and with diastereoselectivities ranging from 12% to 65% de, as shown in Scheme 2.19 [30]. A key factor associated with chiral induction was the incorporation of the reacting olefin in a cycle, indicating the importance of conformational restriction in the reacting double bond.

On the other hand, Satoh et al. have demonstrated that the treatment of chiral α-chlorovinylsulfoxides with N-lithio o-anisidine led to the formation of the corresponding sulfinylaziridines in good yields [31]. The reaction of either (E)-vinylsulfoxide or (Z)-vinylsulfoxide provided the corresponding aziridine as a single diastereomer (Scheme 2.20). These authors proposed the reaction mechanism depicted in Scheme 2.20 in order to explain the chiral induction of the reaction of (E)-vinylsulfoxide with N-lithio o-anisidine. Thus, the lithium cation formed a five-membered chelate between the oxygen of the sulfoxide and the chlorine atom [32]. The nitrogen nucleophile attacked from the less hindered

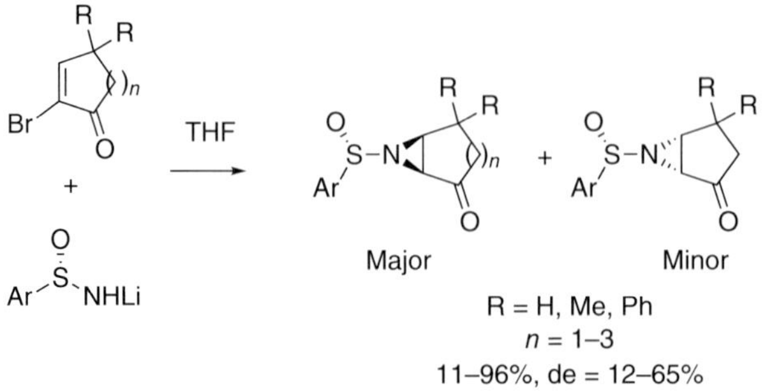

Scheme 2.19 Reaction of cyclic α-bromo enones with lithium salt of a chiral sulfinamide.

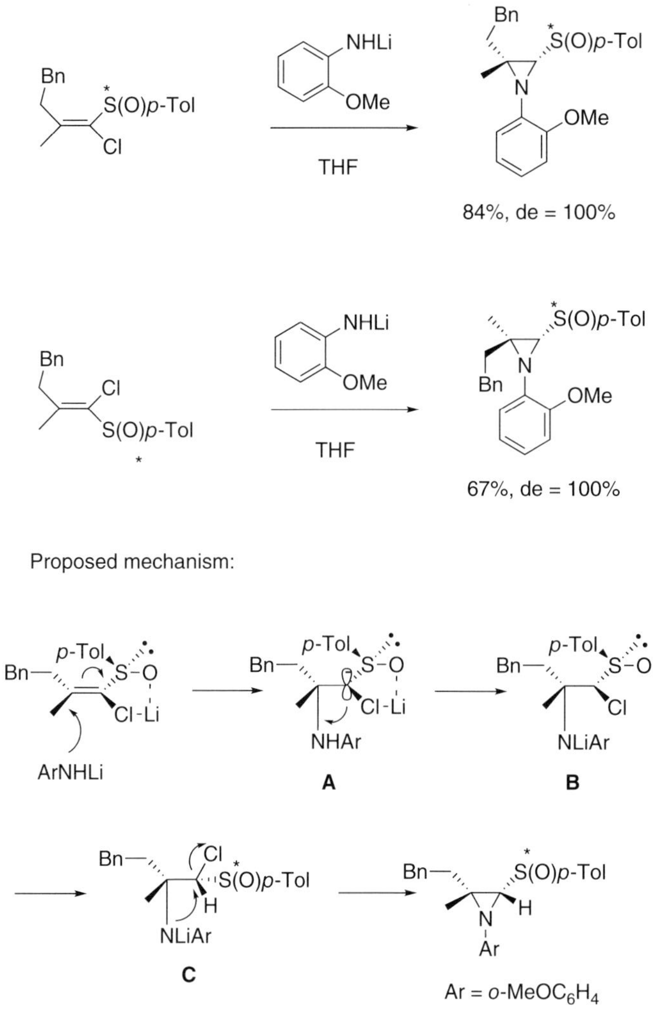

Scheme 2.20 Aziridinations of chiral α-chlorovinylsulfoxides with N-lithio o-anisidine.

Re-face to give the intermediate **A**. The intramolecular proton transfer from the nitrogen to the anionic carbon then took place to give intermediate **B**. The rotation of the carbon–carbon bond at 60° to afford conformer **C** occurred, in which the nitrogen attacked the carbon bearing a chlorine atom, resulting in the formation of the corresponding optically active sulfinylaziridine. This process constituted a novel method for the synthesis of chiral α-quaternary α-amino aldehydes by the treatment of the formed sulfinylaziridines with N-lithio aniline.

In 2005, Ulukanli et al. reported the asymmetric aziridination of 8-phenylmenthol-derived α,β-unsaturated esters by using 3-acetoxyamino-2-ethylquinazolinone, which provided, in good yields, the corresponding aziridines as a single diastereomer for each substrate [33]. The proposed mechanism involved a Michael addition of the exocyclic nitrogen onto the β-position of the α,β-unsaturated ester followed by an S_N2-type displacement of the acetoxy group by the developing partially negative charge at Cα. The high diastereoselectivity of the process could be ascribed to the π-stacking effect [34] between the phenyl group and the double bond of the alkene in the unsaturated ester. This blocked one face of the alkene directing the exocyclic nitrogen in 3-acetoxyamino-2-ethylquinazolinone to attack the other face, as shown in Scheme 2.21. Moreover, these authors observed that the yields of these reactions were greatly improved in the presence of hexamethyldisilazane.

For many years, Pellacani *et al.* have developed an efficient direct aziridination methodology based on the aza-Michael-initiated ring closure (aza-MIRC) reaction promoted by inorganic bases on alkyl nosyloxycarbamates, starting from electron-poor alkenes [35]. An asymmetric version of this methodology was proposed by these authors by using chiral alkyl nosyloxycarbamates as the chiral auxiliaries [36]. Thus, the chiral nosyloxycarbamate derived from Helmchen's auxiliary was reacted via an aza-MIRC pathway with several α,β-unsaturated olefins in the presence of calcium oxide as a base, providing the corresponding highly functionalized aziridines in good yields and with moderate diastereoselectivities (≤46% de), as shown in Scheme 2.22. Similarly, the chiral nosyloxycarbamate derived from menthol was condensed onto ethyl 2-acetylcrotonate to give the corresponding aza-MIRC product in good yield (79%), albeit with moderate diastereoselectivity of 46% de (Scheme 2.22).

Later, these authors reported another asymmetric version of the aza-MIRC reaction based, in this case, on the involvement of chiral (*E*)-nitroalkenes as the chiral auxiliaries, which reacted with various alkyl nosyloxycarbamates, allowing

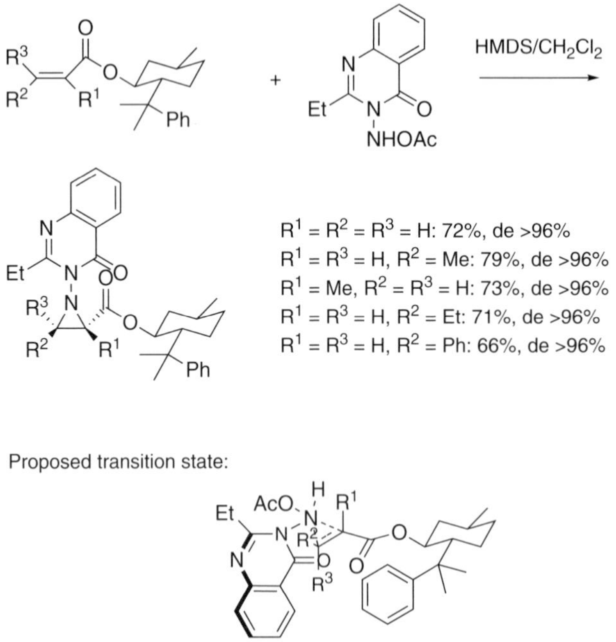

$R^1 = R^2 = R^3 = H$: 72%, de >96%
$R^1 = R^3 = H$, $R^2 = Me$: 79%, de >96%
$R^1 = Me$, $R^2 = R^3 = H$: 73%, de >96%
$R^1 = R^3 = H$, $R^2 = Et$: 71%, de >96%
$R^1 = R^3 = H$, $R^2 = Ph$: 66%, de >96%

Proposed transition state:

Scheme 2.21 Aziridination of 8-phenylmenthol-derived α,β-unsaturated esters with 3-acetoxyamino-2-ethylquinazolinone.

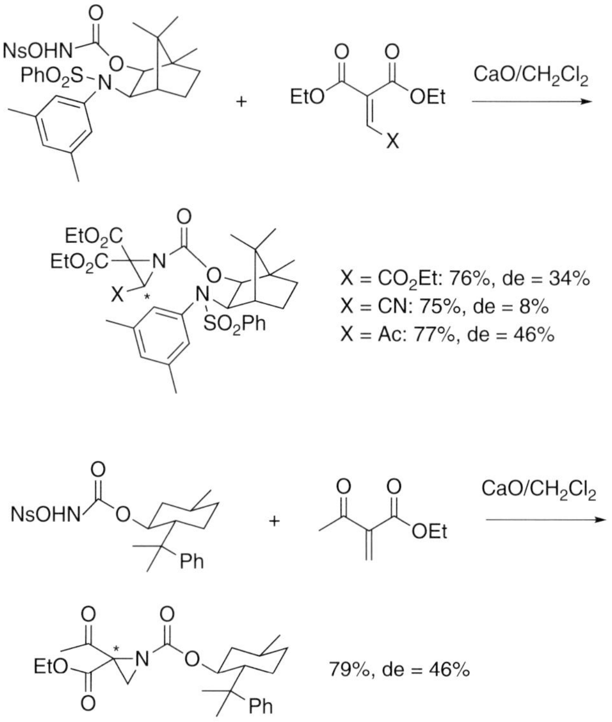

Scheme 2.22 Aziridinations of α,β-unsaturated olefins with chiral alkyl nosyloxycarbamates.

the corresponding chiral aziridines to be synthesized in high yields and with slightly better diastereoselectivities of up to 60% de (Scheme 2.23) [37]. Using this method, optically active (E)-nitroalkenes carrying a 1,3-dioxolane or 1,3-oxazolidine residue underwent stereoselective aziridinations in the presence of alkyl nosyloxycarbamates. Interestingly, it was demonstrated that the stereochemical outcome of the process was strongly influenced by the chiral residue considered, giving stereomers, regardless of the reaction conditions (nature of the base, solvent, or aminating agent). Thus, different stereochemical outcomes were observed according to the nature of the resident chiral residue, chiral 1,3-dioxolane or chiral 1,3-oxazolidine. Indeed, the aziridination reaction of the 1,3-dioxolane derivative occurred with complete retention of stereochemistry of the substrates, while the aziridination of the 1,3-oxazolidine led to the formation of only one of the possible diastereomers of aziridines, which gave in fact a mixture of two diastereomers due to the expected free rotation around the single bond of the reaction intermediates.

In 2009, the same authors employed this methodology to develop a novel asymmetric domino Knoevenagel/aza-MIRC reaction [38]. Thus, a two-step, one-pot synthesis of highly functionalized chiral aziridines, starting from cyanomethylene compounds and (R)-2,2-dimethyl-1,3-dioxolane-4-carboaldehyde or Garner's aldehyde, was successfully developed. Indeed, the first step of the process was the reaction between cyanomethylene compounds and optically active aldehydes mediated by catalytic piperidine, generating the corresponding

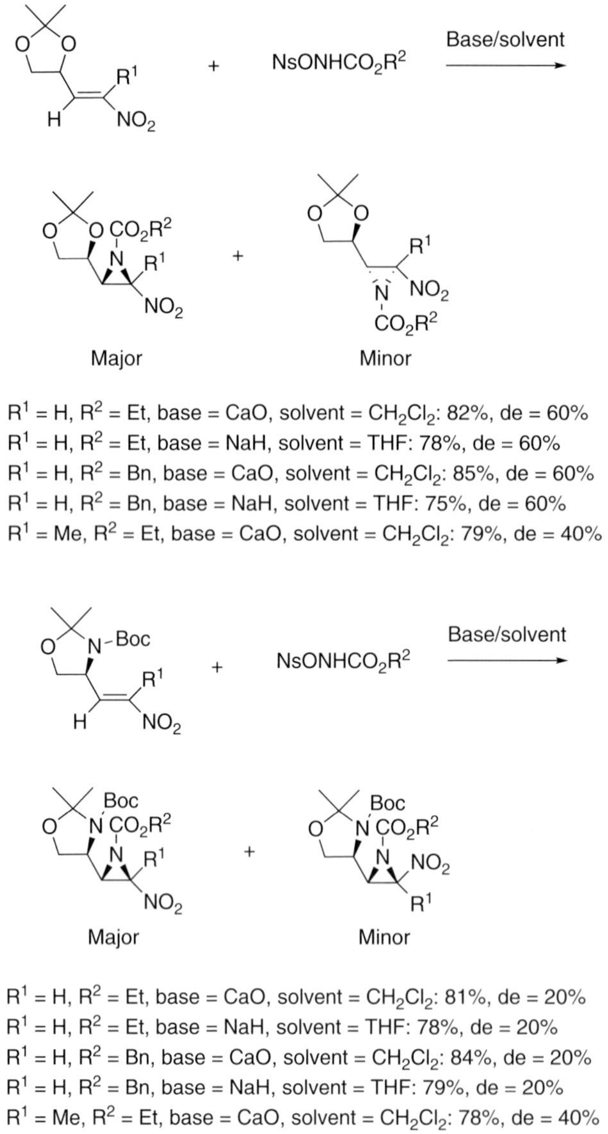

Scheme 2.23 conditions

First reaction (alkene + NsONHCO$_2$R^2 → Major + Minor aziridines):

- R^1 = H, R^2 = Et, base = CaO, solvent = CH$_2$Cl$_2$: 82%, de = 60%
- R^1 = H, R^2 = Et, base = NaH, solvent = THF: 78%, de = 60%
- R^1 = H, R^2 = Bn, base = CaO, solvent = CH$_2$Cl$_2$: 85%, de = 60%
- R^1 = H, R^2 = Bn, base = NaH, solvent = THF: 75%, de = 60%
- R^1 = Me, R^2 = Et, base = CaO, solvent = CH$_2$Cl$_2$: 79%, de = 40%

Second reaction (N-Boc alkene + NsONHCO$_2$R^2 → Major + Minor):

- R^1 = H, R^2 = Et, base = CaO, solvent = CH$_2$Cl$_2$: 81%, de = 20%
- R^1 = H, R^2 = Et, base = NaH, solvent = THF: 78%, de = 20%
- R^1 = H, R^2 = Bn, base = CaO, solvent = CH$_2$Cl$_2$: 84%, de = 20%
- R^1 = H, R^2 = Bn, base = NaH, solvent = THF: 79%, de = 20%
- R^1 = Me, R^2 = Et, base = CaO, solvent = CH$_2$Cl$_2$: 78%, de = 40%

Scheme 2.23 Aziridinations of chiral (E)-nitroalkenes with alkyl nosyloxycarbamates.

chiral alkenes, which were added in the second step of the sequence by ethyl nosyloxycarbamate, yielding the corresponding expected aziridines through addition–elimination process. The latter were obtained in high yields and with moderate-to-high diastereoselectivities of up to >98% de (Scheme 2.24).

2.2.1.3 Miscellaneous Reactions

In 2004, Tanaka *et al.* reported the stereodivergent synthesis of chiral 2-alkenylaziridines on the basis of intramolecular palladium(0)-catalyzed 2,3-*cis*-selective aziridinations and base-mediated 2,3-*trans*-selective aziridinations

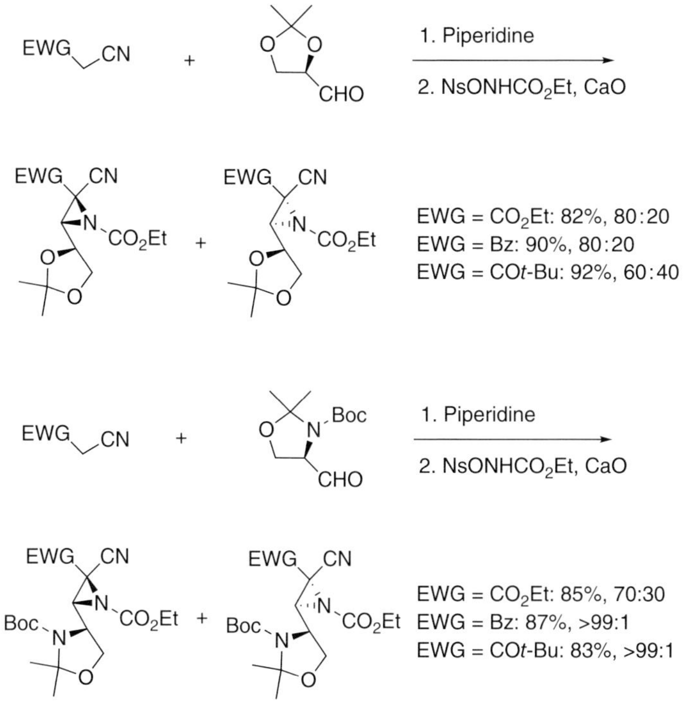

Scheme 2.24 Synthesis of chiral functionalized cyanoaziridines through domino Knoevenagel/aza-MIRC reactions.

[39]. Indeed, whereas the treatment of the chiral allylic mesylates of *N*-protected 2-alkyl-4-amino-(*E*)-2-alken-1-ols with sodium hydride in DMF yielded exclusively the corresponding thermodynamically less stable 2,3-*trans*-2-alkenyl-3-alkylaziridines, the exposure of the chiral methyl carbonates of N-protected 2-alkyl-4-amino-(*E*)-2-alken-1-ols to a catalytic amount of Pd(PPh$_3$)$_4$ in 1,4-dioxane or THF afforded predominantly the corresponding thermodynamically more stable 2,3-*cis*-2-alkenyl-3-alkylaziridines (Scheme 2.25). The exclusive formation of 2,3-*trans*-aziridines from the mesylates could be rationalized by assuming a 1,3-allylic strain in the aza-anionic intermediates. Interestingly, these authors have demonstrated that the introduction of an alkyl group on the double bond of the vinylaziridines improved the thermodynamic preference for the 2,3-*cis*-2-vinylaziridines over the 2,3-*trans*-isomers to up to 98% de, although the alkyl group apparently made the *cis*-isomers more congested as well as the corresponding *trans*-isomers. This novel asymmetric aziridination process provided a powerful methodology for the synthesis of either of the two diastereomers of 2-alkenylaziridines from single intermediates.

2.2.2 Addition to Imines

Among the different methods for achieving asymmetric aziridinations based on the use of chiral auxiliaries, a large area belongs to imine aziridinations, which can be subdivided into four major conceptual categories involving the reactions

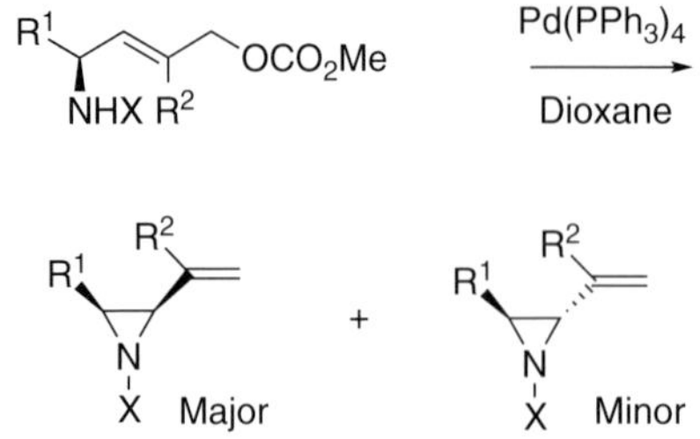

R¹ = i-Pr, R² = Me, X = Mts: 85%, de = 96%
R¹ = i-Pr, R² = Et, X = Mts: 87%, de = 98%
R¹ = i-Bu, R² = Me, X = Ts: 69%, de = 96%
R¹ = Bn, R² = Me, X = Mts: 49%, de = 96%
R¹ = (S)-CH(Me)Et, R² = Me, X = Mts: 68%, de = 94%
R¹ = t-Bu, R² = Me, X = Mts: 68%, de = 74%
R¹ = t-Bu, R² = Me, X = Ts: 64%, de = 82%

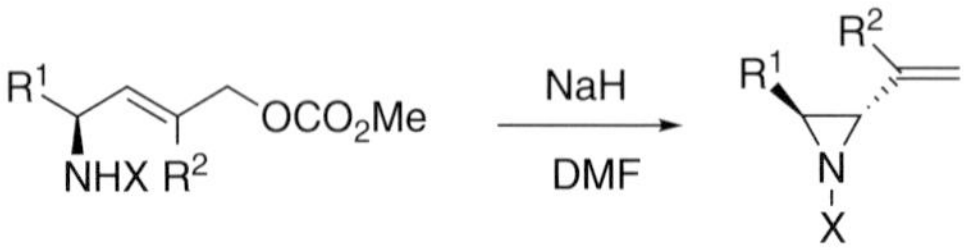

R¹ = i-Pr, R² = Me, X = Mts: 78%, de >98%
R¹ = i-Pr, R² = Et, X = Mts: 95%, de >98%
R¹ = i-Bu, R² = Me, X = Ts: 82%, de >98%
R¹ = Bn, R² = Me, X = Mts: 86%, de >98%
R¹ = (S)-CH(Me)Et, R² = Me, X = Mts: 95%, de >98%
R¹ = t-Bu, R² = Me, X = Mts: 93%, de >98%
R¹ = t-Bu, R² = Me, X = Mes: 92%, de >98%

Scheme 2.25 Palladium-catalyzed intramolecular aziridinations.

of imines with carbenes, ylides, or α-halo enolates (aza-Darzens) and the reactions based on addition/elimination processes. Enantiocontrol can be achieved by using either chiral imines or chiral nucleophiles.

2.2.2.1 Methylidation of Imines

Carbene Methodology The reaction of carbenes with Schiff bases to afford aziridines is well known [40]. In particular, the addition of carbenes generated from diazo compounds to imines is highly attractive, considering the generally good yield, high product stereoselectivity, and ease of procedure. Transition-metal complexes, Lewis acids, Brønsted acids, lithium perchlorate, and ionic liquids have been found to efficiently promote the addition of diazoacetates to imines. The corresponding aziridinecarboxylates are prepared generally in good yields. In addition to diazoacetates, other diazo compounds such as diazoacetoacetates have also been successfully added to imines, providing highly functionalized aziridines. In 2009, Huang *et al.* examined the rhodium-catalyzed addition of diazoacetoacetates derived from various chiral alcohols to imines

[41]. It was shown that the involvement of chiral diazoacetoacetates derived from menthol or borneol did not provide the expected aziridines under the reaction conditions, but complex mixtures of products, from which no substantial amount of aziridines could be isolated. On the other hand, the reaction of (1R)-endo-(+)-fenchol-derived diazoacetoacetate depicted in Scheme 2.26 with N-benzylidene-4-methoxyaniline in the presence of $Rh_2(OAc)_4$ afforded the corresponding cis-aziridine in good yield, albeit with low diastereoselectivity (14% de). A better diastereoselectivity of 50% de was achieved by using (R)-pantolactone-derived diazoacetoacetate, which led to the corresponding cis-aziridine in good yield (Scheme 2.26). According to these results, it seems that the structure of the chiral alcohols had a large effect on the reactivity of the corresponding diazoacetoacetates. For example, the diazoacetoacetates having available C–H bonds gave, preferentially, competitive C–H insertion products instead of aziridination products.

In order to explain the cis stereoselectivity of this process, these authors have proposed the mechanism depicted in Scheme 2.27. Methyl diazoacetoacetate is decomposed by rhodium acetate to generate a metal carbene, which reacts with the imine to provide the azomethine ylide. The consequent intramolecular ring-

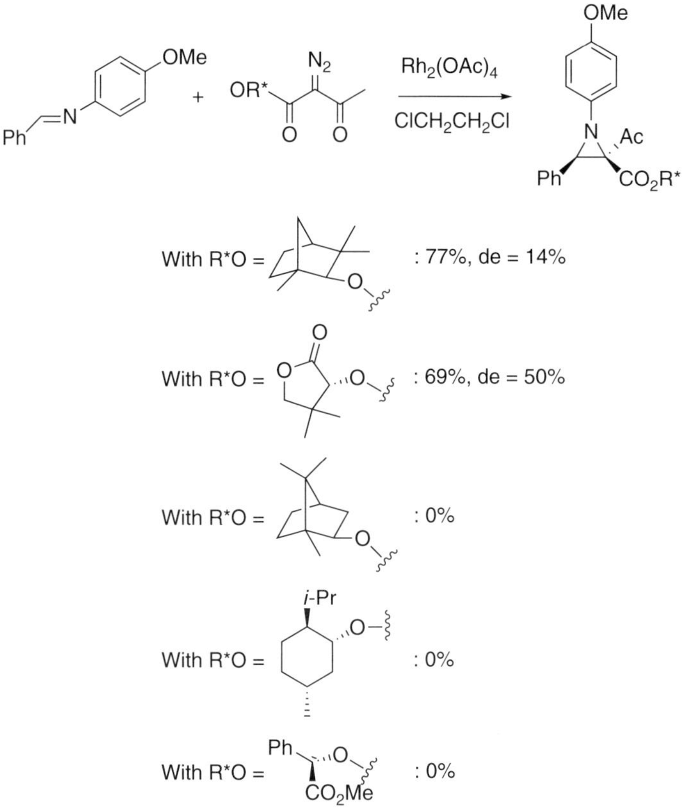

Scheme 2.26 Rh-catalyzed addition of chiral diazoacetoacetates to an aldimine.

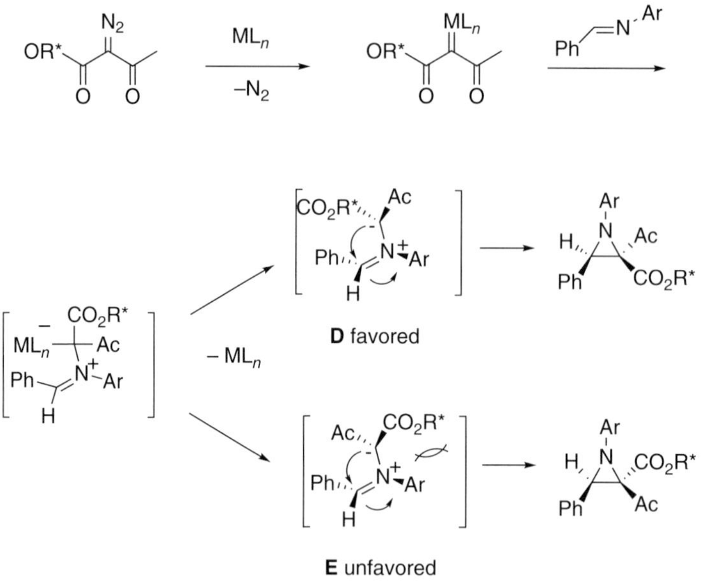

Scheme 2.27 Proposed mechanism for Rh-catalyzed addition of diazoacetoacetates to aldimines.

closing step occurs via conformation **D** to provide the *cis*-aziridine stereoselectively. Conformation **E** is unfavored, due to large steric interactions between the ester group and the *N*-aryl moiety of the imine.

Organofluorine compounds have attracted the attention of synthetic organic chemists, due to their potential activity both in the fields of material sciences and biological sciences [42]. In this context, Akiyama et al. have developed the synthesis of chiral CF_3-substituted aziridines by asymmetric Lewis-acid-mediated aziridination of aldimines with chiral diazoacetates [43]. While the chiral diazoacetates derived from either menthol or binaphthol monomethyl ether were not effective for the $BF_3(Et_2O)$-catalyzed aziridination of imine equivalents, such as *N*,*O*-acetals, the aziridination of the latter substrates performed with a chiral diazo ester derived from (*R*)-pantolactone provided the expected CF_3-substituted aziridines with both high *cis* selectivity (>95:5) and excellent diastereoselectivity (>92% de), as shown in Scheme 2.28.

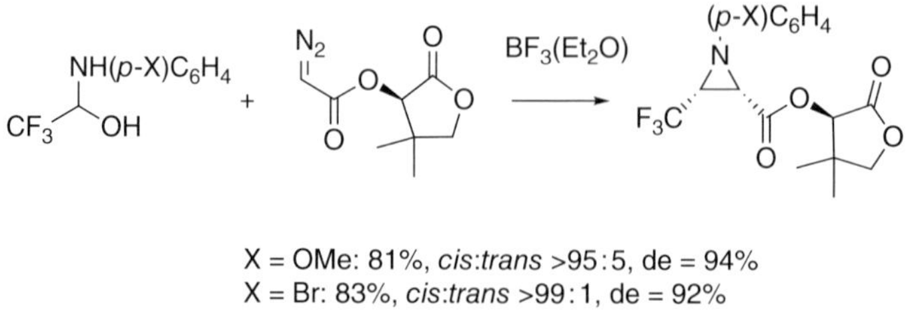

X = OMe: 81%, *cis*:*trans* >95:5, de = 94%
X = Br: 83%, *cis*:*trans* >99:1, de = 92%

Scheme 2.28 $BF_3(Et_2O)$-catalyzed aziridination of aldimine equivalents with chiral (*R*)-pantolactone-derived diazoacetate.

In 2005, Lee and Song reported the asymmetric synthesis of *cis*-aziridines from chiral *N*-benzylimines and ethyl diazoacetate using cobalt(II) as Lewis acid catalyst [44]. A wide range of chiral imines were investigated for this novel process, providing the corresponding chiral *cis*-aziridines in moderate-to-high yields and ratios of the two *cis*-aziridines of up to 2.7, as shown in Scheme 2.29. The best results were observed with imines bearing an electron-withdrawing substituent at the carbon atom, while *p*-methoxybenzaldimine bearing an electron-releasing substituent at carbon gave no reaction. Finally, the ratios of the two *cis*-aziridines produced in these reactions were found to vary from 1.3 to 2.7. These aziridines were further converted into the corresponding arylalanine derivatives by regioselective hydrogenolytic ring opening.

In 2010, Maruoka *et al.* developed an asymmetric synthesis of trisubstituted aziridines based on the Brønsted-acid-catalyzed reaction of α-substituted α-diazocarbonyl compounds bearing camphorsultam as a chiral auxiliary with *N*-alkoxycarbonyl aldimines [45]. As shown in Scheme 2.30, when using triflic acid as catalyst, the corresponding *trans*-aziridines were achieved in good yields (50–74%), with almost complete *trans*-diastereoselectivity (*trans/cis* >95:5) and excellent diastereoselectivity of >90% de in all cases of substrates studied.

As an extension, these authors reported the acid-catalyzed reaction of *N*-α-diazoacetyl camphorsultam bearing a α-hydrogen with ketimines, such as

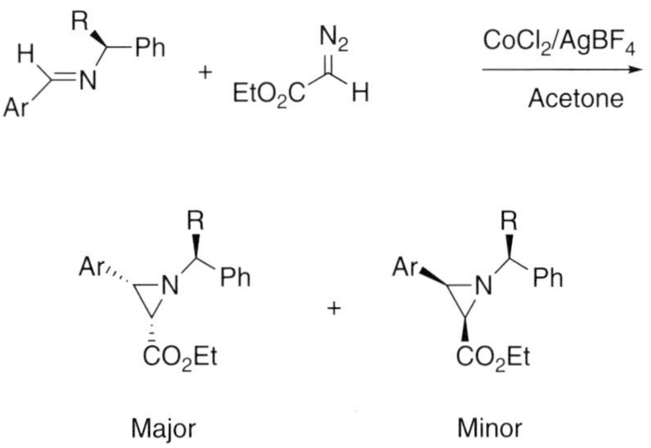

R = H, Ar = Ph: 45%
R = Me, Ar = Ph: 64%, major/minor = 2.1
R = Me, Ar = p-FC$_6$H$_4$: 66%, major/minor = 2.2
R = Me, Ar = o-ClC$_6$H$_4$: 65%, major/minor = 2.2
R = Me, Ar = m-ClC$_6$H$_4$: 71%, major/minor = 2.4
R = Me, Ar = p-ClC$_6$H$_4$: 69%, major/minor = 2.1
R = Me, Ar = p-BrC$_6$H$_4$: 70%, major/minor = 2.0
R = Me, Ar = p-NO$_2$C$_6$H$_4$: 89%, major/minor = 2.7
R = Me, Ar = p-CNC$_6$H$_4$: 85%, major/minor = 2.3
R = Me, Ar = p-Tol: 43%, major/minor = 2.4
R = Me, Ar = p-MeOC$_6$H$_4$: 0%
R = Me, Ar = 2-Naph: 58%, major/minor = 1.3
R = Me, Ar = 2-Py: 34%, major/minor = 1.8

Scheme 2.29 Co-catalyzed aziridination of chiral *N*-benzylaldimines with ethyl diazoacetate.

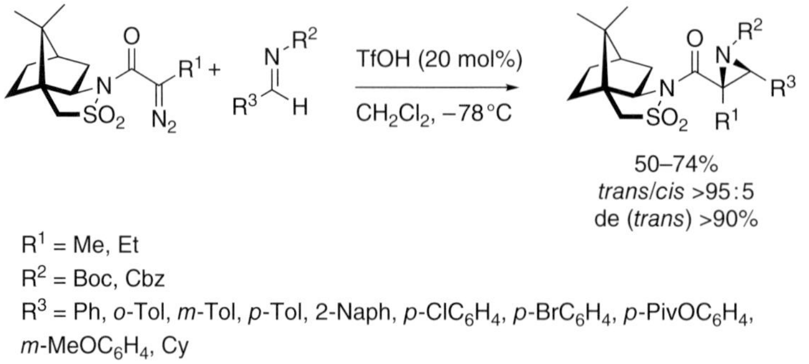

Scheme 2.30

R^1 = Me, Et
R^2 = Boc, Cbz
R^3 = Ph, o-Tol, m-Tol, p-Tol, 2-Naph, p-ClC$_6$H$_4$, p-BrC$_6$H$_4$, p-PivOC$_6$H$_4$, m-MeOC$_6$H$_4$, Cy

Scheme 2.30 TfOH-catalyzed aziridination of aldimines with camphorsultam-derived α-substituted α-diazocarbonyl compounds.

Boc-protected α-ketimino esters depicted in Scheme 2.31 [46]. Among a range of Brønsted acids investigated, Tf$_2$NH was selected as the most efficient to afford the corresponding chiral *trans*-trisubstituted aziridines bearing two carbonyl functionalities. Indeed, these products were achieved as essentially single *trans*-isomers in good yields (51–66%). It must be noted that the process also provided a by-product (Scheme 2.31) derived from the migration of the R group of the ketimine (Scheme 2.31). Moreover, the authors have been unable to expand this methodology to the synthesis of tetrasubstituted chiral aziridines by using α-substituted *N*-α-diazoacyl camphorsultams and ketimines as a route for highly encumbered vicinal quaternary stereogenic centers.

Ylide-Mediated Aziridination In addition to carbenes, carbenoids, and α-haloenolates, ylides have been widely used to achieve asymmetric aziridinations of imines. Indeed, an efficient strategy for the synthesis of chiral aziridines is the reaction of ylides with imines. This reaction forms a betaine, which ring closes to

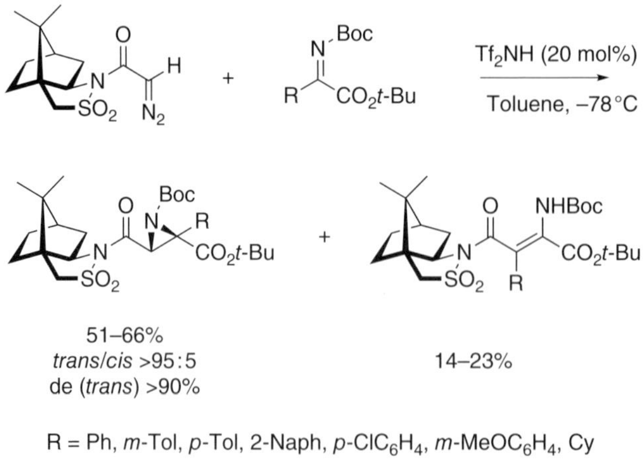

R = Ph, m-Tol, p-Tol, 2-Naph, p-ClC$_6$H$_4$, m-MeOC$_6$H$_4$, Cy

Scheme 2.31 Tf$_2$NH-catalyzed aziridination of ketimines with camphorsultam-derived α-diazocarbonyl compounds.

form an aziridine through elimination of the heteroatom-containing leaving group originating from the ylide. The main class of ylides used in asymmetric aziridination reactions is constituted by sulfur ylides. This asymmetric process can be accomplished using either chiral imines as the chiral auxiliaries, such as chiral sulfinylimines in the reaction with sulfonium ylides [47], or chiral sulfur ylides as the chiral auxiliaries [48]. Concerning the first approach involving chiral sulfinylimines [49], Stockman *et al.* have developed asymmetric reactions of dimethylsulfonium methylide derived from trimethylsulfonium iodide with a wide range of aromatic, heterocyclic, and aliphatic *tert*-butylsulfinylaldimines, providing the corresponding chiral aziridines in good yields and with diastereoselectivities of up to 95% de [50]. As shown in Scheme 2.32, electron-rich, electron-poor, sterically hindered and primary alkyl imines were all successful substrates. Moreover, the scope of this convenient methodology could be extended to some chiral *tert*-butylsulfinylketimines, providing the corresponding chiral highly substituted aziridines with high diastereoselectivity (Scheme 2.32) [51]. In contrast to aldimines, however, it must be noted that ketimines were, in general, found to be not suitable substrates for the reaction with dimethylsulfonium methylide. Indeed, only three of the nine ketimines exposed to these reaction conditions furnished significant amounts of the desired aziridines.

In 2004, the scope of this methodology was extended to another sulfonium ylide, such as that derived from *S*-allyl tetrahydrothiophenium bromide, which was condensed onto a wide variety of chiral *tert*-butylsulfinylaldimines [52]. The expected corresponding chiral vinylaziridines were formed in good yields and *trans*-stereoselectivity combined with excellent diastereoselectivity of up to 95% de. In this case, the use of LiO*t*-Bu as the base in THF exhibited the best results (Scheme 2.33). In order to explain the *trans*-stereoselectivity of the reaction, the authors have proposed a process in two stages. The initial addition step is reversible and is thought to proceed via a quasi [2+2] addition driven by electrostatic attraction, to

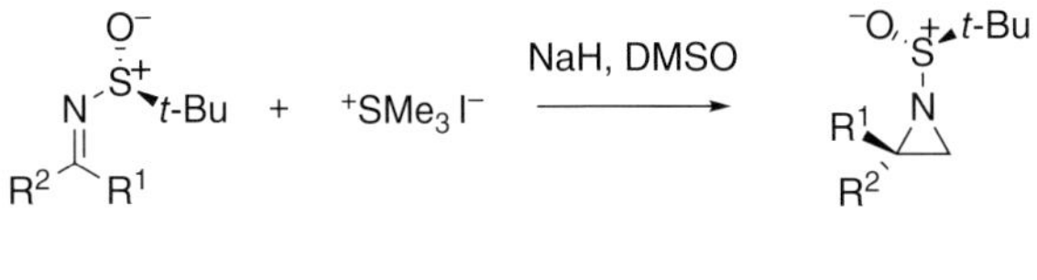

R^1 = H, R^2 = *p*-MeOC$_6$H$_4$: 65%, de = 87%
R^1 = H, R^2 = 2-Py: 77%, de = 90%
R^1 = H, R^2 = 3-Py: 74%, de = 77%
R^1 = H, R^2 = 2-Fu: 72%, de = 91%
R^1 = H, R^2 = Cy: 63%, de >95%
R^1 = H, R^2 = *n*-Pent: 65%, de = 80%
R^1 = R^2 = Ph: 73%, de >95%
R^1 = Et, R^2 = Cy: 47%, de >95%
R^1 = Me, R^2 = *n*-Hex: 36%, de >95%

Scheme 2.32 Reaction of dimethylsulfonium methylide with chiral *tert*-butylsulfinylketimines.

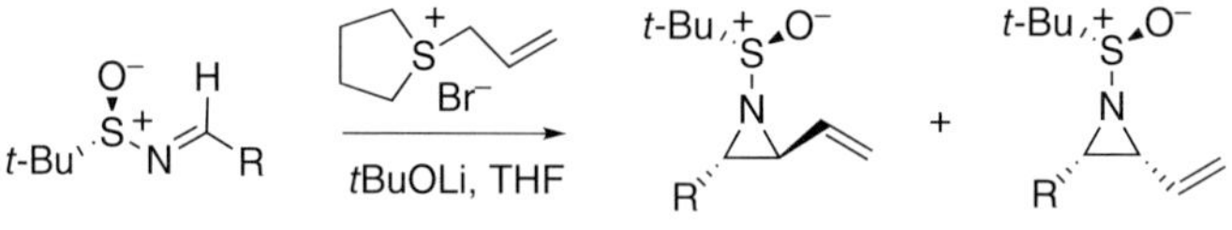

R = Ph: 68%, trans/cis = 71:29, de (trans) = 90%
R = p-MeOC$_6$H$_4$: 76%, trans/cis = 82:18, de (trans) = 92%
R = p-NO$_2$C$_6$H$_4$: 74%, trans/cis = 59:41 de (trans) = 84%
R = 2-Naph: 64%, trans/cis = 80:20, de (trans) >95%
R = Et: 44%, trans/cis = 80:20, de (trans) = 90%
R = n-Pent: 67%, trans/cis = 82:18, de (trans) >95%
R = n-Non: 58%, trans/cis = 81:19, de (trans) >95%
R = i-Bu: 62%, trans/cis = 83:17, de (trans) >95%
R = c-Pr: 61%, trans/cis = 72:28, de (trans) = 86%
R = Cy: 78%, trans/cis = 83:17, de (trans) >95%
R = 2-Py: 54%, trans/cis = 88:12, de (trans) = 88%
R = 2-Fu: 55%, trans/cis = 67:33, de (trans) >95%
R = (E)-CH=CHPh: 82%, trans/cis = 83:17, de (trans) >95%

Proposed transition states:

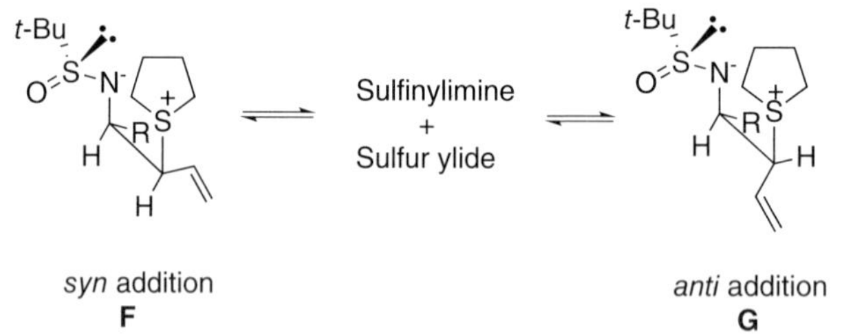

syn addition
F

anti addition
G

Scheme 2.33 Reaction of S-allyl tetrahydrothiophenium bromide with chiral tert-butylsulfinylaldimines.

form the intermediates **F** and **G** (Scheme 2.33). The *anti* addition intermediate **G** is significantly more stable than the *syn* addition product **F**, due to decreased gauche interactions. For ring closure to occur, rotation to an antiperiplanar configuration of the N and S substituents is required. The transition from a high-energy intermediate **F** to a significantly lower-energy rotamer drives the equilibrium to the antiperiplanar configuration, while the rotation of the low-energy intermediate **G** to a sterically more hindered rotamer is disfavored. This is believed to be the rate-determining step. Thus, ring closure gives a predominance of the *trans*-aziridine.

In 2006, the same authors applied this methodology to a range of variously substituted chiral *tert*-butylsulfinylketimines, which allowed a convenient access to a diverse range of highly substituted chiral aziridines to be achieved in up to 78% yield and with >90% de (Scheme 2.34) [51]. Best results were obtained when the reaction was performed in DMSO as solvent.

In addition, more varying substitutions on the alkene of the chiral vinylaziridines could be achieved by these authors by involving a range of other ylides derived

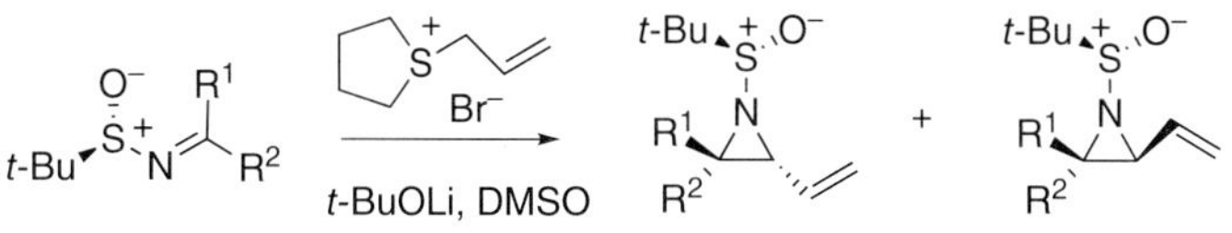

2 Asymmetric Aziridination | 233

R¹ = R² = Ph: 85%, de >95%
R¹ = Me, R² = Ph: 78%, *trans/cis* = 75:25, de (*trans*) >95%
R¹,R² = (CH₂)₄: 55%, de >95%
R¹ = Et, R² = CH₂-CH(Me)Et: 55%, *trans/cis* = 72:28, de (*trans*) >95%
R¹ = Et, R² = 2-thiophene: 72%, *trans/cis* = 60:40, de (*trans*) >95%
R¹ = Et, R² = 2-thiazole: 56%, *trans/cis* = 55:45, de (*trans*) = 90%
R¹ = Me, R² = n-Hex: 55%, *trans/cis* = 69:31, de (*trans*) >95%

Scheme 2.34 Reaction of S-allyl tetrahydrothiophenium bromide with chiral *tert*-butylsulfinylketimines.

from various substituted allyltetrahydrothiophenium salts [53]. In this way, the reaction of substituted sulfur allyl ylides to *tert*-butylphenylsulfinylaldimine gave a range of substituted chiral vinylaziridines in good yields and with moderate-to-excellent diastereoselectivities of up to 99% de, albeit with variable *trans:cis* ratios ranging from 50:50 to 92:8 (Scheme 2.35). In this study, it was found that the use of THF as the solvent tended to give the best compromise between stereoselectivity and conversion, with DMSO giving the best conversion at the expense of *trans:cis* selectivity.

In 2009, Forbes *et al.* reported the employment of another sulfonium salt depicted in Scheme 2.36, which was successfully condensed onto aryl-substituted

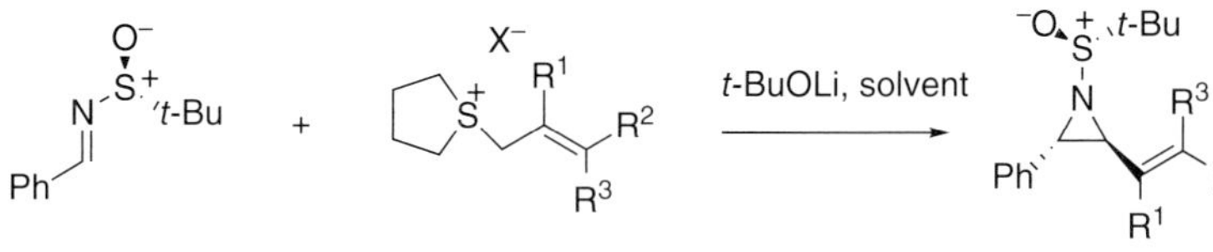

R¹ = R² = R³ = H, X = Br, solvent = THF: 68%, *trans:cis* = 70:30, de (*trans*) = 90%
R¹ = R² = R³ = H, X = Br, solvent = DMSO: 75%, *trans:cis* = 50:50, de (*trans*) = 20%
R¹ = R³ = H, R² = Me, X = Br, solvent = THF: 0%
R¹ = R³ = H, R² = Me, X = Br, solvent = DMSO: 85%, *trans:cis* = 67:23, de (*trans*) = 20%
R¹ = R³ = H, R² = Ph, X = BF₄, solvent = THF: 0%
R¹ = R³ = H, R² = Ph, X = BF₄, solvent = DMSO: 75%, *trans:cis* = 60:40, de (*trans*) = 2%
R¹ = H, R² = R³ = Me, X = Br, solvent = THF: 0%
R¹ = H, R² = R³ = Me, X = Br, solvent = DMSO: 66%, *trans:cis* = 70:30, de (*trans*) = 36%
R¹ = R³ = H, R² = TMS, X = OTs, solvent = THF: 90%, *trans:cis* = 92:8, de (*trans*) = 100%
R¹ = R³ = H, R² = TMS, X = OTs, solvent = DMSO: 70%, *trans:cis* = 55:45, de (*trans*) = 0%
R¹ = Me, R² = R³ = H, X = Br, solvent = THF: 56%, *trans:cis* = 70:30, de (*trans*) = 100%
R¹ = Me, R² = R³ = H, X = Br, solvent = DMSO: 63%, *trans:cis* = 60:40, de (*trans*) = 100%

Scheme 2.35 Reaction of substituted sulfur allyl ylides with chiral *tert*-butylphenylsulfinylaldimine.

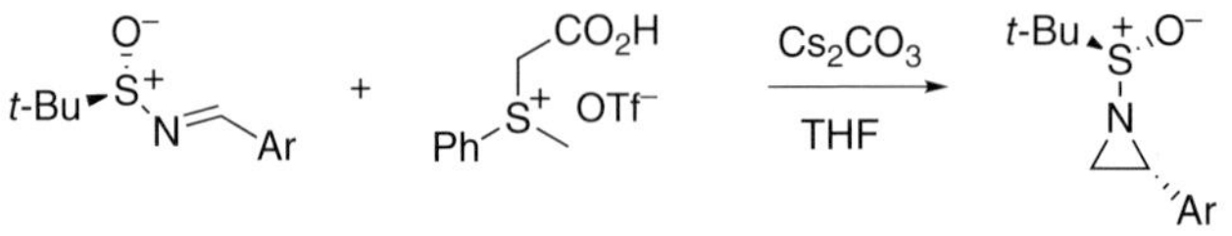

234 | *Asymmetric Synthesis of Three-Membered Rings*

Ar = p-MeOC$_6$H$_4$: 97%, de = 74%
Ar = p-ClC$_6$H$_4$: 96%, de = 56%
Ar = 2,6-Cl$_2$C$_6$H$_3$: 94%, de = 54%
Ar = Ph: 95%, de = 48%
Ar = 3-Ac-4-OHC$_6$H$_3$: 93%, de = 60%
Ar = p-AcC$_6$H$_4$: 89%, de = 32%

Scheme 2.36 Reaction of a sulfonium ylide with chiral aryl-substituted *tert*-butylsulfinylaldimines.

tert-butylsulfinylaldimines in the presence of Cs$_2$CO$_3$ as the base in THF [54]. In this context, a series of chiral aziridines were isolated in excellent yields and with low-to-good diastereoselectivities of up to 74% de, as shown in Scheme 2.36. Indeed, although moderate levels of diastereocontrol were observed, the fact that extremely high levels of conversion were achieved in a process that did not rely on DMSO as solvent and strong bases was encouraging.

In 2005, Khiar *et al.* developed other chiral sulfinylaldimines bearing a less sterically demanding substituent, such as an isopropyl group, presenting, in comparison with a *tert*-butyl group, the advantages of lower molecular weight and higher reactivity, which could be a good alternative to the more popular *tert*-butylsulfinyl group [55]. These authors elaborated a comparative study demonstrating that the isopropylsulfinyl group behaved better than the *tert*-butylsulfinyl and *p*-tolylsulfinyl groups, both in terms of reactivity (time for the reaction to be complete) and stereoselectivity in the reaction of chiral sulfinylaldimines with dimethylsulfonium or dimethyloxosulfonium ylide (Scheme 2.37). As an example, a diastereoselectivity of 70% de was obtained in 48 h in the case of the *tert*-butylsulfinylaldimine, whereas the same diastereoselectivity was achieved with the corresponding isopropylsulfinylaldimine in only 1 h.

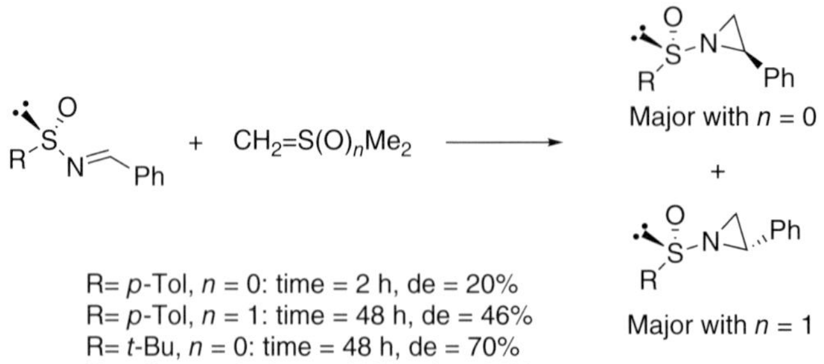

R= p-Tol, n = 0: time = 2 h, de = 20%
R= p-Tol, n = 1: time = 48 h, de = 46%
R= t-Bu, n = 0: time = 48 h, de = 70%
R= t-Bu, n = 1: time = 168 h, de = 90%
R= i-Pr, n = 0: time = 1 h, de = 70%
R= i-Pr, n = 1: time = 48 h, de = 46%

Scheme 2.37 Reaction of dimethylsulfonium or dimethyloxosulfonium ylide with chiral *N*-sulfinylaldimines.

In 2007, Kokotos and Aggarwal reported the asymmetric reaction of chiral *tert*-butylsulfinyl aminals as precursors of imines with aryl-, allyl-, and amide-stabilized sulfur ylides, which led to the corresponding aziridines in high yields and with good stereoselectivities of up to 92:8 *trans/cis* and >90% de (Scheme 2.38) [56].

A closely related methodology was previously used by Aggarwal *et al.* to develop a total synthesis of the protein kinase C inhibitor, (−)-balanol [57]. The key step of this synthesis was constituted by the reaction of diphenyl vinyl sulfonium triflate salt with the same chiral aminal as described in Scheme 2.38, in the presence of NaH as the base, which provided the corresponding aziridine in good yield and with moderate diastereoselectivity (Scheme 2.39).

On the other hand, asymmetric aziridinations have also been achieved by Tang *et al.* by using telluronium allyl ylides instead of sulfur ylides [58]. They showed that the reaction of various allylic telluronium ylides with enantiopure *N-tert*-butylsulfinylimines provided the corresponding optically active *cis*-2-substituted vinylaziridines with excellent diastereoselectivities and in good-to-excellent yields. The best results were obtained in the presence of LiHMDS as the base combined with $Ti(OEt)_4$ as an additive. The procedure was applied to

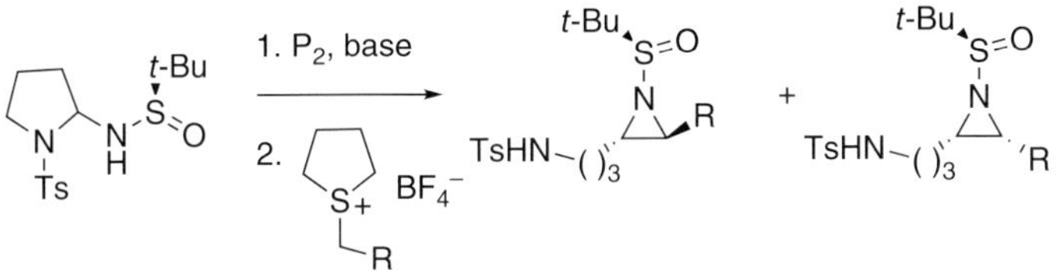

R = *p*-MeOC$_6$H$_4$: 43%, *trans/cis* = 82:18, de (*trans*) = 30%, de (*cis*) >90%
R = *p*-ClC$_6$H$_4$: 66%, *trans/cis* = 92:8, de (*trans*) = 70%, de (*cis*) >90%
R = CH=CH$_2$: 73%, *trans/cis* = 57:43, de (*trans*) >90%, de (*cis*) >90%
R = CONHPh: 63%, *trans/cis* = 52:48, de (*trans*) >90%, de (*cis*) >90%

Proposed mechanism:

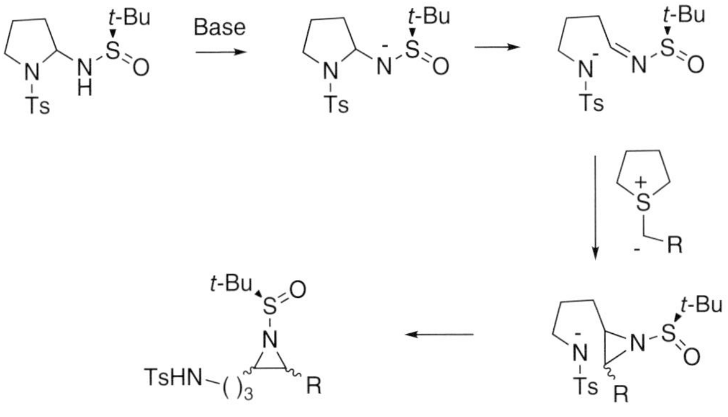

Scheme 2.38 Reaction of sulfur ylides with chiral *tert*-butylsulfinyl aminals.

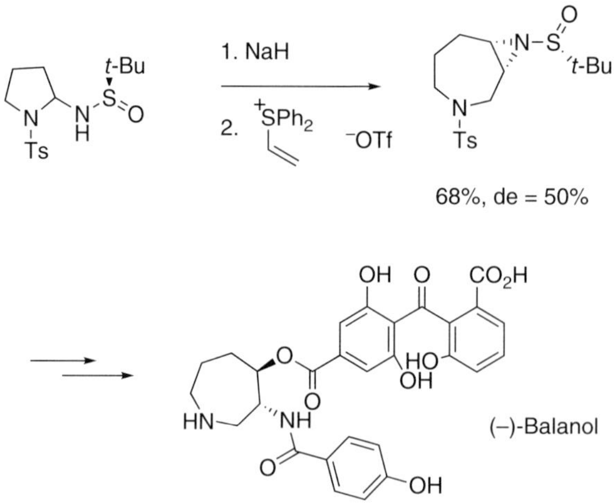

Scheme 2.39 Synthesis of (−)-balanol.

a wide range of imines such as aryl and heteroaryl aldimines as well as alkyl and α,β-unsaturated aldimines, as shown in Scheme 2.40. Furthermore, the reaction was operative with ketimines, giving the corresponding *cis*-aziridines in high yields and with excellent diastereoselectivities. In order to explain the diastereoselectivity of the reaction, these authors have proposed the mechanism depicted in Scheme 2.40, in which the telluronium ylide attacks the imine to form the intermediates **H** and **I**. Transformation of these intermediates into the corresponding intermediates **J** and **K** by rotation was followed by an intramolecular *anti*-elimination to afford the vinylaziridine. Obviously, intermediate **H** was favored over intermediate **I**, due to steric hindrance, and thus, the *cis*-aziridine formed preferentially.

In 2010, Mayer et al. developed an asymmetric aziridination of chiral *N-tert*-butanesulfinylaldimines with benzyl-stabilized sulfur ylides *in situ* generated from rhodium-catalyzed decomposition of phenyldiazomethane in the presence of various sulfides such as tetrahydrothiophene and dibutyl sulfide [59]. It must be noted that these chiral imines are generally unreactive toward diazo compounds, such as ethyl diazoacetate, trimethylsilyldiazomethane, or phenyldiazomethane, in the presence of Lewis acids. As shown in Scheme 2.41, the corresponding chiral aziridines were formed in remarkable quantitative yields in almost all cases of substrates studied, such as alkyl, aryl, as well as ethoxycarbonyl imines. Generally, the 2,3-*trans*-aziridines were found to predominate over the 2,3-*cis*-aziridine diastereomers. The diastereoselectivity of up to 86% de varied significantly, depending upon the solvent and sulfide employed in the reaction. In the case of using tetrahydrothiophene as the sulfide, the reaction gave the best results when performed in toluene as the solvent, providing diastereoselectivities of up to 70% de, while acetonitrile was selected as the most efficient solvent when using dibutylsulfide as the substrate, giving diastereoselectivities of up to 86% de. In general, the *trans/cis*-selectivity of the reactions was not substantially affected by the substituents on the imines with one exception in the case of the imine bearing a 2-pyridyl group. Indeed, in this case, the *cis*-aziridines

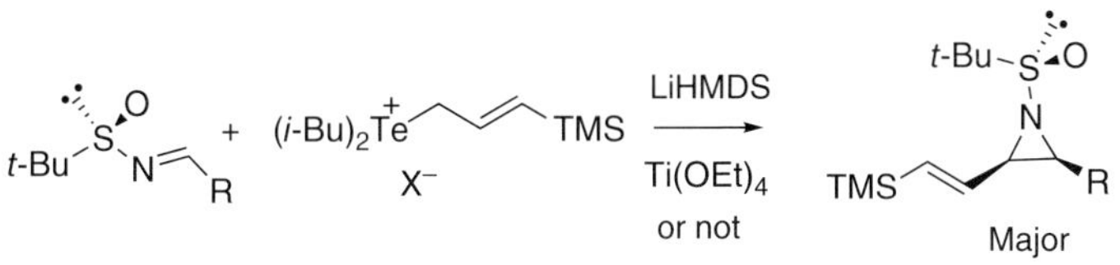

2 Asymmetric Aziridination | 237

with X = Br and with Ti(OEt)$_4$:
R = Ph: 98%, cis/trans = 95/5, de (cis) >98%
R = 1-Naph: 98%, cis/trans = 96/4, de (cis) >98%
R = 2-Fu: 93%, cis/trans = 95/5, de (cis) = 94%
R = p-ClC$_6$H$_4$: 96%, cis/trans = 95/5, de (cis) >98%
R = p-CF$_3$C$_6$H$_4$: 98%, cis/trans = 95/5, de (cis) >98%
R = p-Tol: 98%, cis/trans = 95/5, de (cis) = 97%
R = (E)-CH=CHPh: 88%, cis/trans = 91/9, de (cis) = 88%
R = (E)-CH(Me)=CHPh: 98%, cis/trans = 96/4, de (cis) = 87%
R = (E)-CH(Et)=CH(i-Pr): 91%, cis/trans = 92/8, de (cis) = 86%
with X = BPh$_4$ and without Ti(OEt)$_4$:
R = Cy: 83%, cis/trans = 90/10, de (cis) >98%
R = t-Bu: 53%, cis/trans > 97/3, de (cis) >98%

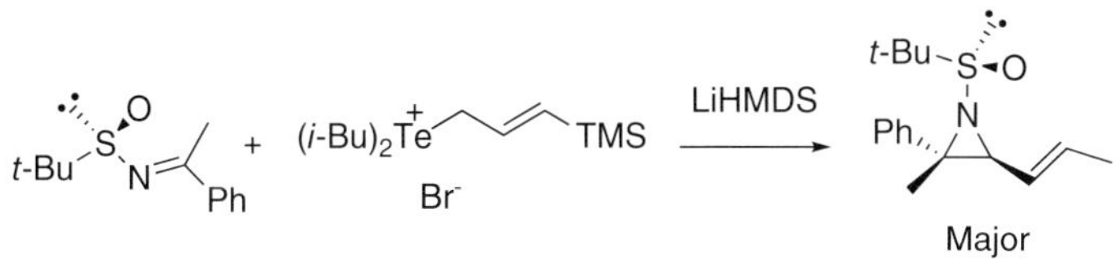

76%, trans/cis = 97/3, de (trans) >98%

Proposed mechanism:

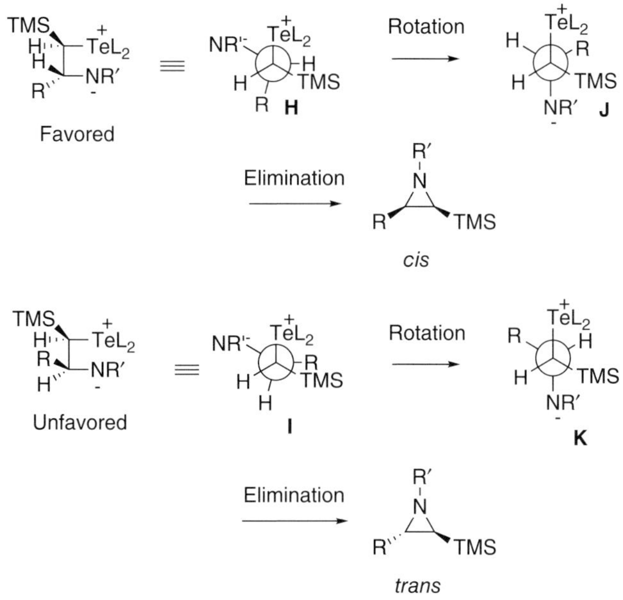

Scheme 2.40 Reactions of allylic telluronium ylides with chiral N-tert-butylsulfinylaldimines.

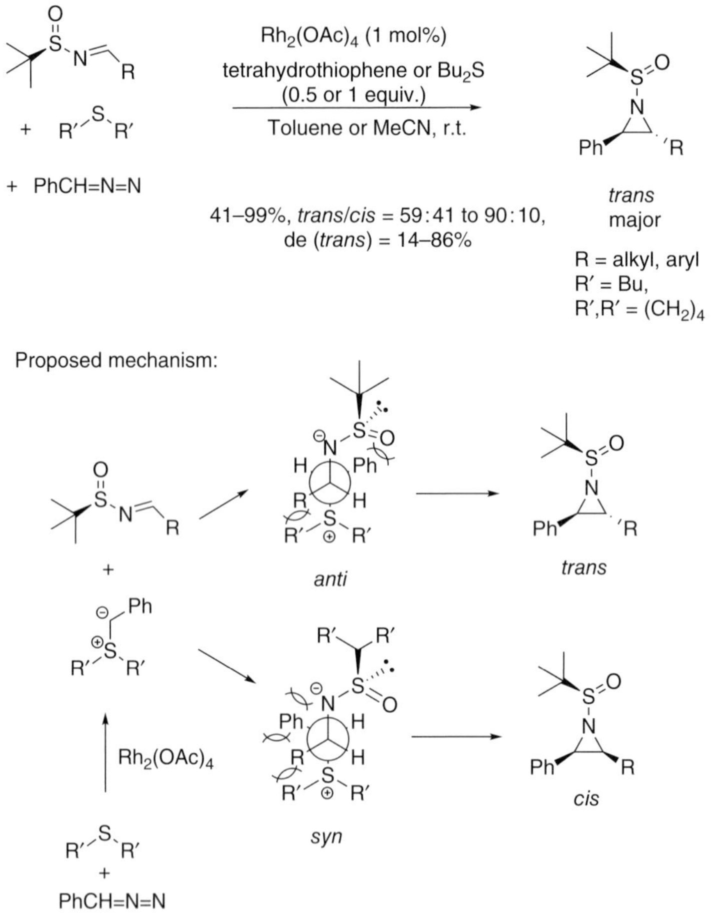

Scheme 2.41 Reaction of chiral *N*-tert-butanesulfinylaldimines with *in situ* generated benzyl-stabilized sulfur ylides.

resulted as the major products, which could be attributable to the involvement of the pyridyl nitrogen in complexation to the rhodium catalyst, thus potentially affecting the steric bulk of the imine substrate and the orientation toward reaction with the sulfur ylide. It has been well established that the addition of a benzyl-stabilized sulfur ylide to an imine is under kinetic control [60]; therefore, the irreversible sulfur ylide addition step determined the stereochemistry of the reaction. If the sulfur ylide adopted a *transoid* approach in its attack upon the imine to minimize steric effects, then the formation of the *anti*-product was favored over the *syn*-product; this rationalization was consistent with the observed preference of formation of the *trans*-aziridines over the *cis*-aziridines in the reaction (Scheme 2.41).

Organofluorine compounds have attracted the attention of synthetic organic chemists, due to their potential activity in both the fields of material sciences and biological sciences [42]. However, although the introduction of the CF_3 group usually results in an improvement of the biological properties of bioactive compounds, its direct introduction into heterocyclic compounds is often difficult. In the past decade, numerous building block approaches for the preparation of trifluoromethylated aziridine derivatives have been reported; however, few examples of their asymmetric synthesis could be found. In this context, Huang et al. recently developed a synthesis of these chiral products on the basis of a reaction between the sulfur ylide derived from trimethyl sulfoxonium iodide and chiral CF_3-substituted N-tert-butanesulfinylketimines [61]. As shown in Scheme 2.42, the corresponding highly strained chiral fluorinated aziridines were obtained in good-to-high yields (45–93%) along with high-to-excellent diastereoselectivities of up to >98% de. Aromatic as well as aliphatic ketimines underwent the reaction smoothly to afford the corresponding aziridines in good yields. In general, the electron-withdrawing groups were beneficial to the construction of trifluoromethylated aziridines with good diastereoselectivities but in lower yields, whereas ketimines bearing an electron-donating group gave the highest yield (93%) and lower diastereoselectivity (78% de). The highest diastereoselectivity of >98% de was reached in the case of formation of acetylenic trifluoromethylated aziridines in moderate-to-good yields (45–65%). A possible mechanism was proposed by the authors, suggesting that the oxygen present in the dimethyloxosulfonium methylide could promote a chelated transition state depicted in Scheme 2.42, in which both the sulfinyl oxygen and the dimethyloxosulfonium methylide oxygen chelated to the sodium ion. In this transition state, the CF_3 group preferred to occupy an equatorial position rather than an axial position due to the electrostatic repulsion between the CF_3 group and the lone pair of the sulfur atom, and the nucleophile attacked the C=N bond from the least hindered face to form a chelated transition state.

The second methodology to prepare chiral aziridines by condensation of ylides onto imines consists of using chiral sulfur ylides as the chiral auxiliaries. Several groups have demonstrated ingenious catalytic asymmetric approaches to chiral aziridines based on the generation of chiral S-ylides. For example, Aggarwal et al., via the reaction of chiral sulfides with rhodium metallocarbenes, have reported on a range of substituted heterocycles, which could be generated in

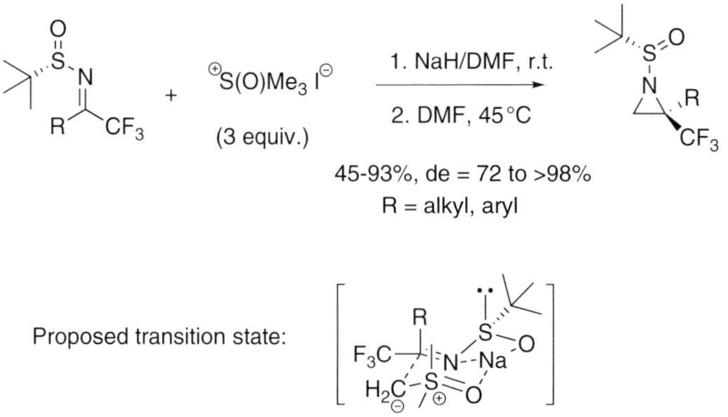

Scheme 2.42 Reaction of chiral CF_3-substituted N-tert-butanesulfinylketimines with sulfur ylide derived from trimethyl sulfoxonium iodide.

high yields and with excellent control of both absolute and relative stereochemistry. In particular, these authors have developed a reliable method that generated diazo compounds *in situ* [62]. This approach was based on the Bamford–Stevens reaction, which uses tosylhydrazone salts as diazo precursors. In the presence of phase-transfer catalysts [63], the tosylhydrazone salts are cleanly converted into the diazo compounds, which are trapped by a transition-metal catalyst, resulting in the formation of a metal–carbene complex. This complex subsequently reacts with a chiral sulfide to form the corresponding chiral sulfur ylide, which finally reacts with an aldimine to furnish the expected aziridine. This methodology was employed as the key step in a synthesis of the taxol side chain [64]. As summarized in Scheme 2.43, the reaction between an N-trimethylsilylethylsulfonyl imine and benzaldehyde tosylhydrazone salt in the presence of a phase-transfer catalyst, $Rh_2(OAc)_4$, and a catalytic quantity of a chiral sulfide, which was recycled during the process, afforded the desired aziridine in good yield as a 89:11 *trans/cis* diastereomeric mixture, in which the major *trans*-diastereomer was obtained with an enantiomeric excess of 98%.

On the other hand, Eliel's oxathiane has been used by Solladié-Cavallo *et al.* as the precursor of a diastereo- and enantiopure sulfur ylide, employed as the chiral auxiliary for the asymmetric synthesis of N-tosyl aziridines [65]. Thus, the aziridines were formed as *trans/cis* diastereomeric mixtures by reaction of the sulfonium salt derived from Eliel's oxathiane with various tosylaldimines in the presence of a phosphazene base (EtP_2) to generate the ylide. Both corresponding *cis*- and *trans*-aziridines were obtained with exceptionally high enantiomeric purities of >99% ee, as shown in Scheme 2.44. In this case, the chiral auxiliary was used in a stoichiometric amount, but was recovered in high yield and reused.

Even though this method was amenable to gram-quantity synthesis, and the chiral auxiliary could easily be recovered and reused, the practicality of this procedure was limited by the use of the sensitive and expensive phosphazene base. In this context, Hamersak *et al.* have developed an application of this method for the asymmetric synthesis of N-Ses, N-Boc, and N-Ts disubstituted aziridines by using sodium hydride as the base [66]. As shown in Scheme 2.45, NaH was successfully used as a substitute for the expensive and sensitive base EtP_2, without any major influence on the yield, enantioselectivity, or diastereoselectivity. Later, this methodology was even more extended to other N-Ts, N-Ses, N-Boc, and N-o-Ns aldimines to allow a range of variously protected chiral aziridines to be achieved in moderate-to-good yields (31–88%) and remarkable enantiomeric excesses for both *cis*- and *trans*-diastereomers of up to >99% ee [67]. The diastereoselectivities of the reactions were variable and influenced by the imine N-protecting group, the aldimine substituent, and the sulfide structure. By comparing the diastereoselectivities, the authors could arrange the protecting groups in the following order of decreasing *trans*-selectivity: Boc > Ses > Ts > o-Ns. The unusual 100% *cis*-selectivity in the formation of N-Ses-2-phenyl-3-(*tert*-butyl) aziridine and N-Ts-2-phenyl-3-*tert*-butylaziridine was explained through the use of computational models.

In 2007, Midura designed a new type of chiral sulfur ylide, containing an enantiopure sulfinyl group bonded to the ylidic carbon atom [68]. Thus, (S)-dimethylsulfonium-(p-tolylsulfinyl)methylide reacted with N-tosylaldimines to

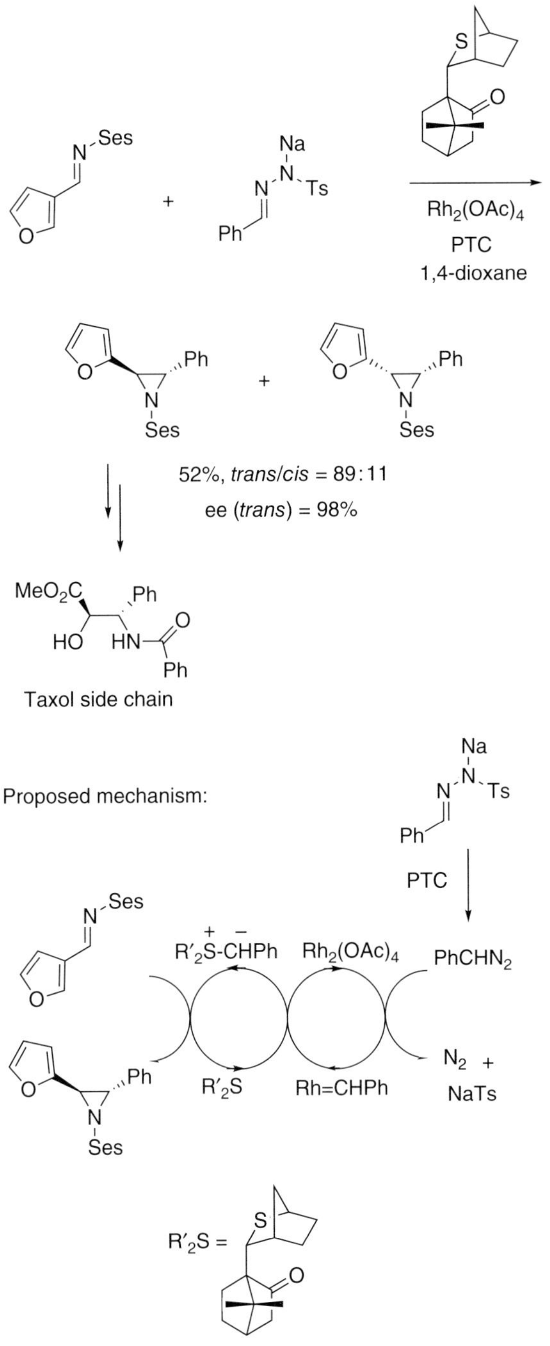

Scheme 2.43 Synthesis of taxol side chain through aziridination of an aldimine with an *in situ* generated chiral sulfur ylide.

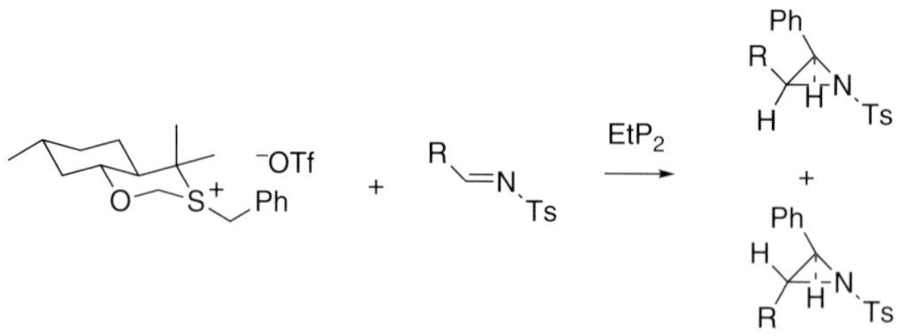

R = 9-Anth: 88%, cis/trans = 55/45, ee (cis) = 100%, ee (trans) = 99%
R = 9-Phen: 75%, cis/trans = 53/47, ee (cis) = 100%, ee (trans) = 100%
R = 1-Naph: 70%, cis/trans = 30/70, ee (cis) = 100%, ee (trans) = 99%
R = t-Bu: 80%, cis/trans = 100/0, ee (cis) = 99%
R = Cy: 60%, cis/trans = 96/4, ee (cis) = 100%, ee (trans) = 100%
R = Ph: cis/trans = 60/40, ee (trans) = 99%
R = p-NO$_2$C$_6$H$_4$: cis/trans = 50/50, ee (cis) = 100%, ee (trans) = 99%
R = p-MeOC$_6$H$_4$: cis/trans = 70/30, ee (cis) = 100%

Scheme 2.44 Reaction of chiral Eliel's oxathiane-derived ylide with tosylaldimines by using EtP$_2$ as base.

give the corresponding *cis*-sulfinyl aziridines with full enantio- and diastereoselectivity and in high yields, as shown in Scheme 2.46. The high facial selectivity observed in the aziridination reactions could be explained by the reasonable assumption that the ylide adopted the conformation depicted in Scheme 2.46, which determined the stereochemical course of the process. The exclusive formation of *cis*-isomers was probably caused by the difference in the ease of betaine formation. Transition state **M** leading to an *anti*-betaine is disfavored due to repulsive interactions between the sulfinyl moiety of the ylide and the tosyl substituent on the nitrogen. This interaction in **M** must be much stronger than the interaction between the sulfinyl group and the R group of the imine in the transition state **L**, leading to a *syn*-betaine (Scheme 2.46).

In 2007, Gais et al. reported the generation of conjugated cyclic and acyclic allyl aminosulfoxonium ylides and their application to the asymmetric aziridination of *N-tert*-butylsulfonyl imino esters, which gave enantioenriched alkenyl- and cycloalkenyl-aziridinecarboxylates for which only a few asymmetric syntheses had been described [69]. When the ylides were generated from aminosulfoxonium-substituted β,γ-unsaturated α-amino acids by treatment with DBU, they provided, by reaction with the *in situ* generated *N-tert*-butylsulfonyl imino ester, the corresponding *cis*-alkenylaziridinecarboxylates as major diastereomers in excellent yields, with good diastereoselectivities, and with moderate-to-high enantioselectivities (Scheme 2.47). The formation of these aziridines was also possible starting from 1-alkenyl aminosulfoxonium salts, which gave, upon treatment with DBU, the corresponding 1-alkenyl aminosulfoxonium ylides, which reacted with the externally added *N-tert*-butylsulfonyl imino ester. The diastereo- and enantioselectivities for the formation of the aziridines from the 1-alkenyl salts differed significantly, however, from those starting from the amino acid derivatives. Indeed, as shown in Scheme 2.47, the aziridines were obtained in good yields, albeit with low *cis/trans* diastereoselectivities, and with moderate

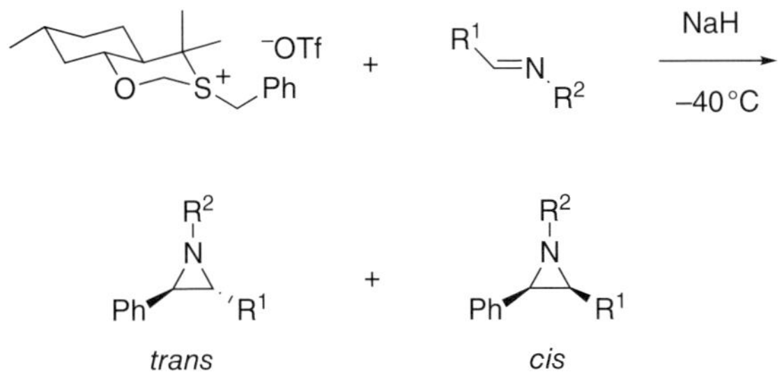

2 Asymmetric Aziridination | 243

R¹ = Ph, R² = Ts: 60%, cis/trans = 60/40, ee (trans) = 98%
R¹ = PMP, R² = Ts: 57%, cis/trans = 58/42, ee (cis) = 98%
R¹ = t-Bu, R² = Ts: 68%, cis/trans = 100/0, ee (cis) = 97%
R¹ = 1-Naph, R² = Ts: 60%, cis/trans = 31/69, ee (cis) = 96%, ee (trans) = 96%
R¹ = 2-Phen, R² = Ts: 76%, cis/trans = 45/55, ee (cis) = 95%, ee (trans) = 96%
R¹ = Ph, R² = Ses: 68%, cis/trans = 44/56, ee (trans) >99%
R¹ = PMP, R² = Ses: 47%, cis/trans = 63/37, ee (cis) >99%, ee (trans) = 98%
R¹ = 1-Naph, R² = Ses: 63%, cis/trans = 24/76, ee (cis) = 97%, ee (trans) = 98%
R¹ = Ph, R² = Boc: 60%, cis/trans = 10/90, ee (trans) = 97%
R¹ = PMP, R² = Boc: 31%, cis/trans = 9/91, ee (trans) = 96%
R¹ = 1-Naph, R² = Boc: 75%, cis/trans = 2/98, ee (trans) = 96%
R¹ = PMP, R² = Ns: 65%, cis/trans = 58:42, ee (cis) = 97%, ee (trans) = 98%
R¹ = 9-phenanthryl, R² = Ses: 80%, cis/trans = 44:56, ee (trans) = 97%, ee (cis) = 97%
R¹ = 9-phenanthryl, R² = Ts: 76%, cis/trans = 45:55, ee (trans) = 99%, ee (cis) = 98%
R¹ = 9-anthryl, R² = Ses: 53%, cis/trans = 9:91, ee (trans) = 98%
R¹ = 9-anthryl, R² = Ts: 88%, cis/trans = 24:76, ee (trans) = 99%, ee (cis) = 99%
R¹ = 9-phenanthryl, R² = Ns: 76%, cis/trans = 70:30, ee (cis) = 98%, ee (trans) = 98%
R¹ = t-Bu, R² = Ses: 61%, cis/trans = 100:0, ee (cis) = 98%
R¹ = t-Bu, R² = Ts: 68%, cis/trans = 100:0, ee (cis) = 97%

Scheme 2.45 Reaction of chiral Eliel's oxathiane-derived ylide with N-protected aldimines by using NaH as base.

enantioselectivities. In addition, this methodology could be applied to the synthesis of cycloalkenylaziridinecarboxylates by the reaction of cyclic allyl aminosulfoxonium ylides with imino ester in good yields, with low diastereoselectivities, and with medium-to-high enantioselectivities (Scheme 2.47).

In 2010, Aggarwal et al. reported the synthesis of a cheap and chiral isothiocineole-derived sulfide, readily available from the solvent-free reaction of elemental sulfur with limonene and γ-terpinene [70]. This rigid benzyl sulfide was employed as chiral precursor of ylide under simple reaction conditions to react with a wide range of aldimines to provide the corresponding chiral aziridines in good yields (63–80%) along with perfect enantioselectivities ranging from 96% to 98% ee, as shown in Scheme 2.48. A notable aspect of the process was the record level of trans-diastereoselectivity observed, which was essentially complete in the case of cinnamaldehyde-derived imine (de >98%). It must be noted that the sulfide could be reisolated in 95% yield. The stereochemical outcome of the reaction was rationalized by the authors by considering the conformation and face selectivity of the ylide intermediate. Of the two conformers, ylide **O** was favored

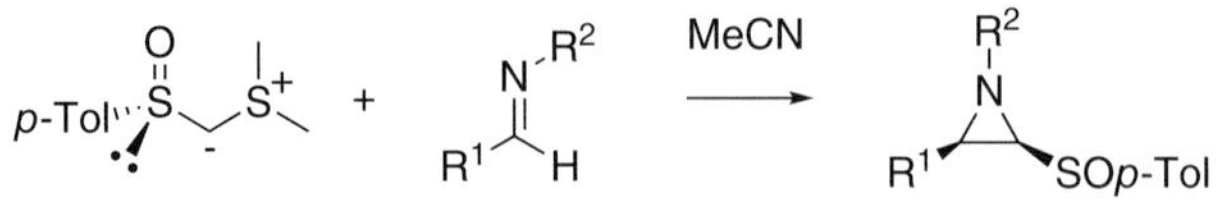

R^1 = Ph, R^2 = Ts: 88%, cis/trans = 100:0, de = 100%
R^1 = p-BrC$_6$H$_4$, R^2 = Ts: 92%, cis/trans = 100:0, de = 100%
R^1 = p-NO$_2$C$_6$H$_4$, R^2 = Ts: 93%, cis/trans = 100:0, de = 100%
R^1 = n-Bu, R^2 = Ts: 72%, cis/trans = 100:0, de = 100%
R^1 = i-Pr, R^2 = Ts: 81%, cis/trans = 100:0, de = 100%
R^1 = Ph, R^2 = p-TolS(O): 76%, cis/trans = 90:10, de = 100%
R^1 = R^2 = Ph: 0%

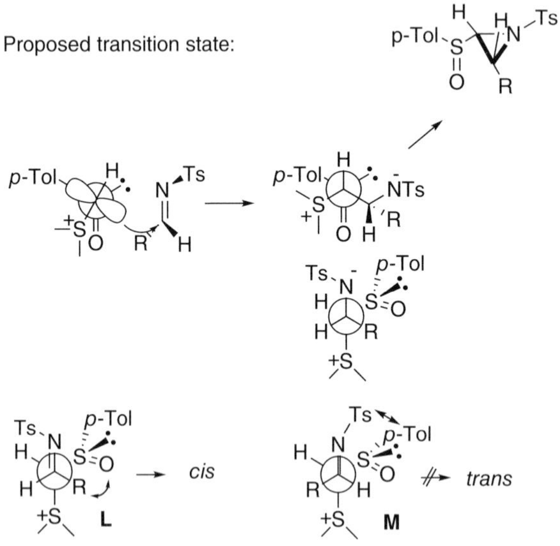

Proposed transition state:

Scheme 2.46 Reaction of (S)-dimethylsulfonium-(p-tolylsulfinyl)methylide with N-tosylaldimines.

over **N** due to nonbonded steric interactions (Scheme 2.48), and this reacted with electrophiles on its *Re* face since the *Si* face was blocked by a flanking methyl group. The minor enantiomer *ent*-aziridine originated from reaction of the minor conformer **1** again reacting with facial selectivity.

Later, the same authors extended the scope of this methodology to related chiral sulfides, such as allyl sulfonium salts, which afforded, under related conditions by reaction with benzaldehyde-derived aldimines, the corresponding chiral aziridines in good yields (72–84%), with moderate *trans*-diastereoselectivities ranging from 78:22 to 86:14, along with generally excellent enantioselectivities of up to >98% ee for both *trans*- and *cis*-diastereomers, as shown in Scheme 2.49 [71]. Different activating groups on nitrogen of imines were tolerated, such as Ts, and P(O)Ph$_2$, whereas Boc-imines were not successful.

In addition, Janardanan and Sunoj have reported a density functional theory investigation on the factors controlling enantio- and diastereoselection in asymmetric aziridination reactions by the addition of chiral bicyclic sulfur ylides

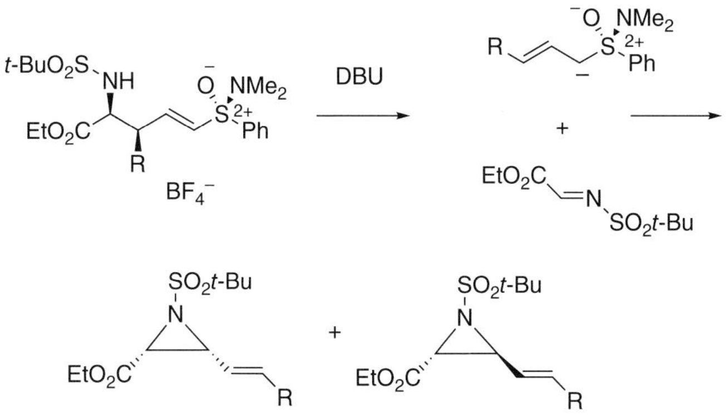

R = Ph: 94%, trans/cis = 93:7, ee (cis) >98%, ee (trans) = 30%
R = i-Pr: 91%, trans/cis = 91:9, ee (cis) = 92%, ee (trans) = 26%
R = Cy: 93%, trans/cis = 90:10, ee (cis) = 71%, ee (trans) = 48%
R = t-Bu: 94%, trans/cis = 91:9, ee (cis) = 50%, ee (trans) = 5%

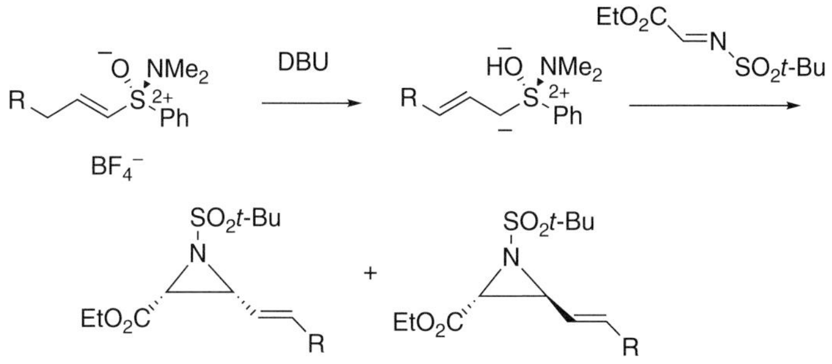

R = i-Pr: 70%, cis/trans = 64:36, ee (cis) = 76%, ee (trans) = 49%
R = Cy: 68%, cis/trans = 70:30, ee (cis) = 47%, ee (trans) = 45%
R = Me: 65%, cis/trans = 60:40, ee (cis) = 65%

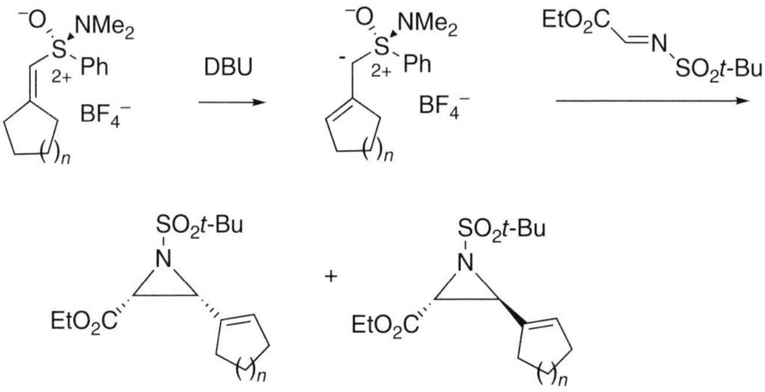

n = 1: 70%, cis/trans = 60:40, ee (cis) = 79%, ee (trans) = 90%
n = 2: 73%, cis/trans = 60:40, ee (cis) = 76%, ee (trans) = 56%
n = 3: 71%, cis/trans = 60:40, ee (cis) = 78%, ee (trans) = 57%
n = 4: 66%, cis/trans = 50:50, ee (cis) = 70%, ee (trans) = 25%

Scheme 2.47 Reactions of chiral allyl aminosulfoxonium ylides with an imino ester.

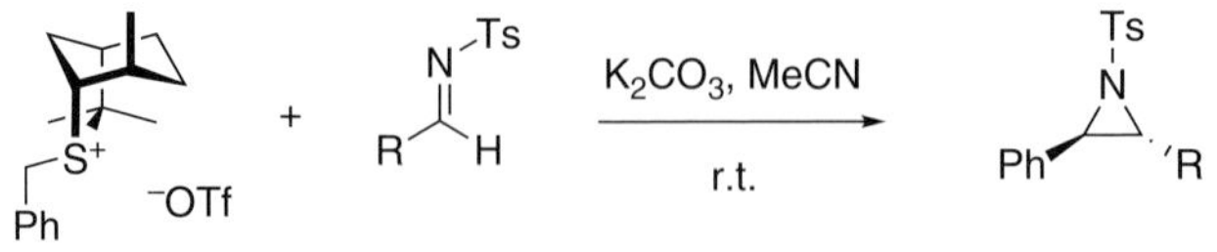

Asymmetric Synthesis of Three-Membered Rings

R = Ph: 72%, trans/cis = 85:15, ee (trans) = 98%
R = p-Tol: 63%, trans/cis = 86:14, ee (trans) = 98%
R = p-ClC$_6$H$_4$: 65%, trans/cis = 75:25, ee (trans) = 98%
R = p-MeOC$_6$H$_4$: 80%, trans/cis = 83:17, ee (trans) = 98%
R = (E)-PhCH=CH: 78%, trans/cis >99:1, ee (trans) = 96%
R = (E)-TMSCH=CH: 78%, trans/cis = 87:13, ee (trans) = 98%
R = t-Bu: 68%, trans/cis = 89:11, ee (trans) = 98%

Rationalization of the reaction selectivity:

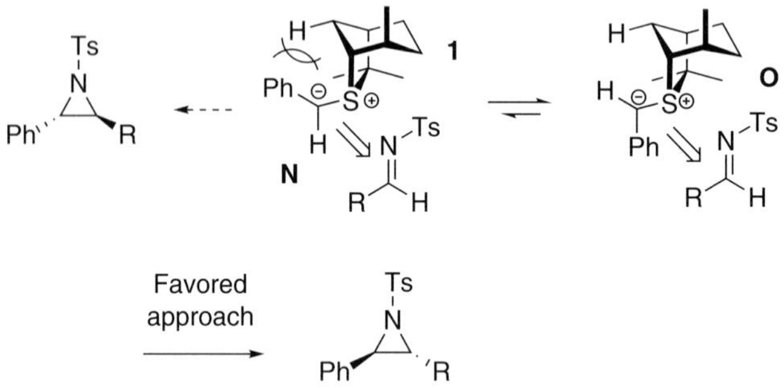

Favored approach

Scheme 2.48 Reaction of a chiral isothiocineole-derived benzyl sulfonium salt with aldimines.

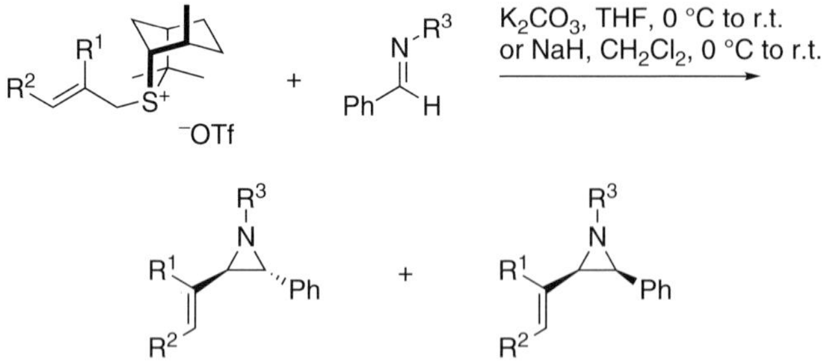

R^1 = Me, R^2 = Ph, R^3 = Ts: 76%, trans/cis = 78:22, ee (trans) = 98%, ee (cis) = 94%
R^1 = Me, R^2 = H, R^3 = Ts: 72%, trans/cis = 83:17, ee (trans) = 98%, ee (cis) = 70%
R^1 = R^2 = H, R^3 = Ts: 73%, trans/cis = 85:15, ee (trans) = 88%, ee (cis) = 70%
R^1 = H, R^2 = Ph, R^3 = Ts: 81%, trans/cis = 80:20, ee (trans) = 90%
R^1 = Me, R^2 = H, R^3 = Ph$_2$PO: 84%, trans/cis = 84:16, ee (trans) = 98%, ee (cis) >98%
R^1 = R^2 = H, R^3 = Ph$_2$PO: 83%, trans/cis = 86:14, ee (trans) = 82%, ee (cis) = 80%

Scheme 2.49 Reaction of chiral isothiocineole-derived allylic sulfonium salts with benzaldehyde-derived aldimines.

to substituted aldimines [72]. High levels of enantioselection were predicted toward the formation of (2S,3S)-cis- and (2R,3S)-trans-aziridines by the addition of stabilized ylide (Ac) to SO_2Me- and CO_2Me-protected aldimines, respectively. Similarly, high enantioselectivity was predicted for the formation of (2S,3R)-cis-aziridines from semistabilized (Ph) ylide. This study highlighted that a correct prediction of the extent of enantioselection requires knowledge of the activation barriers for the elementary steps beyond the initial addition step. In the case of stabilized ylides, the ring closure was found to be crucial in controlling enantio- and diastereoselection. A cumulative effect of electronic and other weak interactions was identified as a factor contributing to the relative energies of transition states, leading to enantio- and diastereomeric products for the stabilized ylide addition to aldimines. On the contrary, steric control appeared to be quite dominant with semistabilized ylide addition.

The addition of stabilized sulfur ylides being nonreversible, the selectivity of this process was determined by the relative rates of formation of the *anti*- and *syn*-betaines. From computational studies, Robiette found that the lowest energy pathway to the *trans*-aziridine occurred via cisoid addition of the ylide to the imine to give the *anti*-betaine intermediate, followed by bond rotation and subsequent ring closure (Scheme 2.50) [60a]. In contrast, the *cis*-aziridine was formed from a transoid addition of the ylide to the imine to give the *syn*-betaine intermediate, followed by direct ring closure. However, the differences between the energies of the barriers of the key transition states leading to the *syn*- and *anti*-betaines and therefore the *cis*- and *trans*-aziridines in the model systems used in the calculations were relatively small, reflecting the low diastereoselectivity generally observed.

Finally, Connon *et al.* have found that a simple chiral salt derived from (2R,5R)-2,5-diisopropylthiolane could mediate the asymmetric methylene transfer to a

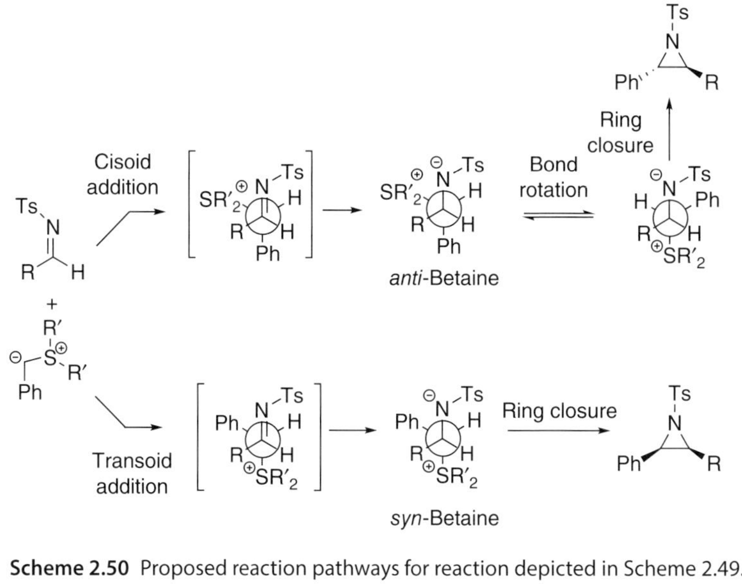

Scheme 2.50 Proposed reaction pathways for reaction depicted in Scheme 2.49.

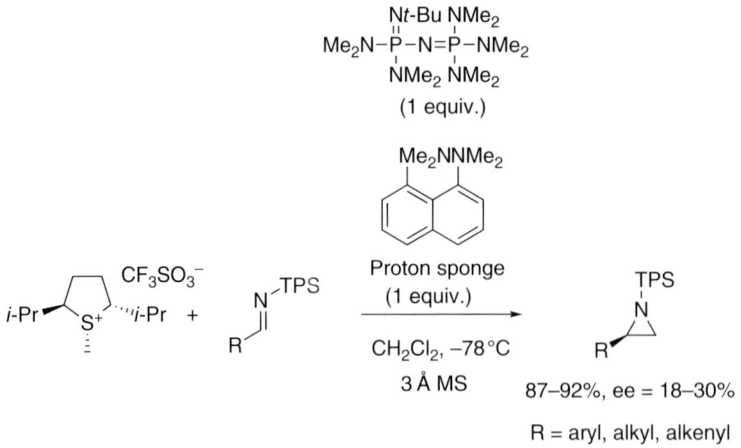

Scheme 2.51 Reaction of a chiral diisopropyl-substituted sulfonium salt with aldimines.

range of aldimines with uniformly excellent yields (87–92%) albeit with moderate enantioselectivities of up to 30% ee, as shown in Scheme 2.51 [73]. The process was carried out in the presence of a strong organic base and a proton sponge in stoichiometric quantities and showed a quite broad scope, since aromatic, aliphatic, and α,β-unsaturated TPS-protected aldimines provided the corresponding enantioenriched aziridines in high yields.

2.2.2.2 Aza-Darzens and Analogous Reactions

The aza-Darzens reaction is analogous to the Darzens synthesis of epoxides but employs imines instead of ketones or aldehydes. The asymmetric version of the aza-Darzens reaction includes the use of either a chiral imine or a chiral α-halo enolate as the chiral auxiliary. As an example, Davis et al. have developed asymmetric aza-Darzens reactions involving the addition of halomethylphosphonate anions to enantiopure sulfinylaldimines, directly providing the corresponding chiral cis-N-sulfinylaziridine 2-phosphonates [74]. Indeed, the reaction between diethyl iodo- or tosyl-methylphosphonate and chiral N-tert-butylsulfinylaldimine derived from benzaldehyde, provided in the presence of LiHMDS, the corresponding aziridines in moderate-to-high yields and as a single diastereomer (Scheme 2.52).

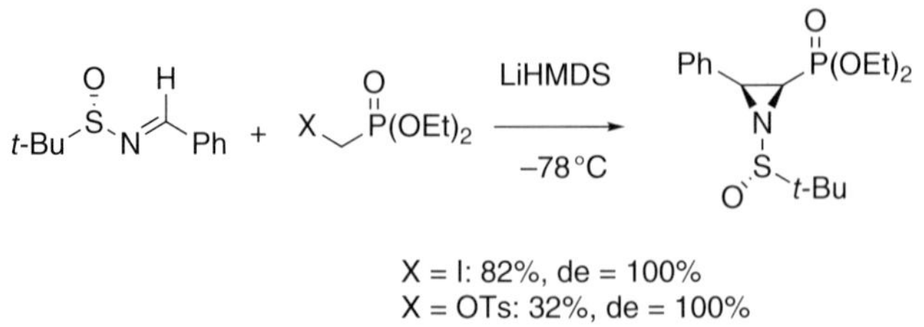

Scheme 2.52 Aza-Darzens reaction of a chiral N-tert-butylsulfinylaldimine with diethyl iodo- or tosylphosphonates.

The scope of the preceding methodology could be extended to the use of chiral *N*-(2,4,6-trimethylphenylsulfinyl)aldimines, which reacted with diethyl iodomethylphosphonate to give the corresponding chiral *cis-N*-sulfinylaziridine 2-phosphonates in high yields and with almost complete diastereoselectivity (>96% de), as shown in Scheme 2.53 [75]. It must be noted, however, that aryl sulfinylaldimines containing electron-attracting groups gave complex mixtures by using this methodology.

Later in 2011, Stockman *et al.* developed the asymmetric synthesis of 2,3-di- and 2,2′,3-tri-substituted aziridines in good yields and with remarkable diastereoselectivities of up to >98% de for the major *syn*-diastereomers on the basis of asymmetric aza-Darzens reactions of a range of chiral *tert*-butanesulfinylaldimines as well as ketimines with ethyl bromoacetate performed in the presence of LiHMDS as a base (Scheme 2.54) [76]. In all cases of chiral *syn*-1,2-disubstituted aziridines arisen from the reaction of sulfinyl aldimines, a generally excellent diastereoselectivity of >98% de was reached along with good yields (49–86%) and *syn/anti* ratios ranging from 71:29 to 98:2 while chiral *syn*-2,2′,3-trisubstituted aziridines were achieved with moderate-to-excellent diastereoselectivities (62% to >98% de) and comparable *syn/anti* ratios albeit in yields generally and predictably lower than those for the aldimine substrates, due to increased steric interactions and lower reactivity of the ketimines. For example, bulky cyclohexylketimine was found not to react, showing the limitations of the method. In 2014, the same authors extended the scope of this methodology to substituted 2-bromoesters, which provided a rapid access to a range of highly substituted chiral aziridines in good yields (63–76%) and with excellent levels of diastereoselectivity (90% to >98% de), as shown in Scheme 2.54 [77]. In the case of aromatic imines, the *anti/syn* selectivity was moderate to good (80:20 to 92:8), whereas an almost complete loss of this selectivity resulted from using aliphatic imines.

In 2008, Kattuboina and Li developed asymmetric aza-Darzens reactions employing novel chiral *N,N*-dibenzyl-*N*-phosphonylaldimines as the electrophiles and the preformed lithium enolate of methyl 2-bromoacetate [78]. This novel methodology allowed the corresponding chiral aziridines to be formed in modest-to-good yields and with good-to-excellent diastereoselectivities, as shown in Scheme 2.55. The electrophilicity of these novel stable chiral *N*-phosphonyl imines could be controlled by introducing a variety of electron-

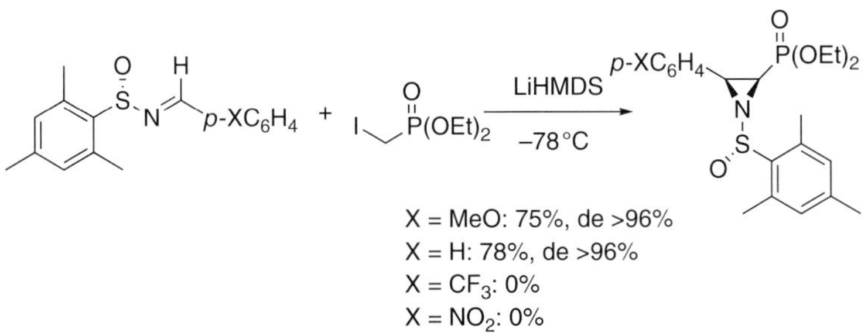

X = MeO: 75%, de >96%
X = H: 78%, de >96%
X = CF$_3$: 0%
X = NO$_2$: 0%

Scheme 2.53 Aza-Darzens reaction of chiral *N*-(2,4,6-trimethylphenylsulfinyl)aldimines with diethyl iodomethylphosphonate.

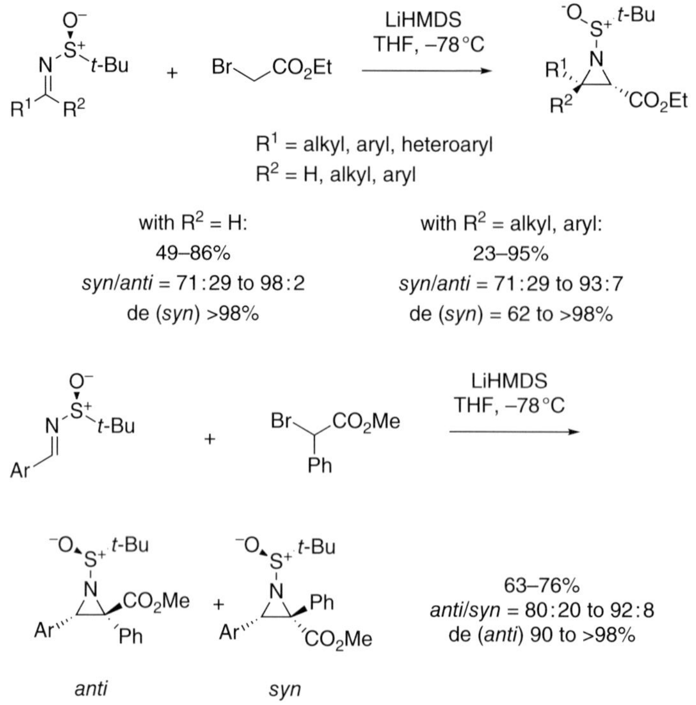

Scheme 2.54 Aza-Darzens reactions of chiral *tert*-butanesulfinylimines with (substituted)ethyl bromoacetates.

donating or -withdrawing groups onto the nitrogen to replace the benzyl group. Interestingly, no *trans*-aziridines were detected for all of the examples studied. In order to explain the stereoselectivity of the reaction, these authors have proposed a cyclic six-membered transition state depicted in Scheme 2.55. The attack of the *E*-configured lithium enolate of methyl 2-bromoacetate onto the *N,N*-dibenzyl-*N*-phosphonyl imine should be directed onto the *Re*-face. The bulky moiety of the chiral auxiliary is pushed away by the sterically hindered side of the six-membered transition state. There are two smaller steric moieties, hydrogen and a lone pair of electrons existing along the enolate-attacking pathway within a widely open space of the *N*-phosphonyl imine template. This asymmetric environment ensures the resulting *S*-chirality on the carbonyl addition center as well as the chirality of the α-position of the *N*-phosphonyl aziridine-2-carboxylic esters.

As an extension of the preceding methodology, the same authors have investigated aza-Darzens reactions of *N,N*-diisopropyl-*N*-phosphonylaldimines as auxiliaries [79]. As shown in Scheme 2.56, their reaction with a pregenerated β-bromo lithium enolate provided the corresponding chiral *cis*-aziridine-2-carboxylic esters in moderate-to-good yields (51–85%) and with generally high diastereoselectivities ranging from 92% to >98% de. The scope of the process was wide since imines bearing both electron-donating and electron-withdrawing groups on the aromatic rings were tolerated as well as heterocyclic aldimines,

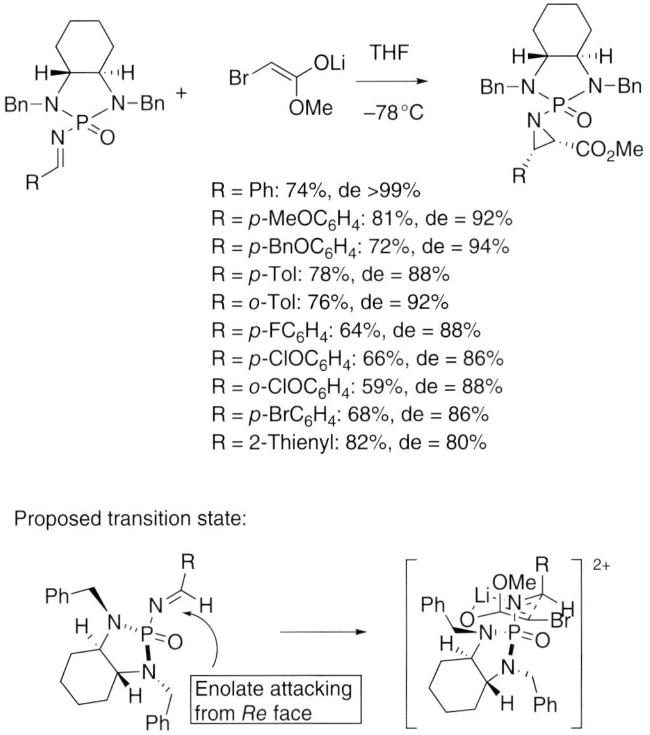
THF
−78 °C
R = Ph: 74%, de >99%
R = p-MeOC6H4: 81%, de = 92%
R = p-BnOC6H4: 72%, de = 94%
R = p-Tol: 78%, de = 88%
R = o-Tol: 76%, de = 92%
R = p-FC6H4: 64%, de = 88%
R = p-ClOC6H4: 66%, de = 86%
R = o-ClOC6H4: 59%, de = 88%
R = p-BrC6H4: 68%, de = 86%
R = 2-Thienyl: 82%, de = 80%
Proposed transition state:
Enolate attacking from Re face

Scheme 2.55 Aza-Darzens reaction of chiral *N,N*-dibenzyl-*N*-phosphonylaldimines with lithium enolate of methyl 2-bromoacetate.

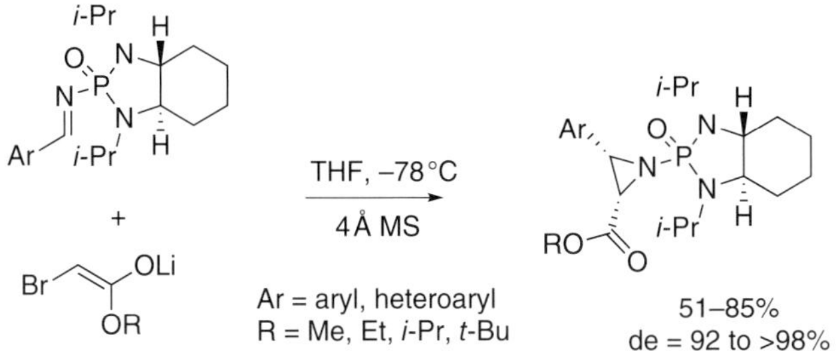
THF, −78 °C
4 Å MS
Ar = aryl, heteroaryl
R = Me, Et, i-Pr, t-Bu
51–85%
de = 92 to >98%

Scheme 2.56 Reaction of chiral *N,N*-diisopropyl-*N*-phosphonylaldimines with pregenerated β-bromo lithium enolates.

providing comparable results. Moreover, several bromo lithium enolates derived from different bromo esters afforded good chemical yields (74–78%) and with generally excellent diastereoselectivity of >98% de.

In 2004, Troisi *et al.* demonstrated that lithiated (α-chloroalkyl)heterocycles, generated by deprotonation with LDA in THF, added to various enantiopure aldimines, affording the corresponding chiral hetero-substituted aziridines in a diastereoselective manner and with satisfactory yields, as shown in Scheme 2.57 [80]. The different steric hindrance and coordinating power of the alkyl group linked to the imine

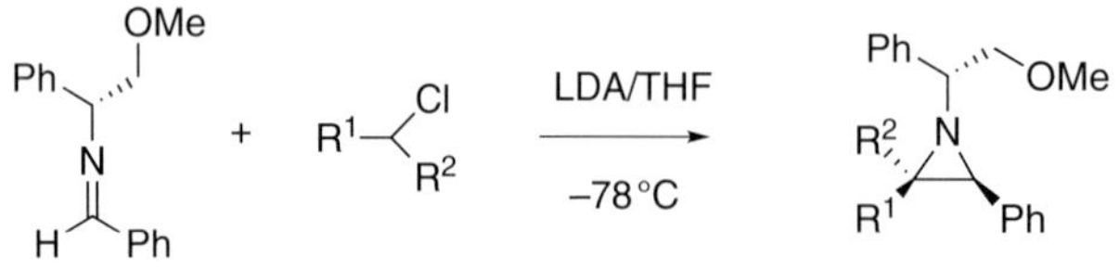

Asymmetric Synthesis of Three-Membered Rings

R^1 = 4-Methylthiazol-2-yl, R^2 = H: 48%, de >96%
R^1 = 4-Methylthiazol-2-yl, R^2 = Me: 90%, de >96%
R^1 = 4-Benzothiazolyl, R^2 = H: 90%, de >96%
R^1 = Me, R^2 = 4-Benzothiazolyl: 60%, de >96%
R^1 = 4,4-Dimethyl-2-oxazolin-2-yl, R^2 = H: 45%, de >96%
R^1 = Me, R^2 = 4,4-Dimethyl-2-oxazolin-2-yl: 35%, de >96%
R^1 = 2-Py, R^2 = H: 60%, de >96%
R^1 = 4-Py, R^2 = H: 30%, de >96%

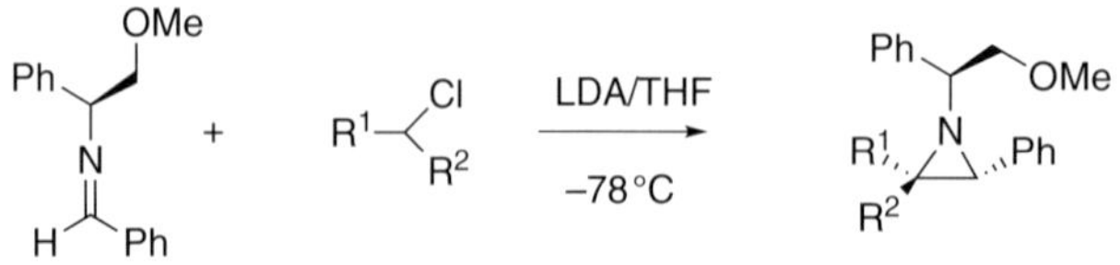

R^1 = 4-Methylthiazol-2-yl, R^2 = H: 45%, de >96%
R^1 = 4,4-Dimethyl-2-oxazolin-2-yl, R^2 = H: 52%, de >96%
R^1 = Me, R^2 = 4,4-Dimethyl-2-oxazolin-2-yl: 31%, de >96%

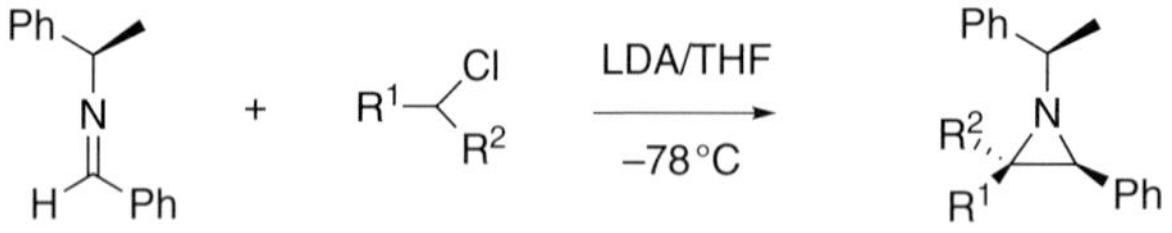

R^1 = 4-Methylthiazol-2-yl, R^2 = H: 50%, de = 60%
R^1 = Me, R^2 = 4,4-Dimethyl-2-oxazolin-2-yl: 55%, de = 34%

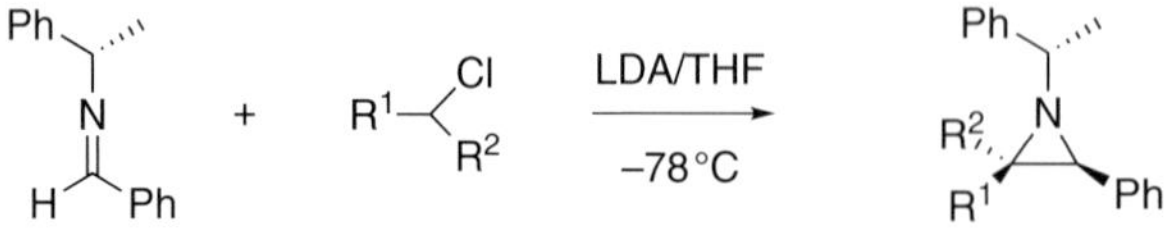

R^1 = 4-Methylthiazol-2-yl, R^2 = H: 53%, de = 60%
R^1 = Me, R^2 = 4,4-Dimethyl-2-oxazolin-2-yl: 52%, de = 34%

Scheme 2.57 Aza-Darzens reactions of chiral aldimines with lithiated (α-chloroalkyl) heterocycles.

nitrogen atom could influence the aziridine ring-closure process and, consequently, the diastereoselectivity of the process. Indeed, aziridines synthesized from chiral aldimines bearing a methoxymethyl group were isolated with a higher diastereoselectivity than those prepared from chiral imines bearing a methyl group.

In 2006, Savoia *et al.* reported a novel asymmetric protocol for the preparation of chiral 2-(2-pyridyl)aziridines from chiral 2-pyridinealdimines [81]. Using this procedure, the addition of chloromethyllithium, generated *in situ* from methyllithium and chloroiodomethane in the presence of lithium bromide, onto the chiral aldimine, derived from 2-pyridinecarboxaldehyde and (S)-valinol protected as its O-trimethylsilyl ether, provided the corresponding 1,2-disubstituted aziridine in good yield and diastereoselectivity, as shown in Scheme 2.58. This protocol was extended to another 2-pyridinealdimine derived from (S)-valine methyl ester, which led to the corresponding aziridine in good yield and with diastereoselectivity of 74% de (Scheme 2.58). Similarly, the reaction of a chiral 2-quinolinealdimine depicted in Scheme 2.58 led to the corresponding aziridine-α-chloroketone in comparable yield and with a diastereoselectivity of 76% de.

Moreover, *in situ* generated iodomethyllithium was condensed by Concellon *et al.* onto aldimines derived from *p*-toluenesulfonamides, yielding the corresponding aziridines [82]. In particular, the reaction of the chiral aldimine derived from phenylalaninal allowed access to the corresponding enantiopure (2R,1'S)-2-(1'-aminoalkyl)aziridine with very high diastereoselectivity (Scheme 2.59). The stereoselectivity of this process was explained by the addition of iodomethyllithium to the chiral aminoimine taking place under nonchelation control, assuming that the energetically more favored transition state had the larger substituent (N,N-dibenzylamino) *anti* to the attack of the iodomethyllithium (Scheme 2.59).

On the other hand, it is also possible to involve a chiral enolate as the chiral auxiliary for the aza-Darzens process instead of a chiral imine. As an example, Sweeney *et al.* have studied the aza-Darzens reaction of a wide range of N-diphenylphosphinylaldimines with chiral camphorsultam-derived α-bromoenolate [83]. In all cases of imines, the corresponding enantiopure

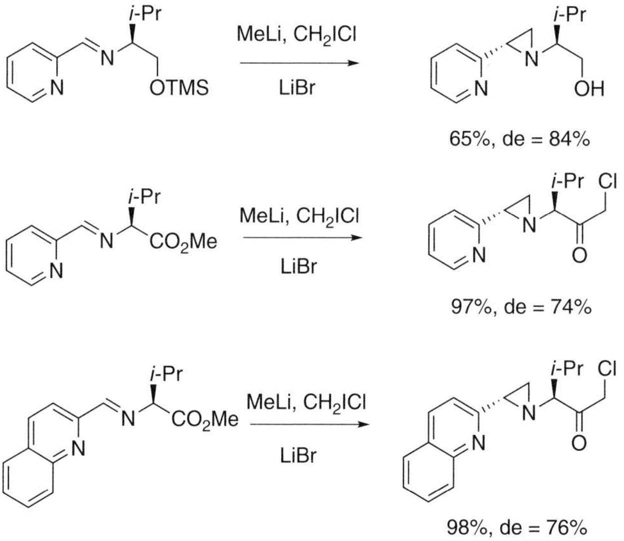

Scheme 2.58 Aza-Darzens reactions of chiral 2-pyridinealdimines and a 2-quinolinealdimine with *in situ* generated chloromethyllithium.

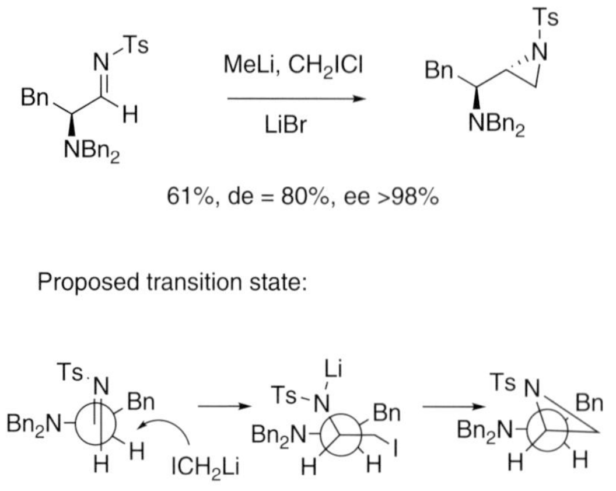

[Scheme showing aziridine synthesis from Bn/NBn₂ aldimine with MeLi, CH₂ICl, LiBr]

61%, de = 80%, ee >98%

Proposed transition state:

[Transition state structures showing Ts-N, Bn₂N, Bn, H, ICH₂Li intermediates]

Scheme 2.59 Aza-Darzens reaction of a chiral phenylalaninal-derived aldimine with *in situ* generated chloromethyllithium.

cis-aziridines were isolated in good yields and with virtually complete diastereo- and enantiocontrol, as shown in Scheme 2.60.

Remarkably, these authors have demonstrated that the stereoselectivity of this process was dependent upon the structure of the aldimine substituent [84]. Indeed, the inherent *cis* selectivity of the reaction could be inverted to exclusively

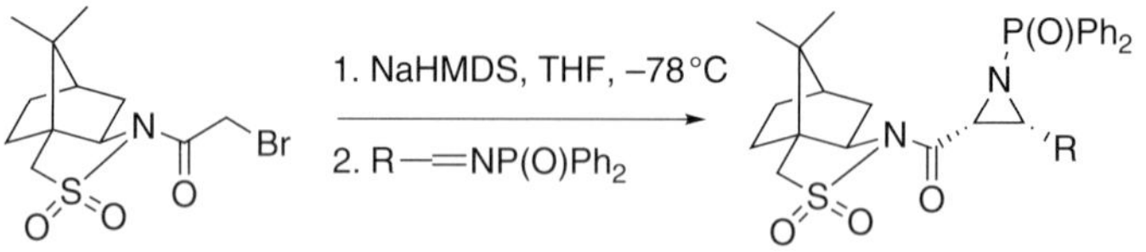

[Scheme showing camphorsultam-derived α-bromoenolate reacting with 1. NaHMDS, THF, −78°C; 2. R−N=P(O)Ph₂ to give aziridine product with P(O)Ph₂ group]

R = Ph: 71%, *cis:trans* = 100:0, de >90%
R = p-NO₂C₆H₄: 75%, *cis:trans* = 100:0, de >90%
R = p-MeOC₆H₄: 78%, *cis:trans* = 100:0, de >90%
R = o-NO₂C₆H₄: 70%, *cis:trans* = 100:0, de >90%
R = p-BrC₆H₄: 60%, *cis:trans* = 100:0, de >90%
R = 2-Naph: 72%, *cis:trans* = 100:0, de >90%
R = CH=CHF: 67%, *cis:trans* = 100:0, de >90%
R = 2-Fu: 68%, *cis:trans* = 100:0, de >90%
R = *t*-Bu: 40%, *cis:trans* = 100:0, de >90%
from (2*S*)-sultam:
R = Ph: 71%, *cis:trans* = 100:0, de >90%
R = p-FC₆H₄: 57%, *cis:trans* = 100:0, de >90%
R = 2,6-Cl₂C₆H₃: 60%, *cis:trans* = 100:0, de >90%
R = m-BrC₆H₄: 60%, *cis:trans* = 100:0, de >90%
R = p-MeOC₆H₄: 60%, *cis:trans* = 100:0, de >90%
R = 2-Py: 67%, *cis:trans* = 100:0, de >90%
R = CH=CH₂: 47%, *cis:trans* = 100:0, de >90%

Scheme 2.60 Aza-Darzens reaction of a camphorsultam-derived α-bromoenolate with *N*-diphenylphosphinylaldimines.

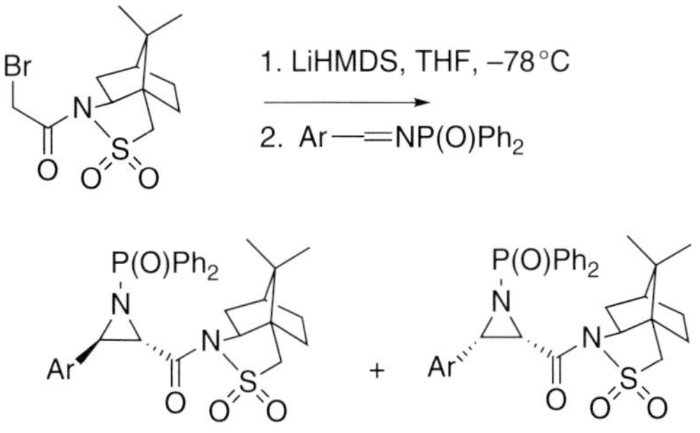

Ar = o-BrC$_6$H$_4$: 67%, *trans:cis* = 100:0, de = 100%
Ar = o-IC$_6$H$_4$: 73%, *trans:cis* = 100:0, de = 100%
Ar = o-MeOC$_6$H$_4$: 65%, *trans:cis* = 100:0, de = 100%
Ar = o-CF$_3$C$_6$H$_4$: 69%, *trans:cis* = 100:0, de = 100%
Ar = o-ClC$_6$H$_4$: 68%, *trans:cis* = 80:20, de = 100%
Ar = o-FC$_6$H$_4$: 84%, *trans:cis* = 50:50, de = 100%
Ar = o-Tol: 87%, *trans:cis* = 50:50, de = 100%
Ar = o-EtC$_6$H$_4$: 93%, *trans:cis* = 37:63, de = 100%
Ar = o-NO$_2$C$_6$H$_4$: 72%, *trans:cis* = 0:100, de = 100%
Ar = Ph: 71%, *trans:cis* = 0:100, de = 100%

Scheme 2.61 Aza-Darzens reaction of a camphorsultam-derived α-bromoenolate with *ortho*-substituted N-diphenylphosphinylarylaldimines.

give the *trans*-configured products, depending on the substitution pattern of the arylaldimine, as demonstrated in Scheme 2.61.

2.2.2.3 Addition/Elimination Processes

In 2006, De Kimpe *et al.* reported a novel stereoselective synthesis of chiral 2-arylated and 2-alkylated aziridines on the basis of the reaction of chiral α-chloro N-(*tert*-butanesulfinyl)aldimines with Grignard reagents, affording chiral β-chloro N-sulfinamides in good yields as nonisolated intermediates [85]. The latter compounds were ring-closed toward the corresponding N-sulfinyl aziridines in a high-yielding, one-pot reaction or after a separate treatment with a base in some cases. As shown in Scheme 2.62, high levels of diastereoselectivities of up to 92% de were observed. This diastereoselectivity was explained via the coordinating ability of the α-chloro atom with magnesium, resulting in an opposite stereochemical outcome to that generally observed for nonfunctionalized N-sulfinyl imines.

Moreover, these authors have attempted to extend this methodology to α-chloro N-(*tert*-butanesulfinyl)ketimines, but, in this case, treatment of the latter imines with a Grignard reagent led to the synthesis of the corresponding chiral N-(1-substituted cyclopropyl)-*tert*-butanesulfinamides in good yields and diastereoselectivities via 1,3-dehydrohalogenation and subsequent addition of the Grignard reagent to the intermediate cyclopropylideneamine [86]. Only in the case of allylmagnesium chloride the reaction did lead to the corresponding aziridines in high yields and with diastereoselectivities of up to 90% de, as shown in Scheme 2.63.

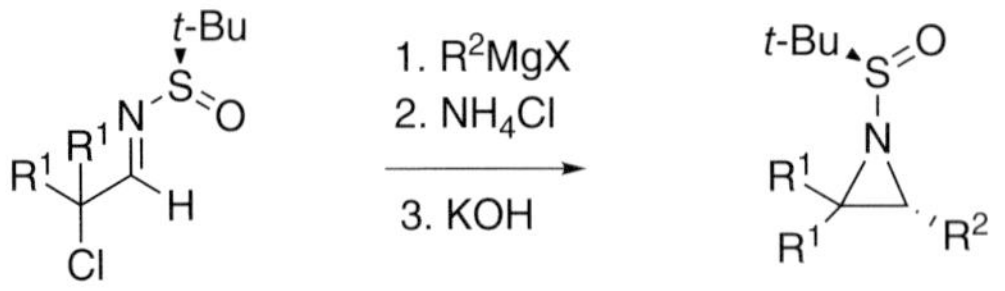

Scheme 2.62 Reaction of chiral α-chloro N-sulfinyl aldimines with Grignard reagents.

$R^1 = R^2 = Et$, $X = Br$: 82%, de = 84%
$R^1 = Et$, $R^2 = Ph$, $X = Cl$: 77%, de = 92%
$R^1 = Et$, $R^2 = CH=CH_2$, $X = Cl$: 83%, de = 86%
$R^1 = Et$, $R^2 = CH_2\text{-}CH=CH_2$, $X = Cl$: 90%, de = 24%
$R^1,R^1 = (CH_2)_5$, $R^2 = Et$, $X = Br$: 82%, de = 84%
$R^1 = Et$, $R^2 = CH_2\text{-}CH=CH_2$, $X = Cl$: 90%, de = 24%
$R^1,R^1 = (CH_2)_5$, $R^2 = Ph$, $X = Cl$: 43%, de = 40%
$R^1 = Et$, $R^2 = CH_2\text{-}CH=CH_2$, $X = Cl$: 90%, de = 24%
$R^1,R^1 = (CH_2)_5$, $R^2 = CH=CH_2$, $X = Br$: 85%, de = 72%
$R^1 = Et$, $R^2 = CH_2\text{-}CH=CH_2$, $X = Cl$: 90%, de = 24%
$R^1,R^1 = (CH_2)_5$, $R^2 = CH_2\text{-}CH=CH_2$, $X = Cl$: 84%, de = 40%

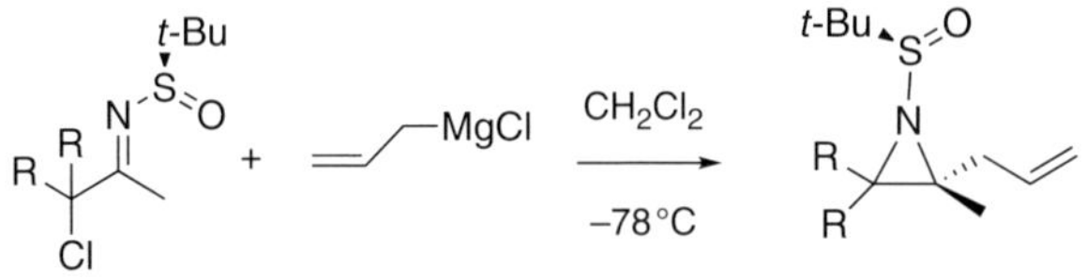

R = Me: 95%, de >90%
R = Et: 91%, de = 90%
R,R = $(CH_2)_5$: 83%, de = 86%

Scheme 2.63 Reaction of chiral α-chloro N-sulfinylketimines with allylmagnesium chloride.

Later in 2010, the preceding methodology could be extended to other α-chloro N-*tert*-butanesulfinylketimines as well as to Grignard reagents, providing a range of chiral N-sulfinyl 2,2-disubstituted aziridines to be synthesized with good-to-excellent diastereoselectivities (72–96% de) via a 1,2-addition across the imino function followed by a subsequent ring closure. These aziridines were obtained almost enantiopure (>98% ee) and in good yields (51–85%), as shown in Scheme 2.64. [87]. It was found that α-chloro N-*tert*-butanesulfinylketimines were the preferred α-halo ketimines for this aziridine formation as the corresponding α-bromo derivatives mainly led to the debrominated ketimines upon treatment with Grignard reagents. The further oxidation of the formed chiral N-sulfinyl 2,2-disubstituted aziridines to the corresponding N-sulfonyl aziridines was possible without loss of enantiomeric purity, allowing the determination of the absolute configuration of the N-sulfinyl aziridines. The stereoselectivity achieved in the Grignard addition was rationalized by the coordinating ability of the α-chloro atom resulting in the opposite stereochemical outcome as observed for nonfunctionalized N-sulfinyl ketimines.

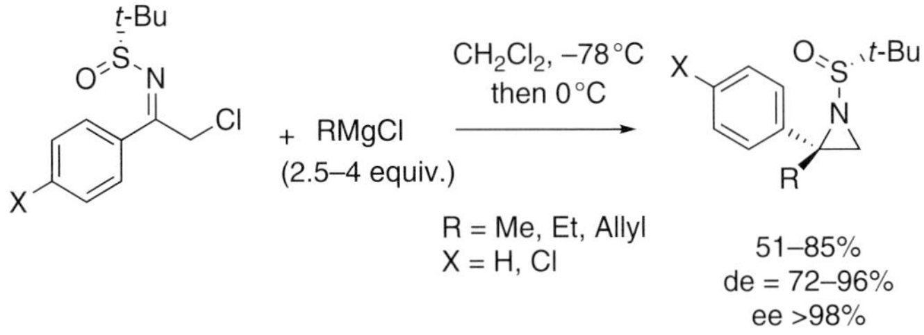

Scheme 2.64 Reaction of chiral α-chloro N-sulfinylketimines with Grignard reagents.

In 2012, Suzuki and Aoyagi reported the first asymmetric total synthesis of a structurally unique alkaloid, chamobtusin A, the key step of which consisted of a novel aziridine formation from the reaction between a chiral tricyclic 1,2-oxazine and allylmagnesium chloride [88]. As shown in Scheme 2.65, the enantiopure key tricyclic aziridine was obtained in 58% yield as a single stereoisomer, presumably via initial abstraction of an allylic hydrogen by an appropriately oriented proximal allyl component on the nitrogen of allylated intermediate complex **T**, followed by simultaneous aziridine formation/N—O bond cleavage. The chiral aziridine was further converted into expected chamobtusin A through seven supplementary steps.

In another context, De Kimpe *et al.* have shown that the reduction of α-chloro N-(*tert*-butylsulfinyl)imines led to the corresponding chiral aziridines in good-to-excellent yields [89]. Therefore, upon reduction of (R_S)-N-(*tert*-butylsulfinyl)imines with $NaBH_4$ in THF, in the presence of MeOH, the intermediate (R_S,S)-β-chloro sulfinamides were formed in excellent yields and diastereoselectivities (>96% de). A simple treatment of the latter sulfinamides with KOH afforded the corresponding (R_S,S)-N-(*tert*-butylsulfinyl)aziridines as almost single diastereomers in quantitative yields, as shown in Scheme 2.66. On the contrary, the epimers, (R_S,R)-N-(*tert*-butylsulfinyl)aziridines, were

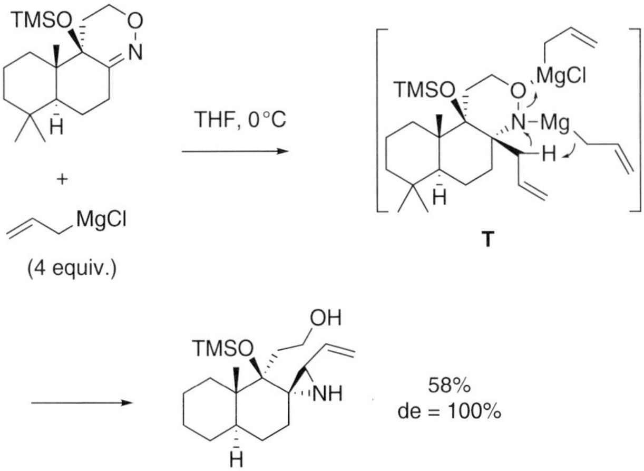

Scheme 2.65 Reaction of a chiral tricyclic 1,2-oxazine with allylmagnesium chloride.

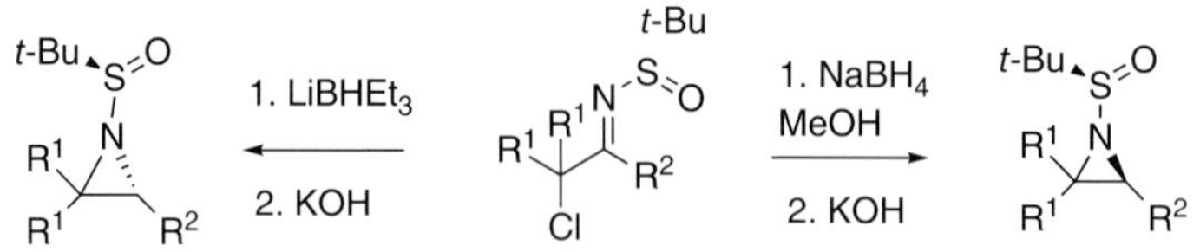

Scheme 2.66 Reductions of chiral α-chloro N-sulfinylimines.

with NaBH$_4$:
R^1 = Me, R^2 = H: 95%, ee = 100%
R^1 = Me, R^2 = Ph: 95%, ee = 100%
R^1 = R^2 = Me: 88%, de = 96%
R^1 = Et, R^2 = Me: 92%, de = 98%
R^1 = H, R^2 = Ph: 90%, de >96%
R^1 = H, R^2 = p-ClC$_6$H$_4$: 89%, de >96%
R^1 = H, R^2 = p-BrC$_6$H$_4$: 91%, de >96%

with LiBHEt$_3$:
R^1 = R^2 = Me: 61%, de = 56%
R^1 = Et, R^2 = Me: 57%, de = 60%
R^1 = H, R^2 = Ph: 62%, de = 74%
R^1 = H, R^2 = p-ClC$_6$H$_4$: 66%, de = 78%
R^1 = H, R^2 = p-BrC$_6$H$_4$: 71%, de = 84%

synthesized by changing the reducing agent from NaBH$_4$ to LiBHEt$_3$. Thus, (R_S,R)-N-(tert-butylsulfinyl)aziridines were achieved in good yields and with diastereoselectivities of up to 84% de by reduction of (R_S)-N-tert-butylsulfinyl α-halo imines performed with LiBHEt$_3$ and subsequent treatment with KOH.

In the same context, Concellon et al. earlier reported the synthesis of a chiral amino aziridine by reduction of its corresponding serine-derived chiral α-amino ketimine with NaBH$_4$ and further treatment with MeLi [90]. This reduction process was performed with total diastereoselectivity, as depicted in Scheme 2.67.

In 2008, Hodgson et al. developed the reaction of N-(2-chloroethylidene)-tert-butylsulfinamide with various organocerium reagents, which provided an efficient and highly diastereoselective access to terminal N-tert-butylsulfinyl aziridines [91]. Alkyl- and allyl-cerium reagents were added with essentially complete diastereocontrol, whereas the reaction was less diastereoselective for aryl-, heteroaryl- and alkynyl-cerium reagents. In order to demonstrate the applicability of this reaction in asymmetric synthesis, n-DecCeCl$_2$ was added to a chiral aldimine to give the corresponding chiral aziridine in good yield and with high diastereoselectivity of 94% de (Scheme 2.68).

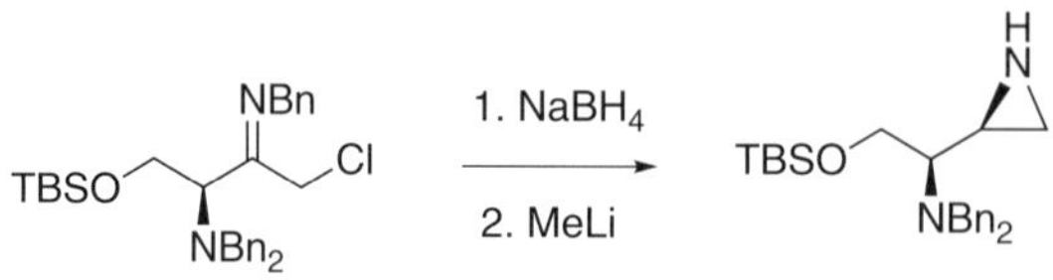

Scheme 2.67 Reduction of a chiral α-amino ketimine.

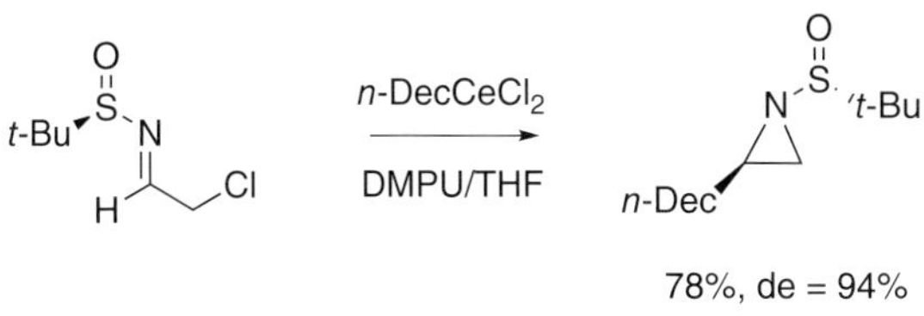
t-Bu
S
O
N
H
Cl
n-DecCeCl2
DMPU/THF
N
S
t-Bu
n-Dec
78%, de = 94%

Scheme 2.68 Reaction of an organocerium reagent with a chiral *N*-(2-chloroethylidene)-*tert*-butylsulfinamide.

Despite their great potential, relatively little investigation has been undertaken so far on the synthesis of chiral alkynylaziridines. In this context, Ferreira and Chemla have developed a concise and efficient synthesis of enantiopure *trans*-ethynyl *N*-*tert*-butylsulfinyl aziridines based on the condensation of the allenylzinc species [92] derived from 3-chloro-1-trimethylsilylpropyne onto *N*-*tert*-butylsulfinyl-aldimines and -ketimines (Scheme 2.69) [93]. The general excellent stereoselectivity of >96% de was shown to result from a high kinetic resolution in the reaction of the racemic allenylzinc reagent with enantiopure *N*-*tert*-butylsulfinylimines. This kinetic resolution was shown to be the consequence of the zinc being coordinated by both the oxygen and the nitrogen atoms of the sulfinylimine in a chelate-type, four-membered transition state (Scheme 2.69).

It was demonstrated that HMPA had a dramatic influence on the stereochemical outcome of the condensation of the allenylzinc reagent onto chiral *N*-*tert*-butylsulfinylaldimines [94]. Indeed, performing the reaction in the presence of 60 equiv. of HMPA in Et_2O allowed the corresponding *cis*-ethynylaziridines to be formed as the major products with good-to-high selectivities (Scheme 2.70). The *cis* selectivity was postulated to result from a high kinetic resolution through a synclinal transition state in a supra- or antarafacial S_E2' process, which has been supported by semiempirical AM1 and MM2 calculations. Furthermore, after

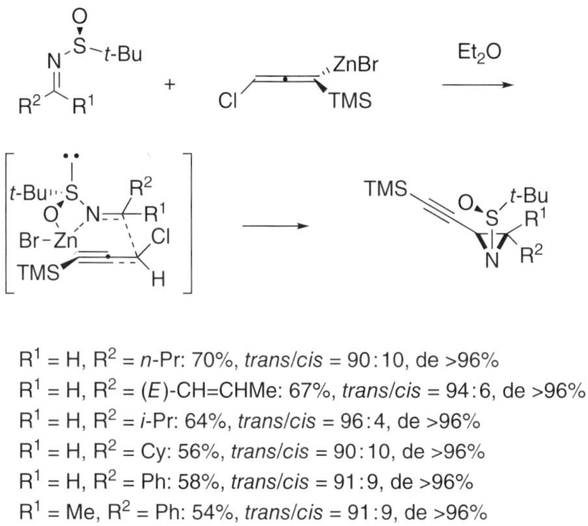
O
S
t-Bu
N
R2
R1
+
ZnBr
Cl
TMS
Et2O
t-Bu
S
R2
O
N
R1
Br-Zn
Cl
TMS
H
TMS
O
t-Bu
S
R1
N
R2
R1 = H, R2 = n-Pr: 70%, trans/cis = 90:10, de >96%
R1 = H, R2 = (E)-CH=CHMe: 67%, trans/cis = 94:6, de >96%
R1 = H, R2 = i-Pr: 64%, trans/cis = 96:4, de >96%
R1 = H, R2 = Cy: 56%, trans/cis = 90:10, de >96%
R1 = H, R2 = Ph: 58%, trans/cis = 91:9, de >96%
R1 = Me, R2 = Ph: 54%, trans/cis = 91:9, de >96%

Scheme 2.69 Reaction of an allenylzinc reagent with chiral *N*-*tert*-butylsulfinylimines.

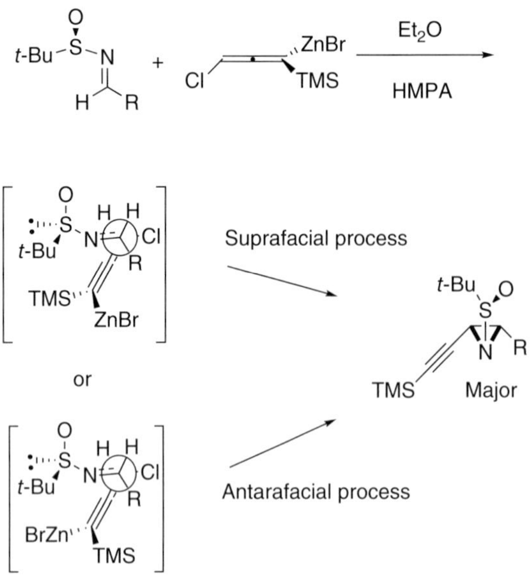

R = Me: 64%, cis/trans = 89:11, de >96%
R = n-Pr: 54%, cis/trans = 84:16, de >96%
R = n-Hept: 56%, cis/trans = 78:22, de >96%
R = (E)-CH=CHMe: 55%, cis/trans = 84:16, de >96%
R = (E)-CH=CH(nPent): 62%, cis/trans = 71:29, de >96%
R = (E)-CH=CHPh: 56%, cis/trans = 87:13, de >96%
R = (CH$_2$)$_2$Ph: 50%, cis/trans = 71:29, de >96%
R = C≡C(nPent): 60%, cis/trans = 73:27, de >96%

Scheme 2.70 Reaction of an allenylzinc reagent with chiral N-tert-butylsulfinylaldimines in the presence of HMPA.

chromatographic separation over silica gel, the major cis-aziridines were obtained as diastereo- and enantiomerically pure products (>96% de and >99% ee).

In 2010, Ruano et al. developed the asymmetric aziridination of a range of (R)-N-sulfinyl aldimines, including aryl, heteroaryl, alkyl, as well as alkenyl derivatives, with (S)-2-(p-tolylsulfinyl)benzyl iodide in the presence of sodium hexamethyl disilazide [95]. As shown in Scheme 2.71, the reaction took place within 1 min with almost complete control of the stereoselectivity, facial- as well as trans-selectivities, and with good-to-very-high yields (61–94%). The only exception was the reaction starting from a cyano aldimine (R^1 = p-Tol, R^2 = p-CNC$_6$H$_4$) containing this strong electron-withdrawing group, which provided a low diastereomeric trans/cis ratio (65:35) along with a low facial diastereoselectivity of 33% de for the trans-isomer. The subsequent simultaneous removal of both C- and N-p-tolylsulfinyl groups from these major trans-aziridines by treatment with t-BuLi afforded the corresponding enantiopure trans-NH aziridines without affecting their optical purity, which was >98% ee.

In 2011, De Kimpe et al. reported the asymmetric synthesis of α-chloro-β-amino-N-sulfinylimidates in good to quantitative yields and with generally high

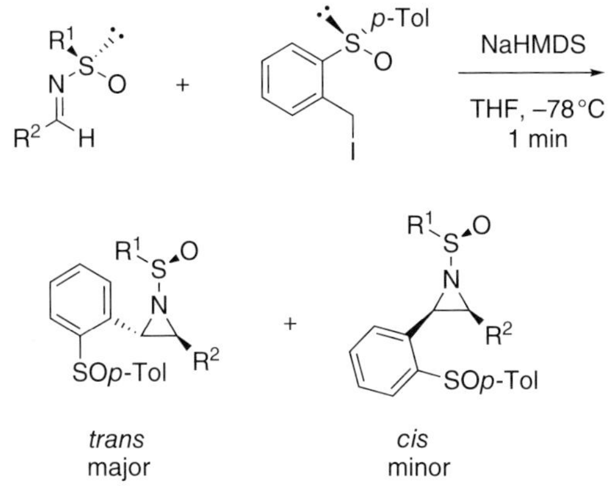

61–94%
trans/cis = 65:35 to >98:2
de (*trans*) >98% except for R^1 = *p*-Tol, R^2 = *p*-CNC_6H_4: de = 33%
de (*cis*) >98% except for R^1 = *p*-Tol, R^2 = Bu: de = 42% and
R^1 = *p*-Tol, R^2 = *i*-Bu: de = 10%

Scheme 2.71 Reaction of chiral *N*-sulfinylaldimines with chiral 2-(*p*-tolylsulfinyl)benzyl iodide.

diastereoselectivities of up to 92% de through a highly *anti*-selective Mannich-type reaction of (R_S)-methyl *N*-*tert*-butanesulfinyl-2-chloro ethanimidate as chiral auxiliary with aromatic aldimines (Scheme 2.72) [96]. The formed isolated chiral α-chloro-β-amino-*N*-sulfinyl imidates proved to be excellent building blocks for the asymmetric synthesis of aziridine-2-carboxylic amides and esters. Indeed, treatment of these chiral α-chloro-β-amino-*N*-sulfinyl imidates by HCl

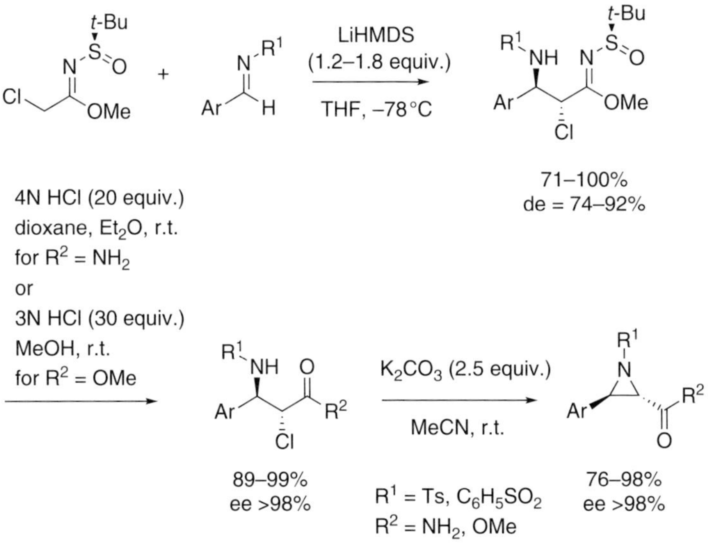

Scheme 2.72 Reaction of chiral methyl *N*-*tert*-butanesulfinyl-2-chloroethanimidate with aromatic aldimines.

provided the corresponding methyl esters (R^2 = OMe) if performed in methanol or the corresponding amides (R^2 = NH$_2$) if performed in dioxane. It must be noted that in both cases of products, they were afforded enantiopure (>98% ee) and in excellent yields (89–99%). In the next step, these amides and esters could be easily cyclized into the corresponding enantiopure aziridines by addition of K$_2$CO$_3$ in MeCN in high yields (76–98%), as shown in Scheme 2.72. It must be noted that the obtained relative *anti*-diastereoselectivity was the opposite of the stereochemical outcome observed for α-methyl-substituted imidates.

In the same area, these authors have also investigated the reaction of a chiral α-chloro-*N*-*p*-toluenesulfinylaldimine with *N*-(diphenylmethylene) glycine esters in the presence of different bases, such as LDA and LiHMDS [97]. The authors have demonstrated that the influence of the base used for the formation of the glycine enolates was of great importance for the *anti*/*syn*-diastereoselectivity of the reaction. As shown in Scheme 2.73, when using LiHMDS, the reaction afforded the corresponding opened *syn*-γ-chloro-α,β-diamino esters in high yields (86–88%) and with generally high diastereoselectivity of >94% de. These products were further cyclized by treatment with K$_2$CO$_3$ in acetone into the corresponding functionalized *syn*-aziridines in high yields (83–99%). On the other hand, the use of LDA as a base allowed the synthesis of the corresponding opened *anti*-γ-chloro-α,β-diamino esters in moderate-to-good yields (52–79%) and diastereoselectivities (44–80% de), which were also subsequently cyclized into the corresponding *anti*-aziridines in good yields (73–79%) by treatment with K$_2$CO$_3$ in acetone. The dramatic influence of the base, LDA or LiHMDS, on the stereochemical outcome of this Mannich-type reaction across the α-chloro-*N*-sulfinylaldimine under kinetic conditions was rationalized on the basis of the enolate geometry of the anions derived from the deprotonation of *N*-(diphenylmethylene) glycine esters. The enolates obtained via deprotonation of *N*-(diphenylmethylene) glycine esters with LDA were expected to have the *Z*-geometry, which was favored by intramolecular chelation. Alternatively, the authors suggested that upon deprotonation of *N*-(diphenylmethylene) glycine esters with the less basic LiHMDS in THF, a shift toward the formation of the *E*-enolate occurred. Unfortunately, the enolate geometry could not be determined via trapping experiments with TMSCl. Reaction of the *Z*- and *E*-enolates via respective transition states **P** and **Q** resulted in the formation of intermediate *anti*- and *syn*-γ-chloro-α,β-diamino esters, respectively (Scheme 2.73).

Furthermore, a related methodology was applied by the same authors to the asymmetric synthesis of *N*-sulfinyl-β,γ-aziridino-α-amino carboxylic amides [98]. As shown in Scheme 2.74, the process involved the Mannich-type reaction between chiral α-chloro-*N*-*p*-toluenesulfinylaldimines and *N*-(diphenylmethylene) glycinamides in the presence of LiHMDS as a base. In contrast to the reaction of glycine esters (Scheme 2.74), the diastereoselectivity of the reaction of the glycinamides was independent of the base used, always providing the corresponding γ-chloro-*syn*-α,β-diaminocarboxylamides, as shown in Scheme 2.74. They were formed in moderate-to-good yields (41–73%) and with complete *syn*-diastereoselectivity of >98% de. The latter were further cyclized into the corresponding *N*-sulfinyl-*syn*-β,γ-aziridino-α-amino carboxylic amides in moderate-to-high yields (36–90%) by treatment with K$_2$CO$_3$ in acetone. The authors have explained the formation of the products

Synthesis of *syn*-aziridines:

86–88%, de >94%

syn 83–99%

Synthesis of *anti*-aziridines:

52–79%, de = 44–80%

anti 73–79%

Transition state models for reaction of *Z*- and *E*-enolates of glycine esters:

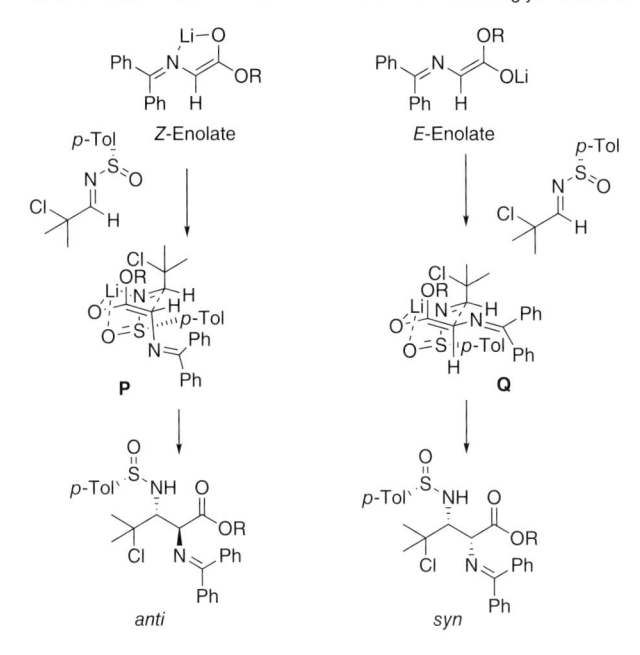

Scheme 2.73 Reactions of chiral α-chloro-*N*-*p*-toluenesulfinylisobutyraldimine with *N*-(diphenylmethylene) glycine esters.

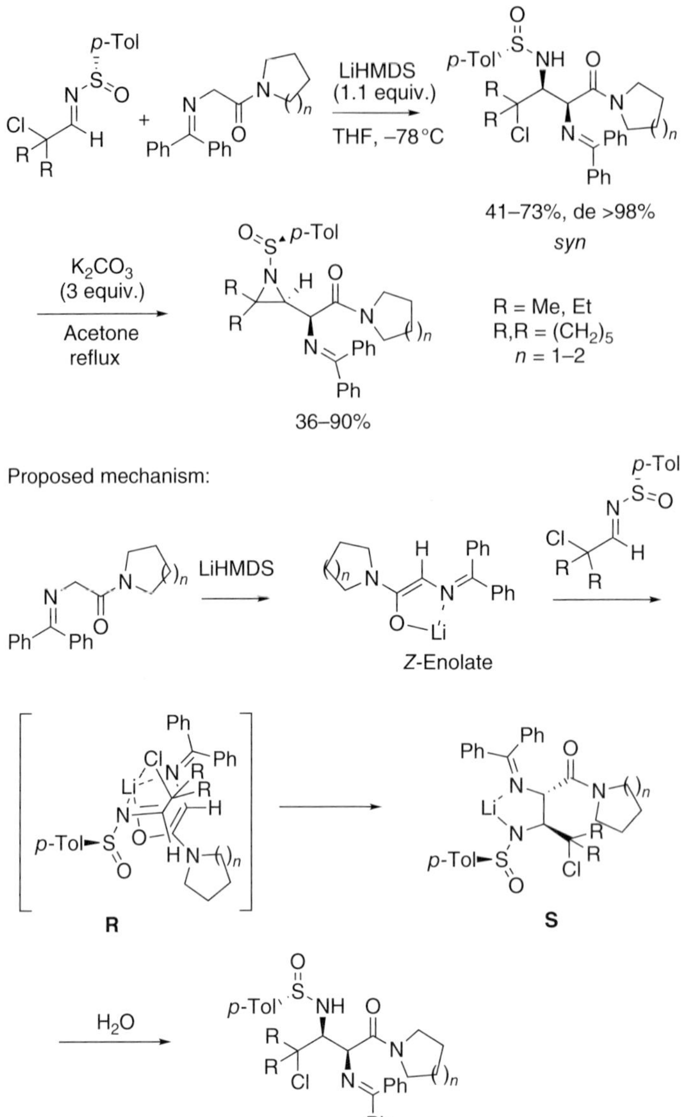

Scheme 2.74 Reaction of chiral α-chloro-N-p-toluenesulfinylaldimines with N-(diphenylmethylene) glycinamides.

by a boat-like transition state model **R** involving the (E)-N-p-toluenesulfinyl aldimines. This less sterically hindered transition state **R**, in which the haloalkyl group (CClR$_2$) occupied the less hindered pseudoequatorial position, and the corresponding Li-adduct **S** were stabilized by the interaction between the Li cation, the diphenylmethyleneamino group, and the sulfinyl imine nitrogen, as shown in Scheme 2.74.

In 2013, Ghorai et al. reported the synthesis of chiral α,β-diamino ester derivatives on the basis of asymmetric imino-aldol reactions between the corresponding protected chiral amino esters and aldimines employing the memory of chirality concept for chiral induction [99]. The imino-aldol reaction of the conformationally chiral enolates of the protected chiral amino esters with aldimines led to a mixture of diastereomers in good yields (62–84%), which were subsequently subjected to Boc-group deprotection by treatment with TFA to give the corresponding α,β-diamino ester derivatives in high yields (70–91%), with moderate diastereoselectivities of up to 66% de, along with enantioselectivities of 56–92% ee for the major diastereomers and 33–88% ee for the minor diastereomers. The major diastereomers of these α,β-diamino ester derivatives were further converted into the corresponding aziridines with conservation of the enantiopurity (80–92% ee). As shown in Scheme 2.75, their ester group was reduced in a first step by treatment with LiBH$_4$ to give the corresponding alcohols in good yields (73–75%). The latter were subsequently submitted to cyclization into aziridines in good yields (65–71%) by treatment with mesyl chloride and triethylamine.

In 2013, Sun et al. reported a facile and efficient approach for the synthesis of chiral vinyl aziridines based on a Zn-mediated reaction of chiral N-tert-butanesulfinylaldimines with 1,3-dibromopropene [100]. As shown in Scheme 2.76, the aziridines were achieved in moderate-to-good yields (25–63%) and with generally high diastereoselectivity of >90% de in all cases of substrates studied. Nitrogen in aza-Barbier product **2** could nucleophilically attack the carbon adjacent to the leaving group and subsequently underwent an S$_N$2-like reaction through a diastereoselective pathway to form the final vinyl aziridine (Scheme 2.76). A six-membered cyclic chair transition state model depicted in

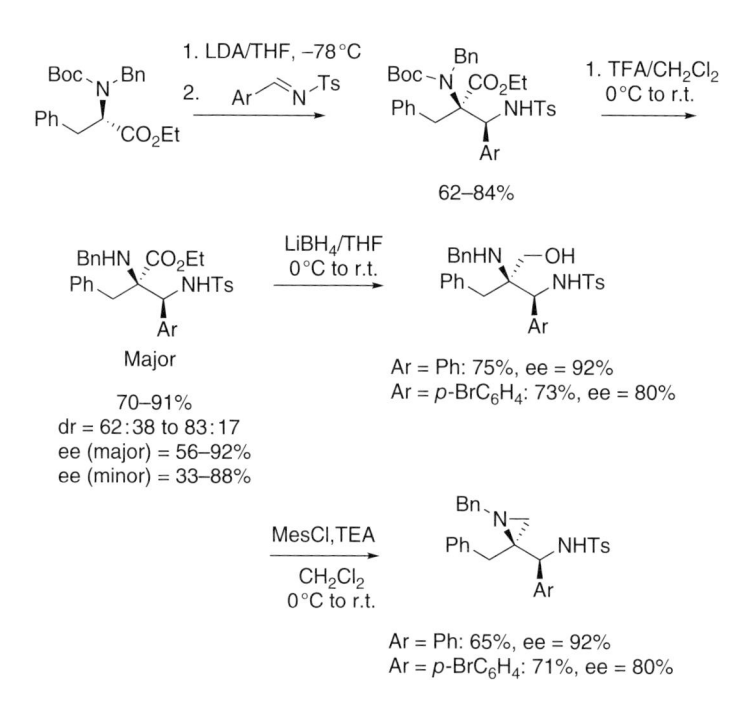

Scheme 2.75 Reaction of a chiral-protected amino ester with aldimines.

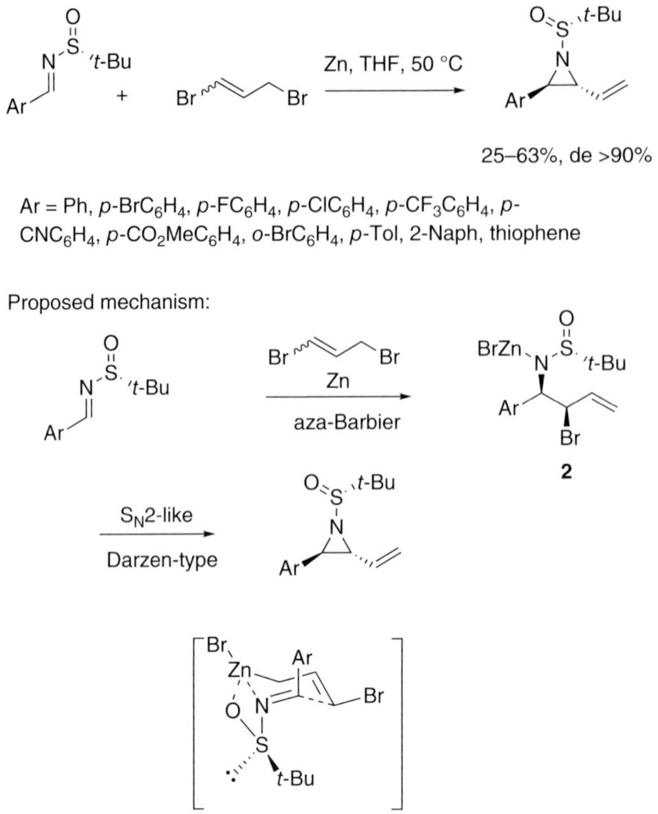

Scheme 2.76 Zn-mediated reaction of chiral N-tert-butanesulfinylaldimines with 1,3-dibromopropene.

Scheme 2.76 was believed to be preferred, and the allylzinc was thought to coordinate both to the imine nitrogen and to sulfinyl oxygen.

In another context, Guijarro et al. recently developed a remarkably efficient one-pot methodology to reach various highly optically enriched, protected, nitrogenated heterocycles with different ring sizes including aziridines [101]. This methodology consisted of an asymmetric ruthenium-catalyzed transfer hydrogenation of N-(tert-butylsulfinyl)haloketimines followed by treatment with a base to promote an intramolecular nucleophilic substitution process. As shown in Scheme 2.77, the one-pot sequential reaction afforded a range of chiral aziridines bearing aromatic, heteroaromatic as well as alkyl substituents in high yields (81–92%) and with excellent diastereoselectivities of 92% to >98% de.

2.2.2.4 Miscellaneous Reactions

In 2010, chiral C-fluoroaziridines were synthesized for the first time by Khlebnikov et al. by reaction of fluorocarbene with N-diphenylmethylidene-substituted natural amino acid esters [102]. As shown in Scheme 2.78, chiral Schiff bases derived from benzophenone imine and amino acid (L-alanine, L-phenylalanine, and L-isoleucine) methyl ester hydrochlorides reacted with

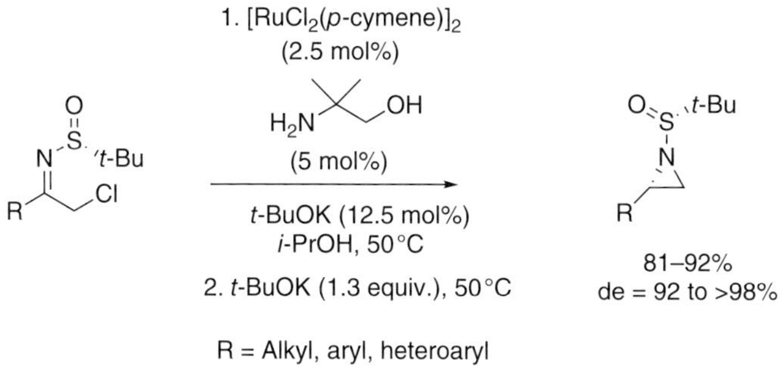

Scheme 2.77 Ru-catalyzed reaction of chiral N-(tert-butylsulfinyl)haloketimines.

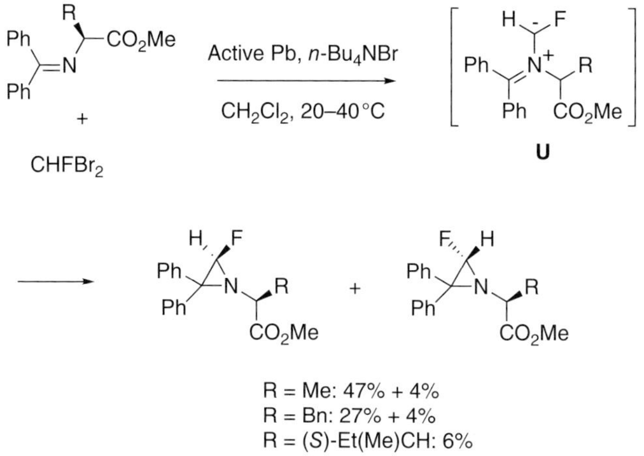

Scheme 2.78 Reaction of chiral ketimines with fluorocarbene.

fluorocarbene *in situ* generated by reduction from CHFBr$_2$ with active lead in the presence of tetrabutylammonium bromide under ultrasonic activation. The reaction involved the electrophilic attack of fluorocarbene to the lone electron pair on the nitrogen atom in the imine, leading to ylide **U**, which underwent further cyclization to provide the final chiral aziridine in moderate yield as a mixture of two diastereomers. It must be noted that the yields decreased when moving from L-alanine to L-isoleucine derivatives as the size of the R substituent at the chiral center increased.

2.2.3 Addition to Azirines

Azirines (three-membered cyclic imines) correspond to the smallest nitrogen-unsaturated heterocyclic system, with two carbon atoms and one double bond in a three-membered ring. Substituted azirines are versatile compounds [103] and have been used for the preparation of various substituted aziridines. The chemistry of these compounds is dominated by processes in which the strain of the three-ring system is relieved. Indeed, the pronounced reactivity of these compounds is due to their ring strain, the electron-rich nature of the C=N bond, and the nitrogen lone pair. Asymmetric nucleophilic addition to azirines is a potentially attractive entry to enantioenriched aziridines. As an example, Alves et al. have developed nucleophilic additions of nitrogen heterocycles to a chiral 2H-azirine-2-carboxylic ester, giving access to optically active aziridine esters [104]. Thus, the relatively nonactivated chiral 2H-azirine-2-carboxylic ester depicted in Scheme 2.79 was used as an electrophile in addition reactions to five-membered ring and five-fused aromatic nitrogen heterocycles, providing the corresponding aziridine esters in moderate-to-good yields and with excellent diastereoselectivity.

A new entry to enantioenriched aziridines has been reported by Somfai et al. on the basis of an asymmetric radical addition of trialkylboranes to various chiral 2H-azirine-3-carboxylates [105]. In particular, high diastereoselectivities of up to 92% de were obtained by using 8-phenylmenthol-derived azirine as the chiral auxiliary, which was reacted with various trialkylboranes (Scheme 2.80). In comparison, another chiral auxiliary, such as Oppolzer's sultam-derived azirine, gave a better yield, albeit combined with a lower diastereoselectivity (Scheme 2.80). It was shown that performing the reaction in the presence of CuCl as a Lewis acid could further increase the diastereoselectivity of the reaction.

The hetero-Diels–Alder reaction is an exceptionally powerful synthetic method for the construction of six-membered heterocycles [106]. In one single transformation, up to four new stereocenters with defined stereochemistry can be formed. By applying azirines as dienophiles, it is possible to form highly functionalized products with a fused [4.1.0] ring system as well as polycyclic structures incorporating the same fused subunit. The aza-Diels–Alder reactions between azirines and many dienes are known to give products with complete regio- and *endo*-selectivities [107]. The azirines are, despite their ring strain, generally poor dienophiles in normal electron-demand Diels–Alder reactions

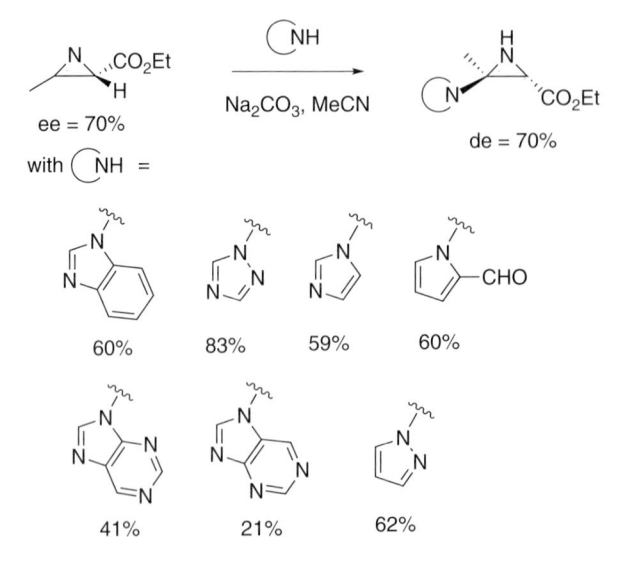

Scheme 2.79 Reaction of nitrogen heterocycles with a chiral 2H-azirine-2-carboxylic ester.

of studying the condensation of enantiopure α-trifluoromethylated aziridinyl anions with various electrophiles, Uneyama et al. have developed the synthesis of chiral α-trifluoromethyl aziridines, starting from the corresponding enantiopure α-trifluoromethyl amino alcohols (Scheme 2.84, eqs 1 and 2) [112]. Similarly, Braga et al. have reported the synthesis of chiral aziridine sulfides to be used as ligands for palladium-catalyzed asymmetric allylic alkylations [113]. Thus, the treatment of α-thioethers of amino alcohols derived from (R)-cysteine with triphenylphosphine combined with DEAD led to the expected enantiopure

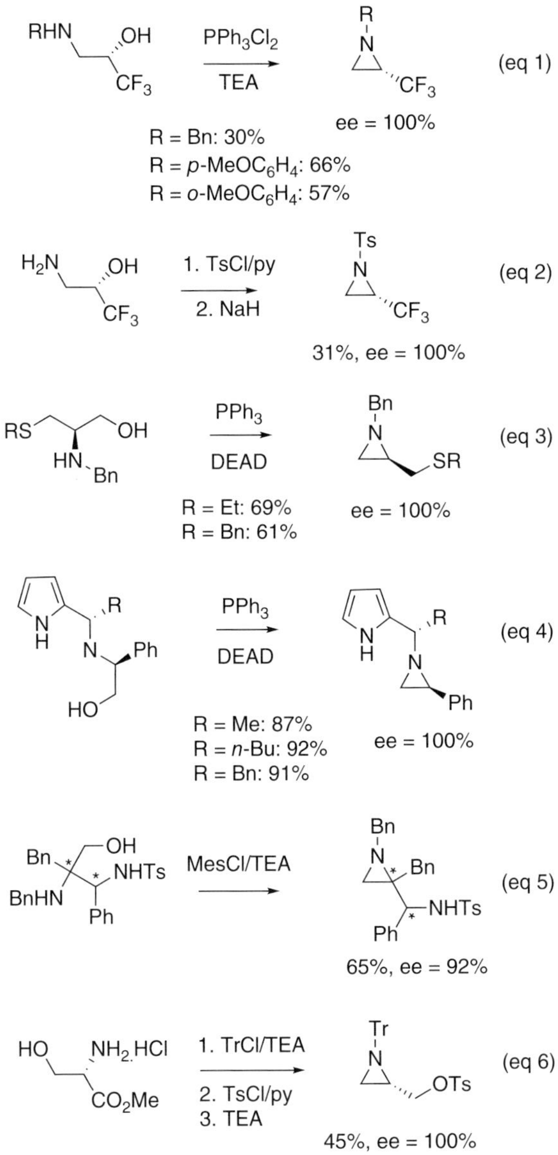

Scheme 2.84 Syntheses of chiral functionalized aziridines from chiral 1,2-amino alcohols.

aziridine sulfides in good yields, as shown in Scheme 2.84, eq 3. These conditions were also applied by Savoia *et al.* to enantiopure β-hydroxyamines derived from (*S*)-phenylglycinol bearing a pyrrole moiety, providing the corresponding chiral pyrrole aziridines in excellent yields (Scheme 2.84, eq 4) [114]. In 2009, Ghorai *et al.* reported the asymmetric synthesis of highly substituted aziridines bearing contiguous quaternary and tertiary stereocenters in good yields and with high enantioselectivities by treating the corresponding amino alcohols with mesyl chloride and triethylamine (Scheme 2.84, eq 5) [115]. In addition, Harrity *et al.* have developed an efficient and practical route to an enantiomerically pure aziridinylmethyl tosylate starting from (*S*)-serine methyl ester hydrochloride [116]. This substrate was successively treated with trityl chloride, tosyl chloride, and then triethylamine to give the expected enantiopure aziridine in 45% overall yield, as shown in Scheme 2.84, eq 6. This aziridine was a key intermediate to prepare chiral triazolylalanine derivatives.

With the aim of preparing chiral β-substituted α-amino phosphonates, Dolence and Roylance have developed a versatile approach for the synthesis of chiral aziridine 2-phosphonates from either (*R*)- or (*S*)-phosphonoserine amino alcohols, which were subsequently submitted to N-tosylation, O-mesylation, and then cyclization with sodium hydride [117]. As shown in Scheme 2.85, the expected chiral aziridine 2-phosphonates were isolated in good yields.

In 2004, Wang *et al.* reported the synthesis of chiral ferrocenyl aziridino alcohols from L-serine and ferrocenecarboxaldehyde in order to employ them as chiral catalysts for the asymmetric addition of diethylzinc to aldehydes [118]. The *N*-alkyl-L-serine ester depicted in Scheme 2.86 was converted upon treatment with triethylamine combined with *p*-toluenesulfonyl chloride into the corresponding *N*-alkylaziridine, which was subsequently transformed into a series of *N*-ferrocenylmethylaziridin-2-ylmethanols by using RMgBr in good yields. In 2015, the utility of this methodology was demonstrated in a synthesis of dendrimer-supported ferrocenylmethyl aziridino alcohol chiral ligands [119].

In 2011, Grellepois *et al.* reported the synthesis of enantiomerically pure *N*-tosyl-2-phenyl- and 2-ethyl-2-trifluoromethylaziridines by Mitsunobu-type cyclization of the corresponding chiral N-protected amino alcohols [120]. As shown in Scheme 2.87, their treatment with diethyl azidocarboxylate and triphenylphosphine in the presence of triethylamine in THF afforded the

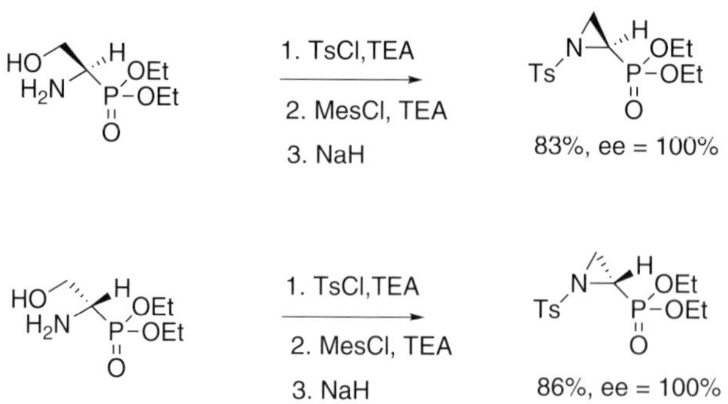

Scheme 2.85 Synthesis of chiral aziridine 2-phosphonates.

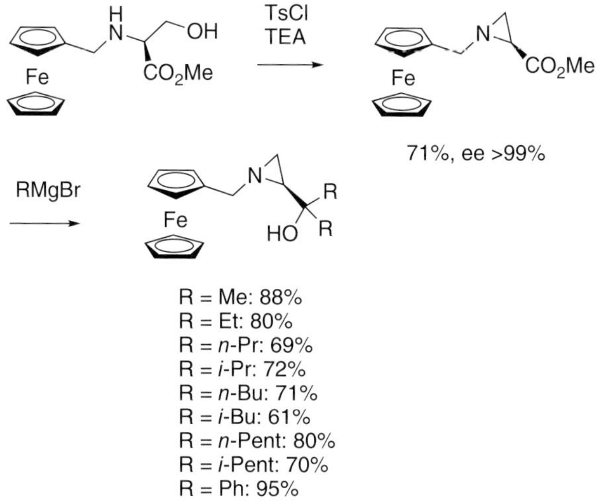

Scheme 2.86 Synthesis of chiral N-ferrocenyl aziridino alcohols.

expected enantiopure aziridines bearing a quaternary stereogenic center in high yields (81–93%). In the same area, related conditions were also applied by Katagiri et al. to the synthesis of enantiopure N-tosyl-2-trifluoromethyl aziridine from the corresponding chiral amino alcohol in 91% yield (Scheme 2.87) [121]. This aziridine was further converted into the corresponding N-tosyl-2-trifluoromethyl-2-alkyloxycarbonylaziridines, which were subsequently submitted to ring opening with various nucleophiles to achieve enantiopure α-trifluoromethyl-α-amino acids after detosylation.

In 2012, a fully protected chiral cis-2-hydroxymethylaziridine was highly efficiently synthesized by Salom-Roig et al. under related Mitsunobu conditions from the corresponding chiral amino alcohol in 77% yield and with high diastereoselectivity of >90% de, as shown in Scheme 2.88 [122].

The asymmetric aziridination based on the use of 1,2-amino alcohols has been applied by several groups for developing total syntheses of various biologically active products. As an example, the key step of a formal enantiospecific synthesis of the antitumor antibiotic, (+)-FR900482, developed by Paleo et al., was based

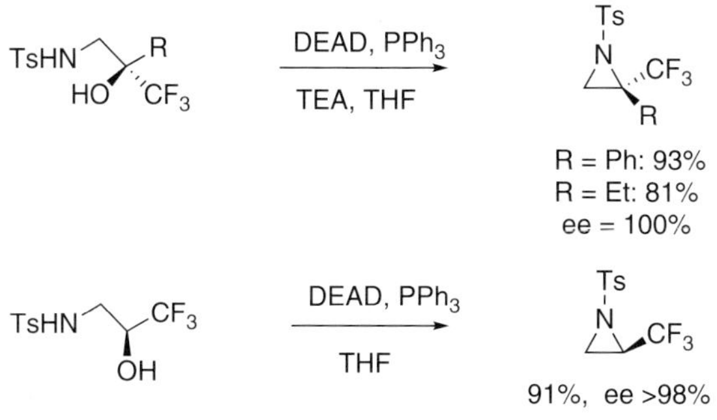

Scheme 2.87 Synthesis of chiral N-tosyl-2-trifluoromethyl aziridines.

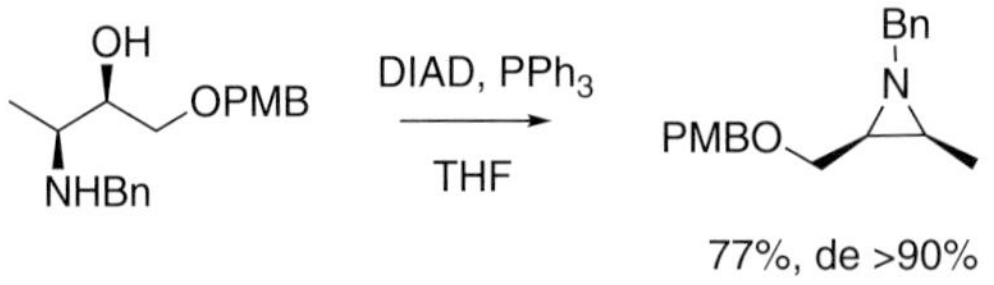

Scheme 2.88 Synthesis of a chiral N-benzyl-2-hydroxymethylaziridine.

on the aziridination of a chiral 1,2-amino alcohol derived from L-vinylglycine mediated by benzenesulfonic anhydride in pyridine (Scheme 2.89) [123].

In 2008, Trost and O'Boyle reported another total synthesis of (+)-FR900482, involving the asymmetric aziridination of a chiral amino diol depicted in Scheme 2.90, which was selectively silylated and mesylated [124]. The mesylate was then exposed to cesium carbonate, affording the expected enantiopure aziridine in good yield, which was further transformed into the final (+)-FR900482.

In 2003, Terashima et al. developed the total synthesis of carzinophilin, an antitumor antibiotic, which involved as a key step in the asymmetric aziridination

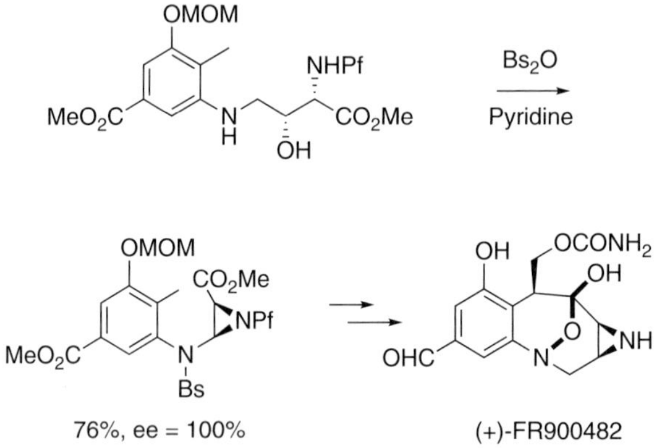

Scheme 2.89 Synthesis of (+)-FR900482.

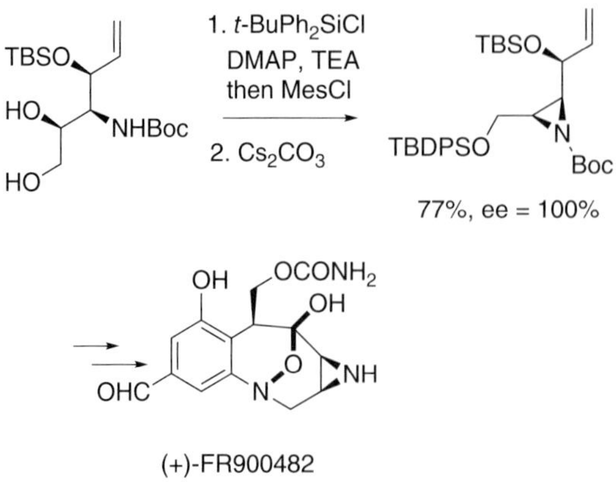

Scheme 2.90 Another synthesis of (+)-FR900482.

of a chiral pyrrolidin-2-ylidenemalonate derived from β-D-arabinofuranose [125]. Thus, the treatment of this pyrrolidin-2-ylidenemalonate with KHMDS led to the expected aziridine in good yield, as shown in Scheme 2.91.

In 2007, Vedejs et al. reported the synthesis of enantiopure aziridinomitosene, which was based on the asymmetric aziridination of a chiral oxazole 1,2-amino alcohol derived from L-serine (Scheme 2.92) [126].

In 2013, Shi and coworkers reported a novel asymmetric azide-free synthesis of oseltamivir phosphate starting from chiral Roche's epoxide [127]. As shown in Scheme 2.93, one of the key steps of this synthesis involved the aziridination of a chiral 1,2-amino mesylate derived from Roche's epoxide performed in a mixed solvent of dichloromethane and dimethyl sulfoxide and in the presence of sodium hydride as a base. This base was selected among a range of other bases, such as sodium bicarbonate, potassium carbonate, potassium *tert*-butoxide, and sodium ethoxide. The reaction afforded the corresponding functionalized enantiopure *cis*-aziridine as single stereoisomer in 91% yield. This aziridine was

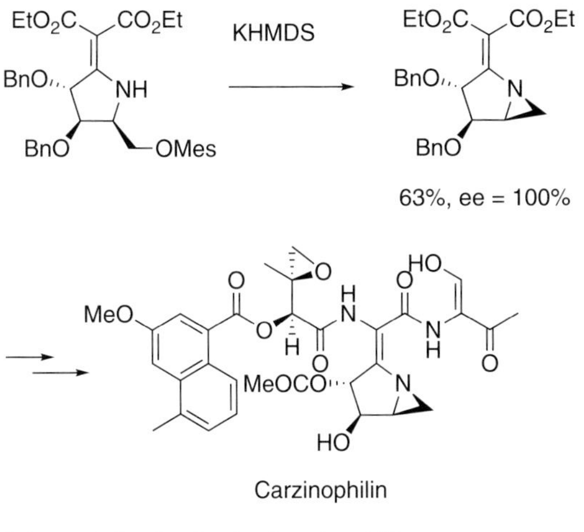

Scheme 2.91 Synthesis of carzinophilin.

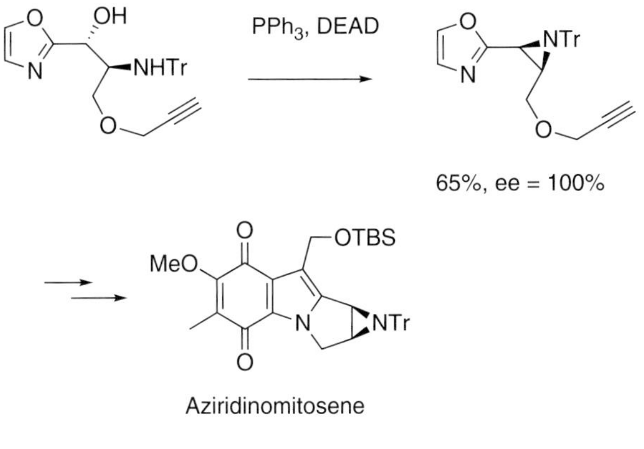

Scheme 2.92 Synthesis of aziridinomitosene.

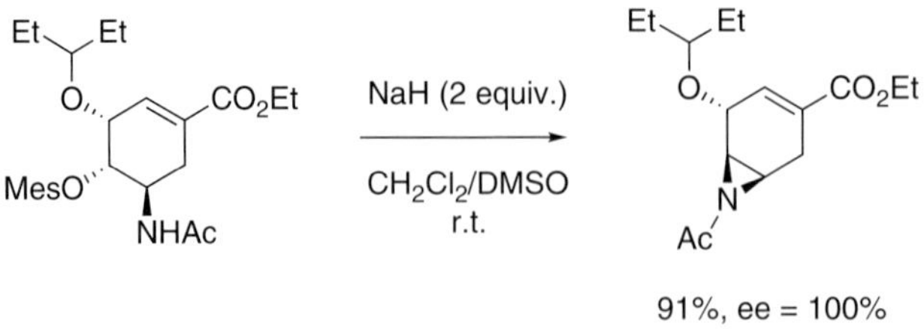

Scheme 2.93 Synthesis of a chiral Roche's epoxide-derived aziridine.

subsequently converted into expected oseltamivir phosphate in 82% yield over two steps.

In the same area, Kongkathip et al. recently reported a novel total synthesis of oseltamivir phosphate, a key step of which was the asymmetric aziridination of a chiral 1,2-amino mesylate derived from D-glucose, as depicted in Scheme 2.94 [128]. The complete strategy gave rise to oseltamivir phosphate in 7.2% overall yield.

In the course of finding a novel route to enantiopure α-amino β-hydroxy acids, Davies et al. investigated the aziridination of chiral N-Boc-protected 2,3-*trans*-β-amino α-hydroxy esters [129]. In a first time, the authors studied the direct aziridination of N-Boc-protected 2,3-*trans*-β-amino α-hydroxy ester, such as that depicted in Scheme 2.95, which was not successful since it led to a mixture of *trans*- and *cis*-aziridines (55:45) in moderate yields of 34% and 28%, respectively. On the other hand, submitting the corresponding mesylate to NaH afforded the corresponding enantiopure *trans*-aziridine as a single stereoisomer in 72% yield, as shown in Scheme 2.95. The scope of the methodology was extended to the synthesis of enantio- and diastereopure *anti*-di-*tert*-butyl *anti*-3-methylaziridine-1,2-dicarboxylate, which was obtained in 45% yield. Subsequent regioselective ring opening of these aziridines with Cl_3CCO_2H followed by hydrolysis led to enantiopure β-amino α-hydroxy acids.

2.2.4.2 From 1,2-Amino Halides

Davis et al. have reported the synthesis of chiral N-sulfinylaziridine 2-phosphonates by cyclization of the corresponding β-amino α-chlorophosphonates by treatment with NaH or n-BuLi as a base [74]. As shown in Scheme 2.96, high yields and enantioselectivities were obtained in all cases of aziridines.

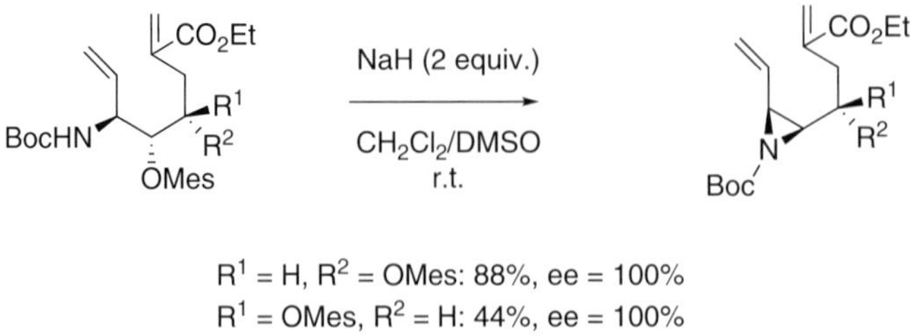

R^1 = H, R^2 = OMes: 88%, ee = 100%
R^1 = OMes, R^2 = H: 44%, ee = 100%

Scheme 2.94 Synthesis of a chiral aziridine intermediate in a total synthesis of oseltamivir phosphate.

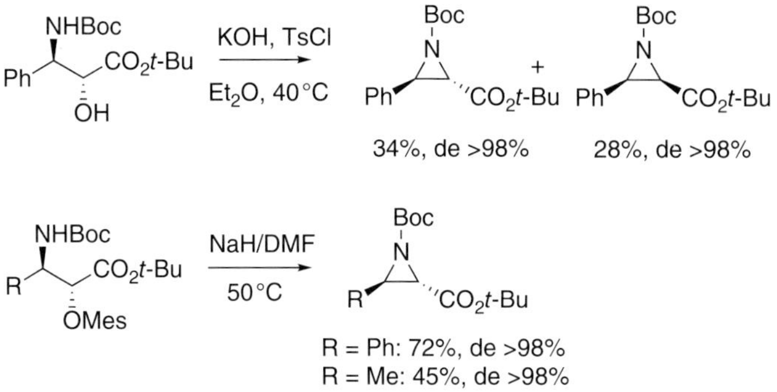

2 Asymmetric Aziridination | 279

Scheme 2.95 Synthesis of chiral *anti*- and *syn*-di-*tert*-butyl *anti*-3-methylaziridine-1,2-dicarboxylates.

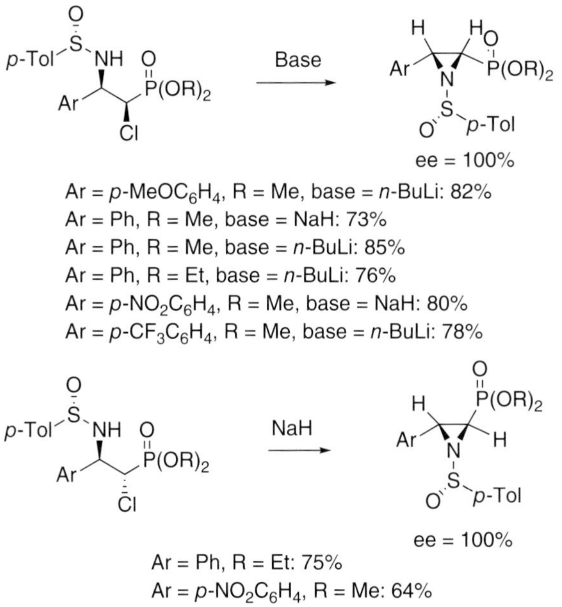

Reaction 1: NHBoc / Ph / CO₂t-Bu / OH → KOH, TsCl, Et₂O, 40°C → *anti* Boc-aziridine Ph, CO₂t-Bu (34%, de >98%) + *syn* Boc-aziridine Ph, CO₂t-Bu (28%, de >98%)

Reaction 2: NHBoc / R / CO₂t-Bu / OMes → NaH/DMF, 50°C → Boc-aziridine R, CO₂t-Bu
R = Ph: 72%, de >98%
R = Me: 45%, de >98%

Scheme 2.96 Syntheses of chiral *N*-sulfinylaziridine 2-phosphonates.

Reaction 1: p-Tol-S(O)-NH / Ar / P(OR)₂(=O) / Cl → Base → N-sulfinyl aziridine with Ar, P(OR)₂, S(O)-p-Tol (ee = 100%)

- Ar = p-MeOC₆H₄, R = Me, base = n-BuLi: 82%
- Ar = Ph, R = Me, base = NaH: 73%
- Ar = Ph, R = Me, base = n-BuLi: 85%
- Ar = Ph, R = Et, base = n-BuLi: 76%
- Ar = p-NO₂C₆H₄, R = Me, base = NaH: 80%
- Ar = p-CF₃C₆H₄, R = Me, base = n-BuLi: 78%

Reaction 2: p-Tol-S(O)-NH / Ar / P(OR)₂(=O) / Cl → NaH → N-sulfinyl aziridine (ee = 100%)

- Ar = Ph, R = Et: 75%
- Ar = p-NO₂C₆H₄, R = Me: 64%

In 2003, Satoh and Fukuda described the synthesis of optically active sulfinylaziridines having a 4-methoxyphenyl group on their nitrogen atom from optically active 1-chloroalkyl *p*-tolyl sulfoxides and an imine derived from benzaldehyde and *p*-anisidine stereoselectively and in good overall yields (Scheme 2.97) [130]. These aziridines were further converted into various α- and β-amino acid derivatives. In the same context, chiral substituted 1,1′-sulfonylbisaziridines were obtained by treatment of the corresponding chiral *N*,*N*′-bis(1-alkyl-2-chloroethyl)sulfamides with K_2CO_3 in high yields (Scheme 2.97) [131].

Later, Kocovsly *et al.* developed an organocatalytic reductive amination of α-chloroketones, yielding chiral α-chloroamines, which were subsequently

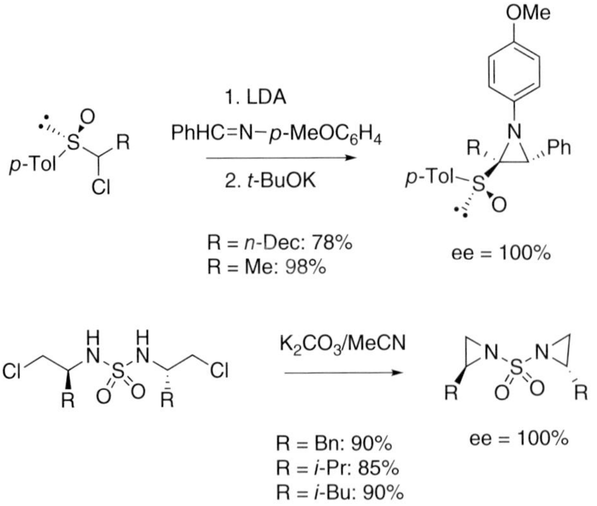

Scheme 2.97 Syntheses of chiral N-sulfinylaziridines and 1,1′-sulfonylbisaziridines.

treated with tBuOK to furnish a wide range of 1,2-diaryl aziridines in excellent yields and enantioselectivities, as shown in Scheme 2.98 [132].

In addition, a new entry to chiral terminal aziridines was reported by Hodgson et al., in 2008, on the basis of ring lithiation of N-tert-butylsulfinyl aziridine in good yield and with excellent enantioselectivity of up to 98% ee (Scheme 2.99)

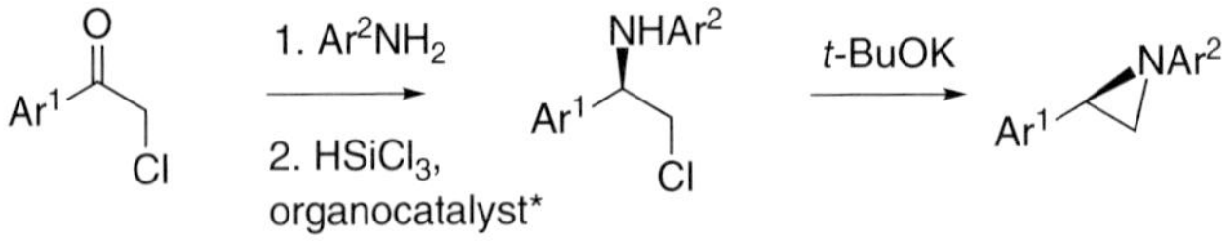

Ar^1 = Ph, Ar^2 = p-MeOC$_6$H$_4$: 98%, ee = 95%
Ar^1 = p-FC$_6$H$_4$, Ar^2 = p-MeOC$_6$H$_4$: 98%, ee = 94%
Ar^1 = p-ClC$_6$H$_4$, Ar^2 = p-MeOC$_6$H$_4$: 94%, ee = 100%
Ar^1 = Ar^2 = Ph: 92%, ee = 90%
Ar^1 = p-FC$_6$H$_4$, Ar^2 = Ph: 94%, ee = 86%
Ar^1 = p-ClC$_6$H$_4$, Ar^2 = Ph: 98%, ee = 100%
Ar^1 = 2-Naph, Ar^2 = Ph: 92%, ee = 90%
Ar^1 = p-Tol, Ar^2 = p-MeOC$_6$H$_4$: 96%, ee = 92%
Ar^1 = Ar^2 = p-MeOC$_6$H$_4$: 76%, ee = 100%
Ar^1 = p-CF$_3$C$_6$H$_4$, Ar^2 = p-MeOC$_6$H$_4$: 73%, ee = 100%
Ar^1 = 2-Naph, Ar^2 = p-MeOC$_6$H$_4$: 91%, ee = 100%
Ar^1 = m-Tol, Ar^2 = p-MeOC$_6$H$_4$: 97%, ee = 92%
Ar^1 = m-MeOC$_6$H$_4$, Ar^2 = p-MeOC$_6$H$_4$: 94%, ee = 92%
Ar^1 = o-ClC$_6$H$_4$, Ar^2 = p-MeOC$_6$H$_4$: 96%, ee = 96%
Ar^1 = o-ClC$_6$H$_4$, Ar^2 = p-ClC$_6$H$_4$: 97%, ee = 100%
Ar^1 = o-ClC$_6$H$_4$, Ar^2 = p-FC$_6$H$_4$: 91%, ee = 100%

Scheme 2.98 Synthesis of chiral 1,2-diaryl aziridines.

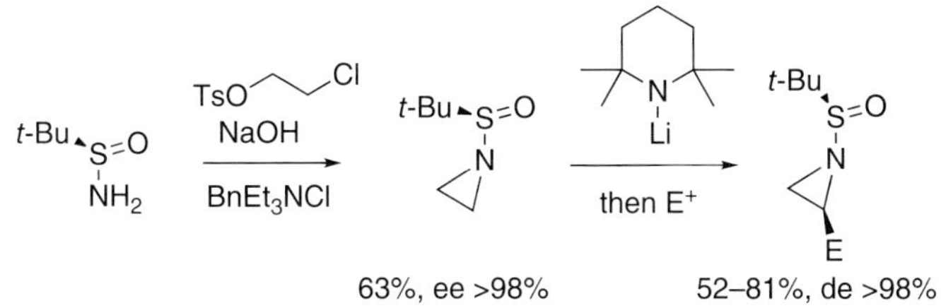

Scheme 2.99 Synthesis of chiral N-tert-butylsulfinyl aziridines.

[133]. The latter chiral aziridine was further submitted to an α-lithiation/electrophile trapping sequence, providing the expected terminal aziridines.

In 2013, chiral aziridinium ions were generated by Chong and Chen by treatment of the corresponding chiral β-bromo amines with a Lewis acid catalyst such as $AlCl_3$ [134]. These intermediate aziridinium bromides were directly submitted to an intramolecular Friedel–Crafts reaction to provide the corresponding chiral 4-substituted tetrahydroisoquinolines in good yields and with enantioselectivities of up to >99% ee, as shown in Scheme 2.100.

Moreover, Shi et al. have reported the transformation of chiral oxazolidinone bromides into the corresponding cis-aziridines in excellent yields of up to 93% and with excellent enantioselectivities of up to 99% ee by simple treatment with K_2CO_3 in methanol at room temperature, as shown in Scheme 2.101 [135].

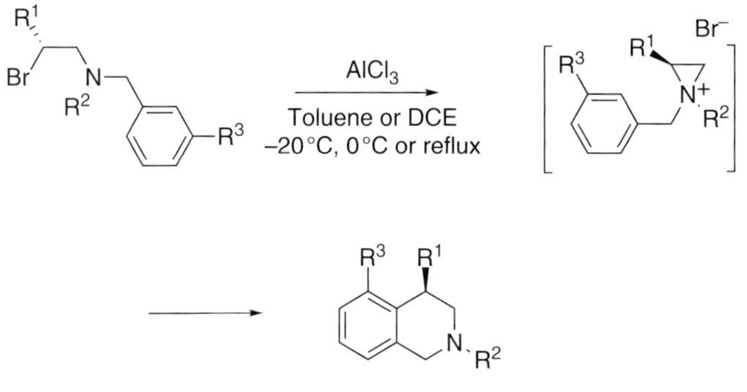

in toluene:
R^1 = Ph, R^2 = Bn, R^3 = H: 81%, ee = 71%
R^1 = Me, R^2 = Bn, R^3 = H: 90%, ee = 97%
R^1 = n-Pr, R^2 = Bn, R^3 = H: 79%, ee >99%
R^1 = Me, R^2 = CH_2(2-Naph), R^3 = H: 49%, ee >99%
R^1 = Me, R^2 = CH_2(m-BrC_6H_4), R^3 = Br: 54%, ee = 78%
in DCE:
R^1 = Ph, R^2 = Allyl, R^3 = H: 41%, ee = 19%

Scheme 2.100 Synthesis of chiral 4-substituted tetrahydroisoquinolines through chiral aziridinium ions.

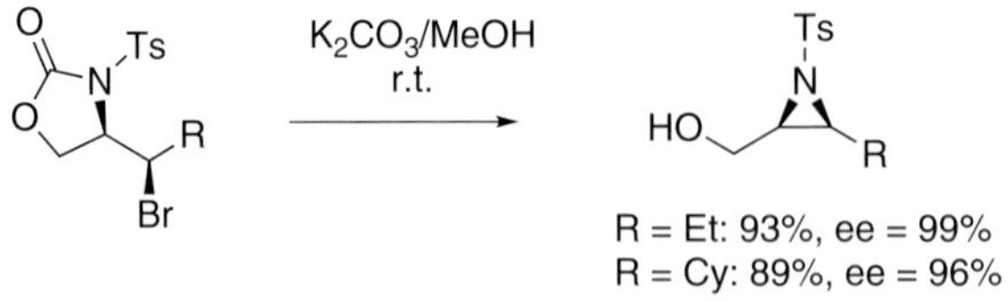

Scheme 2.101 Synthesis of chiral *cis*-aziridines from chiral oxazolidinones.

R = Et: 93%, ee = 99%
R = Cy: 89%, ee = 96%

2.2.4.3 From 1,2-Azido Alcohols

Chiral aziridines can also be prepared by cyclization of enantiopure 1,2-azido alcohols. As an example, Sweeney and Cantrill have developed an efficient synthesis of (2R)-N,O-bis(diphenylphosphinyl)-2-(hydroxymethyl)aziridine, starting from the corresponding 1,2-azido alcohol derived from (R)-glycidol [136]. This reaction consisted of a two-step process summarized in Scheme 2.102. The azide was reacted firstly with triphenylphosphine in refluxing acetonitrile, giving a crude product, which was obtained simply by removal of the solvent. This crude oil was then dissolved in dichloromethane and treated sequentially with triethylamine and diphenylphosphinic chloride, providing the expected enantiopure aziridine in moderate yield.

Glycosidase inhibitors of carbasugar-derived spiroaziridines have been synthesized by Vasella *et al.* by treatment with LiAlH$_4$ in THF of the corresponding chiral azido methanesulfonates prepared from validoxylamine A-derived cyclohexanone [137]. As shown in Scheme 2.103, the expected enantiopure spiroaziridines were obtained in high yields.

In 2003, Schirmeister [138] and Kostyanovsky [139] independently studied the synthesis of enantiopure *trans*-aziridine-2,3-dicarboxylates from the corresponding

29%, ee = 100%

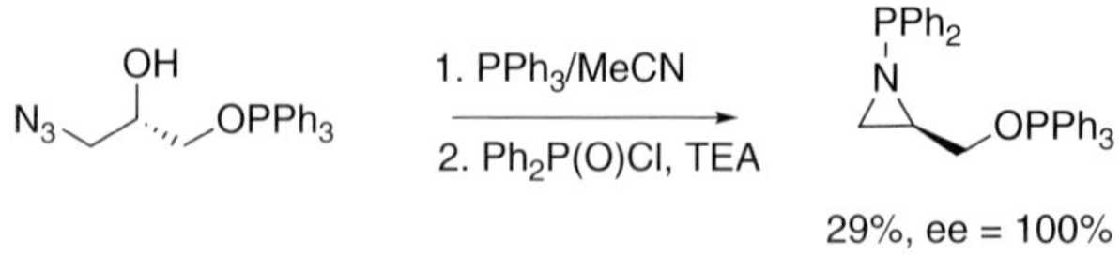

Scheme 2.102 Synthesis of (2R)-N,O-bis(diphenylphosphinyl)-2-(hydroxymethyl)aziridine.

83%, ee = 100%

74%, ee = 100%

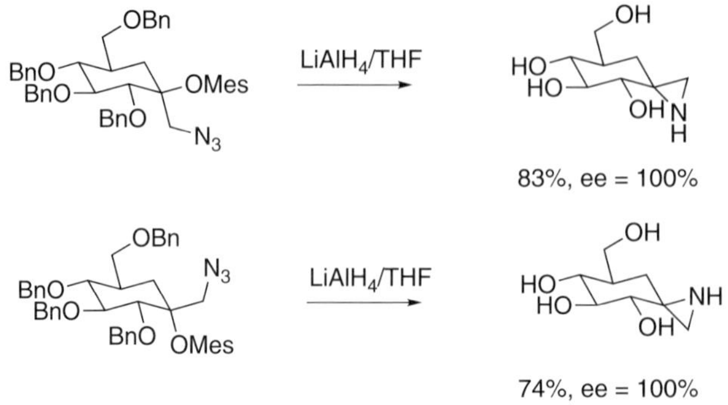

Scheme 2.103 Synthesis of chiral spiroaziridines.

anti-3-azido-2-hydroxysuccinates. The desired dimethyl, diethyl, dibenzyl, and diallyl *trans*-aziridines were obtained in good yields (70–75%) by treatment of these 1,2-azido alcohols with PPh$_3$ in DMF. On the other hand, Voronkov et al. have reported a short, efficient, and scalable route to both enantiopure isomers of limonene aziridines, starting from commercially available mixture of limonene oxides [140]. This methodology was extremely efficient, avoiding the separation of the limonene oxides (and subsequent separate processing of each diastereomer) by either physical or chemical methods. The key to the efficiency of the separation was the exploitation of the differences in rate between the two azido alcohol diastereomers in the ring closure. Indeed, the Staudinger reaction of the secondary azide was much faster, and so it was converted completely into the corresponding *trans*-aziridine at room temperature over a period of 48 h. This resulting *trans*-aziridine was easily separated from the unreacted tertiary azido alcohol by a simple acid–base extraction. Next, the conversion of the tertiary azido alcohol to the corresponding *cis*-aziridine required elevated temperatures and proceeded smoothly in refluxing dioxane over a period of 16 h. Both of the aziridine isomers were obtained in good yields and with enantioselectivities of >98% ee (Scheme 2.104).

The last step of an enantioselective synthesis of an aziridinomitosane, accomplished by Miller et al., was constituted by the cyclization of an enantiopure tricyclic 1,2-azido alcohol depicted in Scheme 2.105 [141]. This reaction was achieved in two steps with resin-bound PPh$_3$, affording the expected aziridinomitosane with the *trans* configuration in good yield (Scheme 2.105).

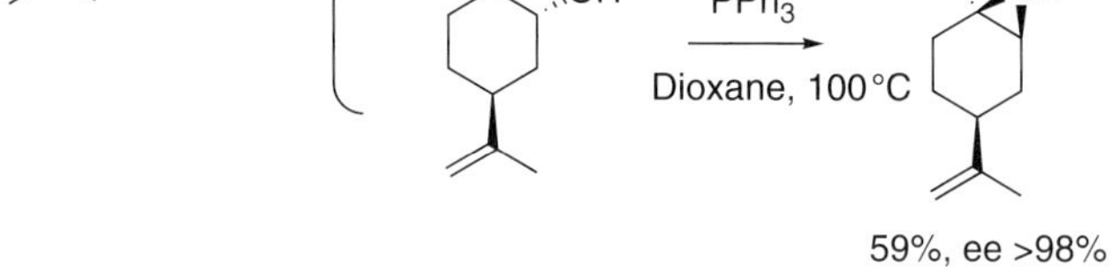

Scheme 2.104 Synthesis of limonene aziridines.

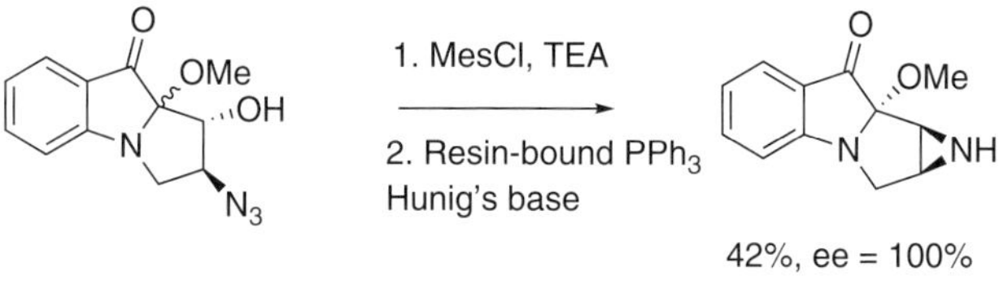

Scheme 2.105 Synthesis of a chiral aziridinomitosane.

In addition, Fürmeier and Metzger have developed the first preparation of chiral fat-derived aziridines, with the twin goals of increasing the variety of interesting fatty compounds with aziridine functions and gaining a deeper insight into their biological properties [142]. The same methodology as described earlier, based on the use of resin-bound PPh$_3$, was applied, for example, to the enantiopure azido alcohol derived from chiral methyl vernolate, depicted in Scheme 2.106. Thus, upon treatment with resin-bound PPh$_3$, the corresponding unsaturated cis-aziridine was isolated in good yield as a single enantiomer, according to the mechanism summarized in Scheme 2.106. This represented the first enantiomerically pure aziridine based on fats and oils.

In 2010, the cyclization of enantiopure 2-azido alcohols was also employed by Coates et al. in the course of synthesizing aziridine analogues of presqualene diphosphate as inhibitors of squalene synthase [143]. As shown in Scheme 2.107, enantiopure 2,3-aziridinofarnesol was prepared as a single stereoisomer in 83% yield from the corresponding azido mesylate by treatment with LiAlH$_4$. The

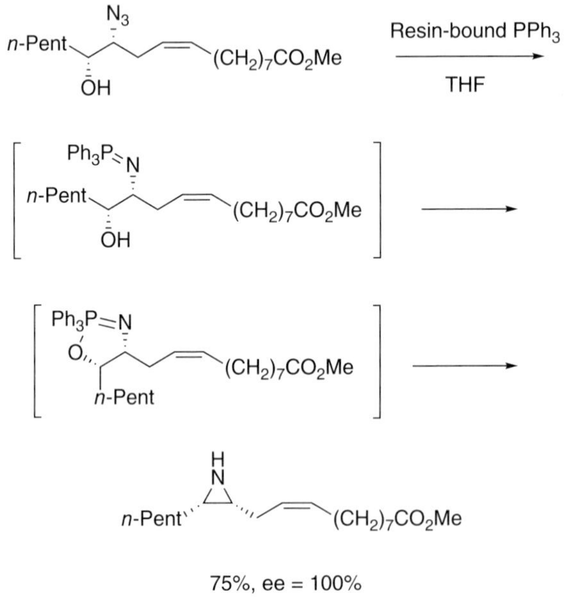

Scheme 2.106 Synthesis of a fat-derived chiral aziridine.

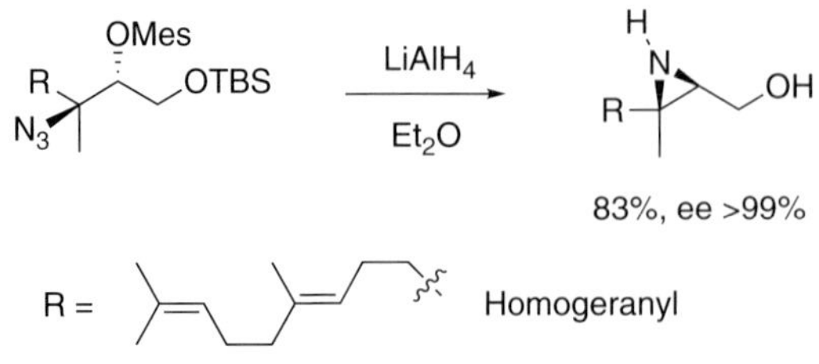

Scheme 2.107 Synthesis of chiral 2,3-aziridinofarnesol.

latter was subsequently converted into diphosphates and methanediphosphonates to be evaluated as squalene synthase inhibitors.

In the same area, other almost enantiopure 2,2-disubstituted unprotected aziridines have been prepared by Molinaro *et al.* through a Staudinger reaction in good-to-high yields (60–94%), as shown in Scheme 2.108 [144]. The starting chiral 2,2-disubstituted azido alcohols were generated through enzymatic resolution of 2-alkyl-2-aryl-disubstituted epoxides using the Codex HHDH P2E2 enzyme and sodium azide.

2.2.4.4 From 1,2-Amino Sulfides and 1,2-Amino Selenides

In 2006, Arroyo *et al.* reported a new entry to optically pure *trans*-2,3-disubstituted N-sulfinyl aziridines starting from 1,2-amino sulfides, involving the formation of a sulfonium salt intermediate followed by intramolecular nucleophilic attack by the sulfinamide nitrogen atom [145]. In this context, a series of chiral aziridines were obtained in high yields and with diastereoselectivities of up to 98% de, as shown in Scheme 2.109.

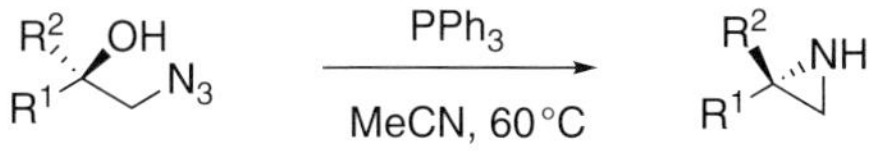

R^1 = 3,5-$F_2C_6H_3$, R^2 = Me: 94%, ee = 94%
R^1 = CH_2Bn, R^2 = Me: 60%, ee = 97%

Scheme 2.108 Synthesis of chiral 2,2-disubstituted unprotected aziridines.

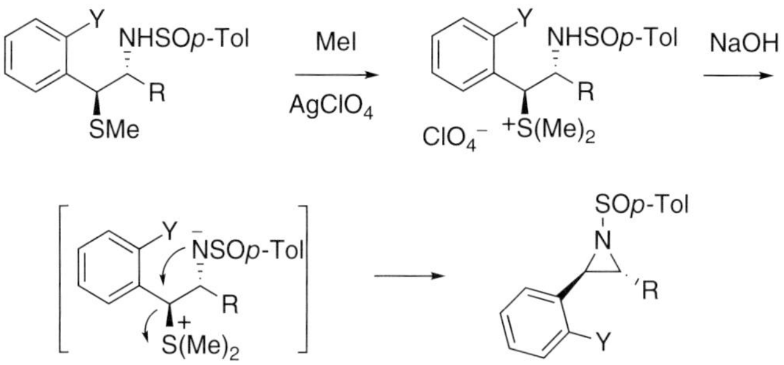

Y = SO*p*-Tol, R = Ph: 78%, de >98%
Y = SO*p*-Tol, R = *o*-BrC$_6$H$_4$: 43%, de >98%
Y = SO*p*-Tol, R = *p*-MeOC$_6$H$_4$: 65%, de >98%
Y = SO*p*-Tol, R = *p*-CNC$_6$H$_4$: 65%, de >98%
Y = SO*p*-Tol, R = 2-Naph: 76%, de >98%
Y = SO*p*-Tol, R = 2-Py: 45%, de >98%
Y = SO*p*-Tol, R = *n*-Bu: 75%, de >98%
Y = H, R = Ph: 77%, de >98%
Y = H, R = *n*-Bu: 76%, de >98%

Scheme 2.109 Synthesis of chiral *trans*-2,3-disubstituted N-sulfinyl aziridines from chiral 1,2-amino sulfides.

Similarly, Tiecco et al. have shown that enantiopure benzoylamino selenides could lead to the formation of the corresponding aziridines upon treatment with *meta*-chloroperbenzoic acid [146]. As shown in Scheme 2.110, the chiral aziridine was submitted to a spontaneous deselenenylation and isolated in moderate yield, along with the corresponding α,β-unsaturated amide resulting from an elimination process.

Moreover, the synthesis of a series of enantiopure *cis*-aziridine esters has been developed by Pannecoucke et al. on the basis of the cyclization of chiral amino selanyl esters induced by the selanyl group activation with either Meerwein's salt or NBS [147]. The best results, collected in Scheme 2.111, were generally obtained by using NBS.

2.2.4.5 From Epoxides

In 2004, Ishikawa et al. demonstrated that it was possible to directly convert chiral epoxides into chiral aziridines by using guanidines as a nitrene source [148]. The reaction of guanidine with the epoxide was supposed to afford a betaine species depicted in Scheme 2.112, which produced the corresponding aziridine via a spiro intermediate. This process proceeded via inversion of configuration at the asymmetric carbon on (*R*)-styrene oxide with high chirality control (96% ee).

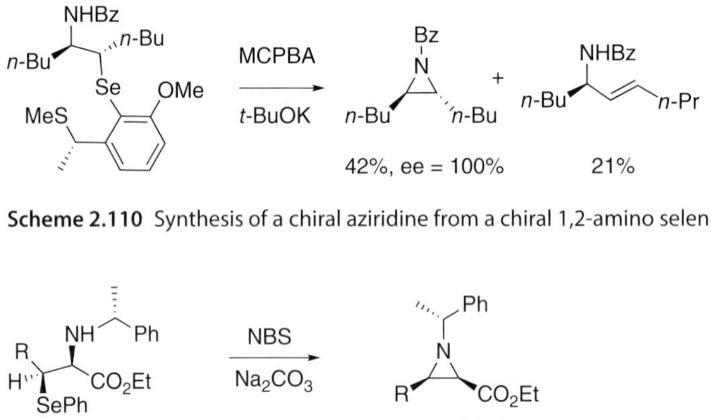

Scheme 2.110 Synthesis of a chiral aziridine from a chiral 1,2-amino selenide.

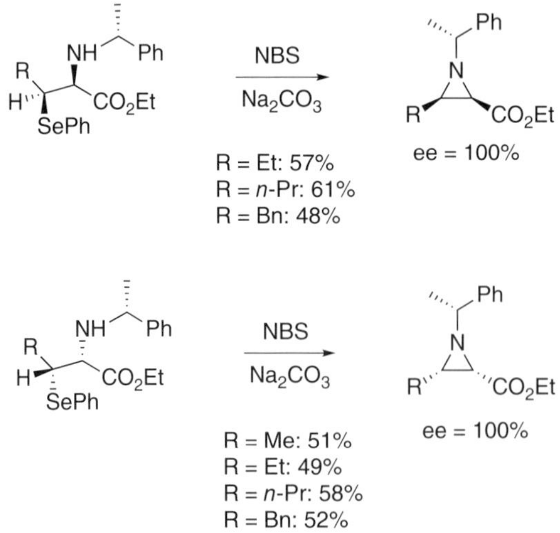

Scheme 2.111 Synthesis of chiral aziridine esters from chiral 1,2-amino selanyl esters.

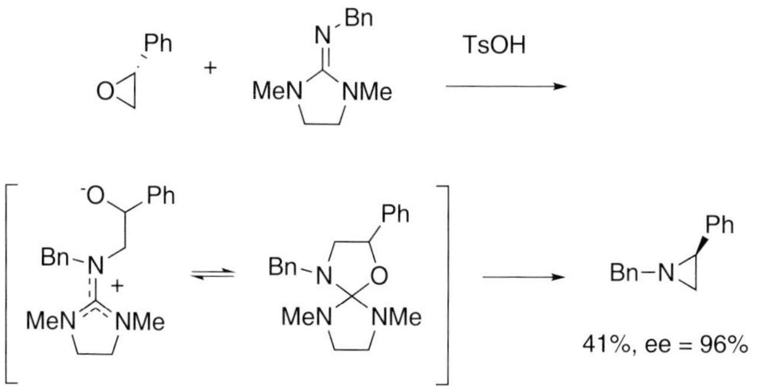

Scheme 2.112 Synthesis of a chiral aziridine from a chiral epoxide by using guanidine.

In another context, Bartoli et al. have reported the asymmetric aminolytic kinetic resolution of racemic terminal epoxides using carbamates as the nucleophiles catalyzed by a chiral (salen)Co(III) complex, which provided a straightforward method for the synthesis of chiral 1,2-amino alcohols [149]. The viability of this novel strategy was proved by the synthesis of a highly enantio-enriched N-Boc-protected aziridine, starting from racemic glycidyl phenyl ether using a practical one-pot procedure shown in Scheme 2.113. The aminolytic kinetic resolution of this epoxide was performed in the presence of tert-butyl carbamate and a chiral cobalt catalyst, affording the corresponding enantiopure N-Boc-protected aziridine in high yield with complete regioselectivity.

2.2.5 Miscellaneous Reactions

In 2001, a novel guanidinium-ylide-mediated procedure was reported by Ishikawa et al. [150], in which guanidinium ylides reacted with aldehydes to form aziridines. The first step was the formation of a C—C bond between a guanidinium salt and an arylaldehyde under basic conditions, in which an initially formed zwitterionic species was in equilibrium with a nonionic spiro compound. The second step of the process was the fragmentation of this intermediate, triggered by Ac$_2$O or SiO$_2$, to afford the expected aziridine and urea. As shown in

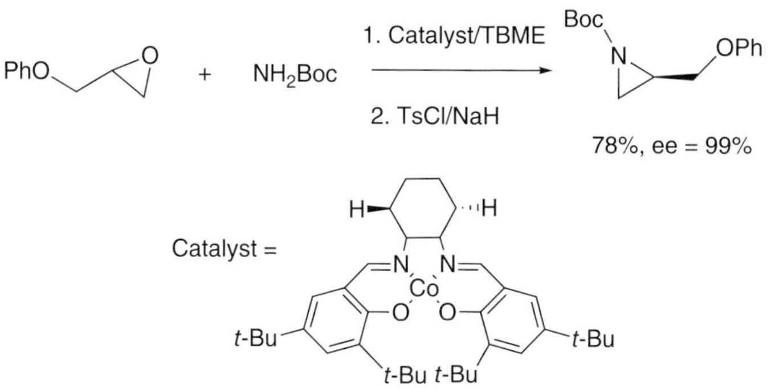

Scheme 2.113 Synthesis of a chiral aziridine through aminolytic kinetic resolution of a racemic epoxide.

Scheme 2.114, eq 1, the reaction of various *p*-substituted benzaldehydes with a chiral guanidinium salt in the presence of TMG as a base led, after subsequent treatment with Ac_2O, to the corresponding aziridines in good yields and with variable diastereo- and enantioselectivities according to the nature of the aldehyde substrate [151]. For example, benzaldehydes bearing a strong electron-donating group allowed the corresponding *trans*-aziridines to be obtained with both excellent diastereo- and enantioselectivities. Later in 2014, these authors extended the scope of this methodology to piperonal, which afforded the corresponding chiral *trans*-aziridine as major product in 82% yield and 97% ee with a

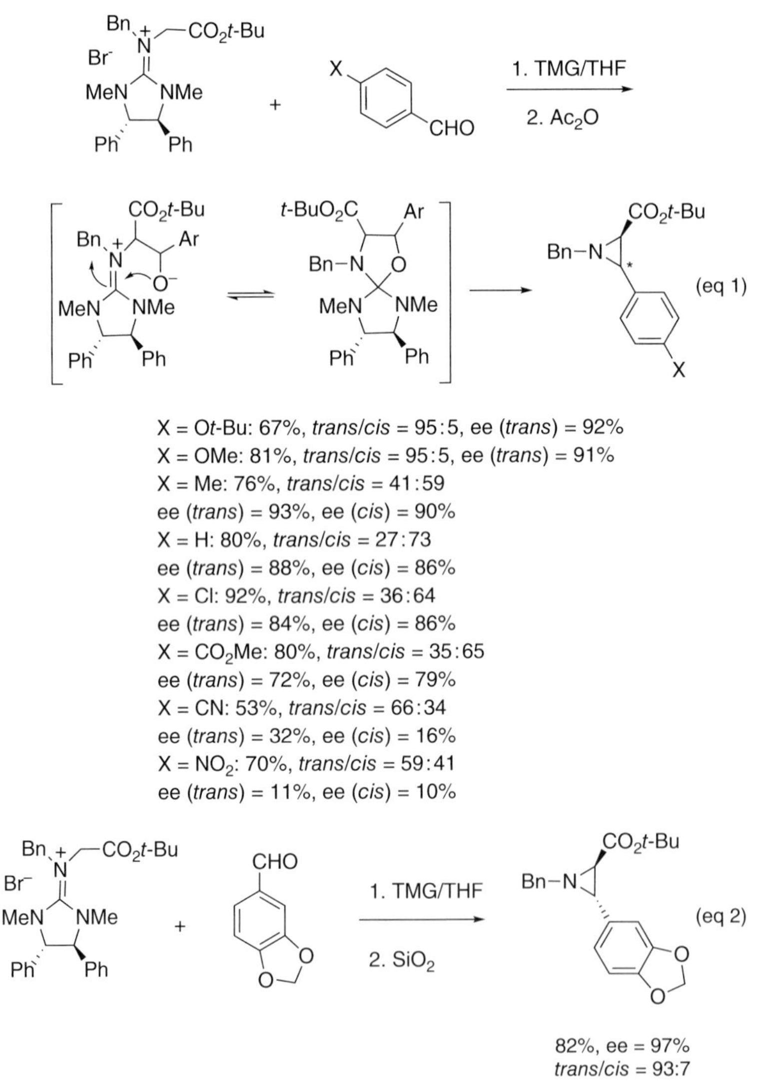
1. TMG/THF
2. Ac2O
(eq 1)
X = Ot-Bu: 67%, trans/cis = 95:5, ee (trans) = 92%
X = OMe: 81%, trans/cis = 95:5, ee (trans) = 91%
X = Me: 76%, trans/cis = 41:59
ee (trans) = 93%, ee (cis) = 90%
X = H: 80%, trans/cis = 27:73
ee (trans) = 88%, ee (cis) = 86%
X = Cl: 92%, trans/cis = 36:64
ee (trans) = 84%, ee (cis) = 86%
X = CO2Me: 80%, trans/cis = 35:65
ee (trans) = 72%, ee (cis) = 79%
X = CN: 53%, trans/cis = 66:34
ee (trans) = 32%, ee (cis) = 16%
X = NO2: 70%, trans/cis = 59:41
ee (trans) = 11%, ee (cis) = 10%
1. TMG/THF
2. SiO2
(eq 2)
82%, ee = 97%
trans/cis = 93:7

Scheme 2.114 Synthesis of chiral aziridines by reaction of aldehydes with a chiral guanidinium salt.

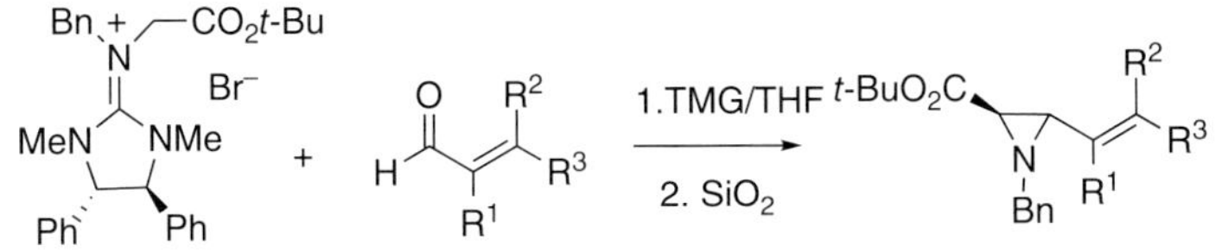

R^1 = Me, R^2 = R^3 = H: 92%, cis/trans = 87:13, ee (cis) = 89%, ee (trans) = 82%
R^1 = R^2 = R^3 = H: 62%, cis/trans = 50:50, ee (cis) = 58%
R^1 = R^2 = H, R^3 = n-$C_{13}H_{27}$: 87%, cis/trans = 53:47, ee (cis) = 95%, ee (trans) = 97%
R^1 = H, R^2 = R^3 = Me: 5% cis/trans = 80:20
R^1,R^3 = $(CH_2)_4$, R^2 = H: 51%, cis/trans = 65:35, ee (cis) = 99%
R^1 = R^2 = H, R^3 = Ph: 82%, cis/trans = 27:73, ee (cis) = 75%, ee (trans) = 65%
R^1 = Me, R^2 = H, R^3 = Ph: 51%, cis/trans = 18:82, ee (cis) = 93%, ee (trans) = 92%
R^1 = Ph, R^2 = Me, R^3 = H: 30%, cis/trans = 93:7, ee (cis) = 98%
R^1 = H, R^2 = R^3 = Ph: 42%, cis/trans = 76:24, ee (cis) = 91%, ee (trans) = 87%

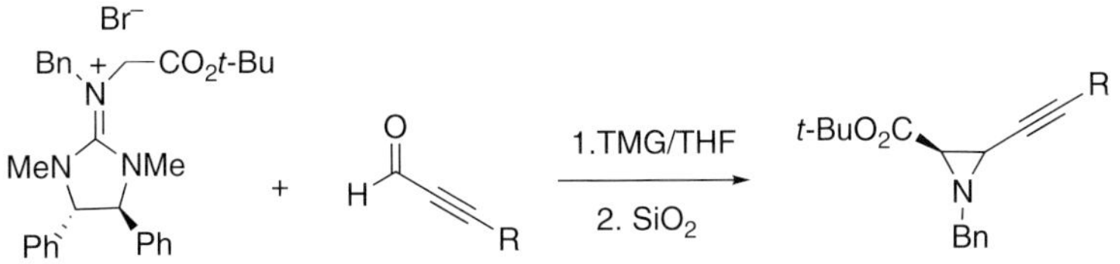

R = Ph: 70%, cis/trans = 31:69, ee (cis) = 91%, ee (trans) = 98%
R = n-$C_{13}H_{27}$: 44%, cis/trans = 41:59, ee (cis) = 18%, ee (trans) = 26%

Scheme 2.115 Synthesis of chiral aziridines by reactions of α,β-unsaturated aldehydes with a chiral guanidinium salt.

trans/cis ratio of 93:7 when the fragmentation step was triggered by SiO_2, as shown in Scheme 2.114, eq 2 [152].

The scope of this methodology could be extended by these authors to α,β-unsaturated aldehydes, which successfully led, under similar conditions, to a variety of chiral α,β-unsaturated aziridine-2-carboxylates in good-to-moderate yields and with the chirality of the guanidinium ylide effectively transferred to the 2- and 3-positions of the azirine products with diastereo- and enantioselectivities of up to 86% de and 99% ee, respectively (Scheme 2.115) [153].

Later, the same authors also performed the reaction of the same chiral guanidinium bromide with a nitroxyl-introduced olefinic aldehyde to provide the corresponding syn-aziridine in 78% yield and with good diastereoselectivity of 84% de (Scheme 2.116) [154]. In this case, the aziridination of the olefinic aldehyde with the chiral guanidinium bromide was performed in the presence of NaH in DMF, followed by stirring the resultant mixture with silica gel in MeCN at room temperature to give the product. It must be noted that this chiral aziridine constituted a candidate for an anticancer drug.

In addition, these authors applied this type of methodology to the synthesis of (−)-benzolactam-V8, an artificially designed cyclic dipeptide with strong tumor-promoter activity [155]. This important product was achieved from benzyl

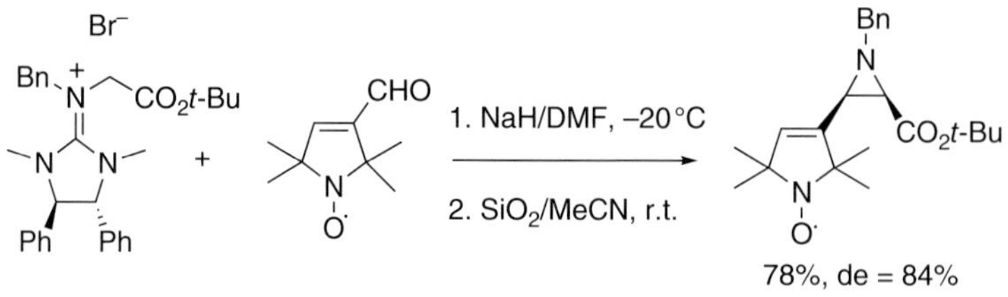

Scheme 2.116 Reaction of a nitroxyl-introduced olefinic aldehyde with a chiral guanidinium bromide.

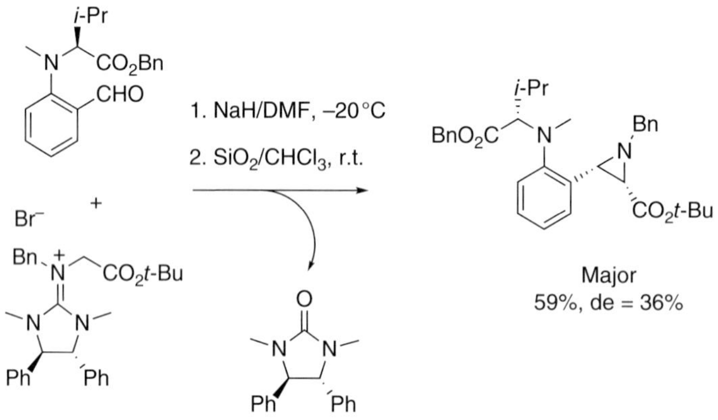

Scheme 2.117 Reaction of a valine-substituted benzaldehyde with a chiral guanidinium bromide.

(S)-N-(2-formylphenyl)-N-methylvalinate through four steps by application of the guanidinium ylide-participated asymmetric aziridination, followed by reductive ring-opening reaction of the 3-arylaziridine-2-carboxylate formed as key steps. The corresponding *syn*-aziridine was afforded by using CHCl$_3$ instead of MeCN for the second step in 59% yield and with a moderate diastereoselectivity of 36% de, as shown in Scheme 2.117.

In another context, Shibasaki *et al.* have prepared chiral α-acylpyrrole aziridines in high yields by treating the corresponding methoxylamines with TiCl$_4$ combined with TEA, as shown in Scheme 2.118, eq 1 [156]. In another study, these authors have achieved other chiral acylaziridines by cyclization of the corresponding chiral methoxylamines performed in the presence of NaO*t*-Bu as base in high yields (Scheme 2.118, eq 2) [157].

In 2006, Bew *et al.* demonstrated that, after O-acylation, the conjugate addition products of (S)-N-(α-methylbenzyl)hydroxylamine underwent an efficient diastereoselective 3-*exo*-tet ring-closure reaction, affording the corresponding substituted (S)-N-(α-methylbenzyl)aziridines in moderate-to-quantitative yields and with moderate-to-good diastereoselectivities of up to 86% de (Scheme 2.119) [158]. The authors have proposed that the process evolved through the conjugate addition of the *in situ* generated lithium salt of O-pivaloyl-(S)-N-(α-methylbenzyl) hydroxylamine to an acrylate ester, providing the corresponding enolate. The

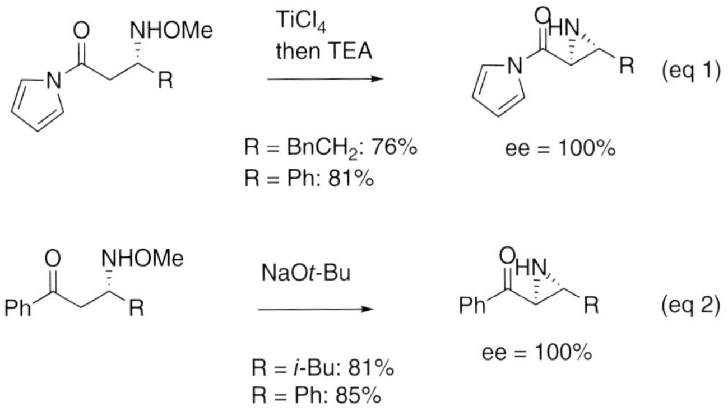

Scheme 2.118 Syntheses of chiral aziridines from chiral methoxylamines.

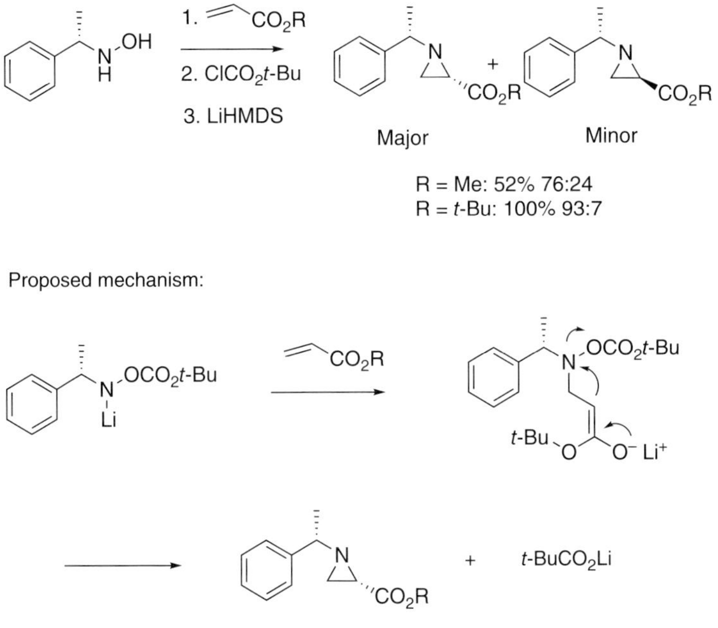

Scheme 2.119 Synthesis of chiral aziridines from (S)-N-(α-methylbenzyl)hydroxylamine.

latter subsequently underwent nucleophilic attack of the O-pivaloyl hydroxylamine to yield the final aziridine with concomitant expulsion of t-BuCO$_2$Li (Scheme 2.119).

Earlier in 2004, Tardella et al. investigated the asymmetric aza-Michael addition of nosyloxycarbamates to 2-(trifluoromethyl)acrylates, demonstrating that it provided either the corresponding α-trifluoromethyl β-amino esters or the corresponding aziridines in high yields by changing the reaction conditions [35]. Indeed, when the amination was performed in the presence of CaO, an aza-Michael 1,4-addition occurred, giving the α-trifluoromethyl β-amino esters, whereas the use of NaH promoted an aza-Michael-initiated ring closure, directly

Scheme 2.120 Synthesis of chiral aziridines from chiral 2-(trifluoromethyl)acrylates.

yielding the aziridines. As shown in Scheme 2.120, the use of (−)-8-phenylmenthol as the chiral auxiliary induced a low diastereoselectivity, while more satisfactory results were obtained by using the bulkier Helmchen's auxiliary.

In 2008, Gade et al. reported the synthesis of chiral aziridinecarboxamides by reaction of enantiopure 2-bromo isocyanates with phenylethylamine in the presence of a base such as t-BuOK [159]. The first reaction intermediates in these reactions were β-bromourea derivatives, which were formed by the nucleophilic attack of the amine and could be detected at low temperature by NMR spectroscopy. Adding the base at low temperature led to the aziridinecarboxamides as single diastereomers in moderate yields (Scheme 2.121).

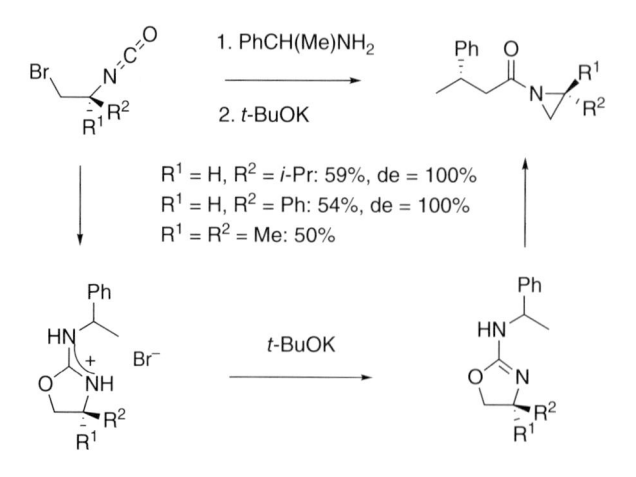

Scheme 2.121 Synthesis of chiral aziridinecarboxamides from chiral 2-bromo isocyanates.

On the other hand, Shipman et al. have observed high levels of diastereocontrol for the lithiation and alkylation of a 2-isopropylidineaziridine bearing an (S)-α-methylbenzyl group on nitrogen [160]. As shown in Scheme 2.122, the chiral aziridinyl anion was alkylated by a series of electrophiles, yielding the corresponding alkylated isopropylidineaziridines in good yields and with diastereoselectivities of up to 90% de.

In another context, Bols et al. have developed the synthesis of various chiral bicyclic aziridines through the reductive cleavage of the N—O bond of chiral 1,2-oxazines carried out with Raney nickel at 1 atm of hydrogen pressure [161]. As shown in Scheme 2.123, the reaction of 1,2-oxazines derived from D-glucose and D-mannose provided the corresponding enantiopure aziridines in moderate-to-good yields.

2,3-Aziridino-γ-lactones represent important precursors for biologically important glutamic acid derivatives. In this context, Kale and Deshmukh have developed an efficient entry to chiral 2,3-aziridino-γ-lactones from azetidin-2-ones [162]. This methodology was based on an acid-catalyzed tandem intramolecular azetidinone ring opening, followed by aziridine ring formation via elimination of a mesylate

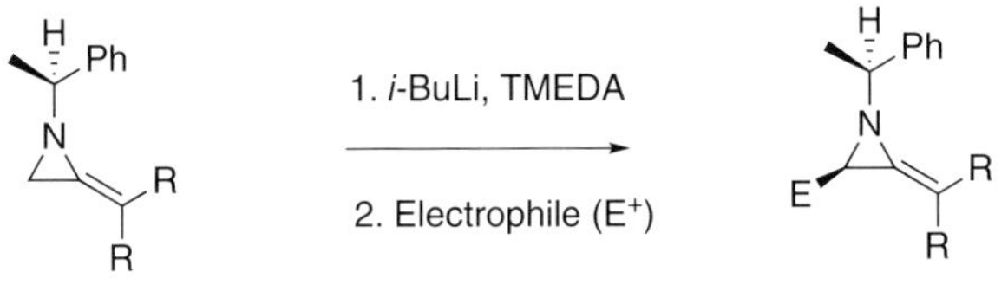

R = H, Electrophile = BnBr: 70%, de = 14%
R = Me, Electrophile = BnBr: 68%, de = 88%
R = Me, Electrophile = MeI: 47%, de = 80%
R = Me, Electrophile = TMSCl: 80%, de = 90%
R = Me, Electrophile = CH$_2$=CH-CH$_2$Br: 63%, de = 84%
R = Me, Electrophile = Ph$_2$CO: 43%, de = 88%

Scheme 2.122 Synthesis of chiral alkylated isopropylidineaziridines.

R = Me: 78%
R = Bn: 48%

ee = 100%

46%, ee = 100%

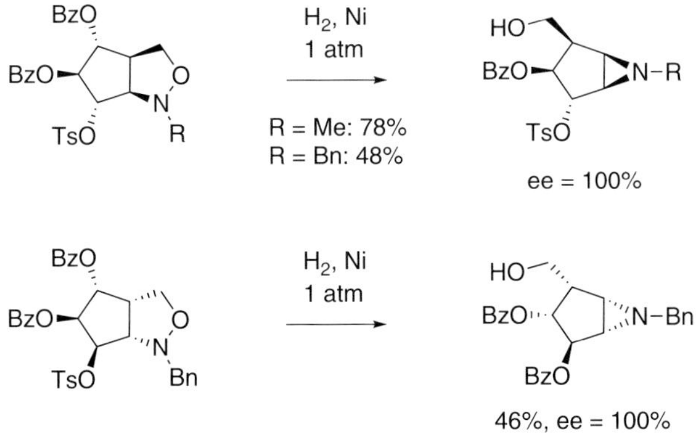

Scheme 2.123 Syntheses of chiral bicyclic aziridines from chiral 1,2-oxazines.

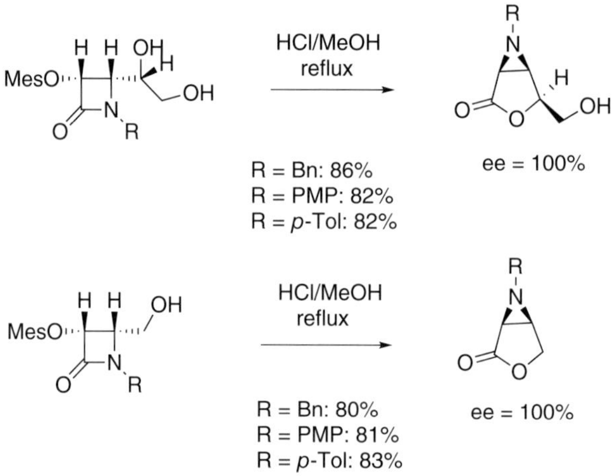

Scheme 2.124 Syntheses of chiral aziridino-γ-lactones.

R = Bn: 86%
R = PMP: 82%
R = p-Tol: 82%

ee = 100%

R = Bn: 80%
R = PMP: 81%
R = p-Tol: 83%

ee = 100%

group. As shown in Scheme 2.124, a series of enantiopure aziridino-γ-lactones could be prepared in good yields using this methodology.

The total synthesis of the potent inhibitor of neuraminidase, (−)-oseltamivir, elaborated by Fukuyama et al. in 2007, included the formation of a bicyclic aziridine by rearrangement of a chiral allyl carbamate [163]. Therefore, treatment of this carbamate with NaOEt resulted in ethanolysis of N-Boc lactam, dehydrobromination, and aziridine formation, which provided the desired aziridine in high yield (Scheme 2.125). This aziridine was further converted into the final (−)-oseltamivir in four steps.

In the course of preparing 1-deoxynojirimycin derivatives, Zhou and Murphy have found that the treatment of a chiral 1,2,3-triazoline, derived from D-glucono-δ-lactone, provided the corresponding aziridine through decomposition, suggesting that the acidity of the middle promoted the loss of nitrogen [164]. As shown in Scheme 2.126, the enantiopure tricyclic aziridine was obtained in moderate yield.

On the other hand, the photochemical decomposition of triazolines derived from D-glucose has been studied by Dahl and Finney [165]. The photolysis of chiral triazolines, as depicted in Scheme 2.127, provided the corresponding enantiopure aziridines in moderate-to-quantitative yields.

Another photocyclization reaction providing chiral aziridines has been developed by Mariano et al., starting from a chiral pyridinium perchlorate derived from D-glucose [166]. Irradiation of this substrate in aqueous $NaHCO_3$ generated a mixture of isomeric N-glycosyl-bicyclic-aziridines, which could be partially separated by silica gel chromatography to yield the major enantiopure aziridine, depicted in Scheme 2.128, in 15% yield. This aziridine was subsequently converted into the aminocyclitol core (trehazolamine) of the potent trehalase inhibitor, trehazolin.

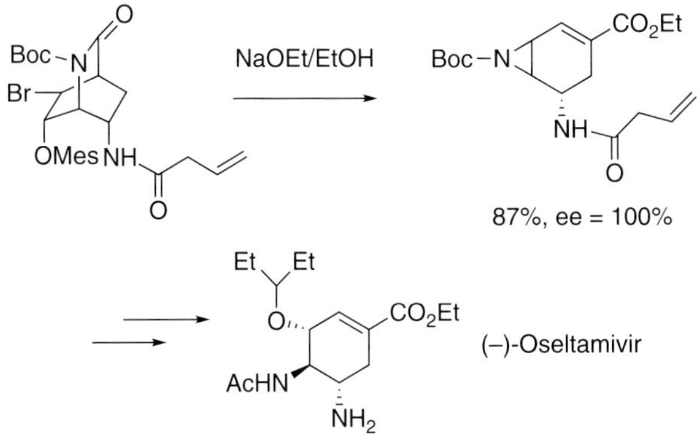

87%, ee = 100%

(−)-Oseltamivir

Proposed mechanism for the formation of the aziridine intermediate:

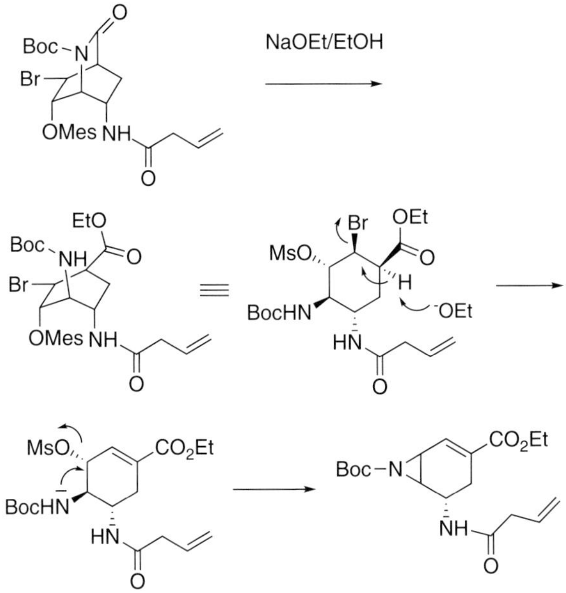

Scheme 2.125 Synthesis of (−)-oseltamivir through rearrangement of a chiral allyl carbamate.

35%, ee = 100%

Scheme 2.126 Synthesis of a chiral tricyclic aziridine through thermal decomposition of a chiral 1,2,3-triazoline.

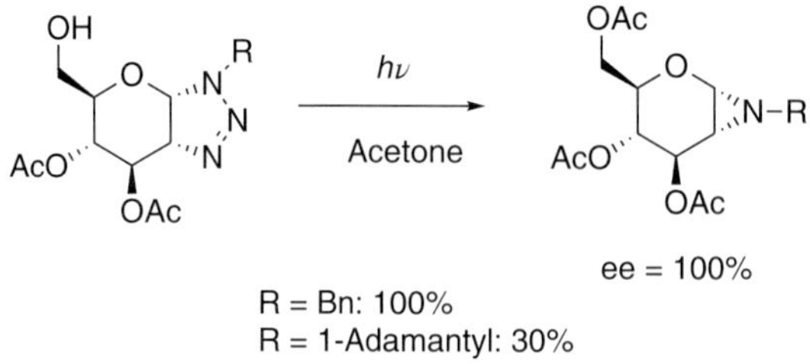

Scheme 2.127 Synthesis of chiral aziridines through photochemical decomposition of D-glucose-derived triazolines.

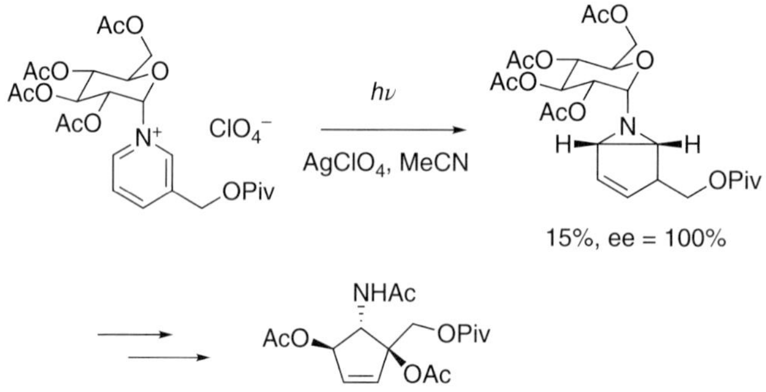

Scheme 2.128 Synthesis of a chiral aziridine through photocyclization of a chiral pyridinium perchlorate.

2.3 Aziridination Based on the Use of Chiral Catalysts

2.3.1 Aziridination via Nitrene Transfer to Alkenes

2.3.1.1 Cu-Catalyzed Aziridination

Although a wide variety of chiral catalysts including copper, rhodium, ruthenium, and other metal complexes have been developed in the past two decades, the field of asymmetric aziridination catalysis has remained for a long time relatively undeveloped. The most commonly employed chiral catalyst systems that have been developed to date for the enantioselective aziridination via nitrene transfer to alkenes are based on copper complexes. The main mechanistic issues of the Cu-catalyzed aziridination of olefins are the transformation of a copper–nitrene intermediate, the oxidation state of the metal, and the nature of the nitrene transfer. It must be noted that experimental evidence for the existence of a metallonitrene has been provided by Jacobsen et al. [167]. The type of chiral ligands that have been most successfully applied to the Cu-catalyzed asymmetric aziridination is constituted by the class of bisoxazolines. More generally, the C_2-symmetric chiral bisoxazolines have emerged as a class of important and efficient ligands in an increasing number of aymmetric transformations over the past decade [168]. The aziridination of alkenes using Cu–bisoxazoline complexes was pioneered by Evans et al. in 1991 [169]. Later in 2007, Cranfill and Lipton

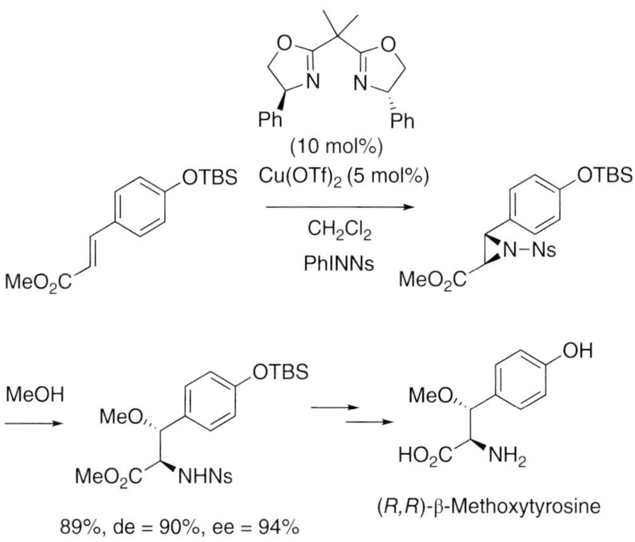

Scheme 2.129 Synthesis of (R,R)-β-methoxytyrosine through Cu-catalyzed aziridination of p-coumarate TBS ether with PhINNs in the presence of Evans's bisoxazoline ligand.

reported the use of Evans' bisoxazoline ligand for the asymmetric aziridination of p-coumarate TBS ether in the presence of Cu(OTf)$_2$ and PhINNs (N-(p-nitrophenylsulfonyl)iminophenyliodinane) as the nitrene source in dichloromethane [170]. This process allowed the corresponding chiral trans-aziridine to be obtained in high yield and with diastereo- and enantioselectivity of 90% de and 94% ee, respectively (Scheme 2.129). In fact, the crude aziridine was directly dissolved in methanol, providing the corresponding methoxy amine by aziridine ring opening. This successful reaction constituted the key step of a total synthesis of (R,R)-β-methoxytyrosine, which is a constituent of several cyclic depsipeptide natural products.

In 2003, Dauban et al. studied the Cu-catalyzed aziridination of allylglycine derivatives, which constituted the first example of the application of this metal-catalyzed nitrene transfer to a nitrogen-containing substrate [171]. It was shown that the Cu-catalyzed and iodosylbenzene-mediated aziridination of chiral N-(9-phenylfluorenyl)allylglycine tert-butyl ester led to the corresponding chiral aziridine in moderate yield and diastereoselectivity. As shown in Scheme 2.130, the reaction was performed in the presence of a combination of tert-butyl Evans' bisoxazoline ligand with [Cu(MeCN)$_4$]PF$_6$. The modest yields observed appeared to be mainly related to the low reactivity of unsubstituted terminal olefins in general. It was not clear, however, whether the nitrogen atom of the substrate had an influence on the course of the reaction.

This ligand has also been employed by Che et al. to develop asymmetric alkene Cu-catalyzed aziridinations mediated by PhI(OAc)$_2$ in the presence of sulfonamides to generate the nitrene precursors (PhI=NR) [172]. This one-pot procedure, employing [Cu(MeCN)$_4$]PF$_6$ as the copper complex, had been optimized using p-nitrobenzenesulfonamide as the nitrene source, allowing the corresponding

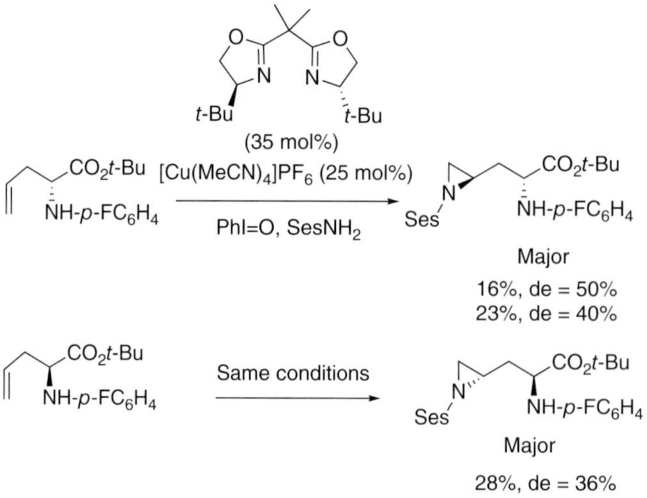

Scheme 2.130 Cu-catalyzed aziridinations of allylglycine derivatives with SesNH$_2$ and PhIO in the presence of Evans' bisoxazoline ligand.

chiral aziridines to be formed in good-to-excellent yields and with enantioselectivities of up to 75% ee. The results obtained from a range of olefins are collected in Scheme 2.131.

In addition, Mayoral et al. demonstrated that Evans' bisoxazoline ligands could be immobilized by electrostatic interactions with anionic supports, furnishing recyclable catalysts for the aziridination of styrene by PhI=NTs as the nitrene precursor [173]. Three different anionic supports were used, namely, laponite, a synthetic clay, SAC-13, a Nafion–silica nanocomposite with 13% Nafion content, and SiO$_2$-CF$_2$SO$_3$H, prepared by grafting a partially fluorinated chain with a sulfonic acid group on silica gel. Although the yields of aziridines were good in all of the cases studied, the enantioselectivity was found to depend on the nature of the chiral ligand (Scheme 2.132). With a bisoxazoline ligand bearing phenyl substituents, the enantioselectivity was always around 25% ee, as in solution. On the other hand, when the chiral ligand had tert-butyl groups attached, the enantioselectivity was noticeably lower than that observed in the homogeneous phase as a consequence of the presence of free copper on the solid.

In the same context, Hutchings et al. have studied the heterogeneous aziridination of styrene using copper-exchanged zeolite HY in the presence of Evans' ligands [174]. Two nitrene donors, PhI=NTs and PhI=NNs, were investigated for each ligand. Excellent enantioselectivities of up to 85% ee were obtained by using PhI=NNs, as shown in Scheme 2.133. It is interesting to note that, in all cases, the heterogeneously catalyzed reaction gave a much higher enantioselection compared to the comparable homogeneously catalyzed reaction performed with Cu(OTf)$_2$. It was possible to even increase the enantioselectivity of the reaction to up to 92% ee by using a 1:3 PhI=NNs:styrene ratio.

This highly efficient methodology could be extended to the heterogeneous aziridination of a series of styrene derivatives, providing the corresponding aziridines in high yields and with enantioselectivities of up to 95% ee, as shown

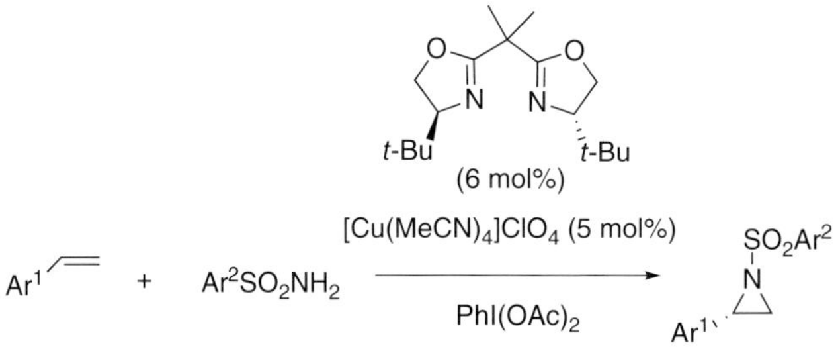

Ar¹ = Ph, Ar² = p-NO$_2$C$_6$H$_4$: 94%, ee = 75%
Ar¹ = Ph, Ar² = p-ClC$_6$H$_4$: 90%, ee = 52%
Ar¹ = p-FC$_6$H$_4$, Ar² = p-NO$_2$C$_6$H$_4$: 95%, ee = 72%
Ar¹ = p-FC$_6$H$_4$, Ar² = p-ClC$_6$H$_4$: 95%, ee = 51%
Ar¹ = p-FC$_6$H$_4$, Ar² = p-Tol: 84%, ee = 40%
Ar¹ = p-CF$_3$C$_6$H$_4$, Ar² = p-NO$_2$C$_6$H$_4$: 64%, ee = 51%
Ar¹ = p-CF$_3$C$_6$H$_4$, Ar² = p-ClC$_6$H$_4$: 68%, ee = 43%
Ar¹ = p-CF$_3$C$_6$H$_4$, Ar² = p-Tol: 43%, ee = 38%
Ar¹ = p-Tol, Ar² = p-NO$_2$C$_6$H$_4$: 78%, ee = 45%
Ar¹ = p-Tol, Ar² = p-ClC$_6$H$_4$: 80%, ee = 43%
Ar¹ = Ar² = p-Tol: 61%, ee = 32%
Ar¹ = m-NO$_2$C$_6$H$_4$, Ar² = p-NO$_2$C$_6$H$_4$: 82%, ee = 52%
Ar¹ = m-NO$_2$C$_6$H$_4$, Ar² = p-ClC$_6$H$_4$: 89%, ee = 48%
Ar¹ = m-NO$_2$C$_6$H$_4$, Ar² = p-Tol: 77%, ee = 45%
Ar¹ = m-Tol, Ar² = p-NO$_2$C$_6$H$_4$: 68%, ee = 57%
Ar¹ = m-Tol, Ar² = p-ClC$_6$H$_4$: 78%, ee = 42%
Ar¹ = m-Tol, Ar² = p-Tol: 76%, ee = 37%

Scheme 2.131 Cu-catalyzed aziridination of alkenes with sulfonamides and PhI(OAc)$_2$ in the presence of Evans' bisoxazoline ligand.

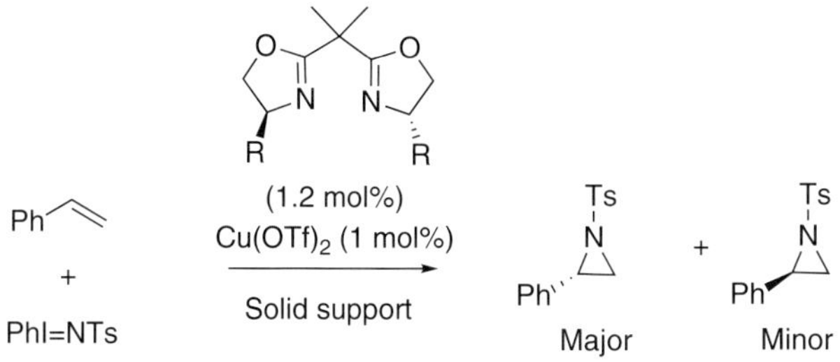

with support = laponite:
R = Ph: 75%, ee = 25%
R = t-Bu: 60%, ee = 2%
with support = SiO$_2$-CF$_2$SO$_3$H:
R = Ph: 58%, ee = 26%
R = t-Bu: 57%, ee = 0%
with support = SAC-13:
R = Ph: 73%, ee = 21%

Scheme 2.132 Heterogeneous Cu-catalyzed aziridination of styrene with PhINTs in the presence of Evans' ligand immobilized by electrostatic interactions.

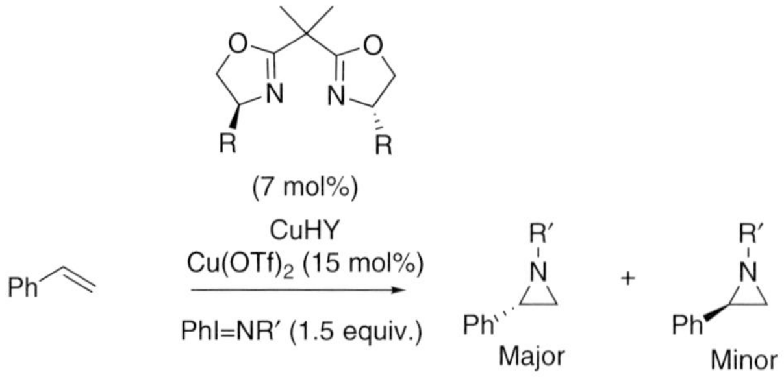

Scheme 2.133 Heterogeneous Cu-catalyzed aziridination of styrene with PhINNs or PhINTs in the presence of Evans' ligand immobilized on zeolite HY.

R = Ph, R' = Ns: 78%, ee = 85%
R = t-Bu, R' = Ns: 68%, ee = 83%
R = Ph, R' = Ts: 78%, ee = 76%
R = t-Bu, R' = Ts: 58%, ee = 24%

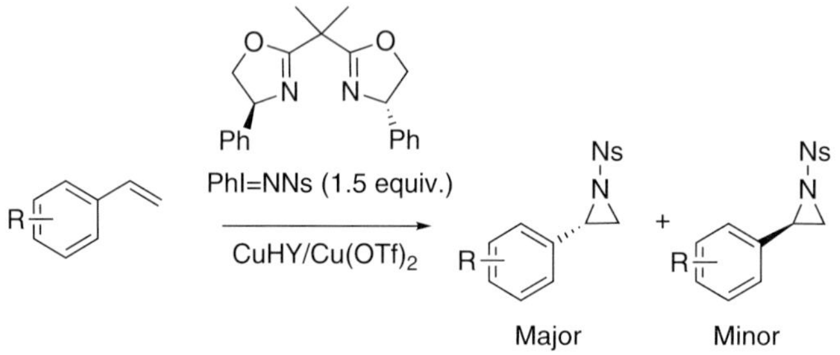

R = 2-Cl: 83%, ee = 88%
R = 3-Cl: 82%, ee = 95%
R = 4-Cl: 85%, ee = 94%
R = 2-F: 82%, ee = 72%
R = 2-Br: 60%, ee = 85%
R = 4-Br: 90%, ee = 74%
R = 2-Me: 60%, ee = 79%
R = 4-OMe: 71%, ee = 64%
R = 3-NO$_2$: 63%, ee = 65%

Scheme 2.134 Heterogeneous Cu-catalyzed aziridination of styrene derivatives with PhINNs in the presence of Evans' ligand immobilized on zeolite HY.

in Scheme 2.134 [175]. In all cases of substrates, higher enantioselection could be achieved with the heterogeneously catalyzed reaction, compared with the homogeneously catalyzed reaction. The effect was considered to be due mainly to the enhanced confinement of the substrate within the pores of the zeolite.

In 2007, Evans' bisoxazoline ligand was used by Dauban et al. to promote the intramolecular Cu-catalyzed aziridination of a wide range of unsaturated sulfamates [176]. This reaction was performed in the presence of PhI=O and [Cu(MeCN)$_4$]PF$_6$, affording the corresponding chiral aziridines in yields of up to 86% and enantiomeric excesses of up to 84%, as shown in Scheme 2.135. In

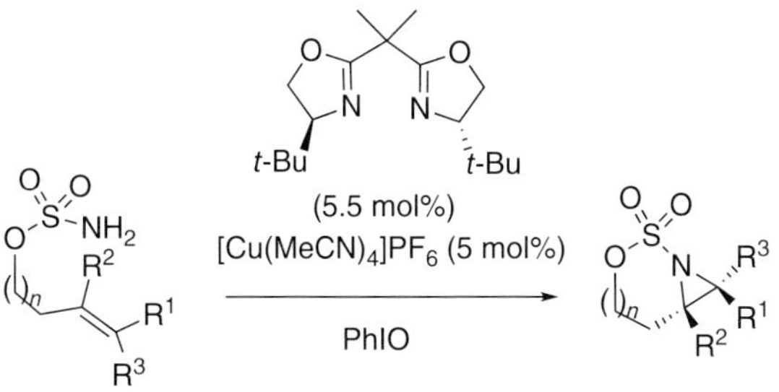

R^1 = Ph, R^2 = R^3 = H, n = 1: 86%, ee = 84%
R^1 = R^2 = R^3 = H, n = 1: 81%, ee = 52%
R^1 = R^3 = H, R^2 = Me, n = 1: 24%, ee = 36%
R^1 = Me, R^2 = R^3 = H, n = 1: 80%, ee = 80%
R^1 = Et, R^2 = R^3 = H, n = 1: 83%, ee = 80%
R^1 = CO_2Me, R^2 = R^3 = H, n = 1: 79%, ee = 62%
R^1 = R^2 = H, R^3 = Et, n = 1: 86%, ee = 72%
R^1 = R^2 = R^3 = H, n = 2: 72%, ee = 47%

Scheme 2.135 Intramolecular Cu-catalyzed aziridination of unsaturated sulfamates with PhIO in the presence of Evans' bisoxazoline ligand.

particular, it is interesting to note that this nitrene transfer occurred with equal success for simple aliphatic olefins and electron-poor alkenes, substrates for which no enantioselective aziridination has been reported to date. Later, one of these products was employed as key intermediate in a total synthesis of spisulosine and its fluoro analog [177].

The size of the chelate in the reactive metal complex of bisoxazolines is an important feature of the catalyst, since it will control the orientation of the substituents on the two oxazolines around the metal ion and the distance between the substituents and the metal ions. This implies that the chelate size of bisoxazolines can tune the chiral environment at the catalytic center and then affect the enantioselectivity of asymmetric catalytic reactions. In order to keep the designed chiral environment at the catalytic center, a series of rigid backbone-linked bisoxazolines were designed and synthesized by Xu et al. As an example, these authors have prepared a series of cyclohexane-linked bisoxazolines (cHBoxes), which have been investigated as chiral ligands for the asymmetric Cu-catalyzed aziridination of chalcones performed in the presence of PhI=NTs as the nitrene source [178]. As summarized in Scheme 2.136, the involvement of (S,S)-1,2-bis-[(S)-(4-phenyl)oxazolin-2-yl]cyclohexane as the chiral ligand in combination with CuOTf allowed the aziridination of a wide range of chalcones to be achieved in good-to-high yields and with enantioselectivities of up to 99% ee. It was found that the enantioselectivity was not substituent dependent with respect to chalcones.

With the aim of finding other efficient bisoxazoline ligands, these authors have developed another novel rigid backbone-containing bisoxazoline ligand (AnBox), in which the two oxazoline rings were attached via the 1,8-positions of a rigid anthracene ring [179]. The use of this ligand for similar reactions to those

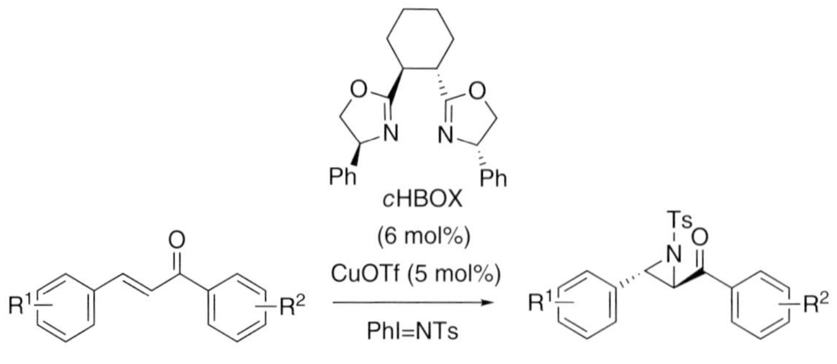

Scheme 2.136 Cu-catalyzed aziridination of chalcones with PhINTs in the presence of cHBOX ligand.

R¹ = R² = H: 56%, ee = 91%
R¹ = p-Me, R² = H: 62%, ee = 94%
R¹ = p-F, R² = H: 62%, ee = 90%
R¹ = p-Cl, R² = H: 80%, ee = 95%
R¹ = R² = p-Me: 50%, ee >99%
R¹ = H, R² = p-Me: 71%, ee >99%
R¹ = H, R² = p-OMe: 73%, ee = 97%
R¹ = R² = p-Cl: 80%, ee = 95%
R¹ = p-Cl, R² = p-Me: 72%, ee = 85%
R¹ = H, R² = p-Br: 63%, ee = 86%
R¹ = m-F, R² = H: 64%, ee = 92%
R¹ = p-CF$_3$, R² = H: 51%, ee = 80%

just described has led to the corresponding *trans*-aziridines with comparably high enantioselectivities of up to 99% ee, but with the opposite enantioselectivity, compared with both the cHBOX ligand and Evans' bisoxazoline ligand (Scheme 2.137). In addition, the results indicated that the enantioselectivity was substituent dependent with respect to the chalcones, in contrast with the results obtained by using cHBOX as the ligand. In general, chalcones bearing electron-donating substituents showed higher enantioselectivities compared to those bearing electron-withdrawing groups. Moreover, it was demonstrated that the coordination of the oxygen atom of the carbonyl group in the chalcones with the copper in the catalyst and the π–π stacking interaction between the aryl group of the nitrene and the aryl substituent attached to the C=C double bond in the chalcones were indispensable for obtaining a high enantioselectivity in the asymmetric aziridination. In 2007, these authors demonstrated that the substituent-dependent enantioselectivity in the asymmetric aziridination of chalcones catalyzed by the AnBOX ligand was rationalized by the π-stacking interaction between the ligand backbone and the substrates, primarily confirmed by the use of bulky substrates and of catalysts without aromatic backbones, such as cHBOX and Evans' bisoxazoline ligand [180].

In a complementary study, these authors have investigated the efficiency of the three bisoxazolines, Evans' ligand, cHBOX ligand, and AnBOX ligand, to induce chirality in the aziridination of 1,3-dienes [181]. When the reaction was

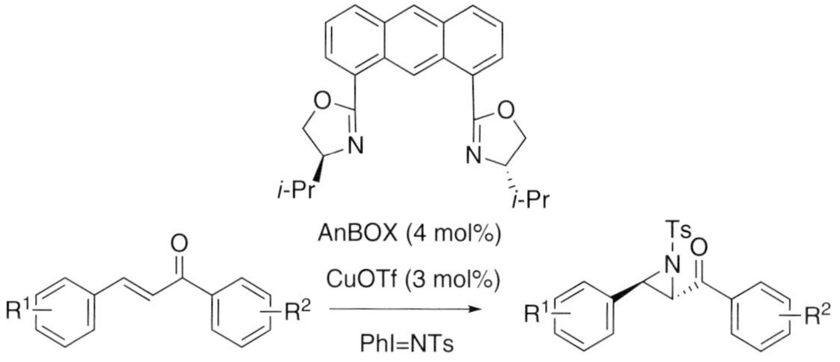

Scheme 2.137 Cu-catalyzed aziridination of chalcones with PhINTs in the presence of AnBox ligand.

$R^1 = R^2 = H$: 80%, ee = 96%
$R^1 = p\text{-Me}$, $R^2 = H$: 86%, ee = 98%
$R^1 = p\text{-Cl}$, $R^2 = H$: 70%, ee = 76%
$R^1 = m\text{-Cl}$, $R^2 = H$: 76%, ee = 84%
$R^1 = o\text{-Cl}$, $R^2 = H$: 91%, ee = 79%
$R^1 = m\text{-F}$, $R^2 = H$: 85%, ee = 71%
$R^1 = H$, $R^2 = p\text{-Me}$: 92%, ee > 99%
$R^1 = R^2 = p\text{-Me}$: 59%, ee > 99%
$R^1 = p\text{-Me}$, $R^2 = p\text{-Cl}$: 51%, ee = 68%
$R^1 = p\text{-Ph}$, $R^2 = H$: 35%, ee = 87%
$R^1 = p\text{-F}$, $R^2 = H$: 72%, ee = 62%
$R^1 = p\text{-Br}$, $R^2 = H$: 58%, ee = 52%
$R^1 = p\text{-OMe}$, $R^2 = H$: 61%, ee = 37%
$R^1 = H$, $R^2 = p\text{-OMe}$: 74%, ee = 62%
$R^1 = m\text{-F}$, $R^2 = H$: 85%, ee = 71%
$R^1 = p\text{-CF}_3$, $R^2 = H$: 69%, ee = 67%

performed in the presence of PhI=NTs combined with CuOTf and the bisoxazoline ligand, the corresponding aziridines were obtained in moderate yields and with enantioselectivities of up to 80% ee, diastereoselectivities of up to 99% de, and regioselectivities of up to 99:1, as shown in Scheme 2.138. It was shown that α,β,γ,δ-unsaturated ketones usually produced *cis*-γ,δ-aziridinated products, while 1,4-diphenyl-1,3-butadiene afforded both the *cis*- and *trans*-aziridine derivatives as major products by the use of the different bisoxazoline ligands.

In 2010, a chiral *gem*-dimethyl-phenylglycine-derived bisoxazoline ligand was used by Lebel and Parmentier to promote the asymmetric copper-catalyzed aziridination of styrenes with tosyloxycarbamates [182]. Best results were obtained when using 2,2,2-trichloroethyl tosyloxycarbamate as the nitrene source. As shown in Scheme 2.139, using a combination of this chiral ligand with Cu(MeCN)$_4$PF$_6$ allowed the aziridination of styrene, 4-nitrostyrene, and 3-nitrostyrene to be achieved in moderate enantioselectivities of up to 75% ee.

With the aim of increasing these enantioselectivities, the same authors performed related reactions using a chiral tosyloxycarbamate reagent, such as (*R*)-Ph-TrocNHOTs [183]. The double stereodifferentiation was observed, and

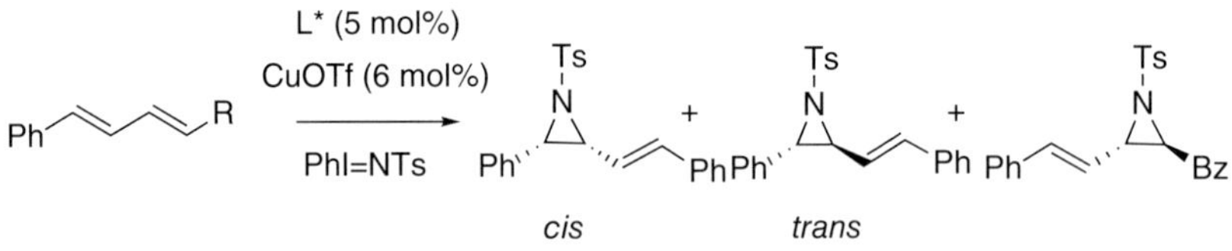

R = Bz, L* = L³: 64%, 26:3:71, ee (cis) = 36% (2S,3R)
R = Bz, L* = L⁴: 59%, 87:13:0, ee (cis) = 13% (2S,3R)
R = Bz, L* = L²: 73%, 99:0:1, ee (cis) = 36% (2R,3S)
R = Bz, L* = L¹: 28%, 100:0:0, ee (cis) = 80% (2R,3S)
R = Ac, L* = L³: 51%, 100:0:0, ee (cis) = 6% (2R,3S)
R = Ac, L* = L²: 25%, 100:0:0, ee (cis) = 19% (2R,3S)
R = Ph, L* = L³: 68%, 26:74, ee (cis) = 12% (2S,3R), ee (trans) = 14% (S,S)
R = Ph, L* = L⁴: 32%, 61:39, ee (cis) = 22% (2S,3R), ee (trans) = 24% (S,S)
R = Ph, L* = L²: 21%, 68:32, ee (cis) = 60% (2R,3S), ee (trans) = 48% (R,R)
R = Ph, L* = L¹: 22%, 18:82, ee (cis) = 21% (2R,3S), ee (trans) = 34% (R,R)

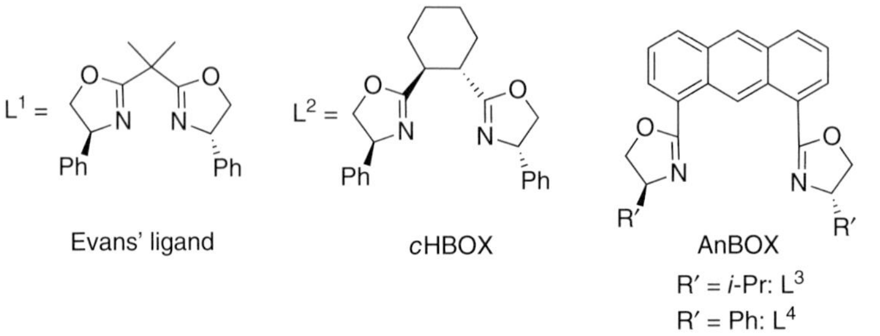

L¹ = Evans' ligand

L² = cHBOX

AnBOX
R' = i-Pr: L³
R' = Ph: L⁴

Scheme 2.138 Cu-catalyzed aziridination of 1,3-dienes with PhINTs in the presence of bisoxazoline ligands.

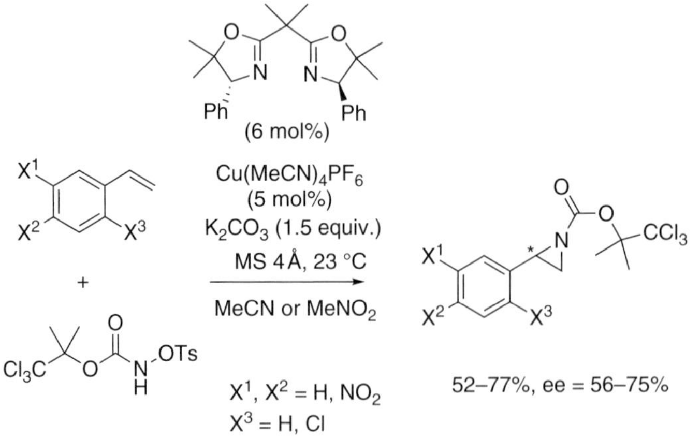

X¹, X² = H, NO₂
X³ = H, Cl

52–77%, ee = 56–75%

Scheme 2.139 Cu-catalyzed aziridination of styrenes with 2,2,2-trichloroethyl tosyloxycarbamate in the presence of a chiral *gem*-dimethyl-phenylglycine-derived bisoxazoline ligand.

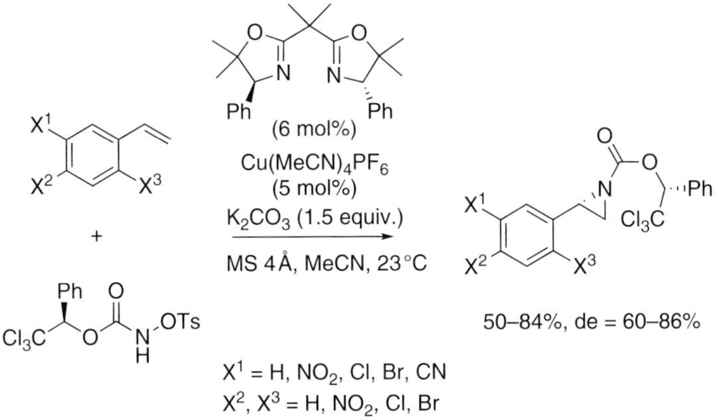

Scheme 2.140 Cu-catalyzed aziridination of styrenes with a chiral tosyloxycarbamate in the presence of a chiral *gem*-dimethyl-phenylglycine-derived bisoxazoline ligand.

both the *N*-tosylcarbamate substituent and the bis(oxazoline) ligand had an effect on the yields and diastereoselectivities of the reaction. Best results were obtained with electron-deficient styrenes, such as nitro-substituted styrenes, with yields of up to 84% and diastereoselectivities of up to 86% de, as shown in Scheme 2.140.

In 2010, Van der Eycken *et al.* reported the synthesis of a novel chiral bisoxazoline ligand bearing a *trans*-(2*R*,3*R*)-diphenylcyclopropane backbone, which was applied to induce chirality in several reactions among which was aziridination of methyl cinnamate with (*N*-(*p*-toluenesulfonyl)imino)phenyliodinane (PhINTs) as nitrene precursor [184]. The corresponding *trans*-aziridine was obtained in moderate yield (25%) albeit with high enantioselectivity of 90% ee (Scheme 2.141).

It must be noted that a number of enantioselective copper-catalyzed aziridinations of olefins performed in the presence of chiral bisoxazoline ligands have been applied to the synthesis of various important chiral products. For example, Hajra

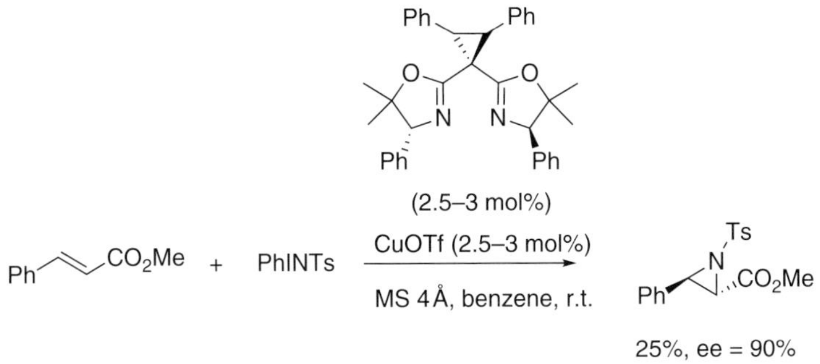

Scheme 2.141 Cu-catalyzed aziridination of methyl cinnamate with PhINTs in the presence of a chiral bisoxazoline ligand bearing a diphenylcyclopropane backbone.

and Sinha have developed one-pot sequential asymmetric aziridination/Friedel–Crafts reactions of aryl cinnamyl ethers, which afforded, in the presence of PhINNs and a copper catalyst based on a chiral indanolamine-derived Box ligand, the corresponding *trans*-3-amino-4-arylchromans in high regio-, diastereo-, and enantioselectivities of up to >98% de and 95% ee, respectively [185]. Related conditions were also successfully applied to the one-pot sequential asymmetric aziridination/Friedel–Crafts reactions of functionalized styrenes, providing the corresponding tetralins with enantioselectivities of up to 95% ee. These products could be further converted into biologically active products, such as dopamine D1 agonist A-86929 [186], dihydrexidine [187], and ecopipam [188].

On the other hand, several chiral ligands other than bisoxazolines have also been investigated in recent years for the Cu-catalyzed asymmetric aziridination of olefins. As an example, Wang and Ding have developed a novel chiral C_2-symmetric diimine ligand derived from D-mannitol, which was found to be highly efficient for the enantioselective control of the Cu-catalyzed asymmetric aziridination of olefins with PhI=NTs as the nitrene source [189]. Indeed, the corresponding *trans*-aziridines were isolated in good-to-excellent yields with enantioselectivities of up to 99% ee, as shown in Scheme 2.142. The catalyst system was also extended to a one-pot enantioselective aziridination of olefins by using $TsNH_2$/$PhI(OAc)_2$ as the nitrene source. In this case, most reactions proceeded smoothly to give the corresponding products in moderate yields and with good-to-excellent enantioselectivities of up to 96% ee (Scheme 2.142).

Earlier in 2003, a chiral binaphthyldiimine ligand, BINIM-DC, was found by Suga *et al.* to be a highly efficient ligand for the Cu-catalyzed asymmetric aziridination of olefins with PhI=NTs as the nitrene precursor [190]. In particular, high levels of enantioselectivity of up to 98% ee were obtained in the aziridination reactions of 3-arylpropenoate and 1,3-disubstituted 2-propen-1-one derivatives, as shown in Scheme 2.143.

A moderate enantioselectivity was obtained by Dauban *et al.* by using Jacobsen's diiminocyclohexane ligand in the intramolecular Cu-catalyzed aziridination of an unsaturated sulfamate. This reaction was performed in the presence of PhI=O and $[Cu(MeCN)_4]PF_6$, affording the corresponding aziridine in good yield, as shown in Scheme 2.144 [176]. It must be noted that better results in terms of yield (86%) and enantioselectivity (84% ee) were reached by the same authors by using Evans' bisoxazoline ligand (Scheme 2.135).

In 2010, Gautun *et al.* investigated the copper-catalyzed aziridination of dihydronaphthalenes in the presence of the same ligand [191]. The process employed aryl imino iodinanes as the nitrene source and afforded the corresponding aziridines in moderate-to-good yields (21–82%) along with moderate-to-high enantioselectivities of up to 87% ee, as shown in Scheme 2.145. The enantioselectivity and the yield were shown to be dependent on the properties of the nitrene precursor. TsNIPh appeared in general to give better results than *p*-NsNIPh. The formed chiral aziridines were further transformed into the respective 2-aminotetralins.

While C_2-symmetric ligands have been extensively used for various metal-mediated enantioselective organic transformations, the analogous C_3-symmetric systems have received much less attention [192]. In this context, Leung *et al.* have reported the synthesis of a chiral C_3-symmetric oxygen tripodal ligand

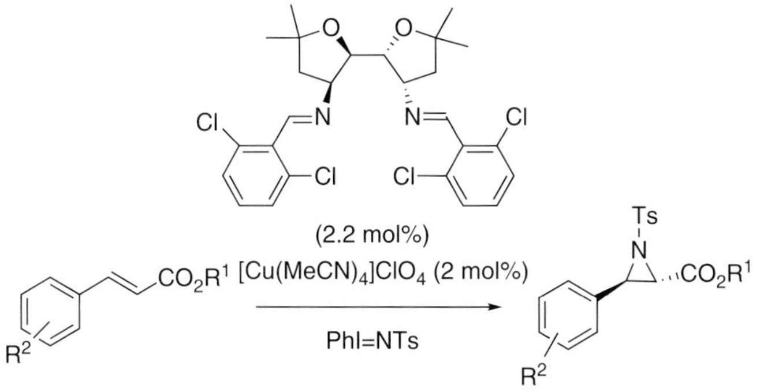

R^1 = Me, R^2 = H: 97%, ee = 88%
R^1 = Ph, R^2 = H: 96%, ee = 87%
R^1 = *t*-Bu, R^2 = H: 99%, ee >99%
R^1 = *t*-Bu, R^2 = 4-F: 99%, ee = 98%
R^1 = *t*-Bu, R^2 = 4-Cl: 97%, ee = 98%
R^1 = *t*-Bu, R^2 = 4-Br: 97%, ee = 98%
R^1 = *t*-Bu, R^2 = 4-Me: 86%, ee = 94%
R^1 = *t*-Bu, R^2 = 4-MeO: 95%, ee = 80%
R^1 = *t*-Bu, R^2 = 2-NO_2: 63%, ee = 99%
R^1 = *t*-Bu, R^2 = 2,3-MeO: 67%, ee = 97%

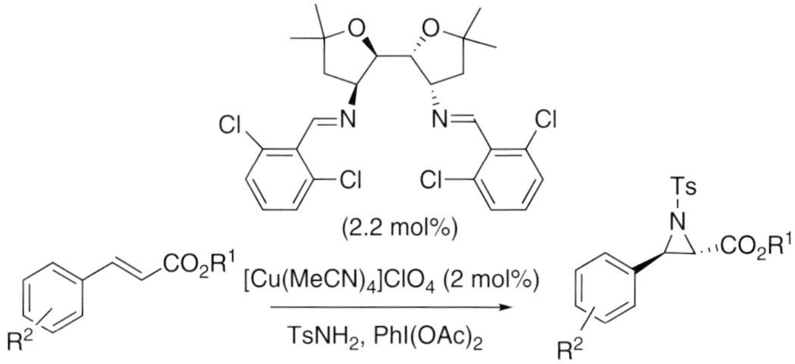

R^1 = *t*-Bu, R^2 = H: 49%, ee = 96%
R^1 = *t*-Bu, R^2 = 4-F: 55%, ee = 95%
R^1 = *t*-Bu, R^2 = 4-Cl: 45%, ee = 95%
R^1 = *t*-Bu, R^2 = 4-Me: 39%, ee = 88%
R^1 = *t*-Bu, R^2 = 4-MeO: 41%, ee = 74%
R^1 = *t*-Bu, R^2 = 2-NO_2: 19%, ee = 97%

Scheme 2.142 Cu-catalyzed aziridinations of olefins with PhINTs or $TsNH_2$/$PhI(OAc)_2$ in the presence of a chiral diimine ligand.

derived from (*S*)-BINOL [193]. The potential of this anionic tris(phosphinite) ligand was evaluated for the Cu-catalyzed aziridination of styrene. Therefore, the use of the corresponding Cu(I) complex prepared *in situ* from $[Cu(MeCN)_4]BF_4$ combined with PhI=NTs as the nitrene source allowed the styrene aziridine to be afforded in 88% yield with moderate enantioselectivity of 43% ee (Scheme 2.146). When this catalytic system was applied to various substituted

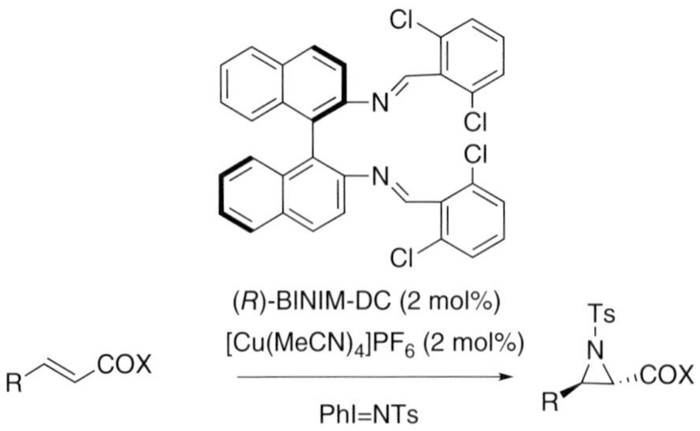

(R)-BINIM-DC (2 mol%)
[Cu(MeCN)₄]PF₆ (2 mol%)

R⌒COX + PhI=NTs → R⌒(N-Ts aziridine)⌒COX

from (R)-BINIM-DC:
R = p-ClC₆H₄, X = OMe: 82%, ee = 81% (2S,3R)
R = 1-Naph, X = OMe: 74%, ee = 77% (2S,3R)
R = 2-Naph, X = OMe: 74%, ee = 68% (2S,3R)
R = Ph, X = OPh: 48%, ee = 89% (2S,3R)
R = Ph, X = Ot-Bu: 57%, ee = 98% (2S,3R)
R = X = Ph: 87%, ee = 84% (2S,3R)

from (S)-BINIM-DC:
R = p-CNC₆H₄, X = OMe: 69%, ee = 90% (2R,3S)
R = p-Tol, X = OMe: 41%, ee = 83% (2R,3S)
R = Me, X = OMe: 13%, ee = 36% (2R,3S)
R = H, X = OMe: 30%, ee = 43% (2R,3S)
R = p-ClC₆H₄, X = Ot-Bu: 64%, ee = 97% (2R,3S)
R = p-CNC₆H₄, X = Ot-Bu: 73%, ee = 95% (2R,3S)
R = X = Ph: 79%, ee = 86% (2R,3S)
R = Ph, X = Me: 73%, ee = 67% (2R,3S)

Scheme 2.143 Cu-catalyzed aziridination of olefins with PhINTs in the presence of BINIM-DC ligand.

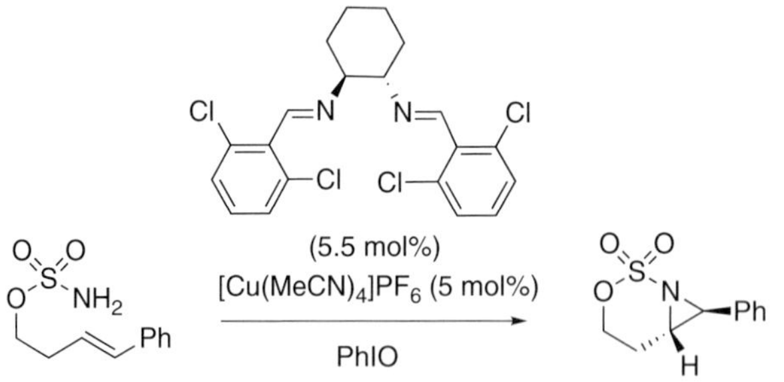

(5.5 mol%)
[Cu(MeCN)₄]PF₆ (5 mol%)
PhIO

75%, ee = 56%

Scheme 2.144 Intramolecular Cu-catalyzed aziridination of an unsaturated sulfamate with PhIO in the presence of Jacobsen's ligand.

styrenes, better enantioselectivities could be obtained, up to 50% ee for the corresponding aziridines, as shown in Scheme 2.146.

In 2008, Gibson et al. investigated the efficiency of a copper(II) complex of a chiral cyclohexyl-fused azamacrocycle, such as (7aR,11aR)-1,4,7-trimethyldodecahydro-1H-1,4,7-benzotriazonine, to induce chirality in the aziridination of

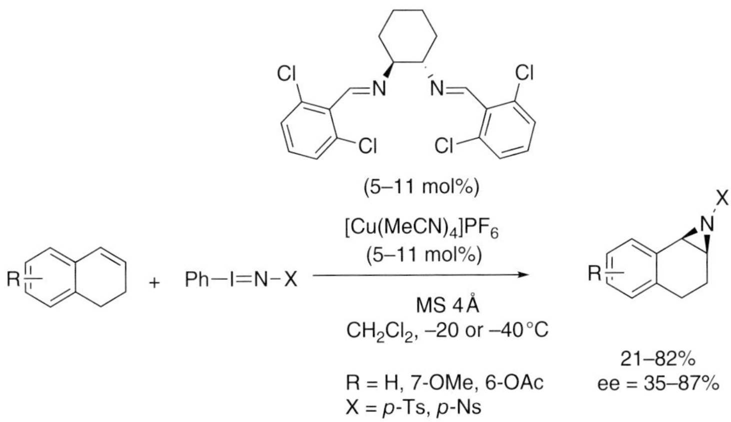

Scheme 2.145 Cu-catalyzed aziridination of dihydronaphthalenes with aryl imino iodinanes in the presence of Jacobsen's ligand.

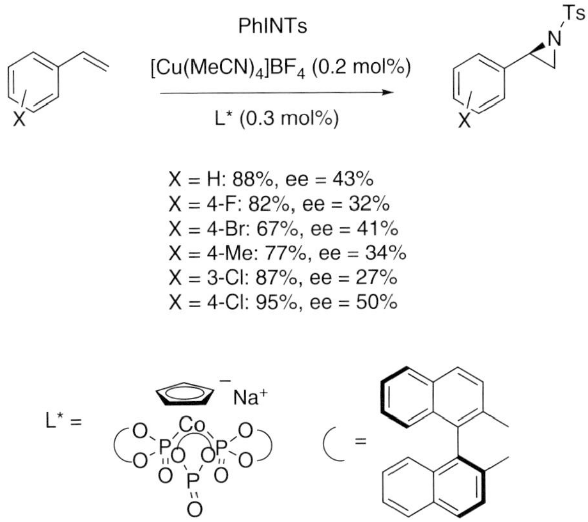

Scheme 2.146 Cu-catalyzed aziridination of styrenes with PhINTs in the presence of a BINOL-derived C_3-symmetric oxygen tripodal ligand.

styrene with PhI=NTs as the nitrene source [194]. Disappointingly, the use of this chiral complex led to the corresponding racemic aziridine. In addition, Chanda et al. have reported the asymmetric synthesis of aziridines using bromamine-T as the nitrene source [195]. Therefore, the Cu-catalyzed aziridination of styrene with bromamine-T was performed in the presence of various chiral cinchona alkaloids such as sparteine and N-benzyl ephedrine as the ligands. Best results were obtained by using the cinchona alkaloid, dihydroquinine, affording the expected aziridine in 39% yield with moderate enantioselectivity of 43% ee in the presence of $CuCl_2$ and bromamine-T in acetonitrile.

2.3.1.2 Rh-Catalyzed Aziridination

Chiral rhodium(II) catalysts are complementary in scope to copper(I) catalysts in asymmetric carbene-transfer reactions. As an example, Che et al. have successfully applied [Rh$_2${(4S)-MEOX}$_4$] as the catalyst of choice for the intramolecular aziridination of a series of unsaturated sulfonamides [196]. In the presence of PhI=O as the oxidant, the corresponding aziridines were obtained in good yields with enantioselectivities of up to 76% ee, as shown in Scheme 2.147.

This rhodium catalyst was also used by Hayes et al. for the asymmetric aziridination of a range of homoallyl-carbamates performed in the presence of PhI=O, in 2006 [197]. In this case, the corresponding aziridines were obtained as major products, but with low enantioselectivities ranging from 1% to 23% ee. Another rhodium catalyst, [Rh$_2${(5S)-MEPY}$_4$], has been successfully employed by Fruit and Müller to induce chirality in the intramolecular aziridination of 2-allyl-substituted aromatic sulfamates [198]. The use of this catalyst in the presence of PhI(OAc)$_2$ combined with MgO allowed the corresponding tricyclic aziridines to be obtained in low-to-moderate yields with moderate enantioselectivities of up to 52% ee (Scheme 2.148).

Another rhodium catalyst, [Rh$_2${(S)-nttl}$_4$], using the 1,8-naphthalimide of L-*tert*-leucine as a bridging ligand, has been successfully employed by Dauban et al. for the asymmetric aziridination of styrene derivatives with chiral sulfonimidamides as iminoiodane precursors [199]. In these conditions, the expected

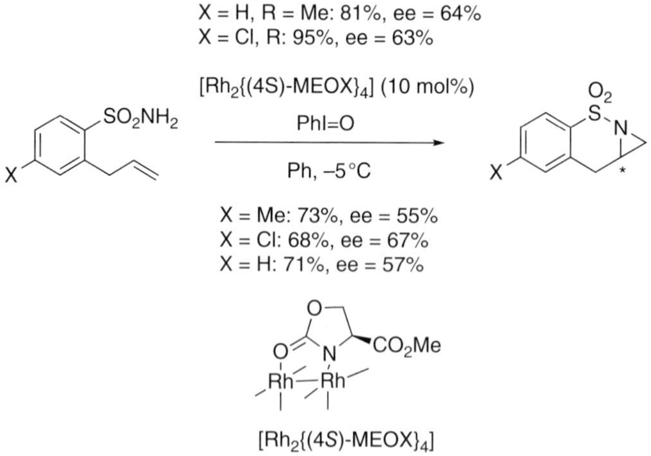

Scheme 2.147 Intramolecular aziridinations of unsaturated sulfonamides with PhIO catalyzed by [Rh$_2${(4S)-MEOX}$_4$].

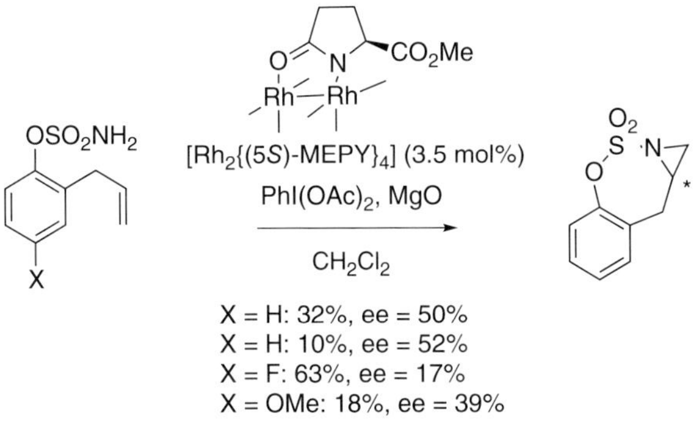

Scheme 2.148 Intramolecular aziridination of 2-allyl substituted aromatic sulfamates with PhI(OAc)$_2$ catalyzed by [Rh$_2${(5S)-MEPY}$_4$].

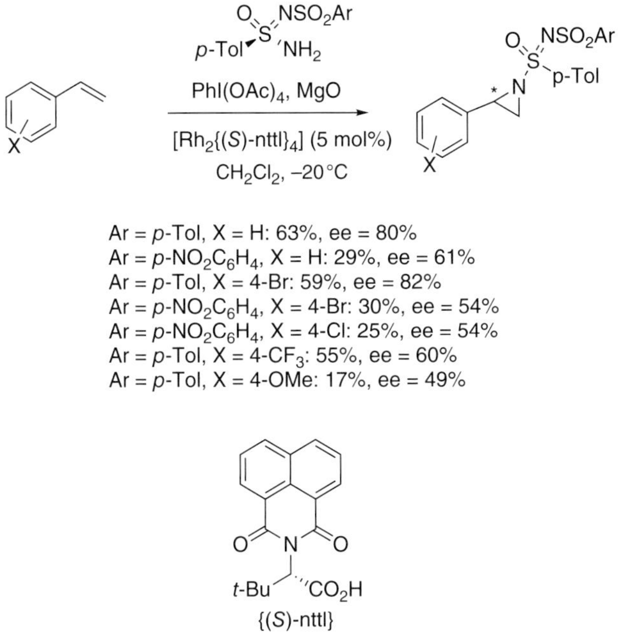

X = H: 32%, ee = 50%
X = H: 10%, ee = 52%
X = F: 63%, ee = 17%
X = OMe: 18%, ee = 39%

Ar = p-Tol, X = H: 63%, ee = 80%
Ar = p-NO$_2$C$_6$H$_4$, X = H: 29%, ee = 61%
Ar = p-Tol, X = 4-Br: 59%, ee = 82%
Ar = p-NO$_2$C$_6$H$_4$, X = 4-Br: 30%, ee = 54%
Ar = p-NO$_2$C$_6$H$_4$, X = 4-Cl: 25%, ee = 54%
Ar = p-Tol, X = 4-CF$_3$: 55%, ee = 60%
Ar = p-Tol, X = 4-OMe: 17%, ee = 49%

{(S)-nttl}

Scheme 2.149 Aziridination of styrenes with sulfonimidamides and PhI(OAc)$_2$ catalyzed by [Rh$_2${(S)-nttl}$_4$].

aziridines were obtained in modest-to-good yields with diastereoselectivities of up to 82% de, as shown in Scheme 2.149.

In addition, Hashimoto et al. have developed enantioselective aziridinations of a range of alkenes with [N-(4-nitrophenylsulfonyl)imino]phenyliodinane catalyzed by [Rh$_2${(S)-TCPTTL}$_4$], which provided the corresponding aziridines in high yields with moderate enantioselectivities, except in the case of 2,2-dimethylchromene, which allowed a high level of enantioselectivity of 94% ee to be achieved (Scheme 2.150) [200].

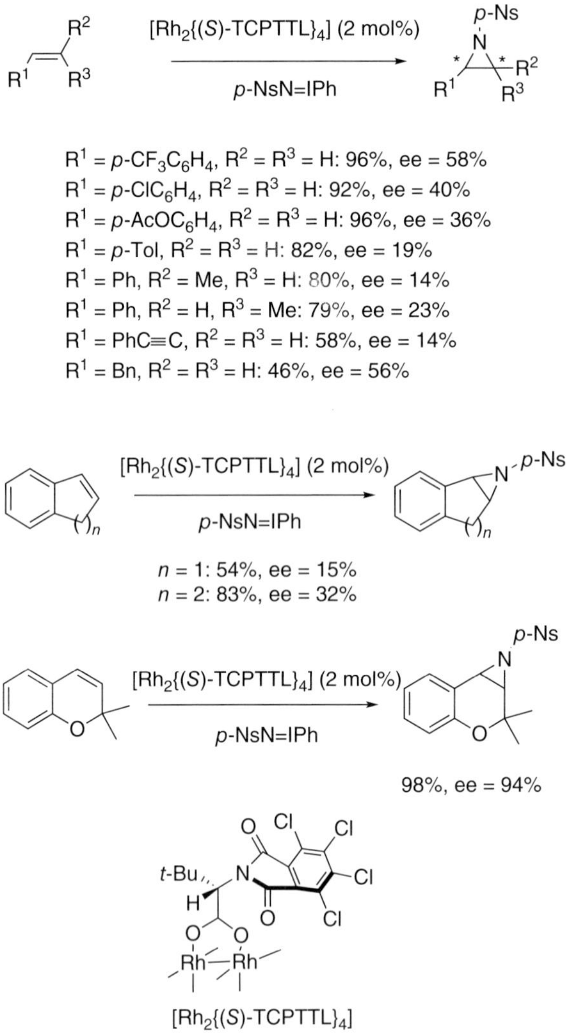

Asymmetric Synthesis of Three-Membered Rings

$R^1 = p\text{-}CF_3C_6H_4$, $R^2 = R^3 = H$: 96%, ee = 58%
$R^1 = p\text{-}ClC_6H_4$, $R^2 = R^3 = H$: 92%, ee = 40%
$R^1 = p\text{-}AcOC_6H_4$, $R^2 = R^3 = H$: 96%, ee = 36%
$R^1 = p\text{-}Tol$, $R^2 = R^3 = H$: 82%, ee = 19%
$R^1 = Ph$, $R^2 = Me$, $R^3 = H$: 80%, ee = 14%
$R^1 = Ph$, $R^2 = H$, $R^3 = Me$: 79%, ee = 23%
$R^1 = PhC\equiv C$, $R^2 = R^3 = H$: 58%, ee = 14%
$R^1 = Bn$, $R^2 = R^3 = H$: 46%, ee = 56%

$n = 1$: 54%, ee = 15%
$n = 2$: 83%, ee = 32%

98%, ee = 94%

Scheme 2.150 Aziridinations of olefins with p-NsNIPh catalyzed by [Rh$_2${(S)-TCPTTL}$_4$].

2.3.1.3 Ru-Catalyzed Aziridination

Excellent enantioselectivities have been reported by Katsuki *et al.* for the asymmetric aziridination of a wide range of alkenes by using chiral ruthenium(salen) (CO) complexes in the presence of azide compounds as the nitrene precursors [201]. Various azide compounds were involved in these reactions, providing the corresponding aziridines in both high yields and with enantioselectivities of up to >99% ee (Scheme 2.151). Best results were obtained by using a fluorinated ruthenium complex or another ruthenium catalyst, which possessed a phenyl substituent bearing chloro and trimethylsilyl groups at its *meta*- and *para*-positions, respectively. Notably, the aziridination of less-reactive α,β-unsaturated esters and

2 Asymmetric Aziridination | 313

$$R^1\text{—CH=CH}_2 + R^2N_3 \xrightarrow{\text{Ru complex}} \underset{R^1}{\overset{R^2}{\text{aziridine}}}$$

R^1 = Ph, R^2 = p-Ns, Ar = 3,5-Cl_2-4-TMSC$_6$H$_2$: 90%, ee = 87%
R^1 = p-BrC$_6$H$_4$, R^2 = p-Ns, Ar = 3,5-Cl_2-4-TMSC$_6$H$_2$: 93%, ee = 83%
R^1 = PhC≡C, R^2 = p-Ns, Ar = 3,5-Cl_2-4-TMSC$_6$H$_2$: 98%, ee = 98%
R^1 = PhC≡C, R^2 = o-Ns, Ar = 3,5-Cl_2-4-TMSC$_6$H$_2$: 58%, ee = 87%
R^1 = Ph, R^2 = o-Ns, Ar = 3,5-Cl_2-4-TMSC$_6$H$_2$: 60%, ee = 81%
R^1 = Ph, R^2 = p-Ts, Ar = Ph: 71%, ee = 87%
R^1 = p-NO$_2$C$_6$H$_4$, R^2 = p-Ts, Ar = Ph: 98%, ee = 92%
R^1 = p-BrC$_6$H$_4$, R^2 = Ses, Ar = 3,5-Cl_2-4-TMSC$_6$H$_2$: 76%, ee = 92%
R^1 = Ph, R^2 = Ses, Ar = 3,5-Cl_2-4-TMSC$_6$H$_2$: 99%, ee = 92%
R^1 = p-BrC$_6$H$_4$, R^2 = Ts, Ar = 3,5-Cl_2-4-TMSC$_6$H$_2$: 90%, ee = 93%
R^1 = PhC C, R^2 = Ses, Ar = 3,5-Cl_2-4-TMSC$_6$H$_2$: 50%, ee >99%
R^1 = 2-$C_{10}H_7$, R^2 = Ts, Ar = 3,5-Cl_2-4-TMSC$_6$H$_2$ 69%, ee = 91%
R^1 = n-Hex, R^2 = Ts, Ar = 3,5-Cl_2-4-TMSC$_6$H$_2$: 64%, ee = 84%
R^1 = CO_2Bn, R^2 = Ses, Ar = 3,5-Cl_2-4-TMSC$_6$H$_2$: 81%, ee >99%
R^1 = CON(OMe)Bn, R^2 = Ses, Ar = 3,5-Cl_2-4-TMSC$_6$H$_2$: 85%, ee >99%

Ru complex = [chiral ruthenium(salen)(CO) complex structure]

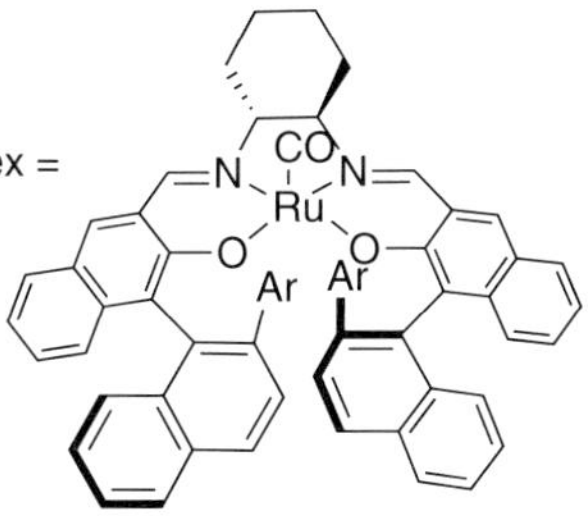

Scheme 2.151 Aziridination of olefins with azides catalyzed by ruthenium(salen)(CO) complexes.

amides performed with 2-(trimethylsilyl)ethanesulfonyl azide (SesN$_3$) also proceeded with excellent enantioselectivities and in good yields (Scheme 2.151).

In addition, Che et al. have developed asymmetric intramolecular aziridinations of sulfonamides in the presence of PhI(OAc)$_2$ and a chiral ruthenium porphyrin [Ru(Por*)(CO)] [202]. As an example, under these conditions, the reaction of 2-vinylbenzenesulfonamide afforded the corresponding tricyclic aziridine in 65% yield, albeit with a low enantioselectivity of 9% ee. On the other hand, remarkable enantioselectivities of up to 99% ee were reported by Katsuki et al. for the ruthenium-catalyzed aziridination of aromatic acyclic as well as cyclic olefins with 2-(trimethylsilyl)ethanesulfonyl (Ses) azide by using very low catalyst loadings as low as 0.5 mol% [203]. The process employed a Jacobsen's type ligand as chiral inducer, providing the corresponding aziridines in generally high yields of up to 99% with enantioselectivities ranging from 89% to 99% ee, as shown in Scheme 2.152. The scope of the methodology was extended using 3 mol% of catalyst loading to a range of aliphatic acyclic olefins, giving the

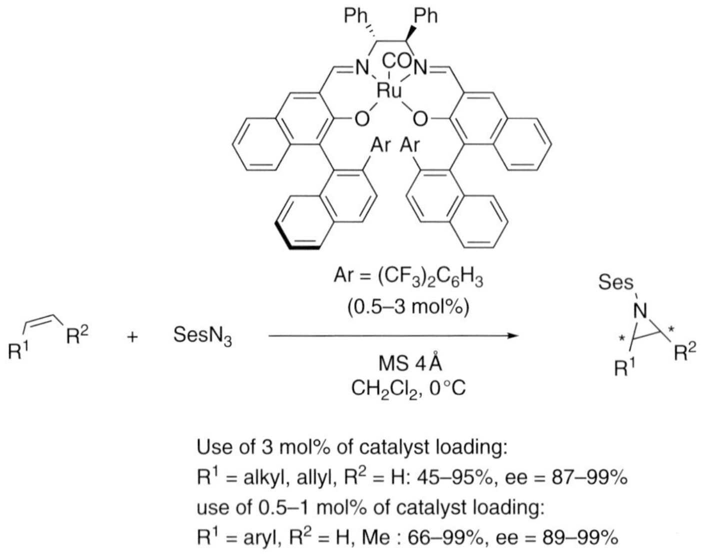

Scheme 2.152 Ru-catalyzed aziridination of acyclic and cyclic olefins with an azide in the presence of a Jacobsen's type ligand.

Use of 3 mol% of catalyst loading:
R^1 = alkyl, allyl, R^2 = H: 45–95%, ee = 87–99%
use of 0.5–1 mol% of catalyst loading:
R^1 = aryl, R^2 = H, Me : 66–99%, ee = 89–99%

corresponding aziridines in moderate-to-very-high yields (45–95%) with high enantioselectivities of 87–99% ee (Scheme 2.152).

A closely related chiral ruthenium catalyst was applied by the same authors to the asymmetric aziridination of vinyl ketones with the same azide [204]. No polymerization of these acid-sensitive ketones was observed, and the reaction provided the expected chiral aziridinyl ketones in both remarkable excellent yields of 92–99% and general enantioselectivity of >99% ee, in all cases of substrates studied. Another important advantage of this nice process was its very low catalyst loading of 0.5 mol%, as shown in Scheme 2.153. The high synthetic utility of this aziridination was demonstrated by its application to a short total synthesis of (+)-PD 128907, which is a potent dopamine D_3 receptor agonist.

In 2014, Lo and Che reported the synthesis of novel chiral bis(NHC) ruthenium(II)–porphyrin complexes and their successful use as catalysts in asymmetric aziridination of acyclic olefins with pentafluorophenyl azide [205]. As shown in Scheme 2.154, the process afforded the expected three-membered products in good-to-high yields with enantioselectivities of up to 98% ee by using only 0.5 mol% of catalyst loading.

2.3.1.4 Catalysis by Other Metals

In 2008, Zhang et al. developed the first cobalt-catalyzed asymmetric aziridination of olefins using diphenylphosphoryl azide (DPPA) as the nitrene source, affording the corresponding N-phosphorylated aziridines [206]. The reaction was carried out in the presence of D_2-symmetric chiral porphyrins, such as that depicted in Scheme 2.155. This novel catalyst system could be applied to a wide variety of aromatic olefins, giving the corresponding aziridines in good yields combined with moderate enantioselectivities of up to 53% ee.

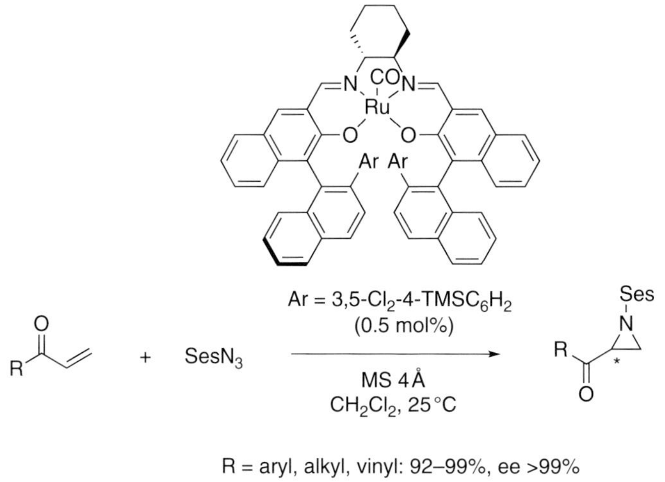

Ar = 3,5-Cl2-4-TMSC6H2
(0.5 mol%)
SesN3
MS 4 Å
CH2Cl2, 25 °C
Ses
R = aryl, alkyl, vinyl: 92–99%, ee >99%

Scheme 2.153 Ru-catalyzed aziridination of vinyl ketones with an azide in the presence of a Jacobsen's type ligand.

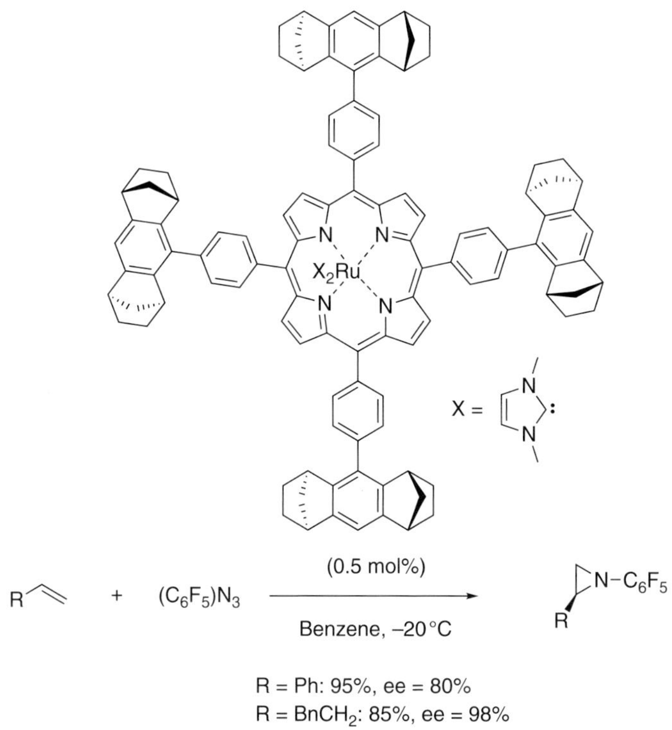

X2Ru
X =
(0.5 mol%)
(C6F5)N3
Benzene, –20 °C
N–C6F5
R = Ph: 95%, ee = 80%
R = BnCH2: 85%, ee = 98%

Scheme 2.154 Ru-catalyzed aziridination of acyclic olefins with an azide in the presence of a porphyrin chiral ligand.

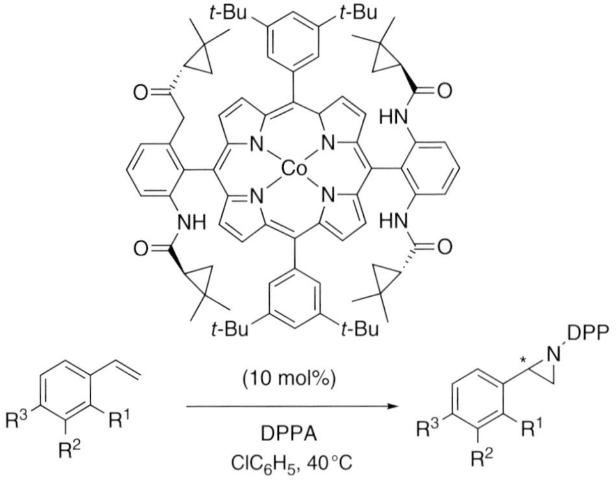

Scheme 2.155 Co-catalyzed aziridination of olefins with DPPA in the presence of a chiral D_2-symmetric amidoporphyrin.

$R^1 = R^2 = R^3 = H$: 88%, ee = 37%
$R^1 = Me$, $R^2 = R^3 = H$: 35%, ee = 46%
$R^1 = R^3 = H$, $R^2 = Me$: 52%, ee = 44%
$R^1 = R^2 = H$, $R^3 = Me$: 58%, ee = 37%
$R^1 = R^2 = H$, $R^3 = t\text{-Bu}$: 77%, ee = 53%
$R^1 = R^2 = H$, $R^3 = Br$: 65%, ee = 28%
$R^1 = Br$, $R^2 = R^3 = H$: 68%, ee = 7%
$R^1 = R^3 = H$, $R^2 = Br$: 58%, ee = 45%
$R^1 = R^2 = H$, $R^3 = Cl$: 64%, ee = 6%
$R^1 = R^2 = H$, $R^3 = F$: 72%, ee = 17%
$R^1 = R^2 = H$, $R^3 = CF_3$: 64%, ee = 44%
$R^1 = R^3 = H$, $R^2 = NO_2$: 58%, ee = 46%

In 2013, the same authors reported a highly efficient asymmetric aziridination of styrenes with fluoroaryl azides induced by a closely related chiral D_2-symmetric amidoporphyrin cobalt catalyst depicted in Scheme 2.156 [207]. The corresponding chiral N-fluoroaryl-containing aziridines were obtained in moderate-to-quantitative yields (52–99%) along with good-to-excellent enantioselectivities (68–96% ee). The scope of the process was wide since mono-*ortho*-fluoro-substituted aryl azides, di-*ortho*-fluoro-substituted aryl azides, as well as 2,6-difluoro-, 2,3,5,6-tetrafluoro-, and pentafluorophenyl azides were shown to be highly effective nitrene sources. Moreover, a wide range of aromatic olefins with varied electronic and steric properties, including heteroaromatic alkenes, such as Boc-protected 3-vinylindole (80% yield, 96% ee), were found suitable to the reaction.

In 2014, another related chiral D_2-symmetric amidoporphyrin cobalt catalyst was employed by these authors in asymmetric aziridinations of styrenes with a phosphoryl azide [208]. This novel catalyst was found suitable for a range of

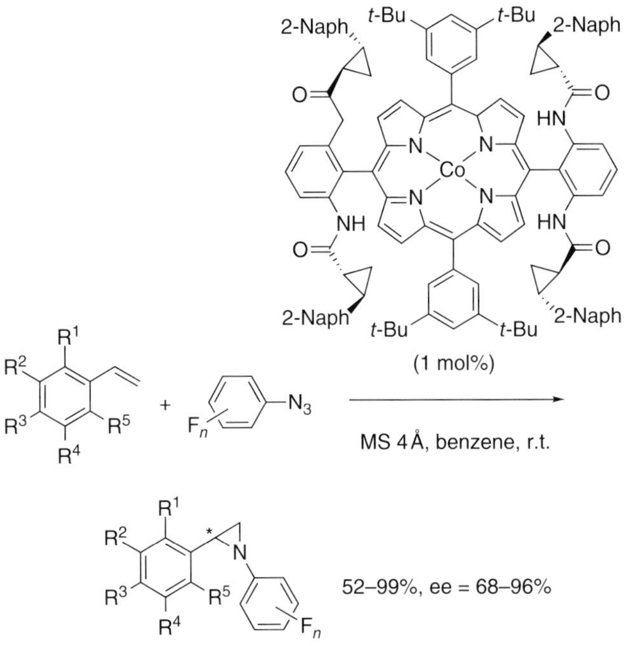

Scheme 2.156 Co-catalyzed aziridination of styrenes with fluoroaryl azides in the presence of a chiral D_2-symmetric amidoporphyrin.

R¹ = H, Me
R² = H, CH=CH₂
R³ = H, Me, CH=CH₂, F, Br, CF₃, *t*-Bu, OMe, CO₂Me
R⁴ = H, Me, Br
R⁵ = H
R¹,R², R²,R³, R⁴,R⁵ = (CH=CH)₂
n = 1–5

52–99%, ee = 68–96%

variously substituted styrenes, which provided by reaction with bis(2,2,2-trichloroethyl)phosphoryl azide the corresponding chiral *N*-phosphorylaziridines in moderate-to-excellent yields (64–99%) along with moderate-to-good enantioselectivities (23–85% ee), as shown in Scheme 2.157.

In 2008, Bolm *et al.* demonstrated that low-cost and nontoxic iron could be applied as catalyst to the asymmetric aziridination of styrene by using PhI=NTs as the nitrene source [209]. Indeed, the combination of iron(II) triflate with chiral ligands was found to induce chirality in the formation of styrene aziridine. Among various tridentate ligands, the (*S*,*S*)-*i*-Pr-py-BOX ligand was shown to be the most efficient, leading to the product with enantioselectivity of 40% ee and 72% yield (Scheme 2.158). Interestingly, 2,6-bis(*N*-pyrazolyl)pyridine ligands were also found to be applied as chiral ligands, giving albeit lower enantioselectivities (≤20% ee).

In 2005, Zhou *et al.* reported the synthesis of a chiral mononuclear complex of Re(I)–NOBIN Schiff base via the reaction of Re(CO)₅Cl and the corresponding tridentate ligand in methanol [210]. This ligand was derived from the reaction

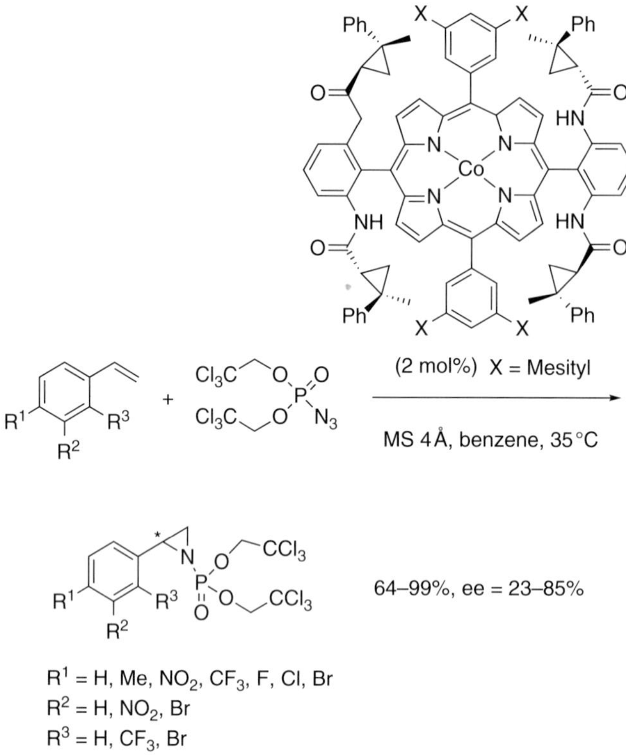

Scheme 2.157 Co-catalyzed aziridination of styrenes with a phosphoryl azide in the presence of a chiral D_2-symmetric amidoporphyrin.

Reaction conditions: (2 mol%) X = Mesityl, MS 4Å, benzene, 35 °C

R^1 = H, Me, NO_2, CF_3, F, Cl, Br
R^2 = H, NO_2, Br
R^3 = H, CF_3, Br

64–99%, ee = 23–85%

between NOBIN and 3,5-dichlorosalicylaldehyde. It was found that this novel rhenium catalyst showed some catalytic ability in the asymmetric aziridination of styrene and *p*-chlorostyrene, which gave the corresponding aziridines in yields of 36% and 30%, respectively, although without chiral induction.

2.3.1.5 Organocatalyzed Aziridination

For a long time, it was not known that organocatalysts could be used to catalyze the aziridination reactions [211]. In recent years, however, several examples of organocatalytic enantioselective aziridination of olefins have been successfully developed, and these have become a topic of interest in asymmetric organocatalysis. Among these reactions are those based on the use of quaternary salts of cinchona alkaloids as the catalysts to induce chirality in the aziridination of 2-(phenylsulfanyl)-2-cycloalkenones [212]. The reaction of these cycloalkenones with ethyl nosyloxycarbamate under phase-transfer conditions provided the corresponding aziridines in low-to-high yields and with enantioselectivities of up to 75% ee (Scheme 2.159).

In the same context, Murugan and Siva have developed other cinchona alkaloids as chiral phase-transfer catalysts for the asymmetric aziridination of a wide range of electron-deficient olefins with *N*-acyl-*N*-aryl hydroxamic acids [213].

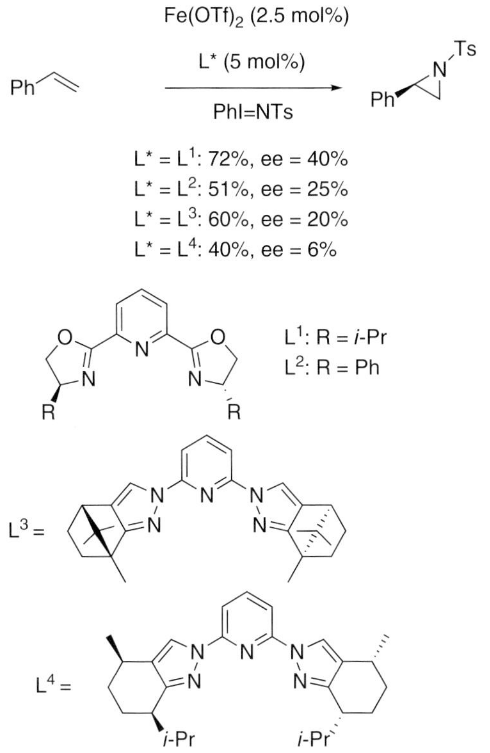

Scheme 2.158 Fe-catalyzed aziridination of styrene.

The corresponding chiral N-arylaziridines were isolated in good yields and with high enantioselectivities of up to 95% ee, as shown in Scheme 2.160. It was shown that the formation of the R- and S-aziridines was solely dependent on chiral transfer between the substrate and the catalyst.

In 2007, Armstrong *et al.* developed the asymmetric aziridination of chalcone using an aminimide generated *in situ* from the treatment of a chiral tertiary amine, such as quinine, with O-(diphenylphosphinyl)hydroxylamine [214]. As shown in Scheme 2.161, a promising level of asymmetric induction of 56% ee was obtained for the aziridination of E-chalcone by using these conditions.

Later in 2014, the same authors extended the scope of this methodology to heteroaromatic-substituted enones [215]. As shown in Scheme 2.162, the corresponding *trans*-aziridines were synthesized in moderate yields (27–47%) and with better enantioselectivities of up to 77% ee.

In 2008, Melchiorre *et al.* reported asymmetric aziridination of α,β-unsaturated ketones by using a chiral primary amine salt, which was prepared by combining the easily available 9-amino(9-deoxy)*epi*-hydroquinine with D-N-Boc-phenylglycine [216]. Therefore, both linear and cyclic α,β-unsaturated ketones reacted with tosylated hydroxycarbamates under these conditions to afford the corresponding aziridines with almost complete diastereocontrol and very high enantioselectivity of up to 99% ee (Scheme 2.163). These results showed the ability of

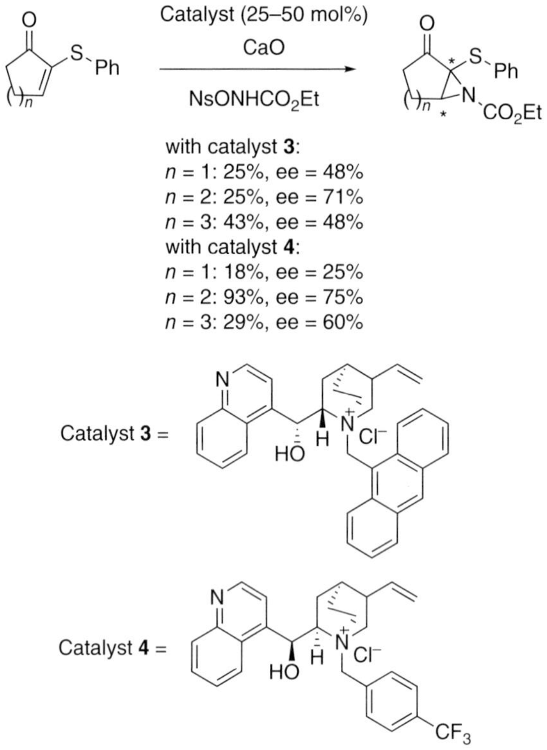
Catalyst (25–50 mol%)
CaO
NsONHCO2Et
with catalyst 3:
n = 1: 25%, ee = 48%
n = 2: 25%, ee = 71%
n = 3: 43%, ee = 48%
with catalyst 4:
n = 1: 18%, ee = 25%
n = 2: 93%, ee = 75%
n = 3: 29%, ee = 60%
Catalyst 3 =
Catalyst 4 =

Scheme 2.159 Aziridination of 2-(phenylsulfanyl)-2-cycloalkenones with ethyl nosyloxycarbamate catalyzed by cinchona alkaloid salts.

this chiral organocatalyst salt to promote an asymmetric domino iminium–enamine intramolecular sequence.

Later in 2010, enantioselectivities of up to 99% ee were also reported by the same authors in the aziridination of cyclic enones with *tert*-butyl tosyloxycarbamate catalyzed by cinchona-derived primary amine salts [217]. As shown in Scheme 2.164, a wide variety of *N*-Boc-protected chiral aziridines could be prepared in moderate-to-excellent yields (33–98%) using this methodology, which could also be successfully applied to indenone derivatives, providing good yields (53–67%) and excellent enantioselectivities (96–98% ee). The efficiency of this method was not only measured by its selectivity and reliability, but also by the possibility of accessing both of the antipodes of the products by simply selecting the appropriate catalyst enantiomer, as shown in Scheme 2.164. It constituted the first catalytic and highly enantioselective aziridination of cyclic enones.

In 2008, Minakata *et al.* reported a new method for the aziridination of electron-deficient olefins based on the use of *N*-chloro-*N*-sodiocarbamate [218]. These reactions were promoted by phase-transfer chiral ammonium salt catalysts derived from cinchona alkaloids, yielding the corresponding aziridines from α,β-unsaturated ketones, esters, sulfones, and amides in good yields and with enantioselectivities of up to 86% ee, as shown in Scheme 2.165.

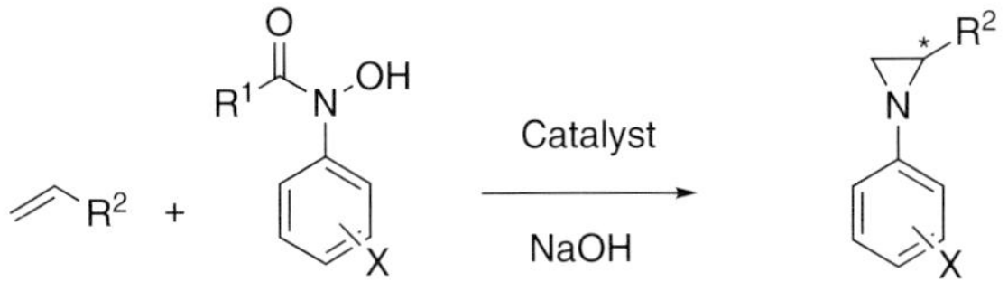

with catalyst 5:
R¹ = t-Bu, X = H, R² = CO₂t-Bu: 79%, ee = 94% (S)
R¹ = t-Bu, X = 3-Br, R² = CO₂t-Bu: 87%, ee = 76% (S)
R¹ = t-Bu, X = 4-Br, R² = CO₂t-Bu: 79%, ee = 87% (S)
R¹ = Ph, X = 4-Me, R² = CO₂Me: 85%, ee = 79% (S)
R¹ = Ph, X = 4-NO₂, R² = CO₂t-Bu: 41%, ee = 43% (S)

with catalyst 6:
R¹ = t-Bu, X = 4-OMe, R² = CO₂t-Bu: 92%, ee = 95% (R)
R¹ = t-Bu, X = 4-Cl, R² = CO₂t-Bu: 86%, ee = 85% (R)
R¹ = t-Bu, X = 4-CO₂H, R² = Ph: 53%, ee = 75% (R)
R¹ = t-Bu, X = 4-OH, R² = SOPh: 77%, ee = 82% (R)
R¹ = t-Bu, X = H, R² = CO₂t-Bu: 56%, ee = 88% (R)
R¹ = t-Bu, X = 4-Br, R² = CO₂Me: 49%, ee = 76% (R)
R¹ = Ph, X = 4-Me, R² = CO₂t-Bu: 69%, ee = 85% (R)
R¹ = t-Bu, X = 4-NO₂, R² = CO₂t-Bu: 92%, ee = 89% (R)
R¹ = t-Bu, X = 4-Me, R² = CO₂t-Bu: 65%, ee = 90% (R)

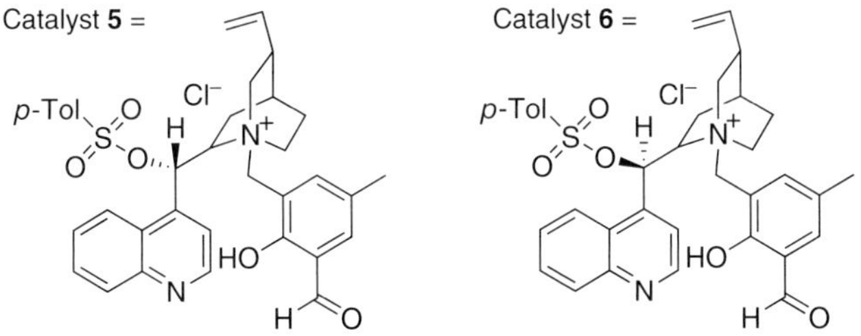

Scheme 2.160 Aziridination of olefins with N-acyl-N-aryl hydroxamic acids catalyzed by cinchona alkaloid salts.

As an extension of the preceding methodology, these authors have investigated the aziridination of chiral electron-deficient olefins, such as oxazolidinone-substituted enones or pyrazole-containing enones, with N-chloro-N-sodiocarbamates [219]. The process afforded the corresponding aziridines in good yields (60–91%) and with diastereoselectivities of up to >98% de, as shown in Scheme 2.166. The key to reach high diastereoselectivity was found to be the employment of the "matching" stereochemical combination of a chiral auxiliary with a chiral ammonium salt catalyst. Furthermore, the preparation of the corresponding enantiomerically pure aziridines by removal of the chiral auxiliary was demonstrated.

Organocatalysts other than cinchona alkaloids have been successfully applied to enantioselective aziridinations of alkenes via nitrene transfer. As an example, Shi et al. have reported the use of a chiral tertiary amine, such as (+)-Tröger's

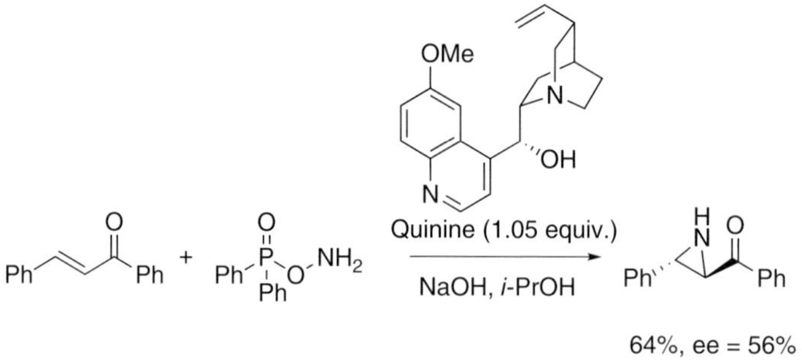

Scheme 2.161 Aziridination of chalcone with an *in situ* generated aminimide derived from quinine.

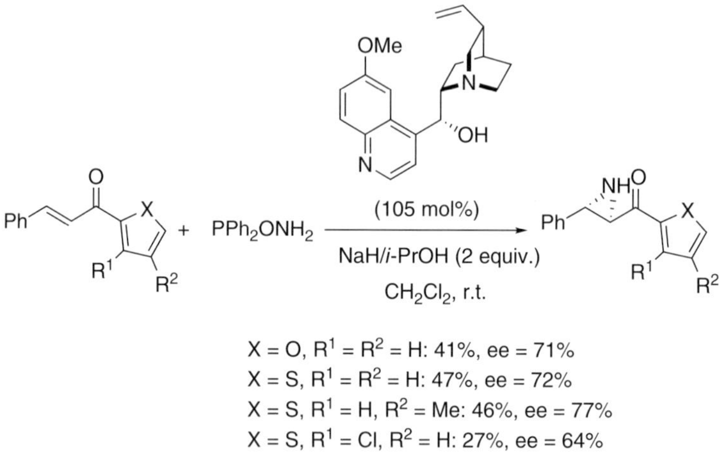

Scheme 2.162 Aziridination of heteroaromatic-substituted enones with diphenylphosphinylhydroxylamine catalyzed by quinine.

X = O, R^1 = R^2 = H: 41%, ee = 71%
X = S, R^1 = R^2 = H: 47%, ee = 72%
X = S, R^1 = H, R^2 = Me: 46%, ee = 77%
X = S, R^1 = Cl, R^2 = H: 27%, ee = 64%

base, in a one-pot process, which involved the *in situ* generation of a hydrazinium salt, deprotonation of the hydrazinium to form an aminimide, and subsequent aziridination by reaction with chalcones [220]. It was shown that O-mesitylenesulfonylhydroxylamine (MSH) could readily aminate the tertiary amine to give the corresponding hydrazinium salt in high yield. The asymmetric aziridination of two chalcones with MSH in the presence of (+)-Tröger's base, and a base such as CsOH·H$_2$O, allowed a reasonable level of enantioselectivity of up to 67% ee to be achieved (Scheme 2.167).

In 2011, Hamada *et al.* also described excellent results for the asymmetric aziridination of cyclic enones with benzyloxyhydroxamate catalyzed by a chiral 1,2-diamine derived from diphenylethylenediamine, such as chiral N-neopentyl 1,2-diphenylethylenediamine [221]. As shown in Scheme 2.168, the reaction was performed in the presence of benzoic acid and afforded the corresponding chiral aziridines in high yields (80–91%) along with high enantioselectivities of

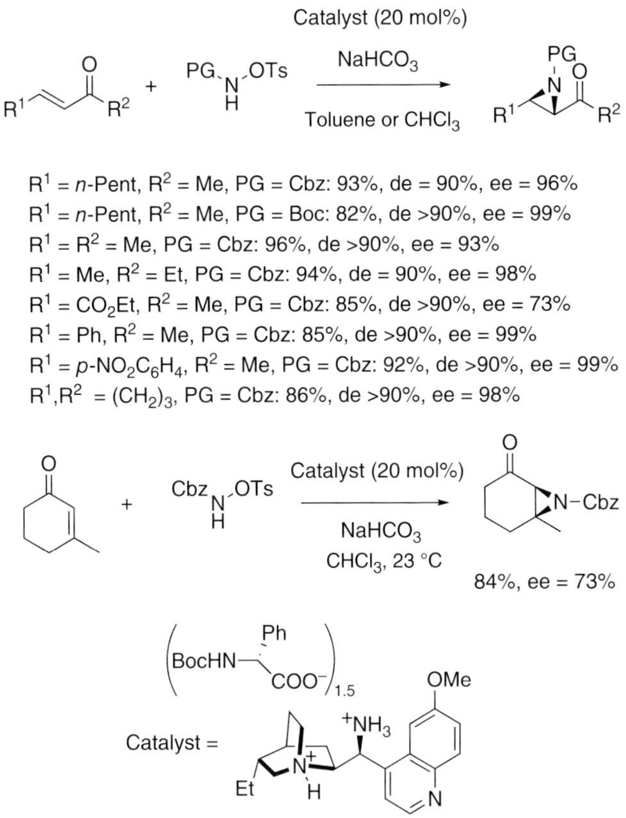

Scheme 2.163 Aziridinations of α,β-unsaturated ketones with tosylated hydroxycarbamates catalyzed by a chiral cinchona-derived primary amine salt.

88–97% ee. The utility of this nice reaction was demonstrated by its application to a total synthesis of (−)-agelastatin A.

Another type of chiral organocatalysts, such as the chiral amino thiourea depicted in Scheme 2.169, was selected by Lattanzi et al. among a range of bifunctional chiral ligands investigated in the asymmetric aziridination of α-acyl acrylates with *tert*-butyl tosyloxycarbamate [222]. The corresponding chiral aziridines bearing a functionalized quaternary stereocenter were formed in good-to-excellent yields (67–98%) and moderate-to-high enantioselectivities ranging from 50% to 82% ee. The synthetic potential of these products was demonstrated by their conversion into α,α-disubstituted-α-amino acid esters.

In addition, a novel chiral binaphthalene-based tertiary amine was applied by Page et al. to mediate the asymmetric aziridination of *trans*-chalcone with diphenylphosphinylhydroxylamine, in 2013 [223]. The corresponding N-unprotected aziridine was afforded, however, in both moderate yield (48%) and enantioselectivity (43% ee) by using a stoichiometric amount of catalyst, as shown in Scheme 2.170.

An unprecedented example of a highly chemo- and enantioselective organocatalytic aziridination of α,β-unsaturated aldehydes with acylated hydroxycarba-

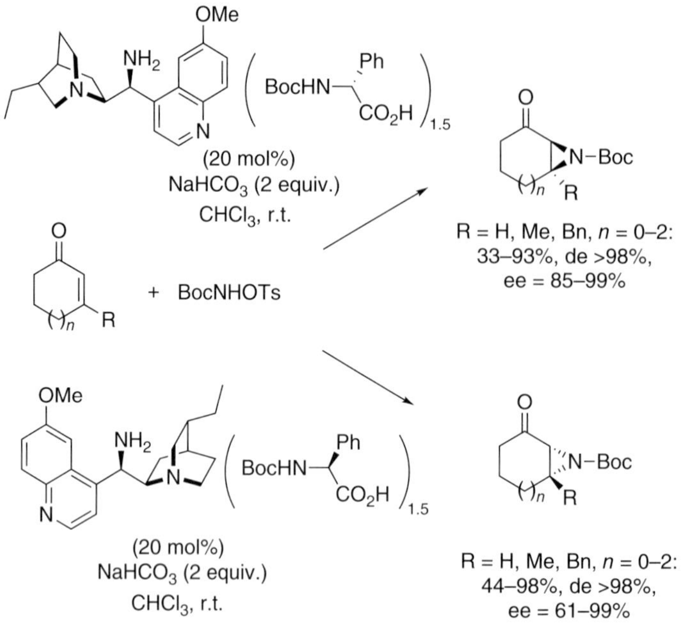

Scheme 2.164 Aziridination of cyclic enones with *tert*-butyl tosyloxycarbamate catalyzed by cinchona-derived primary amine salts.

mates was reported by Cordova et al., in 2007 [224]. This reaction was catalyzed efficiently by simple chiral pyrrolidine derivatives, such as diphenylprolinol trimethylsilyl ether, and led to the corresponding 2-formylaziridines in good-to-high yields with diastereoselectivities of up to 90% de and enantioselectivities of up to 99% ee, as shown in Scheme 2.171. In 2011, Cordova et al. extended the scope of this methodology to a wide range of hydroxylamine derivatives as nitrogen sources and also to a range of aliphatic as well as (hetero)aromatic α,β-unsaturated aldehydes, providing the corresponding chiral aziridines in comparable yields, diastereo-, and enantioselectivities [225]. These authors have proposed a mechanism, depicted in Scheme 2.171, in which an efficient shielding of the *Si* face of the chiral iminium intermediate by the bulky phenyl groups of the organocatalyst led to the stereoselective *Re*-facial nucleophilic conjugate attack on the β carbon atom of the electrophile by the amino group of the acylated hydroxycarbamate. Next, the chiral enamine intermediate generated performed a 3-*exo*-tet nucleophilic attack on the now electrophilic nitrogen atom, and acetic acid was released. The intramolecular ring closure pushed the equilibrium in the forward direction and made this step irreversible.

Moreover, these authors extended the preceding methodology to α,β-disubstituted-α,β-unsaturated aldehydes with a tosylated sulfonamide, such as TsNHOTs [225]. It is noteworthy that the corresponding aziridine products bearing two adjacent tertiary and quaternary stereocenters were formed highly stereoselectively with a diastereoselectivity of 78–92% de along with a remarkable enantioselectivity of 98–99% ee, as shown in Scheme 2.172.

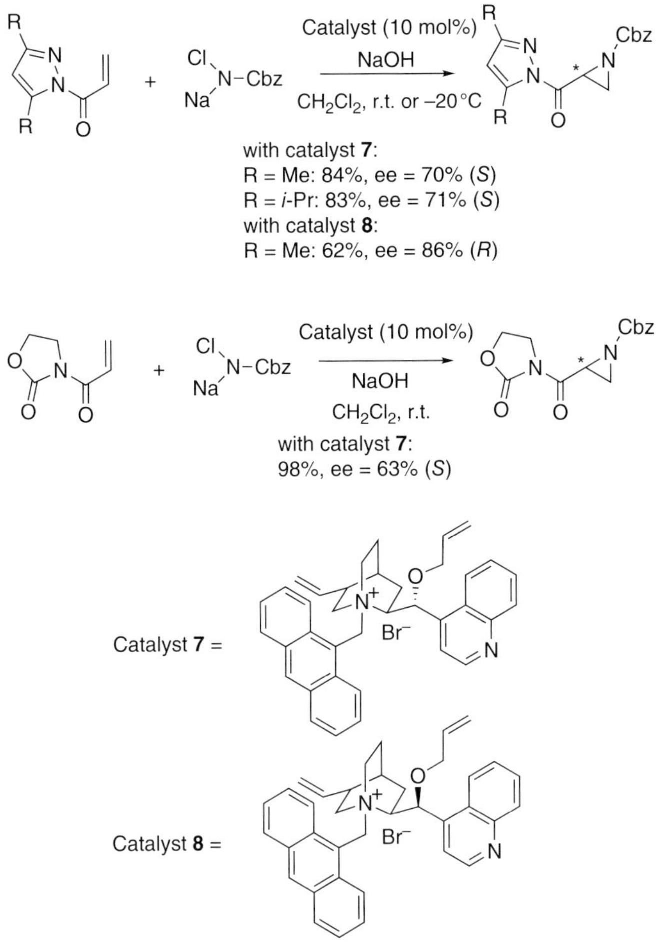

Scheme 2.165 Aziridinations of olefins with N-chloro-N-sodiocarbamate catalyzed by cinchona alkaloid salts.

In 2014, Hamada *et al.* applied this type of methodology to develop a total synthesis of (R)-sumanirole, which exhibits selective dopamine D2 receptor agonist activity (Scheme 2.173) [226]. The aziridination step was catalyzed by chiral diphenylprolinol triethylsilyl ether in the presence of 3 equiv. of base, such as NaOAc or Na_2CO_3, allowing the key intermediate aziridine to be achieved in 94% yield and 97% ee [227].

de Figueiredo *et al.* employed a closely related proline-derived catalyst to induce the asymmetric aziridination of various α-branched α,β-unsaturated aldehydes with a tosylated sulfonamide such as TsNHOTs [228]. The corresponding chiral aziridines bearing a quaternary stereocenter were obtained in good yields (69–86%) with good-to-high enantioselectivities of up to 90% ee, as shown in Scheme 2.174.

In 2011, Jørgensen *et al.* reported the use of the same catalyst as earlier to induce the asymmetric aziridination of structurally rigid α,β-unsaturated aldehydes, such

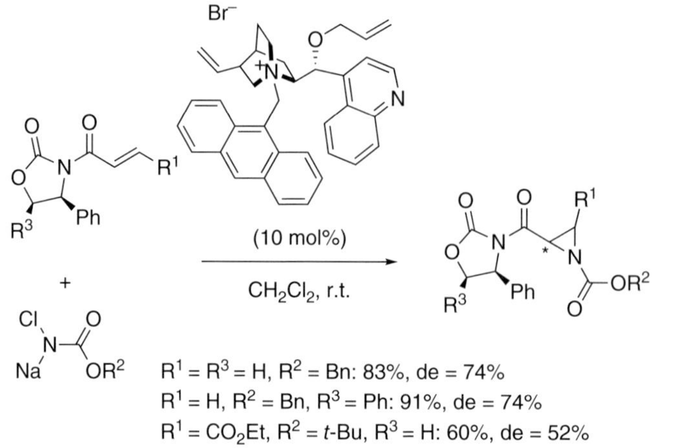
Br⁻
N
O
N
R1
Ph
R3
(10 mol%)
CH2Cl2, r.t.
+
Cl
Na
OR2
R1 = R3 = H, R2 = Bn: 83%, de = 74%
R1 = H, R2 = Bn, R3 = Ph: 91%, de = 74%
R1 = CO2Et, R2 = t-Bu, R3 = H: 60%, de = 52%

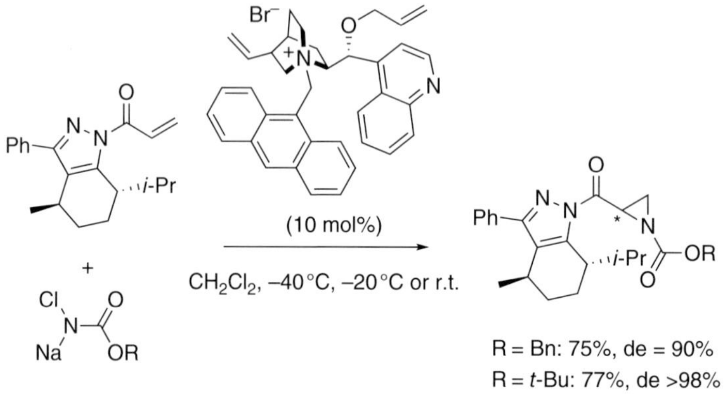
Br⁻
Ph
i-Pr
(10 mol%)
CH2Cl2, –40 °C, –20 °C or r.t.
Cl
Na
OR
R = Bn: 75%, de = 90%
R = t-Bu: 77%, de >98%

Scheme 2.166 Aziridinations of chiral electron-deficient olefins with *N*-chloro-*N*-sodiocarbamates catalyzed by a cinchona alkaloid salt.

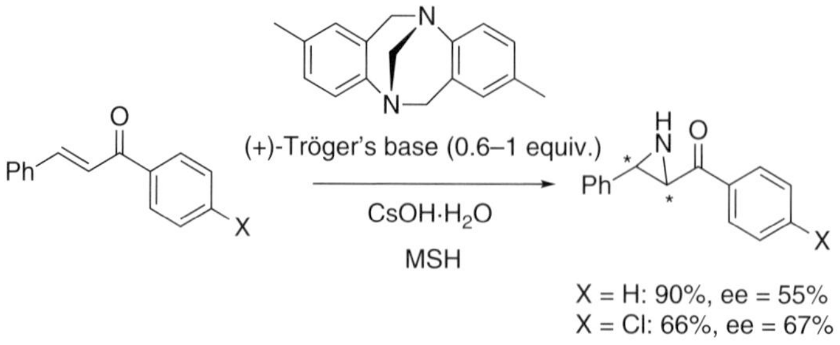
(+)-Tröger's base (0.6–1 equiv.)
CsOH·H2O
MSH
Ph
X
X = H: 90%, ee = 55%
X = Cl: 66%, ee = 67%

Scheme 2.167 Aziridination of chalcones with an aminimide derived from (+)-Tröger's base.

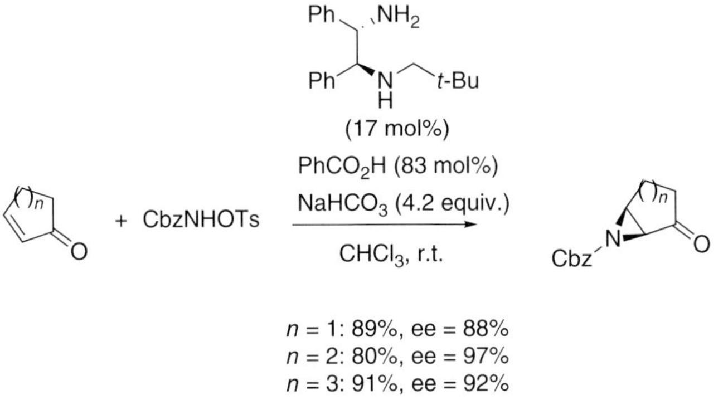

Scheme 2.168 Aziridination of cyclic enones with benzyloxyhydroxamate catalyzed by a chiral 1,2-diamine.

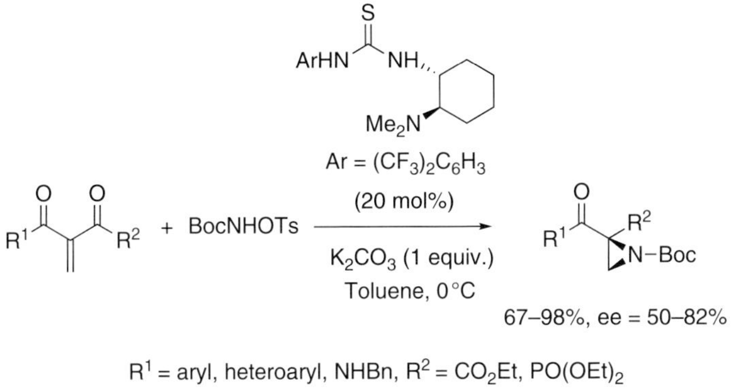

n = 1: 89%, ee = 88%
n = 2: 80%, ee = 97%
n = 3: 91%, ee = 92%

Scheme 2.169 Aziridination of α-acyl acrylates with *tert*-butyl tosyloxycarbamate catalyzed by a chiral thiourea.

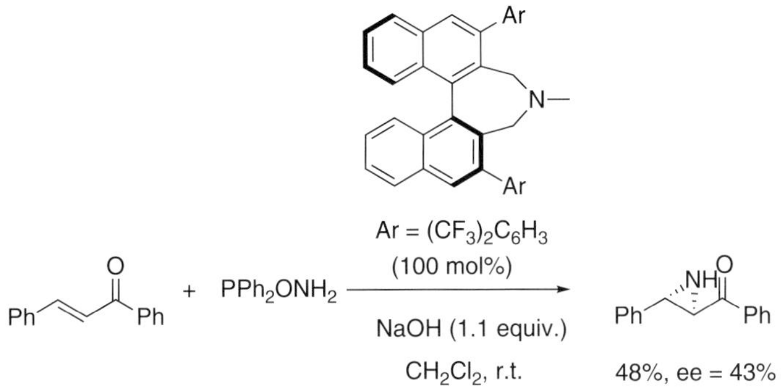

R^1 = aryl, heteroaryl, NHBn, R^2 = CO_2Et, $PO(OEt)_2$

67–98%, ee = 50–82%

Scheme 2.170 Aziridination of *trans*-chalcone with diphenylphosphinylhydroxylamine catalyzed by a binaphthalene-based chiral tertiary amine.

48%, ee = 43%

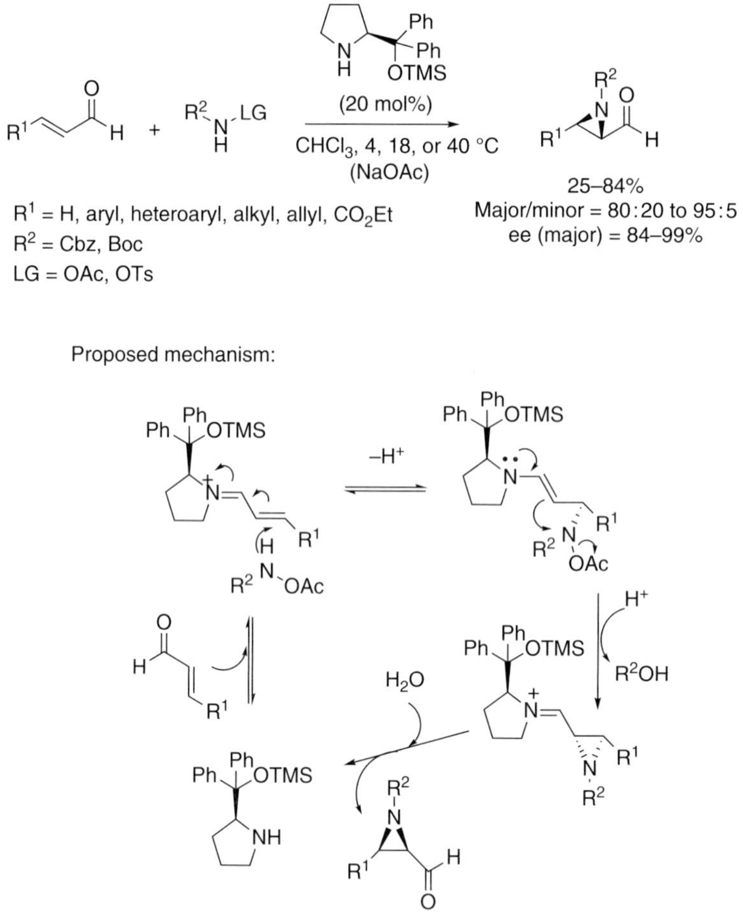

Scheme 2.171 Aziridination of α,β-unsaturated aldehydes with acylated hydroxycarbamates catalyzed by chiral diphenylprolinol trimethylsilyl ether.

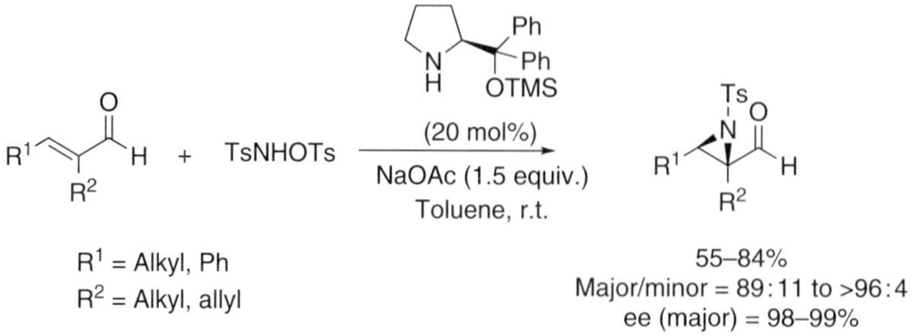

Scheme 2.172 Aziridinations of α,β-disubstituted-α,β-unsaturated aldehydes with TsNHOTs catalyzed by chiral diphenylprolinol trimethylsilyl ether.

as cyclohexane derivatives depicted in Scheme 2.175, with tosyloxy *tert*-butylcarbamate, using a remarkably low catalyst loading as low as 2.5 mol% [229]. The corresponding chiral aziridines were reached in high-to-quantitative yields

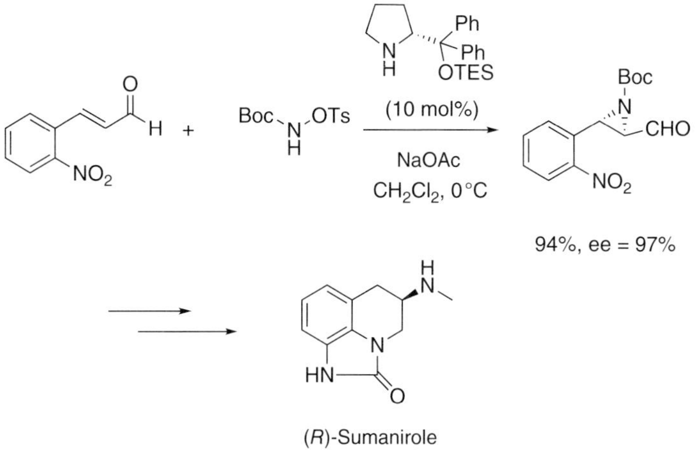

Scheme 2.173 Aziridination of 2-nitrocinnamaldehyde with *tert*-butyl tosyloxycarbamate catalyzed by chiral diphenylprolinol triethylsilyl ether and synthesis of (*R*)-sumanirole.

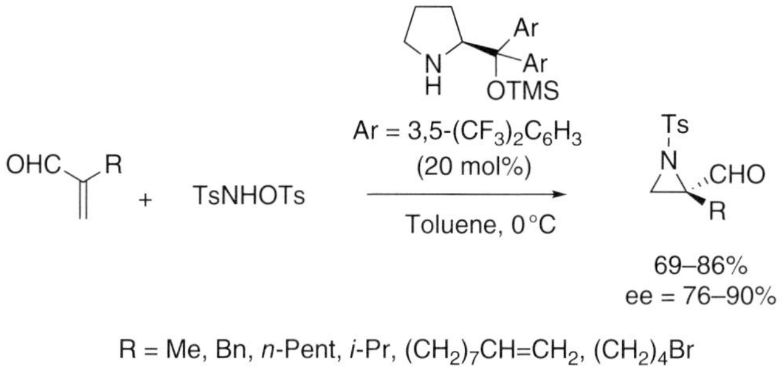

Scheme 2.174 Aziridination of α-branched α,β-unsaturated aldehydes with TsNHOTs catalyzed by a chiral diarylprolinol trimethylsilyl ether.

(85–99%), with moderate-to-high diastereoselectivities of up to 80% de, and with general high enantioselectivities of up to 97% ee.

In 2013, Jørgensen and coworkers reported a novel highly regio- and stereoselective remote aziridination of 2,4-dienals with *tert*-butyl tosyloxycarbamate catalyzed by a chiral diarylprolinol silyl ether, evolving through vinylogous iminium/dienamine catalysis [230]. As shown in Scheme 2.176, the remote aziridinations proceeded in good-to-high yields (53–86%) with excellent regio- and stereoselectivities of up to 95% ee. This reaction concept merged the LUMO-lowering strategy of the vinylogous iminium-ion and the HOMO-raising strategy of the dienamine intermediate in one combined process for the remote enantioselective aziridination of 2,4-dienals. The usefulness of these chiral products was illustrated by their transformation into optically active allylic γ-amino esters and oxazolidinones. It must be noted that a number of enantioselective domino and tandem reactions catalyzed by proline derivatives involving the formation of chiral aziridines as no isolated intermediates have been successfully developed in recent years but will not be detailed in this chapter [231].

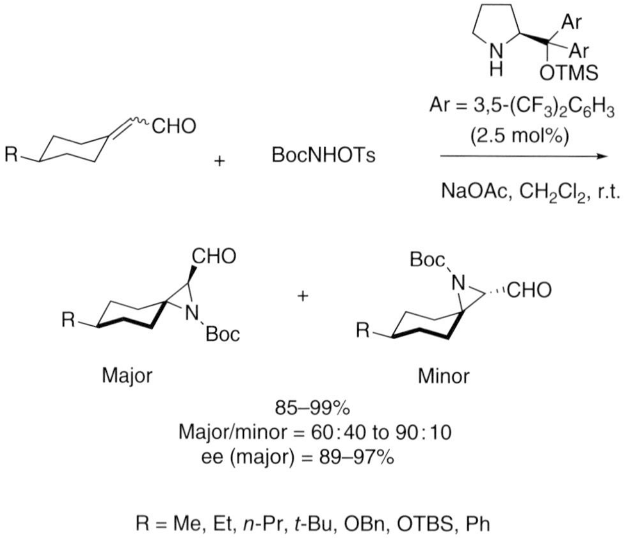

Scheme 2.175 Aziridination of rigid α,β-unsaturated aldehydes with *tert*-butyl tosyloxycarbamate catalyzed by a chiral diarylprolinol trimethylsilyl ether.

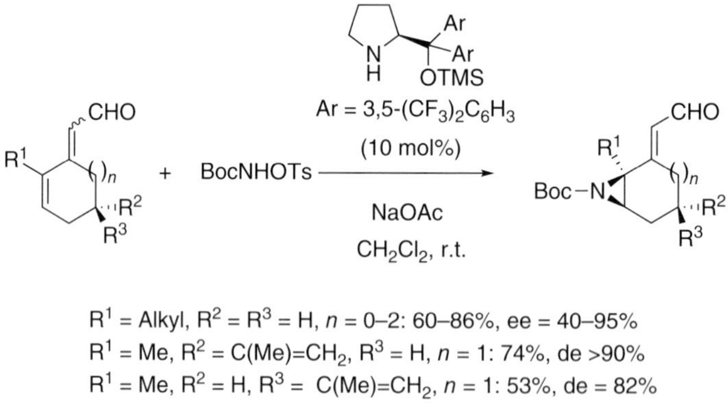

Scheme 2.176 Remote aziridination of 2,4-dienals with *tert*-butyl tosyloxycarbamate catalyzed by a chiral diarylprolinol trimethylsilyl ether.

In contrast with acyclic α,β-unsaturated aldehydes, cyclic enals have been much less studied as substrates in asymmetric aziridination. As a recent example, (S)-2-(fluorodiphenylmethyl)pyrrolidine was used by Gilmour et al. to catalyze the asymmetric aziridination of small, medium, and macro-cyclic enals, such as 5- to 8-, 10-, 12-, and 15-membered rings [232]. As shown in Scheme 2.177, chiral aziridines arisen from small and medium cyclic enals and BocNHOTs were obtained in high yields (78–93%) with remarkable diastereo- and enantioselectivities of >90% de and 91–99% ee, respectively. In the case of macrocyclic enals, the yields in the corresponding aziridines were generally lower (39–62%), while the diastereoselectivities were ranging from 60% to >90% de, along with a general

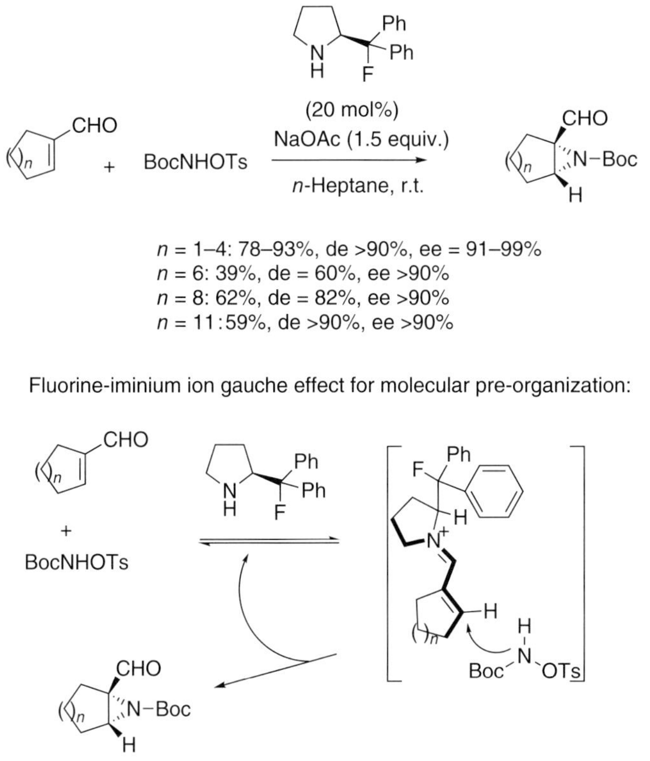

n = 1–4: 78–93%, de >90%, ee = 91–99%
n = 6: 39%, de = 60%, ee >90%
n = 8: 62%, de = 82%, ee >90%
n = 11: 59%, de >90%, ee >90%

Fluorine-iminium ion gauche effect for molecular pre-organization:

Scheme 2.177 Aziridination of cyclic α,β-unsaturated aldehydes with *tert*-butyl tosyloxycarbamate catalyzed by (S)-2-(fluorodiphenylmethyl)pyrrolidine.

excellent enantioselectivity of >90% ee. Central to the reaction design was the reversible formation of a β-fluoroiminium ion intermediate, which was preorganized on account of the fluorine-iminium ion gauche effect (Scheme 2.177). This conformational effect positioned the fluorine substituent *synclinal-endo* to the electropositive nitrogen center, thus benefiting from favorable stereoelectronic and electrostatic interactions. Consequently, one of the shielding groups on the fluorine-bearing carbon atom was positioned above the π-system, forming the basis of an enantioinduction strategy.

In another context, highly efficient enantioselective organocatalytic sequential one-pot reactions were reported by Jørgensen *et al.* in 2011, providing a highly divergent access to a variety of optically active three-, four-, and five-membered bicyclic *N*-heterocycles [233]. Among these, chiral aziridines were prepared through a sequential one-pot Michael addition/[3+2] cycloaddition reaction between α,β-unsaturated aldehydes, malononitrile derivatives, and *N*-aryl dihydroxylamines catalyzed by a chiral diarylprolinol silyl ether. As shown in Scheme 2.178, the scope of the process was broad since a variety of aliphatic α,β-unsaturated aldehydes formed the corresponding aziridines as single diastereomers in all cases of substrates studied. In the cases of unsubstituted alkynes, the products were obtained in good-to-excellent yields (50–89%) along with good-to-high

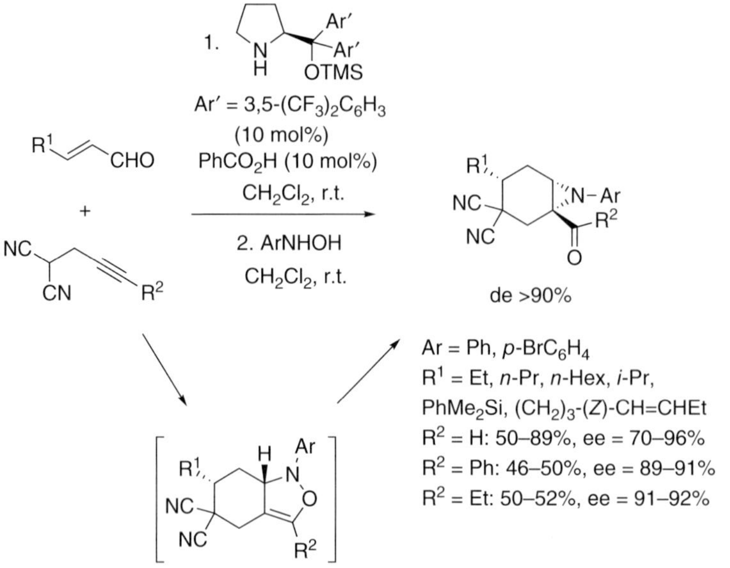

Scheme 2.178 Aziridination through a one-pot sequential Michael addition/[3+2] cycloaddition reaction catalyzed by a chiral diarylprolinol trimethylsilyl ether.

enantioselectivities ranging from 70% to 96% ee, while ethyl- or phenyl-substituted alkynes provided the analogous aziridine ketones in lower yields (46–52%) with high enantioselectivities of 89–92% ee, as shown in Scheme 2.178.

2.3.2 Aziridination via Carbene Transfer to Imines

Although most of the methods for obtaining chiral aziridines by using a chiral catalyst proceeded through the transfer of a nitrogen group to an olefin, a number of methods based on the less-studied enantioselective transfer of a carbenoid to an imine have been successfully developed in recent years [234].

2.3.2.1 Carbene Methodology

The formation of aziridines upon transition metal-catalyzed decomposition of diazo compounds in the presence of imines is well established. In particular, the reaction of ethyl diazoacetate with imines mediated by a Lewis acid is normally selective for the formation of the *cis*-aziridine. In 2003, Tilley et al. studied the enantioselective reaction of ethyl diazoacetate with N-aryl imines catalyzed by a novel chiral benzyl bisoxazoline complex of coordinatively unsaturated monomeric rhodium(II) [235]. This reaction proceeded selectively, giving a 75:25 ratio of the corresponding *cis*- and *trans*-aziridines, albeit with poor enantioselectivity (≤11% ee). On the other hand, the use of a chiral iron–pybox complex to catalyze these reactions was demonstrated by Redlich and Hossain to be more efficient [236]. Indeed, when $AgSbF_6$ was used as an initiator, the reaction of ethyl diazoacetate with phenyl imines afforded the corresponding *cis*-aziridines in enantioselectivity of up to 49% ee in the presence of the *tert*-butyl-pybox iron

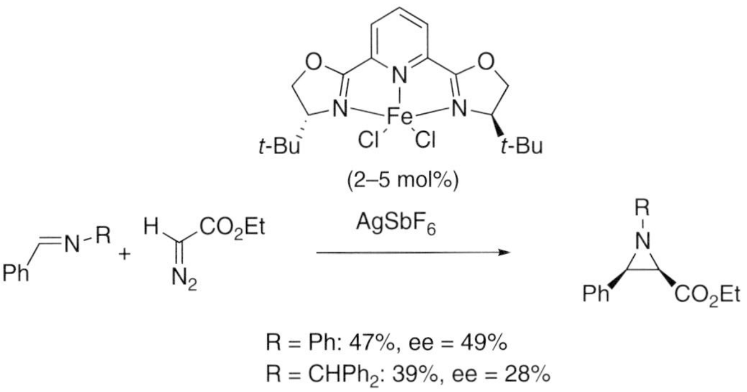

Scheme 2.179 Fe-catalyzed aziridination of phenyl imines with ethyl diazoacetate in the presence of a chiral *tert*-butyl-pybox.

complex depicted in Scheme 2.179. The role of the Ag^+ ion was assumed to create an open site on iron for coordination of the imine to the Lewis acid.

More recently, Mezzetti and coworkers reported a highly enantioselective ruthenium-catalyzed aziridination of aryl aldimines with ethyl diazoacetate, using (1*S*,2*S*)-*N*,*N*′-bis[*o*-(diphenylphosphino)-benzylidene]cyclohexane-1,2-diamine (PNNP) as chiral ligand (Scheme 2.180) [237]. In the presence of a catalytic amount of [RuCl(Et$_2$O)(PNNP)]SbF$_6$, the reaction provided the corresponding *cis*-aziridines in overall low-to-moderate yields (≤34%) albeit the enantioselectivities were unexpectedly good to complete. The authors have not proposed an explanation for the anion effect in particular on the enantioselectivity of the process.

The studies recently developed by Wulff's group based on the use of the vaulted chiral biaryl ligands, VANOL and VAPOL, are among the most successful contributions to date for the enantioselective aziridination of imines with ethyl diazoacetate [238]. Therefore, the asymmetric catalytic aziridination of

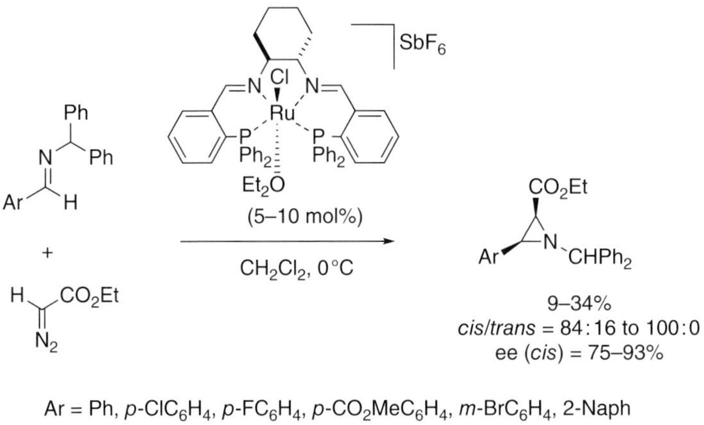

Scheme 2.180 Ru-catalyzed aziridination of aryl aldimines with ethyl diazoacetate in the presence of a PNNP chiral ligand.

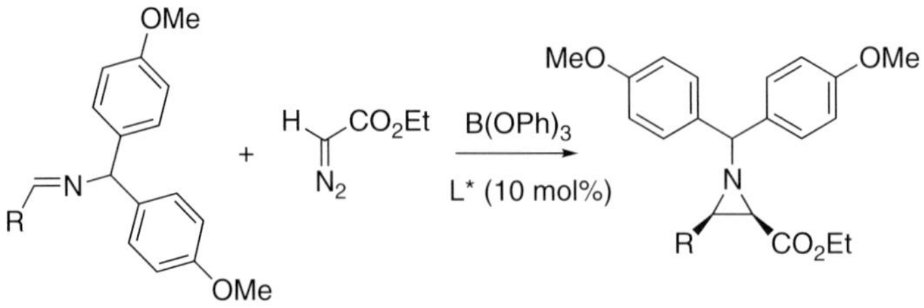

Scheme 2.181 B-catalyzed aziridination of N-dianisylmethylimines with ethyl diazoacetate in the presence of (S)-VANOL and (S)-VAPOL ligands.

with L* = VAPOL:
R = Cy: 86%, cis:trans = 97:3, ee = 84%
R = t-Bu: 70%, cis:trans = 98:2, ee = 75%
R = Ph: 92%, cis:trans = 97:3, ee = 95%
R = o-MeOC$_6$H$_4$: 86%, cis:trans = 98:2, ee = 89%
R = p-BrC$_6$H$_4$: 95%, cis:trans = 98:2, ee = 93%
R = p-NO$_2$C$_6$H$_4$: 97%, cis:trans = 98:2, ee = 97%
R = 1-Naph: 91%, cis:trans = 98:2, ee = 97%

with L* = VANOL:
R = Cy: 69%, cis:trans = 98:2, ee = 77%
R = t-Bu: 77%, cis:trans = 98:2, ee = 87%
R = Ph: 91%, cis:trans = 98:2, ee = 96%
R = p-MeOC$_6$H$_4$: 86%, cis:trans = 97:3, ee = 89%
R = p-BrC$_6$H$_4$: 89%, cis:trans = 97:3, ee = 97%
R = p-NO$_2$C$_6$H$_4$: 88%, cis:trans = 95:5, ee = 96%
R = 1-Naph: 92%, cis:trans = 98:2, ee = 94%

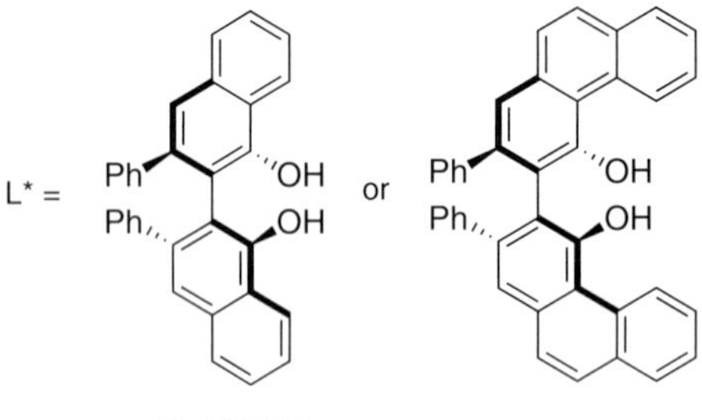

N-dianisylmethylimines with ethyl diazoacetate was developed with chiral catalysts prepared from triphenylborate and both the vaulted binaphthol (VANOL) and vaulted biphenanthrol (VAPOL) ligands [239]. These reactions produced the corresponding N-dianisylmethyl-protected 3-substituted aziridinyl-2-carboxylate esters in high yields and high asymmetric inductions of up to 97% ee combined with a high diastereoselectivity for the cis-aziridines. These two catalysts were both highly enantioselective, and the catalyst loading could be lowered to 0.25 mol% in some cases. Some results have been collected in Scheme 2.181.

In 2008, the scope of this methodology was extended to a wide range of N-benzhydryl imines, including electron-poor aromatic benzhydryl imines as

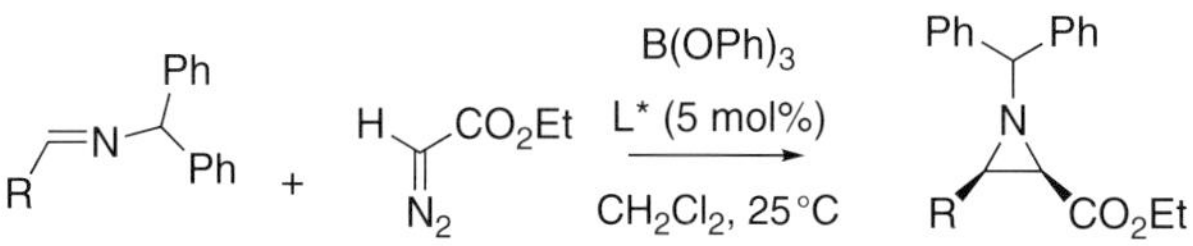

Scheme 2.182 B-catalyzed aziridination of N-benzhydryl imines with ethyl diazoacetate in the presence of (S)-VANOL and (S)-VAPOL ligands.

with L* = (S)-VAPOL:
R = 1-Naph: 76%, cis:trans = 97:3, ee = 93%
R = Ph: 82%, cis:trans = 98:2, ee = 94%
R = o-MeOC$_6$H$_4$: 63%, cis:trans = 91:9, ee = 91%
R = p-Tol: 80%, cis:trans = 98:2, ee = 92%
R = o-BrC$_6$H$_4$: 37%, cis:trans = 62:38, ee = 82%
R = p-BrC$_6$H$_4$: 78%, cis:trans = 95:5, ee = 90%
R = p-NO$_2$C$_6$H$_4$: 79%, cis:trans = 94:6, ee = 79%
R = p-MeOC$_6$H$_4$: 51%, cis:trans = 86:14, ee = 86%
R = 3,4-(OAc)$_2$C$_6$H$_3$: 87%, cis:trans = 100:0, ee = 89%
R = n-Pr: 40%, cis:trans = 93:7, ee = 81%
R = Cy: 73%, cis:trans = 98:2, ee = 81%
R = t-Bu: 72%, cis:trans = 100:0, ee = 87%

with L* = (S)-VANOL:
R = 1-Naph: 80%, cis:trans = 98:2, ee = 93%
R = Ph: 87%, cis:trans = 100:0, ee = 93%
R = o-MeOC$_6$H$_4$: 67%, cis:trans = 92:8, ee = 90%
R = p-Tol: 79%, cis:trans = 98:2, ee = 94%
R = o-BrC$_6$H$_4$: 43%, cis:trans = 66:34, ee = 82%
R = p-BrC$_6$H$_4$: 86%, cis:trans = 95:5, ee = 94%
R = p-NO$_2$C$_6$H$_4$: 86%, cis:trans = 100:0, ee = 89%
R = p-MeOC$_6$H$_4$: 61%, cis:trans = 97:3, ee = 87%
R = 3,4-(OAc)$_2$C$_6$H$_3$: 84%, cis:trans = 100:0, ee = 93%
R = n-Pr: 54%, cis:trans = 93:7, ee = 77%
R = Cy: 79%, cis:trans = 98:2, ee = 82%
R = t-Bu: 89%, cis:trans = 100:0, ee = 85%

well as primary, secondary, and tertiary aliphatic benzhydryl imines [240]. The use of only 5 mol% of catalysts prepared from B(OPh)$_3$ and either the VANOL or VAPOL ligand gave essentially the same profile of asymmetric inductions with enantioselectivities ranging from 77% to 94% for all the substrates (Scheme 2.182).

The synthetic utility of this methodology has been illustrated by the same authors in the asymmetric catalytic synthesis of a leukointegrin LFA-1 antagonist, BIRT-377, which has been developed as an agent for the treatment of inflammatory and immune disorders [241]. As shown in Scheme 2.183, the synthesis was achieved through a *cis*-aziridine, prepared by the reaction of ethyl diazoacetate with an N-benzhydryl imine in the presence of the (S)-VAPOL catalyst.

Later, the same authors investigated the asymmetric boron-catalyzed aziridination reaction of other imines derived from dianisylmethyl (DAM) amine, tetramethyldianisylmethyl (MEDAM) imines, and tetra-*tert*-butyldianisylmethyl (BUDAM) imines with ethyl diazoacetate induced by chiral VANOL and VAPOL ligands [242]. The MEDAM and DAM imines prepared from nine different aryl

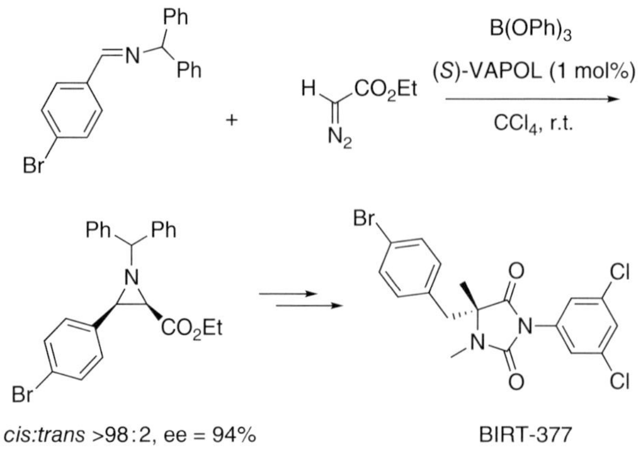

Scheme 2.183 Synthesis of BIRT-377.

87%, cis:trans >98:2, ee = 94%

and aliphatic aldehydes were examined in a first time. The MEDAM imines were superior to the DAM imines in all cases, giving the corresponding cis-aziridines as almost single diastereomers (cis/trans >97:3) in much higher asymmetric inductions of up to 99% ee and higher yields of up to 96%. The MEDAM imines were also found superior to the previously studied and less reactive diphenylmethyl(benzhydryl) imines [240] and to BUDAM imines especially for imines derived from aliphatic aldehydes. The best results obtained with MEDAM imines were reached by using (S)-VAPOL as chiral ligand, as shown in Scheme 2.184. It must be noted that using (S)-VANOL ligand gave closely related results with only slightly lower asymmetric inductions. The authors have elucidated the structure of the active catalyst of these reactions by X-ray diffraction analysis, which provided critical insights into the binding of the substrates with this catalyst [243]. This boroxinate is depicted in Scheme 2.184. In 2011, the authors demonstrated a practical scale-up of these methodologies to gram scale [244].

In 2013, the same authors proposed a stepwise mechanism to explain the origin of the enantio- and diastereoselectivities of these reactions on the basis of relative theoretical energies of the ring-closing transition structures [245]. The authors assumed that the mechanism involved the reversible formation of a diazonium ion intermediate (gauche **I**/*anti* **I**), followed by an irreversible S_N2-like ring closure, being the rate-limiting step, to form the final cis-aziridine with concomitant elimination of N_2 (Scheme 2.185).

In 2011, the highly efficient asymmetric Wulff's aziridination methodology was applied by Chen et al. to develop a total synthesis of antibacterial florfenicol, which is depicted in Scheme 2.186 [246]. In this case, the authors employed (R)-VANOL-boroxinate to aziridinate a benzhydryl aldimine with ethyl diazoacetate, providing the key chiral aziridine in 93% yield as a single cis-diastereomer in 85% ee. The latter was further converted into expected florfenicol in 45% overall yield from commercially available p-(methylsulfonyl)benzaldehyde.

While catalysts derived from linear biaryls such as BINOL gave very-poor-to-moderate enantioselectivity for the aziridination of benzylidene benzhydrylamines

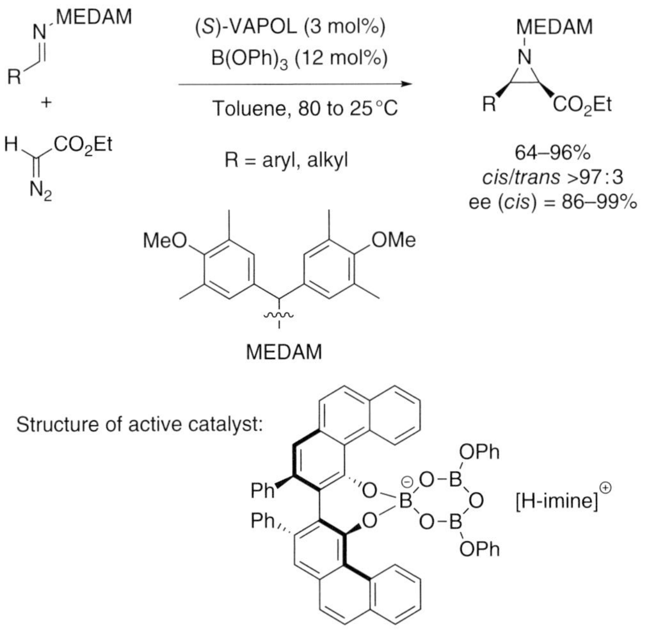

Scheme 2.184 B-catalyzed aziridination of MEDAM aldimines with ethyl diazoacetate in the presence of (S)-VAPOL ligand.

with ethyl diazoacetate [247], Wipf and Lyon have demonstrated that increasing the steric bulk at the 3,3′-positions of BINOL exerted an improved facial control in the reaction [248]. Therefore, these authors have studied a series of 3,3′-disubstituted BINOL derivatives as chiral ligands for the aziridination of benzhydryl imines with ethyl diazoacetate. It was demonstrated that the introduction of bulky arene substituents, in particular, the 2-phenylnaphthalene moiety, into the 3- and 3′-positions of the binaphthol scaffold led to a significant improvement in the level of chiral induction of up to 78% ee, as shown in Scheme 2.187.

Moreover, Wulff and coworkers extended the scope of their highly efficient asymmetric aziridination methodology to diazo compounds other than ethyl diazoacetate, such as diazomethyl vinyl ketones, which were reacted with a series of N-benzhydryl imines in the presence of the VAPOL catalyst [249]. As shown in Scheme 2.188, the corresponding *cis*-vinyl aziridinyl ketones were isolated in high yields and with high degrees of asymmetric induction of up to 100% ee.

Another diazo compound, such as N,N-methylbenzyl diazoacetamide, was shown to react with various aldimines derived from benzaldehyde [242a]. Different diarylmethyl N-substituents for the imines, such as diphenylmethyl, BUDAM, MEDAM, and MEDPM (tetramethyldiphenylmethyl), were investigated in the presence of (S)-VAPOL as borane ligand in toluene or CCl_4 as solvent. Best results were achieved in the case of MEDPM and MEDAM imines, which provided the corresponding *cis*-aziridines as single diastereomers in 66% and 77% yields, respectively, along with enantioselectivities of 97% and 98% ee, respectively (Scheme 2.189).

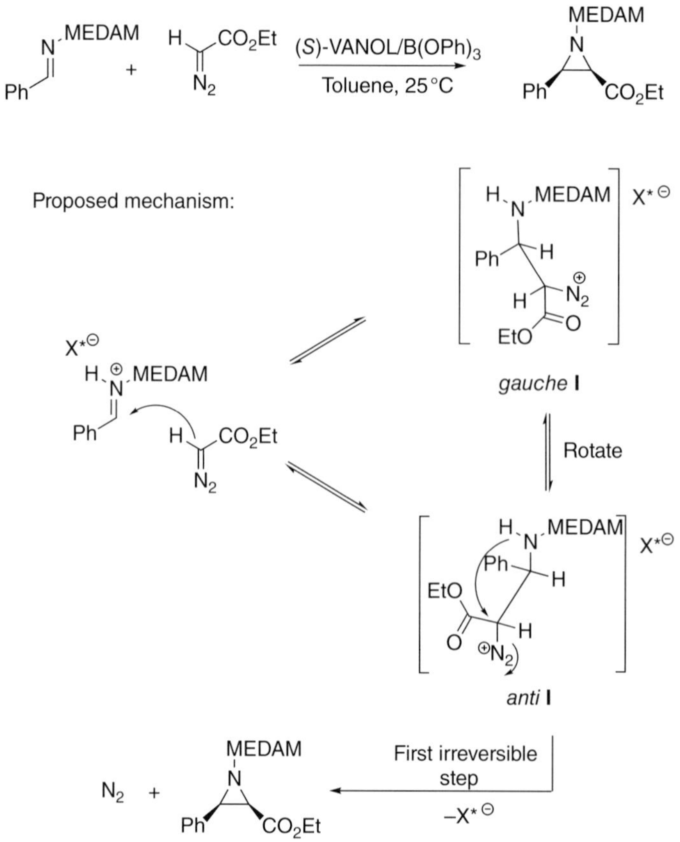

Scheme 2.185 Proposed mechanism for B-catalyzed aziridination of a MEDAM aldimine with ethyl diazoacetate in the presence of (S)-VANOL ligand.

Surprisingly, the same authors have found that using N-monosubstituted diazoacetamides as substrates with various aldimines afforded, when the reaction was catalyzed by (S)-VANOL-boroxinate, the corresponding *trans*-aziridines as major diastereomers (*trans/cis* = 75:25 to >98:2) in good-to-high yields (58–90%) with good-to-excellent enantioselectivities of up to 99% ee, as shown in Scheme 2.190 [250]. In order to explain the origin in the variations of the stereoselectivity of the aziridination reactions of different diazo nucleophiles, the authors have studied the stereochemistry-determining step of the aziridinations of MEDAM imines with three representative diazo nucleophiles, such as ethyl diazoacetate, N-phenyldiazoacetamide, and N,N-methylbenzyl diazoacetamide using ONIOM(B3LYP/6-31G*:AM1) calculations [251]. Thus, the origin of the *cis*-selectivity in the reactions of ethyl diazoacetate and the *trans*-selectivity in reactions of N-phenyldiazoacetamide could be understood on the basis of the difference in specific noncovalent interactions in the stereochemistry-determining transition state. In particular, an H-bonding interaction between the amidic hydrogen of the diazoacetamide and an oxygen atom of the chiral catalyst was

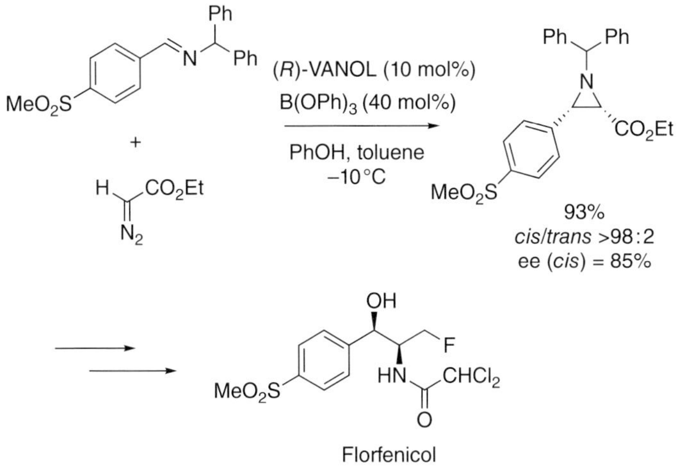

Florfenicol

Scheme 2.186 Synthesis of florfenicol through B-catalyzed aziridination of a benzhydryl aldimine with ethyl diazoacetate in the presence of (*R*)-VANOL ligand.

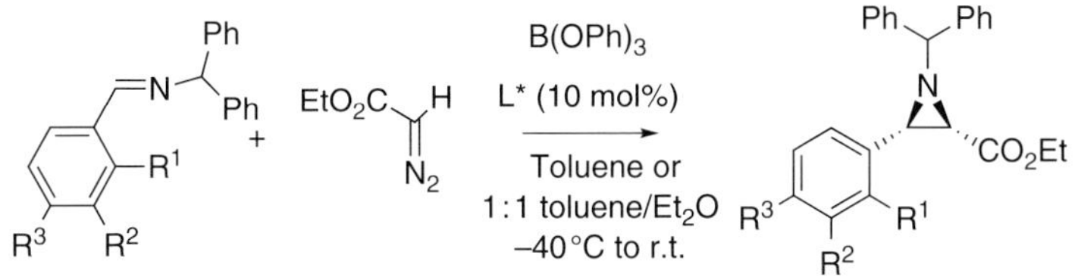

$R^1 = R^2 = R^3 = H$: 55%, *cis:trans* = 99 : 1, ee = 54%
$R^1 = R^2 = H$, $R^3 = Br$: 76%, *cis:trans* = 99 : 1, ee = 78%
$R^1 = R^2 = H$, $R^3 = NO_2$: 66%, *cis:trans* = 99 : 1, ee = 55%
$R^1 = H$, $R^2,R^3 = OCH_2O$: 57%, *cis:trans* = 99 : 1, ee = 55%
$R^1 = OSO_2\text{-}(o\text{-}NO_2)C_6H_4$, $R^2 = R^3 = H$: 39%, *cis:trans* = 60 : 40, ee = 55%

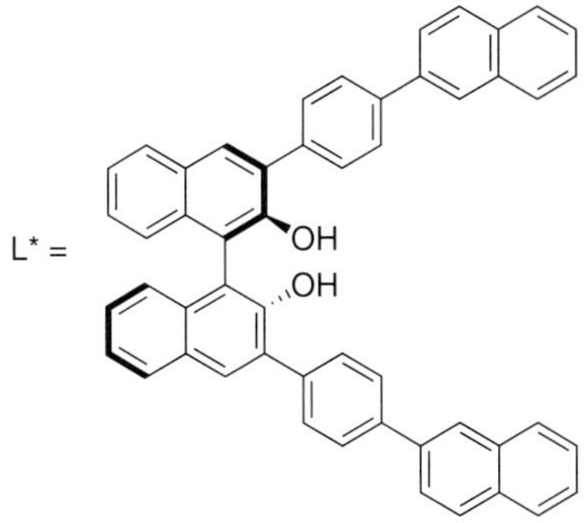

Scheme 2.187 B-catalyzed aziridination of *N*-benzhydryl imines with ethyl diazoacetate in the presence of a bulky BINOL-derived ligand.

Scheme 2.188 B-catalyzed aziridination of N-benzhydryl imines with diazomethyl vinyl ketones in the presence of (S)-VAPOL ligand.

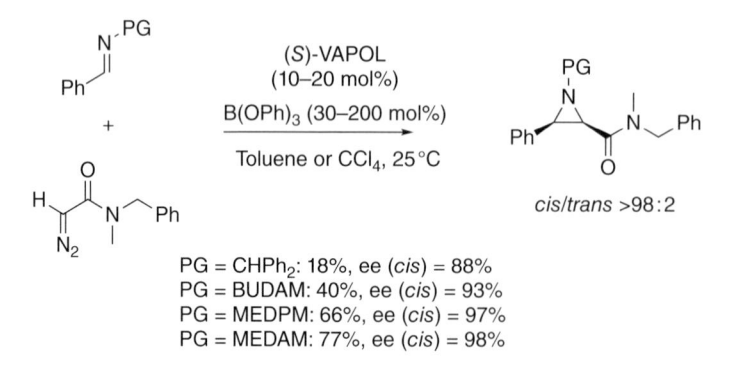

$R^1 = R^3 = H$, $R^2 = R^4 = Ph$: 79%, cis:trans >98:2, ee = 95%
$R^1 = R^3 = H$, $R^2 = Ph$, $R^4 = o$-Tol: 45%, cis:trans >98:2, ee = 91%
$R^1 = R^3 = H$, $R^2 = Ph$, $R^4 = o$-BrC$_6$H$_4$: 55%, cis:trans = 92:8, ee = 93%
$R^1 = R^3 = H$, $R^2 = Ph$, $R^4 = 2$-F-5-C$_6$H$_3$: 64%, cis:trans = 93:7, ee = 95%
$R^1 = R^3 = H$, $R^2 = Ph$, $R^4 = m$-NO$_2$C$_6$H$_4$: 78%, cis:trans >98:2, ee = 94%
$R^1 = R^3 = H$, $R^2 = Ph$, $R^4 = p$-Tol: 71%, cis:trans >98:2, ee = 100%
$R^1 = R^3 = H$, $R^2 = Ph$, $R^4 = p$-BrC$_6$H$_4$: 51%, cis:trans = 93:7, ee = 96%
$R^1 = R^3 = H$, $R^2 = Ph$, $R^4 = p$-NO$_2$C$_6$H$_4$: 80%, cis:trans >98:2, ee = 95%
$R^1 = R^3 = H$, $R^2 = Ph$, $R^4 = 2$-Naph: 84%, cis:trans >98:2, ee = 98%
$R^1 = R^3 = H$, $R^2 = Ph$, $R^4 = Cy$: 90%, cis:trans = 83:7, ee = 93%
$R^1 = H$, $R^2 = R^3 = Me$, $R^4 = Ph$: 76%, cis:trans = 94:6, ee = 98%
$R^1 = H$, $R^2 = R^3 = Me$, $R^4 = p$-Tol: 83%, cis:trans >98:2, ee = 96%
$R^1 = H$, $R^2 = R^3 = Me$, $R^4 = p$-BrC$_6$H$_4$: 76%, cis:trans = 95:5, ee = 98%
$R^1 = H$, $R^2 = R^3 = Me$, $R^4 = p$-NO$_2$C$_6$H$_4$: 67%, cis:trans = 87:13, ee = 96%
$R^1 = H$, $R^2 = R^3 = Me$, $R^4 = Cy$: 75%, cis:trans = 91:9, ee = 94%
$R^1 = R^3 = H$, $R^2 = Me$, $R^4 = Ph$: 85%, cis:trans = 96:4, ee = 96%
$R^1,R^2 = (CH_2)_4$, $R^3 = H$, $R^4 = Ph$: 40%, cis:trans = 86:14, ee = 82%

PG = CHPh$_2$: 18%, ee (cis) = 88%
PG = BUDAM: 40%, ee (cis) = 93%
PG = MEDPM: 66%, ee (cis) = 97%
PG = MEDAM: 77%, ee (cis) = 98%

Scheme 2.189 B-catalyzed aziridination of benzaldehyde-derived aldimines with N,N-methylbenzyl diazoacetamide in the presence of (S)-VAPOL ligand.

identified as the key interaction responsible for the reversal in the diastereoselectivity. This hypothesis was validated when the 3° diazoamide lacking this interaction showed pronounced cis-selectivity both experimentally and theoretically. Actually, the trans-diastereoselectivity of the process could be explained by the involvement of three strong H-bonding interactions between the catalyst and the two substrates, as shown in the transition state proposed in Scheme 2.190.

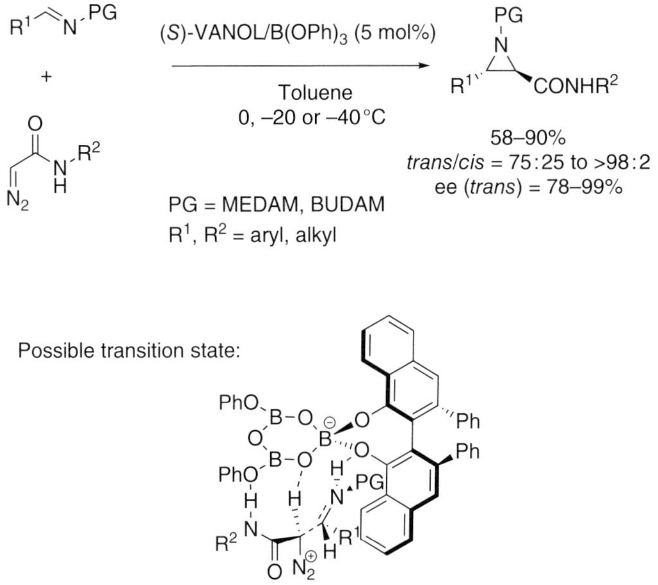

Scheme 2.190 B-catalyzed aziridination of aldimines with N-monosubstituted diazoacetamides in the presence of (S)-VANOL ligand.

The same authors reported a remarkable enantioselective synthesis of trisubstituted aziridines by using another type of diazocompounds, such as α-diazo-N-acyloxazolidinones [252]. As shown in Scheme 2.191, the reaction of N-Boc aldimines with these substrates provided, when catalyzed by (R)-VANOL-boroxinate, the corresponding chiral aziridines in moderate-to-good yields (30–85%), complete *trans*-diastereoselectivity in all cases of substrates studied, along with high-to-excellent enantioselectivities (83–98% ee).

Interestingly, these authors also investigated the influence of using chiral imines in combination with chiral VANOL or VAPOL ligands in boron-catalyzed aziridinations with ethyl diazoacetate [253]. They studied the reactions of chiral aldimines derived from five different chiral disubstituted methyl amines. The strongest matched and mismatched reactions with the two enantiomers of both catalysts (VANOL and VAPOL) were noted with disubstituted methyl amines that had one aromatic and one aliphatic substituent. As shown in Scheme 2.192, it was demonstrated that aldimines derived from (R)-α-methylbenzyl amine underwent the asymmetric aziridination reaction to give moderate-to-good yields of aziridines (35–93%) for which there was a strong matched case between the (R)-imine and (S)-VANOL-derived catalyst for *cis*-aziridines. Curiously, for *trans*-aziridines, the matched case involved the (R)-amine with the (R)-ligand (VAPOL) for aldimines derived from *ortho*-bromo and iodobenzaldehyde. As shown in Scheme 2.192, in the cases of (R)-imines derived from *ortho*-bromo and iodobenzaldehydes, mixtures of all four possible *trans*- and *cis*-diastereomers were achieved with moderate-to-good yields (56–60%), moderate *trans/cis* ratios of up to 74:26, along with good diastereoselectivities for the major *trans*-diastereomers of up to 86% de.

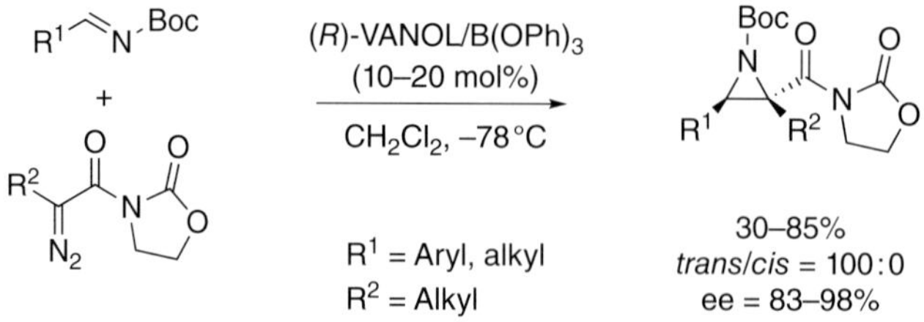

342 | Asymmetric Synthesis of Three-Membered Rings

Scheme 2.191 B-catalyzed aziridination of aldimines with α-diazo-N-acyloxazolidinones in the presence of (R)-VANOL ligand.

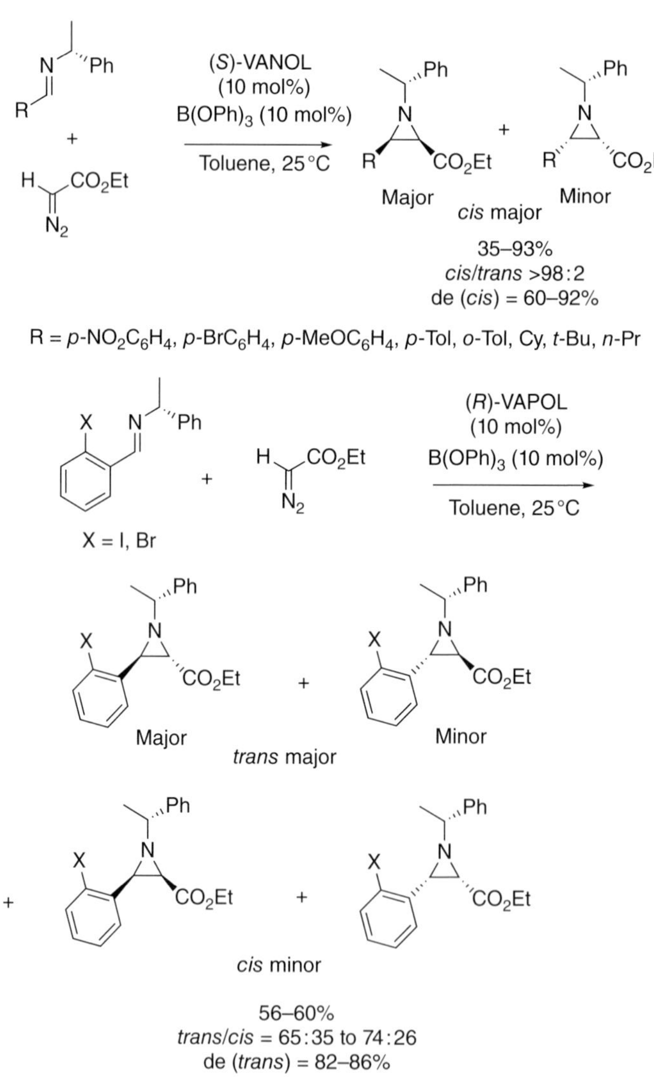

Conditions: (R)-VANOL/B(OPh)$_3$ (10–20 mol%), CH$_2$Cl$_2$, –78 °C

R^1 = Aryl, alkyl
R^2 = Alkyl

30–85%
trans/cis = 100:0
ee = 83–98%

Conditions: (S)-VANOL (10 mol%), B(OPh)$_3$ (10 mol%), Toluene, 25 °C

Major (cis major) + Minor

35–93%
cis/trans >98:2
de (cis) = 60–92%

R = p-NO$_2$C$_6$H$_4$, p-BrC$_6$H$_4$, p-MeOC$_6$H$_4$, p-Tol, o-Tol, Cy, t-Bu, n-Pr

---

X = I, Br

Conditions: (R)-VAPOL (10 mol%), B(OPh)$_3$ (10 mol%), Toluene, 25 °C

Major (trans major) + Minor

cis minor

56–60%
trans/cis = 65:35 to 74:26
de (trans) = 82–86%

Scheme 2.192 B-catalyzed aziridinations of chiral aldimines with ethyl diazoacetate in the presence of (S)-VANOL and (R)-VAPOL ligands.

In 2013, Maruoka and coworkers developed a novel chiral Brønsted acid catalyst, which was composed of two independent organic molecules, a chiral diol and 2-boronobenzoic acid [254]. The *in situ* formation of the corresponding chiral boronate ester was employed as a key process to generate an active chiral catalyst. This boronate-ester-assisted chiral carboxylic acid catalyst was successfully applied to the *trans*-aziridination of *N*-Boc and *N*-benzyl aldimines with *N*-phenyldiazoacetamide. As shown in Scheme 2.193, in both cases of imines, the corresponding *trans*-aziridines were achieved as almost single diastereomers (*trans*/*cis* >95:5) in good yields and with good-to-high enantioselectivities of up to 98% ee. It must be noted that this novel catalyst constituted the first one to achieve high enantioselectivities using *N*-benzyl imines.

Even though the history of multicomponent reactions dates back to the second half of nineteenth century with the reactions of Strecker, Hantzsch, and Biginelli, it was only in the last decades with the work of Ugi *et al.* that the concept of the multicomponent reaction has emerged as a powerful tool in synthetic chemistry [255]. This strategy, which consists of achieving multibond formation in one operation and in which the subsequent reactions result as a consequence of the functionality formed by bond formation or fragmentation in the previous step, is atom economical and avoids the necessity of protecting groups and isolation of intermediates [256]. The chemists have devoted more and more effort in the development of new and powerful strategies in these powerful one-pot reactions that avoid the use of costly and time-consuming protection–deprotection processes, as well as purification procedures of intermediates. Furthermore, they help lower the risk in the storage, transportation, and handling of toxic, unstable, or explosive intermediates. The use of one-pot multicomponent reactions [257] in organic synthesis is increasing constantly, since they allow the synthesis of a wide range of complex molecules, including natural products and biologically active compounds in an economically favorable way [258]. In 2010, Ren and Wulff reported a very interesting sequential highly diastereo- and enantioselective multicomponent reactions using (*R*)-VAPOL- or (*R*)-VANOL-boroxinate as chiral catalyst [259]. The sequence begun with the *in situ* formation of a functionalized diazomethyl ketone from the corresponding carboxylic acid via its *in situ* generated acyl chloride and trimethylsilyldiazomethane. The formed diazomethyl ketone subsequently underwent aziridination with protected imines in the presence of the catalytic system $B(OPh)_3$/(*R*)-VANOL or (*R*)-VAPOL ligand. The sequence afforded the corresponding *cis*-aziridines as single diastereoisomers (*cis*/*trans* >98:2) in all cases of substrates studied in good-to-excellent yields (58–95%) and with generally high-to-excellent enantioselectivities ranging from 87% to 99% ee, as shown in Scheme 2.194. This work demonstrated that the acylation of trimethylsilyldiazomethane with aliphatic acid chlorides can be coupled to the catalytic asymmetric aziridination of aldimines with a chiral polyborate Brønsted acid catalyst derived from the biaryl ligands VANOL and VAPOL.

Later, the same authors reported the first sequential multicomponent catalytic asymmetric aziridination reaction to give chiral aziridine-2-carboxylic esters with very high diastereo- and enantioselectivity from aromatic and aliphatic aldehydes [260]. This novel method pushed the boundary of the aziridination reaction to substrates that failed with preformed aldimines, such as unbranched

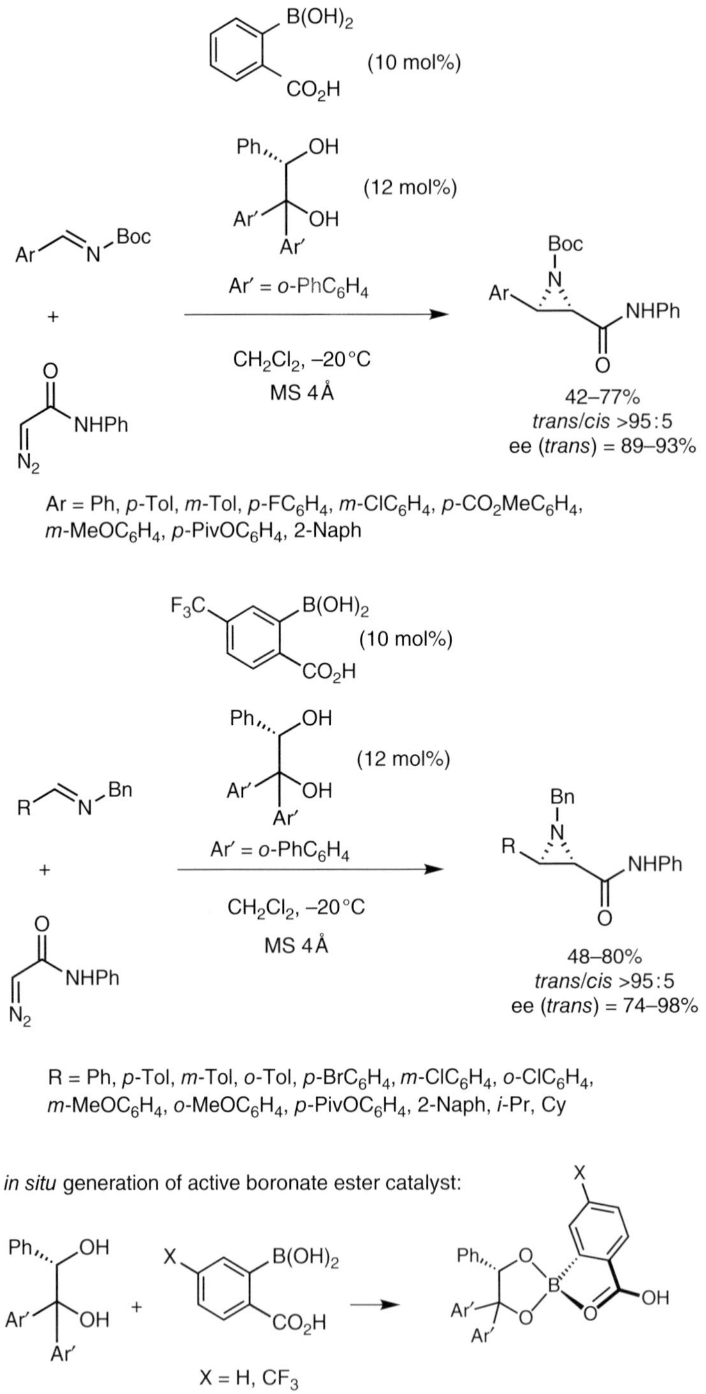

Scheme 2.193 Aziridinations of aldimines with N-phenyldiazoacetamide catalyzed by in situ generated chiral boronate esters.

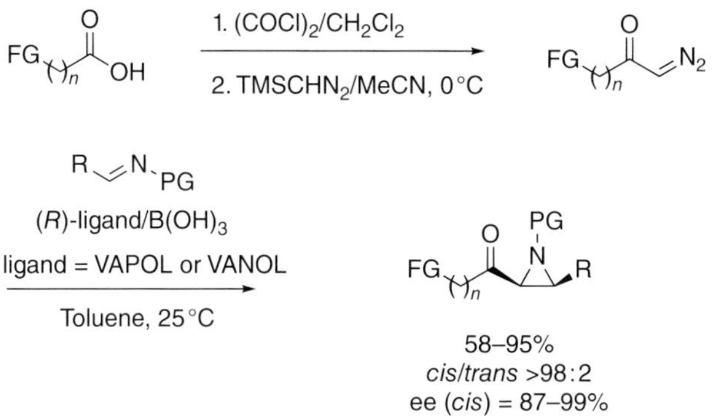

FG = ester, alkene, alkyne, halide, imide, ketal
PG = MEDAM, $CHPh_2$
R = Ph, Cy
n = 2–3

Scheme 2.194 Sequential multicomponent reaction including a B-catalyzed aziridination of aldimines with *in situ* generated diazomethyl ketones in the presence of (*R*)-VANOL/(*R*)-VAPOL ligand.

aliphatic aldehydes. As shown in Scheme 2.195, in a first time, the VAPOL ligand was mixed with 3 equiv. of $B(OPh)_3$ and 20 equiv. of the primary amine and stirred in toluene to ensure the complete formation of the boroxinate catalyst. This was followed by the addition of 4 Å MS and the aldehyde. After subsequent addition of ethyl diazoacetate, the resulting mixture was stirred to give the final *cis*-aziridine as single diastereomer (*cis/trans* >98 : 2) in good-to-high yield of up to 97% and with enantioselectivity of up to 98% ee. It must be noted that very similar results were obtained if the diazo compound was added before the aldehyde, which revealed that this was a true multicomponent reaction since imine formation could occur in the presence of all other compounds. Generally excellent asymmetric inductions were observed for the aziridination of a number of functionalized unbranched, α-branched, as well as α,α-branched aliphatic aldehydes, including problematical aldehydes that have failed to give aziridines via *in*

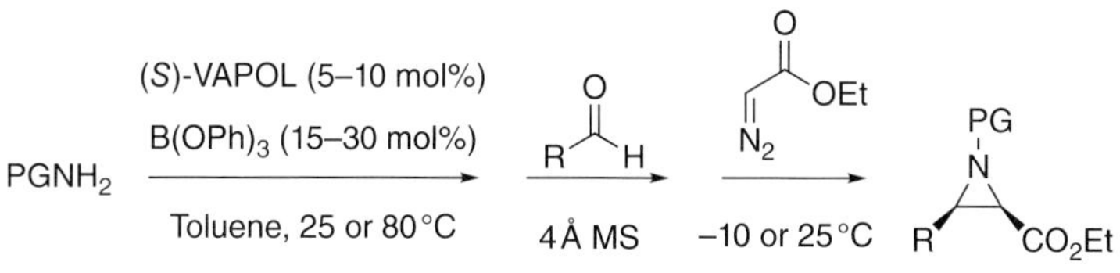

with R = *t*-Bu, Cy, Bn, CH_2Bn, $(CH_2)_2Bn$, $(CH_2)_3CO_2Et$, $(CH_2)_7OTBS$, $(CH_2)_{14}Me$ and PG = MEDAM: 50–97%, *cis/trans* >98 : 2, ee (*cis*) = 90–98%
with R = Ph, PG = MEDAM or $CHPh_2$: 70–92%, *cis/trans* >98 : 2, ee (*cis*) = 92–98%

Scheme 2.195 Sequential multicomponent reaction including a B-catalyzed aziridination of *in situ* generated aldimines with ethyl diazoacetate in the presence of (*S*)-VAPOL ligand.

situ generated imines, such as dihydrocinnamaldehyde, phenyl acetaldehyde, and ethyl 5-oxopentanoate. Later, the scope of this methodology was extended to an aromatic aldehyde, such as benzaldehyde, which also provided the corresponding diastereopure *cis*-aziridines in comparable good-to-high yields (70–92%) and with excellent enantioselectivities of up to 98% ee, as shown in Scheme 2.195 [261].

In 2014, the authors demonstrated the utility of this remarkable sequential multicomponent reaction by its application to an asymmetric synthesis of all four stereoisomers of sphinganine, an important biologically active sphingolipid involved in cell structure and regulation [262]. As shown in Scheme 2.196, the syntheses all started from hexadecanal, MEDAM amine, and ethyl diazoacetate as the three substrates. Using either (R)-VANOL- or (S)-VAPOL-boroxinate catalyst, the multicomponent reaction of hexadecanal with MEDAM amine and ethyl diazoacetate gave rise to almost enantiopure aziridine-2-carboxylates. Access to all four stereoisomers of sphinganine was achieved upon ring opening of the enantiopure aziridine-2-carboxylate at the C-3 position through direct S_N2 attack of an oxygen nucleophile, which occurred with inversion of configuration and by ring expansion of an *N*-acyl aziridine to an oxazolidinone followed by

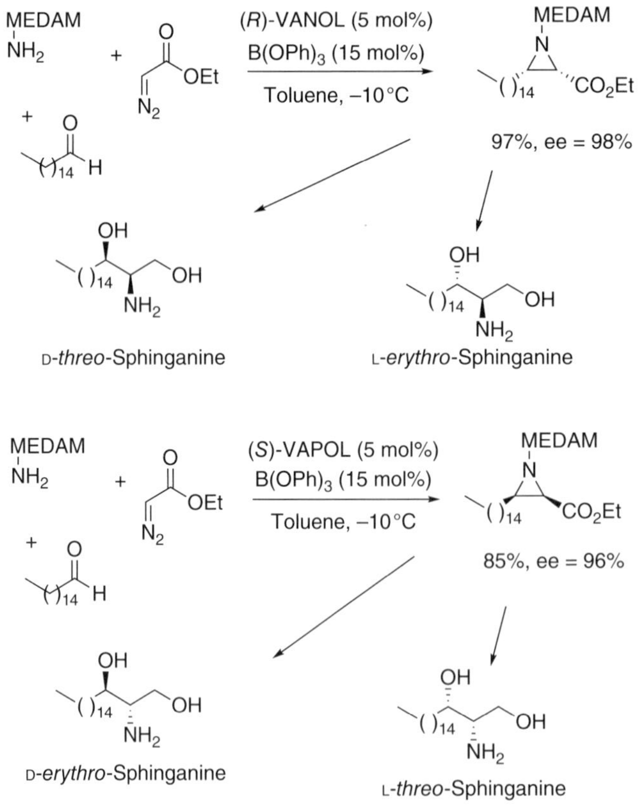

Scheme 2.196 Synthesis of four stereoisomers of sphinganine.

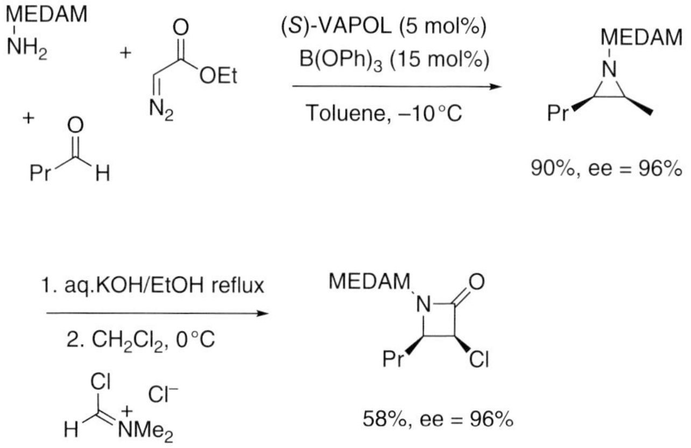

Scheme 2.197 Synthesis of chiral β-lactams.

hydrolysis. Overall, this process resulted in the formal ring opening of the aziridine with an oxygen nucleophile with retention of configuration.

The scope of this methodology was also recently included in an asymmetric synthesis of β-lactams on the basis of a cascade three-step reaction [263]. As shown in Scheme 2.197, the intermediate *cis*-aziridines formed from the multicomponent reaction between *n*-butanal, MEDAM amine, and ethyl diazoacetate was obtained in 90% isolated yield and 96% ee without purification. The crude was directly hydrolyzed into the corresponding carboxylic acid by treatment with aqueous KOH in ethanol. This crude product was finally submitted to ring expansion by treatment with the Vilsmeier reagent to give the corresponding final β-lactam, which was isolated in 58% overall yield with excellent enantioselectivity of 96% ee, as shown in Scheme 2.197.

For a long time, it was not known that organocatalysts could be used to catalyze aziridination reactions [211]. In recent years, however, a range of examples of organocatalytic enantioselective aziridination of olefins have been successfully developed as demonstrated in Section 2.3.1.5, and these have become a topic of interest in asymmetric organocatalysis. In contrast, much less examples of enantioselective organocatalytic aziridinations of imines have been reported so far. Among these, an early highly efficient *trans*-selective asymmetric aziridination of diazoacetamides with *N*-Boc imines was successfully achieved in 2008 by Maruoka et al. [264]. In contrast to all the preceding aziridinations, which are *cis*-selective, this *trans*-selective asymmetric aziridination of diazoacetamides with *N*-Boc imines was organocatalyzed by an axially chiral dicarboxylic acid. Screening of the reaction between benzaldehyde *N*-Boc imine and *N*-phenyldiazoacetamide using various axially chiral dicarboxylic acids having 3,3′-diaryl substituents led to the identification of the 3,3′-dimesityl-substituted dicarboxylic acid depicted in Scheme 2.198 as the optimal catalyst, providing the corresponding *trans*-aziridine exclusively in good yield and with high enantioselectivity (97% ee). The scope of this *trans*-selective aziridination was further extended to a wide range of substrates with generally excellent enantioselectivi-

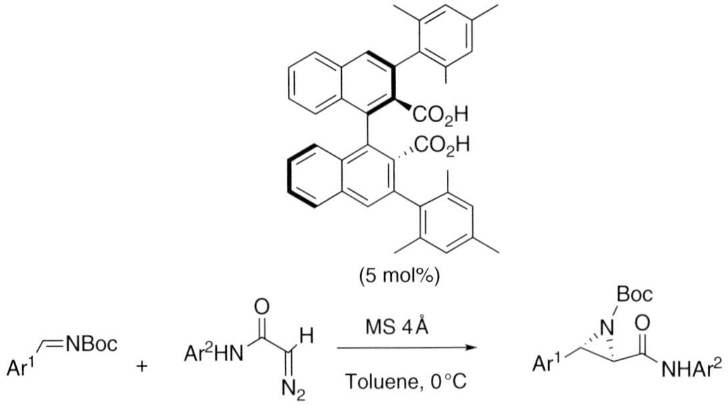

348 | Asymmetric Synthesis of Three-Membered Rings

Ar^1 = Ar^2 = Ph: 61%, ee = 97%
Ar^1 = p-Tol, Ar^2 = Ph: 51%, ee = 99%
Ar^1 = m-Tol, Ar^2 = Ph: 52%, ee = 98%
Ar^1 = 2-Naph, Ar^2 = Ph: 66%, ee = 99%
Ar^1 = p-PivOC$_6$H$_4$, Ar^2 = Ph: 49%, ee = 96%
Ar^1 = m-MeOC$_6$H$_4$, Ar^2 = Ph: 55%, ee = 96%
Ar^1 = p-FC$_6$H$_4$, Ar^2 = Ph: 31%, ee = 89%
Ar^1 = m-ClC$_6$H$_4$, Ar^2 = Ph: 50%, ee = 91%
Ar^1 = Ph, Ar^2 = p-MeOC$_6$H$_4$: 61%, ee = 97%
Ar^1 = 2-Naph, Ar^2 = p-MeOC$_6$H$_4$: 71%, ee = 99%
Ar^1 = p-PivOC$_6$H$_4$, Ar^2 = p-MeOC$_6$H$_4$: 57%, ee = 97%
Ar^1 = Ph, Ar^2 = p-ClC$_6$H$_4$: 60%, ee = 97%
Ar^1 = 2-Naph, Ar^2 = p-ClC$_6$H$_4$: 70%, ee = 99%

Proposed transition state:

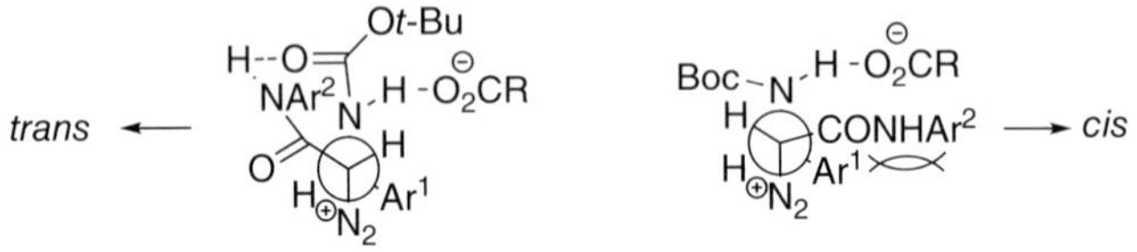

Scheme 2.198 Organocatalyzed *trans*-aziridination of N-Boc aryl aldimines with diazoacetamides.

ties (89–99% ee), good yields, and complete *trans* selectivity, as shown in Scheme 2.198. In order to explain the *trans* selectivity, the authors have speculated that it arose from the preference of a rotamer in which the carboxamide group and the aryl group of the N-Boc imine adopted an antiperiplanar orientation. Synclinal orientation could be destabilized by the steric repulsion (Scheme 2.198). The hydrogen bonding between the amide N—H bond and the Boc group could act as a secondary factor.

Later, the same authors established a remarkable and general procedure for the enantioselective synthesis of trisubstituted aziridines bearing three different

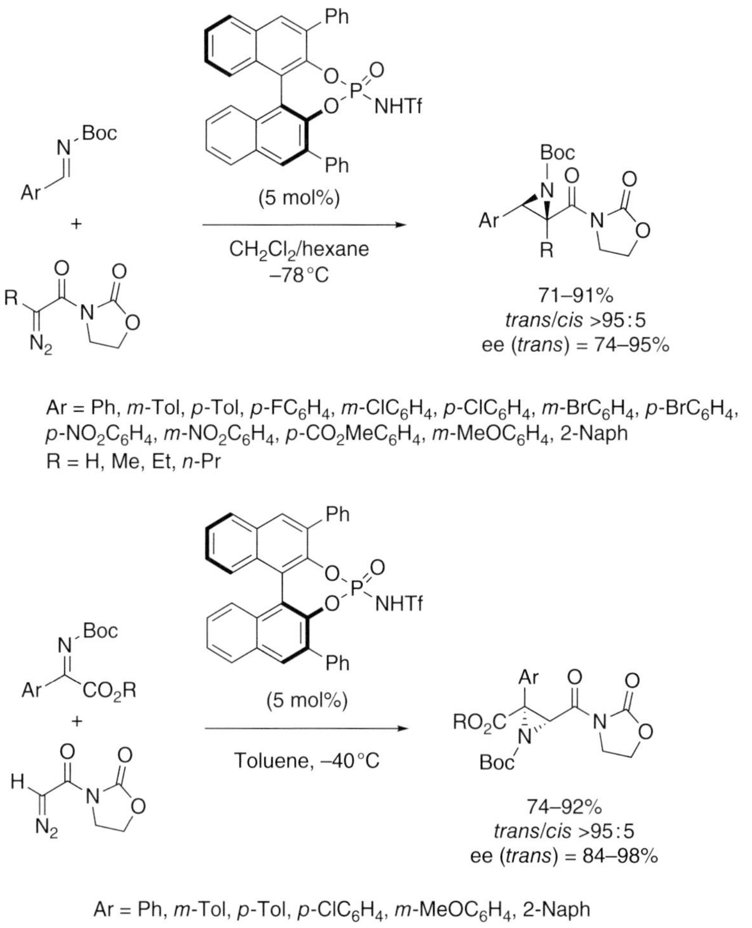

Scheme 2.199 Organocatalyzed aziridinations of aryl aldimines with α-substituted α-diazocarbonyl compounds and of aryl ketimines with a α-unsubstituted α-diazocarbonyl compound.

substituents, which had rarely been accessible directly by any kind of catalytic asymmetric transformation [265]. Catalyzed by a chiral N-triflyl phosphoramide, a range of α-substituted α-diazocarbonyl compounds reacted with various aryl aldimines to provide the corresponding chiral *trans*-aziridines in good-to-high yields (71–91%), with generally excellent diastereoselectivity of >90% de, along with high enantioselectivities ranging from 74% to 95% ee, as shown in Scheme 2.199. Applying related conditions to the reaction of an α-unsubstituted α-diazocarbonyl compound with various ketimines afforded comparable excellent results since the corresponding *trans*-aziridines were reached in even higher yields (74–92%), with the same general diastereoselectivity of >90% de, and even with higher enantioselectivities of 84–98% ee, as shown in Scheme 2.199.

In 2012, Cahard *et al.* reported a remarkable multicomponent organocatalyzed highly diastereo- and enantioselective synthesis of CF_3-substituted aziridines

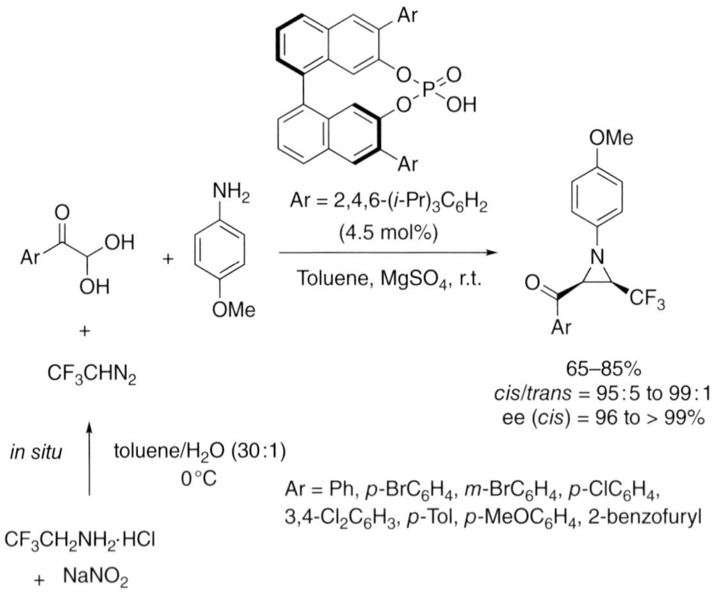

Scheme 2.200 Organocatalyzed multicomponent reaction of *in situ* generated aldimines with *in situ* generated CF$_3$CHN$_2$.

[266]. This reaction involved the *in situ* generation of CF$_3$CHN$_2$ and also that of aryl aldimines from the corresponding aryl glyoxal monohydrates and *p*-anisidine in the presence of a chiral phosphoric acid as organocatalyst. It provided the corresponding *cis*-aziridines in good-to-high yields (65–85%) as almost single stereoisomers since the diastereoselectivities and enantioselectivities were ranging from 90% to 98% de and 96% to >99% ee, respectively, as shown in Scheme 2.200. It must be noted that this novel methodology constituted the first catalytic asymmetric synthesis of CF$_3$-substituted aziridines. It was remarkable by its simplicity, low catalyst loading, and very high enantioselectivity. The utility of these chiral products was illustrated by easy access to chiral β-CF$_3$ isocysteine and aziridine-containing dipeptides.

2.3.2.2 Sulfur-Ylide-Mediated Aziridination

As an extension of his methodology dealing with epoxide synthesis, Aggarwal developed an aziridination procedure based on asymmetric carbene transfer via a chiral *in situ* generated sulfonium ylide [267]. This procedure consists of the generation of a carbene via a diazo decomposition with [Rh$_2$(OAc)$_4$], its association to a chiral sulfide, and subsequent transfer to an appropriate imine [268]. This highly efficient sulfur ylide methodology has been used to construct the taxol side chain with a high degree of enantioselectivity via a *trans*-aziridine [64]. Therefore, the reaction of the *N*-Ses imine depicted in Scheme 2.201 with the tosylhydrazone salt derived from benzaldehyde in the presence of a phase-transfer catalyst (PTC), Rh$_2$(OAc)$_4$, and catalytic quantities (20 mol%) of a chiral sulfide led to the corresponding aziridine in good yield and with an 89:11

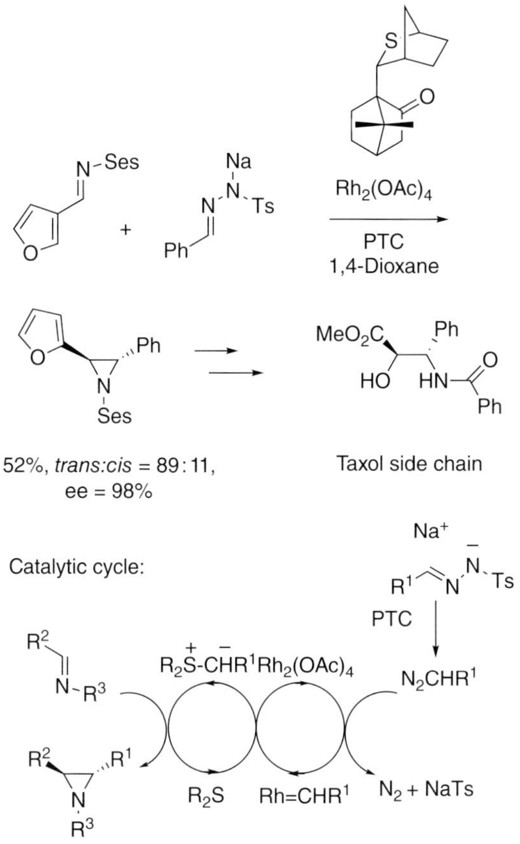

Scheme 2.201 Synthesis of taxol side chain.

trans/cis diastereoisomeric ratio (Scheme 2.201). The expected *trans*-aziridine was obtained with an enantiomeric excess of 98% ee and was further converted into the desired final taxol side chain. A catalytic cycle was proposed involving the decomposition of the diazo compound in the presence of the rhodium complex to yield the metallocarbene. This was then transferred to the chiral sulfide, forming a sulfur ylide, which underwent a reaction with the imine to give the expected aziridine, returning the sulfide to the cycle to make it available for further catalysis (Scheme 2.201).

In 2007, Huang et al. reported the synthesis of a novel chiral C_2-symmetric sulfide from L-tartaric acid and its application to an asymmetric tandem reaction between benzyl bromide and *N*-tosyl imines to form the corresponding aziridines [269]. In this case, the sulfide reacted with benzyl bromide in the presence of potassium carbonate to generate the corresponding chiral sulfonium ylide *in situ*, which then reacted with an *N*-tosyl imine to yield the aziridine. In the presence of a catalytic amount of the sulfide (0.5 equiv.), the aziridine was isolated in moderate yield as a dominant *trans*-isomer in high enantioselectivity, as shown

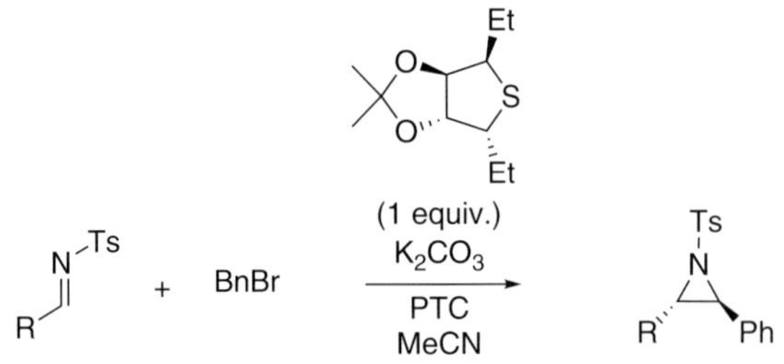

Scheme 2.202 Sulfur-ylide-mediated aziridination of N-tosyl imines with benzyl bromide.

with 1 equiv. of sulfide:
R = Ph: 72%, trans:cis = 75:25, ee = 96%
R = p-ClC$_6$H$_4$: 62%, trans:cis = 80:20, ee = 87%
R = p-FC$_6$H$_4$: 65%, trans:cis = 75:25, ee = 93%
R = p-NO$_2$C$_6$H$_4$: 50%, trans:cis = 65:35, ee = 85%
R = p-MeOC$_6$H$_4$: 60%, trans:cis = 60:40 ee = 80%
R = p-Tol: 68%, trans:cis = 80:20, ee = 91%
R = PhCH=CH: 75%, trans:cis = 90:10, ee = 90%

with 50 mol% of sulfide:
R = Ph: 43%, trans:cis = 75:25, ee = 87%

in Scheme 2.202. Increasing the amount of sulfide from 0.5 to 1 equiv. allowed both better yields and enantioselectivities of up to 96% ee.

The downside of this methodology was that a stoichiometric amount of the chiral sulfide was required to achieve high enantioselectivities. In the context of finding a more economical methodology, Chein et al. very recently developed related aziridinations using only 20 mol% of chiral (thiolan-2-yl)diphenylmethanol benzyl ether as organocatalyst [270]. As shown in Scheme 2.203, a range of N-diphenylphosphinic aryl aldimines could be converted into the corresponding

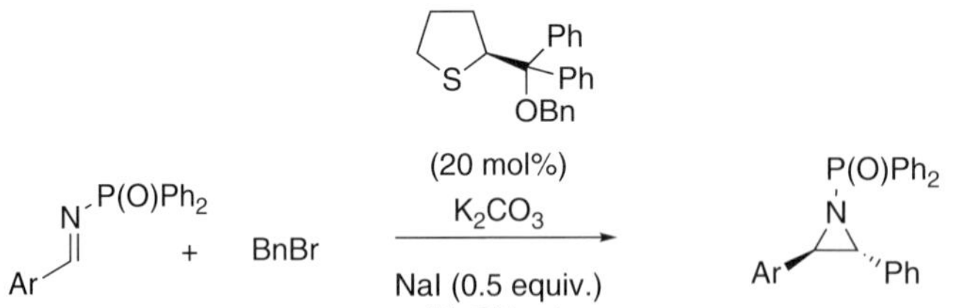

73–93%
trans/cis = 54:46 to 83:17
ee (trans) = 95–98%

Ar = Ph, p-CF$_3$C$_6$H$_4$, p-ClC$_6$H$_4$, p-FC$_6$H$_4$, p-MeOC$_6$H$_4$, m-MeOC$_6$H$_4$, o-MeOC$_6$H$_4$, p-CNC$_6$H$_4$, p-Tol, m-Tol, o-Tol, p-PhC$_6$H$_4$, o-BrC$_6$H$_4$, 3,4-MeO$_2$C$_6$H$_3$, 2-Naph

Scheme 2.203 Organocatalyzed sulfur-ylide-mediated aziridination of N-diphenylphosphinic aryl aldimines with benzyl bromide.

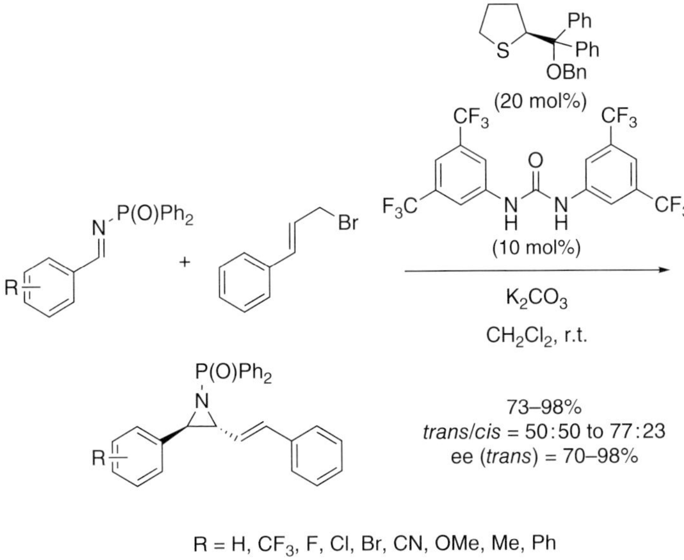

Scheme 2.204 Organocatalyzed sulfur-ylide-mediated aziridination of N-diphenylphosphinic aryl aldimines with cinnamyl bromide.

major *trans*-aziridines in high yields and with general excellent enantioselectivities ranging from 95% to 98% ee. These chiral products could be further transformed into chiral β-amino alcohols, making this novel methodology the most efficient route to this important class of products.

The same organocatalyst was later applied by these authors to induce the asymmetric aziridination of variously substituted *N*-diphenylphosphinic aryl aldimines with cinnamyl bromide [271]. As shown in Scheme 2.204, the reaction required the use of urea as cocatalyst to afford a range of *trans*-aryl cinnamyl aziridines in good-to-excellent yields (73–98%), moderate diastereoselectivities of up to 54% de, along with moderate-to-excellent enantioselectivities of 70–98% ee.

2.3.3 Miscellaneous Reactions

In 2003, Timén and Somfai developed the first catalytic enantioselective aza-Diels–Alder reactions involving azirines as dienophiles [272]. In this study, a wide range of Lewis-acidic metals and chiral ligands were screened together with benzyl-2*H*-azirine-3-carboxylate as dienophile and cyclopentadiene. Of all the Lewis acids screened, AlMe$_3$, together with, especially, oxygen-containing ligands, such as BINOL, TADDOL, or bisoxazoline, but also with nitrogen-containing ligands, such as bis-sulfonamides, proved to be the most successful, providing the corresponding aziridine with moderate enantioselectivity of up to 52% ee (Scheme 2.205).

In 2004, Kim and Jacobsen developed a new and practical route to highly enantioenriched aziridines from racemic epoxides, which was actually constituted by three successive reactions, with the intermediate products isolated by simple fil-

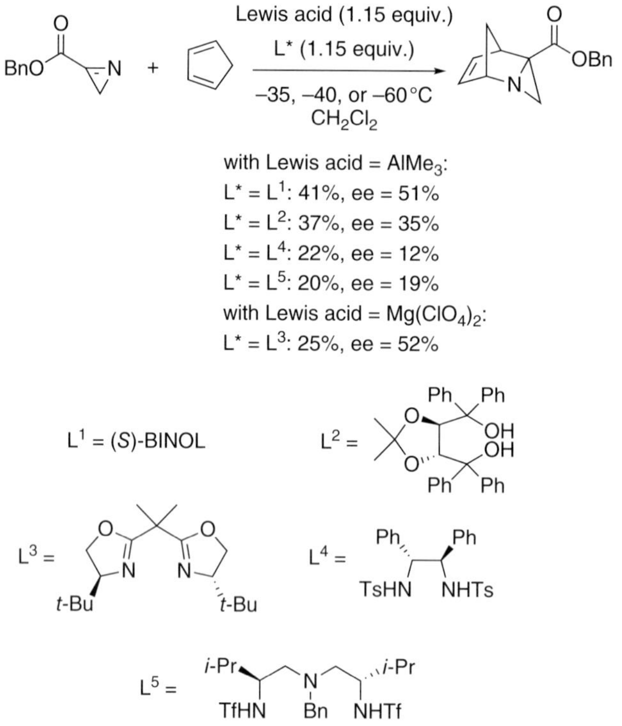

Scheme 2.205

Lewis acid (1.15 equiv.)
L* (1.15 equiv.)

BnO–C(O)–N=CH–... + cyclopentadiene →(−35, −40, or −60 °C, CH$_2$Cl$_2$) bicyclic aziridine-CO$_2$Bn adduct

with Lewis acid = AlMe$_3$:
L* = L^1: 41%, ee = 51%
L* = L^2: 37%, ee = 35%
L* = L^4: 22%, ee = 12%
L* = L^5: 20%, ee = 19%

with Lewis acid = Mg(ClO$_4$)$_2$:
L* = L^3: 25%, ee = 52%

L^1 = (S)-BINOL

L^2 = TADDOL-type (Ph, Ph, OH, OH, Ph, Ph)

L^3 = bis(oxazoline) with t-Bu, t-Bu

L^4 = (Ph, Ph, TsHN, NHTs) diamine

L^5 = (i-Pr, i-Pr, TfHN, Bn, NHTf)

Scheme 2.205 Lewis-acid-catalyzed aza-Diels–Alder reaction of an azirine with cyclopentadiene.

tration of the crude reaction mixtures [273]. The first step of the sequence consisted of the kinetic resolution of a racemic epoxide with a sulfonamide in the presence of a chiral [(salen)Co] complex, yielding the corresponding enantiopure 1,2-amino alcohol derivative, which was isolated by filtration of the crude material through silica gel. This product was then transformed into the corresponding N-nosyl aziridine by successive removal of the Boc group, conversion into the O-mesylate, and then cyclization with K$_2$CO$_3$ or Cs$_2$CO$_3$. A series of chiral aziridines was thus synthesized in almost complete enantioselectivity and in high yields, as shown in Scheme 2.206.

In 2003, Vedejs et al. studied the stereochemistry of aziridine borane lithiation, demonstrating that it occurred *syn* to the borane subunit [274]. Moreover, the enantioselective lithiation of stable borane complexes was performed in the presence of (−)-sparteine as the chiral ligand of lithium, providing the corresponding lithioaziridines, which were subsequently trapped with electrophiles. The corresponding free aziridines were isolated with good enantioselectivities of up to 72% ee (Scheme 2.207).

The conventional methods to prepare enantiomerically aziridines have mostly provided these products in their N-protected forms. In 2004, Yudin et al. explored routes from unprotected aziridines to their N-allylated derivatives [275]. In this context, these authors found that unprotected aziridines underwent facile

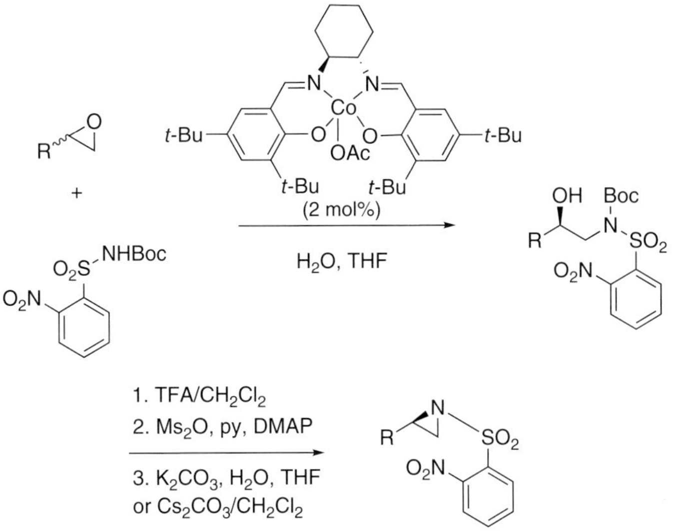

Scheme 2.206 Four-step synthesis of chiral aziridines from racemic epoxides.

R = Cy: 86%, ee >99%
R = n-Bu: 79%, ee >99%
R = t-Bu: 67%, ee >99%
R = CO$_2$Me: 69%, ee >99%
R = CH$_2$Cl: 75%, ee >99%
R = (CH$_2$)$_2$CH=CH$_2$: 58%, ee >99%
R = Ph: 81%, ee = 99%
R = m-ClC$_6$H$_4$: 82%, ee >99%
R = o-ClC$_6$H$_4$: 81%, ee >99%
R = p-ClC$_6$H$_4$: 85%, ee >99%
R = m-MeOC$_6$H$_4$: 80%, ee >99%
R = m-NO$_2$C$_6$H$_4$: 72%, ee >99%

palladium-catalyzed allylic amination with allyl acetates in the presence of K$_2$CO$_3$, providing the corresponding aziridines in high yields. When the reaction was carried out in the presence of (R)-BINAP as the chiral ligand of [Pd(η^3-C$_3$H$_5$)Cl]$_2$, it yielded the corresponding aziridines with excellent enantioselectivity of up to 98% ee, as shown in Scheme 2.208.

In 2010, Lindsley and coworkers reported a three-step, one-pot protocol involving an enantioselective organocatalyzed α-chlorination of aldehydes, a subsequent reductive amination reaction with a primary alkyl amine, and an S$_N$2 displacement to afford the corresponding chiral N-alkyl terminal aziridines [276]. As shown in Scheme 2.209, these final aziridines were obtained in moderate-to-high yields (40–86%) and with enantioselectivities of up to 96% ee, when using proline-derived Jørgensen catalyst [277]. It must be noted that this novel methodology provided access to aziridines that were previously difficult to prepare, using aldehydes and amines for which thousands are commercially available. In 2015, the scope of this methodology was extended to various other aliphatic aldehydes in comparable results [278].

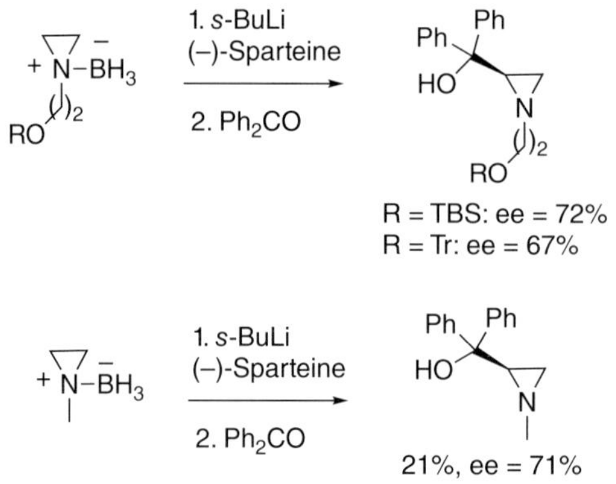

Scheme 2.207 Synthesis of chiral aziridines through enantioselective borane lithiations.

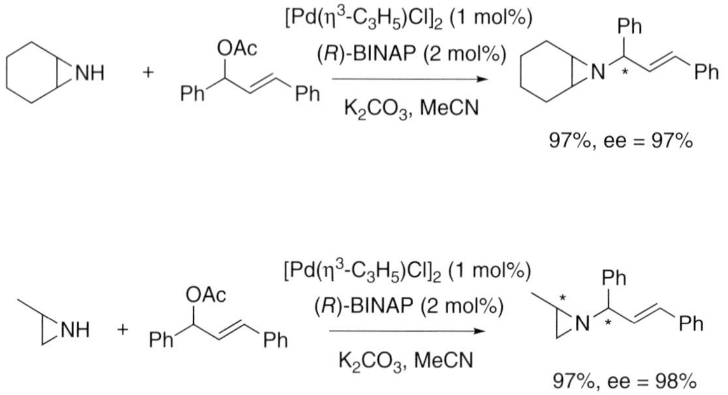

Scheme 2.208 Synthesis of chiral aziridines through Pd-catalyzed allylic aminations.

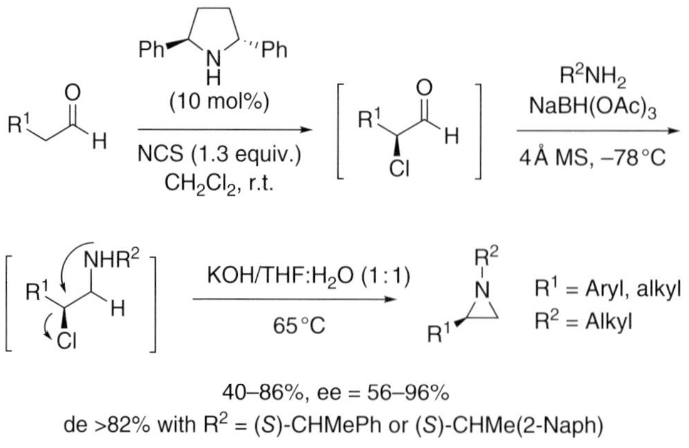

40–86%, ee = 56–96%
de >82% with R^2 = (S)-CHMePh or (S)-CHMe(2-Naph)

Scheme 2.209 Organocatalyzed aziridination through a one-pot sequential α-chlorination/reductive amination/S_N2 reaction.

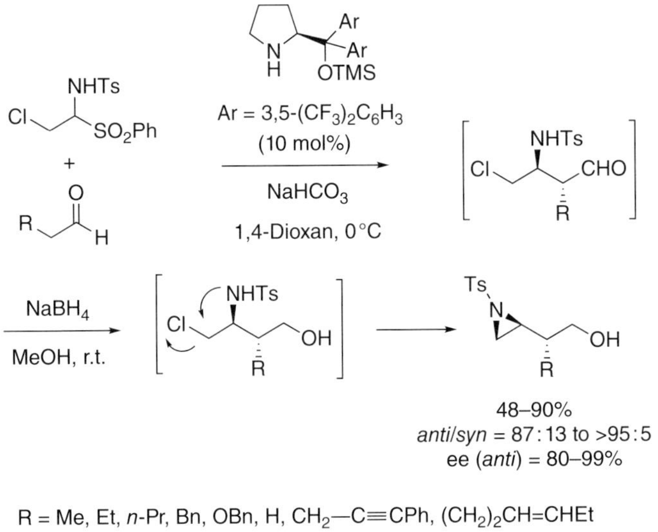

Scheme 2.210 Organocatalyzed aziridination through a one-pot sequential imine formation/Mannich reaction/reduction/S_N2 reaction.

On the other hand, another proline-derived organocatalyst was applied by Hayashi et al. to induce a one-pot synthesis of chiral aziridines [279]. It involved the synthesis of chiral 1-substituted-2-hydroxyethylaziridine derivatives through uninterrupted sequential reactions, including a desulfonylative formation of an N-tosyl imine from the corresponding aldehyde and a α-amido sulfone, an asymmetric direct Mannich reaction organocatalyzed by a (S)-diarylprolinol silyl ether, a reduction, and a final aziridine formation through intramolecular substitution. As shown in Scheme 2.210, these synthetically useful chiral aziridines bearing a hydroxyethyl moiety were prepared in a single flask with good *anti/cis* ratios of up to >95:5, in good-to-high yields (48–90%), along with high-to-excellent enantioselectivities of up to 99% ee.

In addition, Antilla et al. have demonstrated the utility of a chiral VAPOL magnesium phosphate chiral salt catalyst to induce an enantioselective aza-Darzens aziridination reaction between N-benzoyl aryl aldimines and a α-chloro 1,3-diketone [280]. The corresponding chiral aziridines were obtained in good yields (52–78%) and with good-to-high enantioselectivities of up to 92% ee, as shown in Scheme 2.211. A theoretical study of the active catalytic species was undertaken by the authors to explain the stereoselectivity of the process. It showed coordination of the magnesium to the carbonyls of the imine and diketone's enol form. Additionally, the enol could hydrogen bond to the oxygens on the catalyst. The catalyst could simultaneously stabilize the nucleophile and electrophile, while providing the chiral requirement for asymmetric induction.

2.3.4 Kinetic Resolutions of Aziridines

Despite the increased industrial demand for enantiomerically pure compounds, to date, only a few catalytic asymmetric processes have found commercial

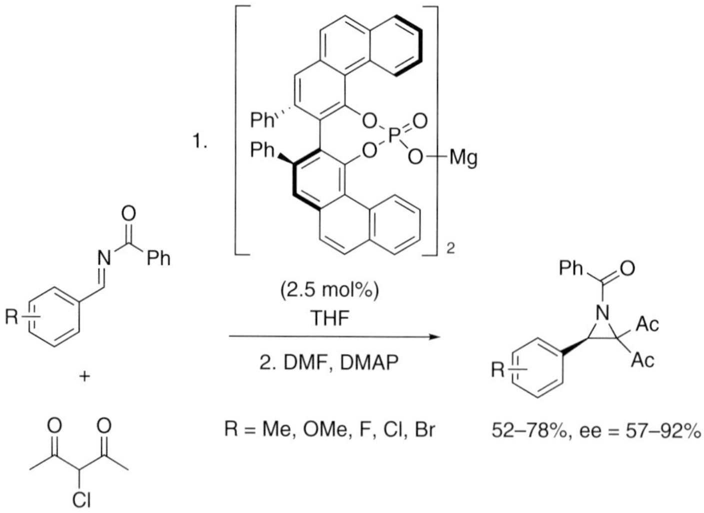

Scheme 2.211 Mg-catalyzed aza-Darzens aziridination reaction.

application [281], among them, rare exceptions are catalytic kinetic resolutions [282]. Indeed, kinetic resolution as one of the most powerful tools in asymmetric catalysis has found wide applications in both academics and industry, complementing approaches such as asymmetric synthesis and classical resolution. The use of enzymes for the kinetic resolution of racemic substrates to afford enantiopure compounds with high enantioselectivity and in good yield has long been a popular strategy in synthesis. When an asymmetric route to an enantiomerically pure aziridine cannot be achieved, there still exists the possibility of performing a kinetic resolution of the racemic aziridine mixture. This can be achieved either chemically or enzymatically. Despite the synthetic utility of chiral aziridines, it must be recognized that their production by kinetic resolution remains undeveloped. In recent years, however, several examples of the enzymatic resolution of aziridines have been successfully developed. As an example, Kumar et al. have demonstrated that methyl N-arylaziridine-2-carboxylates could be enzymatically resolved using the lipase from *Candida rugosa* to afford optically active aziridine carboxylates in moderate-to-high enantiomeric purity, as shown in Scheme 2.212 [283].

In 2007, Wang et al. demonstrated that, catalyzed by the *Rhodococcus erythropolis* AJ270 whole-cell catalyst under very mild conditions, biotransformations of racemic N-arylaziridine-2-carbonitriles proceeded efficiently and enantioselectively to produce highly enantiopure (S)-N-arylaziridine-2-carboxamides and (R)-N-arylaziridine-2-carboxylic acids in excellent yields (Scheme 2.213) [284]. The latter carboxylic acids were converted by treatment with diazomethane into the corresponding methyl esters.

As an extension of this methodology, several racemic *trans*-3-arylaziridine-2-carbonitriles were efficiently transformed into their corresponding enantiopure (2R,3S)-3-arylaziridine-2-carboxamides under similar conditions (Scheme 2.214) [285].

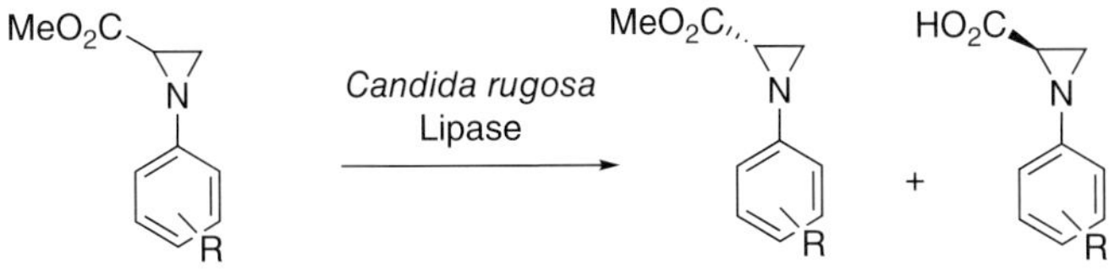

Scheme 2.212 Enzymatic kinetic resolution of methyl N-arylaziridine-2-carboxylates.

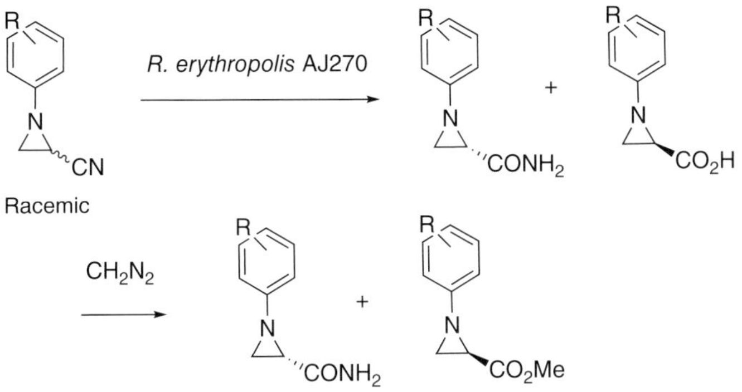

R = H: 48%, ee = 84%
R = 4-Br: 50%, ee = 99%
R = 4-NO$_2$: 45%, ee = 12%
R = 4-Me: 48%, ee = 70%
R = 4-F: 44%, ee = 7%
R = 4-Me,3-Br: 46%, ee = 15%
R = 4-MeO: 46%, ee = 79%

R = H: 45% (carboxamide), ee (carboxamide) >99% + 47%, (ester), ee (ester) = 91%
R = 4-F: 48% (carboxamide), ee (carboxamide) >99% + 50% (ester), ee (ester) = 94%
R = 4-Cl: 46% (carboxamide), ee (carboxamide) = 95% + 49% (ester), ee (ester) = 87%
R = 4-Br: 25% (carboxamide), ee (carboxamide) >99% + 74% (ester), ee (ester) = 67%
R = 4-MeO: 28% (carboxamide), ee (carboxamide) = 96% + 22% (ester), ee (ester) >99%
R = 4-Me: 50% (carboxamide), ee (carboxamide) >99% + 50% (ester), ee (ester) >99%
R = 3-Me: 45% (carboxamide), ee (carboxamide) = 97% + 50% (ester), ee (ester) = 90%

Scheme 2.213 Kinetic resolution of N-arylaziridine-2-carbonitriles by biotransformation.

While the nitrile hydratase exhibited low selectivity against nitrile substrates, the amidase was highly enantioselective toward 3-arylaziridine-2-carboxamides. Indeed, the conditions described earlier could be applied to the resolution of a number of racemic *trans*-3-arylaziridine-2-carboxamides, as shown in Scheme 2.215 [285].

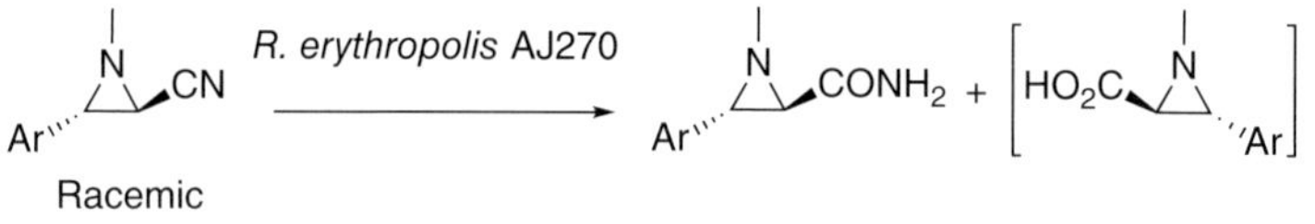

Ar⋯△(N-CN) **R. erythropolis AJ270** → Ar⋯△(N-CONH$_2$) + [HO$_2$C-△(N)⋯Ar]

Racemic

Ar = Ph: 45% (carboxamide), ee (carboxamide) >99%
Ar = p-Tol: 48% (carboxamide), ee (carboxamide) >99%
Ar = p-MeOC$_6$H$_4$: 47% (carboxamide), ee (carboxamide) <5%
Ar = p-BrC$_6$H$_4$: 26% (carboxamide), ee (carboxamide) = 24%

Scheme 2.214 Kinetic resolution of *trans*-3-arylaziridine-2-carbonitriles by biotransformation.

Ar⋯△(N-R)(CONH$_2$) **R. erythropolis AJ270** → Ar⋯△(N-R)(CONH$_2$) + [HO$_2$C-△(N-R)⋯Ar]

Racemic

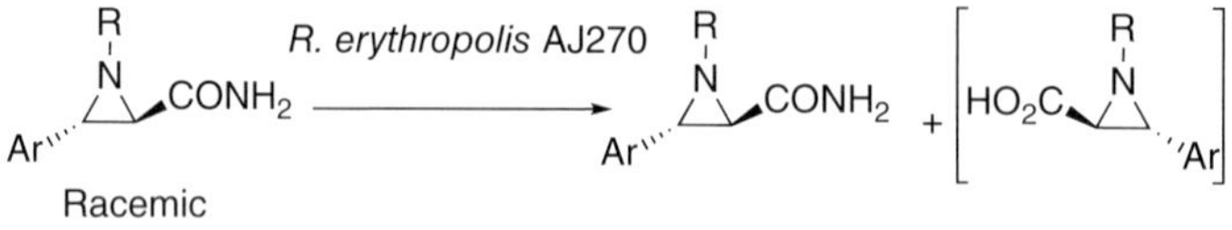

Ar = Ph, R = Me: 48%, ee >99%
Ar = p-MeOC$_6$H$_4$, R = Me: 40%, ee <5%
Ar = p-BrC$_6$H$_4$, R = Me: 48%, ee >99%
Ar = p-FC$_6$H$_4$, R = Me: 49%, ee >99%
Ar = p-ClC$_6$H$_4$, R = Me: 48%, ee >99%
Ar = o-ClC$_6$H$_4$, R = Me: 50%, ee = 12%
Ar = m-ClC$_6$H$_4$, R = Me: 46%, ee >99%
Ar = Ph, R = H: 47%, ee >99%
Ar = Ph, R = Me: 48%, ee >99%

Scheme 2.215 Kinetic resolution of *trans*-3-arylaziridine-2-carboxamides by biotransformation.

△(N-R)(CONH$_2$) **Rhodococcus rhodochrous IFO 15564** → △(N-R)(CONH$_2$) + △(N-R)(CO$_2$H)

Racemic

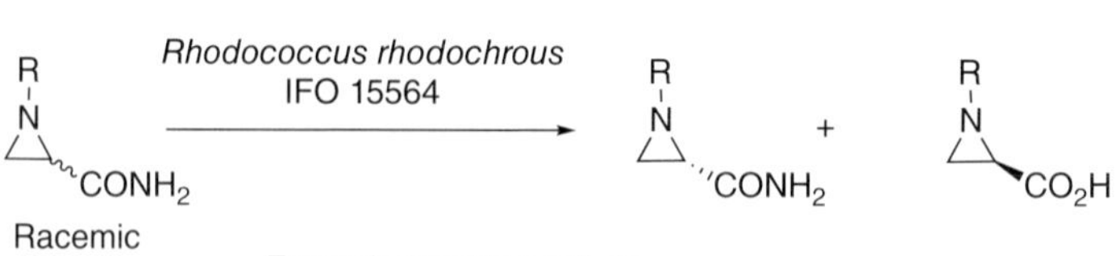

For carboxamide products:
R = Bn: 45%, ee >99%
R = Ph: 47%, ee >99%
R = p-MeOC$_6$H$_4$: 46%, ee >99%
R = p-CF$_3$C$_6$H$_4$: 45%, ee >99%

Scheme 2.216 Kinetic resolution of *N*-benzyl- and *N*-arylaziridine-2-carboxamides by biotransformation.

In the same context, enantiopure *N*-benzyl- and *N*-arylaziridine-2-carboxamides were obtained by Gotor *et al.* through kinetic resolution of their corresponding racemates by treatment with *Rhodococcus rhodochrous* IFO 15564 (Scheme 2.216) [286].

In 2010, these authors extended the scope of the preceding methodology to a series of N-unsubstituted *trans*-3-arylaziridine-2-carboxamides, which provided the recovered enantiopure aziridines in 40–47% yields (Scheme 2.217) [287].

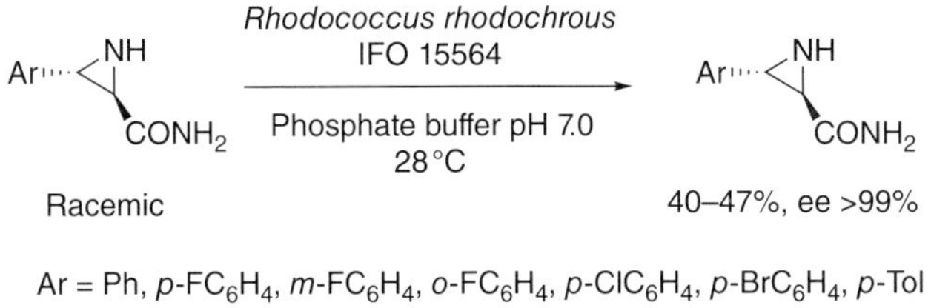

Scheme 2.217 Kinetic resolution of N-unsubstituted *trans*-3-arylaziridine-2-carboxamides by biotransformation.

Among a range of substrates investigated, only the reaction of 3-(*p*-anisyl) substrate failed, leading to a complex mixture of products.

In addition, Sakai *et al.* have reported the lipase-catalyzed resolution of (2*R**,3*S**)- and (2*R**,3*R**)-3-methyl-3-phenyl-2-aziridinemethanols [288]. Therefore, upon treatment with lipase PS-C II, (2*R**,3*S**)-3-methyl-3-phenyl-2-aziridinemethanol gave enantioenriched (2*R*,3*S*)-3-methyl-3-phenyl-2-aziridinemethanol and its corresponding acetate, while a similar reaction of (2*R**,3*R**)-3-methyl-3-phenyl-2-aziridinemethanol afforded enantioenriched (2*R*,3*R*)-3-methyl-3-phenyl-2-aziridinemethanol and its corresponding acetate (Scheme 2.218).

Kinetic resolutions of racemic substrates performed with enzymes or microorganisms to afford chiral compounds with high enantioselectivities and in good yields constitute common strategies in synthesis. However, transition-metal-mediated and more recently organocatalyzed kinetic resolutions have gained popularity within the synthetic community over the past two decades due to the progress made in the development of chiral catalysts for asymmetric reactions [289]. Many procedures of catalytic nonenzymatic kinetic resolution have been developed, providing high enantioselectivity and yield for both products and recovered starting materials. Indeed, the nonenzymatic kinetic resolution of racemic compounds based on the use of a chiral catalyst is presently an area of great importance in asymmetric organic synthesis. The first examples of catalytic nonenzymatic kinetic resolution were reported by Fajans and Bredig in 1908, dealing with the decarboxylation of camphor-3-carboxylic acid in the

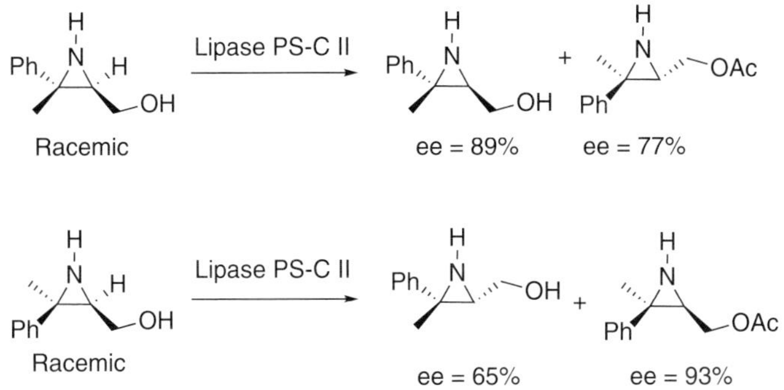

Scheme 2.218 Enzymatic kinetic resolution of 3-methyl-3-phenyl-2-aziridinemethanol.

presence of chiral alkaloids [290]. A recent and nice example of organocatalytic kinetic resolution of aziridines was described by Ooi et al., using a chiral 1,2,3-triazolium salt as catalyst [291]. For example, the reaction of vinylcyclohexane-derived aziridine with TMSCl in the presence of 5 mol% of this organocatalyst in toluene at −40 °C gave rise to the corresponding ring-opened product in 51% yield with 80% ee along with the recovered aziridine in 44% yield with 94% ee (selectivity factor [282] of 23). Even higher selectivity factors of up to 78 were observed with aziridines bearing sterically demanding *tert*-butyl and trimethylsilyl substituents, as shown in Scheme 2.219. Another feature of this approach was an unprecedented kinetic resolution of different 2,2-disubstituted aziridines, which exhibited impressive levels of selectivity factors of up to 470, enabling the preparation of enantiomerically pure aziridines **9**, otherwise not readily accessible through conventional asymmetric methodologies. When racemic aziridines **9** were exposed to the same reaction conditions as earlier, a regioisomeric mixture of the corresponding chlorinated products were obtained in 48–53% yields and the enantiomeric excesses of the major β-chloro-*tert*-amine derivatives were of 90–97% ee. Importantly, the starting aziridines were recovered in 41–51% yields in an essentially enantiopure form (ee = 96–99%).

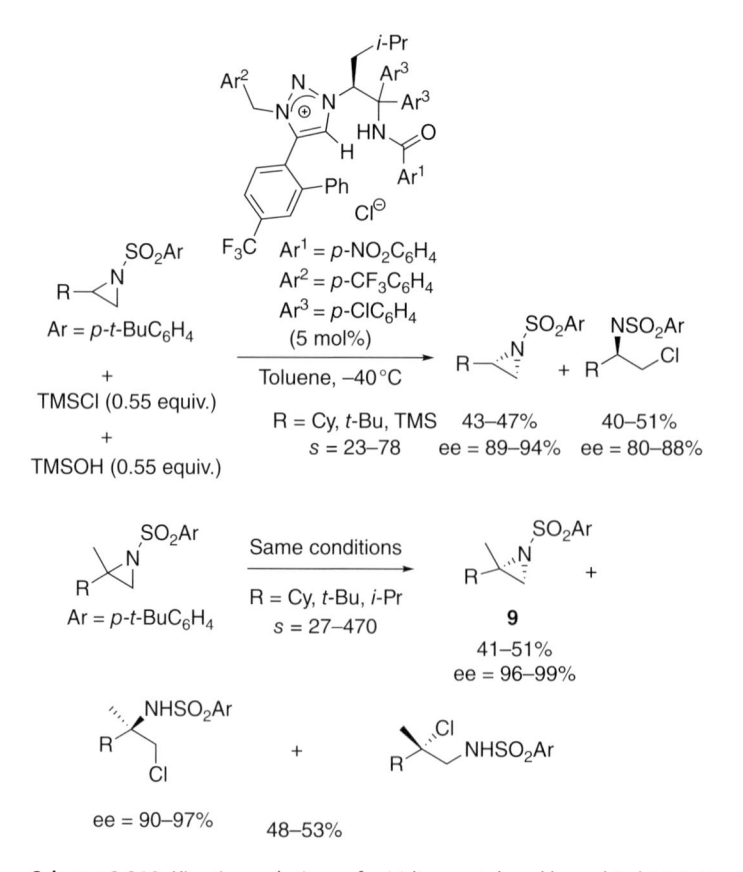

Scheme 2.219 Kinetic resolutions of aziridines catalyzed by a chiral 1,2,3-triazolium salt.

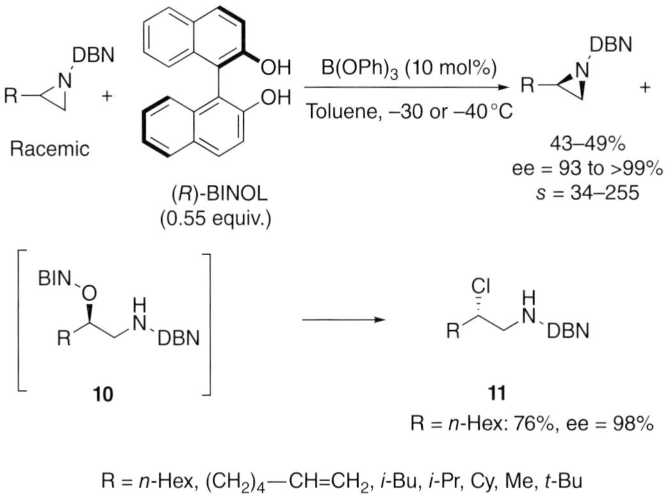

Scheme 2.220 B-catalyzed kinetic resolution of N-acylaziridines using (R)-BINOL as resolution agent.

In 2012, Morgan *et al.* reported a boron-catalyzed kinetic resolution of aziridines, dealing with the resolution of N-acylaziridines with a broad scope, using (R)-BINOL as the resolution agent (Scheme 2.220) [292]. Best results were obtained with 3,5-dinitrobenzoyl(DBN)-protected aziridines, which underwent kinetic resolution in the presence of triphenylborate and (R)-BINOL, the latter functioning as a nucleophile. Steric bulk was shown to play an important role in both reactivity and selectivity, with aziridines bearing tertiary substituents found as the most selective substrates, producing nearly enantiopure recovered aziridines. Moreover, the BINOL-derived by-product **10** arisen from the consumed enantiomer of aziridine was further processed to recover BINOL and produce an enantiomerially pure 1,2-chloroamide **11**.

2.4 Conclusions

This chapter collects the recent developments in asymmetric aziridination covering the literature since 2003. For a long time, asymmetric approaches toward chiral aziridines appeared to be less developed compared to the analogous reactions leading to other three-membered cycles such as epoxides or cyclopropanes. In recent years, however, great advances have been made, which now allow the preparation of chiral aziridines to be achieved by various synthetic strategies in an efficient manner, affording products in excellent yields and enantioselectivities. In addition to describing the large number of highly efficient processes based on the use of various chiral auxiliaries or substrates, this chapter demonstrates that the most important achievements in asymmetric aziridination are the spectacular expansion of novel chiral catalysts, including the especially attractive chiral organocatalysts, which have been recently applied to this type of reaction. Indeed, a collection of new chiral Lewis-acid catalysts and organocatalysts have provided new opportunities for these enantioselective reactions and widely expanded their scope. Although most of the methods for obtaining chiral aziridines by using a chiral catalyst proceed through the transfer of a nitrogen group to an olefin, a number of highly efficient methods based on the previously less-studied enantioselective transfer of a carbenoid to an imine have been successfully developed in the last few years. For example, special mention has been made to the VANOL and VAPOL catalysts, which have been applied to the borane-catalyzed carbene transfer to imines. On the other hand, a range of chiral organocatalysts have been successfully applied to both carbene transfer to imines and nitrene transfer to alkenes, including the first catalytic and highly enantioselective aziridination of cyclic enones recently reported. Furthermore, the last few years have witnessed the first one-pot multicomponent catalytic asymmetric aziridination reactions being described. Indeed, a number of these powerful one-pot reactions, avoiding the use of costly and time-consuming protection–deprotection processes as well as the purification procedures of intermediates, have allowed a range of chiral aziridines to be achieved in remarkable yields and enantioselectivities. This chapter well demonstrates that the asymmetric aziridination reaction constitutes an important tool in organic synthesis, still attracting a considerable interest due to the potential use of enantiopure aziridines as useful intermediates in the synthesis of complex molecules and to the intriguing biological activities of numerous aziridine-containing products including natural important compounds. The development of new catalytic systems for the synthesis of certain highly useful synthetic intermediates is, however, awaited in the coming few years.

References

1 (a) Gabriel, S. (1888) *Chem. Ber.*, 21, 1049–1057; (b) Gabriel, S. (1888) *Chem. Ber.*, 21, 2664–2669.
2 (a) Padwa, A. (2008) in *Comprehensive Heterocyclic Chemistry III*, Chapter 1 (eds A. Ramsden, E.F.V. Scriven, and R.J.K. Taylor), Elsevier, Oxford, pp. 1–104; (b) Beck Bisol, T. and Mandolesi Sà, M. (2007) *Quim. Nova*, 30, 106–115; (c) Singh, G.S., D'hooghe, M., and De Kimpe, N. (2007) *Chem. Rev.*, 107, 2080–2135; (d) Sweeney, J.B. (2006) in *Aziridines and Epoxides in Organic Synthesis*, Chapter 4 (ed. A. Yudin), Wiley-VCH Verlag GmbH, Weinheim, pp. 117–144; (e) Aggarwal, V.K., Badine, M., and Moorthie, V. (2006) in *Aziridines and Epoxides in Organic Synthesis*, Chapter 1 (ed. A. Yudin), Wiley-VCH Verlag GmbH, Weinheim, pp. 1–35; (f) Zhou, P., Chen, B.-C., and Davis, F.A. (2006) in *Aziridines and Epoxides in Organic Synthesis*, Chapter 3 (ed. A. Yudin), Wiley-VCH Verlag GmbH, Weinheim, pp. 73–115; (g) Padwa, A. and Murphree, S.S. (2006) *Arkivoc*, 2006 (3), 6–33; (h) Cardillo, G., Gentilucci, L., and Tolomelli, A. (2003) *Aldrichimica Acta*, 36, 39–50; (i) Lee, W.K. and Ha, H.-J. (2003) *Aldrichimica Acta*, 36, 57–63; (j) Padwa, A. and Murphree, S. (2003) *Prog. Heterocycl. Chem.*, 15, 75–99; (k) Padwa, A. and Murphree, S.S. (2000) in *Progress in Heterocyclic Chemistry*, vol. 12, Chapter 4.1 (eds G.W. Gribble and T.L.

Gilchrist), Elsevier Science, Oxford, p. 57; (l) Padwa, A., Pearson, W.H., Lian, B.N., and Bergmeier, S.C. (1996) in *Comprehensive Heterocyclic Chemistry II*, vol. 1A (eds A.R. Katritzky, C.W. Reese, and E.F. Scriven), Pergamon, Oxford, p. 1; (m) Kemp, J.E.G. (1991) in *Comprehensive Organic Synthesis*, vol. 7, Chapter 3.5 (eds B.M. Trost and I. Fleming), Pergamon, Oxford, p. 469; (n) Padwa, A. and Woolhouse, A.D. (1984) in *Comprehensive Heterocyclic Chemistry*, vol. 7 (ed. W. Lwowski), Pergamon Press, Oxford, pp. 47–93.

3 (a) Padwa, A. (1991) in *Comprehensive Organic Synthesis*, vol. 4, Chapter 4.9 (eds B.M. Trost and I. Fleming), Pergamon, Oxford, p. 1069; (b) Tanner, T. (1993) *Pure Appl. Chem.*, 65, 1319–1328; (c) Tanner, D. (1994) *Angew. Chem. Int. Ed. Engl.*, 33, 599–619; (d) Atkinson, R.S. (1999) *Tetrahedron*, 55, 1519–1559; (e) Stamm, H. (1999) *J. Prakt. Chem.*, 341, 319–331; (f) McCoull, W. and Davis, F.A. (2000) *Synthesis*, 2000, 1347–1365; (g) Righi, G. and Bonini, C. (2000) *Targets Heterocycl. Syst.*, 4, 139–165; (h) Zwanenburg, B. and ten Holte, P. (2001) in *Stereoselective Heterocyclic Synthesis III*, Topics in Current Chemistry, vol. 216 (ed. P. Metz), Springer, Berlin, pp. 93–124; (i) Sweeney, J.B. (2002) *Chem. Soc. Rev.*, 31, 247–258; (j) Aires-de-Sousa, J., Prabhakar, S., Lobo, A.M., Rosa, A.M., Gomes, M.J.S., Corvo, M.C., Williams, D.J., and White, A.J.P. (2002) *Tetrahedron: Asymmetry*, 12, 3349–3365; (k) Hu, X.E. (2004) *Tetrahedron*, 60, 2701–2743; (l) Pineschi, M. (2006) *Eur. J. Org. Chem.*, 2006, 4979–4988; (m) Yudin, A. (2006) *Aziridines and Epoxides in Organic Synthesis*, Wiley-VCH Verlag GmbH, Weinheim; (n) Florio, S. and Luisi, R. (2010) *Chem. Rev.*, 110, 5128–5157; (o) Lu, P. (2010) *Tetrahedron*, 66, 2549–2560.

4 Zalialov, I.A. and Dahanubar, V.H. (2002) *Curr. Opin. Drug Discovery Dev.*, 5, 918–927.

5 Lesniak, S., Rachwalski, M.I., and Pieczonka, A.M. (2014) *Curr. Org. Chem.*, 18, 3045–3065.

6 (a) Osborn, H.M.I. and Sweeney, J.B. (1997) *Tetrahedron: Asymmetry*, 8, 203–238; (b) Müller, P. and Fruit, C. (2003) *Chem. Rev.*, 103, 2905–2919; (c) Mössner, C. and Bolm, C. (2004) in *Transition Metals for Organic Synthesis*, 2nd edn (eds M. Beller and C. Bolm), Wiley-VCH Verlag GmbH, Weinheim, pp. 389–402; (d) Pellissier, H. (2010) *Tetrahedron*, 66, 1509–1555; For reviews concentrating on not especially asymmetric aziridination (and epoxidation), see: (e) Aggarwal, V.K., McGarrigle, E.M., and Shaw, M.A. (2010) *Sci. Synth.*, 37, 311–347; (f) Muchalski, H. and Johnston, J.N. (2011) *Sci. Synth.*, 1, 155–184; (g) Karila, D. and Dodd, R.H. (2011) *Curr. Org. Chem.*, 15, 1509–1540; (h) Chang, J.W.W., Ton, T.M.U., and Chan, P.W.H. (2011) *Chem. Rec.*, 11, 331–357; (i) Jung, N. and Bräse, S. (2012) *Angew. Chem. Int. Ed.*, 51, 5538–5540; (j) Chawla, R., Singh, A.K., and Yadav, L.D.S. (2013) *RSC Adv.*, 3, 11385–11403; (k) Wang, P.-A. (2013) *Beilstein J. Org. Chem.*, 9, 1677–1695; (l) Charette, A.B., Lebel, H., and Roy, M.-N. (2014) in *Copper-Calalyzed Asymmetric Synthesis* (eds A. Alexakis, N. Krause, and S. Woodward), Wiley-VCH Verlag GmbH, Weinheim; (m) Pellissier, H. (2014) *Adv. Synth. Catal.*, 356, 1899–1935; (n) Degennaro, L., Trinchera, P., and Luisi, R. (2014) *Chem. Rev.*, 114, 7881–7929; (o) Callebaut, G., Meiresonne, T., De Kimpe, N., and Mangelinckx, S. (2014) *Chem. Rev.*, 114, 7954–8015.

7 Dauban, P. and Dodd, R.H. (2003) *Synlett*, 11, 1571–1586.

8 (a) Evans, D.A., Faul, M.M., Bilodeau, M.T., Anderson, B.A., and Barnes, D.M. (1993) *J. Am. Chem. Soc.*, 115, 5328–5329; (b) Li, Z., Conser, K.R., and Jacobsen, E.N. (1993) *J. Am. Chem. Soc.*, 115, 5326–5327.
9 Trost, B.M. and Dong, G. (2006) *J. Am. Chem. Soc.*, 128, 6054–6055.
10 Flock, S. and Frauenrath, H. (2007) *Arkivoc*, 10, 245–259.
11 Dauban, P. and Dodd, R.H. (1999) *J. Org. Chem.*, 64, 5304–5307.
12 Di Chenna, P.H., Dauban, P., Ghini, A., Baggio, R., Garland, M.T., Burton, G., and Dodd, R.H. (2003) *Tetrahedron*, 59, 1009–1014.
13 (a) Södergren, M.J., Alonso, D.A., Bedekar, A.V., and Andersson, P.G. (1997) *Tetrahedron Lett.*, 38, 6897–6900; (b) Stang, P.J. and Zhdankin, V.V. (1996) *Chem. Rev.*, 96, 1123–1178.
14 (a) Dauban, P., Sanière, L., Tarrade, A., and Dodd, R.H. (2001) *J. Am. Chem. Soc.*, 123, 7707–7708; (b) Duran, F., Leman, L., Ghini, A., Burton, G., Dauban, P., and Dodd, R.H. (2002) *Org. Lett.*, 4, 2481–2483.
15 Di Chenna, P.H., Robert-Peillard, F., Dauban, P., and Dodd, R.H. (2004) *Org. Lett.*, 6, 4503–4505.
16 Sanière, L., Leman, L., Bourguignon, J.-J., Dauban, P., and Dodd, R.H. (2004) *Tetrahedron*, 60, 5889–5897.
17 Yu, X.-Q., Huang, J.-S., Zhou, X.-G., and Che, C.-M. (2000) *Org. Lett.*, 2, 2233–2236.
18 (a) When, P.M., Lee, J., and Du Bois, J. (2003) *Org. Lett.*, 5, 4823–4826; (b) Guthikonda, K., When, P.M., Caliando, B.J., and Du Bois, J. (2006) *Tetrahedron*, 62, 11331–11342.
19 Keaney, G.F. and Wood, J.L. (2005) *Tetrahedron Lett.*, 46, 4031–4034.
20 (a) Trost, B.M. and Zhang, T. (2008) *Angew. Chem. Int. Ed.*, 47, 3759–3761; (b) Trost, B.M. and Zhang, T. (2011) *Chem. Eur. J.*, 17, 3630–3643.
21 Trost, B.M., Malhotra, S., Olson, D.E., Maruniak, A., and Du Bois, J. (2009) *J. Am. Chem. Soc.*, 131, 4190–4191.
22 Diaper, C.M., Sutherland, A., Pillai, B., James, M.N.G., Semchuk, P., Blanchard, J.S., and Vederas, J.C. (2005) *Org. Biomol. Chem.*, 3, 4402–4411.
23 Duan, P.-W., Chiu, C.-C., Lee, W.-D., Pan, L.S., Venkatesham, U., Tzeng, Z.-H., and Chen, K. (2008) *Tetrahedron: Asymmetry*, 19, 682–690.
24 Katsuki, T. (2005) *Chem. Lett.*, 34, 1304–1309.
25 Yoshimisu, T., Ino, T., and Tanaka, T. (2008) *Org. Lett.*, 10, 5457–5460.
26 Mendlik, M.T., Tao, P., Hadad, C.M., Coleman, R.S., and Lowary, T.L. (2006) *J. Org. Chem.*, 71, 8059–8070.
27 Sureshkumar, D., Maity, S., and Chandrasekaran, S. (2006) *J. Org. Chem.*, 71, 1653–1657.
28 Barros, M.T., Matias, P.M., Maycock, C.D., and Rita Ventura, M. (2003) *Org. Lett.*, 5, 4321–4323.
29 Tarrade, A., Dauban, P., and Dodd, R.H. (2003) *J. Org. Chem.*, 68, 9521–9524.
30 Bonifacio, V.D.B., Gonzalez-Bello, C., Rzepa, H.S., Prabhakar, S., and Lobo, A.M. (2010) *Synlett*, 1, 145–149.
31 Satoh, T., Endo, J., Ota, H., and Chyouma, T. (2007) *Tetrahedron*, 63, 4806–4813.
32 Sugiyama, S., Kido, M., and Satoh, T. (2005) *Tetrahedron Lett.*, 46, 6771–6775.
33 Ulukanli, S., Karabuga, S., Celik, A., and Kazaz, C. (2005) *Tetrahedron Lett.*, 46, 197–199.

34 Jones, G.B. and Chapman, B.J. (1995) *Synthesis*, 1995, 475–497.
35 Colantoni, D., Fioravanti, S., Pellacani, L., and Tardella, P.A. (2004) *Org. Lett.*, 6, 197–200.
36 (a) Fioravanti, S., Morreale, A., Pellacani, L., and Tardella, P.A. (2003) *Tetrahedron Lett.*, 44, 3031–3034; (b) Fioravanti, S., Morreale, A., Pellacani, L., and Tardella, P.A. (2003) *Eur. J. Org. Chem.*, 2003, 4549–4552.
37 Fioravanti, S., Marchetti, F., Pellacani, L., Ranieri, L., and Tardella, P.A. (2008) *Tetrahedron: Asymmetry*, 19, 231–236.
38 Fioravanti, S., Morea, S., Morreale, A., Pellacani, L., and Tardella, P.A. (2009) *Tetrahedron*, 65, 484–488.
39 Ohno, H., Takemoto, Y., Fujii, N., Tanaka, T., and Ibuka, T. (2004) *Chem. Pharm. Bull.*, 52, 111–119.
40 (a) Kirmse, W. (1971) *Carbene Chemistry*, 2nd edn, McGraw-Hill, New York, p. 412; (b) Badea, F., Condeiu, C., Gherghiu, M., Iancu, A., Iordache, A., and Simion, C. (1992) *Rev. Roum. Chim.*, 37, 393–405.
41 Zhang, X.-J., Yan, M., and Huang, D. (2009) *Org. Biomol. Chem.*, 7, 187–192.
42 Resnati, G., Soloshonok, V.A. (1996) *Tetrahedron*, Tetrahedron Symposium-in-Print N°58 on Fluoroorganic Chemistry. Synthetic Challenges and Biomedicinal Rewards., 52, 1–330.
43 Akiyama, T., Ogi, S., and Fuchibe, K. (2003) *Tetrahedron Lett.*, 44, 4011–4013.
44 Lee, S.-L. and Song, I.-W. (2005) *Bull. Korean Chem. Soc.*, 26, 223–224.
45 Hashimoto, T., Nakatsu, H., Watanabe, S., and Maruoka, K. (2010) *Org. Lett.*, 12, 1668–1671.
46 Hashimoto, T., Nakatsu, H., Yamamoto, K., Watanabe, S., and Maruoka, K. (2011) *Chem. Asian J.*, 6, 607–613.
47 (a) Davis, F.A., Zhou, P., Liang, C.-H., and Reddy, R.E. (1995) *Tetrahedron: Asymmetry*, 6, 1511–1514; (b) Garcia Ruano, J.L., Fernandez, I., and Hamdouchi, C. (1995) *Tetrahedron Lett.*, 36, 295–298; (c) Garcia Ruano, J.L., Fernandez, I., del Prado Catalina, M., and Cruz, A.A. (1996) *Tetrahedron: Asymmetry*, 7, 3407–3414.
48 (a) Li, A.-H., Dai, L.-X., and Aggarwal, V.K. (1997) *Chem. Rev.*, 97, 2341–2372; (b) Dai, L.-X., Hou, X.-L., and Zhou, Y.-G. (1999) *Pure Appl. Chem.*, 71, 369–376.
49 Morton, D. and Stockman, R.A. (2006) *Tetrahedron*, 62, 8869–8905.
50 Morton, D., Pearson, D., Field, R.A., and Stockman, A. (2003) *Synlett*, 13, 1985–1988.
51 Morton, D., Pearson, D., Field, R.A., and Stockman, R.A. (2006) *Chem. Commun.*, 17, 1833–1835.
52 Morton, D., Pearson, D., Field, R.A., and Stockman, R.A. (2004) *Org. Lett.*, 6, 2377–2380.
53 Chigboh, K., Morton, D., Nadin, A., and Stockman, R.A. (2008) *Tetrahedron Lett.*, 49, 4768–4770.
54 Forbes, D.C., Bettigeri, S.V., Patrawala, S.A., Pischek, S.C., and Standen, M.C. (2009) *Tetrahedron*, 65, 70–76.
55 (a) Fernandez, I., Valdivia, V., Gori, B., Alcudia, F., Alvarez, E., and Khiar, N. (2005) *Org. Lett.*, 7, 1307–1310; (b) Fernandez, I., Gori, B., Alcudia, F., and Khiar, N. (2005) *Phosphorus, Sulfur Silicon Relat. Elem.*, 180, 1511–1512.

56 Kokotos, C.G. and Aggarwal, V.K. (2007) *Org. Lett.*, 9, 2099–2102.
57 Unthank, M.G., Hussain, N., and Aggarwal, V.K. (2006) *Angew. Chem. Int. Ed.*, 45, 7066–7069.
58 (a) Zheng, J.-C., Liao, W.-W., Sun, X.-X., Sun, X.-L., Tang, Y., Dai, L.-X., and Deng, J.-G. (2005) *Org. Lett.*, 7, 5789–5792; (b) Sun, X.-L. and Tang, Y. (2008) *Acc. Chem. Res.*, 41, 937–948.
59 Xue, Z., Dee, V.M., Hope-Weeks, L.J., Whittlesey, B.R., and Mayer, M.F. (2010) *Arkivoc*, 7, 65–80.
60 (a) Robiette, R. (2006) *J. Org. Chem.*, 71, 2726–2734; (b) Aggarwal, V.K., Charmant, J.P.H., Ciampi, C., Hornby, J.M., O'Brien, C.J., Hynd, G., and Parsons, R. (2001) *J. Chem. Soc., Perkin Trans.*, 23 (1), 3159–3166.
61 Yang, Y., Huang, Y., and Qing, F.-L. (2013) *Tetrahedron Lett.*, 54, 3826–3830.
62 Fulton, J.R., Aggarwal, V.K., and de Vicente, J. (2005) *Eur. J. Org. Chem.*, 2005, 1479–1492.
63 (a) Ooi, T. and Maruoka, K. (2007) *Angew. Chem. Int. Ed.*, 46, 4222–4266; (b) Maruoka, K. (2008) *Org. Process Res. Dev.*, 12, 679–697.
64 Aggarwal, V.K. and Vasse, J.-L. (2003) *Org. Lett.*, 5, 3987–3990.
65 Solladié-Cavallo, A., Roje, M., Welter, R., and Sunjic, V. (2004) *J. Org. Chem.*, 69, 1409–1412.
66 Stipetic, I., Roje, M., and Hamersak, Z. (2008) *Synlett*, 20, 3149–3152.
67 Dokli, I., Matanovic, I., and Hamersak, Z. (2010) *Chem. Eur. J.*, 16, 11744–11752.
68 Midura, W.H. (2007) *Tetrahedron Lett.*, 48, 3907–3910.
69 Reddy Iska, V.B., Gais, H.-J., Tiwari, S.K., Surendra Babu, D., and Adrien, A. (2007) *Tetrahedron Lett.*, 48, 7102–7107.
70 Illa, O., Arshad, M., Ros, A., McGarrigle, E.M., and Aggarwal, V.K. (2010) *J. Am. Chem. Soc.*, 132, 1828–1830.
71 Illa, O., Namutebi, M., Saha, C., Ostovar, M., Chen, C.C., Haddow, M.F., Nocquet-Thibault, S., Lusi, M., McGarrigle, E.M., and Aggarwal, V.K. (2013) *J. Am. Chem. Soc.*, 135, 11951–11966.
72 Janardanan, D. and Sunoj, R.B. (2008) *J. Org. Chem.*, 73, 8163–8174.
73 Kavanagh, S.A., Piccinini, A., and Connon, S.J. (2013) *Org. Biomol. Chem.*, 11, 3535–3540.
74 Davis, F.A., Wu, Y., Yan, H., McCoull, W., and Prasad, K.R. (2003) *J. Org. Chem.*, 68, 2410–2419.
75 Davis, F.A., Ramachandar, T., and Wu, Y. (2003) *J. Org. Chem.*, 68, 6894–6898.
76 Solà, T.M., Churcher, I., Lewis, W., and Stockman, R.A. (2011) *Org. Biomol. Chem.*, 9, 5034–5035.
77 Moragas, T., Churcher, I., Lewis, W., and Stockman, R.A. (2014) *Org. Lett.*, 16, 6290–6293.
78 Kattuboina, A. and Li, G. (2008) *Tetrahedron Lett.*, 49, 1573–1577.
79 Kattamuri, P.V., Xiong, Y., Pan, Y., and Li, G. (2013) *Org. Biomol. Chem.*, 11, 3400–3408.
80 De Vitis, L., Florio, S., Granito, C., Ronzini, L., Troisi, L., Capriati, V., Luisi, R., and Pilati, T. (2004) *Tetrahedron*, 60, 1175–1182.
81 Savoia, D., Alvaro, G., Di Fabio, R., Gualandi, A., and Fiorelli, C. (2006) *J. Org. Chem.*, 71, 9373–9381.

82 (a) Concellon, J.M., Rodriguez-Solla, H., and Simal, C. (2008) *Org. Lett.*, 10, 4457–4460; (b) Concellon, J.M., Rodriguez-Solla, H., Bernad, P.L., and Simal, C. (2009) *J. Org. Chem.*, 74, 2452–2459.
83 Sweeney, J.B., Cantrill, A.A., McLaren, A.B., and Thobhani, S. (2006) *Tetrahedron*, 62, 3681–3693.
84 Sweeney, J.B., Cantrill, A.A., Drew, M.G.B., McLaren, A.B., and Thobhani, S. (2006) *Tetrahedron*, 62, 3694–3703.
85 Denolf, B., Mangelinckx, S., Tornroos, K.W., and De Kimpe, N. (2006) *Org. Lett.*, 8, 3129–3132.
86 Denolf, B., Mangelinckx, S., Tornroos, K.W., and De Kimpe, N. (2007) *Org. Lett.*, 9, 187–190.
87 Colpaert, F., Mangelinckx, S., Leemans, E., Denolf, B., and De Kimpe, N. (2010) *Org. Biomol. Chem.*, 8, 3251–3258.
88 Suzuki, H. and Aoyagi, S. (2012) *Org. Lett.*, 14, 6374–6376.
89 Denolf, B., Leemans, E., and De Kimpe, N. (2007) *J. Org. Chem.*, 72, 3211–3217.
90 Concellon, J.M., Riego, E., Rivero, I.A., and Ochoa, A. (2004) *J. Org. Chem.*, 69, 6244–6248.
91 Hodgson, D.M., Kloesges, J., and Evans, B. (2008) *Org. Lett.*, 10, 2781–2783.
92 Botuha, C., Chemla, F., Ferreira, F., Pérez-Luna, A., and Roy, B. (2007) *New J. Chem.*, 31, 1552–1567.
93 Chemla, F. and Ferreira, F. (2004) *J. Org. Chem.*, 69, 8244–8250.
94 Ferreira, F., Audouin, M., and Chemla, F. (2005) *Chem. Eur. J.*, 11, 5269–5278.
95 Arroyo, Y., Meana, A., Sanz-Tejedor, M.A., Alonso, I., and Garcia Ruano, J.L. (2010) *Chem. Eur. J.*, 16, 9874–9883.
96 Colpaert, F., Mangelinckx, S., De Brabandere, S., and De Kimpe, N. (2011) *J. Org. Chem.*, 76, 2204–2213.
97 Callebaut, G., Mangelinckx, S., Kiss, L., Sillanpää, R., Fülöp, F., and De Kimpe, N. (2012) *Org. Biomol. Chem.*, 10, 2326–2338.
98 Callebaut, G., Mangelinckx, S., Van der Veken, P., Törnroos, K.W., Augustyns, K., and De Kimpe, N. (2012) *Beilstein J. Org. Chem.*, 8, 2124–2131.
99 Ghorai, M.K., Ghosh, K., Yadav, A.K., Nanaji, Y., Halder, S., and Sayyad, M. (2013) *J. Org. Chem.*, 78, 2311–2326.
100 Liu, W.-J., Zhao, Y.-H., and Sun, X.-W. (2013) *Tetrahedron Lett.*, 54, 3586–3590.
101 Pablo, O., Guijarro, D., and Yus, M. (2013) *J. Org. Chem.*, 78, 9181–9189.
102 Konev, A.S., Tehrani, K.A., Khlebnikov, A.F., Novikov, M.S., and Magull, J. (2010) *Russ. J. Org. Chem.*, 46, 976–986.
103 Palacios, F., Ochoa de Retana, A.M., Martinez de Marigorta, E., and de los Santos, J.M. (2001) *Eur. J. Org. Chem.*, 2001, 2401–2414.
104 Alves, M.J., Fortes, A.G., and Gonçalves, L.F. (2003) *Tetrahedron Lett.*, 44, 6277–6279.
105 (a) Risberg, E., Fischer, A., and Somfai, P. (2004) *Chem. Commun.*, 18, 2088–2089; (b) Risberg, E., Fischer, A., and Somfai, P. (2005) *Tetrahedron*, 61, 8443–8450.
106 Pellissier, H. (2009) *Tetrahedron*, 65, 2839–2877.
107 Gilchrist, T.L. (2001) *Aldrichimica Acta*, 34, 51–55.
108 Timen, A.S., Fischer, A., and Somfai, P. (2003) *Chem. Commun.*, 10, 1150–1151.

109 Garrier, E., Le Gac, S., and Jabin, I. (2005) *Tetrahedron: Asymmetry*, 16, 3767–3771.
110 Vicario, J.L., Badia, D., and Carrillo, L. (2007) *Arkivoc*, 4, 304–311.
111 Choi, J.Y. and Borch, R.F. (2007) *Org. Lett.*, 9, 215–218.
112 Yamauchi, Y., Kawate, T., Itahashi, H., Katagiri, T., and Uneyama, K. (2003) *Tetrahedron Lett.*, 44, 6319–6322.
113 Braga, A.L., Paixao, M.W., Milani, P., Silveira, C.C., Rodrigues, O.E.D., and Alves, E.F. (2004) *Synlett*, 7, 1297–1299.
114 Alvaro, G., Di Fabio, R., Gualandi, A., and Savoia, D. (2007) *Eur. J. Org. Chem.*, 2007, 5573–5582.
115 Ghorai, M.K., Ghosh, K., and Yadav, A.K. (2009) *Tetrahedron Lett.*, 50, 476–479.
116 Jamookeeah, C.E., Beadle, C.D., and Harrity, J.P.A. (2009) *Synthesis*, 1, 133–137.
117 Dolence, E.K. and Roylance, J.B. (2004) *Tetrahedron: Asymmetry*, 15, 3307–3322.
118 (a) Wang, M.-C., Wang, D.-K., Zhu, Y., Liu, L.-T., and Guo, Y.-F. (2004) *Tetrahedron: Asymmetry*, 15, 1289–1294; (b) Wang, M.-C., Liu, L.-T., Zhang, J.-S., Shi, Y.-Y., and Wang, D.-K. (2004) *Tetrahedron: Asymmetry*, 15, 3853–3859.
119 Zhao, W.-X., Liu, N., Li, G.-W., Chen, D.-L., Zhang, A.-A., Wang, M.-C., and Liu, L. (2015) *Green Chem.*, 17, 2924–2930.
120 Grellepois, F., Nonnenmacher, J., Lachaud, F., and Portella, C. (2011) *Org. Biomol. Chem.*, 9, 1160–1168.
121 Katagiri, T., Katayama, Y., Taeda, M., Ohshima, T., and Iguchi, N. (2011) *J. Org. Chem.*, 76, 9305–9311.
122 Géant, P.-Y., Martinez, J., Rocard, L., and Salom-Roig, X.J. (2012) *Synthesis*, 44, 1247–1252.
123 Paleo, M.R., Aurrecoechea, N., Jung, K.-Y., and Rapoport, H. (2003) *J. Org. Chem.*, 68, 130–138.
124 Trost, B.M. and O'Boyle, B.M. (2008) *Org. Lett.*, 10, 1369–1372.
125 Hashimoto, M., Matsumoto, M., and Terashima, S. (2003) *Tetrahedron*, 59, 3041–3062.
126 (a) Bobeck, D.R., Warner, D.L., and Vedejs, E. (2007) *J. Org. Chem.*, 72, 8506–8518; (b) Vedejs, E., Naidu, B.N., Klapars, A., Warner, D.L., Li, V.-S., Na, Y., and Kohn, H. (2003) *J. Am. Chem. Soc.*, 125, 15796–15806.
127 Nie, L.-D., Wang, F.-F., Ding, W., Shi, X.-X., and Lu, X. (2013) *Tetrahedron: Asymmetry*, 24, 638–642.
128 Kongkathip, B., Akkarasamiyo, S., and Kongkathip, N. (2015) *Tetrahedron*, 71, 2393–2399.
129 Davies, S.G., Fletcher, A.M., Frost, A.B., Lee, J.A., and Roberts, P.M. (2013) *Tetrahedron*, 69, 8885–8898.
130 Satoh, T. and Fukuda, Y. (2003) *Tetrahedron*, 59, 9803–9810.
131 Khettache, N., Bendjeddou, A., Berredjem, M., Regainia, Z., Montero, V., Menut, C., Aouf, N.-E., and Winum, J.-Y. (2006) *Synth. Commun.*, 36, 2299–2305.
132 Malkov, A.V., Stoncius, S., and Kocovsly, P. (2007) *Angew. Chem. Int. Ed.*, 46, 3722–3724.
133 Hodgson, D.M., Hughes, S.P., Thompson, A.L., and Heightman, T.D. (2008) *Org. Lett.*, 10, 3453–3456.

134 Chong, H.-S. and Chen, Y. (2013) *Org. Lett.*, 15, 5912–5915.
135 Huang, D., Liu, X., Li, L., Cai, Y., Liu, W., and Shi, Y. (2013) *J. Am. Chem. Soc.*, 135, 8101–8104.
136 Sweeney, J.B. and Cantrill, A.A. (2003) *Tetrahedron*, 59, 3677–3690.
137 Kapferer, P., Birault, V., Poisson, J.-F., and Vasella, A. (2003) *Helv. Chim. Acta*, 86, 2210–2218.
138 Breuning, A., Vicik, R., and Schirmeister, T. (2003) *Tetrahedron: Asymmetry*, 14, 3301–3312.
139 Kostyanovsky, R.G., Krutius, O.N., Stankevich, A.A., and Lyssenko, K.A. (2003) *Mendeleev Commun.*, 13, 223–225.
140 Voronkov, M.V., Gontcharov, A.V., Kanamarlapudi, R.C., Richardson, P.F., and Wang, Z.-M. (2005) *Org. Process Res. Dev.*, 9, 221–224.
141 Papaioannou, N., Blank, J.T., and Miller, S.J. (2003) *J. Org. Chem.*, 68, 2728–2734.
142 Fürmeier, S. and Metzger, J.O. (2003) *Eur. J. Org. Chem.*, 2003, 649–659.
143 Koohang, A., Bailey, J.L., Coates, R.M., Erickson, H.K., Owen, D., and Poulter, C.D. (2010) *J. Org. Chem.*, 75, 4769–4777.
144 Molinaro, C., Guilbault, A.-A., and Kosjek, B. (2010) *Org. Lett.*, 12, 3772–3775.
145 Arroyo, Y., Meana, A., Rodriguez, J.F., Santos, M., Sanz-Tejedor, M.A., and Garcia-Ruano, J.L. (2006) *Tetrahedron*, 62, 8525–8532.
146 Tiecco, M., Testaferri, L., Santi, C., Tomassini, C., Marini, F., Bagnoli, L., and Temperini, A. (2003) *Angew. Chem. Int. Ed.*, 42, 3131–3133.
147 Miniejew, C., Outurquin, F., and Pannecoucke, X. (2006) *Tetrahedron*, 62, 2657–2670.
148 Tsuchiya, Y., Kumamoto, T., and Ishikawa, T. (2004) *J. Org. Chem.*, 69, 8504–8505.
149 Bartoli, G., Bosco, M., Carlone, A., Locatelli, M., Melchiorre, P., and Sambri, L. (2004) *Org. Lett.*, 6, 3973–3975.
150 Ishikawa, T., Hada, K., Watanabe, T., and Isobe, T. (2001) *J. Am. Chem. Soc.*, 123, 7705–7706.
151 (a) Haga, T. and Ishikawa, T. (2005) *Tetrahedron*, 61, 2857–2869; (b) Ishikawa, T. (2006) *Arkivoc*, 7, 148–168.
152 Oda, Y., Hada, K., Miyata, M., Takahata, C., Hayashi, Y., Takahashi, M., Yajima, N., Fujinami, M., and Ishikawa, T. (2014) *Synthesis*, 46, 2201–2219.
153 Disadee, W. and Ishikawa, T. (2005) *J. Org. Chem.*, 70, 9399–9406.
154 Kumamoto, T., Suzuki, K., Kim, S.-k., Hoshino, K., Takahashi, M., Sato, H., Iwata, H., Ueno, K., Fukuzumi, M., and Ishikawa, T. (2010) *Helv. Chim. Acta*, 93, 2109–2113.
155 Khantikaew, I., Takahashi, M., Kumamoto, T., Suzuki, N., and Ishikawa, T. (2012) *Tetrahedron*, 68, 878–882.
156 Yamagiwa, N., Qin, H., Matsunaga, S., and Shibasaki, M. (2005) *J. Am. Chem. Soc.*, 127, 13419–13427.
157 Yamagiwa, N., Matsunaga, S., and Shibasaki, M. (2003) *J. Am. Chem. Soc.*, 125, 16178–16179.
158 Bew, S.P., Hughes, D.L., Savic, V., Soapi, K.M., and Wilson, M.A. (2006) *Chem. Commun.*, 33, 3513–3515.
159 Foltz, C., Bellemin-Laponnaz, S., Enders, M., Wadepohl, H., and Gade, L.H. (2008) *Org. Lett.*, 10, 305–308.

160 Hayes, J.F., Prévost, N., Prokes, I., Shipman, M., Slawin, A.M.Z., and Twin, H. (2003) *Chem. Commun.*, 12, 1344–1345.
161 Lopez Lopez, O., Fernandez-Bolanos, J.G., Lillelund, V.H., and Bols, M. (2003) *Org. Biomol. Chem.*, 1, 478–482.
162 Kale, A.S. and Deshmukh, A.R.A.S. (2005) *Synlett*, 15, 2370–2372.
163 (a) Satoh, N., Akiba, T., Yokoshima, S., and Fukuyama, T. (2007) *Angew. Chem. Int. Ed.*, 46, 5734–5736; (b) Satoh, N., Akiba, T., Yokoshima, S., and Fukuyama, T. (2009) *Tetrahedron*, 65, 3239–3245.
164 Zhou, Y. and Murphy, P.V. (2008) *Org. Lett.*, 10, 3777–3780.
165 Dahl, R.S. and Finney, N.S. (2004) *J. Am. Chem. Soc.*, 126, 8356–8357.
166 Feng, X., Duesler, E.N., and Mariano, P.S. (2005) *J. Org. Chem.*, 70, 5618–5623.
167 Li, Z., Quan, R.W., and Jacobsen, E.N. (1995) *J. Am. Chem. Soc.*, 117, 5889–5890.
168 (a) Pfaltz, A. (1993) *Acc. Chem. Res.*, 26, 339–345; (b) Ghosh, A.K., Mathivanan, P., and Cappiello, J. (1998) *Tetrahedron: Asymmetry*, 9, 1–45; (c) McManus, H.A. and Guiry, P.J. (2004) *Chem. Rev.*, 104, 4151–4202; (d) Desimoni, G., Faita, G., and Jorgensen, K.A. (2006) *Chem. Rev.*, 106, 3561–3651.
169 Evans, D.A., Woerpel, K.A., Hinman, M.M., and Faul, M.M. (1991) *J. Am. Chem. Soc.*, 113, 726–728.
170 Cranfill, D.C. and Lipton, M.A. (2007) *Org. Lett.*, 9, 3511–3513.
171 Leman, L., Sanière, L., Dauban, P., and Dodd, R.H. (2003) *Arkivoc*, 6, 126–134.
172 Kwong, H.-L., Liu, D., Chan, K.-Y., Lee, C.-S., Huang, K.-H., and Che, C.-M. (2004) *Tetrahedron Lett.*, 45, 3965–3968.
173 Fraille, J.M., Garcia, J.I., Lafuente, G., Mayoral, J.A., and Salvatella, L. (2004) *Arkivoc*, 4, 67–73.
174 (a) Taylor, S., Gullick, J., McMorn, P., Bethell, D., Bulman Page, P.C., Hancock, F.E., King, F., and Hutchings, G.J. (2003) *Top. Catal.*, 24, 43–50; (b) Taylor, S., Gullick, J., Galea, N., McMorn, P., Bethell, D., Bulman Page, P.C., Hancock, F.E., King, F., Willock, D.J., and Hutchings, G.J. (2003) *Top. Catal.*, 25, 81–88; (c) Gullick, J., Taylor, S., Ryan, D., McMorn, P., Coogan, M., Bethell, D., Bulman Page, P.C., Hancock, F.E., King, F., and Hutchings, G.J. (2003) *Chem. Commun.*, 22, 2808–2809; (d) Gullick, J., Ryan, D., McMorn, P., Bethell, D., King, F., Hancock, F.E., and Hutchings, G.J. (2004) *New J. Chem.*, 28, 1470–1478.
175 Ryan, D., McMorn, P., Bethell, D., and Hutchings, G. (2004) *Org. Biomol. Chem.*, 2, 3566–3572.
176 Estéoule, A., Duran, F., Retailleau, P., Dodd, R.H., and Dauban, P. (2007) *Synthesis*, 8, 1251–1260.
177 Malik, G., Estéoule, A., Retailleau, P., and Dauban, P. (2011) *J. Org. Chem.*, 76, 7438–7448.
178 Ma, L., Du, D.-M., and Xu, J. (2005) *J. Org. Chem.*, 70, 10155–10158.
179 (a) Xu, J., Ma, L., and Jiao, P. (2004) *Chem. Commun.*, 14, 1616–1617; (b) Ma, L., Jiao, P., Zhang, Q., and Xu, J. (2005) *Tetrahedron: Asymmetry*, 16, 3718–3734.
180 Ma, L., Jiao, P., Zhang, Q., Du, D.-M., and Xu, J. (2007) *Tetrahedron: Asymmetry*, 18, 878–884.
181 Ma, L., Du, D.-M., and Xu, J. (2006) *Chirality*, 18, 575–580.
182 Lebel, H. and Parmentier, M. (2010) *Pure Appl. Chem.*, 82, 1827–1833.

183 Lebel, H., Parmentier, M., Leogane, O., Ross, K., and Spitz, C. (2012) *Tetrahedron*, 68, 3396–3409.
184 Gök, Y., Noël, T., and Van der Eycken, J. (2010) *Tetrahedron: Asymmetry*, 21, 2275–2280.
185 Hajra, S. and Sinha, D. (2011) *J. Org. Chem.*, 76, 7334–7340.
186 Hajra, S. and Bar, S. (2011) *Chem. Commun.*, 47, 3981–3982.
187 Hajra, S. and Bar, S. (2011) *Tetrahedron: Asymmetry*, 22, 775–779.
188 Hajra, S. and Bar, S. (2012) *Tetrahedron: Asymmetry*, 23, 151–156.
189 Wang, X. and Ding, K. (2006) *Chem. Eur. J.*, 12, 4568–4575.
190 Suga, H., Kakehi, A., Ito, S., Ibata, T., Fudo, T., Watanabe, Y., and Kinoshita, Y. (2003) *Bull. Chem. Soc. Jpn.*, 76, 189–199.
191 Aaseng, J.E., Melnes, S., Reian, G., and Gautun, O.R. (2010) *Tetrahedron*, 66, 9790–9797.
192 Moberg, C. (1998) *Angew. Chem. Int. Ed. Engl.*, 37, 248–268.
193 Lam, T.C.H., Mak, W.-L., Wong, W.-L., Kwong, H.-L., Sung, H.H.Y., Lo,S.M.F., Williams, I.D., and Leung, W.-H. (2004) *Organometallics*, 23, 1247–1252.
194 Stones, G., Tripoli, R., McDavid, C.L., Roux-Duplâtre, K., Kennedy, A.R., Sherrington, D.C., and Gibson, C.L. (2008) *Org. Biomol. Chem.*, 6, 374–384.
195 Chanda, B.M., Vyas, R., and Landge, S.S. (2004) *J. Mol. Catal. A: Chem.*, 223, 57–60.
196 Liang, J.-L., Yuan, S.-X., Hong Chan, P.W., and Che, C.-M. (2003) *Tetrahedron Lett.*, 44, 5917–5920.
197 Hayes, C.J., Beavis, P.W., and Humphries, L.A. (2006) *Chem. Commun.*, 43, 4501–4502.
198 Fuit, C. and Müller, P. (2004) *Tetrahedron: Asymmetry*, 15, 1019–1026.
199 Fruit, C., Robert-Peillard, F., Bernardinelli, G., Müller, P., Dodd, R.H., and Dauban, P. (2005) *Tetrahedron: Asymmetry*, 16, 3484–3487.
200 Yamawaki, M., Tanaka, M., Abe, T., Anada, M., and Hashimoto, S. (2007) *Heterocycles*, 72, 709–721.
201 (a) Omura, K., Murakami, M., Uchida, T., Irie, R., and Katsuki, T. (2003) *Chem. Lett.*, 32, 354–355; (b) Uchida, T., Tamura, Y., Ohba, M., and Katsuki, T. (2003) *Tetrahedron Lett.*, 44, 7965–7968; (c) Omura, K., Uchida, T., Irie, R., and Katsuki, T. (2004) *Chem. Commun.*, 18, 2060–2061; (d) Kawabata, H., Omura, K., and Katsuki, T. (2006) *Tetrahedron Lett.*, 47, 1571–1574; (e) Kawabata, H., Omura, K., Uchida, T., and Katsuki, T. (2007) *Chem. Asian J.*, 2, 248–256.
202 Liang, J.-L., Yuan, S.-X., Huang, J.-S., and Che, C.-M. (2004) *J. Org. Chem.*, 69, 3610–3619.
203 Kim, C., Uchida, T., and Katsuki, T. (2012) *Chem. Commun.*, 48, 7188–7190.
204 Kukunaga, Y., Uchida, T., Ito, Y., Matsumoto, K., and Katsuki, T. (2012) *Org. Lett.*, 14, 4658–4661.
205 Chan, K.-H., Chan, X., Kar-Yan Lo, V., and Che, C.-M. (2014) *Angew. Chem. Int. Ed.*, 53, 2982–2987.
206 Jones, J.E., Ruppel, J.V., Gao, G.-Y., Moore, T.M., and Zhang, X.P. (2008) *J. Org. Chem.*, 73, 7260–7265.
207 Jin, L.-M., Xu, X., Lu, H., Cui, X., Wojtas, L., and Zhang, X.P. (2013) *Angew. Chem. Int. Ed.*, 52, 5309–5313.

208 Tao, J., Jin, L.-M., and Zhang, X.P. (2014) *Beilstein J. Org. Chem.*, 10, 1282–1289.
209 Nakanishi, M., Salit, A.-F., and Bolm, C. (2008) *Adv. Synth. Catal.*, 350, 1835–1840.
210 Li, Z.-K., Li, Y., Lei, L., Che, C.-M., and Zhou, X.-G. (2005) *Inorg. Chem. Commun.*, 8, 307–309.
211 (a) Pellissier, H. (2007) *Tetrahedron*, 63, 9267–9331; (b) McGarrigle, E.M., Myers, E.L., Illa, O., Shaw, M.A., Riches, S.L., and Aggarwal, V.K. (2007) *Chem. Rev.*, 107, 5841–5883.
212 Fioravanti, S., Mascia, M.G., Pellacani, L., and Tardella, P.A. (2004) *Tetrahedron*, 60, 8073–8077.
213 Murugan, E. and Siva, A. (2005) *Synthesis*, 12, 2022–2028.
214 Armstrong, A., Baxter, C.A., Lamont, S.G., Pape, A.R., and Wincewicz, R. (2007) *Org. Lett.*, 9, 351–353.
215 Armstrong, A., Pullin, R.D.C., Jenner, C.R., Foo, K., and White, A.J.P. (2014) *Tetrahedron: Asymmetry*, 25, 74–86.
216 Pesciaioli, F., De Vincentiis, F., Galzerano, P., Bencivenni, G., Bartoli, G., Mazzanti, A., and Melchiorre, P. (2008) *Angew. Chem. Int. Ed.*, 47, 8703–8706.
217 De Vincentiis, F., Bencivenni, G., Pesciaioli, F., Mazzanti, A., Bartoli, G., Galzerano, P., and Melchiorre, P. (2010) *Chem. Asian J.*, 5, 1652–1656.
218 Minakata, S., Murakami, Y., Tsuruoka, R., Kitanaka, S., and Komatsu, M. (2008) *Chem. Commun.*, 47, 6363–6365.
219 Murakami, Y., Takeda, Y., and Minakata, S. (2011) *J. Org. Chem.*, 76, 6277–6285.
220 Shen, Y.-M., Zhao, M.-X., Xu, J., and Shi, Y. (2006) *Angew. Chem. Int. Ed.*, 45, 8005–8008.
221 Menjo, Y., Hamajima, A., Sasaki, N., and Hamada, Y. (2011) *Org. Lett.*, 13, 5744–5747.
222 De Fusco, C., Fuoco, T., Croce, G., and Lattanzi, A. (2012) *Org. Lett.*, 14, 4078–4081.
223 Page, P.C.B., Bordogna, C., Strutt, I., Chan, Y., and Buckley, B.R. (2013) *Synlett*, 24, 2067–2072.
224 Vesely, J., Ibrahem, I., Zhao, G.-L., Rios, R., and Cordova, A. (2007) *Angew. Chem. Int. Ed.*, 46, 778–781.
225 Deiana, L., Dziedzic, P., Zhao, G.-L., Vesely, J., Ibrahem, I., Rios, R., Sun, J., and Cordova, A. (2011) *Chem. Eur. J.*, 17, 7904–7917.
226 Nemoto, T., Hayashi, M., Xu, D., Hamajima, A., and Hamada, Y. (2014) *Tetrahedron: Asymmetry*, 25, 1133–1137.
227 Arai, H., Sugaya, N., Sasaki, N., Makino, K., Lectard, S., and Hamada, Y. (2009) *Tetrahedron Lett.*, 50, 3329–3332.
228 (a) Desmarchelier, A., Pereira de Sant'Ana, D., Terrasson, V., Campagne, J.M., Moreau, X., Greck, C., and de Figueiredo, R.M. (2011) *Eur. J. Org. Chem.*, 2011, 4046–4052; (b) Desmarchelier, A., Coeffard, V., Moreau, X., and Greck, C. (2012) *Chem. Eur. J.*, 18, 13222–13225.
229 Jiang, H., Halskov, K.S., Johansen, T.K., and Jørgensen, K.A. (2011) *Chem. Eur. J.*, 17, 3842–3846.
230 Halskov, K.S., Naicker, T., Jensen, M.E., and Jørgensen, K.A. (2013) *Chem. Commun.*, 49, 6382–6384.

231 For examples, see:(a) Albrecht, L., Jiang, H., Dickmeiss, G., Gschwend, B., Hansen, S.G., and Jørgensen, K.A. (2010) *J. Am. Chem. Soc.*, 132, 9188–9196; (b) Albrecht, L., Ransborg, L.K., Gschwend, B., and Jørgensen, K.A. (2010) *J. Am. Chem. Soc.*, 132, 17886–17893; (c) Albrecht, L., Albrecht, A., Ransborg, L.K., and Jørgensen, K.A. (2011) *Chem. Sci.*, 2, 1273–1277; (d) Albrecht, L., Ransborg, L.K., Albrecht, A., Lykke, L., and Jørgensen, K.A. (2011) *Chem. Eur. J.*, 17, 13240–13246.

232 Molnar, I.G., Tanzer, E.-M., Daniliuc, C., and Gilmour, R. (2014) *Chem. Eur. J.*, 20, 794–800.

233 Worgull, D., Dickmeiss, G., Jensen, K.L., Franke, P.T., Holub, N., and Jørgensen, K.A. (2011) *Chem. Eur. J.*, 17, 4076–4080.

234 (a) Jacobsen, E.N. (1999) in *Comprehensive Asymmetric Catalysis II*, Chapter 17 (eds E.N. Jacobsen, A. Pfaltz, and H. Yamamoto), Springer, Berlin, p. 607; (b) Katsuki, T. (2003) in *Comprehensive Coordination Chemistry II*, vol. 9, Chapter 9.4 (ed. J. McCleverty), Elsevier Science Ltd., Oxford, p. 207.

235 Krumper, J.R., Gerisch, M., Suh, J.M., Bergman, R.G., and Tilley, T.D. (2003) *J. Org. Chem.*, 68, 9705–9710.

236 Redlich, M. and Hossain, M.M. (2004) *Tetrahedron Lett.*, 45, 8987–8990.

237 Egloff, J., Ranocchiari, M., Schira, A., Schotes, C., and Mezzetti, A. (2013) *Organometallics*, 32, 4690–4701.

238 Desai, A.A., Moran-Ramallal, R., and Wulff, W.D. (2011) *Org. Synth.*, 88, 224–237.

239 Lu, Z., Zhang, Y., and Wulff, W.D. (2007) *J. Am. Chem. Soc.*, 129, 7185–7194.

240 Zhang, Y., Desai, A., Lu, Z., Hu, G., Ding, Z., and Wulff, W.D. (2008) *Chem. Eur. J.*, 14, 3785–3803.

241 Patwardhan, A.P., Pulgam, V.R., Zhang, Y., and Wulff, W.D. (2005) *Angew. Chem. Int. Ed.*, 44, 6169–6172.

242 (a) Mukherjee, M., Gupta, A.K., Lu, Z., Zhang, Y., and Wulff, W.D. (2010) *J. Org. Chem.*, 75, 5643–5660; (b) Zhang, Y., Lu, Z., Desai, A., and Wulff, W.D. (2008) *Org. Lett.*, 10, 5429–5432.

243 Hu, G., Gupta, A.K., Huang, R.H., Mukherjee, M., and Wulff, W.D. (2010) *J. Am. Chem. Soc.*, 132, 14669–14675.

244 Desai, A.A., Ren, H., Mukherjee, M., and Wulff, W.D. (2011) *Org. Process Res. Dev.*, 15, 1108–1115.

245 Vetticatt, M.J., Desai, A.A., and Wulff, W.D. (2013) *J. Org. Chem.*, 78, 5142–5152.

246 Wang, Z., Li, F., Zhao, L., He, Q., Chen, F., and Zheng, C. (2011) *Tetrahedron*, 67, 9199–9203.

247 Yu, S., Rabalakos, C., Mitchell, W.D., and Wulff, W.D. (2005) *Org. Lett.*, 7, 367–369.

248 Wipf, P. and Lyon, M.A. (2007) *Arkivoc*, 12, 91–98.

249 Deng, Y., Lee, Y.R., Newman, C.A., and Wulff, W.D. (2007) *Eur. J. Org. Chem.*, 2007, 2068–2071.

250 Desai, A.A. and Wulff, W.D. (2010) *J. Am. Chem. Soc.*, 132, 13100–13103.

251 Vetticatt, M.J., Desai, A.A., and Wulff, W.D. (2010) *J. Am. Chem. Soc.*, 132, 13104–13107.

252 Huang, L. and Wulff, W.D. (2011) *J. Am. Chem. Soc.*, 133, 8892–8895.

253 Huang, L., Zhang, Y., Staples, R.J., Huang, R.H., and Wulff, W.D. (2012) *Chem. Eur. J.*, 18, 5302–5313.

254 Hashimoto, T., Galvez, A.O., and Maruoka, K. (2013) *J. Am. Chem. Soc.*, 135, 17667–17670.

255 (a) Ugi, I., Meyr, R., Fetzer, U., and Steinbrückner, C. (1959) *Angew. Chem. Int. Ed. Engl.*, 71, 386–388; (b) Orru, R.V.A. and de Greef, M. (2003) *Synthesis*, 2003, 1471–1499.

256 (a) Tietze, L.F. and Modi, A. (2000) *Med. Res. Rev.*, 20, 304–322; (b) Zhu, J. and Bienaymé, H. (eds) (2005) *Multicomponent Reactions*, Wiley-VCH Verlag GmbH, Weinheim; (c) Ramon, D.J. and Yus, M. (2005) *Angew. Chem. Int. Ed.*, 44, 1602–1634; (d) Guillena, G., Ramon, D.J., and Yus, M. (2007) *Tetrahedron: Asymmetry*, 18, 693–700; (e) Orru, R.V.A. and Ruijter, E. (eds) (2010) *Synthesis of Heterocycles via Multicomponent Reactions*, Topics in Heterocyclic Chemistry, vols. 1 and 2, Springer, Berlin.

257 (a) Posner, G.H. (1986) *Chem. Rev.*, 86, 831–844; (b) Tietze, L.F. and Beifuss, U. (1993) *Angew. Chem. Int. Ed. Engl.*, 32, 131–163; (c) Tietze, L.F. (1996) *Chem. Rev.*, 96, 115–136; (d) Tietze, L.F., Brasche, G., and Gericke, K. (2006) *Domino Reactions in Organic Synthesis*, Wiley-VCH Verlag GmbH, Weinheim; (e) Pellissier, H. (2006) *Tetrahedron*, 62, 2143–2173; (f) Pellissier, H. (2006) *Tetrahedron*, 62, 1619–1665; (g) Enders, D., Grondal, C., and Hüttl, M.R.M. (2007) *Angew. Chem. Int. Ed.*, 46, 1570–1581; (h) D'Souza, D.M. and Müller, T.J.J. (2007) *Chem. Soc. Rev.*, 36, 1095–1108; (i) Alba, A.-N., Companyo, X., Viciano, M., and Rios, R. (2009) *Curr. Org. Chem.*, 13, 1432–1474; (j) Cowen, B.J. and Miller, S.J. (2009) *Chem. Soc. Rev.*, 38 (11), 3102–3116, Special Issue on Rapid Formation of Molecular Complexity in Organic Synthesis; (k) Ruiz, M., Lopez-Alvarado, P., Giorgi, G., and Menéndez, J.C. (2011) *Chem. Soc. Rev.*, 40, 3445–3454; (l) Albrecht, L., Jiang, H., and Jørgensen, K.A. (2011) *Angew. Chem. Int. Ed.*, 50, 8492–8509; (m) Pellissier, H. (2012) *Adv. Synth. Catal.*, 354, 237–294; (n) De Graaff, C., Ruijter, E., and Orru, R.V.A. (2012) *Chem. Soc. Rev.*, 41, 3969–4009; (o) Pellissier, H. (2013) *Chem. Rev.*, 113, 442–524.

258 (a) Zhang, Z. and Antilla, J.C. (2012) *Angew. Chem. Int. Ed.*, 51, 11778–11782; (b) Vaxelaire, C., Winter, P., and Christmann, M. (2011) *Angew. Chem. Int. Ed.*, 50, 3605–3607; (c) Nicolaou, K.C. and Chen, J.S. (2009) *Chem. Soc. Rev.*, 38, 2993–3009; (d) Touré, B.B. and Hall, D.G. (2009) *Chem. Rev.*, 109, 4439–4486; (e) Colombo, M. and Peretto, I. (2008) *Drug Discovery Today*, 13, 677–684; (f) Hulme, C. and Gore, V. (2003) *Curr. Med. Chem.*, 10, 51–80; (g) Padwa, A. and Bur, S.K. (2007) *Tetrahedron*, 63, 5341–5378.

259 Ren, H. and Wulff, W.D. (2010) *Org. Lett.*, 12, 4908–4911.

260 Gupta, A.K., Mukherjee, M., and Wulff, W.D. (2011) *Org. Lett.*, 13, 5866–5869.

261 Gupta, A.K., Mukherjee, M., Hu, G., and Wulff, W.D. (2012) *J. Org. Chem.*, 77, 7932–7944.

262 Mukherjee, M., Zhou, Y., Gupta, A.K., Guan, Y., and Wulff, W.D. (2014) *Eur. J. Org. Chem.*, 2014 (7), 1386–1390.

263 Huang, L., Zhao, W., Staples, R.J., and Wulff, W.D. (2013) *Chem. Sci.*, 4, 622–628.

264 Hashimoto, T., Uchiyama, N., and Maruoka, K. (2008) *J. Am. Chem. Soc.*, 130, 14380–14381.

265 Hashimoto, T., Nakatsu, H., Yamamoto, K., and Maruoka, K. (2011) *J. Am. Chem. Soc.*, 133, 9730–9733.
266 Chai, Z., Bouillon, J.-P., and Cahard, D. (2012) *Chem. Commun.*, 48, 9471–9473.
267 (a) Aggarwal, A.K., Thompson, A., Jones, R.V.H., and Standen, M.C.H. (1996) *J. Org. Chem.*, 61, 8368–8369; (b) Aggarwal, V.K. (1998) *Synlett*, 4, 329–336.
268 Aggarwal, V.K. and Winn, C.L. (2004) *Acc. Chem. Res.*, 37, 611–620.
269 Gui, Y., Shen, S., Wang, H.-Y., Li, Z.-Y., and Huang, Z.-Z. (2007) *Chem. Lett.*, 36, 1436–1437.
270 Huang, M.-T., Wu, H.-Y., and Chein, R.-J. (2014) *Chem. Commun.*, 50, 1101–1103.
271 S.-H. Wang, R.-J. Chein, *Tetrahedron* 2016, 72, 2607–2615.
272 Timén, A.S. and Somfai, P. (2003) *J. Org. Chem.*, 68, 9958–9963.
273 Kim, S.K. and Jacobsen, E.N. (2004) *Angew. Chem. Int. Ed.*, 43, 3952–3954.
274 Vedejs, E., Bhanu Prasad, A.S., Kendall, J.T., and Russel, J.S. (2003) *Tetrahedron*, 59, 9849–9856.
275 (a) Watson, I.D.G., Styler, S.A., and Yudin, A.K. (2004) *J. Am. Chem. Soc.*, 126, 5086–5087; (b) Watson, I.D.G. and Yudin, A.K. (2005) *J. Am. Chem. Soc.*, 127, 17516–17529; (c) Watson, I.D.G., Yu, L., and Yudin, A.K. (2006) *Acc. Chem. Res.*, 39, 194–206.
276 Fadeyi, O.O., Schulte, M.L., and Lindsley, C.W. (2010) *Org. Lett.*, 12, 3276–3278.
277 Halland, N., Braunton, A., Bachmann, S., Marigo, M., and Jørgensen, K.A. (2004) *J. Am. Chem. Soc.*, 126, 4790–4791.
278 Senter, T.J., O'Reilly, M.C., Chong, K.M., Sulikowski, G.A., and Lindsley, C.W. (2015) *Tetrahedron Lett.*, 56, 1276–1279.
279 Hayashi, Y., Urushima, T., Sakamoto, D., Torii, K., and Ishikawa, H. (2011) *Chem. Eur. J.*, 17, 11715–11718.
280 Larson, S.E., Li, G., Rowland, G.B., Junge, D., Huang, R., Woodcock, H.L., and Antilla, J.C. (2011) *Org. Lett.*, 13, 2188–2191.
281 Blaser, H.U. and Schmidt, E. (2004) *Asymmetric Catalysis on Industrial Scale*, Wiley-VCH Verlag GmbH, Weinheim.
282 (a) Kagan, H.B. and Fiaud, J.C. (1988) *Top. Stereochem.*, 18, 249–330; (b) Hoveyda, A.H. and Didiuk, M.T. (1998) *Curr. Org. Chem.*, 2, 489–526; (c) Cook, G.R. (2000) *Curr. Org. Chem.*, 4, 869–885; (d) Keith, M., Larrow, J.F., and Jacobsen, E.N. (2001) *Adv. Synth. Catal.*, 343, 5–26; (e) Robinson, D.E.J.E. and Bull, S.D. (2003) *Tetrahedron: Asymmetry*, 14, 1407–1446; (f) Vedejs, E. and Jure, M. (2005) *Angew. Chem. Int. Ed.*, 44, 3974–4001; (g) Jarvo, E.R. and Miller, S.J. (2004) in *Comprehensive Asymmetric Catalysis, Supplement* (eds E.N. Jacobsen, A. Pfaltz, and H. Yamamoto), Springer, Berlin, pp. 189–206.
283 Kumar, H.M.S., Rao, M.S., Chakravarthy, P.P., and Yadav, J.S. (2004) *Tetrahedron: Asymmetry*, 15, 127–130.
284 Wang, J.-Y., Wang, D.-X., Zheng, Q.-Y., Huang, Z.-T., and Wang, M.-X. (2007) *J. Org. Chem.*, 72, 2040–2045.
285 Wang, J.-Y., Wang, D.-X., Pan, J., Huang, Z.-T., and Wang, M.-X. (2007) *J. Org. Chem.*, 72, 9391–9394.
286 Moran-Ramallal, R., Liz, R., and Gotor, V. (2007) *Org. Lett.*, 9, 521–524.

287 Moran-Ramallal, R., Liz, R., and Gotor, V. (2010) *J. Org. Chem.*, 75, 6614–6624.
288 Sakai, T., Liu, Y., Ohta, H., Korenaga, T., and Ema, T. (2005) *J. Org. Chem.*, 70, 1369–1375.
289 Pellissier, H. (2011) *Adv. Synth. Catal.*, 353, 1613–1666.
290 (a) Bredig, G. and Fajans, K. (1908) *Ber. Dtsch. Chem. Ges.*, 41, 752–763; (b) Fajans, K. (1910) *Z. Phys. Chem.*, 73, 25–96.
291 Ohmatsu, K., Hamajima, Y., and Ooi, T. (2012) *J. Am. Chem. Soc.*, 134, 8794–8797.
292 Cockrell, J., Wilhelmsen, C., Rubin, H., Martin, A., and Morgan, J.B. (2012) *Angew. Chem. Int. Ed.*, 51, 9842–9845.

3
Asymmetric Epoxidation

3.1 Introduction

Epoxides are strained three-membered heterocycles of extraordinary importance as versatile synthetic intermediates and in the Nature. Calculations and experimental data enabled the estimation of high strain energy associated with differently substituted epoxides ranging from 24 to 28 kcal mol^{-1} [1]. This intrinsic property justifies easy ring opening of epoxides, a process generally proceeding in an S_N2 fashion with inversion of configuration. The regioselectivity of the nucleophilic attack is nevertheless highly dependent on the structural features of epoxides and reaction conditions. According to the nature of nucleophiles employed, a countless examples of stereo- and regioselective ring-opening reactions of epoxides have been reported in the literature to afford 1,2-functionalized products such as the most common amino alcohols, diols, hydroxy sulfides, hydrazino alcohols, 1,2-halohydrins, 1,2-cyanohydrins, alcohols, and pharmaceuticals in stereodefined manner [2]. Given the considerable number of asymmetric methodologies accessible to synthesize chiral epoxides, chemists have judiciously applied the combination of asymmetric epoxidation/ring-opening reactions, including powerful intermolecular cascade ring-opening versions to accomplish several total syntheses of natural and biologically active molecules [3]. Asymmetric ring-opening reactions of achiral and readily available *meso*-epoxides have been the focus of several studies, expanding the possibility of producing chiral 1,2-amino alcohols, 1,2-diols, and 1,2-hydroxy sulfides, which are also recognized among the most important ligands used in asymmetric metal-based catalysis [4]. Kinetic resolution of racemic epoxides has been used to obtain challenging chiral terminal epoxides and 1,2 functionalized alcohols, both of which are privileged structural motives for industrial applications [4a, 5]. Noteworthy examples of commonly used pharmaceuticals, such as β-adrenergic blocking agents used for the treatment of hypertension and angina pectoris, can be easily obtained from ring opening of chiral terminal epoxides including (*S*)-propranolol [6], (*R*)-dichloroisoproterenol [7]. Similarly, in living organisms, the highly selective ring-opening reaction of chiral epoxides is the key reaction associated with a biological activity. Carboxypeptidase A, a zinc-containing proteolytic enzyme of physiological importance, has been found to be irreversibly

Asymmetric Synthesis of Three-Membered Rings, First Edition.
Hélène Pellissier, Alessandra Lattanzi and Renato Dalpozzo.
© 2017 Wiley-VCH Verlag GmbH & Co. KGaA. Published 2017 by Wiley-VCH Verlag GmbH & Co. KGaA

inactivated, with the highest activity among all the possible stereoisomers, by terminal epoxide (2R,3S)-2-benzyl-3,4-epoxybutanoic acid via S_N2-type ring cleavage [8]. Moreover, [3+2] asymmetric cycloaddition reactions of epoxides have also recently emerged as a valuable tool to obtain chiral dihydro- or tetrahydrofuran derivatives, important core structures present in natural products [9]. Besides the relevance of chiral epoxides in synthesis, a great number of natural products and bioactive compounds contain the epoxide subunit in their structure as exemplified by sex pheromone for the gypsy moth (+)-disparlure [10], antibiotic agent monocillin I [11], potent oral hypoglycemic and antiketogenic agent in mammals (R)-methyl palmoxirate [12], anticancer agents ovalicin, fumagillin, and epothilones A and B [13].

In this chapter, advances in the asymmetric synthesis of epoxides using chiral substrates (auxiliaries) as well as chiral catalysts will be illustrated by collecting the literature published in the past 11 years.

This chapter is divided into eight sections, dealing successively with asymmetric synthesis of epoxides based on the use of chiral auxiliaries, different methods employed to prepare them, for instance, asymmetric epoxidation of alkenes, subdivided into two sections, dealing with metal-based-catalyzed epoxidation and organocatalyzed epoxidation of alkenes. Other sections illustrate the kinetic resolution of racemic epoxides followed by asymmetric sulfur-ylide-mediated epoxidation of carbonyl compounds, asymmetric Darzens reaction, and finally biocatalyzed synthesis of epoxides. Selected examples of the methodologies applied to synthesis of epoxides to prepare pharmaceuticals, natural or biologically active compounds, are also included.

3.2 Asymmetric Epoxidations Based on the Use of Chiral Auxiliaries

The epoxidation of alkenes is undoubtedly the most investigated and convenient approach to obtain epoxides, thanks to the large variety of commercially available alkenes or readily prepared via carbon–carbon double-bond-forming reactions [14]. A stereoselective epoxidation can be approached according to two strategies: (i) reagent-controlled process, where a chiral oxidant is used; (ii) substrate-controlled process, where a chiral alkene undergoes the epoxidation by an achiral oxidant. In the latter case, the epoxidation can proceed on an alkene, with preexisting chiral center previously inserted in the scaffold, which will direct the preferential approach of the oxidant over one of the two faces to give a preferential diastereoisomer. This intermediate can maintain all previously installed stereocenters as it generally occurs in an impressive number of synthetic examples reported in the literature [15]. When employing the chiral auxiliary approach [16], after the diastereoselective epoxidation, cleavage of the chiral auxiliary ends up with an enantiomerically enriched epoxide useful as a building block for further elaboration. Although both strategies require a stoichiometric amount of the chiral reagent/substrate, the main drawback of the chiral auxiliary approach

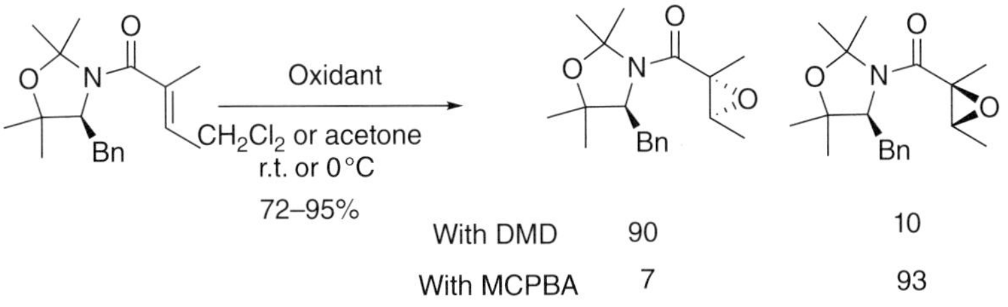

	With DMD	With MCPBA
	90	7
	10	93

Scheme 3.1 Chiral auxiliary approach in the epoxidation of epoxides with DMD and MCPBA.

concerns the poor step economy, as two more steps are necessary to install and remove the auxiliary from the reagent. General methods based on the chiral auxiliary approach met with poor success during the past decade [17]. In 2005, Pastor and coauthors illustrated the diastereoselective epoxidation of optically active tiglic acid derivatives incorporating the 2,2-dimethyloxazolidine chiral auxiliary using dimethyl dioxirane (DMD) and MCPBA as the oxidants (Scheme 3.1) [18].

A reversed sense of diastereoselectivity has been achieved by using the two oxidants, which was ascribed to stereoelectronic effect in the approach of DMD to the nonplanar conformational preference of the carbon–carbon double bond relative to the carbonyl, whereas hydrogen-bonding interactions would be involved in the MCPBA-mediated epoxidation. Epoxidation of various chiral *N*-enoylsultams has been reported by Zhang and Chen using urea–H_2O_2 complex (UHP) and trifluoroacetic anhydride [19]. However, modest control of the diastereoselectivity has been observed (dr up to 85:15).

3.3 Asymmetric Metal-Catalyzed Epoxidations

Over the past decade, a vast amount of research has been focused on the development of asymmetric metal-catalyzed procedures for the epoxidation of functionalized and unfunctionalized alkenes [20]. Most of the efforts were directed toward the expansion of metal/chiral ligand systems suitable for the process as well as on the substrate scope. Moreover, more attention has been paid to the environmental concerns with an increasing number of processes using H_2O_2 as the oxidant, low catalyst loading, or supported catalysts suitable for recycling.

3.3.1 Ti-, Zr-, Hf-Catalyzed Epoxidations

The Katsuki–Sharpless asymmetric epoxidation (KSAE) of allylic alcohols mediated by Ti/L-DET/*tert*-butyl hydroperoxide (TBHP) represents a milestone discovery in the area of asymmetric synthesis [21]. The reaction can be successfully applied to a wide range of (*E*)-2,3-substituted and trisubstituted allylic alcohols to yield the corresponding epoxyalcohols with complete chemoselectivity, good-to-high yield, and high enantioselectivity (Scheme 3.2) [22].

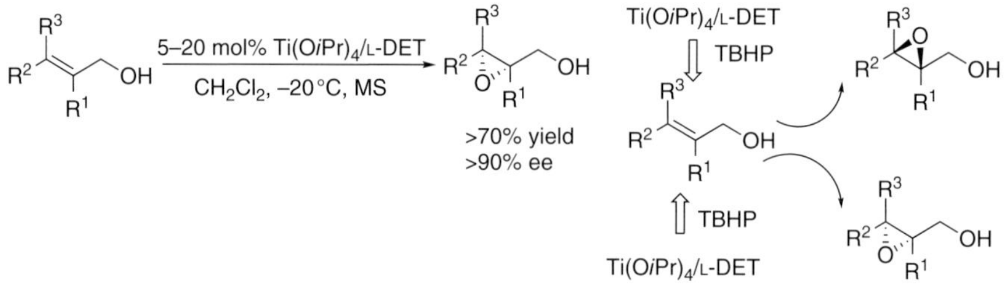

Scheme 3.2 Katsuki–Sharpless asymmetric epoxidation of allylic alcohols.

Typical reaction conditions require catalytic loading of Ti(O*i*Pr)$_4$/L-DET in the presence of activated molecular sieves in dichloromethane as solvent at −20 °C. Lower degrees of conversion and enantiocontrol have been observed when reacting the (Z)-3-substituted allylic alcohols. An empirical rule, tested on a wide variety of allylic alcohols, helps to predict the final absolute configuration of the epoxy alcohol when using L- or D-tartrate esters (Scheme 3.2). The presence of different functional groups is tolerated in the allylic alcohol, although easily oxidizable amines, mercaptans, phosphanes or carboxylic esters, and phenols are not tolerated. Another successful application of the KSAE involves the kinetic resolution of racemic secondary allylic alcohols to afford unreactive allylic alcohols [23] and epoxides, with three contiguous stereogenic centers, with high-to-excellent diastereo- and enantioselectivity [5c, 24]. The desymmetrization of prochiral dialkenylcarbinols and *meso*-secondary diallylic alcohols, under KSAE oxidative conditions, has been successfully demonstrated [25]. Spectroscopic and theoretical studies have been carried out to elucidate the mechanism of the KSAE, which led to the identification of the crucial species as a dinuclear oxo-bridged complex [26]. These investigations showed that dynamic equilibria among different species are involved. Homoallylic alcohols are epoxidized under the KSAE conditions at lower rates, with modest and inverted sense of enantiofacial control (up to 55% ee) [27]. Since its discovery, the KSAE has found continuous application as the key step in the total synthesis of biologically active and natural compounds [3a, 28]. Recently, Echavarren and coauthors illustrated a short asymmetric synthesis of three natural aromadendrene sesquiterpenes, such as (−)-epiglobulol, (−)-4α,7α-aromadendranediol and (−)-4β,7α-aromadendranediol, using as the first step the KSAE of low cost and commercially available (E,E)-farnesol (Scheme 3.3) [29]. These compounds, widespread in plant species, are endowed with a variety of antiviral, antibacterial, antifungal activities [30]. The reaction carried out with 5 mol% of Ti(O*i*Pr)$_4$/L-DIPT at −48 °C afforded the epoxide in high yield with good enantioselectivity. A following stereodivergent gold-catalyzed reaction was applied to install four new stereogenic centers into the bicyclic structures.

Muthukrishnan and coworkers recently illustrated a simple synthesis of (R)-2-benzylmorpholine, an appetite-suppressant agent, via Sharpless asymmetric epoxidation of commercially available E-cinnamyl alcohol as the first key step to set up the stereocenter (Scheme 3.4) [31]. The epoxy alcohol was isolated in high

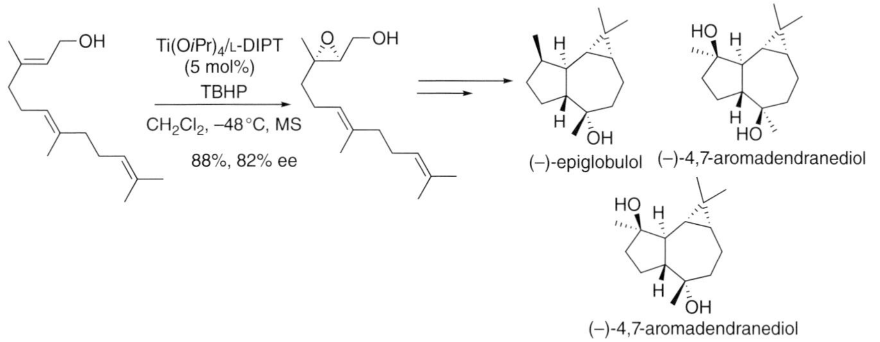

Scheme 3.3 Katsuki–Sharpless asymmetric epoxidation of (E,E)-farnesol as key step to sesquiterpene derivatives.

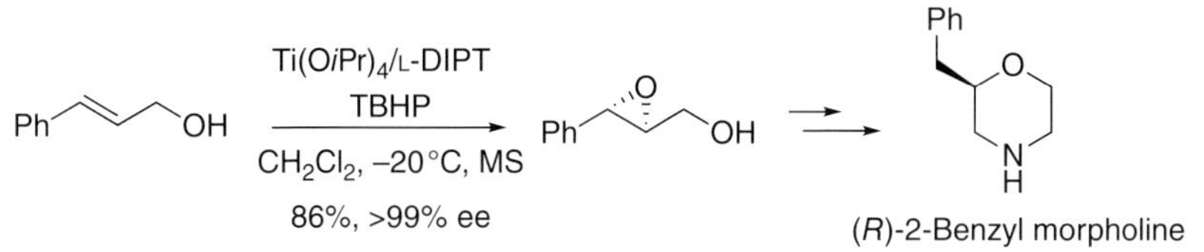

Scheme 3.4 Katsuki–Sharpless asymmetric epoxidation of E-cinnamyl alcohol as the key step to appetite suppressant agent (R)-2-benzylmorpholine.

yield with excellent enantioselectivity under standard KSAE conditions. The final (R)-2-benzylmorpholine was obtained in 24% overall yield.

In the total enantioselective synthesis of potent antibiotic GSK966587, KSAE has been used to install the sole stereogenic center of the drug (Scheme 3.5) [32]. The epoxidation smoothly proceeded with a catalytic loading of the metal complex, in the presence of cumyl hydroperoxide (CHP), to give the epoxy alcohol in 81% yield with 90% ee. The synthesis afforded the target compound in eight steps and 25% overall yield.

Yamamoto recently developed a conceptually new optically active ligand that can accommodate two closely positioned independent titanium centers able to activate TBHP and primary/tertiary homoallylic alcohols or 2-allylic phenols for the epoxidation (Scheme 3.6) [33]. The reaction showed to be very practical since it can be performed at room temperature using aqueous TBHP, in contrast to KSAE conditions, over a variety of di-, trisubstituted challenging homoallylic primary and tertiary alcohols to give the corresponding epoxides in moderate-to-high yield and enantioselectivity according to the type of substitution. The presence of the OH group was found necessary to coordinate to one Ti center as attested by the lack of reactivity observed for styrene. Interestingly, the first asymmetric epoxidation of 2-allylic phenols afforded epoxides in good yield and enantioselectivity. The same system has been found to be useful in the kinetic resolution of secondary homoallylic alcohols and asymmetric sulfoxidation of γ-hydroxypropyl sulfides.

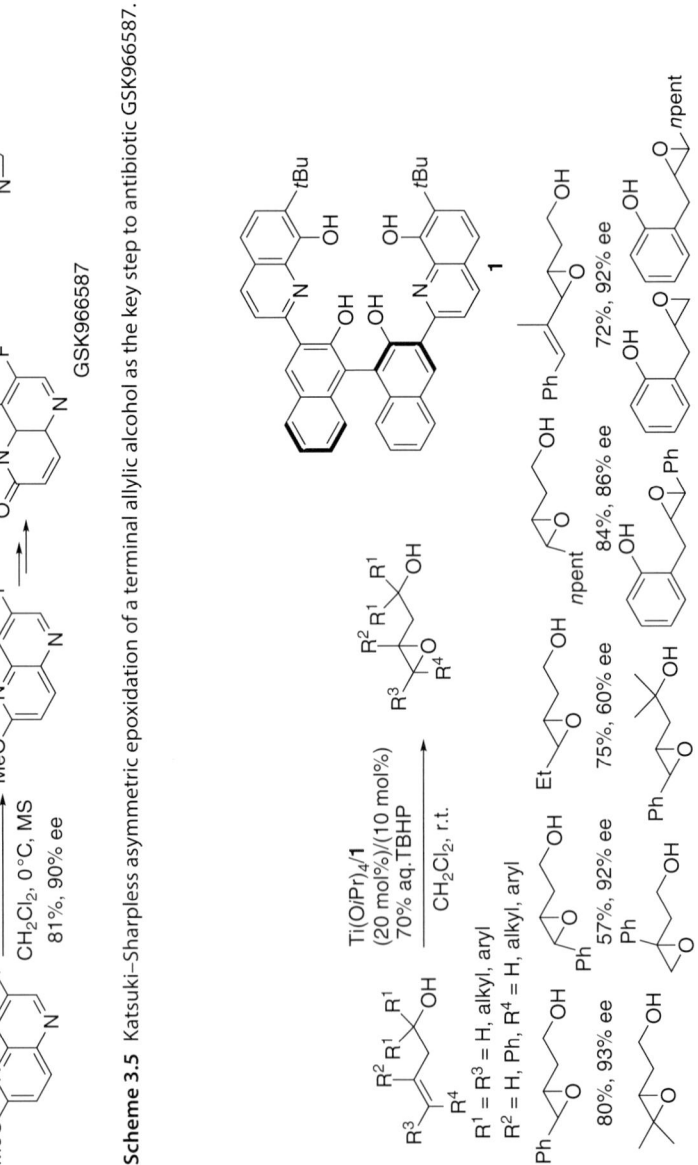

Scheme 3.5 Katsuki–Sharpless asymmetric epoxidation of a terminal allylic alcohol as the key step to antibiotic GSK966587.

Scheme 3.6 Ti-catalyzed asymmetric epoxidation of homoallylic alcohols and 2-allylic phenols.

In the past decade, more efforts have been put to develop sustainable methodologies, requiring mild reaction conditions, safe reagents, minimum waste, reduction, whenever possible, of reaction steps, thus saving cost and time. In the area of asymmetric epoxidation, green and cheap H_2O_2 replaced classical alkyl hydroperoxides or NaOCl in several processes, with the main advantage of producing water as environmentally friendly by-product [34]. In 2005, Katsuki and coworkers paved the way for the development of novel Ti-based catalysts suitable for the asymmetric epoxidation of challenging olefins under mild and sustainable conditions [35]. For example, chiral salalen–di-μ-oxo titanium complex **2** was found to efficiently catalyze, at only 1 mol% (up to 3 mol%) loading, the epoxidation of unfunctionalized alkenes using aqueous H_2O_2 in dichloromethane or ethyl acetate at room temperature (Scheme 3.7). This system is particularly suited for the oxidation of challenging aliphatic and acyclic Z-disubstituted, trisubstituted, and terminal alkenes, whose epoxides were recovered in high yield with good-to-excellent enantioselectivity, without observing any other isomeric epoxide. A concerted mechanism was suggested, where a monomeric peroxotitanium species, activated by an intramolecular hydrogen bond with the secondary amine, would be involved. Rather surprisingly, Katsuki observed that the corresponding titanium(salen) complexes found to react with H_2O_2 to form peroxo complexes responsible for sulfide oxidation [36] were inert to olefin epoxidation. Moreover, the enantiomeric catalyst, *ent-*2, was used in the highly enantioselective epoxidation of aromatic Z-alkenylsilanes. These products can be regarded as precursors of chiral monosubstituted or 1,1-disubstituted epoxides, via simple desilylation or alkylation and desilylation procedures [37]. A simple di-μ-oxo titanium–salan complex **3**, of lower molecular weight, was developed by the same group and checked under the same reaction conditions, achieving nearly similar results compared to complex **2** (Scheme 3.7) [38]. A more appealing procedure with *in situ* generation of the catalytically active species by mixing Ti(O*i*Pr)$_4$ and similar salan ligands enabled work at 1 mol% loading of the catalyst, achieving excellent conversion and level of enantioselectivity [39].

In 2007, Berkessel developed a rapid access to a variety of simplified salalen ligands, lacking the axial chirality motif. The corresponding Ti complexes, such as **4**, *in situ* generated using Ti(O*i*Pr)$_4$, were tested at 10 mol% loading in the epoxidation of unfunctionalized alkenes (Scheme 3.8) [40]. They proved to be less effective and enantioselective compared to complex **2**, although cyclic epoxides were isolated in high yield and ee values. A rationale for this behavior has been achieved by investigations using electrospray ionization–mass spectrometry (ESI-MS) and kinetic studies, which confirmed the involvement of a mononuclear active complex in solution [41]. Moreover, an oxidative degradation of the secondary amine in the ligand was observed, which would explain the limited catalyst activity observed toward less reactive terminal olefins. Further contributions from Katsuki's group in mechanistic investigations on the catalysis provided by Ti–salan complexes helped to synthesize and characterize a novel μ-oxo-μ-$η^2$:$η^2$-peroxo titanium–salan complex, which would act as a reservoir of the real active species, showing even in this case, a certain complexity of the species involved in the Ti-based epoxidizing systems [42].

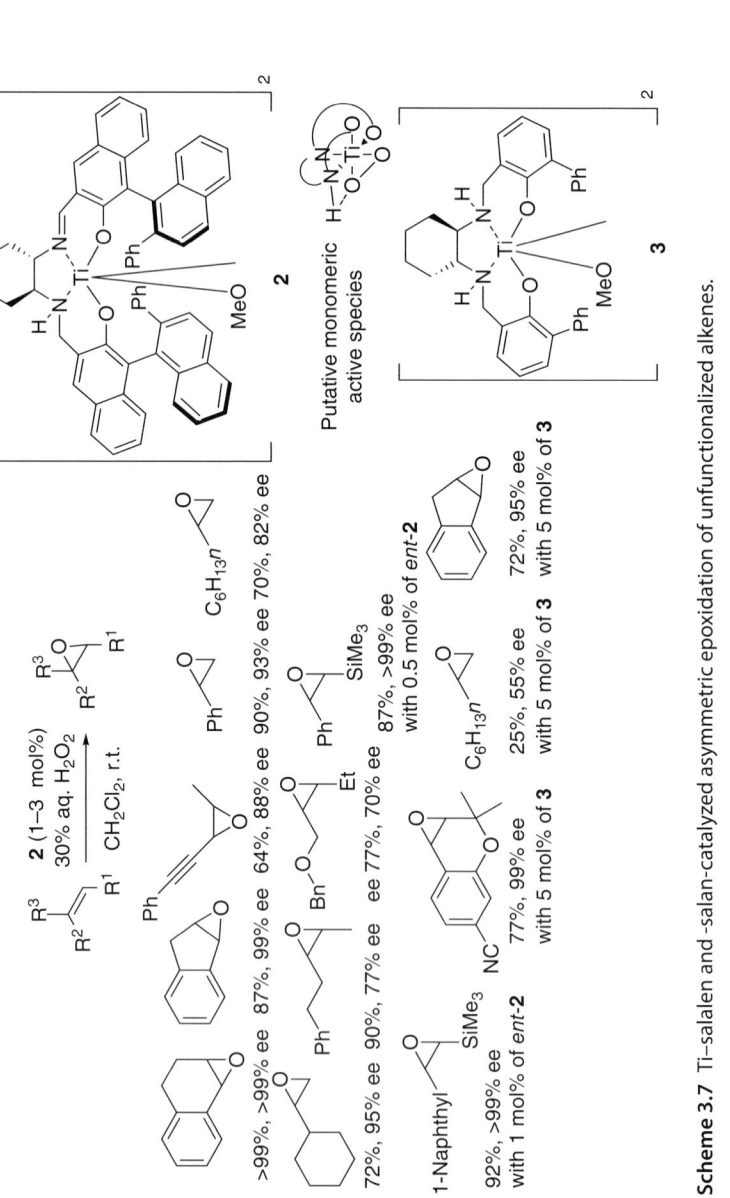

Scheme 3.7 Ti-salalen and -salan-catalyzed asymmetric epoxidation of unfunctionalized alkenes.

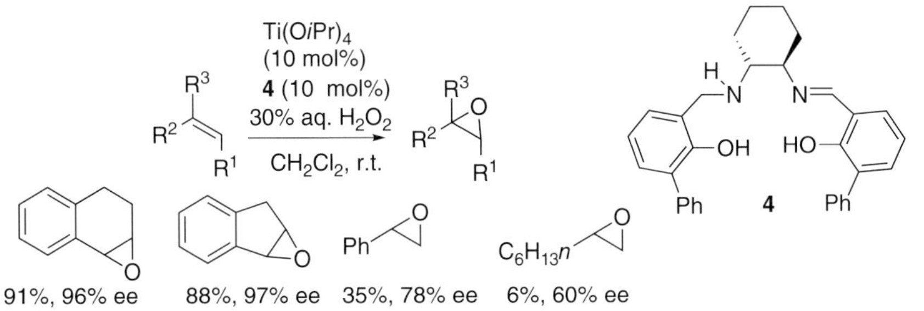

Scheme 3.8 *In situ* generated Ti–salalen complexes for the asymmetric epoxidation of unfunctionalized alkene.

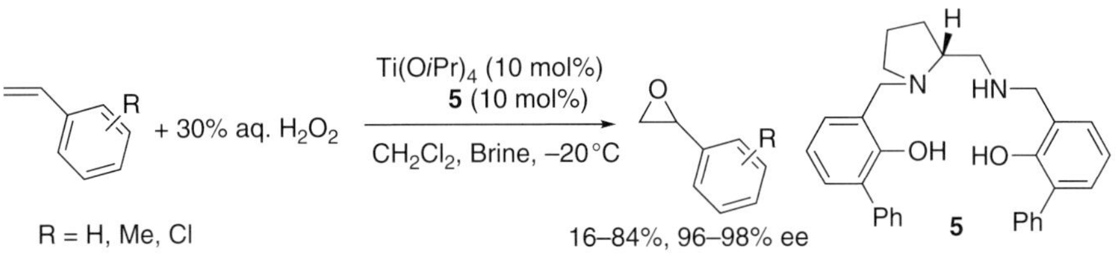

Scheme 3.9 *In situ* generated Ti–salan complexes for the asymmetric epoxidation of styrenes.

Katsuki and coworkers synthesized a C_1-symmetric proline derived diamine as useful ligand for a titanium-catalysed asymmetric epoxidation of styrenes (Scheme 3.9) [43]. The idea was born from the necessity of generating a unique species of Ti complex, where the pyrrolidine ring would make the complex adopt a *cis*-β-configuration. This simple chiral scaffold enabled formation of a highly enantioselective Ti complex, suitable for the epoxidation of challenging styrenes, whose epoxides were isolated in good yield and with excellent enantioselectivity. When the N-methylated ligand was used in the reaction, no conversion to the epoxides was observed, which attested the importance of H-bonding in the activation of a peroxotitanium species (Scheme 3.7).

Zirconium catalysts have been of limited use in the area of asymmetric epoxidation, with a particular focus on homoallylic alcohols. In the first example of Zr(O*i*Pr)$_4$/tartramide/TBHP system, reported by Katsuki in 1987 [44], higher enantioselectivity was achieved with respect to the corresponding titanium–tartrate system, when starting with *Z*-homoallylic alcohols as the substrates (up to 77% ee). This result was rationalized on the basis of longer metal–oxygen bond of Zr-based catalysts that would help to better accommodate the folded conformation of the chelated alcohol in the transition state postulated for the epoxidation. However, this system proved to be difficult to improve over time, especially in terms of metal loading and efficiency [45]. The conversions were generally low at stoichiometric loading of the chiral complex. Under catalytic conditions (20 mol% of Zr(O*t*Bu)$_4$), a dependence of the enantiofacial selection on the amount of the tartramide used was observed, which is likely ascribed to the presence of monomeric and/or polymeric active species. However, moderate to fairly good enantioselectivity was observed working at −40 °C in the presence of molecular sieves [46]. In 2010, a significant improvement of this challenging reaction was illustrated by the work of Yamomoto, who employed a chiral C_2-symmetric bishydroxamic acid **6** as ligand, with a catalytic loading of $Zr(OtBu)_4$ (2 mol%), employing CHP as the oxidant in toluene (Scheme 3.10) [47]. The presence of additives such as DMPU and molecular sieves was necessary to preserve and maintain a monomeric catalytically active chiral complex. The group supposed that a large Zr metal would create a better cavity-like arrangement for the ligand, to induce higher enantioselectivity in the epoxidation step. Monohydroxamic acids derived from α-amino acid were reported in 2003 by the same group in the vanadium-catalyzed asymmetric epoxidation of homoallylic alcohols [48]. Under these conditions, the epoxidation proceeded smoothly on a variety of terminal, Z-disubstituted and trisubstituted homoallylic alcohols in satisfactory yield with generally good-to-high enantioselectivity. This protocol has been recently applied as the key step in the synthesis of the tricyclic polar segment of fusarisetin A, a fungal metabolite with anticancer activity, due to potent inhibition against acinar morphogenesis, cell migration, and cell invasion in MDAMB-231 cells [49]. The synthesis commenced with the epoxidation of commercially available (Z)-pent-3-en-1-ol, which when treated under Yamamoto conditions gave the (3R,4S)-epoxyalcohol in 88% yield with 84% ee (Scheme 3.10) [50]. The same group introduced an asymmetric epoxidation using $Hf(OtBu)_4$ and ligand **6**, which having a big metal center would generate, even better than its Zr counterpart, a doubly coordinated complex with the sterically bishydroxamic acid, thus preventing catalyst oligomerization [48]. As a general remark, the epoxidation proceeded with better yield and slightly better level of enantioselectivity when compared to the same process catalyzed by Zr/**6** complex. The $Hf(OtBu)_4$/ligand **6**/CHP system proved to be more effective with less reactive and highly challenging terminal *bis*-homoallylic alcohols, which were converted into the corresponding epoxides in moderate yield but with excellent enantioselectivity (Scheme 3.10). In some examples, the tetrahydrofuranol derivative was isolated at the end of the reaction, which derived from metal-catalyzed intramolecular epoxide ring-opening process.

The pioneering work of Sharpless and the following asymmetric metal-catalyzed epoxidations of allylic alcohols pointed out that the crucial factor to the success of these reactions is the presence of the OH group, which assures a directing link to the metal, thus regulating the regio- and stereoselectivity of the oxidation. Yamamoto and coworkers disclosed a further interesting application of the $Hf(OtBu)_4$/ligand **6**/CHP for the asymmetric epoxidation of N-alkenyl amine derivatives. This novel type of epoxidation would proceed via sulfonyl-oxygen-directing group, giving synthetically useful allylic and homoallylic epoxy amines in moderate-to-good yield and enantioselectivity when working under mild reaction conditions (Scheme 3.11) [51]. The positive role of MgO additive in enhancing the yield and enantioselectivity was unclear. Structural modification of the reacting compounds indicated that the sulfonyl oxygen played the role of a directing group. Interestingly, when this system was applied to the oxidation of N-tosyl aldimines and ketimines, the corresponding enantioenriched oxaziridines were obtained. The long-standing problem of asymmetric

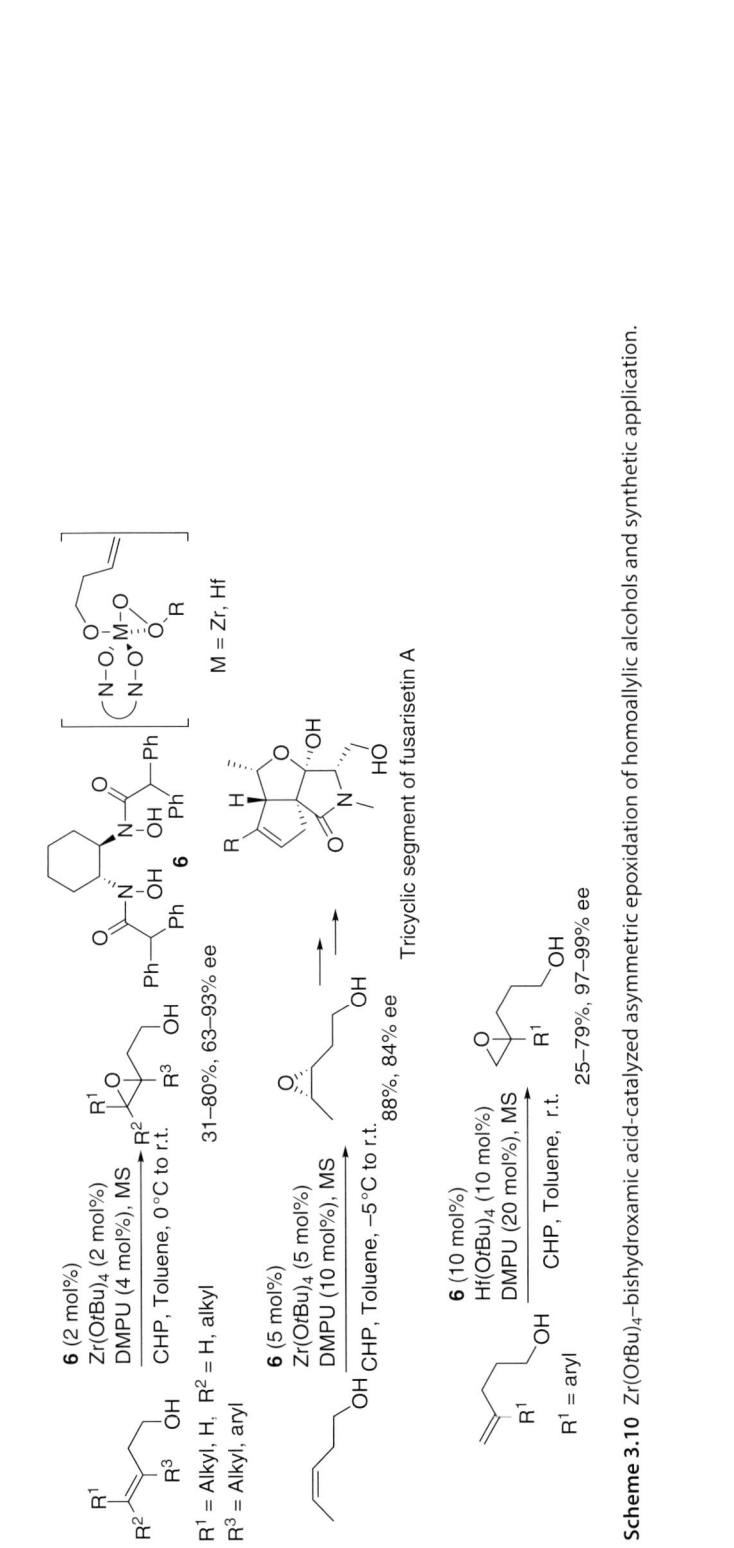

Scheme 3.10 Zr(OtBu)₄–bishydroxamic acid-catalyzed asymmetric epoxidation of homoallylic alcohols and synthetic application.

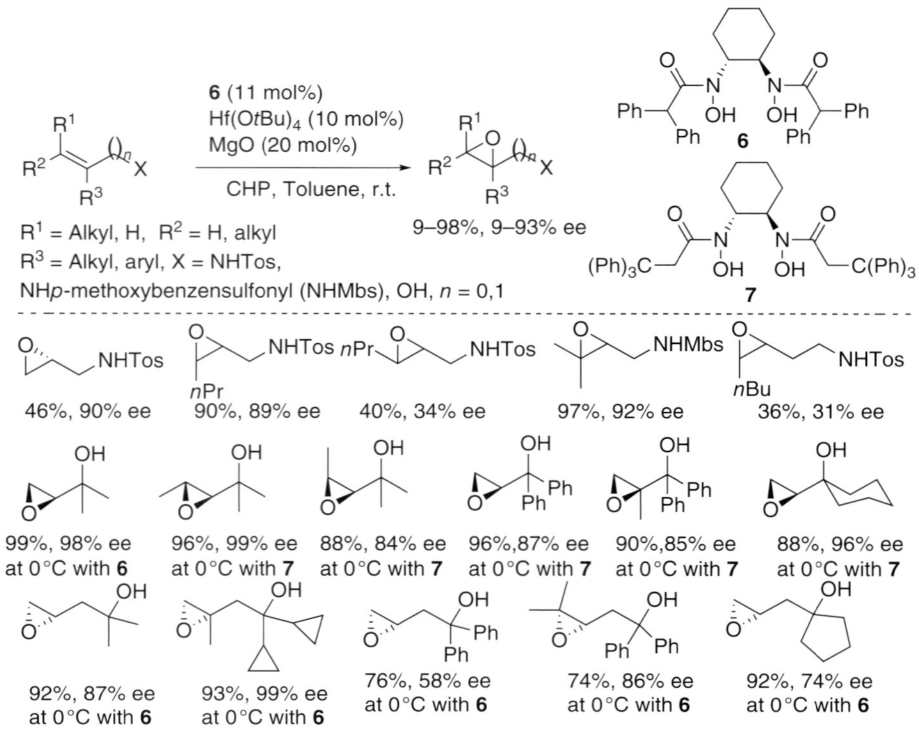

Scheme 3.11 Zr(OtBu)$_4$–bishydroxamic acid catalyzed asymmetric epoxidation of N-alkenyl amine derivatives and tertiary allylic and homoallylic alcohols.

epoxidation of tertiary allylic and homoallylic alcohols has been solved by the same group, who revealed the versatility of the Hf(OtBu)$_4$/ligands **6**/CHP system in addressing this issue (Scheme 3.11) [52]. A variety of substitution is tolerated in the starting allylic and homoallylic reagents, and the employment of more sterically demanding ligand **7** helped to improve, in some examples, the level of enantiocontrol, which was generally very high. Additionally, this system appears to be suitable for performing the kinetic resolution and desymmetrization of tertiary allylic and homoallylic alcohols. The combined kinetic resolution in the asymmetric epoxidation of secondary allylic alcohol followed by regioselective W-catalyzed epoxide aminolysis has been set up by the Yamamoto group to obtain 3-amino-1,2-diols with three consecutive stereocenters with excellent enantioselectivity [53].

3.3.2 V-, Nb-, Ta-Catalyzed Epoxidations

Chiral ligand/vanadium-based-catalyzed epoxidations represent another important area of investigation, which slowly flourished over the years, since Sharpless and coworkers' discovery in 1977 (Scheme 3.12) [54]. They found that some chiral hydroxamic acids, such as **8**, would be suitable ligands for VO(acac)$_2$ in the asymmetric epoxidation of allylic alcohols mediated by TBHP, achieving up to 50% ee (Scheme 3.12). Although vanadium complexes are more appealing than titanium complexes, thanks to higher moisture stability, a drawback of the vanadium-catalyzed oxidation was the excess of ligand used. The excess of ligand was necessary to improve the enantiomeric excess of the product, at the expense of conversion. This ligand deceleration effect was ascribed to the presence of different equilibria, where only species **II** assures an enantioselective pathway. Species **I** and **III** have to be limited in the reaction mixture as they decrease the conversion to the product and the asymmetric induction.

Sharpless postulated the involvement of complexes of type **a** and **b**, which were later detected by Bryliakov *et al.* via NMR studies [55]. Over the years, we have assisted in the development of this methodology with the introduction of hydroxamic acids with planar chirality [56], those derived from BINOL [57] or based on α-amino acids [58]. A wider substrate scope employing a V/ligand ratio close to 1/1 has been demonstrated, achieving moderate-to-high level of enantioselectivity. The employment of other chiral ligands [59] or enantioenriched hydroperoxides in the presence of achiral hydroxamic acids was also investigated as an alternative approach to an asymmetric epoxidation reaction of allylic alcohols, achieving the epoxides with up to 72% ee [60]. However, a concrete advancement in the asymmetric epoxidation of allylic alcohol to be considered competitive with the Ti/tartrate system has been reported by Yamamoto and coworkers in 2005. They designed the first C_2-symmetric bishydroxamic acids based on 1,2-cyclohexanediamine core to form V-based catalysts **9** and **10**. They planned to be able to avoid ligand deceleration, thanks to a second binding site and sterically hindered substituents that would help to disfavor the intermolecular coordination (Scheme 3.13) [61].

A 1/1 vanadium/ligand complex displayed high activity at only 1 mol% loading, using 70% aqueous TBHP. High-to-excellent level of enantioselectivity

Scheme 3.12 VO(acac)$_2$/bishydroxamic acid-catalyzed asymmetric epoxidation of allylic alcohols.

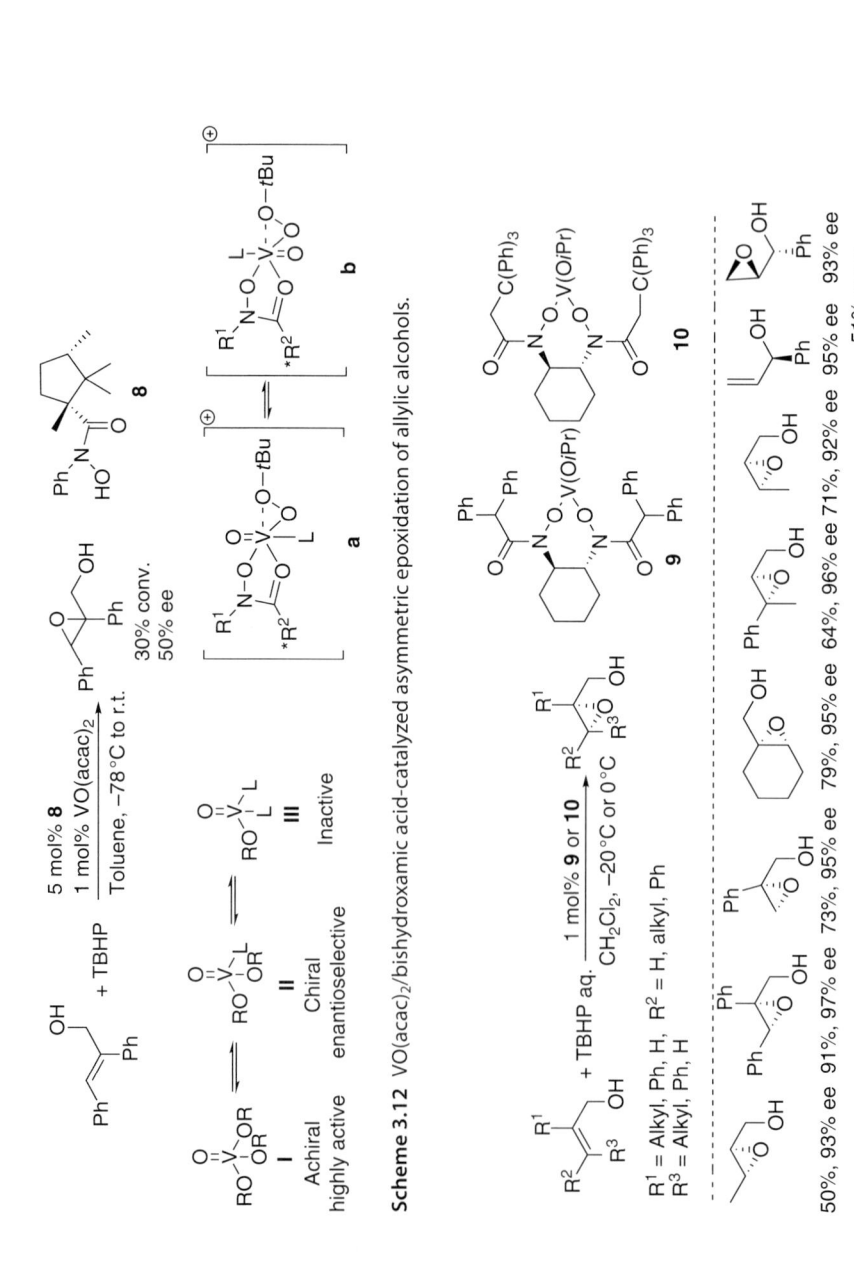

Scheme 3.13 V/bishydroxamic acid-catalyzed asymmetric epoxidation of allylic alcohols.

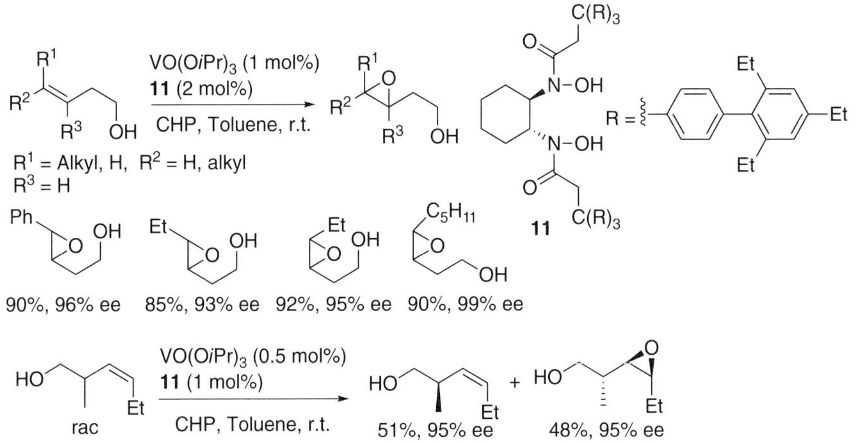

Scheme 3.14 V/bishydroxamic acid-catalyzed asymmetric epoxidation of disubstituted homoallylic alcohols.

starting from a broad range of *E*- or *Z*-disubstituted, terminal, or trisubstituted allylic alcohols was demonstrated. This moisture-insensitive system was also investigated in the kinetic resolution of allylic alcohol and desymmetrization of dialkenylcarbinols, showing an excellent performance [62]. Yamamoto *et al.* reported the VO(O*i*Pr)$_3$/α-amino-acid-derived hydroxamic ligand/CHP system in the epoxidation of homoallylic alcohols, achieving modest-to-satisfactory yields and enantioselectivity [48]. Unfortunately, *E*- and *Z*-disubstituted homoallylic alcohols yielded unsatisfactory results under these conditions. A highly improved catalytic methodology for the epoxidation of these alkenes has been reported by the same group using VO(O*i*Pr)$_3$ and sterically demanding bishydroxamic acid **11** working at room temperature (Scheme 3.14) [63]. These challenging epoxy alcohols were isolated in high yield with excellent ee values. Interestingly, a useful process of kinetic resolution can be performed at very low catalytic loading, recovering hydroxy alkene and epoxy alcohol in perfect yield with excellent enantioselectivity. Recently, the efficacy of bishydroxamic acids as ligands has been further illustrated by Noji and coworkers, who introduced a binaphthylbishydroxamic acid used at 7.5 mol% loading with VO(acac)$_2$ (5 mol%) in the asymmetric epoxidation of a variety of allylic alcohols with TBHP, achieving good-to-high yield and enantioselectivity (up to 98% ee) [64].

Domino processes involve the formation of complex molecules in the same reaction vessel, starting from two or more reagents via a sequence of bond-forming reactions occurring on the functional groups created in the previous steps [65]. This approach has several advantages in terms of reduced costs, time saving, and minimum waste. At present, ever-growing synthetic efforts are focused on the development of convenient domino processes to obtain functionalized complex targets with different stereocenters in a highly efficient manner. In this context, You *et al.* recently designed a domino sequence, involving a VO(acac)$_2$/ligand **6** catalyzed epoxidation of tryptophols followed by

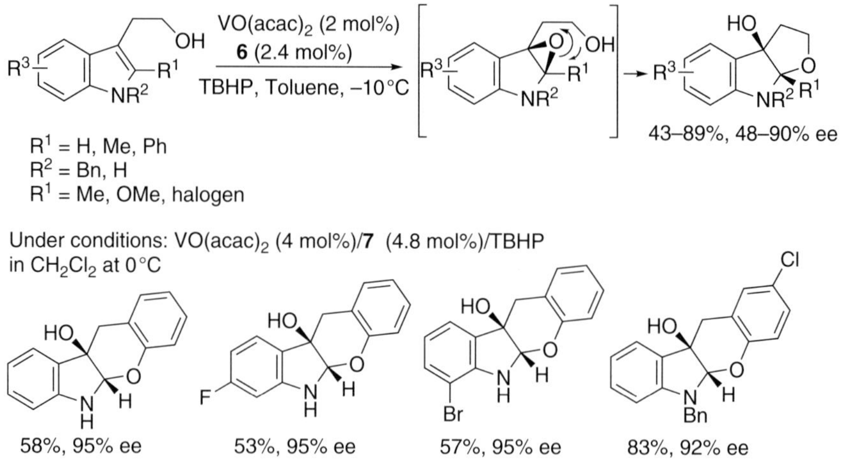

Scheme 3.15 V–bishydroxamic acids catalyzed asymmetric domino epoxidation/ring-opening of indole derivatives.

intramolecular ring opening to yield hydroxyfuroindolines, a scaffold found in some natural products, in good yield and enantioselectivity (Scheme 3.15) [66]. This enantioselective dearomatization was also extended, using higher catalyst loading and ligand **7**, to phenol-linked derivatives to obtain tetrahydrochromene indoles in satisfactory yield with up to 98% ee (Scheme 3.15) [67].

Malkov and coworkers reported a new class of amino acid sulfonamide-derived hydroxamic acids, suitable for the vanadium-catalyzed epoxidation of allylic alcohols in water as benign medium [68]. From the initial studies, it was found that $VOSO_4/H_2O$ and 70% aqueous solution of TBHP were the reagents of choice. Interestingly, a 1 : 1 ligand/vanadium salt ratio afforded the best conversion to the epoxide at 1 mol% metal loading. With respect to organic solvent, where the process was found ligand-decelerated, in water it turned out into a ligand-accelerated catalysis. The epoxidation was suggested to proceed via a soluble vanadium complex species in the organic phase. Although hydrolysis of the epoxide was problematic at room temperature, the epoxidation could be performed at 0 °C, achieving up to 72% ee. A significant improvement of the enantioselectivity was achieved by employing new sulfonamide-derived hydroxamic acids containing two stereogenic centers (Scheme 3.16) [69]. A variety of epoxy alcohols could be isolated in good yield with up to 94% ee working under environmentally friendly conditions.

A heterogeneous asymmetric version of the vanadium-catalyzed epoxidation has been developed by He *et al.* by using inorganic nanosheets modified with α-amino acids as ligands [70]. The nanosheets were positively charged brucite-like layers of layered double hydroxides (LDHs) containing vanadium metal centers. LDHs are generally intercalated with anions and water molecules. The authors intercalated different α-amino acids into the interlayer regions of LDHs and tested the catalysts in the epoxidation of cinnamyl alcohol and a terminal homoallylic alcohol (Scheme 3.17). Best results were achieved with the intercalated

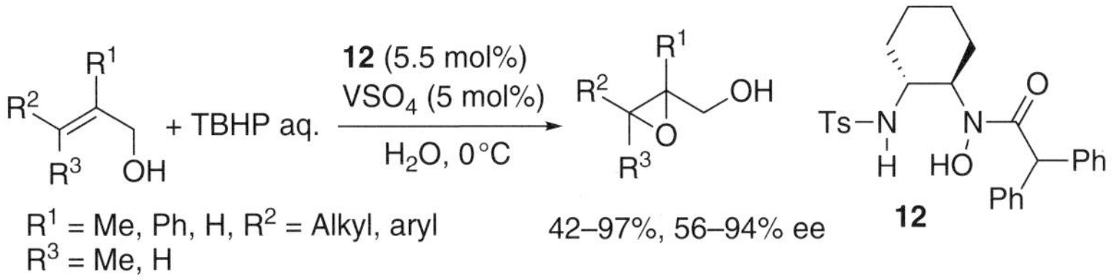

Scheme 3.16 V/sulfonamide-derived hydroxamic acid catalyzed asymmetric epoxidation of allylic alcohols in water.

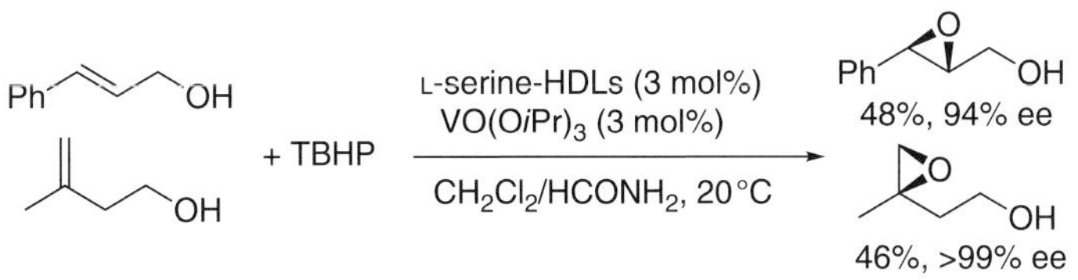

Scheme 3.17 Asymmetric epoxidation with inorganic nanosheets attached to L-serine anions as ligands.

L-serine ligand, which afforded the epoxides in moderate yield with excellent ee values. Positively charged layer interacts with the α-amino acid anions through electrostatic attraction with the carboxylate groups, making a steric planar environment, which greatly influenced the approaching directions of the reagents, eventually improving the level of enantioselectivity. The methodology was then extended to other allylic alcohols, whose *trans-* and *cis-*diastereoisomeric epoxides were isolated in excellent yield, at variable ratios (22–66% de) and with highest enantiocontrol (>90% ee) for the *trans-*epoxy alcohols [71]. According to density functional theory (DFT) calculations, the LDH layer would improve the enantiocontrol by assisting and controlling the formation of stable transition states via steric and hydrogen bonding interactions. Recycling and reuse of the heterogeneous catalysts were feasible as investigated on the epoxidation of model cinnamyl alcohol, whose epoxide was obtained with comparable ee value although in slightly reduced yield.

An example of the synthetic utility of the vanadium-catalyzed asymmetric epoxidations has been demonstrated in the preparation of the fragrance α-bisabolol [48]. The key step in installing the second stereocenter was an asymmetric epoxidation of the (S)-limonene-derived homoallylic alcohol using the vanadium/ligand **13**/CHP system. The required epoxide intermediate was isolated in 84% yield with 90% de (Scheme 3.18). An enantioselective synthesis of florfenicol, a fluorinated derivative of thiamphenicol but with superior antibacterial spectrum, has been developed in 37% overall yield from commercially available 4-methylthiobenzaldehyde [72]. The key epoxidation step was performed on the *trans-*cinnamyl alcohol derivative with 5 mol% loading of the Yamamoto vanadium complex **9** and TBHP to give the epoxide in 75% yield with 90.5% ee (Scheme 3.18).

In 2008, Egami and Katsuki reported the first example of an asymmetric epoxidation of allylic alcohols catalyzed by a chiral Nb–salan complex [73]. Ti–salan

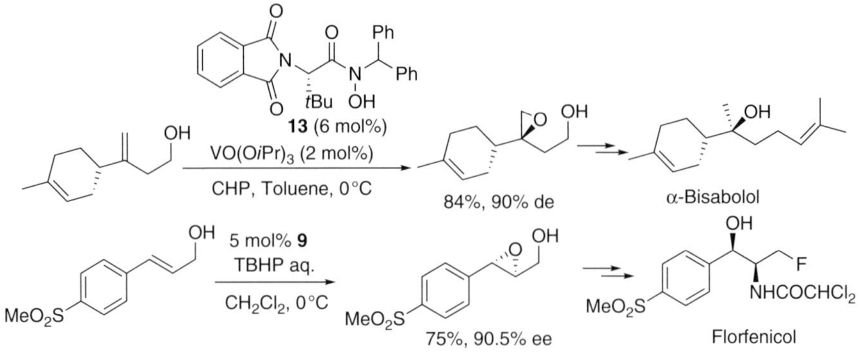

Scheme 3.18 V/hydroxamic acid-catalyzed asymmetric epoxidation of homoallylic and allylic alcohols as the key step for the synthesis of a fragrance and an antibacterial agent.

based catalysts, reported by the same group, proved to be optimal for the asymmetric epoxidation of unfunctionalized alkenes, but they could not be applied in the epoxidation of functionalized alkenes such as allylic alcohols. The authors envisioned that metals with higher valency and coordination number could accommodate tetradentate salan ligands, the allylic alcohol, and the oxidant in pentagonal–bipyramidal complexes. Preliminary experiments carried out with 2 mol% of a dimeric μ-oxo Nb(salan) complex **14**, prepared from Nb(O*i*Pr)$_5$ and a salan ligand, proved that it could serve as suitable catalyst for the epoxidation of some allylic alcohols with UHP as the oxidant, achieving up to 80% ee (Scheme 3.19).

Further mechanistic studies showed that the μ-oxo dimer dissociates into monomeric species, which are involved in the epoxidation [74]. The monomeric

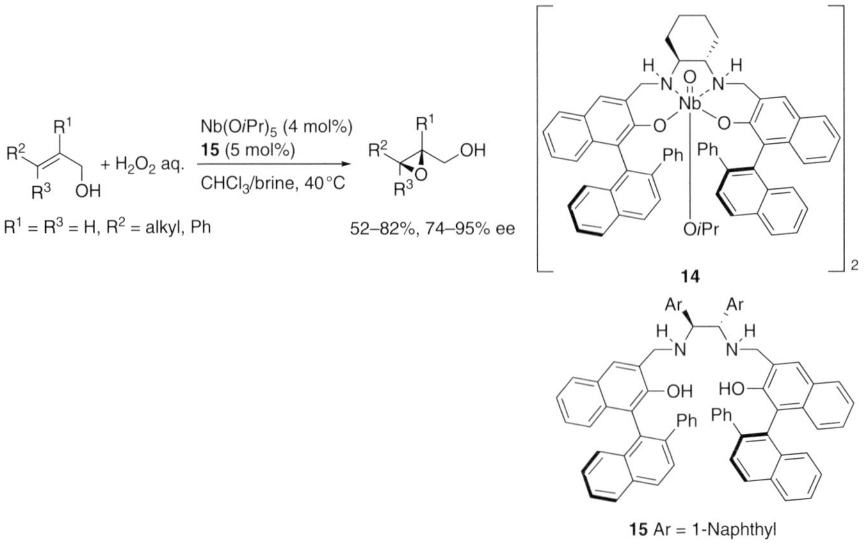

Scheme 3.19 Nb/salan/H$_2$O$_2$-catalyzed asymmetric epoxidation of allylic alcohols.

Nb(salan) complexes, *in situ* prepared from Nb(O*i*Pr)$_5$ and compound **15**, were among one of most effective salan ligands tested, able to catalyze the epoxidation of allylic alcohols using aqueous H$_2$O$_2$ in CHCl$_3$/brine, attaining enantioselectivity ranging from 74% to 95% ee. The authors suggested that the catalytically active species is a monomeric peroxoniobium heptacoordinated species. The asymmetric epoxidation mediated by chiral tantalum complexes was poorly investigated. Heterogeneous silica supported tantalum catalysts, pretreated with tartrate esters, were investigated in the epoxidation of two representative allylic alcohols with TBHP as the oxidant to give epoxides in modest yield with up to 96% ee [75].

3.3.3 Cr-, Mo-, W-Catalyzed Epoxidations

Over the past decades, chiral Cr(III)(salen) complexes were far less investigated in the epoxidation of olefins compared to the most famous Mn–salen catalysts [76]. Optimization of the substituents, counteranions, additives, nature of the oxidant enabled Gilheany and coauthors to find the most striking results with Cr–salen complexes, which turned out to be more effective than Mn–salen complexes in the enantioselective epoxidation of *E*-alkenes over *Z*-alkenes (Scheme 3.20) [77]. The reaction could be performed under catalytic conditions, using PhIO as the oxidant that is able to transform precatalyst Cr(III)salen complex into the catalytically active oxo-Cr(V)salen complex. Contributions from Katsuki's group showed the important role played by the solvent in affecting the asymmetric induction [78].

During the 1970s, molybdenum(VI) complexes, derived from MoO$_2$(acac)$_2$ and chiral aminoalcohols [79] and chiral oxodiperoxo molybdenum complexes [80], were reported as the first systems useful for the epoxidation of functionalized and unfunctionalized alkenes. However, only modest level of enantiocontrol was achieved (up to 35% ee), likely ascribed to weak coordinating ability of the ligands for the metal center [14d, e, 81]. In 2006, a remarkable improvement in this area has been illustrated by Yamamoto and coauthors, who developed an active complex by mixing catalytic loadings of MoO$_2$(acac)$_2$ with bishydroxamic acids using different alkyl hydroperoxides as the oxidant at room temperature

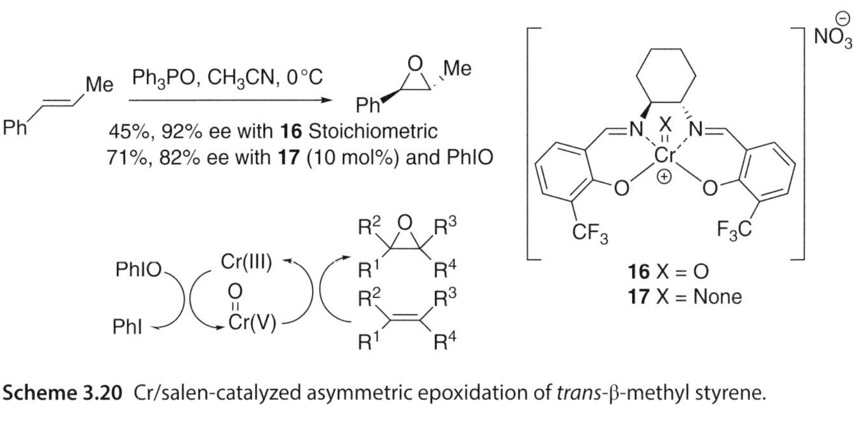

Scheme 3.20 Cr/salen-catalyzed asymmetric epoxidation of *trans*-β-methyl styrene.

(Scheme 3.21) [82]. A variety of Z-disubstituted, trisubstituted, and terminal alkenes were smoothly converted into the corresponding epoxides in good-to-high yield and enantioselectivity. According to the substitution pattern of the alkene, an appropriate ligand and sterically hindered trityl (THP) or CHPs were used to achieve the highest level of asymmetric induction. Interestingly, although this system has a strong oxidizing power, it is able to oxidize the most electron-rich double bond in polyenes (Scheme 3.21). Squalene, a polycyclic terpenoid and biogenetic precursor of steroids, was selectively converted into the 2,3-epoxysqualene in 53% yield with 76% ee. The mechanism proposed for the molybdenum-catalyzed epoxidation, originally suggested by Sharpless and coauthors, would involve the attack of the alkene on the chiral Mo(VI) alkyl peroxide, then leading to a butterfly transition state complex evolving to epoxide and alkoxide products [83].

In 2009, Wang and coworkers developed a $MoO_2(acac)_2$/L-diphenyl prolinol/TBHP system for the asymmetric epoxidation of styrenes (Scheme 3.22) [84]. This simple protocol based on commercially available reagents and ligand was effective after mixing and aging at room temperature 5 mol% of metal and 5.5 mol% loadings of aminoalcohol to give terminal epoxides in good yield and enantioselectivity. The highest ee values were achieved when the aromatic ring was substituted with an electron-withdrawing group. Further developments of this system and mechanistic studies were not reported.

Over the past years, other chiral dioxomolybdenum catalysts were applied in the asymmetric epoxidation of olefins, although poor level of asymmetric induction was observed [85]. A possible explanation of the difficulty in achieving high stereocontrol might be the generation of diastereoisomeric species active in the epoxidation of the alkene. Peroxotungstates have been studied in the area of epoxidation, thanks to their ability to easily transfer oxygen atom, without showing significant catalase activity when using H_2O_2 as the terminal oxidant. However, the development of asymmetric versions with this type of catalysts has met with poor success. The first synthesis and characterization of chiral dioxo tungsten(VI)-pyridinyl alcoholate complexes were reported by Herrmann [86]. The chiral W-catalyzed epoxidation of trans-β-methylstyrene with TBHP afforded a poor result (24% ee). An analogous complex was used with H_2O_2 as the oxygen source in the epoxidation of Z-1-propenylphosphonic acid in CH_2Cl_2 at 0 °C leading to a complete conversion to the epoxide with 74% ee [87].

In 2014, a breakthrough discovery in this area was illustrated by Yamamoto and coworkers with the first highly enantioselective epoxidation of allylic and homoallylic alcohols mediated by $WO_2(acac)_2$/bishydroxamic acid **20**/H_2O_2 in the presence of NaCl as additive (Scheme 3.23) [88]. Catalytic loadings of the metal complex and a sterically demanding ligand guaranteed, for both epoxide classes and substitution pattern, high yield and enantioselectivity. The presence of NaCl was necessary to prevent ring opening of the epoxide [89]. By competing experiments, it was observed that this environmentally friendly system can assure a high degree of chemoselectivity for primary alcohol over secondary and tertiary ones in the epoxidation. Moreover, it appears applicable to the kinetic resolution of secondary allylic alcohols.

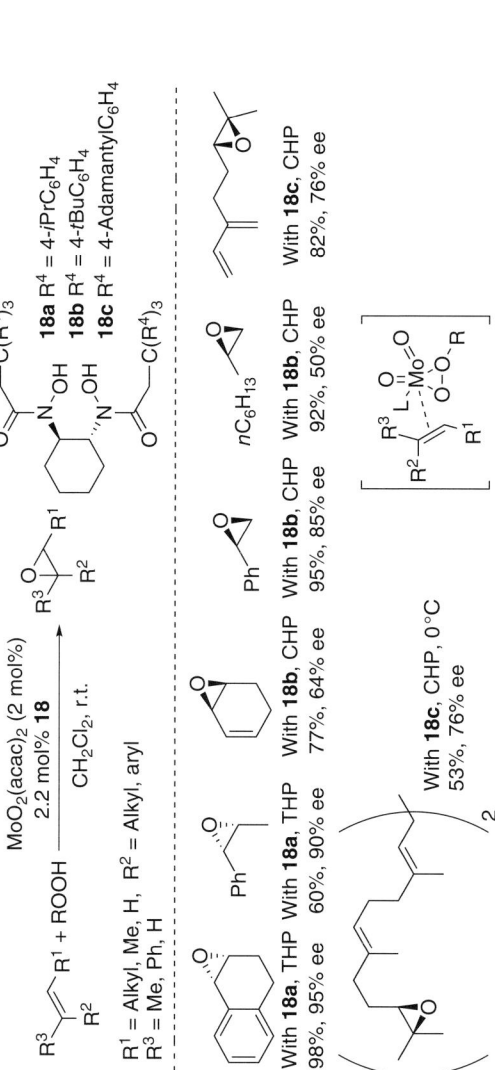

Scheme 3.21 Mo/bishydroxamic acid-catalyzed asymmetric epoxidation of unfunctionalized alkenes.

Scheme 3.22 Mo/amino alcohol-catalyzed asymmetric epoxidation of styrenes.

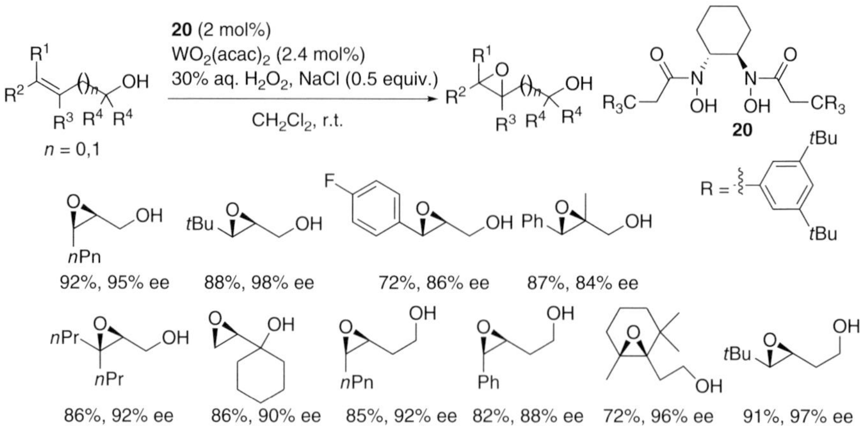

Scheme 3.23 WO$_2$(acac)$_2$/bishydroxamic acid-catalyzed asymmetric epoxidation of allylic and homoallylic alcohols.

3.3.4 Mn-, Re-, Fe-, Ru-Catalyzed Epoxidations

In 1990, a pivotal discovery was made by the groups of Jacobsen [90] and Katsuki [91], who independently reported the asymmetric epoxidation of a variety of unfunctionalized alkenes with optically pure Mn(III)–salen complexes using readily available PhIO, bleach, H$_2$O$_2$, Oxone as the terminal oxidants, achieving good level of enantioselectivity. A lot of investigations followed these reports to improve the original systems as this methodology has particularly appealing features: (i) it is based on easily accessible chiral ligands of large tunable stereoelectronic properties, (ii) manganese is a nontoxic metal with a favorable environmental impact, (iii) different oxygen sources are useful for the process comprising environmentally benign and atom-economic H$_2$O$_2$ [34c–e], and (iv) the substrate applicability covers different classes of alkenes comprising the most challenging ones (Scheme 3.24) [14f, 76, 92].

The majority of the asymmetric epoxidation protocols are based on C_2-symmetric chiral ligand, since they are straightforwardly prepared by chiral diamines and substituted salicylaldehydes. The Mn–salen complexes can adopt planar or folded conformations according to the chiral ligand used, their valency, and the presence of apical ligands, thus influencing their catalytic performance and the stereocontrol [92a]. From a mechanistic point of view, a lot of studies have been carried out to identify the species involved in the oxidation, initially suggested by Kochi [93]. In the simplified catalytic oxidative cycle, the formation of O=Mn^V(salen)$^+$ species has been proposed as the reactive intermediate responsible for the epoxidation. Concerning the oxygen delivery, radical intermediates after the oxygen attack are formed, which would justify the experimentally detected *trans*-epoxides from *cis*-alkenes, especially in the case of enynes and conjugate dienes as the starting reagents as well as by-products of allylic oxidation (Scheme 3.24). As a general remark, the Mn–salen complexes/oxygen source represents the method of choice for the catalytic asymmetric epoxidation of mono-, di-, tri-, and tetrasubstituted alkenes. Indeed, some of the catalysts are

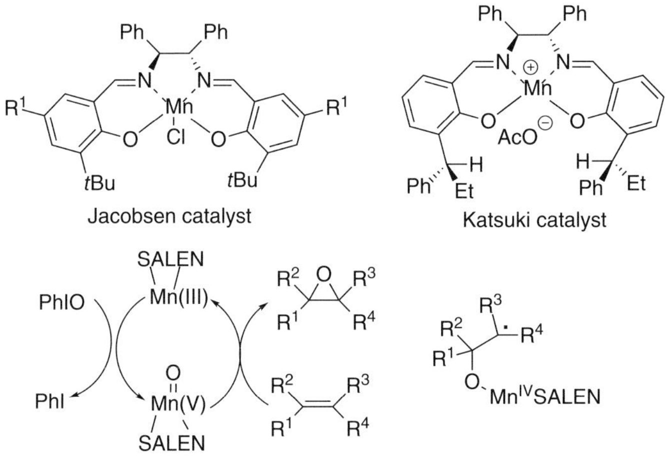

Scheme 3.24 The Jacobsen and Katsuki Mn–salen catalysts.

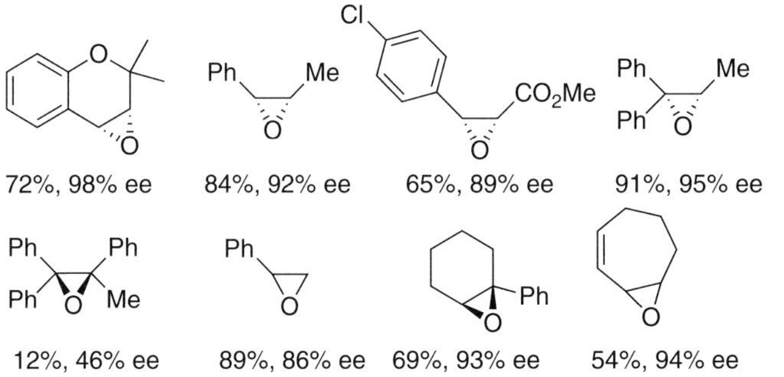

Scheme 3.25 Overview of chiral epoxides accessible via Mn–salen epoxidation.

commercially available, and the loading required is low (<10 mol%). The best level of enantioselectivity has been achieved on *cis*-disubstituted and trisubstituted acyclic and cyclic unfunctionalized and functionalized alkenes, whereas *trans*-alkenes react sluggishly, and modest ee values are observed. An overview of chiral epoxides achievable in the asymmetric epoxidation of the corresponding alkenes mediated by these systems is illustrated in Scheme 3.25 [91, 94].

In the past decade [95], the investigations of the Mn–salen-catalyzed asymmetric epoxidation were directed toward the design of different chiral salen scaffolds, introducing macrocycles, which imparted more stability and also recyclability via precipitation of the catalyst [96] (Scheme 3.26). Low-cost sugars, accessible from the chiral pool, were also incorporated in the construction of flexible libraries of modular Mn–salen complexes, which showed good level of stereocontrol [97]. Interestingly, classical Jacobsen's Mn–salen catalysts showed the ability to work in water as a medium in the presence of surfactants affording results comparable to the organic media [98]. Additionally, the epoxidation carried out with the Jacobsen's catalyst in fluorinated solvents, instead of commonly

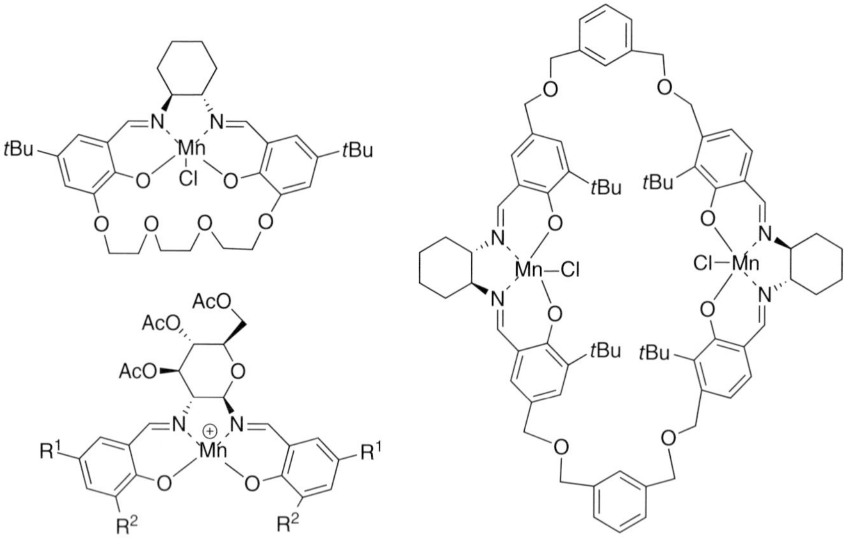

Scheme 3.26 Modified chiral Mn–salen catalysts.

used dichloromethane, gave slightly better results in terms of enantioselectivity [99]. Katsuki's initial studies on the enantiomeric conformational isomerism of achiral Mn–salen in solution showed that this equilibrium can be shifted to one diastereisomerically more active conformer by adding a chiral N-oxide neutral donor ligand, thus generating a nonracemic catalyst suitable for asymmetric epoxidation [92a, 100].

Taking advantage of these observations, List and coworkers recently reported a highly enantioselective epoxidation of alkenes catalyzed by a chiral ion-pair consisting of an achiral Mn(III)–salen cation and a chiral phosphate counteranion (Scheme 3.27) [101]. Notably, the role of the anion is to block one of the enantiomorphic conformations of the achiral cationic Mn–salen to provide a

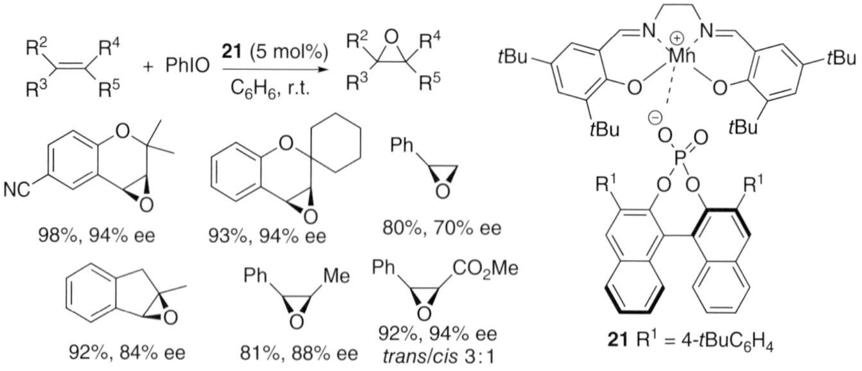

Scheme 3.27 Achiral Mn(III)–salen cation/chiral phosphate counteranion-catalyzed asymmetric epoxidation.

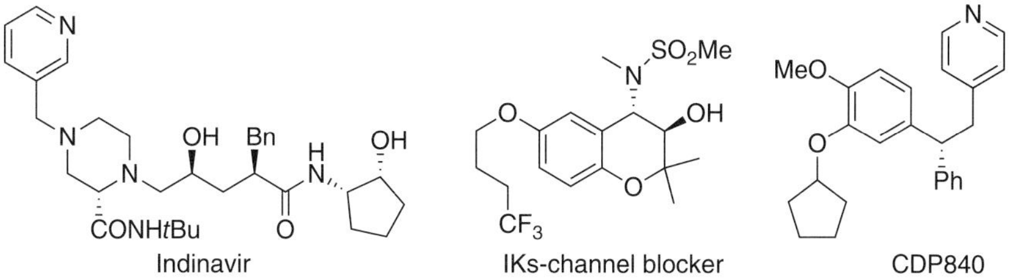

Scheme 3.28 Important biologically active compounds synthesized exploiting Mn–salen-catalyzed epoxidation as the key step.

catalyst that, in the presence of PhIO as the oxidant, led to the epoxidation of different alkenes in high yield with good-to-high enantioselectivity.

To improve the environmental concerns and the practicality of the Mn–salen-catalyzed epoxidation, several efforts were devoted to set up immobilized versions of the catalyst via heterogenization within inorganic matrices [102]. Different strategies have been adopted to link the catalyst via functional groups on the surface of mesoporous supports [103], or axial coordination of the apical ligand [104], or through covalent grafting [105]. Catalyst anchoring to organic polymers [106] and employment of ionic liquid [107] have also been pursued. Studies have shown confinement effects of pore size on the catalytic performance and enantioselectivity of the heterogeneous catalysts immobilized onto mesoporous matrices [108]. Moreover, the linkage used to connect Mn–salen complexes to the support surface and its length influenced the catalytic performance [104, 108c, 109]. For certain alkenes, the level of asymmetric induction achieved was comparable to homogeneous Jacobsen–Katsuki catalysts. Clearly, advantages in the use of these supported catalysts reside in their suitability for recycling and easy separation of the products. Soon after its discovery, the application of the Mn–salen-catalyzed asymmetric epoxidation as key step in the synthesis of biologically active has been actively reported. Relevant examples include the Merck synthesis of anti-HIV compound indinavir (Crixivan®) [110] or Iks-channel blockers [111] and phosphodiesterase IV inhibitor CDP840 [112] (Scheme 3.28).

In more than a decade, Fuchs and coauthors applied and finely optimized this methodology for the asymmetric epoxidation of cyclic dienyl sulfones and triflates to the corresponding monoepoxides [113]. These compounds proved to be very useful intermediates to access a variety of natural products as exemplified in Scheme 3.29 for a large-scale, concise approach to (+)-pretazettine alkaloid core **22** starting from dienyl triflate **23** as the starting reagent, to obtain the intermediate epoxide **24** in good yield with high enantiomeric excess by using commercially available catalyst **25** and H_2O_2 as terminal oxidant.

Biologically inspired catalysts based on Mn(II) complexes with tetradentate nitrogen (N_4)-based ligands have attracted attention in the area of asymmetric epoxidation, after the discovery of Mn–salen catalysts. The initial studies were reported by Stack and coworkers [114], who developed ligands of type **26** complexed with $Mn(CF_3SO_3)_2$ using peracetic acid as the oxidant, followed by more sterically congested ligand **27** developed by Costas [115]. Although being

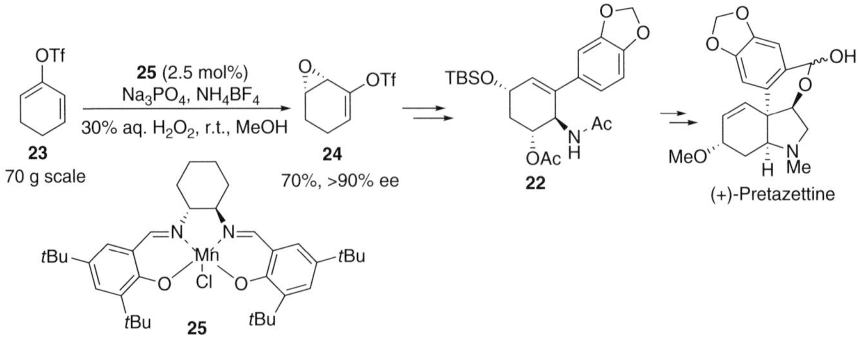

Scheme 3.29 Efficient large-scale asymmetric epoxidation of a dienyl triflate to prepare an intermediate of the (+)-pretazettine alkaloid core.

efficient in the epoxidation of styrenes, they were moderately enantioselective (ee up to 46%) (Scheme 3.30). The same group replaced the nature of the oxidant with a more convenient introduction of the H$_2$O$_2$ solution and acetic acid as necessary additive in order to eliminate the deleterious manganese-catalyzed disproportion of H$_2$O$_2$ [116]. In 2009, Sun, Xia, and coauthors inserted additional chiral centers to the previous ligand scaffolds; thus, a precatalyst **28** was formed and used for the epoxidation at only 1 mol% loading in CH$_3$CN at room temperature with H$_2$O$_2$ and acetic acid [117]. For the comparative purposes, substrates typically epoxidized via Mn–salen methodology, namely 6-cyano-2,2-dimethylchromene and styrene, were obtained with moderate enantioselectivity. However, *trans*-chalcones were converted into epoxides in high yield with fairly good level of enantioselectivity.

Notable improvements of this epoxidation were achieved by Bryliakov's groups [118] and soon after Costas [119] and Sun [120], who synthesized and applied highly reactive complexes **29**, **30**, and **31** in the epoxidation of unfunctionalized

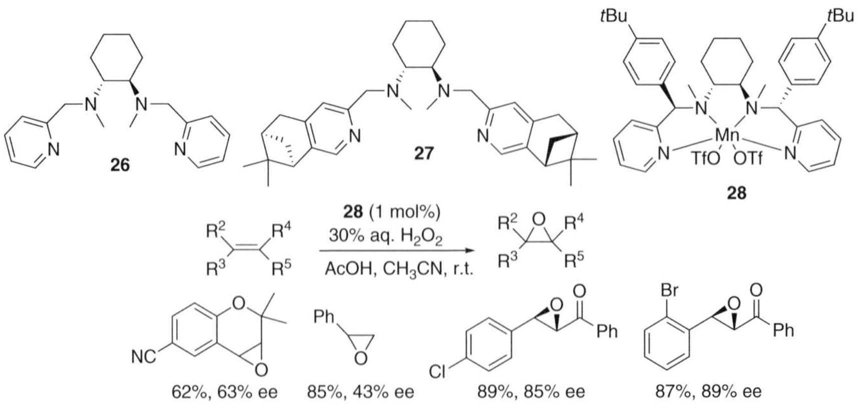

Scheme 3.30 Tetradentate nitrogen-based ligands and precatalyst for Mn(II)-catalyzed asymmetric epoxidation.

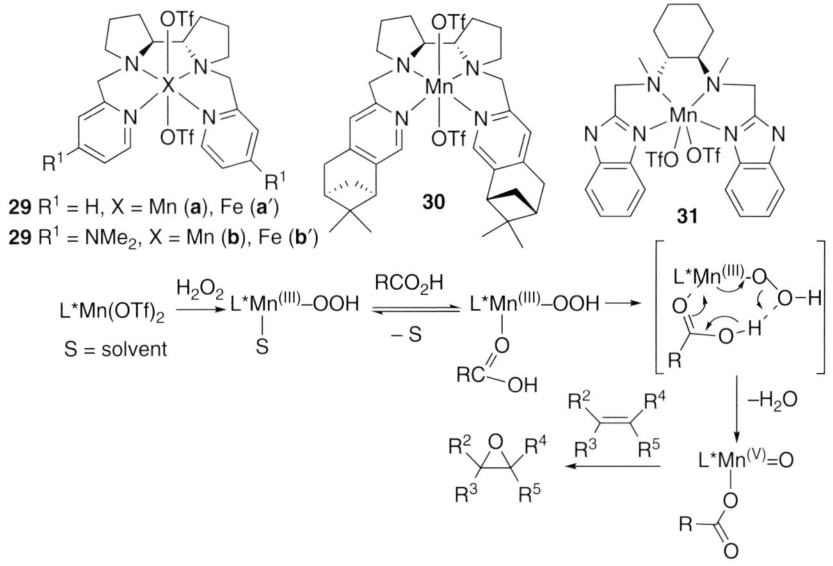

Scheme 3.31 Mn/chiral tetradentate nitrogen ligand complexes for asymmetric epoxidation and catalytic cycle.

and functionalized alkenes. As low as 0.1 mol% loading of the catalysts, in the presence of H_2O_2 and an excess of carboxylic acid as additive were employed in acetonitrile at low temperatures, achieving moderate-to-high enantioselectivity (up to 98% ee) (Scheme 3.31). Former mechanistic investigations, performed on the achiral version of the system, gave some useful insights: (i) epoxidation using $H_2{}^{18}O_2$ showed high incorporation of ^{18}O into the epoxide; (ii) stereoretention of geometrically pure alkenes pointed to a concerted oxygen transfer, without intermediate radical species; and (iii) Hammett plot analysis led to a negative ρ value indicative of the involvement of an electrophilic oxidative species. Bryliakov studied complex **29a** and the corresponding Fe(II)-based complex, varying the steric nature of the carboxylic acid additive [121]. EPR spectroscopy helpfully identified the species formed in the Fe-catalyzed epoxidation. A significant enhancement of the enantioselectivity was observed (up to 93% ee) when using hindered carboxylic acids, a result that could be justified with the acid incorporation into the active species in the enantioselectivity-determining step. Indeed, all the data could be framed under the oxidative pathway illustrated in Scheme 3.31, where the relevant catalytic oxygen-transferring species is a highly electrophilic chiral metal(V)-oxo complex that is able to deliver the oxygen to electron-rich and electron-poor alkenes.

In 2013, Gao and coauthors reported a porphyrin-inspired system exploiting an *in situ* generated catalyst obtained after mixing $Mn(OTf)_2$ and a new sterically tunable ligand **32**, bearing chiral oxazoline groups [122] (Scheme 3.32) [123].

The methodology showed to be competitive with classical Jacobsen–Katsuki Mn–salen catalysts in terms of easily accessible ligands, low catalyst loading, green and atom-economic oxidant, facile generation of the active complex,

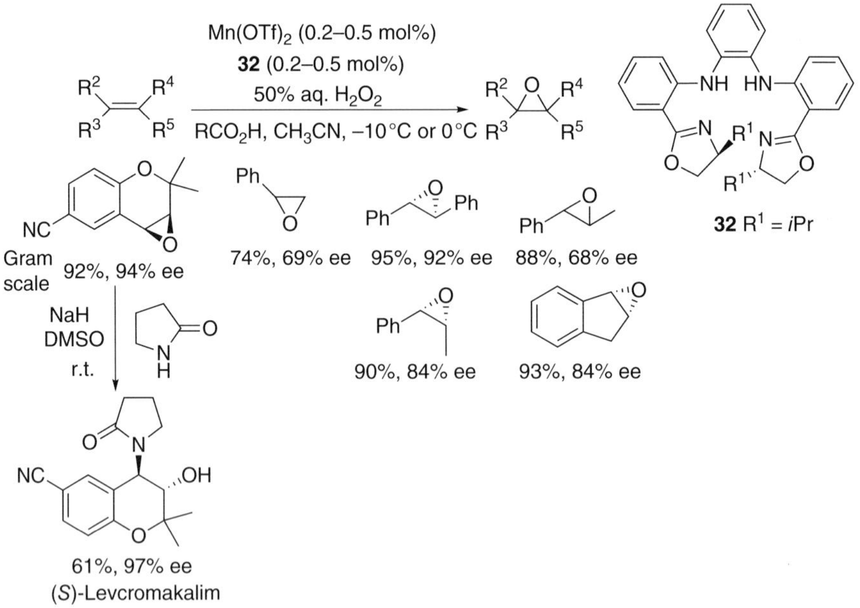

Scheme 3.32 Porphyrin-inspired Mn complexes for asymmetric epoxidation.

good-to-high level of enantiocontrol for a variety of epoxides, excluding *trans*-chalcones, which were isolated with up to 50% ee. The epoxidation could be scaled up to produce gram amounts of an intermediate epoxide easily ring-opened to give the antihypertensive drug (*S*)-levcromakalim (Scheme 3.32). In 2014, Bryliakov reported an in-depth investigation on the performances of catalysts of type **29a,b** by varying the substitution at the pyridine ring, achieving the best results when using electron-donating substituents at the *para* position of the pyridine ring (Scheme 3.31) [124]. Further remarkable improvements on catalyst loading, used as low as 0.01–0.05 mol%, were achieved in the epoxidation of *trans*-chalcones and *trans*-α,β-unsaturated esters in the presence of H_2O_2 and 2-ethylhexanoic acid as additive in acetonitrile at −30 °C. Under these conditions, the epoxides were isolated in excellent yield and enantioselectivity (>95% ee). In 2016, Sun et al. modified precatalysts of type **31**, by introducing strong electron-donating dimethyl amino groups at the *para* position of the pyridine moieties [125]. This modified class of Mn(II)-based catalysts was able to work at 0.1 mol% loading, providing the best enantioselectivity in the epoxidation of styrenes (up to 93% ee). It is fair to say that asymmetric epoxidation mediated by chiral Mn-based catalysts occupies a preeminent position in this area, thanks to the significant developments attained in the past decade in terms of practicality, low catalyst loading, green oxidants employed, wide substrate scope including unfunctionalized and functionalized olefins, high yield, and good-to-excellent control of the enantioselectivity.

It is important to mention that ReO_3Me has been actively investigated with great success under homogeneous and heterogeneous versions as a catalyst in the epoxidation of alkenes using H_2O_2 or UHP as the terminal oxidant [81b, 126].

However, it has been found problematic to develop effective asymmetric protocols mostly because of the unselective formation of highly reactive monoperoxo- and diperoxo-renium species. To date, only modest results have been achieved in the epoxidation of terminal alkenes and methyl styrenes by using chiral Re complexes (up to 41% ee) [127].

The increasing demand for low-cost and environmentally friendly methods to produce fine chemicals and biologically active compounds has only recently prompted several groups to focus their research on chiral iron-complex-catalyzed reactions, with iron being a widespread metal, with very low toxicity, hence appealing for industrial applications [128]. Heme-based enzymes such as cytochrome P450s and peroxidases have been considered as models to construct catalytic chiral iron-based complexes suitable for epoxidation [129]. In 1999, Jacobsen and Francis demonstrated that $FeCl_2$ with peptides as ligands in the presence of H_2O_2 could promote the epoxidation of *trans*-β-methyl styrene to give the epoxide with 20% ee [130]. However, it was only in 2007 that Beller and coworkers disclosed a simple and practical biomimetic nonheme system, based on commercially available $FeCl_3 \cdot 6H_2O$ in combination with pyridine-2,6-dicarboxylic acid, chiral 1,2-diphenyl-ethylene-1,2-diamine, and H_2O_2 in 2-methyl-butano-2-ol [131]. This system afforded, when working at room temperature, *trans*-stilbene epoxides in good-to-high yield and enantioselectivity (Scheme 3.33).

The most successful iron catalyst for the asymmetric epoxidation of alkenes reported so far are of chiral bis-amine-bis-pyridine, strictly similar to those illustrated for manganese-catalyzed epoxidations (Schemes 3.30 and 3.31). Octahedral metal centers of *cis*-α-topology are preferentially formed, as attested by spectroscopic studies. Preliminary investigations on the achiral version were useful to establish the nature of oxidant, ligand, and additives required to achieve the highest efficiency [132]. The analogous iron(II)-based precatalyst of type **28**, stirred in the presence of H_2O_2 and acetic acid as additive in CH_3CN at −15 °C, was reported to give enantioselectivities up to 87% exclusively in the epoxidation of *trans*-chalcones as substrates [133]. As previously mentioned, 1 mol% iron(II)-based chiral complex of type **29a′**, under similar reaction conditions, except the use of ethylhexanoic acid as additive at −30 °C, led to ee values up to 86% for epoxides deriving from *trans*-chalcones, although for the epoxidation of chromene, *trans*-cinnamic acid esters, and styrene derivatives, moderate enantioselectivity

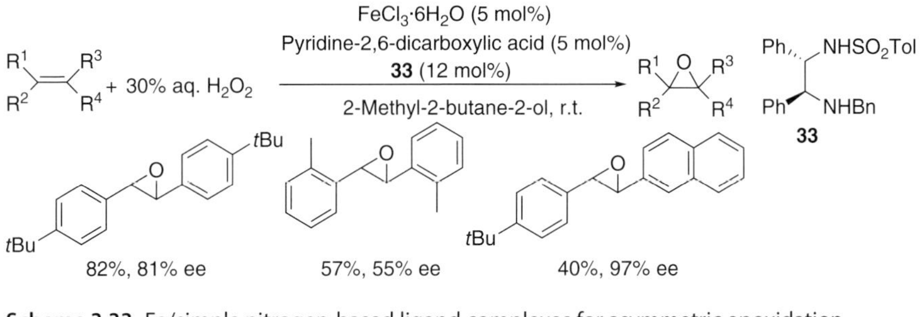

Scheme 3.33 Fe/simple nitrogen-based ligand complexes for asymmetric epoxidation.

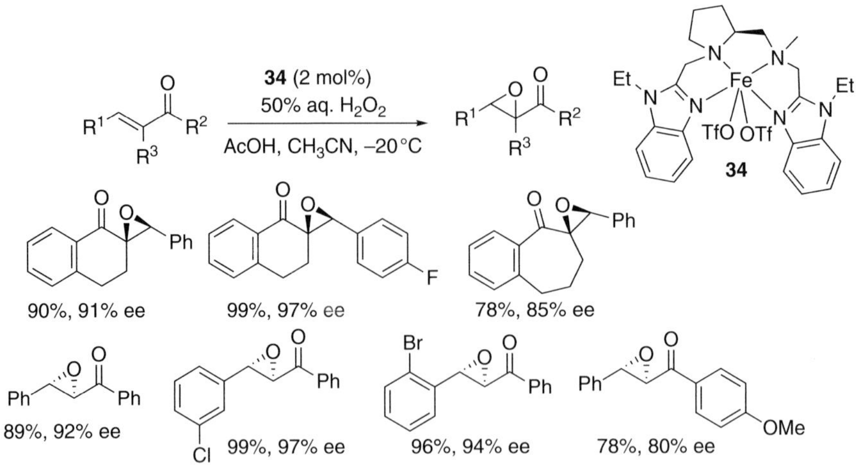

Scheme 3.34 Fe/chiral tetradentate nitrogen ligand complex in asymmetric epoxidation of α,β-unsaturated ketones.

was observed (Scheme 3.31) [121a]. A proline-derived C_1-symmetric diamine ligand for FeCl$_2$ salt, synthesized by Sun and coworkers, gave a highly efficient complex **34**, used at 2 mol% loading under usual conditions, in the asymmetric epoxidation of *trans*-chalcones and tetralones (Scheme 3.34) [120a]. The epoxides were rapidly obtained in excellent yield and enantioselectivity. Further notable improvements in this area have been achieved by Costas and coauthors, who developed a really catalytic system of wide applicability based on the Fe(II)-based **29b′** complex (1 mol%) and of (*S*)-ibuprofen or 2-ethylhexanoic (at only 3 mol%) as additive, using 1.6 equiv. of H$_2$O$_2$ at −30 °C in CH$_3$CN [134]. The epoxidation of *cis*-β-methyl styrene, *cis*- and *trans*-cinnamic acid esters or amides, *trans*-enones, *trans*-tetralones, chromene derivatives smoothly afforded the epoxides in high yield with high-to-excellent ee values (up to 99% ee).

Complex **29b′** was also applied in the epoxidation of highly challenging terminal α-alkyl substituted styrenes, by adding N-protected amino acid **35** as co-ligand (Scheme 3.35) [135]. Irrespective of the substitution pattern in the aromatic ring, good yield and good-to-high enantioselectivity were achieved for the corresponding epoxides. Absolute configuration of the epoxides was controlled by

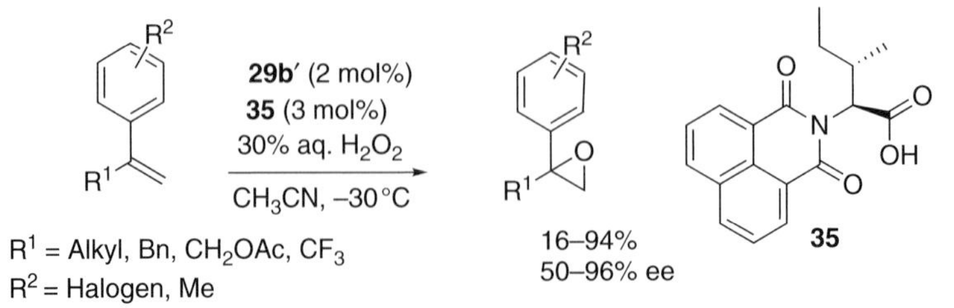

Scheme 3.35 Fe/chiral tetradentate nitrogen ligand complex and N-protected amino acid co-ligand-catalyzed epoxidation of α-alkyl substituted styrenes.

the chirality of the iron complex, and only minor matching and mismatching effects were observed, changing the aminoacid chirality and enantiomeric iron catalyst. Unfortunately, the epoxidation of α,α′-dialkyl-substituted substrates such as 2-methylhept-1-ene proceeded with low stereocontrol (16% ee). However, this system represents the most effective and sustainable protocol reported up to date for the asymmetric epoxidation of terminal α-alkyl substituted styrenes.

A remarkable development in this area concerned the introduction of a new phenantroline ligand, derivatized with binaphthyl moieties, combined with Fe(OTf)$_2$ in the presence of peracetic acid as oxygen source [136]. Yamamoto and coauthors found this system to be suitable for the first asymmetric epoxidation of acyclic β,β-disubstituted enones, which are difficult substrates, because of being sterically encumbered at the β-position for a classical nucleophilic epoxidation approach (Weitz–Scheffer epoxidation). An octahedral, mononuclear pseudo-C_2-symmetric complex with two homochiral phenanthroline ligands coordinated to the iron center in a *cis* topology was identified via X-ray analysis of crystals grown after mixing Fe(OTf)$_2$ and a ligand similar to compound **36**. The epoxidation reactions, irrespective of the different electronic features on the phenacyl group of the enones, proceeded smoothly to give the epoxides in good yield, with complete control of the diastereoselectivity and excellent enantioselectivity (Scheme 3.36). Low conversion and enantiocontrol were achieved when reacting β,β-dialkyl-substituted enones. Some experiments revealed the electrophilic nature of the active oxidant, which is able to attack relatively electron-poor alkenes. Interestingly, the same system proved to be efficient in the epoxidation of *trans*-α-methyl cinnamic acid esters, another class of highly challenging alkenes [137]. The corresponding trisubstituted epoxides were isolated in moderate yield, with complete diastereocontrol and excellent level of enantioselectivity (Scheme 3.36). Gao and coauthors reported that ligand **32** was also useful when combined with Fe(OTf)$_2$ to generate an *in situ* chiral complex at 10 mol% loading, which in the presence of peracetic acid or MCPBA at −10 °C in CH$_3$CN served as an effective epoxidizing system [138]. With this protocol, a large variety of *trans*-tetralones were converted into the epoxides in excellent yield and enantioselectivity (up to 99% ee). The electrophilic nature of the oxidant was

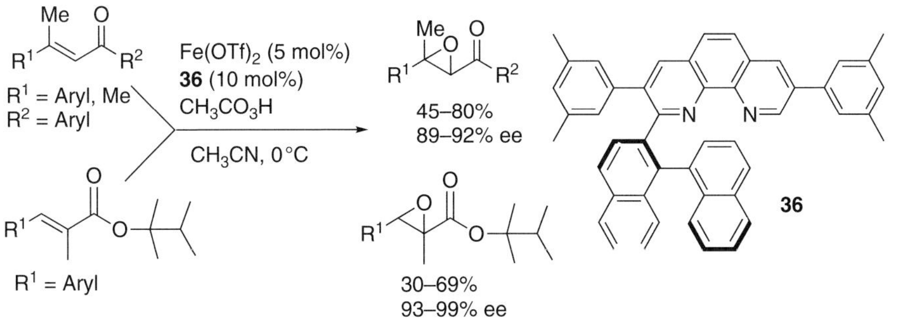

Scheme 3.36 Fe/catalyzed asymmetric epoxidation of trisubstituted α,β-unsaturated enones and esters.

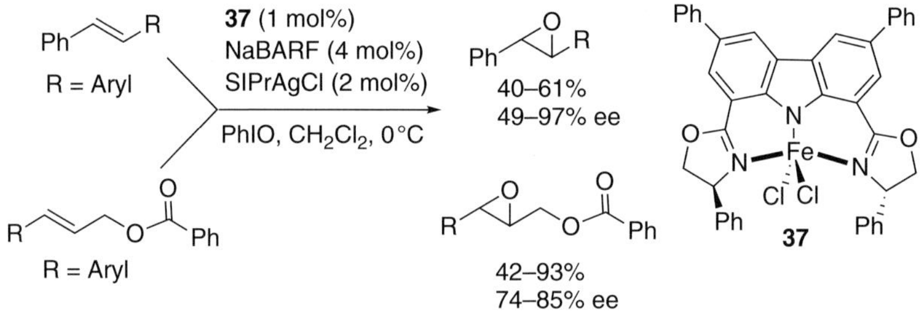

Scheme 3.37 Fe/catalyzed asymmetric epoxidation of *trans*-stilbenes and protected cinnamic alcohols.

ascertained by Hammet correlation mechanistic study. A diluted solution of peracetic acid is employable, which renders the protocol safer.

In 2012, Nakada and coauthors developed an iron(III) complex **37** containing a new chiral carbazole-based tridentate ligand, obtained by combination with FeCl$_2$·4H$_2$O, NaBARF (sodium tetrakis[3,5-bis(trifluoromethyl)phenyl] borate) and SIPrAgCl (SIPr = *N,N'*-bis(2,6-diisopropylphenyl)-4,5-dihydroimidazol-2-ylidene) served as additives, in the presence of PhIO as the oxidant. This system was particularly effective at only 1 mol% loading for the epoxidation of *trans*-stilbenes and cinnamyl alcohol derivatives (Scheme 3.37) [139].

Both classes of epoxides were isolated in moderate-to-good yield and good-to-high enantioselectivity. The presence of additives was necessary to improve yield and asymmetric induction. By employing UV–vis and EPR analyses, it was demonstrated the involvement of an intermediate Fe(IV)-oxo complex bearing a π-cation radical, in analogy to Fe porphyrin-mediated epoxidations. Although highly desirable from sustainability and atom-economy points of view, epoxidations mediated by chiral Fe complexes with molecular oxygen were scarcely reported [140]. Indeed, metal activation of O$_2$ is a rather difficult task, also coupled with scarce product selectivity, which significantly hampered the use of this terminal oxidant.

Ruthenium metal complexes would be advantageous in asymmetric epoxidations thanks to low cost, large arena of chiral ligands suitable for coordination, and several oxidation numbers for the metal [141]. High oxidative activity of the RuO$_4$ species with alkenes has been demonstrated, affording mainly cleavage of the carbon–carbon double bond or diols as prevalent products [142]. However, in the presence of proper ligands, such as pyridine derivatives, the oxidative power of RuO$_4$ species, *in situ* formed via oxidation of RuCl$_3$ with NaIO$_4$, could be lowered by complexation, thus generating a selective epoxidizing system for unfunctionalized alkenes [143]. Early examples on asymmetric epoxidation using Ru(II)–Schiff bases with iodosylbenzene [144] and Ru(III)–salen chiral complexes with molecular oxygen and isobutyraldehyde or 2,4-dichloropyridine *N*-oxide as the terminal oxidants [145] afforded styrene epoxides in variable yields and up to 89% ee. The formation of high-valent Ru-oxo species was postulated to be involved in the process. In 2006, a ruthenium(III)–salen complex bearing a glucosamine moiety was reacted in the presence of TBHP, and a variety of alkenes

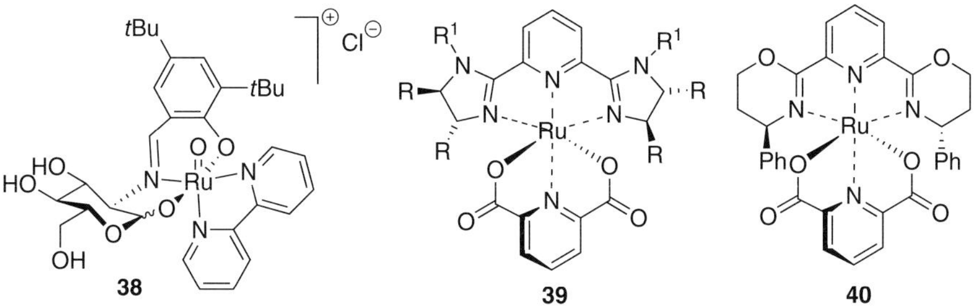

Scheme 3.38 Ru/oxo complex and precatalysts involved in asymmetric epoxidations.

were converted into epoxides with modest enantioselectivity (39–47% ee) [146]. Involvement of high-valent Ru(V)-oxo species **38** was suggested by investigation via IR and UV analysis of the reaction mixture (Scheme 3.38).

Chiral bisoxazoline-containing ligands, firstly introduced by Pfaltz and coworkers, proved to be a useful motif in the Ru-catalyzed asymmetric epoxidation of stilbenes achieving moderate enantiocontrol [147]. Interesting and more general results were displayed by the catalytic systems introduced by Beller and coauthors in the epoxidation of terminal, di-, and trisubstituted electron-rich alkenes. The chiral tridentate N,N,N-pyridinebisimidazoline ligand Ru complex **39** and chiral tridentate N,N,N-pyridinebisoxazine Ru complex **40** were used at 5 mol% loading in t-amyl alcohol as solvent at room temperature with a slow addition of H_2O_2 as terminal oxidant. The incorporation of pyridinedicarboxylate ligand within complexes **39** and **40** served to suppress the catalase activity of the ruthenium (Scheme 3.38) [148]. Under these conditions, high-to-excellent yield and up to 71% ee values were obtained for aromatic epoxides when using catalyst **39** and up to 84% ee when using catalyst **40**. Different spectroscopic studies to intercept the catalytically active species proved to be unsuccessful. ESI-MS suggested the presence of mono- and dioxo-Ru species. Computational studies pointed to the involvement of a mono-oxo-Ru complex as the most plausible catalytically relevant species. The authors suggested that π–π interactions between the ligand and the substrate are responsible for the asymmetric induction. In 2010, Katsuki synthesized Ru(II)(NO)–salen complex **41a**, incorporating BINOL moieties, as catalyst for the asymmetric aerobic oxygen atom transfer epoxidation in the presence of water under visible light irradiation [149]. *trans*-β-Methyl styrenes and *cis*-β-methyl styrene were epoxidized with molecular oxygen as terminal oxidant in satisfactory yield, with complete diastereoselectivity and good-to-high ee (Scheme 3.39). Experimental data indicated that an aqua ligand coordinated with the ruthenium ion in complex **41a** serves as a proton transfer source for the oxygen activation process, and it is recycled and used as proton transfer mediator during the process. These aerobic oxygen atom transfer reactions do not need a special proton and electron transfer system, similarly in oxygen atom transfer biological reactions, which is an interesting example of a sustainable process. This system was further improved by employing catalyst **41b**, synthesized under photoirradiation conditions, in the asymmetric epoxidation of Z- and E-alkenes under air or O_2 atmosphere, without photoirradiating

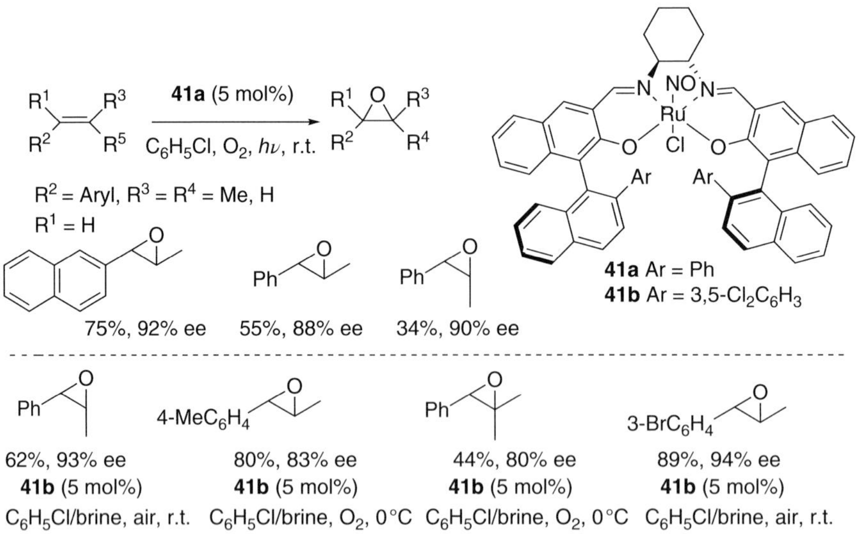

Scheme 3.39 Ru/salen-complex-catalyzed asymmetric epoxidation with molecular oxygen.

the reaction mixture (Scheme 3.39) [150]. Comparable or better results were generally observed under these conditions.

Another class of useful complexes is found in chiral ruthenium–porphyrin-based catalysts, which have been used in the presence of 2,6-dichloropyridine N-oxide (2,6-DCPNO), PhIO, or molecular oxygen as the terminal oxidant. Initial studies showed low yields for aromatic epoxides (deriving from styrene, E-stilbene) and moderate level of enantioselectivity [151]. Improvement of the enantiocontrol has been achieved by Berkessel and coworkers, who reported the epoxidation at room temperature of styrene and 1,2-dihydronaphthalene with 0.1 mol% of catalyst **42** giving epoxides with up to 83% ee (Scheme 3.40) [152]. In 2010, Bach and coworkers, inspired by fundamental interactions typical of enzymatic catalysis, designed the Ru(CO)–porphyrin **43**, incorporating a stereodirecting lactam unit close to the metal center, with a view to preferentially direct the alkene via hydrogen bonding toward the active Ru center [153]. This system was successfully applied for the epoxidation of some 3-vinylquinolones, bearing a complementary H-bonding accepting moiety (Scheme 3.40). Proof of concept on the involvement of directing effects in the epoxidation was clearly evidenced by the isolation in lower yield and in racemic form of the epoxide deriving from the reaction of N-methylquinolone as the substrate. The formation of a Ru(V) oxo species during the oxidative process after decarbonylation and oxidation has been suggested [154].

3.3.5 Pt-, Zn-, Lanthanoid-Catalyzed Epoxidations

In 2006, a peculiar example in the area of asymmetric epoxidation has been reported by Strukul and coworkers, who developed a chiral Pt(II)–diphosphane complex/H_2O_2 system for the asymmetric epoxidation of challenging terminal alkenes [155]. This second generation system was developed by the same group

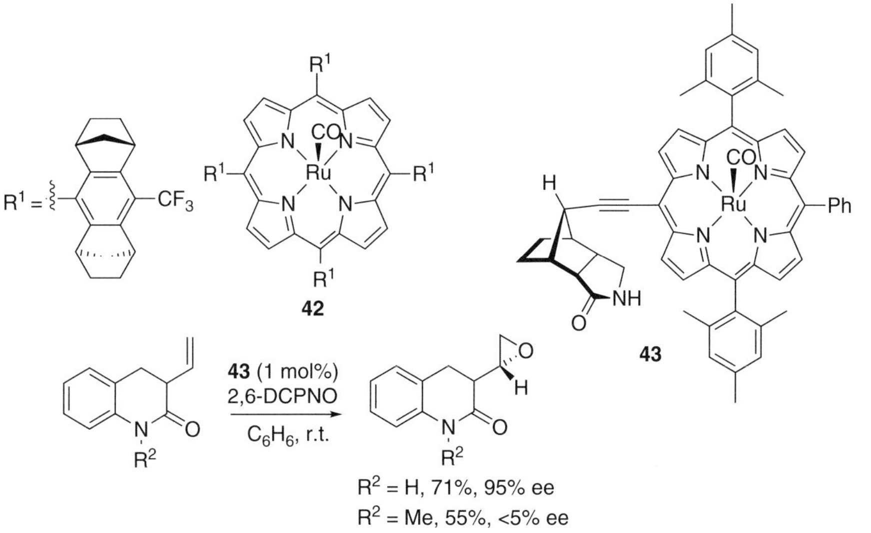

Scheme 3.40 Ru(CO)/porphyrin used in asymmetric epoxidation.

on the grounds of electron-poor Pt(II) diphosphane complexes amenable of efficient activation of aqueous H_2O_2 toward racemic epoxidation of terminal alkenes [156]. A pentafluorophenyl-substituted ligand was incorporated within the chiral complexes obtained by using commercially or easily available ligands. Among the various complexes tested, **44** gave the best results (Scheme 3.41). Aliphatic and aromatic substituted epoxides were isolated in moderate-to-good yield and up to 98% ee. It is interesting to note that a completely regioselective epoxidation of the terminal carbon–carbon double bond was observed when reacting dienes. The proposed mechanism does not involve classical metal activation of H_2O_2 to generate electrophilic oxo, peroxo, or hydroperoxo species. On the contrary, the cationic platinum center would coordinate the alkene, whereas H_2O_2 would be hydrogen-bonded and nucleophilically activated by the closely located fluorine atoms (Scheme 3.41) [157]. The high regioselectivity observed for terminal, less electron-rich double bonds of not sterically encumbered alkenes can be framed under this mechanistic picture.

To date, a few examples have been reported on stereoselective epoxidations mediated by chiral zinc complexes. In the 1990s, Enders and coworkers discovered the epoxidation of *trans*-α,β-unsaturated ketones and *trans*-nitroalkenes mediated by overstoichiometric amounts of a complex formed by Et_2Zn and (1R,2R)-N-methylpseudoephedrine in the presence of molecular oxygen as a benign and highly atom-economic oxidant [158]. This protocol has been recently applied as the final step in the total synthesis of citrinadin A and an analog epoxide citrinadin B, two alkaloids isolated from marine-derived fungus *Penicillium citrinum* (Scheme 3.42) [159].

Citrinadin A and B showed activity against murine leukemia L1210 and human carcinoma KB cells [160]. Although an excess of easily available (1R,2R)-N-methylpseudoephedrine ligand is required, the process afforded the epoxides in

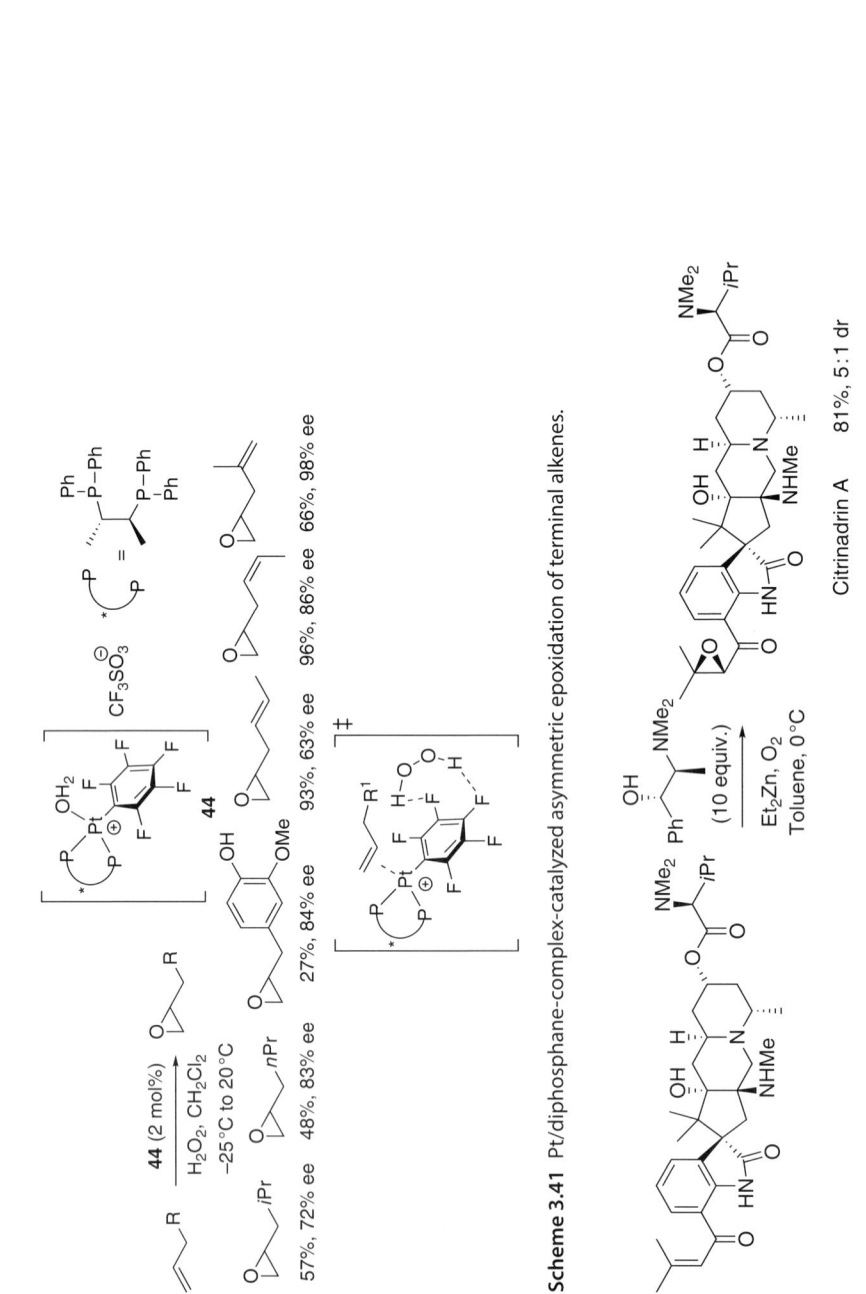

Scheme 3.41 Pt/diphosphane-complex-catalyzed asymmetric epoxidation of terminal alkenes.

Scheme 3.42 Ender's diastereoselective epoxidation as final step in the synthesis of alkaloid citrinadin A.

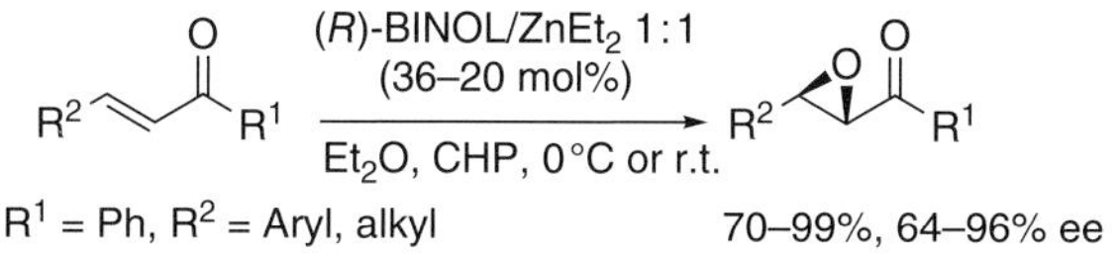

Scheme 3.43 Asymmetric epoxidation of *trans*-enones with ZnEt$_2$/(R)-BINOL/CHP system.

good yield and diastereoselectivity in favor of the epoxide isomer of interest. In this epoxidation, a chiral nucleophilic zinc alkyl peroxide intermediate adds to the electron-poor alkene according to a Weitz–Scheffer mechanism [161]. More recent efforts were directed to render the process catalytic. A first catalytic version, where (R)-binaphthyl polymer was combined with diethyl zinc and molecular oxygen or TBHP, enabled the epoxidation of some *trans*-enones with moderate-to-good enantioselectivity (up to 81% ee) [162]. The polymeric structure of this incorporated ligand helped to increase the enantioselectivity with respect to monomeric BINOL. In 2006, Dötz and coauthors developed an optimized monomeric substoichiometric loading zinc/(R)-BINOL catalyst for the asymmetric epoxidation of *trans*-chalcones and β-alkyl phenyl enones using CHP as the oxidant. Diastereoisomerically pure *trans*-epoxides were isolated in good-to-excellent yield and up to 96% ee (Scheme 3.43) [163].

A similarly performing recyclable heterogeneous catalyst of this system has been reported by Ding and coworkers where the immobilization of BINOL was built up through assembly of chiral multitopic ligands and metal ions without the use of any support [164]. An interesting and synthetically useful diastereo- and enantioselective one-pot methodology to obtain acyclic epoxy alcohols and allylic epoxy alcohols has been reported by Walsh and coauthors [165]. They combined an enantioselective carbon–carbon bond-forming reaction employing enantiopure ligand **45**, to give an allylic zinc alkoxide intermediate, *in situ* epoxidized using either dioxygen or TBHP in the presence of a titanium tetraalkoxide as co-catalyst for the epoxidation. This route gave access to a great variety of epoxy alcohols difficult to prepare via Sharpless asymmetric epoxidation of allylic alcohols. Moreover, a high diastereo and enantiocontrol has been generally observed independently of the substitution pattern of the starting aldehyde (Scheme 3.44). An alternative initial vinyl zinc asymmetric addition to

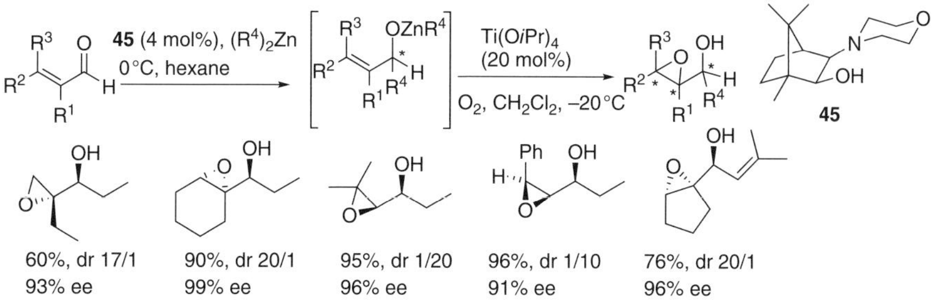

Scheme 3.44 One-pot asymmetric alkyl zinc addition/epoxidation route to epoxy alcohols.

α,β-unsaturated aldehydes followed by the epoxidation step allows for further expansion of the substrate scope of the one-pot approach to allylic epoxy alcohols (dr up to 20:1, ee up to 96%).

The most notable and versatile examples of lanthanoid-based catalysts reported in asymmetric epoxidation reactions have been developed in the late 1990s by the Shibasaki group. *In situ* generation of lanthanum and ytterbium complexes after mixing (*R*)-BINOL and Ln(*i*PrO)$_3$ was firstly found to be suitable for the highly enantioselective epoxidation of *trans*-α,β-unsaturated ketones with TBHP or CHP as the oxidants [166]. Afterward, the system, also applied to other challenging classes of electron-poor alkenes, has been improved in terms of efficiency, nature of the lanthanum complex, and presence of additives. Indeed, the first asymmetric epoxidation reactions of *cis*-α,β-unsaturated ketones [167], *trans*-α,β-unsaturated amides [168] were developed, achieving remarkable levels of diastereo- and enantioselectivity. Interestingly, a heterogeneous recyclable version (up to six cycles) of the Shibasaki's catalyst was also reported for epoxidation of different α,β-enones affording the epoxides in excellent yield (91–99%) with high enantioselectivity (85–97% ee) [169]. In 2005, Shibasaki and coauthors developed a catalytic complex (<10 mol%), generated in the presence of MS from Y(O*i*-Pr)$_3$, (Ph)$_3$As=O additive and ligand **46** in 1:1:1 ratio, suitable for the asymmetric epoxidation of β-aromatic and aliphatic α,β-unsaturated esters (Scheme 3.45) [170].

This method can be considered the first and until now the only general protocol for the asymmetric epoxidation of *trans*-disubstituted α,β-unsaturated esters, which are known for being low-reactive and challenging olefins for a nucleophilic epoxidation. A slightly different biphenyl diol ligand **47** and achiral triaryl phosphane oxide **48** additive were employed under the same conditions, in the yttrium-catalyzed epoxidation of aromatic and aliphatic *trans*-α,β-unsaturated phosphane oxides [171]. The corresponding *trans*-α,β-epoxy phosphane oxides were obtained in high yield and enantioselectivity. A general mechanistic cycle for the epoxidation reactions mediated by lanthanoid complexes has been proposed [172]. The additive would stabilize the active lanthanum species via coordination. Bifunctional activation of the reagents is provided by this complex, via Brønsted base to *in situ* generate a lanthanum peroxide and then as a Lewis acid, activating the Michael acceptor. 1,4-Addition of the nucleophilic Ln peroxide to the Michael acceptor follows according to the general mechanism of the Weitz–Scheffer epoxidation [161]. Shibasaki's systems have been successfully employed as key step for the synthesis of different biologically active compounds, such as anticancer agent (+)-decursin [173], antifeedant (−)-marmesin [173], antifungal agent (+)-strictifolione [174], and antidepressant drug fluoxetin [168d] (Scheme 3.46).

Recently, Zhao, Yao, and coauthors reported the synthesis of a heterobimetallic yttrium–lithium complex stabilized by a chiral phenoxy-functionalized prolinolate ligand that is able to catalyze, at 10 mol% loading, the asymmetric epoxidation of *trans*-chalcones in CH$_3$CN at 0 °C with TBHP as the oxidant. The epoxides were recovered in satisfactory yield and with good-to-excellent enantioselectivity (up to 99% ee) [175]. However, aliphatic enones proved to be unreactive. The same group developed a similar system, based on a chiral ytterbium–lithium

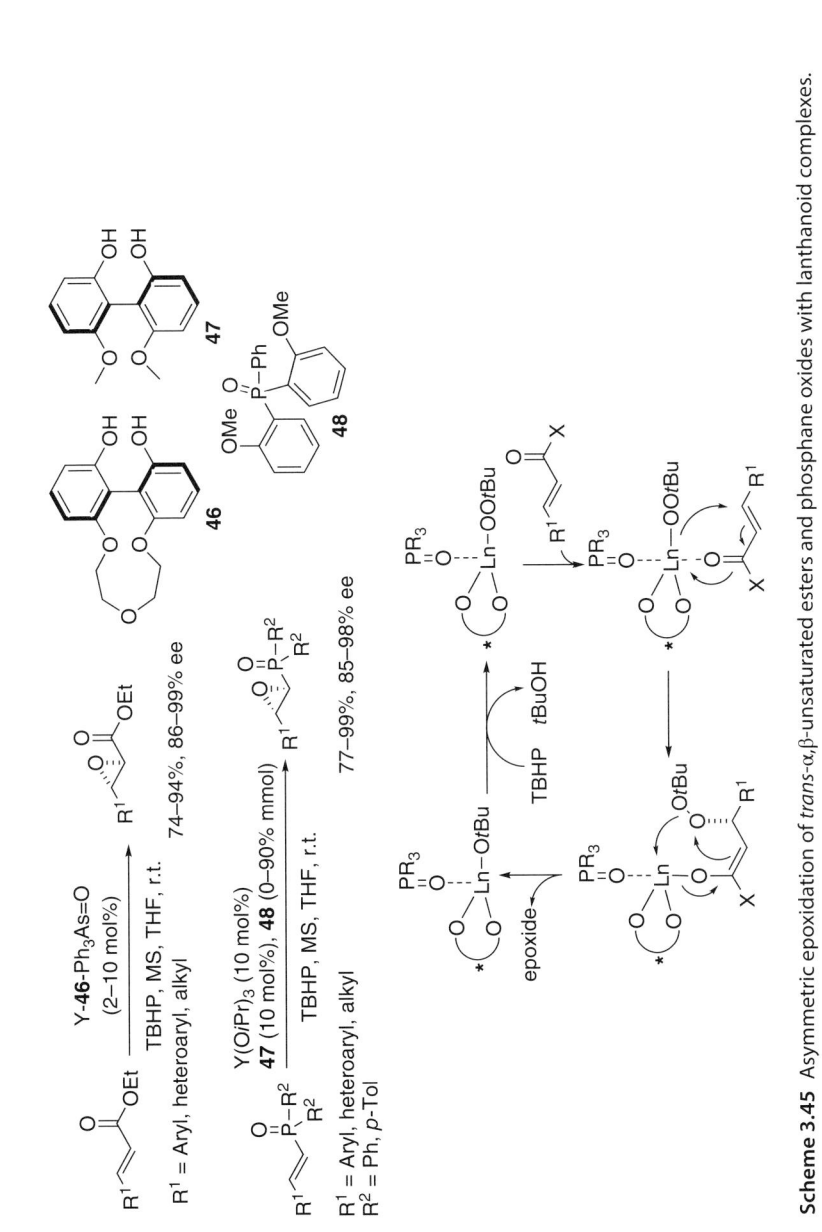

Scheme 3.45 Asymmetric epoxidation of *trans*-α,β-unsaturated esters and phosphane oxides with lanthanoid complexes.

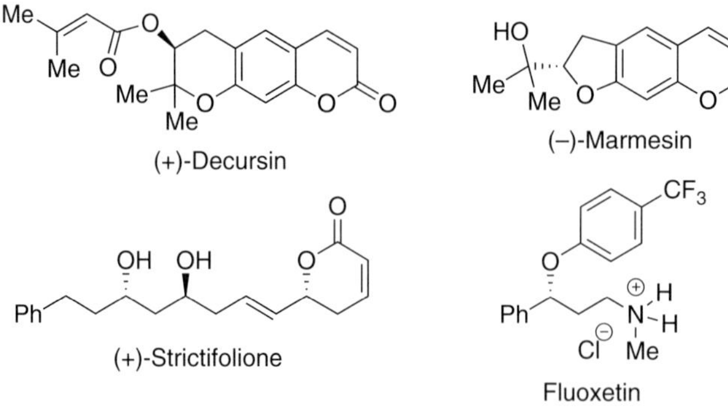

Scheme 3.46 Natural and biologically active molecules synthesized via Shibasaki's epoxidizing systems as key step.

complex, suitable for an asymmetric epoxidation of *trans*-chalcones and enone derivatives with an exocyclic double bond, at 4 mol% loading of the chiral catalyst [176]. Significant improvements in epoxides yield and enantioselectivity were observed. In 2012, Feng and coworkers reported an innovative and environmentally friendly system for the epoxidation of *trans*-chalcones and other electron-poor alkenes based on a chiral *N,N'*-dioxide-Sc(III) complex and H_2O_2 (Scheme 3.47) [177]. Under simple and mild reaction conditions, a good variety of enones, prevalently *trans*-chalcones, were epoxidized in high yield and with excellent level of enantioselectivity using only 5 mol% of the catalyst. $Y(OTf)_3$ and $La(OTf)_3$ proved to be unreactive or much less efficient than $Sc(OTf)_3$ when employed under the same conditions. The authors postulated that the scandium complex would activate the enone as a Lewis acid, which was then attacked by nucleophilic H_2O_2 according to typical Weitz–Scheffer mechanism. An interesting feature of this reaction is high water tolerance of the catalytic system, as demonstrated by addition of water up to 100 times the volume of the oxidant, still maintaining the same conversion and level of asymmetric induction, while working under diluted and safer oxidative conditions. They demonstrated that the

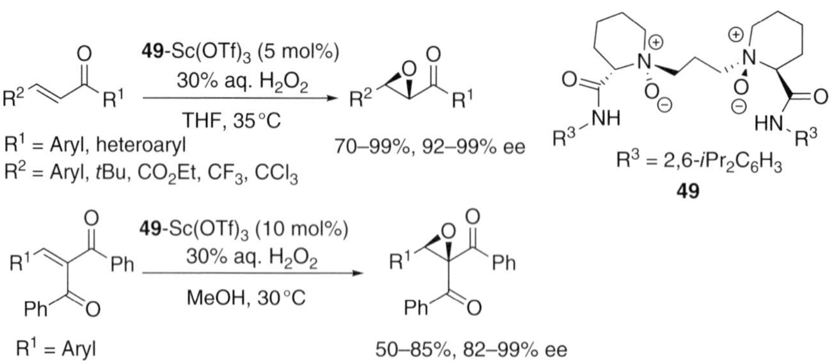

Scheme 3.47 Asymmetric epoxidation of *trans*-di- and trisubstituted enones with $Sc(OTf)_3$/ligand/H_2O_2 system.

catalyst could be recovered, reused, and also employable for gram-scale epoxidation. The same group extended the application of this system to the epoxidation of sterically demanding and more challenging trisubstituted alkenes, such as 2-arylidene-1,3-diketones (Scheme 3.47) [178]. Under slightly modified conditions, the same catalyst, used at 10 mol% loading in MeOH as solvent, afforded a variety of differently substituted aromatic epoxides in moderate-to-good yield and with good-to-high enantioselectivity.

3.4 Asymmetric Organocatalyzed Epoxidations

Asymmetric organocatalysis has increasingly become one of the most useful and practical tools for performing a synthetic transformation [179]. A chiral organic compound serves as a catalyst working under simple reaction conditions, which generally exclude the use of dry solvents and inert atmosphere. Long reaction times and significantly high loading of the organocatalysts are still the main drawbacks, although improvements have been achieved with the development of novel, more effective promoters catalytically active at low loadings [180]. In the area of asymmetric epoxidation, some of the organocatalytic systems appeared before 2000, which signed the formal birth of organocatalysis. A great number of methodologies encompassing nucleophilic and electrophilic types of asymmetric epoxidation of alkenes have been reported in the past decade [181]. Their potential in the synthesis of chiral epoxides to complement and enlarge the rich synthetic repertoire available under metal catalysis has been demonstrated.

3.4.1 Phase-Transfer Catalyst

Asymmetric phase-transfer catalysis (PTC) includes highly valuable methodologies to be used for several carbon–carbon and carbon–heteroatom bond formation [182]. The mildness and often environmentally friendly reaction conditions of asymmetric PTC reactions cope well for large-scale and industrial applications [183]. The first example of phase-transfer-catalyzed methodology for the asymmetric epoxidation of *trans*-chalcones and naphthoquinones was reported in 1976 by Wynberg, using quaternary ammonium salts derived from cinchona alkaloids quinine and quinidine, H_2O_2 as the oxidant under biphasic conditions (NaOH/toluene) achieving up to 45% ee [184]. Over the past decades, different groups modified the structure of PTCs and reaction conditions to improve the level of asymmetric induction and substrate scope. The most notable results have been achieved by the groups of Arai [185], Corey [186], and Lygo [187], who synthesized differently nitrogen substituted cinchona-alkaloid-derived quaternary ammonium salts. These catalysts demonstrated to be compatible with alternative oxidants such easily available NaClO, KOCl, or trichloroisocyanuric acid [188] in the epoxidation of *trans*-enones, affording epoxides with 70–90% ee (Scheme 3.48). The asymmetric epoxidation of α,β-unsaturated sulfones has been developed by employing the Lygo–Corey catalysts, achieving up to 82% ee for *trans*-phenyl styrylsulfone epoxide [189]. Very recently, this epoxidizing

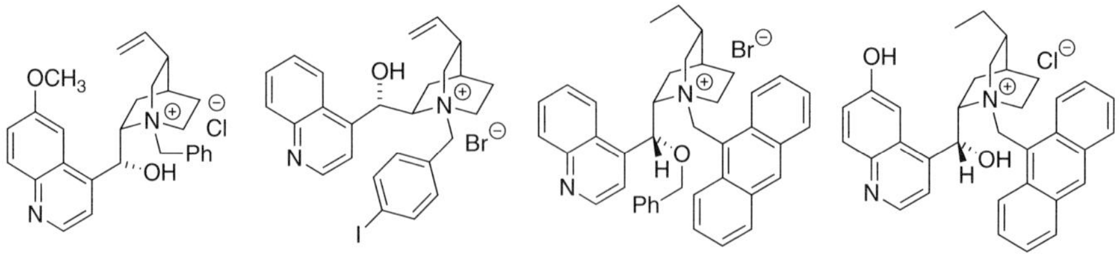

Scheme 3.48 Cinchona-alkaloid-derived PTCs used in the asymmetric epoxidation of enones and naphthoquinones.

system has been found useful for the asymmetric epoxidation of *trans*-β-alkyl-β-nitrostyrenes to the corresponding epoxides achieving good-to-high level of enantioselectivity (up to 92% ee) [190]. Berkessel reported the best level of asymmetric induction in the epoxidation of vitamin K_3, a particularly challenging naphthoquinone, with a PTC derived from quinine (85% ee for the recovered epoxide) [191].

In 2005, Jew, Park, and coworkers reported a highly enantioselective epoxidation of *trans*-chalcones using only 1 mol% of dimeric cinchona-alkaloid quaternary ammonium salts **50** in the presence of surfactant Span 20, H_2O_2 and aqueous KOH [192]. The surfactant dramatically increased the reaction rate and the enantioselectivity (Scheme 3.49).

Soluble ammonium salts of dimeric cinchonidine and quinine, supported on long linear polyethylene glycole chains (PEG), ease to separate and potentially recyclable, have been also reported [193]. However, these catalysts resulted only moderately enantioselective in the epoxidation of *trans*-chalcones (33–57% ee). PTCs with different chiral shapes have been synthesized starting from classical BINOL ligand as those reported by Maruoka [194] and by Hori, where the quaternary ammonium salt was embedded in the azacrown ether [195], or, reported by Bakó, where chiral crown ethers were derived from D-mannose and D-glucose (Scheme 3.50) [196]. Simple chiral crown ethers differ in the mechanism of action from quaternary ammonium salts, which function as surfactants in the liquid–liquid or liquid–solid two-phase reaction system [197]. Indeed, in their case, the whole inorganic salt is transferred into the organic phase without any preliminary ion-pair exchange.

PTCs have been applied in the epoxidation of *trans*-α,β-unsaturated ketones, mainly chalcones, achieving good-to-excellent control of the asymmetric induction. Shibata and coauthors recently illustrated the first asymmetric epoxidation of a sterically hindered class of electron-poor alkenes, the β,β-disubstituted enones bearing a β-trifluoromethyl group [198a]. The PTC **51** derived from cinchona alkaloids was used in the presence of methylhydrazine, an inorganic base, under air (Scheme 3.51). This peculiar and mild system worked well at room temperature, affording challenging and pharmaceutically interesting epoxides, bearing a quaternary stereocenter, in excellent yield, with high diastereoselectivity and excellent enantioselectivity (up to >99% ee). From a mechanistic point of view, the authors proposed a single electron-transfer radical pathway with the oxidation of methylhydrazine by molecular oxygen, leading to an *in situ*

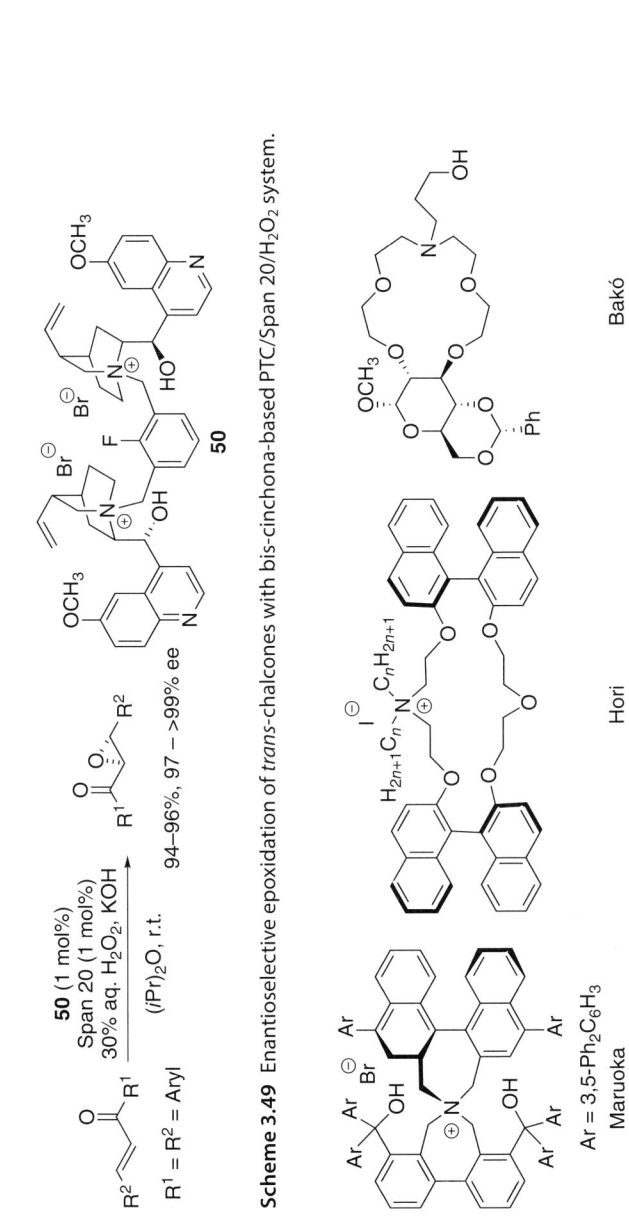

Scheme 3.49 Enantioselective epoxidation of *trans*-chalcones with bis-cinchona-based PTC/Span 20/H$_2$O$_2$ system.

Scheme 3.50 PTCs used in the asymmetric epoxidation of *trans*-α,β-unsaturated ketones.

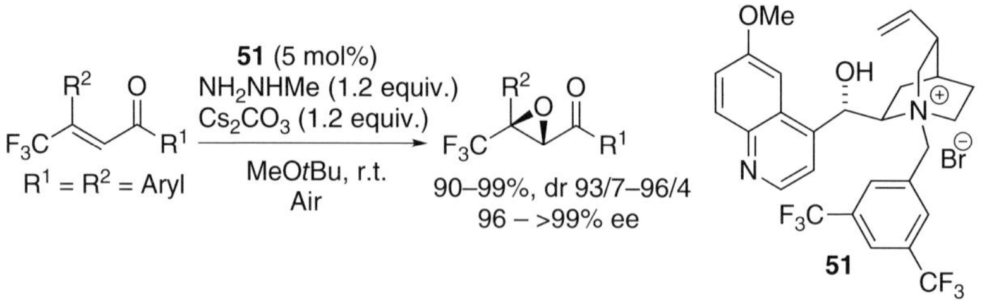

Scheme 3.51 PTC-catalyzed asymmetric epoxidation of β,β-disubstituted enones under air.

generation of H_2O_2 as the real oxidant. This assumption was verified by performing the epoxidation with 50% H_2O_2 as the oxidant, thus achieving a lower yield but almost the same enantiocontrol. Experiments, performed with isotopically labeled ^{18}O as well as TEMPO radical scavenger, supported the mechanistic hypothesis. Concurrently, Chen and coauthors reported a similar more practical system, where the process was catalyzed by 3 mol% of a quinidine-derived PTC in the presence of 30% H_2O_2 and 50% KOH in $CHCl_3$ as solvent at 0 °C [198b]. Under these conditions, better diastereoselectivity and comparable yields and enantioselectivity were observed. The catalyst could be recovered and reused twice, giving similar results. Bis-quaternary cinchona-alkaloid-derived ammonium bromides have been recently synthesized as multifunctional phase-transfer catalysts and investigated in the asymmetric epoxidation of chalcones with 30% H_2O_2 at room temperature [199]. Epoxides were recovered in high yield and with good-to-high ee values.

The first two encouraging examples reported by Murphy's group on a novel class of C_2-symmetric guanidinium salts in the asymmetric epoxidation of *trans*-chalcones, which proceeded with up to 93% ee [200], inspired Nagasawa *et al.* to synthesize modified C_2-symmetric guanidinium salts of different cyclic structures. These catalysts used at 10 mol% loading in the epoxidation of *trans*-enones yielded the products with moderate enantioselectivity (up to 60% ee) [201]. The same group recently introduced a new modular class of bifunctional C_2-symmetric guanidinium–urea catalysts derived from α-amino acids as the chiral spacers [202]. The epoxidation of *trans*-chalcones smoothly gave the epoxides with high enantioselectivity when using PTC **52** (5 mol%) with H_2O_2 as the oxidant and NaOH as the base (Scheme 3.52). A quantitative recovery of PTC **52** from the reaction mixture enabled its employment without any loss of catalytic activity and enantioselectivity for five runs. Synergistic H-bonding interactions of the guanidinium and urea moieties with the peroxide anion and the carbonyl of the enone, respectively, were proposed to be involved in the activation of the reactive partners according to two possible interaction types (Scheme 3.52). Supportive data for this hypothesis derived from complete inactivity displayed by catalysts were devoid of either the urea or the guanidinium groups. These organocatalysts and their thiourea derivatives proved to be highly efficient promoters in a variety of asymmetric processes including nitroaldol, Mannich type, Friedel–Crafts reactions [203].

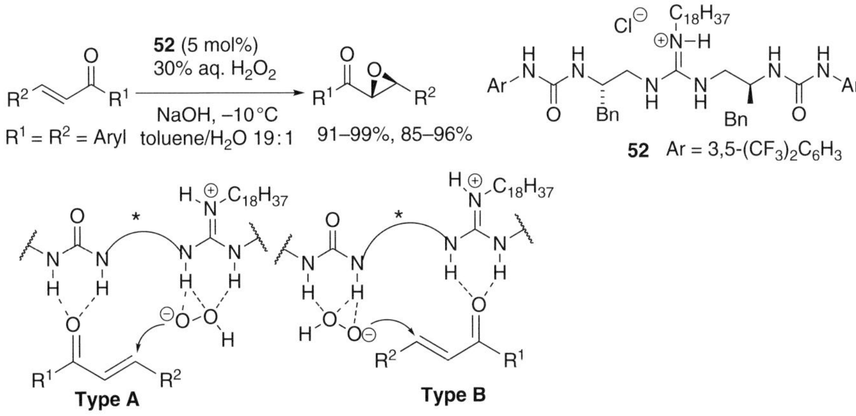

Scheme 3.52 Guanidinium/urea/H$_2$O$_2$-catalyzed asymmetric epoxidation of *trans*-chalcones.

An interesting application of the cinchona-alkaloid-derived PTC-catalyzed diastereo- and enantioselective epoxidation to access potent naturally occurring cysteine protease inhibitor, the epoxysuccinyl peptide E-64c, has been illustrated by Lygo and coworkers (Scheme 3.53) [204]. The epoxidation proceeded with satisfactory yield and good diastereoselectivity, similar to the one required in the final target.

3.4.2 Polyamino Acids and Aspartate-Derived Peracids

Peptides offer a great structural variety, and their straightforward synthesis stimulated an increasing interest toward their application as organocatalysts for asymmetric synthesis [205]. Although catalyst design is difficult, due to the presence of many functional groups potentially involved in the catalysis and numerous achievable conformations of peptides in solution, combinatorial techniques and DFT calculations proved to be helpful for determining the most active peptide catalyst in its most plausible conformation [206]. In the early 1980s, Juliá and Colonna disclosed that homo-oligopeptides as poly-L-alanine (PLA) or poly-L-leucine (PLL) catalyzed the asymmetric epoxidation of *trans*-chalcones, dienones, unsaturated ketoesters in a three-phase mixture composed of toluene/basic H$_2$O$_2$ water solution/gel-like homo-oligopeptide as the third phase [207]. The best level of asymmetric induction (up to 98% ee) was obtained with an optimized number of 30 amino acidic residues. Under these conditions, *trans*-chalcones, dienones, and unsaturated ketoesters were epoxidized in good yield and with generally good-to-high ee values (Scheme 3.54).

Long reaction times, problematic enolizable enones, and the difficulty of catalyst recovery prompted some research groups to improve this reaction. An improvement of the epoxidation under three phase conditions was achieved by adding tetrabutylammonium bromide as phase-transfer additive to increase the concentration of the peroxide in the organic phase, as demonstrated by Geller and coworkers. A rapid conversion to the epoxides was achieved also at 100 g substrate scale, maintaining the level of asymmetric induction [208]. Anhydrous

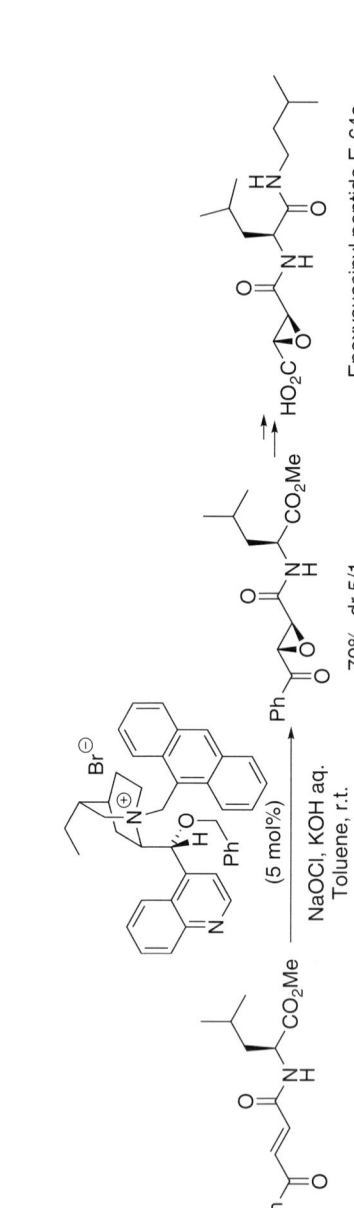

Scheme 3.53 PTC-catalyzed epoxidation to access cysteine protease inhibitor peptide E-64c.

Scheme 3.54 Enantioselective epoxidation of *trans*-α,β-enones with triphasic system polyamino acids/H_2O_2.

solvents, urea–H_2O_2 complex as oxygen source, organic or inorganic bases in polar medium turned the epoxidation system into a biphasic one with reduction of reaction times and catalyst loading [181e, 209]. Morever, recovery of PLL was succeeded by catalyst absorption or grafting to silica gel, although reactivation of the catalyst was necessary prior to use [210]. Recently, a recyclable imidazolium-modified poly(amino acid) catalyst was synthesized and successfully used in the heterogeneous asymmetric epoxidation of *trans*-chalcones achieving up to 99% ee. The insoluble powder catalyst greatly reduced the reaction times, and it was easily separated by filtration and reused for seven cycles without decrease in catalytic efficiency [211]. Nanohybrid materials based on polyamino acids immobilized onto rehydrated hydrotalcite were also developed to employ as recyclable catalysts in the Juliá–Colonna epoxidation [212]. Their performance was highly satisfactory in both epoxide conversion and asymmetric induction. The nanohybrid catalyst could be separated from the reaction mixture by simple centrifugation. In this way, the catalyst could be recovered and reused, keeping the same activity and enantioselectivity for at least five consecutive runs. Scrimin and coworkers showed that Cα-tetrasubstituted amino-acid-based peptides can promote the epoxidation under biphasic conditions, but the level of enantioselectivity was poor (ee <20%) [213]. The Roberts group incorporated the PLL onto a soluble polymer PEG, leading to a copolymer, which formed a homogeneous monophasic system for the asymmetric epoxidation with UHP/diazabicycloundecene (DBU) system [214]. Another interesting feature of the Juliá–Colonna reaction was that a minimum of four leucine residues in an α-helical structure are sufficient to achieve high conversion and enantioselectivity. The Roberts and Berkessel groups elucidated by experimental molecular modeling and, more recently, NMR studies the role of amino acids in the oligomers, disclosing that those close to the N-terminus amino acid determined the stereochemical outcome of the epoxidation [215]. Mechanistically, it has been proposed that the sense of asymmetric induction in the epoxidation is determined by the helicity of the catalyst, through carbonyl oxygen of the enone hydrogen bonded to the NH of the amino acid at the N-terminus of position $n-2$, whereas the hydroperoxide anion would be delivered face-selectively to the β-carbon atom of chalcone by the NH group of amino acid $n-1$. Recent application of chiral cyclic α-amino-acid-containing oligopeptides confirmed the importance of the α-helical secondary structure of the peptide catalysts to achieve high stereocontrol in the asymmetric epoxidation of chalcones with DBU and UHP as the oxidant [216]. Kinetic studies reported by the groups of Colonna, Roberts, and Blackmond on the Juliá–Colonna epoxidation showed a complex mechanistic scenario, suggesting an enzyme-like catalysis provided by PLL (Scheme 3.55) [217]. Reaction progress kinetic analysis helped to demonstrate that the epoxidation proceeds via reversible addition of the enone to PLL-bound hydroperoxide (PLL:HOO⁻), forming a transitory hydroperoxy enolate that would evolve to the epoxide according to pathway (a), pathway (b) being kinetically negligible.

Besides the most studied *trans*-chalcones, some acyclic enones, tetralones, *trans*-vinyl sulfones were efficiently epoxidized using PLL as a catalyst, whereas particularly challenging trisubstituted alkenones and less-reactive α,β-unsaturated

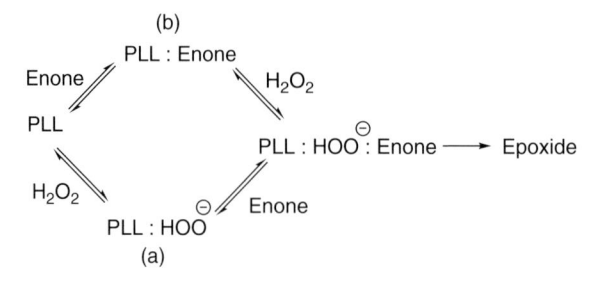

Scheme 3.55 Proposal of consecutive-parallel reactions network for the Juliá–Colonna epoxidation.

esters are not suitable compounds for this readily available epoxidizing system. It can be affirmed that the PLL-catalyzed epoxidation is one of the most applied methods for the asymmetric synthesis of chalcone-derived epoxides. A recent investigation by Kudo and Akagawa illustrated the first application of an amphiphilic resin-supported polyamino acid catalyst in the diastereo- and enantioselective epoxidation of α,β-unsaturated aldehydes using H_2O_2 in a water medium (Scheme 3.56) [218]. It has to be pointed out that this class of alkenes is unsatisfactorily epoxidized by the chiral metal-based systems.

The terminal sequence D-Pro-Ach-[Ala(1-Pyn)]$_3$ for the supported catalyst with the introduction of the unnatural amino acid Ala(1-Pyn), bearing hydrophobic and bulky aromatic ring, was particularly effective for enhancing reaction rate and stereocontrol in aqueous medium. In order to have compounds easy to analyze, the epoxy aldehydes were *in situ* reduced to epoxy alcohols. Interestingly, high level of diastereoselectivity and good-to-high enantioselectivity were generally achieved. The recyclability of the resin-supported peptide was tested up to three epoxidation cycles, observing a decrease in the conversion, albeit the enantioselectivity was maintained. Although substrate scope was limited to alkene with β-aromatic moiety bearing electron-withdrawing groups and improvement in the conversion to the epoxide was necessary, the protocol offers an environmentally friendly procedure for the asymmetric epoxidation of a rather challenging class of electron-poor alkene. Two of the most notable synthetic applications of the Juliá–Colonna epoxidation as key step concerned the preparation of the potent blood-pressure-lowering agent diltiazem [219] and the side chain of anticancer agent taxol (Scheme 3.57) [219].

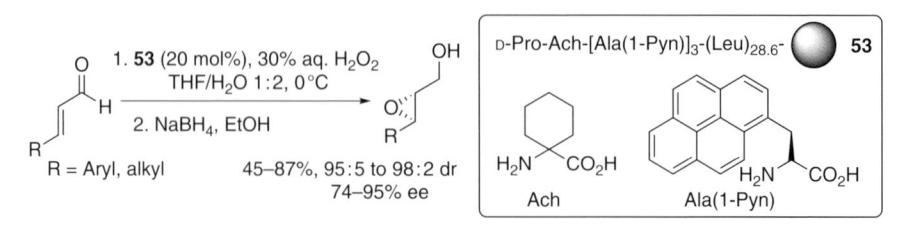

Scheme 3.56 Stereoselective epoxidation of *trans*-α,β-unsaturated aldehydes using a resin-supported peptide catalyst/H_2O_2 system.

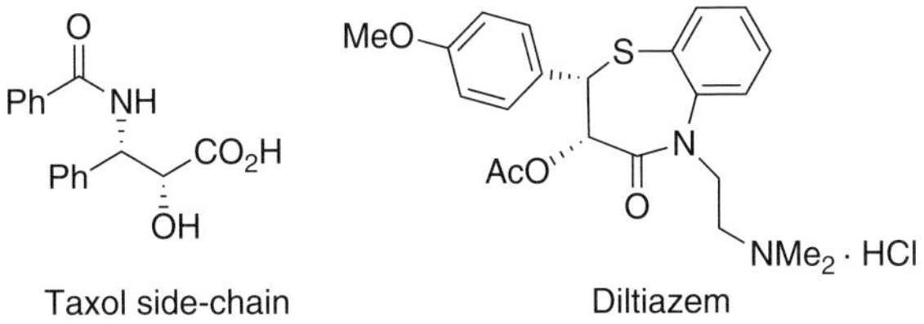

Scheme 3.57 Pharmaceuticals or their molecular fragments synthesized via Juliá–Colonna epoxidation.

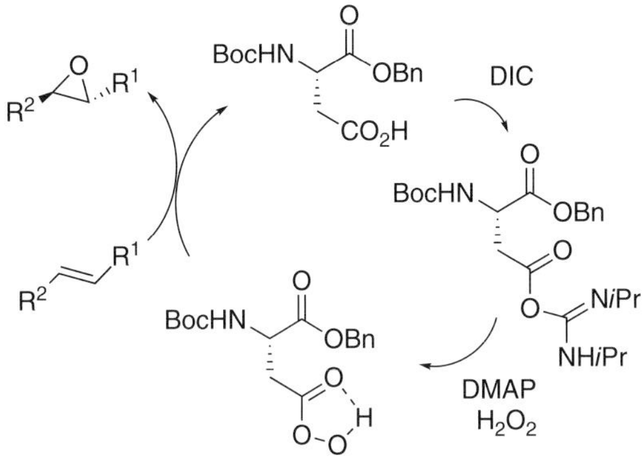

Scheme 3.58 Asymmetric epoxidation with Asp-derivative/carbodiimide/H$_2$O$_2$ system.

A fascinating approach to the asymmetric epoxidation of alkenes has been illustrated by the work of Miller and coauthors. In 2007, they developed the first highly enantioselective electrophilic epoxidation using a peptide catalyst and H$_2$O$_2$ as the oxidant, a system that complements the Juliá–Colonna peptide-based epoxidation [220]. The idea was to form a transient peroxyacid within the peptide chiral scaffold. The pioneering approach, which met with scarce success, has been based on the use of stoichiometric simple, small chiral percarboxylic acids in the epoxidation of alkenes [221]. Initially, the alkene was epoxidized by a model chiral peracid, *in situ* generated from diisopropylcarbodiimide (DIC) as a stoichiometric activator of an Asp-derivative as the catalyst, whereas *N,N*-dimethyl-4-aminopyridine (DMAP) acted as co-catalyst enhancing catalytic turnover for the generation of the peroxyacid (Scheme 3.58).

The real asymmetric protocol of the process was developed by inserting the catalytically active aspartate in a more complex β-turn forming a tripeptide catalyst **54** (Scheme 3.59). A carbamate moiety was introduced in the allylic alcohols with a view to establish hydrogen bonding interactions with the peptide catalyst. Indeed, the presence of the carbamate group helped to achieve high enantioselectivity in the epoxidation. Electron-withdrawing and electron-donating substituents on the aromatic ring of the carbamate moiety did not affect the yield and enantioselectivity, which were generally high. Different Henbest-type peroxyacid-substrate ensembles are plausible to explain the origin

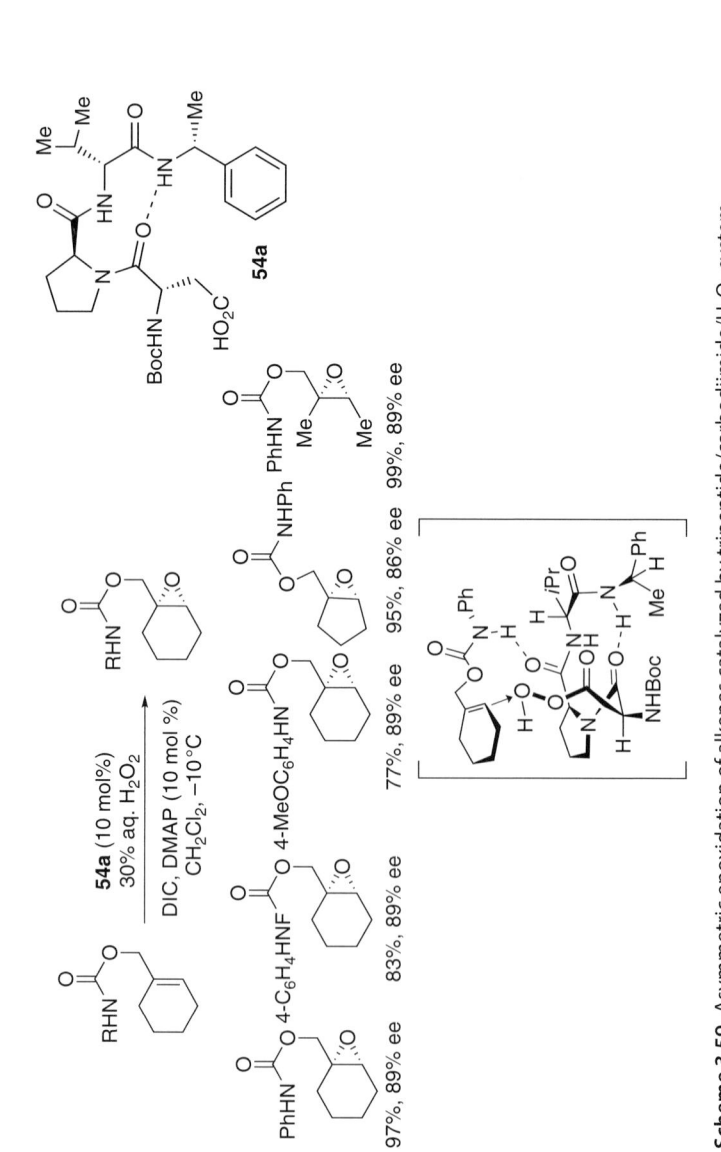

Scheme 3.59 Asymmetric epoxidation of alkenes catalyzed by tripeptide/carbodiimide/H_2O_2 system.

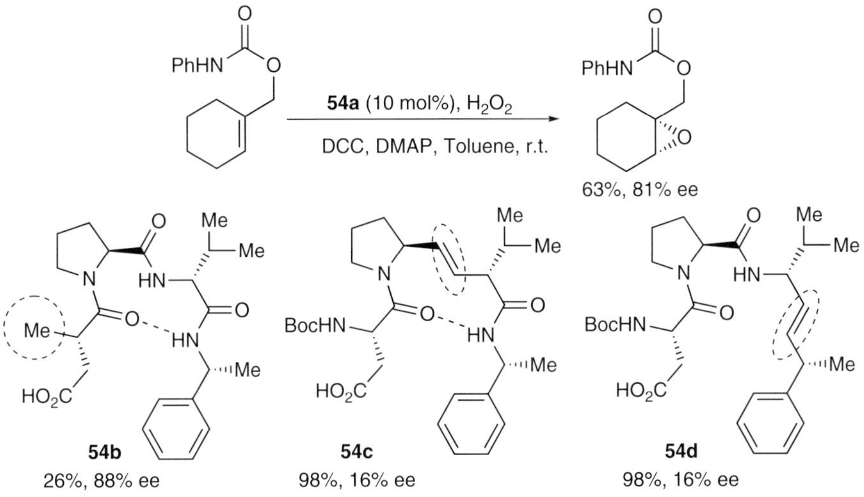

Scheme 3.60 Asymmetric epoxidation of model alkene with modified tripeptide/carbodiimide/H_2O_2 system.

of enantioselectivity. A transition state has been proposed, consistent with the absolute configuration of the epoxides.

The same group synthesized tripeptides with specific-site modifications to shed light on the stereochemical information transfer in the model epoxidation (Scheme 3.60) [222].

Substitution of the NHBoc moiety with a methyl group as in catalyst **54b** did not affect the enantioselectivity, thus showing that NHBoc group was not involved in the orientation of the substrate with a specific interaction. Hence, catalysts **54a** and **54b** might adopt the same three-dimensional structure. The functional role of the Pro-D-Val amide portion was studied by replacing this portion with an isosteric alkene, able to mimic the steric characteristics of the amide bond as in catalyst **54c**. The epoxidation carried out with compound **54c** yielded the epoxide with dramatically reduced ee. Results obtained with catalyst **54d** were not unequivocal since substitution of the C-terminal amide with a double bond prevents formation of the intramolecular hydrogen bond and β-turn; therefore, the erosion of the selectivity could be due to lack of functional or structural motif. An interesting and challenging application of this catalytic system was reported by the same authors to effect the asymmetric oxidation of indole derivatives (Scheme 3.61) [223]. The oxidation turned out to be remarkably chemoselective, a result that was not obvious.

Several catalysts in which the Asp residue was incorporated at the N-terminal of β-turn-shaped peptides were tested, and the most promising compound was peptide **54e**. A variety of indole derivatives were treated under optimized conditions in $CHCl_3$ at 0°C with catalyst **54e** (Scheme 3.61). Unexpectedly, *ortho*-substituted compounds afforded the 3-hydroxy-indolenines with the highest enantioselectivity. Recently, the same group reported an interesting behavior of these peptide-based catalysts in the epoxidation of farnesol, where a notable

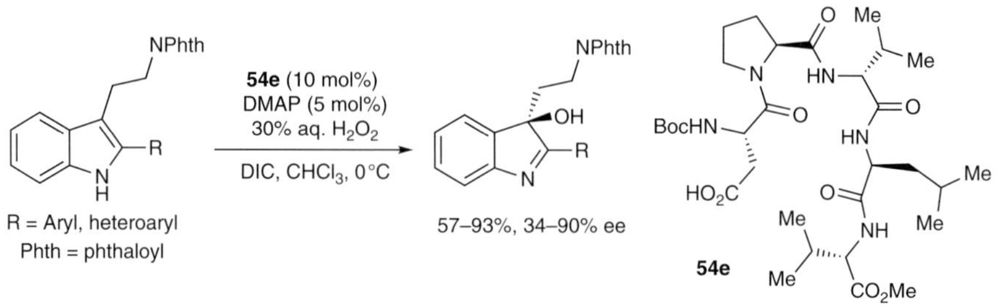

Scheme 3.61 Asymmetric oxidation of indole derivatives with peptide-based catalyst.

selectivity has been displayed by peptides in site epoxidation of different carbon–carbon double bonds of the polyunsaturated molecule (Scheme 3.62) [224]. Peptide **54f** would operate according to a Sharpless-type epoxidizing system via a hydroxyl-directed mechanism and with comparable level of asymmetric induction. Catalyst **54g** preferentially afforded the internal epoxide of farnesol although with poor enantioselectivity. Better insight into this atipycal selectivity was achieved by further modification of peptides and NMR studies, which help to clarify that, even in the epoxidation with catalyst **54g**, the process is hydroxyl directed and that the catalyst exhibits propensity for β-turn formation [225]. These preliminary studies underline the intriguing possibility of achieving high site- and enantioselective epoxidation of complex molecules by using short peptides, which are reasonably simple to obtain by split-pool combinatorial synthesis [226].

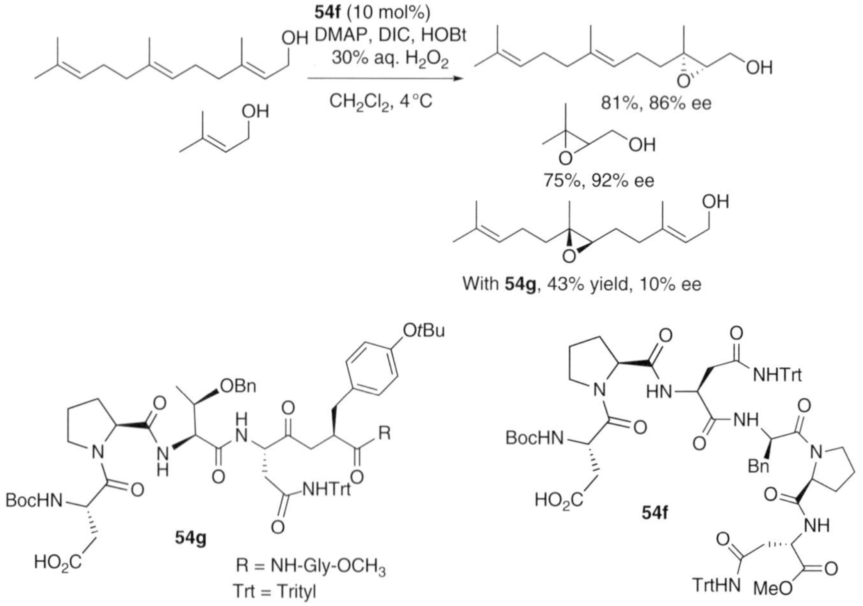

Scheme 3.62 Asymmetric epoxidation of allylic alcohols with peptide-based catalysts.

3.4.3 Chiral Dioxiranes, Iminium Salts, and Alkyl Hydroperoxides

The most efficient class of oxidants to effect the epoxidation of a generic carbon–carbon double bond are likely dioxiranes [227]. In contrast to similar oxidizing agents, such as percarboxylic acids, the development of chiral dioxiranes for asymmetric epoxidation reactions met with great success over the past decades [228]. Since the pioneering discovery by Curci and coauthors in 1984, who used simple chiral (+)-isopinocamphone as the stoichiometric precursor of an *in situ* generated chiral dioxirane with peroxides in the epoxidation of *trans*-β-methyl styrene achieving 12% ee [229], remarkable progress in terms of asymmetric induction and practicality has been attained (Scheme 3.63).

The first striking example was reported by Yang and coworkers in 1996, who disclosed a C_2-symmetric ketone with an axially chiral binaphthyl backbone substituted at the 3,3′-positions (Scheme 3.63) [230]. In the presence of potassium peroxomonosulfate, commercially available as Oxone (mixture of $2KHSO_5$ $KHSO_4$ K_2SO_4), a relevant feature of the system was that only 10 mol% loading of the ketone was required, thanks to a satisfactory regeneration of the chiral dioxirane, whose involvement was convincingly demonstrated through ^{18}O-labeling. A spiro transition state for the epoxidation has been suggested. *para*-Substituted *trans*-stilbenes and trisubstituted alkenes were epoxidized at room temperature with enantioselectivity up to 95% ee. The same system was also useful for the asymmetric epoxidation of challenging *trans*-cinnamates [231]. Denmark's group reported another modified class of axially chiral ketones bearing fluorine atoms to increase the carbonyl electrophilicity and then the dioxirane formation (Scheme 3.63) [232]. This promoter was used at 30 mol% loading, giving the best enantioselectivity in the epoxidation of stilbenes and *trans*-β-methyl styrene. The same group developed a series of chiral cyclic ammonium ketones to employ in the asymmetric epoxidation with Oxone, although moderate level of asymmetric induction was achieved [233]. Armstrong and coauthors developed α-fluorinated tropinone-based ketones, which enabled the achievement of good level of enantioselectivity in the epoxidation of representative phenylstilbene [234]. Contributions from different groups highlighted the possibility of exploiting cyclic six-membered chiral ketones bearing halogen substituents, which showed the importance of polar interactions to effect the asymmetric induction [235]. Noncyclic chiral ketones based on oxazolidinone [236], peptide [237], and cyclodextrin [238] scaffolds were also checked in the asymmetric epoxidation, achieving modest-to-good level of asymmetric induction. In the late 1990s, Shi and coworkers introduced a variety of pseudo C_2-symmetric six-membered carbocyclic ketones derived from quinic acid suitable for the asymmetric epoxidation of *trans*-disubstituted, trisubstitued, and terminal alkenes and electron-poor alkenes. Oxone was used in DME under basic buffered conditions at 0 °C in the presence of only 5–10 mol% of ketone **55** (Scheme 3.64) [239]. The system showed fairly good applicability, and the results were very promising, with the epoxides being recovered in satisfactory-to-high yield and enantioselectivity.

Fructose-derived ketone **56** is the most effective and reliable ketone reported by Shi and coworkers in the Oxone-mediated asymmetric epoxidation of alkenes [240]. The notable features of this system are as follows: (i) two-step synthesis of

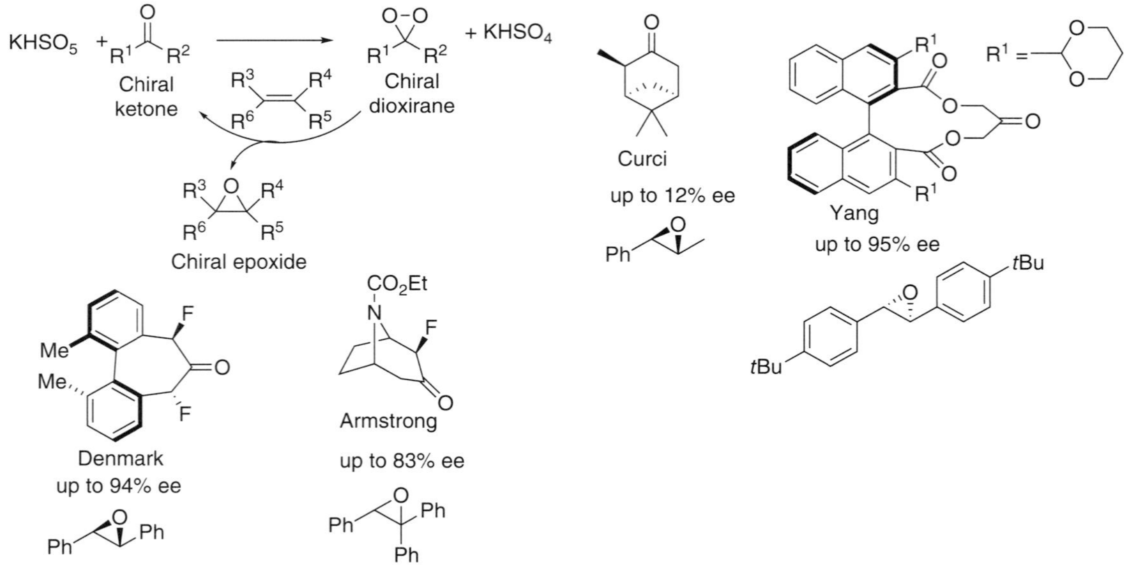

Scheme 3.63 Chiral ketones precursors of dioxirane-catalyzed asymmetric epoxidation.

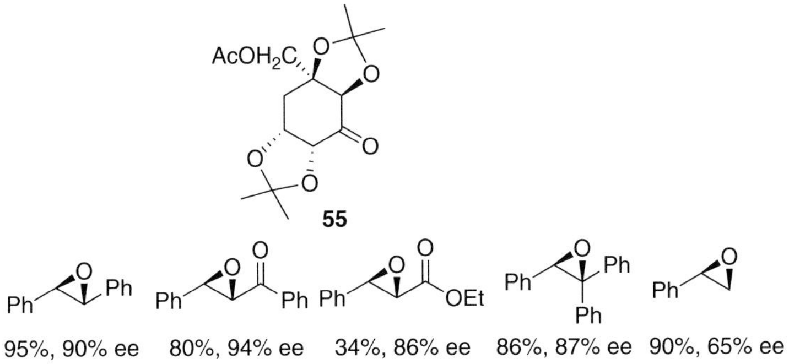

Scheme 3.64 Shi's quinic-acid-derived ketone-catalyzed asymmetric epoxidation.

compound **56** from readily available D-fructose; (ii) epoxides can be prepared in both absolute configurations as the enantiomeric catalyst is accessible from L-fructose; and (iii) fast reaction times. A detailed and fine study of the reaction parameters enabled to set up an efficient and rapid highly enantioselective epoxidation of trans-disubstituted and trisubstituted alkenes using 20–30 mol% of ketone **56** at 0 °C in acetonitrile/dimethoxymethane under buffered conditions, which secured pH > 10 (Scheme 3.65) [240, 241]. Indeed, this ketone proved to be of general employment as it was demonstrated by its application in the highly regio- and stereoselective epoxidation of a great variety of unfunctionalized and functionalized alkenes, such as allylic alcohols [242], dienes [243], silyl enol ethers and enolesters [244], vinyl silanes [245], fluoro alkenes [246]. Whenever possible, no isomerization in the epoxide product was observed, suggesting that the oxidation is stereospecific. The stereochemical outcome of the reaction when using ketone **56** was highly predictable, and the epoxidation has been suggested to proceed via spiro transition state, a hypothesis also reinforced by studies on the kinetic resolution of trisubstituted cyclic alkenes [247].

Hydrogen peroxide can be used in CH_3CN as an alternative terminal oxidant of comparable performance to Oxone [248]. In this case, intermediate peroxyimidic acid is thought to be involved in the catalytic cycle as the real oxidant. Further improvement of the epoxidation was achieved by the Shi group, who planned to reduce the catalyst loading by limiting Baeyer–Villiger oxidation of ketone **56**, likely occurring during the epoxidation. To this end, they synthesized ketone **57**, replacing the fused ketal moiety with a more robust oxazolidinone group (Scheme 3.66) [249]. Under usual conditions, catalyst **57** performed well at only 1–5 mol% loading, achieving high stereocontrol in the epoxidation of disubstituted and trisubstituted alkenes. Diacetate ketone **58** was particularly effective in the asymmetric epoxidation of low-reactive and highly challenging trans-cinnamates and other α,β-unsaturated esters, although cis-cinnamates afforded modest results [250]. The same group directed further efforts to tackle the difficult task of the asymmetric epoxidation of cis-alkenes. A modified chiral ketone **59** proved to be a suitable promoter at 10–30 mol% loading, under usual conditions, giving notable results in the epoxidation of unfunctionalized and functionalized acyclic and cyclic cis-alkenes and terminal alkenes [251].

Scheme 3.65 L-Fructose-derived ketone-catalyzed asymmetric epoxidation.

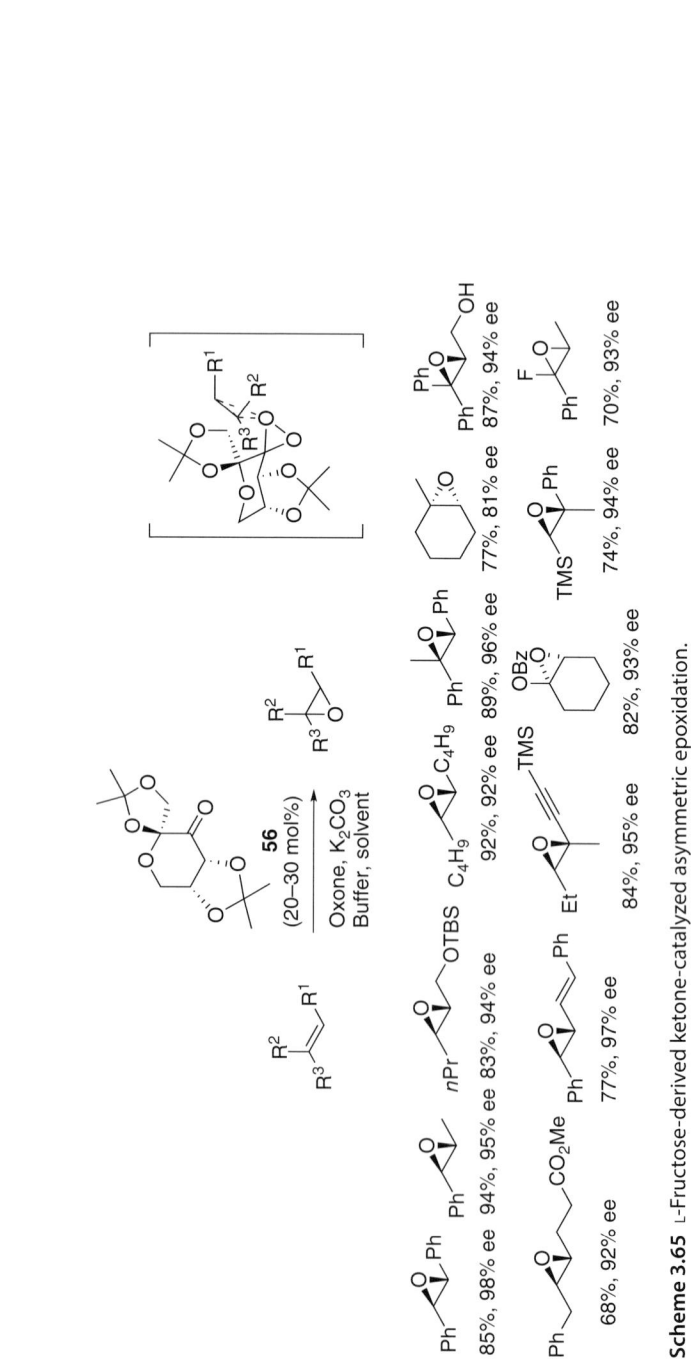

Scheme 3.66 Structurally modified Shi's ketone-catalyzed asymmetric epoxidations.

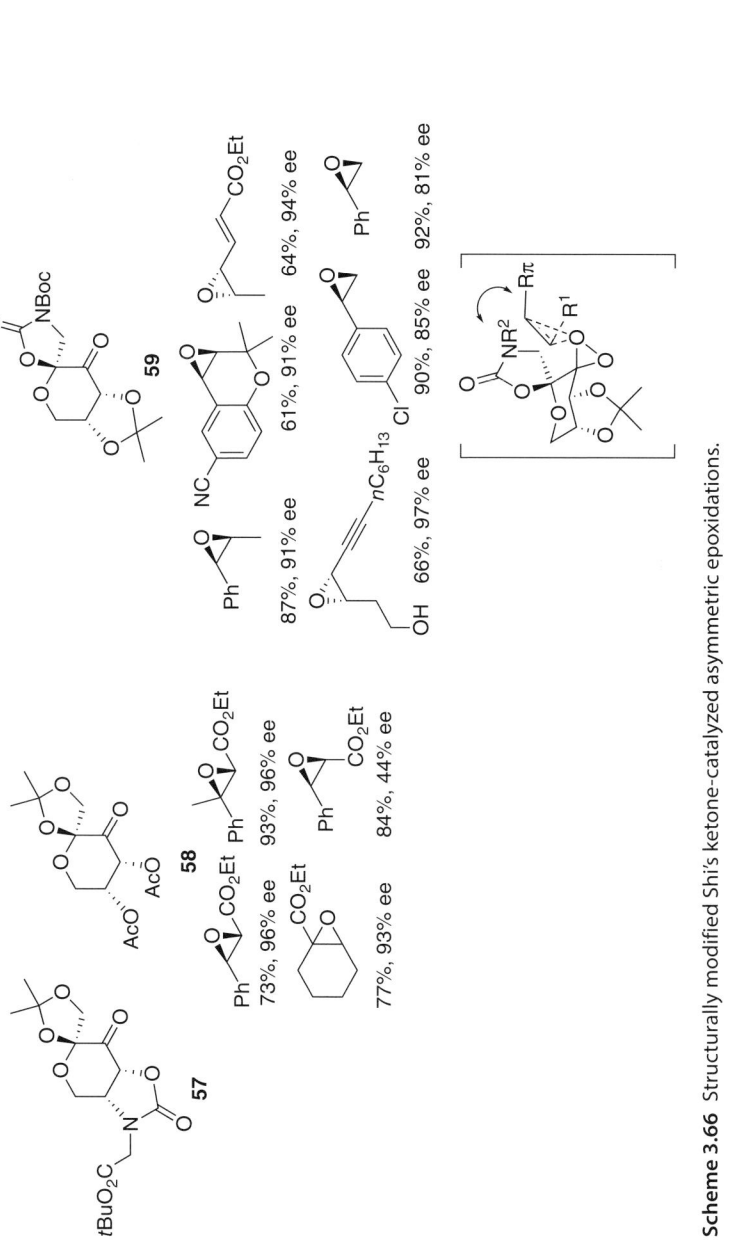

Concerning the stereochemical outcome, a spiro transition state, where the Rπ groups with π systems of the alkene are involved in attractive interactions with the oxazolidinone group of the ketone, was proved with N-aromatic substitution in the ketone structure [252]. Higher ee and predictable absolute configuration of epoxides derived from *cis*-alkenes bearing aromatic groups are in line with this hypothesis. Van der Waals forces and hydrophobic interactions have also been invoked to affect the enantiocontrol of the epoxidation.

In the past decade, the Shi group largely expanded the application of ketone of type **60** to achieve the asymmetric epoxidation of hitherto not accessible chiral epoxides, exemplified by conjugated *cis*-dienes [253], *cis*-enynes [254], nonconjugate *cis*-alkenes [255], styrenes (Scheme 3.67) [256]. Interestingly, when starting from tri- or tetrasubstituted benzylidenecyclobutanes, the corresponding epoxides were rearranged with Lewis acid to give difficult-to-access 2-arylcyclopentanones with inversion or retention of configuration, according to the nature of the Lewis acid and reaction conditions employed [257]. This is an example of an asymmetric epoxidation of tetrasubstituted alkenes [257b]. The level of enantioselectivity in the cyclopentanone was affected by the nature of the substituent in the phenyl ring, indicative of a concerted or S_N1 stepwise rearrangement. In the case of tri- or tetrasubstituted benzylidenecyclopropanes, a sequential asymmetric epoxidation, ring expansion, and Baeyer–Villiger oxidation enabled the synthesis of enantioenriched γ-butyrolactones with tertiary or quaternary stereocenters with up to 91% ee [258]. Working under an increased ketone **60** to Oxone ratio at lower pH could be exploited to synthesize chiral 2-aryl cyclobutanones. A morpholinone-based ketone **61** proved to be the most effective promoter for the asymmetric epoxidation of challenging 1,1-disubstituted terminal alkenes, which were isolated in good yield and with up to 88% ee (Scheme 3.67) [259].

Davis and coworkers studied the performance of conformationally blocked ketone derived from N-acetyl-D-glucosamine **62** in the epoxidation of different alkenes with Oxone at room temperature achieving up to 81% ee in the epoxidation of styrene at stoichiometric loading (Scheme 3.68) [260]. The carboxamide at the chiral α-position was thought be useful to tune the reactivity and the stereoselectivity of the process via stereoelectronic effects.

Jäger and coworkers recently introduced ketone **63**, derived from N-acetyl-D-glucosamine, which afforded interesting results (up to 80% ee) in the epoxidation of cinnamates (Scheme 3.68) [261]. Recently, Vega-Pérez, Iglesias-Guerra, and coworkers described the epoxidation with glucose-derived seven-membered-ring ketones, achieving modest enantiocontrol in the epoxidation of di- and trisubstituted alkenes [262]. More recently, they also reported mannose-derived ketone **64**, which, among other derivatives, enabled, when used in stoichiometric amounts, to achieve up to 90% ee for trisubstituted epoxides [263]. Undoubtedly, according to the results previously illustrated, the sugar-derived ketone organocatalyzed epoxidation with Oxone represents the most general system to produce chiral epoxides from alkenes, irrespective of the degree of substitution at the carbon–carbon double bond and the presence or absence of functional groups. Moreover, carbohydrates and Oxone are readily available low-cost compounds, which is another added value for the application of this protocol as key

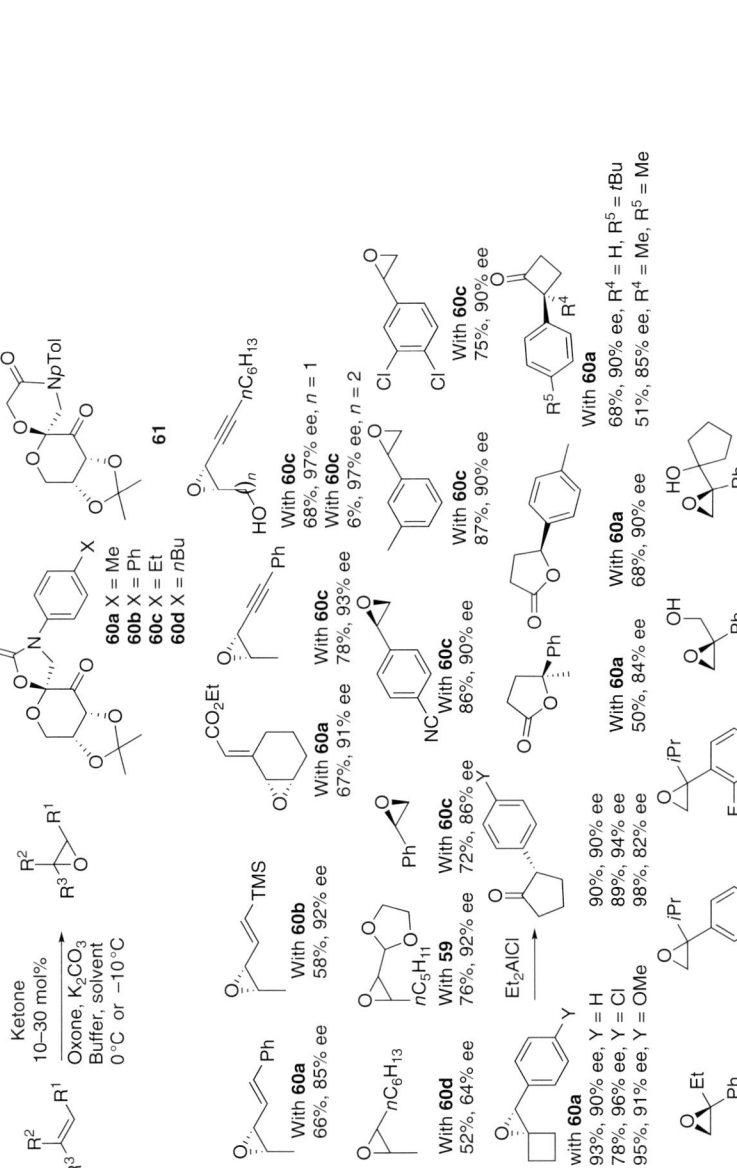

Scheme 3.67 Shi's ketone-catalyzed asymmetric epoxidations.

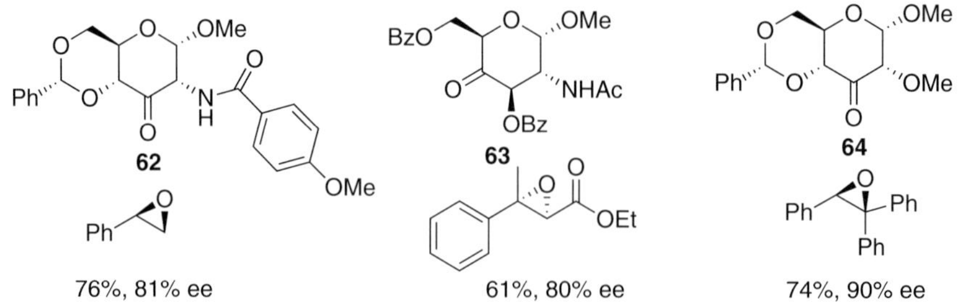

Scheme 3.68 Carbohydrate-based organocatalysts for asymmetric epoxidations.

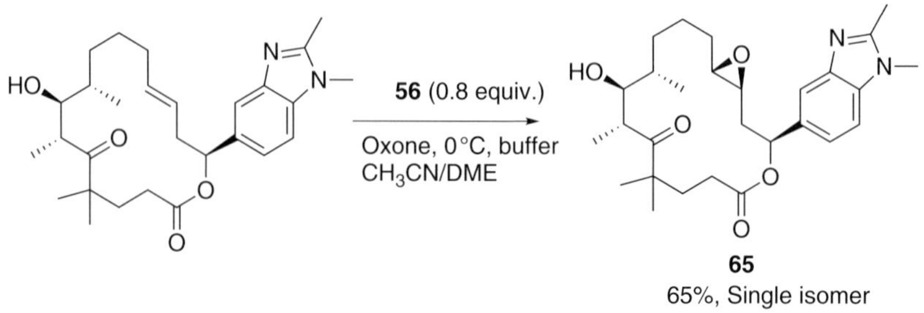

Scheme 3.69 Shi's epoxidation as key step to access epothilone analog.

step in organic synthesis [264]. Indeed, it has been successfully used for the preparation of a vast number of small chiral targets and intermediates or in the synthesis of complex biologically active molecules bearing the epoxide group. Altmann and coworkers illustrated an important application of ketone **56** in the final synthetic step to obtain macrocyclic epoxide **65** from the corresponding highly functionalized alkene [265]. The reaction afforded the epoxide as single diastereoisomer in 65% yield (Scheme 3.69). Compound **65** is an analogous of highly potent epothilones, anticancer agents. Interestingly, tests conducted with product **65** in tubulin-polymerizing and antiproliferative activity showed virtually identical behavior to those of epothilone A and taxol.

Recently, Chandrasekhar and coworkers reported the synthesis of pladienolide B, a natural anticancer macrolide, [266]. The homoallylic alcohol fragment was epoxidized with overstoichiometric amounts of ketone **56** and Oxone to give epoxide **66** in 64% yield and with 95% de (Scheme 3.70). It is interesting to note the recurrent importance of asymmetric epoxidation reactions in total synthesis. The second fragment that was coupled with compound **66** via Stille reaction to give pladienolide **B**, has been constructed starting with the Sharpless asymmetric epoxidation of geraniol.

The examples reported on the epoxidation mediated by sugar-derived ketones/Oxone system to approach small epoxidic intermediates, which were further inter- or intramolecularly ring-opened to final targets, are impressive [228c, 264]. A particularly interesting example has been illustrated by DSM Pharma Chemicals, who performed the Shi epoxidation on an industrial scale to prepare

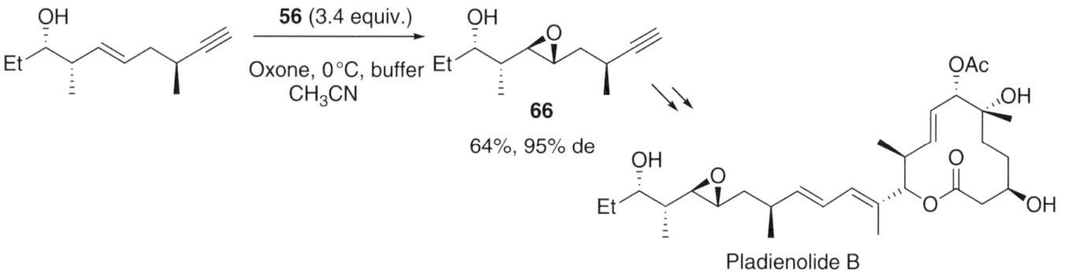

Scheme 3.70 Shi's epoxidation as key step in the synthesis of pladienolide B.

hydroxy lactone **67**, key intermediate for the synthesis of different protease inhibitors (Scheme 3.71) [267]. The unsaturated potassium salt was epoxidized, without buffer, with ketone **56** and sequential slow additions of Oxone solutions to give directly the hydroxy lactone **67** as the ring-opened product in 63% yield and useful level of enantioselectivity. The production of compound **67** was set up to >100 kg scale working under similar laboratory-scale conditions.

Shi and coauthors recently reported a gram-scale simple synthesis of optically pure (+)-ambrisentan, a drug clinically used for the treatment of pulmonary arterial hypertension, starting from ethyl 3,3-diphenylacrylate (Scheme 3.72) [250c]. The electron-poor alkene was epoxidized at gram scale with the most effective ketone **58**, at substoichiometric loadings, to give the epoxide in 90% estimated yield and with 85% ee. The reaction crude was then directly subjected to the other steps to afford (+)-ambrisentan in 53% overall yield and with >99% ee.

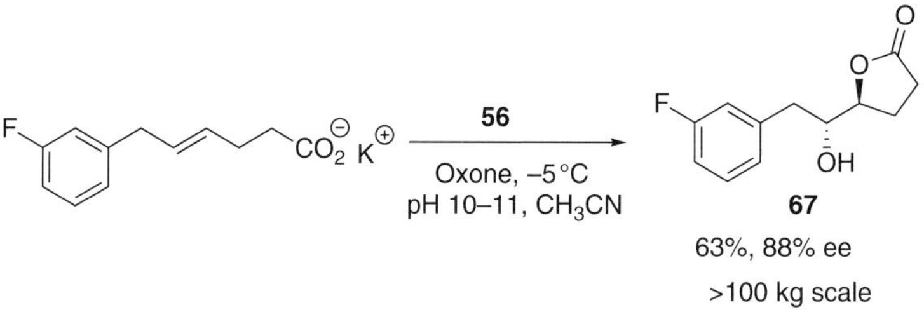

Scheme 3.71 Industrial synthesis of hydroxy lactone via Shi's epoxidation.

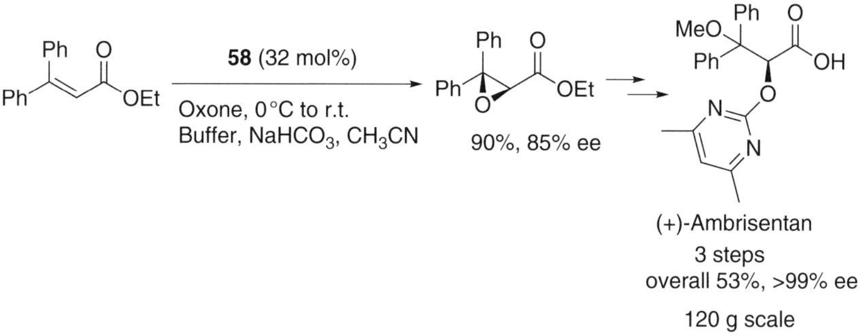

Scheme 3.72 Synthesis of pure (+)-ambrisentan via Shi's epoxidation.

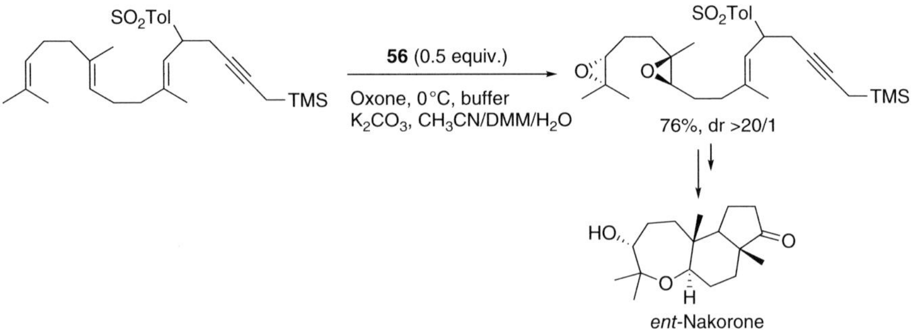

Scheme 3.73 Synthesis of *ent*-nakorone via Shi's epoxidation.

Biomimetic synthesis exploiting intramolecular cascade ring-opening reaction of chiral epoxides has also been reported as illustrated by McDonald and coworkers (Scheme 3.73) [268]. With a view to develop a synthetic pathway to abudinols A and B, oxygenated triterpenoid marine natural products, the triene was selectively epoxidized at two of the three carbon–carbon double bonds in good yield and with high diastereoselectivity when using ketone **56** as the organocatalyst. The allylic electron-withdrawing group proximal to the other carbon–carbon double bond prevented its epoxidation. *ent*-Nakorone, an oxidative degradation product of abudinols, was then obtained via TMSOTf-promoted cyclization with the propargylsilane nucleophile, followed by ozonolysis.

Strictly related to the ketone/dioxirane system, chiral iminium salts/oxaziridinium salts system has been investigated in the asymmetric epoxidation of unfunctionalized alkenes [269]. The oxaziridinium salts were originally reported by Lusinchi and Hanquet via treatment of iminium salts with nucleophilic oxidants (peracids under basic conditions) and demonstrated to be highly electrophilic oxidants [270]. Hence, the asymmetric epoxidation of electron-rich alkenes can be mediated by *in situ* generated chiral oxaziridinium salts from chiral iminium salts in a catalytic cycle analogous to that reported for chiral ketones (Scheme 3.74). In 1993, Lusinchi and coworkers firstly attested this hypothesis using Oxone and $NaHCO_3$ in acetonitrile [271]. Although the enantioselectivity was poor in the epoxidation of *trans*-stilbene (33% ee), they demonstrated the same enantiocontrol and comparable yield either by isolating and reacting the exclusively prepared diastereoisomer of an oxaziridinium salt derived from (1*S*,2*R*)-norephedrine or by its *in situ* generation from catalytic loadings of the corresponding iminium salt with Oxone/$NaHCO_3$ in acetonitrile. This result is indicative of the identical oxaziridinium salt intermediate involved as the real oxidant in the process, which is able to stereospecifically transfer the electrophilic oxygen, as only the *trans*-epoxide was detected. Over the years, a variety of chiral iminium salts were tested to improve the level of enantiocontrol in the epoxidation. However, a first challenge to face is assuring complete diastereoselectivity in the generation of the oxaziridinium salt along the catalytic cycle. Aggarwal and Wang reported a binaphthyl-derived iminium salt achieving up to 71% ee in the epoxidation of 1-phenyl cyclohexene (Scheme 3.74) [272]. Although the oxaziridinium salt was not stable enough to be isolated, an important

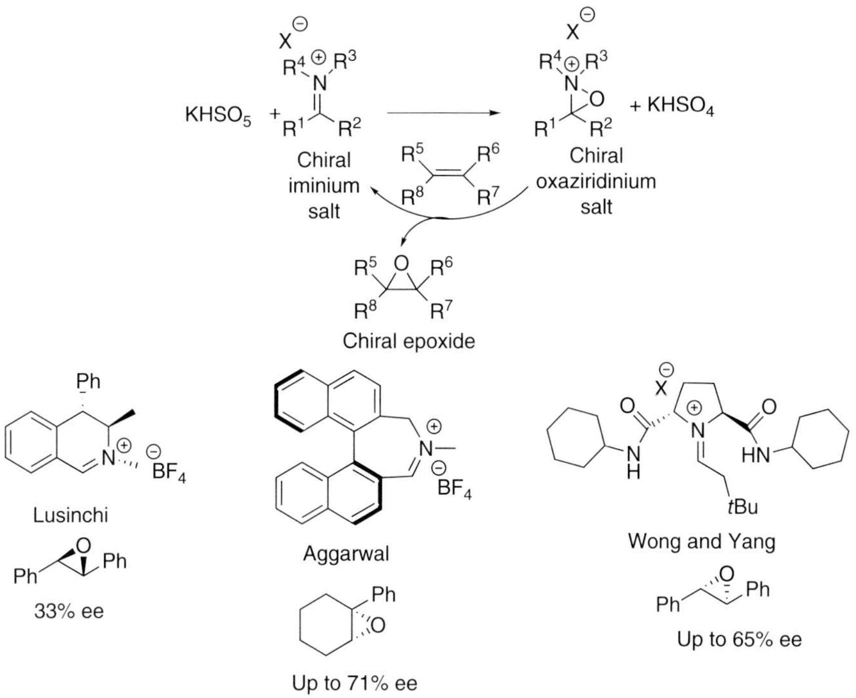

Scheme 3.74 First examples of chiral iminium salts reported in the asymmetric epoxidation.

indication of its intermediacy derived from testing the oxidation of the precursor imine, which resulted completely diastereoselective.

Stoichiometric and catalytic (20–50 mol%) acyclic chiral iminium salts, *in situ* obtained by mixing substoichiometric loadings of aldehydes and a chiral amine (pyrrolidine-based), were reported to give modest enantiocontrol in the epoxidation of unfunctionalized alkenes as demonstrated by Wong, Yang, and coworkers (Scheme 3.74) [273]. However, this approach enables a rapid modular screening of amines and aldehydes in the epoxidation. The presence of a chiral anion in the chiral oxaziridinium-salt-mediated epoxidation has also been studied by Lacour and coworkers, demonstrating a synergic effect in the chiral ion pair of the anion chirality to improve the level of enantioselectivity in the epoxidation [274]. In 2004, a significant improvement and efficiency of the catalytic protocol for the epoxidation have been illustrated by the work of Page's group, who developed more elaborated benzodiazepinium salts with asymmetric centers in the exocyclic substituent on the iminium nitrogen atom (Scheme 3.75) [275].

The enantioselectivity was found to be greatly dependent on the substitution pattern and ring size for cyclic alkenes. However, the low catalytic loading applied and high enantiomeric excess achieved in the case of trisubstituted alkenes are noteworthy. The same group demonstrated that the process can be performed at low temperatures working in organic solvents by introducing a more soluble oxidant in organic solvents such as tetraphenylphosphonium monoperoxybisulfate

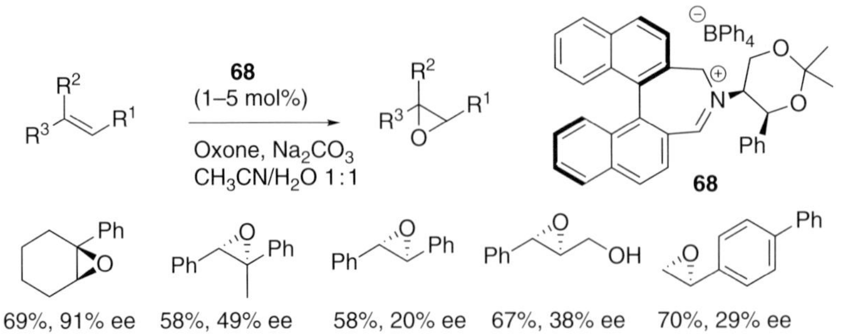

Scheme 3.75 Page's chiral iminium salt working at catalytic loading in the asymmetric epoxidation.

(TPPP), which did not require a base as additive [276]. Under anhydrous reaction conditions, similar or better ee values were generally obtained. In particular, organocatalyst **69** satisfactorily performed at 10 mol% loading in the asymmetric epoxidation of some Z-alkenes to the corresponding epoxides, which were isolated in good yield and with good-to-high enantioselectivity (Scheme 3.76) [277]. The epoxidation of benzopyran substrates occurred with the highest efficiency, and the protocol was used to synthesize the antihypertensive agent (−)-cromakalim with high stereoselectivity via ring opening of the epoxide unit with pyrrolidone. Alternative stoichiometric oxidants have also been used in the asymmetric epoxidation with chiral iminium salts, such as H_2O_2 [278], sodium hypochlorite [279], and electrochemically generated oxidants [280]. Organocatalyst **69** was also used in the key epoxidation step for a concise asymmetric synthesis of natural coumarin derivatives (−)-lomatin and (+)-*trans*-khellactone [281]. Moreover, compound **69** was recently employed in the kinetic resolution of racemic 2-substituted chromenes via epoxidation, achieving encouraging results for both major and minor diastereoisomeric epoxides, although the unreacted alkenes were isolated with low ee values [282].

Page and coauthors reported the application of chiral iminium-salt-catalyzed epoxidation as key step for the first highly enantioselective synthesis of (+)-scuteflorin A, one of the numerous and pharmacological active compounds, recently isolated from *Scutellaria* species (Scheme 3.77) [283]. A previously reported chiral iminium salt **70** [284] served as an excellent catalyst at catalytic loading, under nonaqueous conditions, to epoxidize the starting Z-alkene to the corresponding epoxide in 97% yield and enantiopure form. Epoxide hydrolysis/oxidation and esterification completed the synthesis.

Further modification of the iminium scaffold has been reported by Lacour and coauthors introducing diastereomeric doubly bridged biphenyl azepine-derived iminium salts of type **71**, with a chiral moiety at the nitrogen atoms (Scheme 3.78) [285]. By exploiting atropisomerism at different temperatures (*atropos* vs *tropos*), they were able to improve at 80 °C the amount of the kinetically and also thermodynamically most abundant (S,P)-configured diastereoisomer. The iminium salts were investigated at 5 mol% loading against the monobridge iminium salt, observing in some selected epoxidation reactions a slight improvement of the

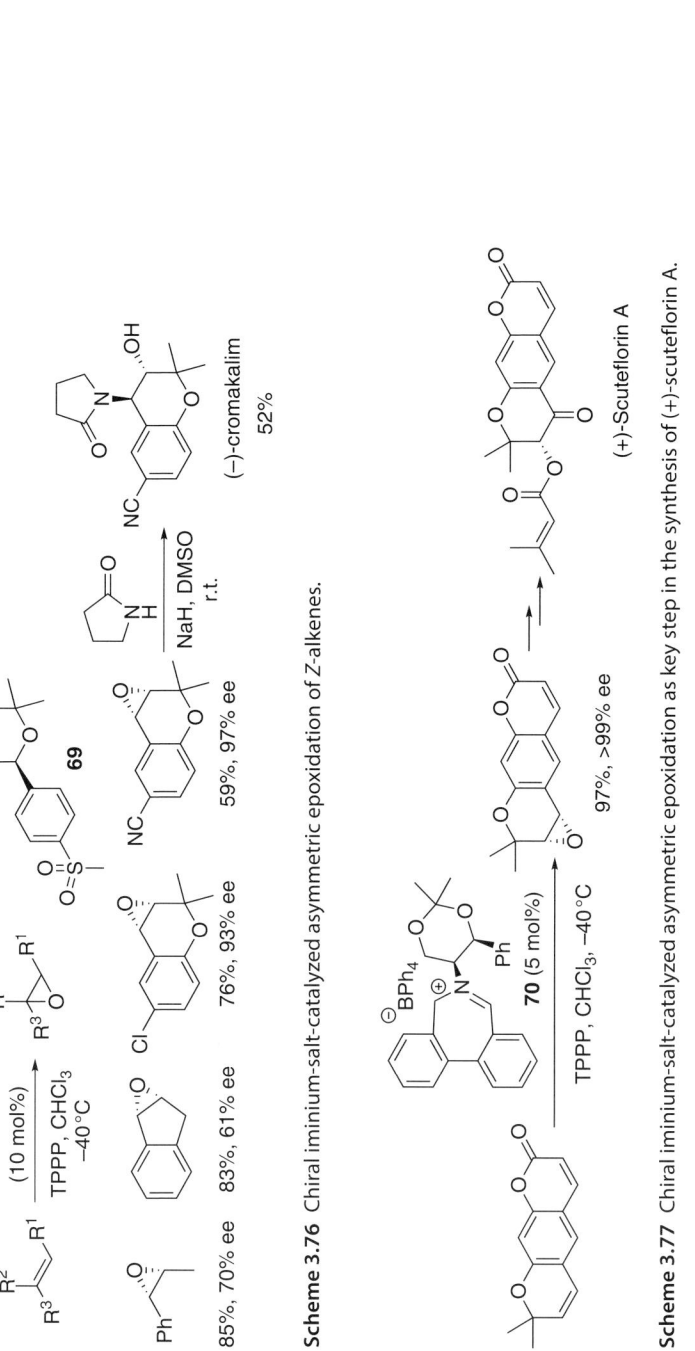

Scheme 3.76 Chiral iminium-salt-catalyzed asymmetric epoxidation of Z-alkenes.

Scheme 3.77 Chiral iminium-salt-catalyzed asymmetric epoxidation as key step in the synthesis of (+)-scuteflorin A.

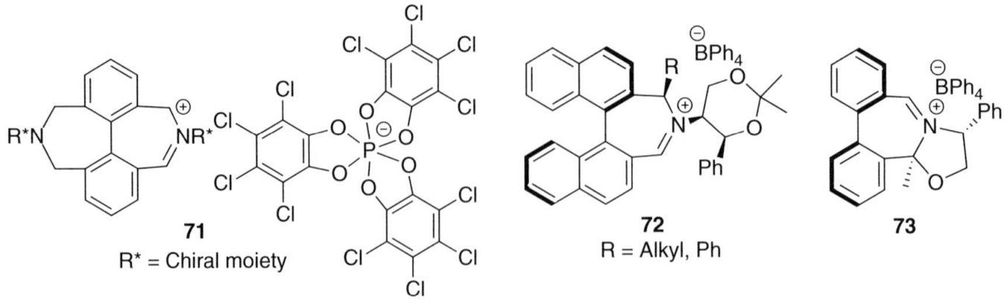

Scheme 3.78 Modified chiral iminium salts for asymmetric epoxidation of alkenes.

enantiocontrol (up to 85% ee). Page and coauthors recently observed that the stereocontrolled introduction of a pseudoaxial substituent adjacent to the nitrogen atom in binaphthyl- and biphenyl-derived azepinium salt of type **72** improved enantioselectivities and yields when used as organocatalysts in the epoxidation of unfunctionalized alkenes (Scheme 3.78) [286]. The new chiral center controlled both the atropisomerism at the biphenyl axis and the regiochemistry of introduction of the iminium double bond. In particular, high yield and enantioselectivity were observed in the epoxidation of chromenes, whose epoxides were isolated with >85% ee. The same group recently developed a range of new biphenylazepinium salts incorporating an additional substituted oxazolidine ring, starting from enantiopure aminoalcohols in a four-step sequence [287]. The tetracyclic iminium salt representative **73** (Scheme 3.78) showed comparable activity to or slightly better activity than previously reported similar organocatalyst **70**, affording epoxides with up to 76% ee.

It is useful to point out that chiral ketones and iminium salts with Oxone as terminal oxidant are conceptually similar systems to effect the asymmetric epoxidation of unfunctionalized alkenes. However, chiral ketones/Oxone has been established as the most reliable and effective system to obtain highly enantioenriched epoxides from a large number of alkenes. Indeed, the generation of a single chiral dioxirane species assured good-to-high control of the stereoselectivity. On the contrary, single *in situ* generation of a diastereoisomeric oxaziridinium ion involved in the epoxidation of alkenes revealed to be much more difficult task to succeed. Hence, the design of precursor chiral iminium ions of general use and high stereocontrol will likely be the subject of further efforts.

Optically pure alkyl hydroperoxides can be employed as chiral reagents in a Weitz–Scheffer epoxidation to generate a chiral peroxyalkyl anion involved in the 1,4-addition to an electron-poor alkene, followed by ring closure to the epoxide. This route to chiral epoxides met with poor success in comparison to other epoxidation reactions, because of the rare efficient approaches disclosed for the synthesis of enantiopure alkyl hydroperoxides to be employed at stoichiometric loadings. Moreover, alkyl hydroperoxides, according to their structure, show a variable degree of chemical and thermal stability, a drawback that hampered the development of this alternative route for the enantioselective epoxidation of α,β-unsaturated carbonyl compounds. The first study on the use of chiral alkyl hydroperoxides has been illustrated by Adam's group in the epoxidation of α,β-enones mediated by a secondary optically pure hydroperoxide using KOH at −40 °C (Scheme 3.79) [288]. The stereocontrol was dependent on the structure of the alkene achieving up to 90% ee. The authors demonstrated the involvement of "templating effects" of the alkaline metal cation with the peroxy anion and the carbonyl group of the alkene in regulating the π-facial differentiation. In 2005, Taylor and coworkers introduced optically pure sugar-derived hydroperoxides/DBU system, working in toluene at room temperature for the enantioselective epoxidation of naphthoquinones, including vitamin K$_3$. Good conversion to the epoxides and moderate-to-good enantioselectivity of up to 82% ee were achieved (Scheme 3.79) [289]. Interestingly, this system performed much better for *cis*-alkenes as the epoxidation of *trans*-chalcone proceeded with poor enantiocontrol (13% ee). The intervention of a template effect was supported by molecular modeling studies of the reaction. The first tertiary enantiopure alkyl hydroperoxide, obtained from commercially available (−)-TADDOL, was synthesized by Seebach and coworkers [290]. The epoxidation of *trans*-chalcone with *n*BuLi in THF at −30 °C afforded the corresponding epoxide in 90% ee. A model to account for the enantioselectivity observed in the epoxidation of *trans*-chalcone has been suggested (Scheme 3.79). Assuming the conjugate addition was the rate- and stereoselectivity-determining step, the approach of the lithium peroxide to the enone would lead to the preferential formation of the (α*S*,β*R*)-configured epoxide. Other enones afforded the epoxides with variable level of enantioselectivity. (−)-TADDOL, by-product of the reaction, could be isolated during the purification process and recycled for the synthesis of the oxidant.

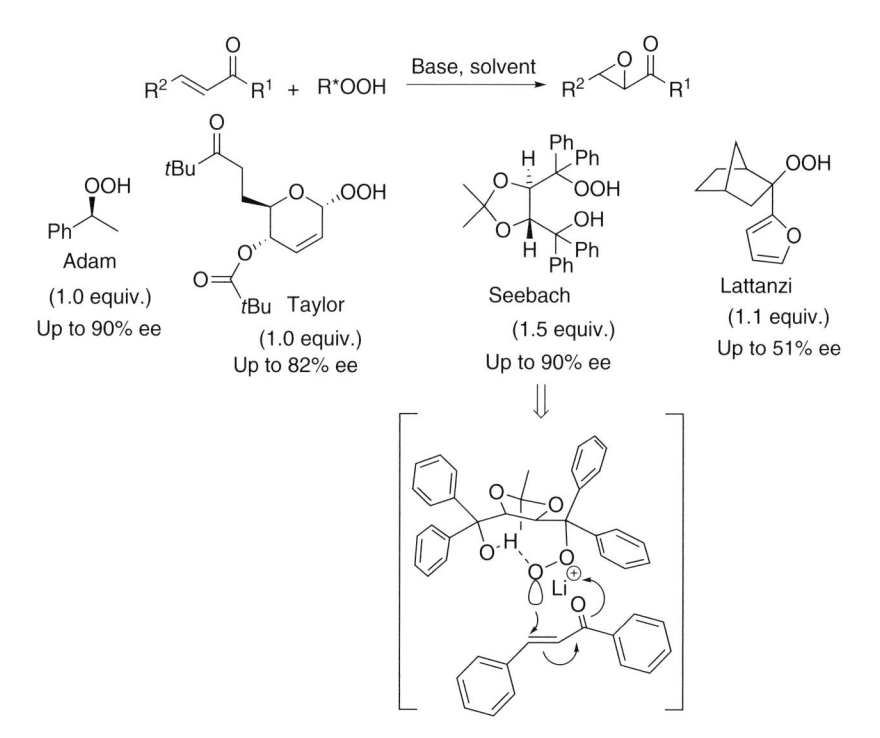

Scheme 3.79 Optically pure alkyl hydroperoxides used as reagents in nucleophilic epoxidation.

Lattanzi and coworkers efficiently synthesized, in a two-step sequence, a (S)-norcamphor-based tertiary alkyl hydroperoxide [291] employed with *n*BuLi in the asymmetric epoxidation of *trans*-chalcones and naphthoquinones. Good yield and moderate level of enantiocontrol (up to 51% ee) were observed (Scheme 3.79) [292]. During the purification process, the chiral alcohol could be isolated and reoxidized to the hydroperoxide, making the process more convenient.

Recently, Chiemlewski and coworkers synthesized alkyl hydroperoxides derived from 2-deoxygalactose, which were checked in the epoxidation of some *trans*-α,β-enones with different bases at room temperature [293]. The α-anomeric hydroperoxide **74** with NaOH as the base was found to be the most efficient oxidant (Scheme 3.80). High conversions to the epoxides were observed as well as good-to-high enantioselectivity. The *s-cis*-conformation of the enone was a prerequisite important to achieve high stereocontrol, as the epoxidation of vitamin K_3, showing an s-*trans*-conformation, proceeded with modest asymmetric induction (40% ee). A remarkable effect of the cation on the stereochemical outcome of the reaction was observed, which was interpreted in terms of cation coordination of both reagents during the epoxidation reaction and was confirmed by DFT calculations. Hemiacetal **75** could be recovered during the purification and reoxidized to hydroperoxide **74**.

It is interesting to highlight in this context the strategy of the double activation of an achiral alkyl hydroperoxide by lithiation and subsequent chelate formation with a chiral ligand developed by Tomioka and coworkers [294]. After checking different bidentate and tridentate ligands and various alkyl hydroperoxides in a model epoxidation, tridentate compound **76** and CHP turned out to be the best combination in toluene as the solvent for the epoxidation (Scheme 3.81). The tridentate ligand **76** was useful to form the bicyclo[3.3.0] complex **77** by coordination to the lithium cation. The hemilabile methoxyphenyl side chain allowed a coordinating atom exchange with the carbonyl oxygen of the enone. Enones bearing a bulky substituent on the carbonyl group afforded the epoxides with the

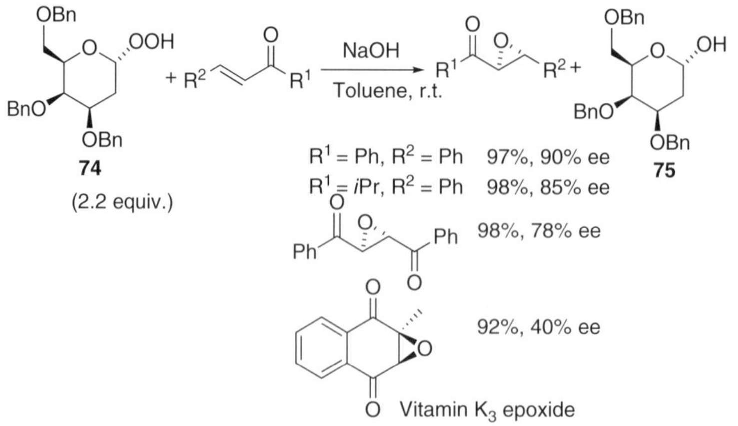

R^1 = Ph, R^2 = Ph   97%, 90% ee
R^1 = *i*Pr, R^2 = Ph   98%, 85% ee

Ph-CO-CH-CH-CO-Ph 98%, 78% ee

Vitamin K_3 epoxide   92%, 40% ee

Scheme 3.80 Optically pure 2-deoxygalactose hydroperoxide-mediated nucleophilic epoxidation.

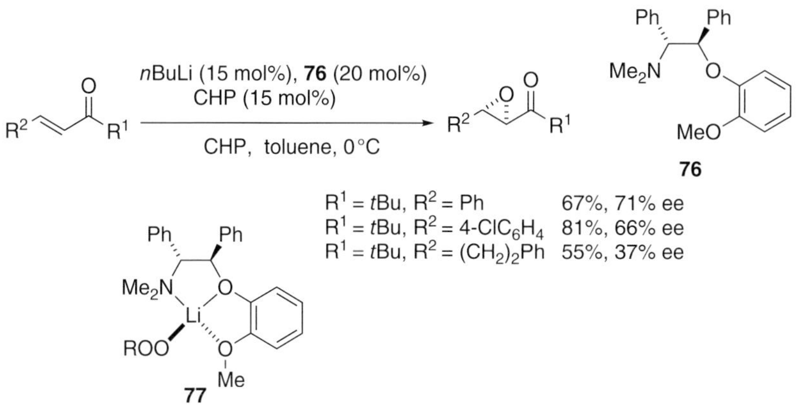

Scheme 3.81 Asymmetric epoxidation of *trans*-α,β-unsaturated ketones with CHP/nBuLi/ligand system.

highest control of the asymmetric induction (up to 71% ee). Ligand **76** could be quantitatively recovered after the reaction and reused.

3.4.4 Chiral Amines

Chiral neutral guanidines are strong nitrogen bases recently introduced as tools in asymmetric synthesis for carbon–carbon and carbon–heteroatom bond formation [295]. Besides being strong bases, they can act as nucleophilic catalysts to effect, among others, conjugate additions, carbonylations, silylations, aldol reactions [296]. In the 1980s, it was demonstrated that alkyl hydroperoxides and organic bases such as DBU are useful combinations to perform the Weitz–Scheffer epoxidation [297]. In 2003, the first asymmetric version using enantiopure guanidine as bases was illustrated by Taylor and coworkers [298]. Optically pure cyclic or acyclic guanidines were synthesized and used in stoichiometric amounts in the epoxidation of quinone **78** with TBHP in toluene/5% of isopropanol mixture (Scheme 3.82). After several days, conversions to epoxide **79** were at best moderate, although the epoxidation proceeded with encouraging levels of asymmetric induction. Linear or cyclic guanidines were of comparable efficiency, and steric effects could be responsible for the decreased ee values, most likely due to destabilization of the peroxide anion–guanidinium cation ion pair. Interestingly, O-methylated C_2-symmetric pyrrolidine-based catalyst was less enantioselective than the corresponding hydroxyl-substituted compound. The best enantioselectivity was achieved with hydroxyl-free guanidines (up to 60% ee).

Ishikawa and coauthors designed a variety of cyclic guanidines to employ as organocatalysts at 20 mol% loading in the epoxidation of *trans*-chalcone with TBHP as oxygen source (Scheme 3.83) [299]. The most effective catalysts **80** and **81** afforded the epoxide in modest enantioselectivity. It is interesting to note that the presence of the hydroxyl group in catalyst **81** greatly enhanced the conversion to the epoxide.

Scheme 3.82 Asymmetric epoxidation of quinones with chiral guanidines.

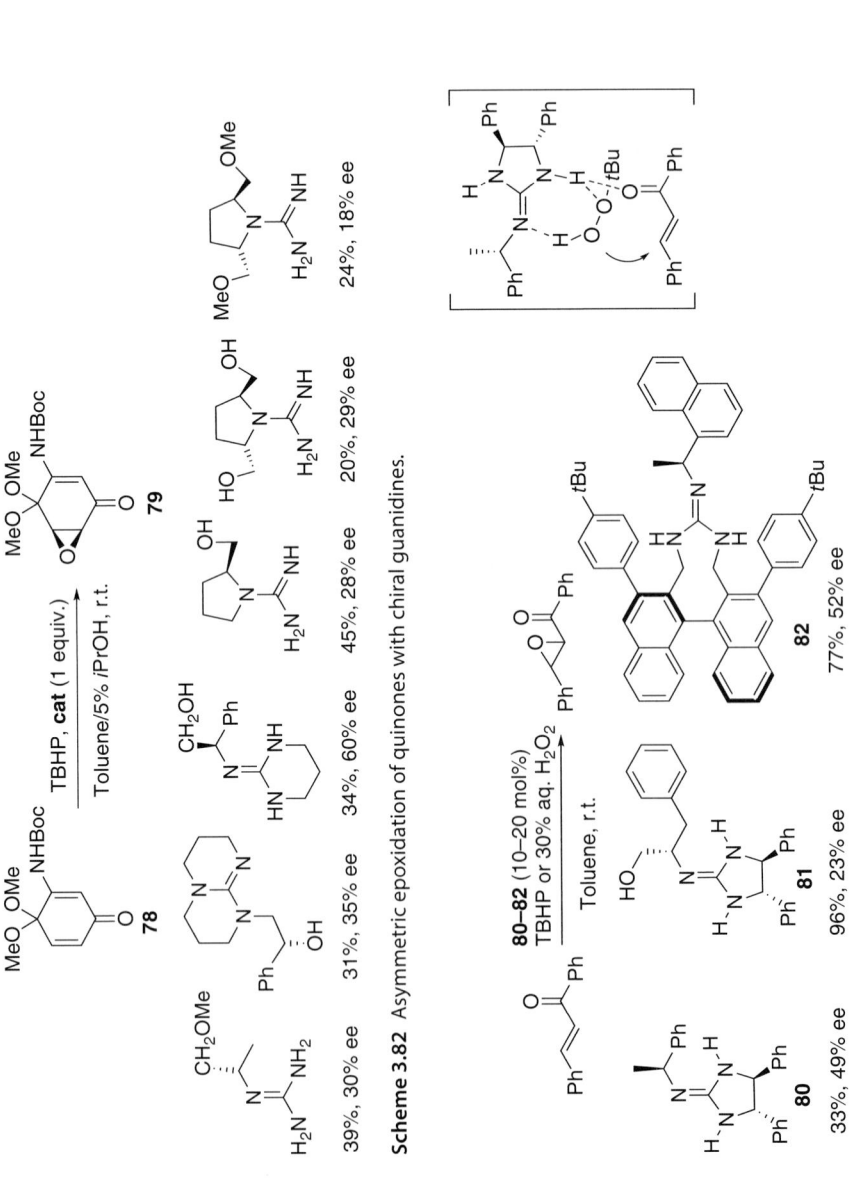

Scheme 3.83 Asymmetric epoxidation of *trans*-chalcone with chiral guanidines.

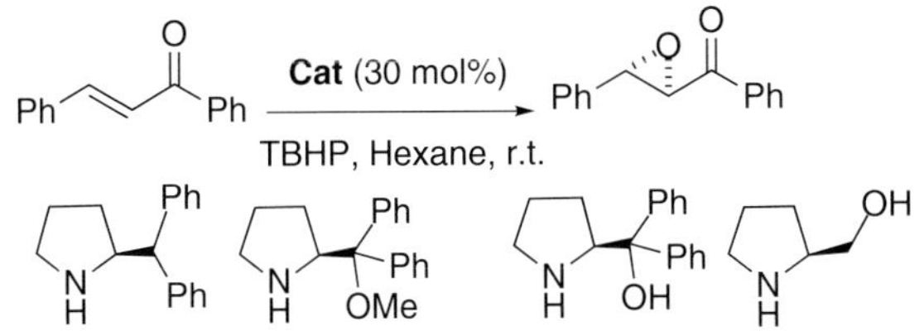

Scheme 3.84 Asymmetric epoxidation of *trans*-chalcone with L-proline-based catalysts.

83	84	19	85
22%, 23% ee	14%, 63% ee	72%, 76% ee	82%, *rac*

They suggested that the acid–base interaction of the catalyst with TBHP would give rise to a H-bonded complex, where the approach of the enone toward the nucleophilic oxidant would be directed by H-bonding with the NH group of the catalyst. Terada and coworkers synthesized BINOL-derived guanidine **82** with sterically hindered aromatic groups at 3,3′-positions and a chiral moiety with central chirality, achieving up to 52% ee in the epoxidation of *trans*-chalcone (Scheme 3.83) [300]. The bifunctionality of the organocatalyst has been found to be important to improve the conversion to the epoxide, likely due to H-bonding coordination of the enone and deprotonation of the pronucleophile ROOH by the Brønsted basic group, which would increase the rate of the oxa-Michael addition step. Along this line and with a view to improve the stereocontrol, Lattanzi and coworkers thought to investigate easily available diaryl-L-prolinols as organocatalysts in the nucleophilic epoxidation of *trans*-enones. Catalysts **83** and **84** proved to be inefficient, whereas the conversion and the enantioselectivity significantly improved up to 76% ee when commercially available α,α-L-diphenyl prolinol **19** was used in the epoxidation of *trans*-chalcone working at 30 mol% catalyst loading in hexane at room temperature with TBHP as the oxygen donor (Scheme 3.84) [301]. Moreover, catalyst **85** with the simplest carbinolic group, although active, proved to be completely unselective.

When employing diaryl prolinols bearing electron-donating groups in the phenyl rings such as commercially available prolinol **86** at 20 mol% loading in the epoxidation of a variety of aryl and alkyl *trans*-α,β-unsaturated ketones, a significant improvement of the efficiency and enantiocontrol was attained (Scheme 3.85) [302].

Further investigations by the same group on aromatic pattern substitution and ring size of the secondary cyclic amine core of the organocatalyst enabled identification of the best structural features necessary to reduce the catalyst loading to 10 mol%, while working at room temperature, maintaining the asymmetric induction (up to 90% ee) [303]. Moreover, it was demonstrated that commercially available primary β-amino alcohols or easily available from α-amino acid esters were also able to promote the asymmetric epoxidation of *trans*-chalcone with encouraging level of asymmetric induction (up to 52% ee). Nucleophilic epoxidations of enones, proceeding according to the Weitz–Scheffer mechanism in the presence of a primary amine and an acid co-catalyst, can give rise to the formation of an iminium ion, which then evolves, after ring closure to the final epoxide (b) (Scheme 3.86). A noncovalent general acid and base activation of

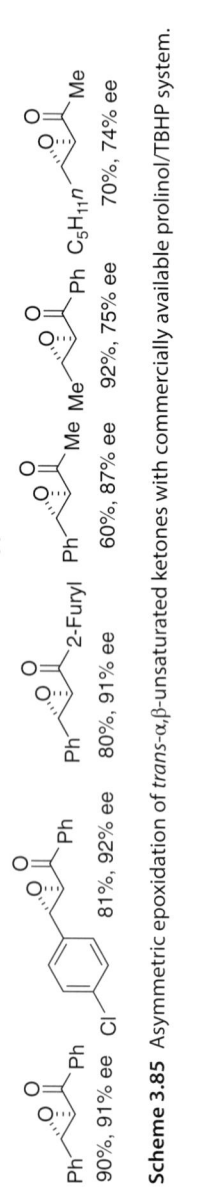

Scheme 3.85 Asymmetric epoxidation of *trans*-α,β-unsaturated ketones with commercially available prolinol/TBHP system.

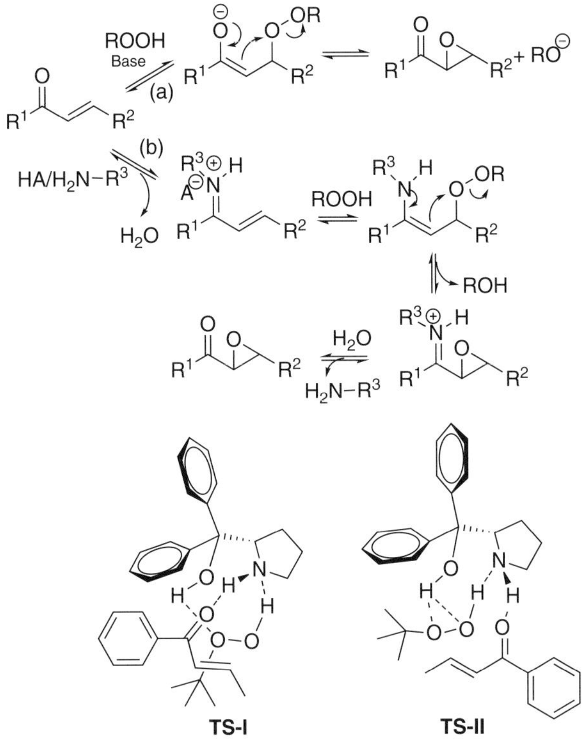

Scheme 3.86 Mechanistic picture of base- and prolinol-catalyzed nucleophilic epoxidation of enones.

enone and alkyl hydroperoxide has been considered plausible when using a sterically encumbered secondary amine of type **19**, **85**–**86** and enones less prone to form iminium ion than enals (a), which is in line with the experimental and computational studies performed on this reaction [304].

The oxa-Michael addition of the nucleophilic peroxide was found to be the rate- and stereoselectivity-determining step, followed by a fast ring-closure step to the epoxide, which accounts for the complete control of the diastereoselectivity observed in the epoxidation. The most stable transition states for the oxa-Michael addition of *trans*-1-phenylbut-2-en-1-one with TBHP using prolinol **19** were computed to be **TS-I** and **TS-II** (Scheme 3.86). The lowest-energy transition state **TS-II**, in which a more staggered conformation for the C—O forming bond and a more effective H-bonding network are present, would lead to the formation of 2R,3S-configured epoxide, as experimentally observed.

Diaryl prolinol modified at the proline scaffold, such as compound **87**, was reported by Zhao and coauthors as successful promoter in the asymmetric epoxidation of disubstituted *trans*-enones with TBHP, achieving up to 96% ee (Scheme 3.87) [305]. They also studied the epoxidation of a variety of *trans*-enones using recyclable diaryl prolinols, by synthesizing dendrimeric catalyst **88** [306] and fluoro-derivative **89** [307], used at 30 mol% loading in CCl$_4$ as solvent (Scheme 3.87). Although catalysts **88** and **89** are not effective as catalysts

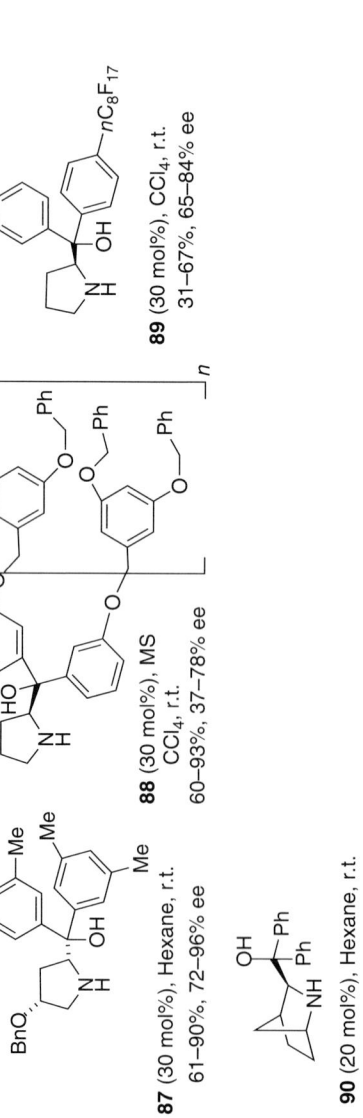

Scheme 3.87 L-Prolinols and other amino alcohols used in asymmetric homogeneous or heterogeneous epoxidation of α,β-unsaturated ketones.

86 and **87**, they could be recovered and reused up to five runs with little loss of efficiency. Loh and coworkers demonstrated that a secondary bicyclic amine **90**, bearing the chiral diphenyl carbinol unit, is able to promote the epoxidation of *trans*-α,β-enones with TBHP isolating the products in good yield and enantioselectivity [308].

The diaryl prolinol/TBHP system has been further investigated by Zhao, developing the first asymmetric epoxidation reactions of a variety of *trans*-disubstituted electron-poor alkenes, whose epoxides were recovered in good yield, with complete diastereocontrol and good-to-high enantioselectivity (Scheme 3.88) [309]. Interesting and useful elaborations were demonstrated starting from enantioenriched halogenated epoxyketones or epoxyketoesters. Moreover, a short synthesis of natural products (−)-(5R,6S)-norbalasubramide and (−)-(5R,6S)-balasubramide was also demonstrated.

Lattanzi and coworkers illustrated the applicability of L-prolinols **91** or **92**/TBHP system in the asymmetric epoxidation of scarcely investigated trisubstituted electron-poor alkenes (Scheme 3.89) [310]. The enantioselective epoxidation protocol has been successfully applied to acyclic 2-arylidene-1,3-diketones, achieving a good level of enantioselectivity, which could be improved up to 99% ee by single crystallization. Cyclic alkenes derived from 1,3-indandione afforded the spiroepoxides in excellent yield and with up to 80% ee. These spiroepoxides are the precursors of a potent inhibitors series of the human papillomavirus HPV11 E1-E2 protein–protein interaction [311]. The same group investigated the stereoselective epoxidation of *trans*-2-aroyl-3-substituted acrylonitriles, which led to the formation of epoxides containing two contiguous tertiary and quaternary stereocenters in excellent yield with complete control of the diastereoselectivity and up to 84% ee [310b]. Gasperi and coworkers investigated the asymmetric epoxidation of α-ylideneoxindole esters by using **19**/TBHP system (Scheme 3.89) [312]. The spirocyclic oxiranes bearing two contiguous stereocenters were obtained as nearly equivalent mixture of *trans* and *cis* spirooxirane-2-carboxylates in good yield and with modest-to-good enantioselectivity. In this case, the peroxyenolate would enable ring closure at significantly lower rate as justified by the presence of *cis*-epoxide.

A breakthrough in the epoxidation of aliphatic α,β-unsaturated ketones has been achieved by List and coworkers, who developed the first efficient and environmentally friendly method for the asymmetric epoxidation of highly challenging cyclic enones [313]. Quinine-derived diamine salt was used as catalyst at 10 mol% loading with 50% H_2O_2 in dioxane at 30 °C. According to the mechanistic picture illustrated in Scheme 3.86, these bifunctional catalysts would activate both enone via iminium ion formation and H_2O_2 under general base catalysis (Scheme 3.90). 9-Amino-9-deoxyepiquinine salt **93** proved to be the most effective organocatalyst to obtain different substituted cyclic enones in good yield and with excellent ee values.

The same activation strategy was independently applied by Deng [314] and List [315] groups for the asymmetric epoxidation of acyclic aliphatic *trans*-α,β-unsaturated ketones mediated by organocatalyst **93** under different reaction conditions (Scheme 3.91). Deng employed 10 mol% of catalyst **93** at 23° or 55 °C in toluene and using CHP as the oxidant. The epoxides were isolated in

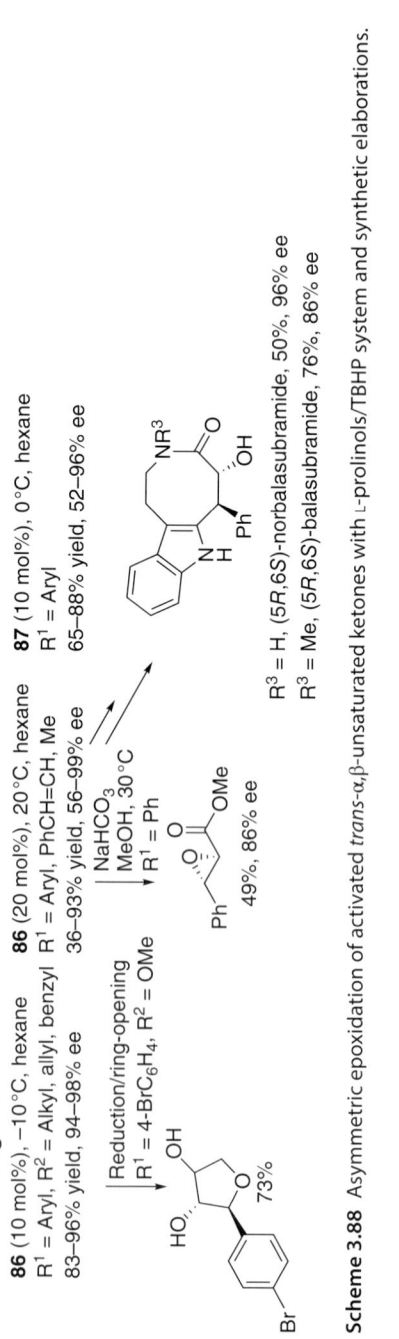

Scheme 3.88 Asymmetric epoxidation of activated *trans*-α,β-unsaturated ketones with L-prolinols/TBHP system and synthetic elaborations.

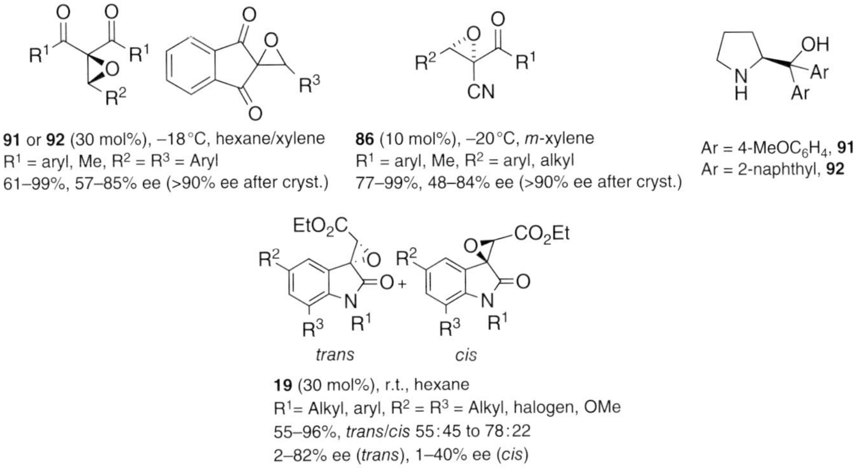

Scheme 3.89 Asymmetric epoxidation of trisubstituted electron-poor alkenes with diaryl prolinols/TBHP system.

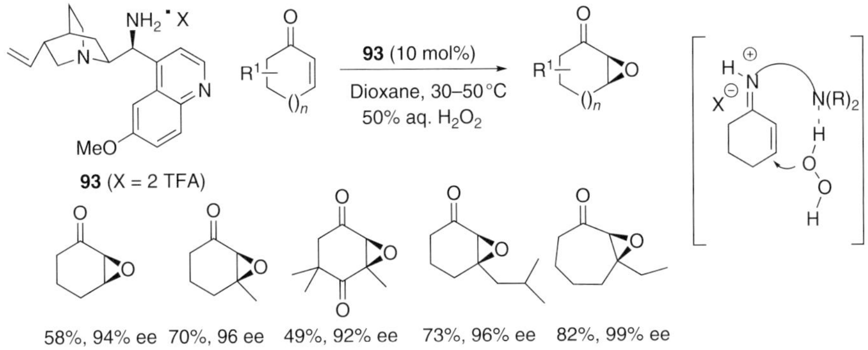

Scheme 3.90 Asymmetric epoxidation of cyclic enones by cinchona-derived amine salt/H_2O_2 system.

moderate-to-high yield with excellent enantioselectivity. The presence of the intermediate β-peroxide **94** was also observed, although reaction conditions were optimized to minimize their formation. List and coauthors applied the conditions previously reported in Scheme 3.90, observing the preferential formation of cyclic peroxyhemiketal intermediates **95**. Basic treatment of the crude reaction mixture led to quantitative transformation of the peroxyhemiketals **95** to the epoxides in generally good-to-high yields with excellent enantioselectivity. Enones bearing an aromatic group at the carbon–carbon double bond or with the trisubstituted double bond turned out to be unreactive under these conditions, whereas *E*- or *Z*-enones afforded *trans*-epoxides, likely via a common dienamine intermediate (Scheme 3.91).

The catalytic asymmetric epoxidation of *trans*-α,β-unsaturated aldehydes has been a highly challenging transformation since 2005, when the first direct and

Deng's system
93 (10 mol%)
CHP, Toluene
23–55 °C

List's system
93 (10 mol%)
50% aq. H$_2$O$_2$
Dioxane, 50 °C

NaOH, Et$_2$O
r.t.
95

R^1 = Me, R^2 = (CH$_2$)$_2$Ph   **Epoxide/94** 99/1, 88%, 97% ee
R^1 = Me, R^2 = nC$_5$H$_{11}$ **Epoxide/94** 99/1, 91%, 97% ee
R^1 = Me, R^2 = Me   **Epoxide/94** 68/32, 55%, 97% ee
R^1 = nBu, R^2 = Me   **Epoxide/94** 87/13, 71%, 97% ee
R^1 = Et, R^2 = nC$_7$H$_{13}$ **Epoxide/94** 67/33, 54%, 96% ee

R^1 = Me, R^2 = (CH$_2$)$_2$Ph   85%, 97% ee
R^1 = Me, R^2 = nC$_6$H$_{13}$ 72%, 97% ee
R^1 = Me, R^2 = Cyclohexyl   83%, 97% ee
R^1 = Me, R^2 = ⁓⁓   76%, 97% ee
R^1 = Et, R^2 = Me   55%, 97% ee
R^1 = iBu, R^2 = nC$_5$H$_{11}$ 81%, 97% ee

Scheme 3.91 Asymmetric epoxidation of acyclic aliphatic *trans*-α,β-unsaturated ketones by cinchona-derived amine salt/ROOH systems.

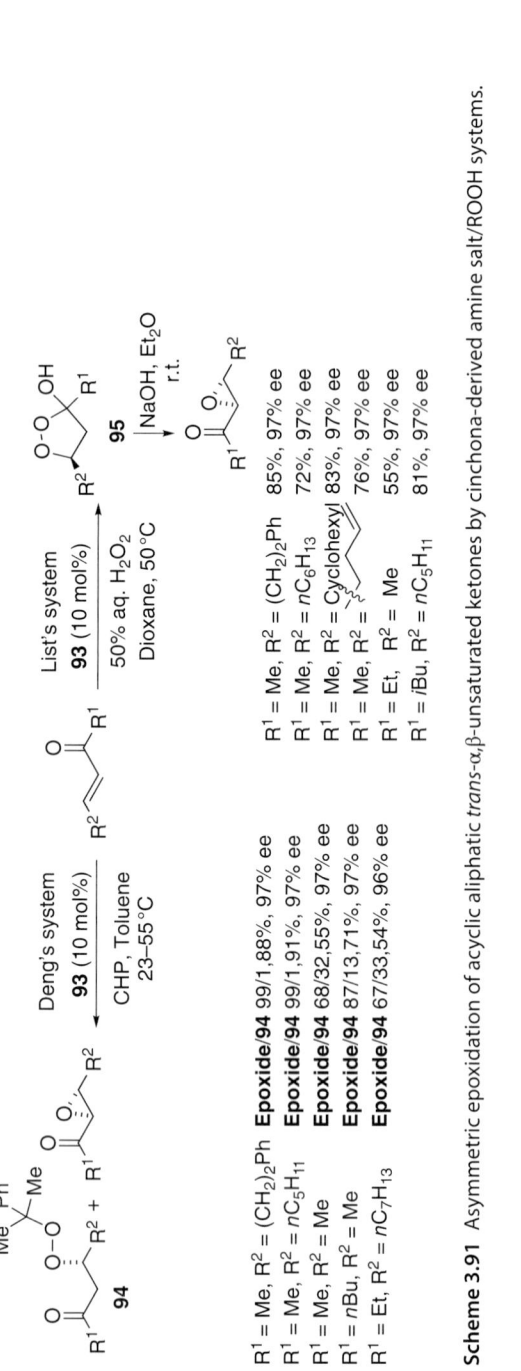

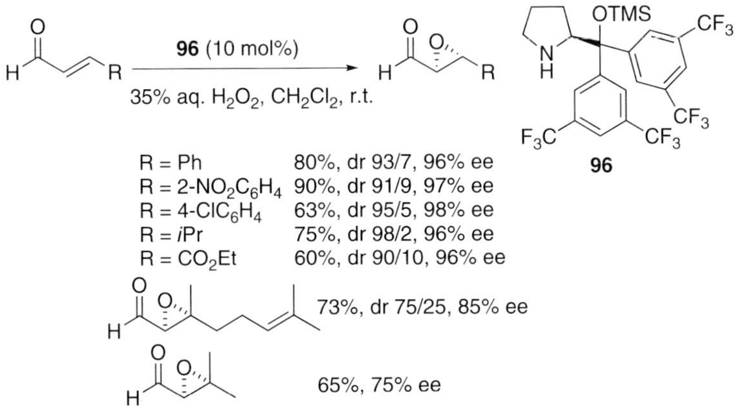

Scheme 3.92 Asymmetric epoxidation of trans-α,β-unsaturated aldehydes with TMS-prolinol/H_2O_2 system.

R	Yield, dr, ee
R = Ph	80%, dr 93/7, 96% ee
R = 2-$NO_2C_6H_4$	90%, dr 91/9, 97% ee
R = 4-ClC_6H_4	63%, dr 95/5, 98% ee
R = iPr	75%, dr 98/2, 96% ee
R = CO_2Et	60%, dr 90/10, 96% ee

73%, dr 75/25, 85% ee

65%, 75% ee

general approach to epoxyaldehydes has been reported by Jørgensen and coworkers. They illustrated an efficient and environmentally friendly epoxidation of trans-enals, using H_2O_2 and 10 mol% of O-trimethylsilyl-protected L-diaryl pyrrolidinemethanol **96** as the catalyst in dichloromethane at room temperature [316]. Different oxygen donors, such as UHP, TBHP, CHP, proved to be of comparable efficiency, and the reaction could be carried out in nonpolar, protic, and halogenated solvents (Scheme 3.92).

A variety of aromatic and alkyl-substituted epoxides were obtained in good-to-high yield, with high diastereoselectivity in favor of the trans-epoxide, which showed excellent ee values. The epoxidation has been significantly improved in terms of sustainability working in an environmentally benign EtOH/H_2O 3/1 solvent mixture [317]. Under these conditions, the epoxides could be obtained with slightly reduced performance (dr up to 86:14, ee up to 96%). The formation of enantioenriched β-disubstituted epoxides from the corresponding trisubstituted alkenes proceeded with good yield and slightly reduced control of the diastereo- and enantioselectivity. A recent investigation on the mechanism of the epoxidation of trans-enals led to the discovery of an autoinductive behavior with in situ formed peroxydrate species **98**, involved as a phase-transfer catalyst, increasing the reaction rate (Scheme 3.93) [318].

Consequently, the original procedure could be conveniently performed at significantly reduced catalyst loading (up to 1 mol% of **96**), in the presence of 5 mol% of chloral hydrate as an additive in CH_2Cl_2 as solvent, maintaining similar level of stereocontrol. Mechanistically, experimental findings and a computational study [319] supported the covalent activation of the enal by the organocatalyst **96** to give a more reactive iminium ion species **97**, which would undergo a face-selective Michael addition of H_2O_2 from the less-hindered face of the carbon–carbon double bond to form nucleophilic enamine **99**. Ring closure would generate the epoxy iminium ion **100**, which after hydrolysis would afford epoxide and catalyst **96** suitable for a new catalytic cycle. Córdova and coworkers disclosed that commercially available trimethyl silyl protected

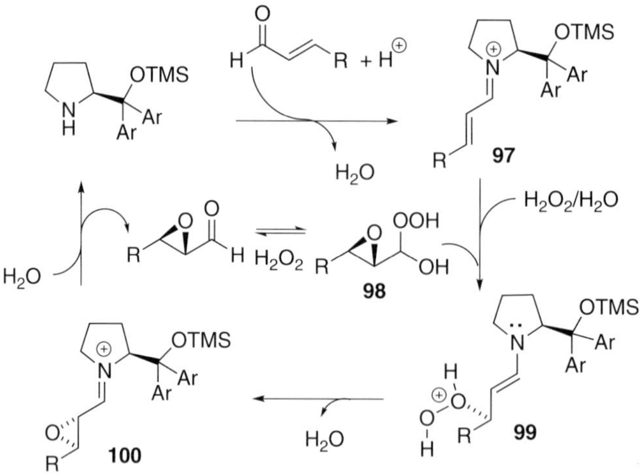

Scheme 3.93 Mechanism suggested for the asymmetric epoxidation of trans-α,β-unsaturated aldehydes with TMS-prolinol/H_2O_2 system.

diphenyl prolinol **101** is an effective catalyst for the asymmetric epoxidation of *trans*-enals with H_2O_2 or sodium percarbonate (SPC) working in $CHCl_3$ as the solvent (Scheme 3.94) [320]. They set up an interesting application of the epoxidation reaction by developing a tandem organocatalytic asymmetric synthesis of 1,2,3-triols, via reduction of the aldehyde- and acid-catalyzed epoxide hydrolysis [321]. The same group developed one-pot organocatalytic direct asymmetric tandem epoxidation–Wittig reaction and cascade epoxidation–Mannich reaction sequences (Scheme 3.94) [322]. The catalytic enantioselective cascade reaction assembled 2-keto-4-amino-5-epoxides in an asymmetric manner with high diastereo- (up to 19:1 dr) and enantioselectivity (>95% ee). According to the H_2O_2/ylide ratio used, the synthesis of complex mono- or diepoxyaldehydes was accomplished.

Schore and coworkers investigated the applicability of polymer-supported diphenylprolinols in the solid-state asymmetric epoxidation of *trans*-cinnamaldehyde (Scheme 3.95) [323]. Poorer yield and stereoselectivity for the epoxide compared to the homogeneous version were achieved. In order to obtain a better yield with the cross-link-functionalized catalyst **102**, it was necessary to seed the resin with the α,β-unsaturated aldehyde for 24 h prior to the addition of trimethylorthoformate (TMOF), which promoted imine formation. Under these conditions, moderate conversion, diastereo-, and enantioselectivity were observed.

The asymmetric epoxidation of α-substituted acroleins still represents one of the synthetic challenges in this area, together with all classes of terminal olefins. In 2010, Hayashi and coworkers developed the first successful asymmetric epoxidation of aliphatic α-substituted acroleins, catalyzed by diphenylprolinol silyl ether **103**, using H_2O_2 as an oxidant (Scheme 3.96) [12c]. Terminal epoxides, bearing quaternary stereocenters, have been obtained in good yield with excellent enantioselectivity (up to 94% ee). The synthetic versatility of

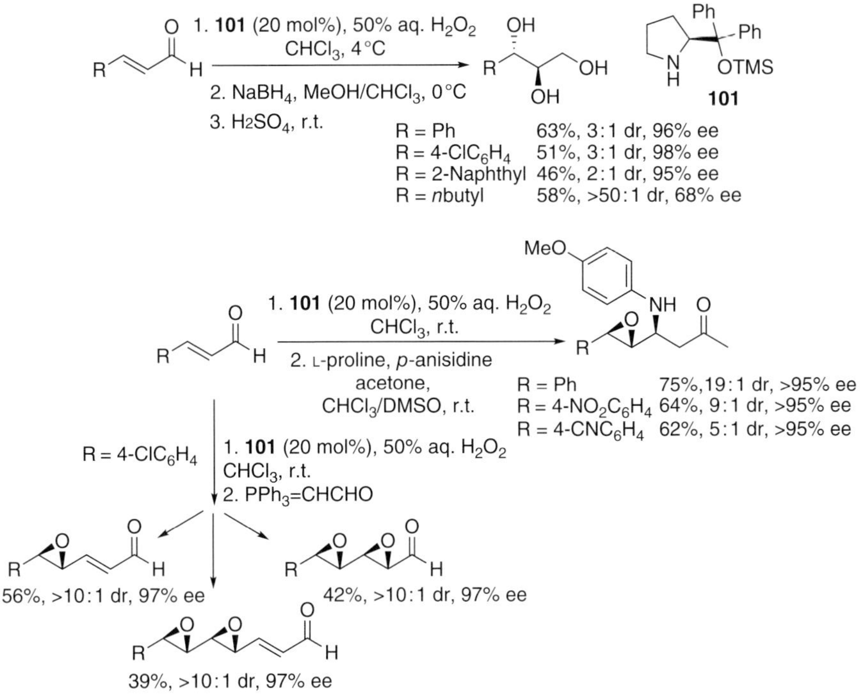

Scheme 3.94 Tandem asymmetric epoxidation of trans-α,β-unsaturated aldehydes with TMS-prolinol/H_2O_2 system followed by reduction/hydrolysis or Mannich and Wittig reactions.

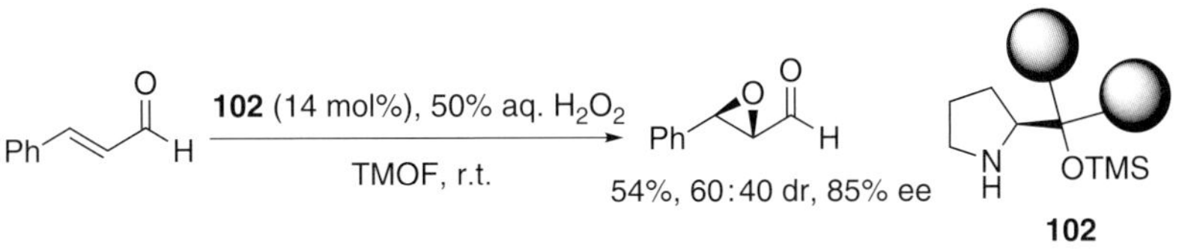

Scheme 3.95 Asymmetric epoxidation of cinnamic aldehyde by a polymer-supported prolinol.

α-substituted-β,β-unsubstituted-α,β-epoxyaldehydes was illustrated in a short synthesis of (R)-methyl palmoxirate, a potent oral hypoglycemic agent.

The asymmetric epoxidation of α,β-unsaturated aldehydes with O-protected diaryl prolinols/H_2O_2 system proved to be a synthetically useful process, which has been applied as key step in the preparation of functionalized compounds. In 2010, Jørgensen's group developed a novel organocatalytic strategy for a formal asymmetric trans-dihydroxylation of α,β-unsaturated aldehydes (Scheme 3.97) [324]. Addition of NaOMe to the carbonyl group in epoxy aldehydes promoted a rearrangement leading to the formation of 2,3-dihydroxyaldehydes with the aldehyde functionality protected as the corresponding dimethyl acetal. Using only 2.5 mol% of organocatalyst **96**, in a one-pot sequence, yielded the diols in more than 20:1 dr with excellent level of enantiocontrol. Payne rearrangement of the epoxyaldehyde

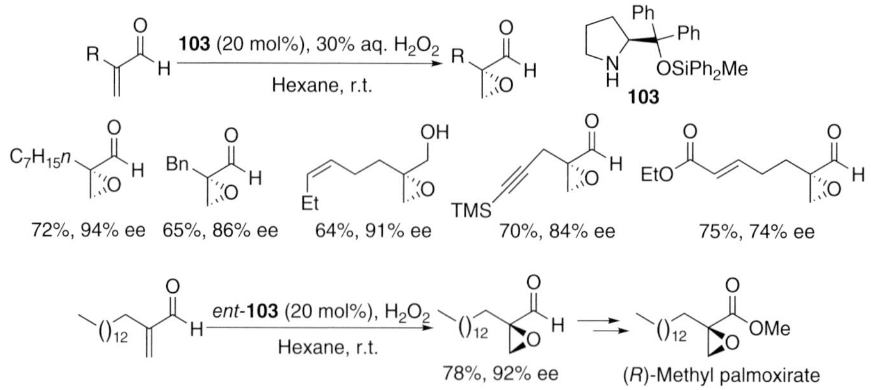

Scheme 3.96 Asymmetric organocatalytic epoxidation of α-substituted acroleins and synthetic application to (R)-methyl palmoxirate.

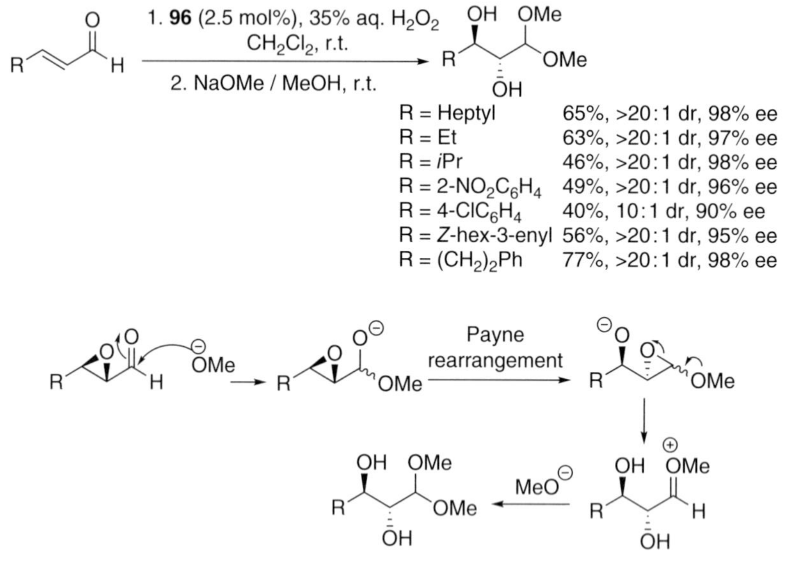

Scheme 3.97 Asymmetric organocatalytic epoxidation/ring opening/protection of trans-α, β-unsaturated aldehydes.

initiated by addition of the methoxide to the carbonyl group was proposed to afford the final products. The intermolecular 1,2-addition of methoxide to the oxocarbenium ion intermediate led to trans-2,3-dihydroxyaldehyde dimethyl acetals. Optically active epoxyaldehydes were further functionalized to synthesize carbohydrates with up to five continuous stereogenic centers, introduced with complete diastereo- and enantioselectivity.

The same group explored the functionalization of the epoxyaldehyde synthons, by combining amino- and N-heterocyclic carbene (NHC)-catalyzed reactions in a one-pot manner to access interesting compounds such as enantioenriched

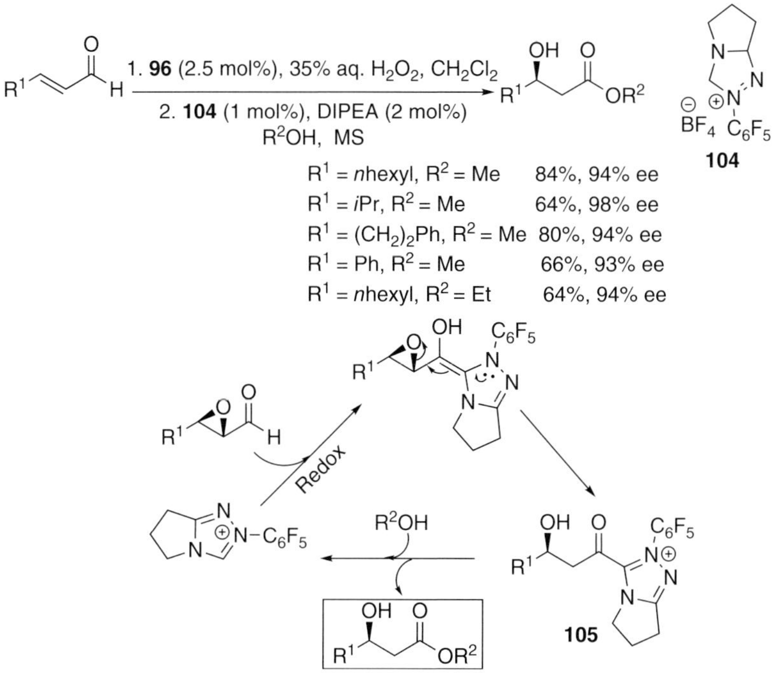

Scheme 3.98 Synthesis of β-hydroxy esters using combined amino- and NHC-catalysis.

β-hydroxy esters (Scheme 3.98) [325]. The desired products were obtained in high yield with excellent enantioselectivity, and the process could be scaled up to gram scale. Mechanistically, the firstly formed enantioenriched epoxyaldehyde undergoes attack by the *in situ* generated diisopropyl ethyl amine (DIPEA)–carbene NHC catalyst from **104**. An intramolecular redox reaction to the activated carboxylate **105** occurs, followed by the acyl transfer in the presence of nucleophilic alcohols to give the β-hydroxy esters.

Jørgensen *et al.* developed an innovative one-pot enantioselective cascade reaction for the preparation of electron-poor 2-hydroxyalkyl furanes (Scheme 3.99) [326].

A wide variety of α,β-unsaturated aldehydes and different 1,3-dicarbonyl compounds were transformed into furanes, with good-to-high level of enantioselectivity, under mild reaction conditions using low loading of commercially available catalyst **96**. It has been postulated that the epoxyaldehyde was involved in the base-promoted interrupted Feist–Bénary reaction [327], initiated by nucleophilic addition of 1,3-dicarbonyl enolate to the carbonyl group of the aldehyde. 7-Methyl-1,5,7-triazabicyclo[4.4.0]dec-5-ene (MTBD) was found to be the most effective base to use in the process (Scheme 3.100). In the absence of camphorsulfonic acid (CSA), the reaction could be stopped at 2-hydroxyalkyl-2,3-dihydrofuranes **106**, bearing three contiguous stereogenic centers as the products, obtained with good diastereoselectivity and excellent ee values.

In addition to O-protected α,α-diaryl prolinols, further studies were devoted to modify this scaffold to obtain new secondary chiral amines useful in the

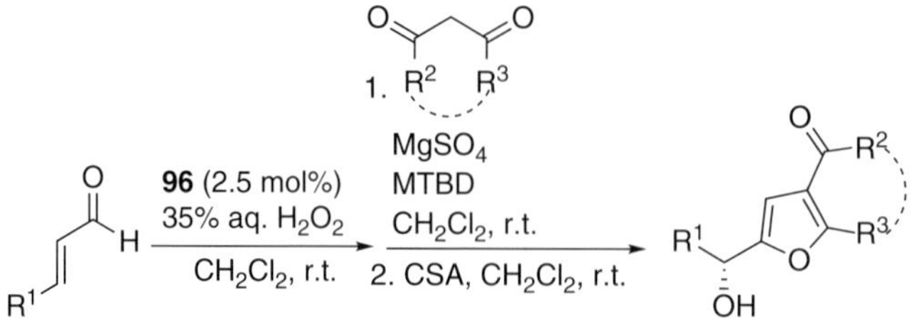

Scheme 3.99 One-pot asymmetric synthesis of electron-poor 2-hydroxyalkyl furans via organocatalyzed epoxidation of *trans*-enals.

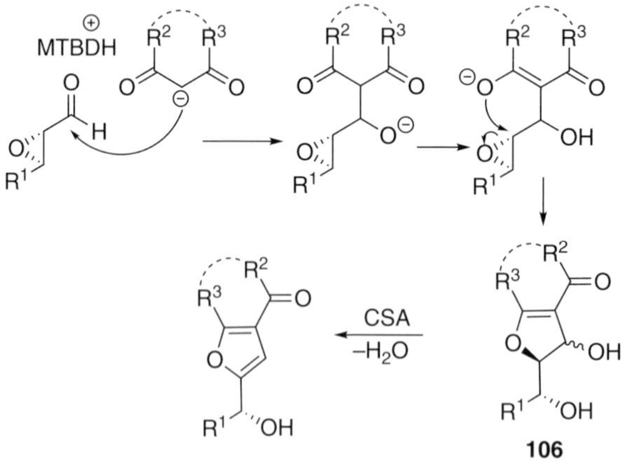

R¹ = Pr, R² = OMe, R³ = Me — 72%, 92% ee
R¹ = Ph, R² = OMe, R³ = Me — 56%, 88% ee
R¹ = Pr, R² = OEt, R³ = Et — 58%, 86% ee
R¹ = Pr, R² = Me, R³ = Me — 75%, 89% ee
R¹ = Pr, R² = R³ = CH$_2$C(CH$_3$)$_2$CH$_2$ — 62%, 94% ee
R¹ = CH$_2$OBn, R² = OMe, R³ = Me — 73%, 92% ee
R¹ = CO$_2$Et, R² = OMe, R³ = Me — 53%, 88% ee

106
60–87%
2:1 to 7:1 dr
92–98% ee

Scheme 3.100 Mechanism of Feist–Bénary reaction starting from epoxyaldehydes.

asymmetric epoxidation of enals. Gilmour and coworkers synthesized an effective fluorinated pyrrolidine-based organocatalyst **107** (Scheme 3.101) [328]. They disclosed a fluorine–iminium ion gauche effect as valuable conformational tool to achieve stereocontrol in the asymmetric epoxidation of β-aryl and β-alkylsubstituted enals. The *gauche* effect, which is induced upon reversible formation of an iminium ion, provides a powerful method for the preorganization of the transient intermediates, providing an extra degree of torsional rigidity that could assist the asymmetric induction (Scheme 3.101). This turned out to be true as the epoxides were recovered in high yield, with good diastereoselectivity and excellent enantiocontrol. This effect has been corroborated by performing the

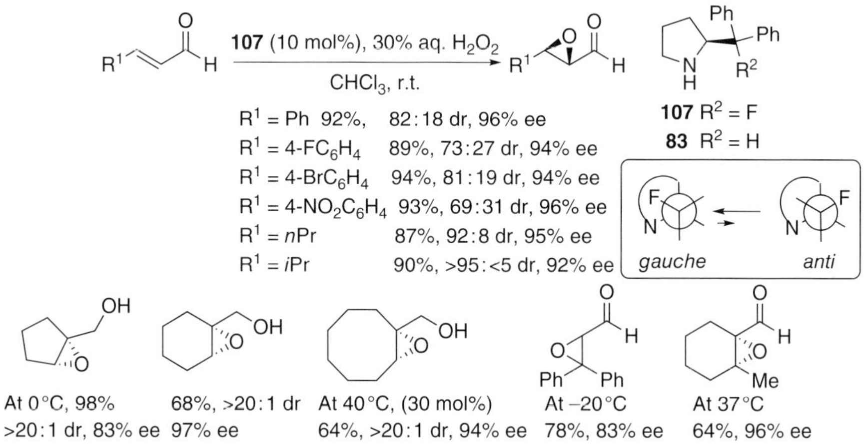

Scheme 3.101

R¹-CH=CH-CHO + 107 (10 mol%), 30% aq. H₂O₂, CHCl₃, r.t. → epoxy aldehyde

Catalyst 107: (S)-2-(fluorodiphenylmethyl)pyrrolidine, R² = F
Catalyst 83: R² = H

R¹ = Ph   92%, 82:18 dr, 96% ee
R¹ = 4-FC₆H₄   89%, 73:27 dr, 94% ee
R¹ = 4-BrC₆H₄ 94%, 81:19 dr, 94% ee
R¹ = 4-NO₂C₆H₄ 93%, 69:31 dr, 96% ee
R¹ = nPr   87%, 92:8 dr, 95% ee
R¹ = iPr   90%, >95:<5 dr, 92% ee

gauche ⇌ anti

At 0 °C, 98%, >20:1 dr, 83% ee | 68%, >20:1 dr, 97% ee | At 40 °C, (30 mol%) 64%, >20:1 dr, 94% ee | At −20 °C, 78%, 83% ee | At 37 °C, 64%, 96% ee

Scheme 3.101 Asymmetric epoxidation of *trans*-α,β-unsaturated aldehydes with fluorinated pyrrolidine organocatalyst/H₂O₂ system.

epoxidation under identical conditions of *trans*-cinnamaldehyde with the non-fluorinated catalyst **83**. In this case, the epoxide was obtained with 85% ee compared to 96% ee achieved with catalyst **107**. DFT calculations predicted the preference of the *gauche* over the *anti* conformation of the β-fluoroiminium ions to be 16.5 kJ mol^{-1} for the *E* geometry and 18.1 kJ mol^{-1} for the *Z* geometry, whereas in the nonfluorinated congeners, the *gauche* conformer was favored by only 5–6 kJ mol^{-1}. The Gilmour's group developed a high-yielding, five-step synthesis of (S)-2-(fluorodiphenylmethyl)pyrrolidine **107** from L-proline for an easier application of this catalyst [329], which was found to be effective in the asymmetric epoxidation of more challenging cyclic and acyclic trisubstituted α,β-unsaturated aldehydes and a tetrasubstituted enal (Scheme 3.101) [330]. Very recently, novel chiral spiro-pyrrolidine-derived organocatalysts was found to be effective in the asymmetric epoxidation of α,β-unsaturated aldehydes [331].

MacMillan and coworkers reported that chiral imidazolidinone salt **108** is a suitable organocatalyst for the epoxidation of *trans*-α,β-unsaturated aldehydes with [(nosylimino)iodo]benzene (NsNIPh) as the terminal oxidant in a solvent mixture of dichloromethane or chloroform/acetic acid at −30 °C [332]. Different α,β-unsaturated aldehydes were converted into the corresponding epoxides in good-to-high yield, with complete diastereocontrol for the *trans*-epoxides and high enantioselectivity (Scheme 3.102). The formation of a covalent iminium species between the catalyst and the aldehyde, in analogy with proline-based catalysts **96**, **101**, **103**, has been suggested to justify the high diastereo- and enantioselectivity achieved. The *t*-butyl and benzyl groups in compound **108** regulate the iminium geometry through nonbonding steric interactions and face selectivity for the nucleophilic attack at β-position by effective face shielding.

List and coworkers disclosed a new concept in asymmetric synthesis termed "asymmetric counteranion-directed catalysis" (ACDC)) [333]. Catalytic reactions, where cationic intermediates are involved, can proceed in enantioselective way when a chiral counteranion is incorporated into the catalyst. This strategy

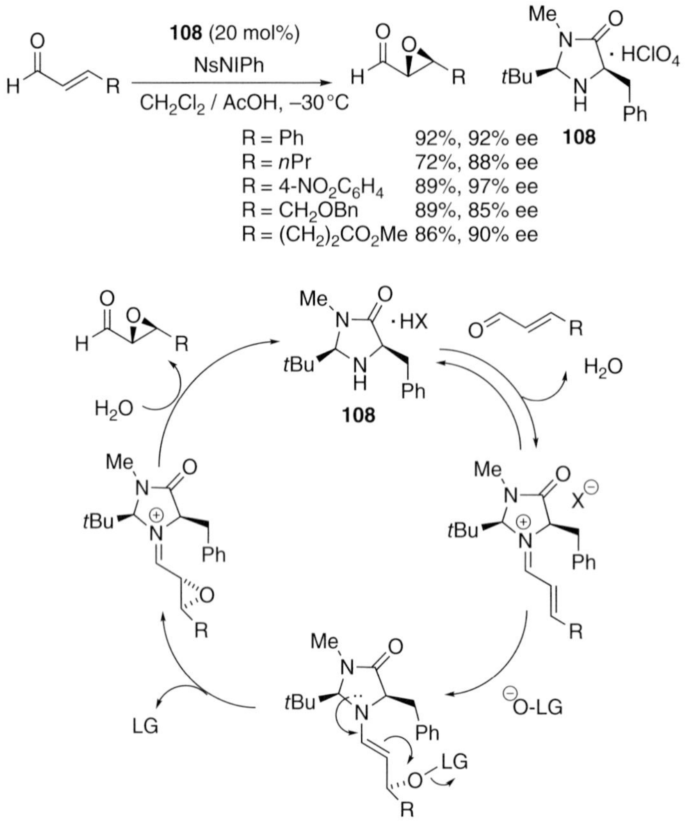

Scheme 3.102 Asymmetric epoxidation of *trans*-α,β-unsaturated aldehydes with imidazolidinone salt/NsNIPh system.

has been applied to develop a highly effective epoxidation of *trans*-α,β-unsaturated aldehydes using 10 mol% of the dibenzylammonium salt of 3,3′-bis-(2,4,6-triisopropyl-phenyl)-1,1′-binaphthyl-2,2′-diyl hydrogen phosphate (TRIP) **109** in dioxane or *tert*-butyl methyl ether with TBHP at 35 °C (Scheme 3.103) [334]. Aromatic and aliphatic *trans*-disubstituted enals gave the epoxides in good yield, with excellent diastereoselectivity and good-to-high enantiocontrol. Moreover, β,β-disubstituted α,β-unsaturated aldehydes were also obtained with high ee values. The epoxidation of citral afforded the *trans*-epoxide with comparable stereocontrol achieved when using Jørgensen's system (Scheme 3.92), and the *cis*-epoxide was obtained with 92% ee. From a mechanistic point of view, the intriguing point of this reaction was the control of the asymmetric induction by salt **109**. According to the general mechanism proposed for amino-catalyzed epoxidation, in this case, the stereocenter is not created in the 1,4-addition step, as intermediate **110** is an achiral species. The stereocenter is created in the cyclization to iminium ion **111**, thus implying that the chiral TRIP anion is involved in the carbon–oxygen bond formation. This is an uncommon example of an enantioselective, organocatalyzed Michael addition reaction promoted by a

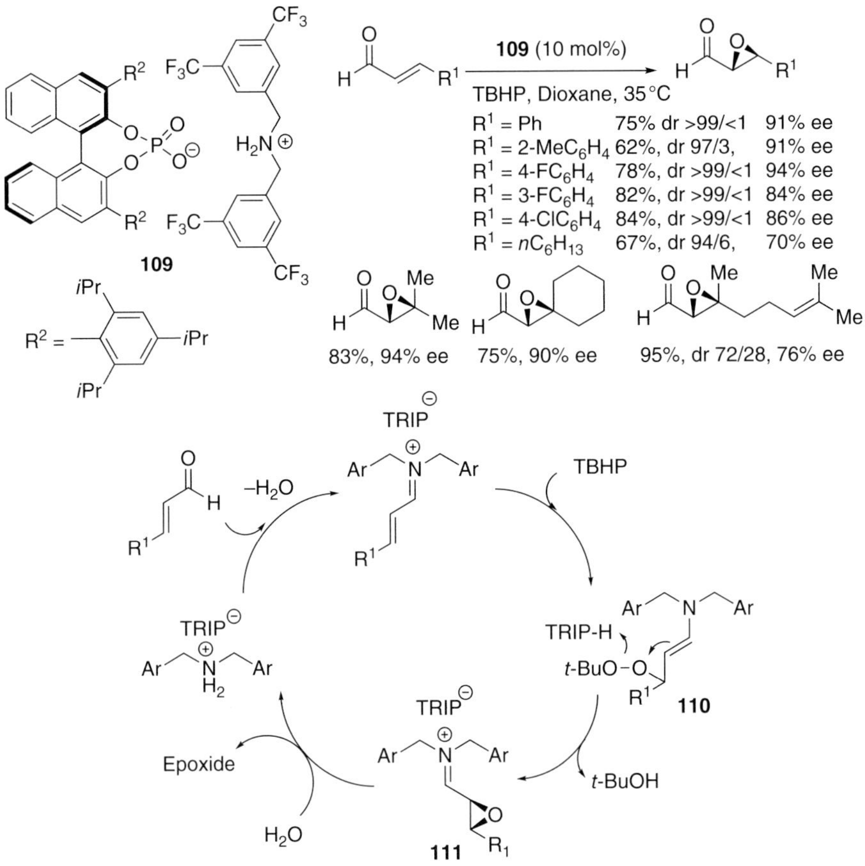

Scheme 3.103 Asymmetric epoxidation of trans-α,β-unsaturated aldehydes via ACDC.

secondary achiral amine with a chiral Brønsted acid, in contrast to commonly reported iminium activation via employment of secondary or primary chiral amines with achiral Brønsted acids [335].

More recently, a combination of the previous concepts has been demonstrated by the work of List and coauthors, who reported a highly diastereo- and enantioselective organocatalytic epoxidation of α,β-substituted α,β-unsaturated aldehydes and α-substituted acroleins under environmentally friendly conditions (Scheme 3.104) [336]. The combined 9-amino-9-deoxyepiquinine and (R)-BINOL-based phosphoric acid salt **112** performed at best with H_2O_2 as the oxidant in THF at 50 °C, giving the α,β-epoxyaldehydes in good yield with high-to-excellent diastereo- and enantiocontrol. According to previously shown mechanism (Scheme 3.103), this chiral catalyst provided additional enantiodiscrimination within the ion pair formed by the phosphoric acid anion of the Brønsted acid and the chiral iminium ion generated between enal and the primary amine. Indeed, dramatic matching/mismatching effects were observed when using the (S)-enantiomer of phosphoric acid in salt **112**, with a significant

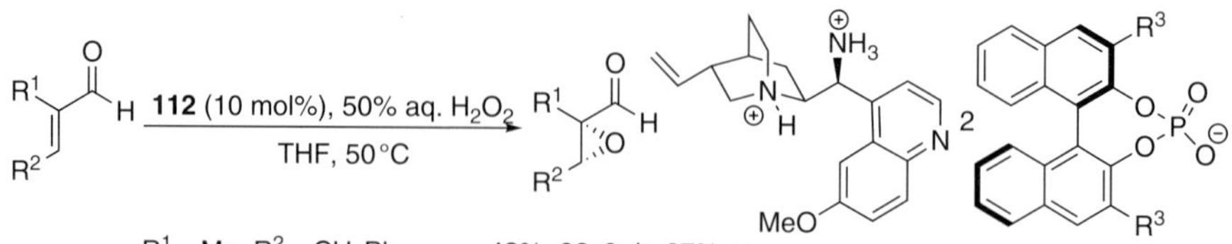

R¹ = Me, R² = CH2Ph   43%, 92:8 dr, 97% ee
R¹ = (CH₂)₂Ph, R² = CH₂Ph   77%, 90:10 dr, 98% ee
R¹ = R² = −(CH₂)₄−   70%, 97% ee
R¹ = Me, R² = (CH₂)₂Ph   75%, 95:5 dr, 97% ee
R¹ = CH₂Ph, R² = H   78%, 98% ee with **112**
R¹ = (CH₂)₇CH₃, R² = H   78%, 98% ee with **113**

112 R³ = Ph
113 R³ = 2,4,6-*i*Pr₃C₆H₂

Scheme 3.104 Asymmetric epoxidation of α,β-unsaturated aldehydes with chiral cation and anion salt.

decrease in the diastereoselectivity and formation of the opposite enantiomer of the epoxide with moderate enantioselectivity. The process is another interesting example of ACDC.

In the past decade, noncovalent bifunctional organocatalysis has undoubtedly demonstrated great potential as a tool to construct carbon–carbon and carbon–heteroatom stereocenters. Among the structurally different organocatalysts, cinchona-alkaloid-derived thioureas and squaramides proved to be particularly useful in different transformations [337]. Cinchona thioureas are able to generate enolates of 1,3-dicarbonyl compounds in chiral environments as illustrated in Michael-type additions by the pivotal reports of Takemoto's group [338]. The noncovalent bifunctional catalysis demonstrated to be a useful approach to develop asymmetric epoxidation reactions of disubstituted and less explored trisubstituted electron-poor alkenes bearing electron-withdrawing groups unable to undergo activation via iminium ion strategy, but proceeding via general acid and base catalysis. On these grounds, Lattanzi and coworkers reasoned that they could be suitable catalysts to promote the enantioselective nucleophilic epoxidation of terminal electron-poor α-aroyl acrylamides and acrylates with TBHP as the oxidant. A first deprotonation of TBHP would form the peroxyanion, followed by its conjugate addition to the alkene to give the prochiral peroxyenolate hydrogen-bonded by the thiourea group of the chiral catalyst [339]. Indeed, a highly selective ring closure was observed as the corresponding epoxides bearing a quaternary stereocenter were isolated in excellent yield and enantioselectivity when using only 5 mol% of easily available catalyst **114** (Scheme 3.105). Interestingly, the presence of the secondary amide bond was important to preorganize the alkene, through hydrogen bonding, as attested by the unreactive nature of the alkene bearing a tertiary amide portion. Taking advantage of high regioselectivity in ring-opening reactions of terminal epoxides, epoxidation/ring-opening reactions were exploited to afford highly functionalized hydroxyl sulfides or *N*-Boc-α-hydroxy-β-amino acid derivatives, bearing a quaternary stereocenter, in good-to-high yield and with high enantioselectivity.

The same group has recently disclosed the first asymmetric epoxidation of alkylidene malononitriles, a challenging class of Michael acceptors due to their

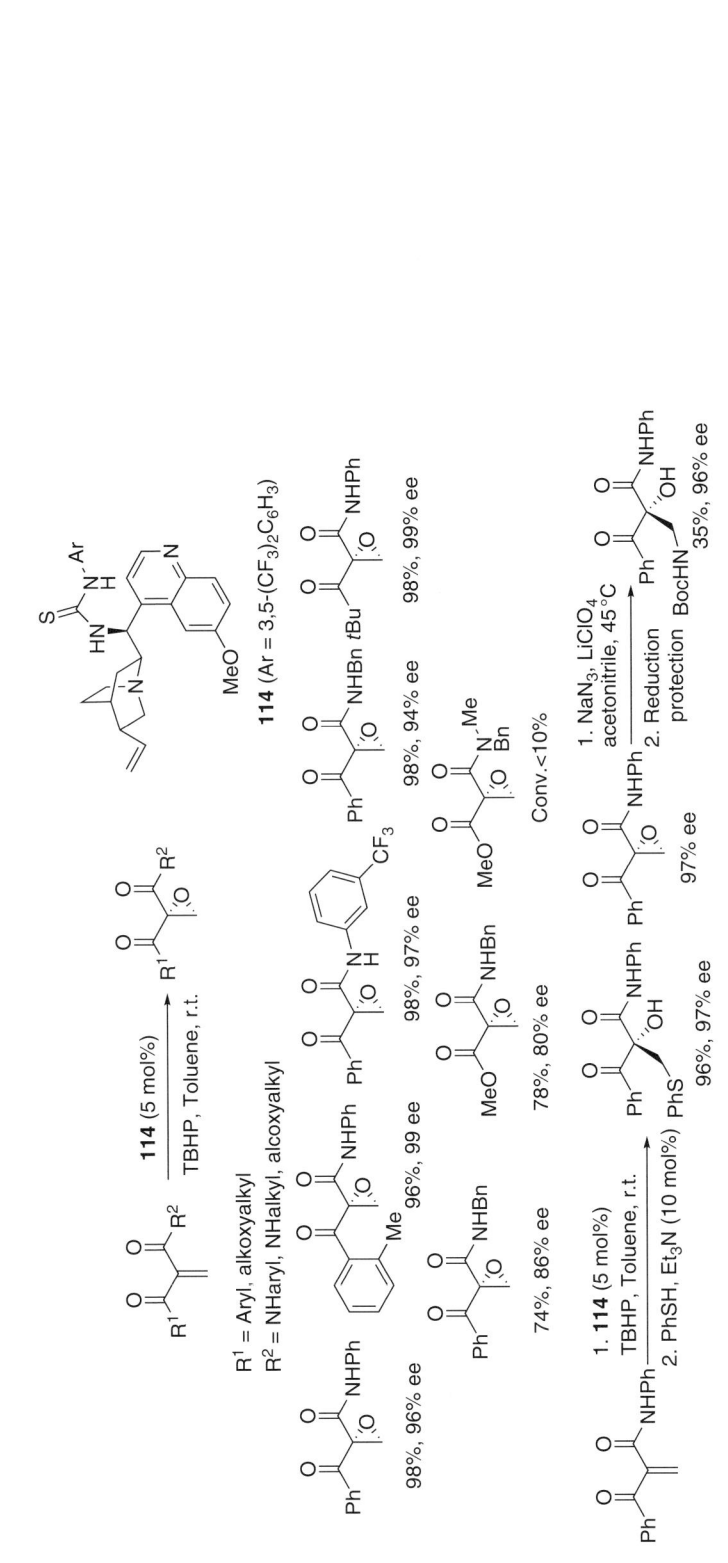

Scheme 3.105 Asymmetric epoxidation of α-aroyl acrylamides and acrylates with cinchona thiourea/TBHP system and synthetic elaboration.

high reactivity and weak H-bonding acceptor ability of the cyano group. The best results were achieved by using a multifunctional quinine-derived thiourea **115**, bearing additional stereocenters and hydrogen-bonding donor (Scheme 3.106) [340]. Matching and mismatching effects of the cinchona alkaloid and β-aminoalcohol moieties affected the stereocontrol. Aryl and alkyl substituted alkenes were epoxidized in good yield and with moderate-to-satisfactory level of enantioselectivity when using CHP as the oxygen source in toluene at −20 °C. *gem*-Dicyanoepoxides behave formally as synthetic equivalent of dication ketenes, when subjected to the presence of binucleophilic reagents to give heterocyclic compounds of interest such as imidazoles, 1,4-benzoxazin-2-ones, 3-substituted piperazin-2-ones.

Few approaches are known to prepare optically active 3-substituted piperazin-2-ones, which are endowed with a wide range of important biological activities. Hence, a one-pot methodology based on asymmetric epoxidation/ring-opening reaction with diamines to access enantioenriched 3-substituted piperazin-2-ones was established (Scheme 3.106). A transition state for the oxa-Michael step, involving a network of hydrogen-bonding interactions among the catalyst and the reagents, has been suggested. Takemoto and coworkers recently illustrated the application of organocatalysts such as benzothiadiazine **116** bearing a stronger H-bonding donating group, easily derived from (*R*,*R*)-1,2-cyclohexanediamine, in the asymmetric epoxidation of low-reactive *trans*-α-cyano-α,β-unsaturated amides with TBHP as the oxidant (Scheme 3.107) [341]. The reaction proceeded with β-aryl-substituted amides in excellent yields with complete control of the diastereoselectivity and moderate-to-good level of enantiocontrol, which was significantly affected by the electronic properties of the groups on nitrogen. The enantioselectivity could be eventually increased over 90% by a single crystallization. A neutral transition state has been suggested, where the carbonyl group is hydrogen-bonded with the benzothiadiazine group, and the TBHP is engaged in H-bonding interaction with the tertiary amine. The epoxides, bearing quaternary and tertiary stereocenters, proved to be useful building blocks to access biologically important 2-quinolone scaffolds with complete stereocontrol.

3.5 Kinetic Resolution of Racemic Epoxides

Besides reactions mediated by metal complexes, enzymes, and organocatalysts, another important tool in asymmetric synthesis to obtain optically active compounds is represented by the kinetic resolution of racemic molecules [5c, 342]. In general, enzymatic kinetic resolutions can be likely considered the most synthetically useful and effective examples of kinetic resolution applied in academia and industrially, given the excellent stereocontrol associated with the highest yield of the unreactive product achievable (50% maximum theoretical yield), satisfying the criteria of an ideal process of kinetic resolution. The first pivotal and highly effective example of kinetic resolution in asymmetric catalysis was illustrated in the 1980s by the Ti/tartrate-catalyzed Katsuki–Sharpless epoxidation of secondary allylic alcohols, which enabled to obtain both unreacted enantioenriched secondary allylic alcohols and epoxyalcohol products [23a]. Nucleophilic kinetic

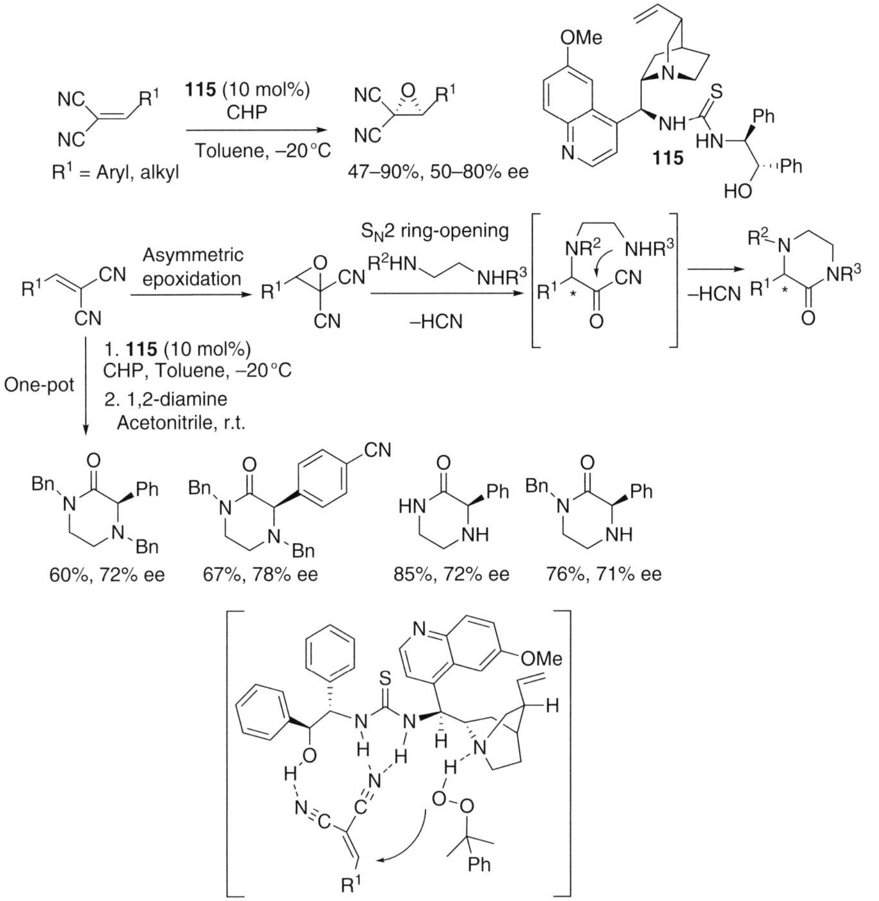

Scheme 3.106 Asymmetric epoxidation of alkylidenemalononitriles with cinchona thiourea/CHP system.

resolution of racemic epoxides appears to be a fundamental strategy when chiral epoxides cannot be obtained through classical asymmetric methods. To serve as an efficient and practical tool, the two enantiomers have to differ significantly in reaction rates, as measured by the stereoselectivity factor $(S = k_{rel} = k_{fast}/k_{slow})$, which according to Kagan's equations can be estimated by experimental parameters such as conversion of the reagent and the enantiomeric excess of the unreacted compound and/or the chiral product [5c, 24c]. Another important issue for a kinetic resolution to be of practical use is the highest possible control of the regioselectivity in the ring opening of the oxirane. As this section is focused on the methodologies of kinetic resolution developed to obtain optically active epoxides, dynamic kinetic resolution of racemic epoxides to obtain ring-opened products is outside the scope of this section. According to the results previously illustrated, significant progress in the synthesis of structurally diverse and substituted epoxides in enantioenriched form has been achieved via asymmetric epoxidation. However, it is easy to recognize that only a few methodologies enable the

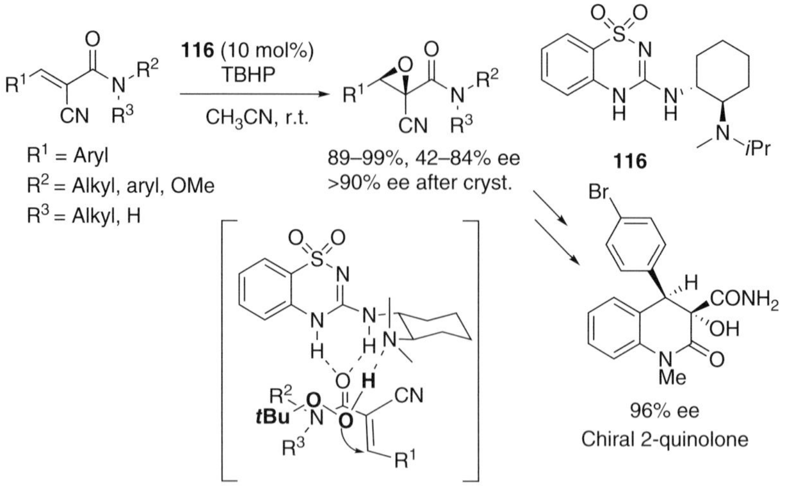

Scheme 3.107 Asymmetric epoxidation of α-cyano-α,β-unsaturated amides with benzothiadiazines/TBHP system.

access to terminal epoxides, which are among the most useful synthetic precursors to give functionalized products such as 1,2-amino alcohols, 1,2- diols, 2-hydroxy sulfides, 1,2-halohydrins, and 1,2-cyanohydrins. In the mid-1990s, Jacobsen and coworkers demonstrated the exceptional catalytic performance of Cr(III)–salen in the kinetic resolution of racemic terminal monosubstituted and 2,2-disubstituted epoxides by TMSN$_3$, which occurred with complete regioselectivity for the terminal position and very high stereoselectivity factors $44 < S < 280$ [343]. Although this method was a first impressive example, a more practical, environmentally friendly, and less expensive method to obtain in close-to-theoretical yield a broad variety of functionalized and simple terminal epoxides in essentially enantiopure form has been disclosed by the same group using water as the nucleophile in the presence of Co(III)–salen, generally working under solvent-free conditions (Scheme 3.108) [5d]. The active catalyst was formed by single-electron air oxidation of inactive Co(II)–salen precursor.

The level of stereoselectivity was found to be exceptionally high for a chemical process to also give diols as highly valuable by-products with excellent ee values

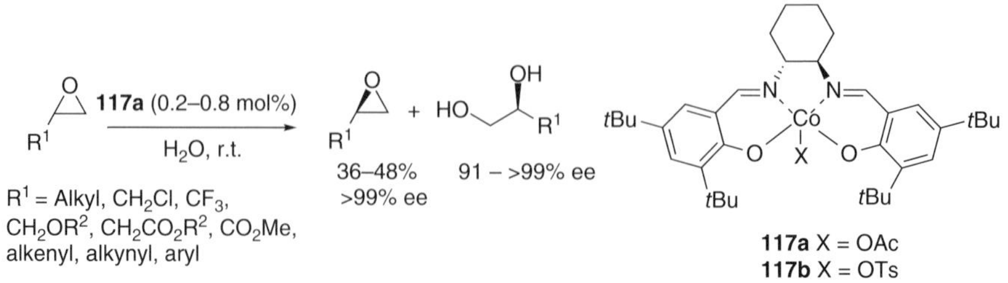

Scheme 3.108 Hydrolytic kinetic resolution (HKR) of terminal epoxides catalyzed by Jacobsen's Co–salen complex.

($S > 50$ and up to $S = 500$). Both products are easily separated from the reaction mixture by distillation. All these favorable features, including a significantly low catalyst loading, clearly showed the feasibility of the HKR as a first powerful tool to synthesize, among others, propylene oxide, styrene oxide, and epichlorohydrin, fundamental chiral building blocks in the industry [344]. Catalyst **117a** was also useful for the inter-and intramolecular KR of terminal epoxide by phenols, although with slightly less efficiency than that observed in HKR [345]. Studies also focused on mechanistic aspects of the HKR for an implementation of the reaction parameters necessary to work at best for large-scale applications. A second-order dependence on catalyst concentration indicated that two catalyst molecules participate in the ring-opening rate-determining step, and the reaction proceeds via cooperative bimetallic mechanisms [346]. An important role in regulating both reaction rate and ring opening of the epoxide was displayed by counteranion X in complex **117**. One Co(salen) unit serves as Lewis acid species responsible for binding and activation of the epoxide, whereas the nucleophilic (salen)Co–OH species attacks the activated epoxide to give the diol. The most effective complex was found to be **117b** bearing the tosylate counteranion [346b]. To overcome limitations such as reaction rate decrease by reduction of catalyst loading due to the second-order kinetics, efforts were focused to develop linked multi-(salen) metal complexes via dendritic [347], oligomeric frameworks [348]. Enhanced reactivity was observed with oligomeric (prevalently dimeric) catalysts **118** in HKR of terminal epoxides ascribed to cooperative reaction within the linked catalysts, attested by first-order kinetic dependence on catalyst concentration (Scheme 3.109) [349]. Indeed, commercially available catalyst **118a** demonstrated to be the most general and effective catalyst in ring-opening reactions of terminal epoxides with water, phenols, and primary alcohols as nucleophiles [350]. A recent synthetic application of kinetic resolution using primary alcohols has been illustrated by Jacobsen and coworkers in the total synthesis of (+)-reserpine via intermediate enone **120** [351]. The kinetic resolution of racemic epoxide performed in acetonitrile with 4.5 mol% of catalyst **118a** and benzyl alcohol as the nucleophile afforded the secondary alcohol in 41% yield and with 96% ee, which was elaborated to enone **120**. In turn, this compound was coupled with dihydro-β-carboline **121** to access antipsychotic and antihypertensive drug (+)-reserpine. The HKR of racemic terminal epoxides performed under solvent-free conditions afforded the recovered epoxides with perfect enantioselectivity and in high yield, working at very low catalyst loadings (Scheme 3.109). Enhancement in reactivity and enantioselectivity relative to monomeric catalyst **117** has mainly ascribed to the flexible tethering of the Co(salen) units, thus allowing their reciprocal optimal relative position for activation of the reagents. Catalysts **118** have been recycled by removing the volatile reactants followed by charging the reaction vessel with starting materials. Jones, Weck, and coworkers reported poly-(styrene)-supported Co(salen) catalysts by the free-radical polymerization of a salen monomer, which gave, among others, flexible catalyst **119** performing in an excellent way in the HKR of racemic epichlorohydrin (Scheme 3.109) [352]. Recycling of copolymer-bound Co(salen) complex **119** was realized by precipitation after HKR of epichlorohydrin by the addition of diethyl ether. The catalyst after being recycled three times proved to give comparable ee values with only a slight decrease in the

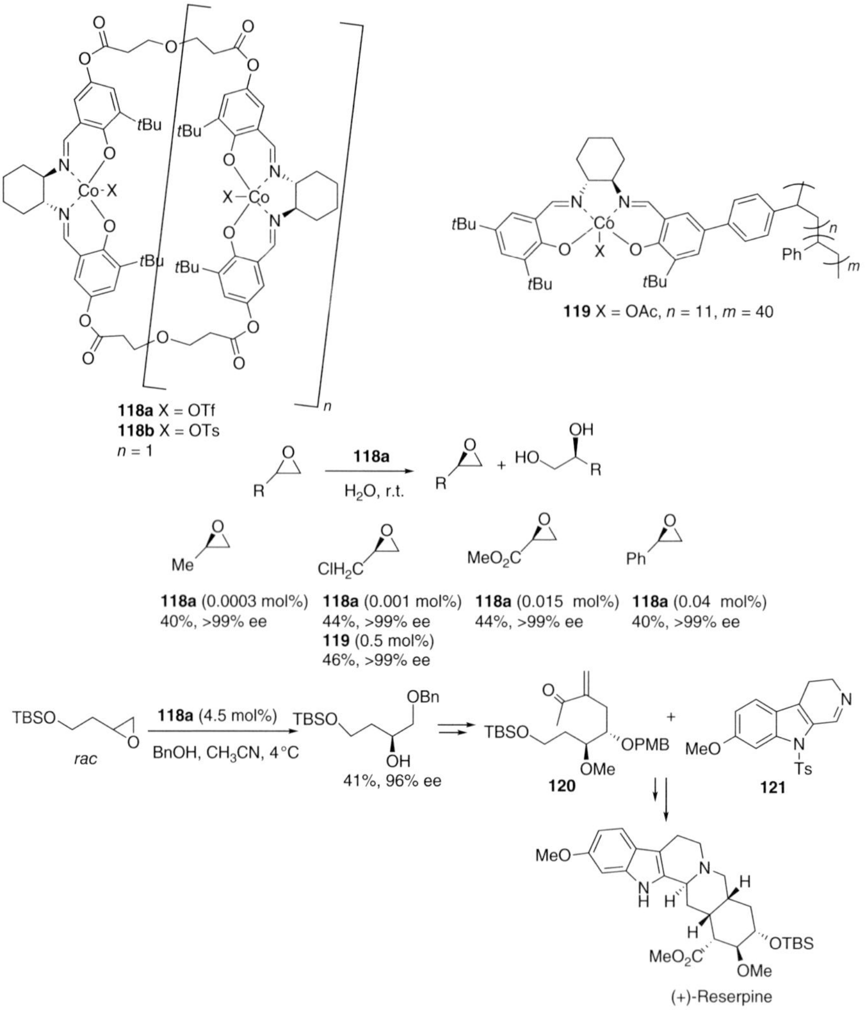

Scheme 3.109 Cyclic oligomeric and polymeric Co–salen catalysts for HKR of racemic terminal epoxides.

conversion with respect to the first run. Several polymer-supported catalysts for HKR of terminal epoxides were developed by the same group [353]. It has been demonstrated that the precise amount of water (0.5–0.7 equiv.) to be used in the HKR of epoxide as limiting reagent can be overcome by using a Co(salen) complex supported on an amphiphilic, water-soluble block copolymer [354]. In this case, excellent performance in the HKR of terminal epoxides has been achieved working in pure water as medium and reagent. The formation of micellar aggregates would create nanoreactors with high local concentration of catalyst in the hydrophobic core, and limited amount of water can penetrate into the micellar core during the catalysis. Further examples of Co(salen)-(BF$_3$) immobilized on a meso/macroporous carbon monolith as effective heterogeneous recyclable catalysts [355] and Co(salen) immobilized via HO$_3$S-linkers on ordered mesoporous SBA-16 silica by electrostatic interactions [356] have been successfully developed for the HKR of terminal epoxides. Examples of immobilization of Co(salen) catalyst on various supports, such as polymers [354, 357], gold colloids [358], mesoporous silica [353c], or zeolite [352a], allow recovery and reuse at the end of reactions.

Recently, Kleij and coworkers developed dinuclear Co(salen) complex **122** and dinuclear Co(salen) complex calix[4]arene hybrid **123**, both tested in the HKR of a few racemic terminal epoxides (Scheme 3.110) [359]. The precatalyst of complex **122** was synthesized via olefin metathesis of precursors bearing a *para*-allylic moiety in the aromatic rings.

The performances in the HKR of (*rac*)-1,2-epoxyhexane showed that besides the catalytic activity, which is significantly different for the two catalysts, the proximity of the catalytic Co(salen) moieties plays a significant role in the control of the stereoselectivity. Dimeric Co(salen) complexes **124** activated by GaCl$_3$, InCl$_3$, TlCl$_3$, and, more recently, Y(OTf)$_3$ played efficiently as catalysts in the HKR of terminal epoxides [360]. The linking of two Co(salen) moieties through the Ga and other metal additives enabled proper proximity and orientation for reagent activation, reinforcing the reactivity and selectivity (Scheme 3.111). The reactions proceeded with low catalyst loading, affording unreacted epoxides in high yield and with excellent enantioselectivity. Bimetallic monomeric [Co(salen)]-type catalysts **125** complexed with common transition-metal salts also proved to be useful at low loading for the HKR of terminal epoxides [361].

Rare examples of HKR of functionalized racemic epoxides bearing two stereocenters have been reported. The groups of Tae and Sudalai studied the HKR of racemic *syn-* and *anti-*2-hydroxy-1-oxiranes, in order to obtain both enantiomers of *syn-* and *anti-*epoxides depending on the chirality of the commercially available Co(salen) catalysts **117** used (Scheme 3.112) [362]. The approach is alternative to using enantiomerically enriched hydroxy ketones or epoxyxalcohols as the starting materials to access synthetically useful 1,3-diol chiral units. Racemic *syn-*O-protected terminal epoxide **126** underwent HKR with catalyst **117a** to give enantioenriched *syn-*epoxide **126** and the 1,2-diol in excellent yield and ee value. Epoxide **126** was employed as the starting reagent for the total synthesis of the opposite enantiomer of natural product (5*S*,7*R*)-kurzilactone. HKR of racemic *syn-* or *anti-* alkoxy- and azido epoxides **127** can be catalyzed by both enantiomeric Co(salen) complexes **117a** to afford highly enantioenriched *syn-* or *anti-* alkoxy- and azido epoxides and the corresponding 1,2-diols. Optically active *trans-***128** was employed in a concise enantioselective synthesis of the cytokine modulator (+)-*epi*-cytoxazone.

Sudalai *et al.* have recently applied the HKR of racemic *anti-* and *syn-*3-alkyl or aryl substituted epoxy esters **129–130** to obtain in excellent yield and enantioselectivity *cis-* or *trans-*3,4-substituted-γ-butyrolactones and unreacted epoxides by using 0.5 mol% of *ent-***117a** (Scheme 3.113) [363]. This simple reaction is an appealing process, which can yield *cis-* or *trans-*3,4-disubstituted γ-butyrolactones, important scaffolds endowed with several biological activities. Moreover, the authors illustrated a concise synthesis of natural product (+)-eldanolide, a long-range sex attractant, via regioselective ring opening of the chiral epoxide with an organometallic reagent.

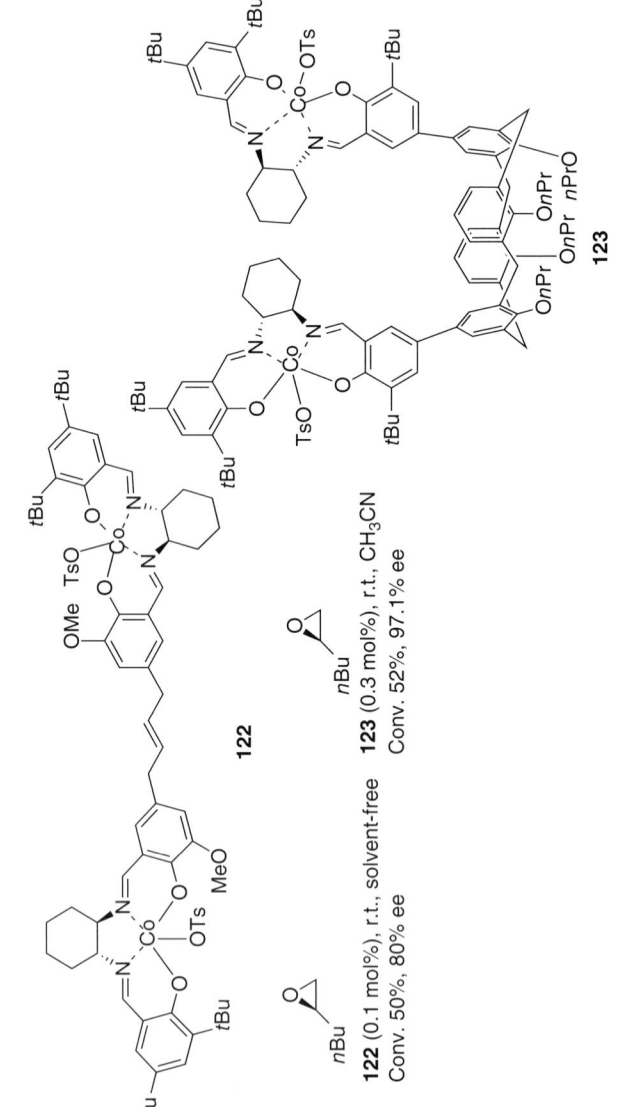

Scheme 3.110 Dinuclear Co(salen) catalysts for HKR of racemic terminal epoxides.

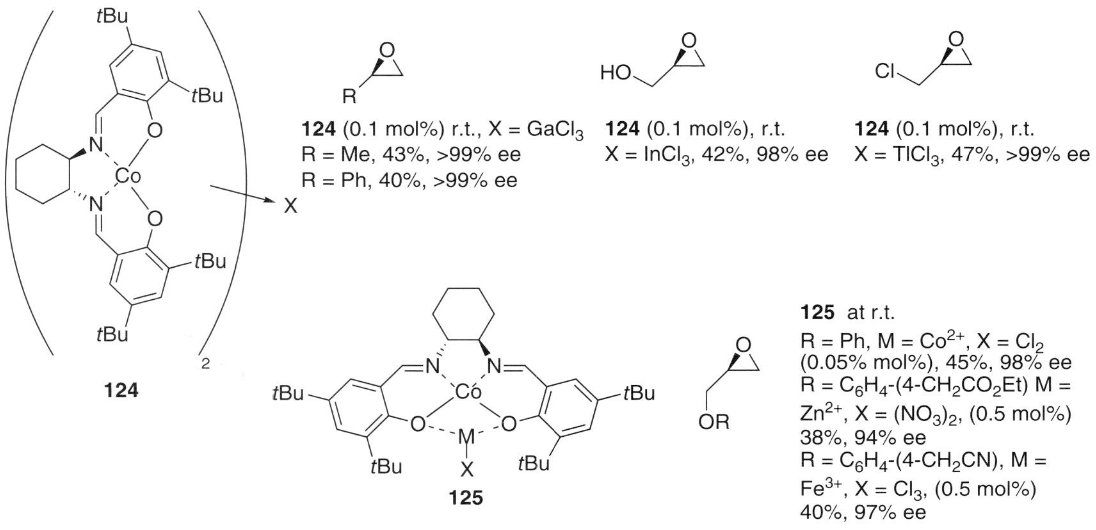

124 (0.1 mol%) r.t., X = GaCl₃
R = Me, 43%, >99% ee
R = Ph, 40%, >99% ee

124 (0.1 mol%), r.t.
X = InCl₃, 42%, 98% ee

124 (0.1 mol%), r.t.
X = TiCl₃, 47%, >99% ee

125 at r.t.
R = Ph, M = Co^{2+}, X = Cl₂ (0.05% mol%), 45%, 98% ee
R = C₆H₄-(4-CH₂CO₂Et) M = Zn^{2+}, X = (NO₃)₂, (0.5 mol%) 38%, 94% ee
R = C₆H₄-(4-CH₂CN), M = Fe^{3+}, X = Cl₃, (0.5 mol%) 40%, 97% ee

Scheme 3.111 Bimetallic Co(salen) catalysts for HKR of racemic terminal epoxides.

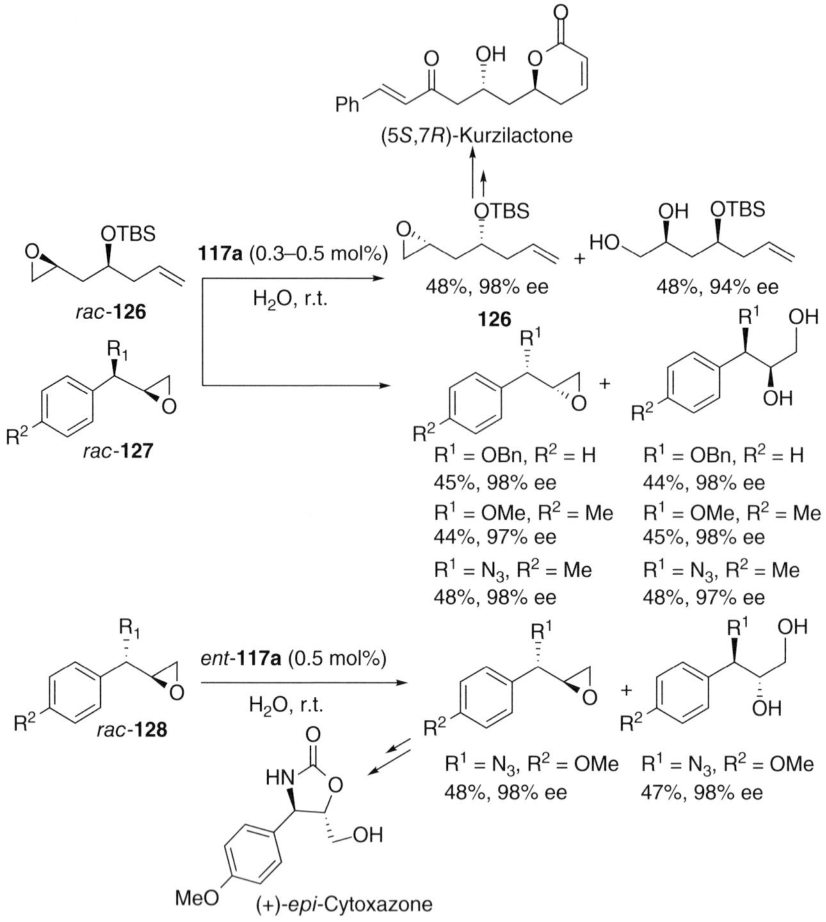

Scheme 3.112 HKR of racemic terminal epoxides with two stereocenters using Jacobsen's catalyst and synthetic application.

The HKR of terminal epoxides was recently applied in the total synthesis of the anti-Parkinson agent safinamide, starting from commercially available benzyl glycidyl ether (Scheme 3.114) [364]. Safinamide undergoes phase III clinical trials as an add-on therapy to dopamine agonists and L-Dopa for patients with Parkinson's disease. The chiral building block epoxide **131** has been obtained in excellent yield and enantioselectivity via HKR of the corresponding racemic epoxide, which secured a highly enantioselective approach to safinamide via simple elaboration steps. In the past decade, several examples attested the importance of HKR of racemic epoxides in the synthesis of natural and bioactive compounds such as (S)-timolol [365], decarestrictine D [366], (+)-isolaurepan [367], (R)-tuberculostearic acid [368], (R)-mexiletine [369], neocarazostatin [370], amprenavir [371].

Amines, carbamates, azides, imides, and, to a lesser extent, carbon nucleophiles have been used as the nucleophiles in the kinetic resolution of racemic epoxides

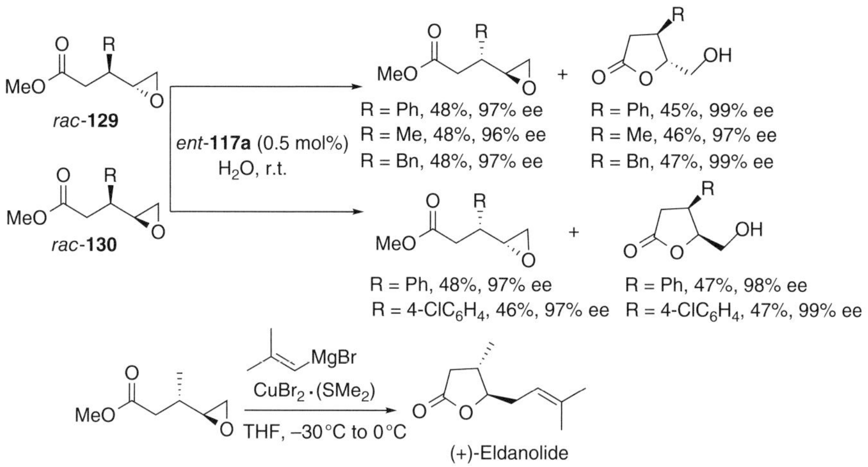

Scheme 3.113 HKR of functionalized terminal epoxides using Jacobsen's catalyst and synthetic application to γ-butyrolactones.

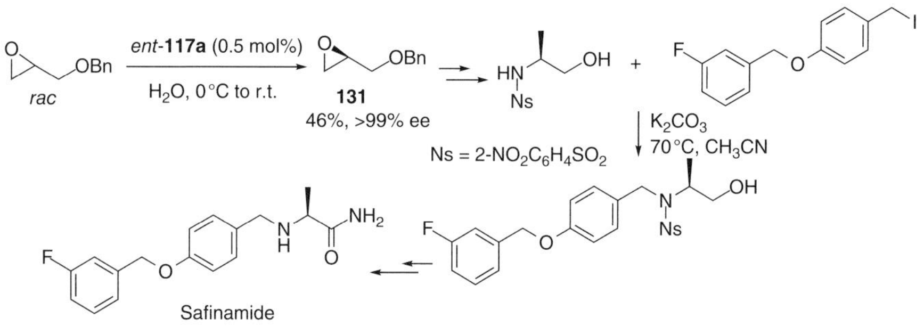

Scheme 3.114 HKR of benzyl glycidyl ether using Jacobsen's catalyst as building block for the synthesis of safinamide.

to give ring-opened products of undiscussed value, namely chiral 1,2-difunctionalized alcohols [4]. In particular, 1,2-amino alcohols are useful scaffolds in the pharmaceutical industry and in synthetic organic chemistry. Bartoli, Melchiorre, and coworkers demonstrated the utility of an aminolytic kinetic resolution (AKR) of *trans*-epoxides by using commercially available Cr(salen) complex **132** and anilines in CH_2Cl_2 (Scheme 3.115) [372]. The reaction proceeded in *anti*-stereoselective manner, giving the *trans*-aminoalcohol with complete regiocontrol in good yield and with good-to-high enantioselectivity, whereas the epoxides were recovered with moderate-to-good ee values. An improvement in terms of practicality and environmental concerns of AKR has been illustrated by Kureshi and coworkers, who carried out the same process in ionic liquids as medium achieving faster reaction rates and enabling recycling of the catalyst (Scheme 3.115) [373]. Indeed, up to six cycles with no loss in performance have been attested. The same group developed a recyclable polymeric Cr(salen) catalyst used in the AKR of *trans*-stilbene oxide and *trans*-β-methylstyrene oxide with anilines [374].

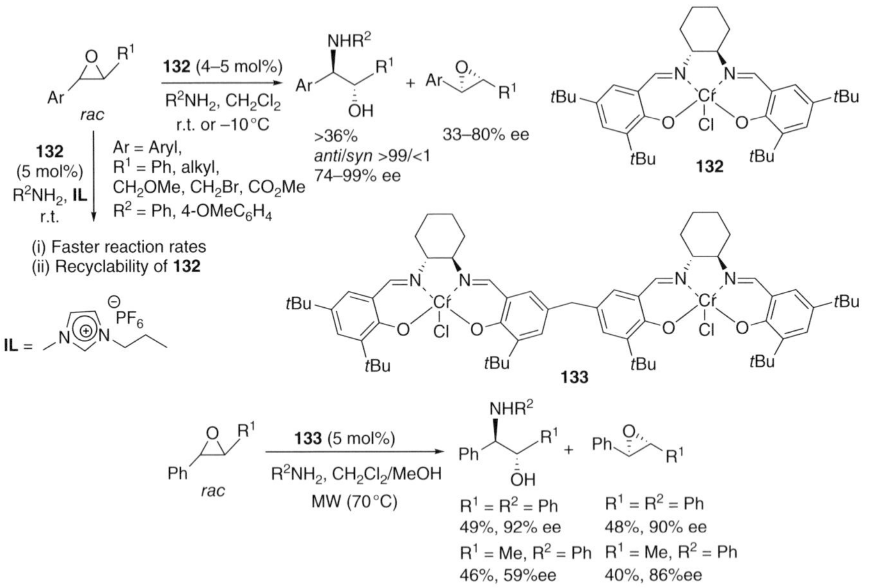

Scheme 3.115 AKR of racemic *trans*-epoxides with anilines using Cr(salen) catalysts.

Interestingly, the AKR of these epoxides was efficiently carried out with dimeric Cr(salen) catalyst **133** under microwave irradiation affording the *anti*-β-amino alcohols with up to 94% ee and the recovered epoxides with up to 92% ee (Scheme 3.115) [375]. The reactions were very rapid to finish in a few minutes, and catalyst **133** could be reused for different runs after precipitation from hexane:diethyl ether mixture without loss of efficiency.

Bartoli, Melchiorre *et al.* showed that Co(salen) catalyst **117a** could very efficiently catalyze the highly regioselective kinetic resolution of terminal aliphatic and aromatic epoxides with carbamates as nucleophiles and 4-nitrobenzoic acid as an additive to obtain in a straightforward way β-amino alcohols in excellent ee values as demonstrated by impressive values of stereoselectivity factors $S > 500$ (Scheme 3.116) [376]. This approach has been exploited to prepare antihypertensive drug (β-blocker) (*S*)-propanolol in enantiopure form. Moreover, basic treatment of the β-amino alcohols afforded chiral 5-substituted oxazolidinones [377]. The AKR of aryloxy terminal epoxides with carbamates can be conducted with excellent performance in ionic liquids showing faster rates than in the ethereal solvent and allowing recycling of the catalyst for different runs [378].

Rare examples of kinetic resolution of racemic epoxides by using soft carbon nucleophiles have been reported in the literature [379] with the most notable one concerning the addition of 2-methyl indole to racemic *trans*- and *cis*-epoxides by using Cr(salen) catalyst **134** exemplified in Scheme 3.117 [380]. Although the stereoselectivity factors were not particularly high ($S = 10–30$), it was possible to obtain the unreacted epoxides with high level of enantioselectivity forcing the conversion of the epoxide to the ring-opened products. Similarly, operating with an excess of epoxide guaranteed the isolation of the ring-opened products with ee values >90%.

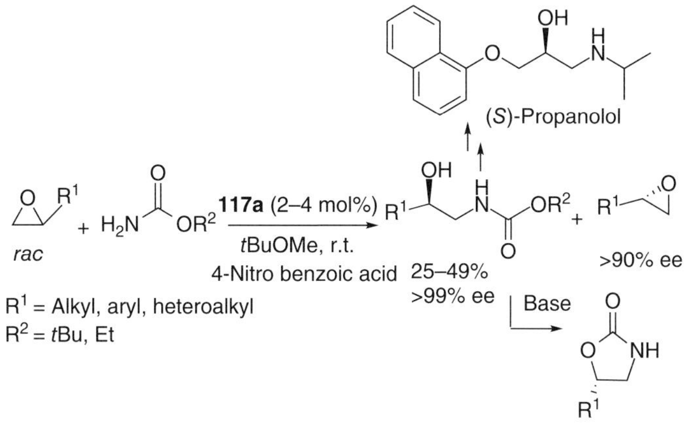

Scheme 3.116 Application of Jacobsen's catalyst in AKR of racemic terminal epoxides with carbamates for the asymmetric synthesis of β-amino alcohols and oxazolidinones.

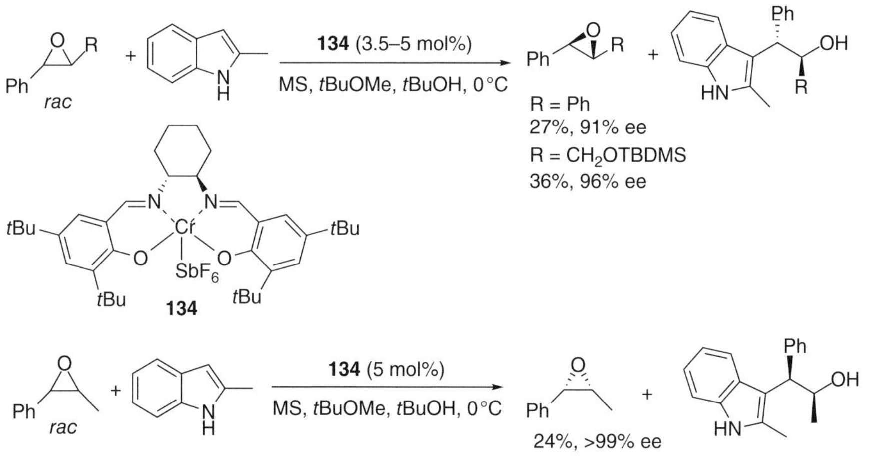

Scheme 3.117 Kinetic resolution of *trans*- and *cis*-aromatic epoxides with 2-methylindole catalyzed by a Cr(salen) catalyst.

The kinetic resolution of racemic epoxides mediated by organocatalysts has received less attention. An early example can be recognized in the work of Andersson and coworkers, who exploited a β-elimination reaction of cyclohexene epoxides catalyzed by optically pure diamine **135** in the presence of LDA and DBU as stoichiometric bases to give enantioenriched allylic alcohols and unreacted epoxides with up to 99% ee values (Scheme 3.118) [381]. However, this interesting process was successful only for the resolution of cyclic α-substituted and β-disubstituted epoxides. One strictly related example of kinetic resolution of cyclohexanone epoxide was recently reported by Jørgensen and coworkers by using thiourea-amine **136**, derived from cinchona alkaloids, as the organocatalyst (Scheme 3.118) [382]. This bifunctional promoter was thought to be able to enhance via hydrogen bonding of the thiourea group with the carbonyl moiety,

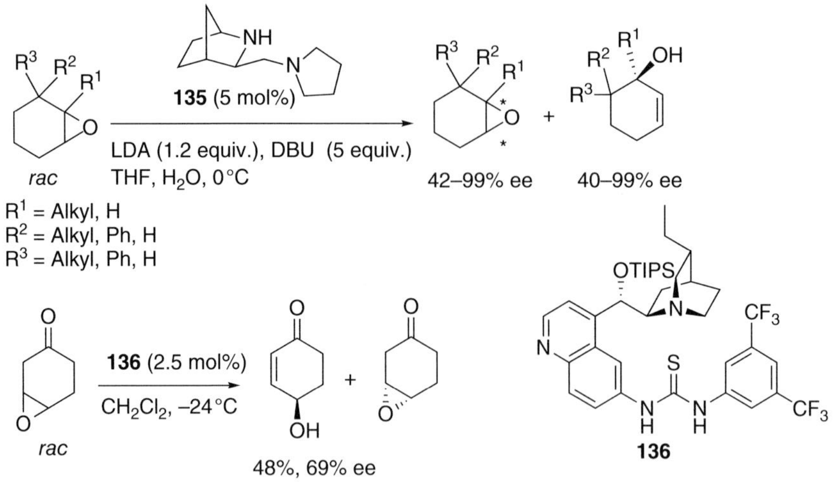

Scheme 3.118 Organocatalytic kinetic resolution of racemic cyclic epoxides via β-elimination to allylic alcohols.

the acidity of the α-protons, thus easily and selectively deprotonated by the quinuclidine nitrogen of **136**. Although the enantioselectivity of the epoxide was not mentioned, synthetically useful (R)-4-hydroxy-cyclohexenone was isolated with an encouraging level of enantioselectivity.

It has been known since the 1970s that ring-opening reactions of racemic β-aryl-substituted α-nitroepoxides by different nucleophiles proceed with complete regioselectivity to give α-substituted ketones or heterocyclic compounds [383]. Taking advantage of this selectivity, Lattanzi and coworkers developed an asymmetric AKR of racemic β-aryl-substituted α-nitroepoxides AKR with aniline as the nucleophile catalyzed by thiourea-amine **137** (Scheme 3.119) [384]. Working in excess of aniline enabled recovery of β-aryl-substituted α-nitroepoxides in acceptable yield and with good-to-high enantioselectivity. The kinetic resolution of aliphatic α-nitroepoxides proved to be inefficient. Moderate values of selectivity factors ($3 < S < 7$) were calculated, although underestimated by the contribution of background ring-opening pathway, partially taking place in the absence of the catalyst. The organocatalyst is thought to activate the reagents by general acid–base catalysis. This represented a first approach to prepare enantioenriched β-aryl-substituted α-nitroepoxides, which can be synthetically exploited to access optically active *trans*-β-amino alcohols via one-pot ring-opening/reduction sequence (Scheme 3.119).

3.6 Asymmetric Sulfur-Ylide-Mediated Epoxidations

An alternative nonoxidative preparation of epoxides relies on the use of chiral sulfur ylides and aldehydes as reactive partners, with the contemporary generation of a carbon–oxygen and carbon–carbon bonds. This process has been the focus of significant interest to produce enantioenriched epoxides in recent years [385].

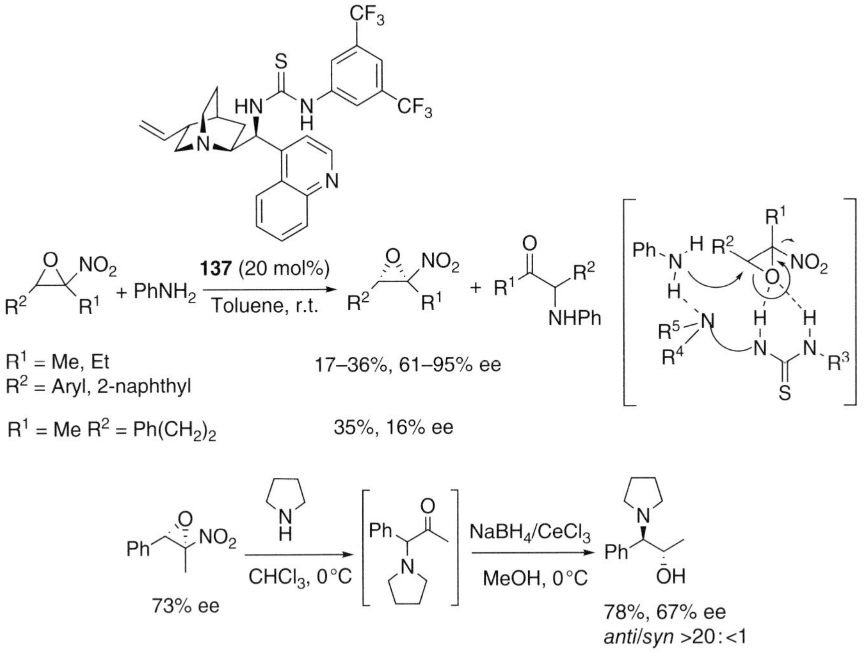

Scheme 3.119 Amino-thiourea organocatalyzed AKR of α-nitroepoxides.

The first and representative examples, reported by different groups, employed enantiopure cyclic sulfides to be transformed via alkylation to sulfonium salts, which treated under basic conditions gave sulfur ylides, ready to combine with aldehydes to form an intermediate betaine ready to evolve to epoxides (Scheme 3.120) [386]. Sulfides shown in Scheme 3.120, mostly derived from easily available terpenes, have been prevalently used at stoichiometric loadings, but according to the catalytic cycle, the sulfide being regenerated, some of them were successfully investigated as catalyst at substoichiometric amounts. The reactions generally proceeded with high control of the diastereo- and enantioselectivity for the synthesis of *trans*-disubstituted aromatic epoxides.

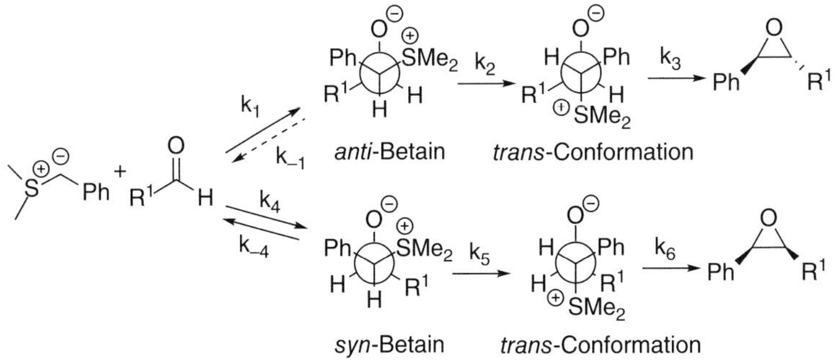

Scheme 3.120 Catalytic cycle for the epoxidation via sulfur ylides under basic conditions.

As an example, the stoichiometric process was carried out via alkylation of sulfide **142**, generating a single diastereoisomeric salt. The salt was then deprotonated at −40 °C to give *trans*-diaryl, vinyl–aryl/heteroaryl epoxides with high diastereocontrol and enantioselectivity (ee >90%) [386e, 387]. Similarly, sulfonium salt derived from sulfide **140** was effective with a broader substrate scope to produce *trans*-diaryl, aryl–alkyl, and aryl–vinyl epoxides with high stereocontrol, and, for the first time, trisubstituted epoxides were obtained starting from challenging ketones (ee >70%) [388]. Indeed, sulfide **140** proved to be useful in the asymmetric sulfur-ylide-mediated key epoxidation step in the total syntheses of naturally compounds, for instance, D-*erythro*-sphingosine [389] and arenastatin A, a highly potent cytotoxic cyclic depsipeptide from marine sponge [390]. Achieving high control of the diastereo- and enantioselectivity in sulfur-ylide-mediated epoxidation is a rather challenging task when compared to the epoxidation of geometrically defined alkenes, where only the control of absolute configuration is generally the prevalent issue to address throughout the process. Differently, in the asymmetric epoxidation via sulfur ylides, both relative and absolute configurations need to be controlled. Mechanistically, the level of diastereo- and enantioselectivity is related to the fine control and balance of key steps: (i) 1,2-addition, (ii) bond-rotation in the betaine to achieve the right conformation for the cyclization, and (iii) rate of ring closure (Scheme 3.121) [391]. Several mechanistic studies mostly performed by Aggarwal and coworkers helped to shed light on the structural requirement of the sulfide, reaction parameters to adopt in order to achieve the highest stereocontrol in the epoxidation. Some general clues have been provided: high reversibility in the betaine formation leads to high diastereoselectivity but low enantioselectivity, whereas irreversible betaine formation leads to low diastereoselectivity and high enantioselectivity. High diastereo- and enantioselectivities are attainable when the formation of the *anti*-betaine is nonreversible, whereas the formation of the *syn*-betaine is reversible. It has been observed that enhancing the *syn*-betaine reversibility, via introduction of bulky groups in the aldehyde and ylide, improves the diastereoselectivity, thanks to a higher energy barrier for the carbon–carbon bond rotation to the reactive *trans*-conformer. Protic solvents or metals can help to solvate the alkoxide lowering the barrier for the carbon–carbon bond rotation, thus reducing the reversibility of *syn*-betaine, hence lowering the diastereoselectivity. Crucial to the control of the enantioselectivity is that only one face of one ylide isomer is exposed to the incoming aldehyde and that this step is irreversible and consequently enantio-determining.

All the data showed that the stereocontrol is significantly affected by reaction conditions, nature and stability of ylides, and sulfide structure. The asymmetric sulfur-ylide-mediated epoxidation has demonstrated the possibility to access a restricted number of terminal epoxides, when reacting aromatic ylides of oxathiane **142** with paraformaldehyde at low temperature achieving moderate-to-good yield and up to 96% ee [392]. Disappointingly, employment of Simmons–Smith-type Zn-carbenoid with stoichiometric loading of chiral sulfides gave modest results in terms of conversion and enantioselectivity (ee <60%)[393].

Another class of accessible optically active epoxides are those bearing an electron-withdrawing group (ester, amide, sulfonyl, ketone) nearby the oxirane ring.

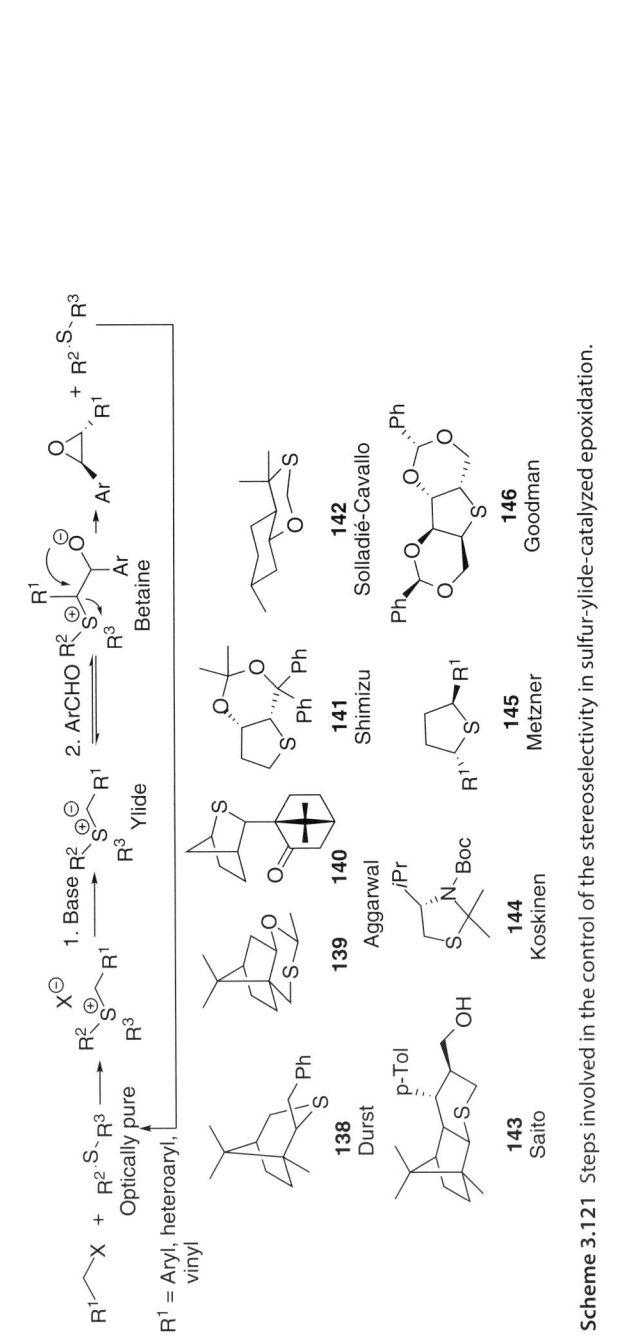

Scheme 3.121 Steps involved in the control of the stereoselectivity in sulfur-ylide-catalyzed epoxidation.

In this case, the stability of the ylide plays a crucial role in the reactive pathway toward the epoxide as demonstrated by the nature and ability of the group to stabilize the carbanion. Ylides with the amide group, of low electron-withdrawing ability, demonstrated to react with aldehydes to give epoxides, whereas the presence of an ester group is too stabilizing for the ylide to react. The first asymmetric synthesis of glycidic amides used stoichiometric amount of easily available camphor-derived sulfonium salt **147** under basic conditions to *in situ* form the ylide, which reacted with aromatic aldehydes to give *trans*-glycidic amides exclusively with up to 72% ee (Scheme 3.122) [394]. An improvement of the reaction was attained by employing the diastereoisomerically pure O-protected sulfonium salt **148**, which, when treated with KOH at a lower temperature, afforded the *trans*-epoxides in good-to-high yield with high enantioselectivity for aromatic aldehydes [395]. Modest ee values were observed for sterically unhindered alkyl aldehydes. Crossover experiments and DFT calculations suggested that the rate- and selectivity-determining step was the ring closure. The reaction has been applied as key step for the synthesis of SK&F 104353, a leukotriene D_4 inhibitor. Recently, Xiao, Lu, and coworkers developed a highly diastereo- and enantioselective synthesis of spiroepoxyindoles, a class of important compounds to access therapeutic agents, by treating sulfonium salt **148** with DBU in the presence of isatins (Scheme 3.122) [396]. The reaction allowed an efficient access to *trans*-spiroepoxyoxindoles, obtained in high yields and with excellent enantioselectivities, representing the best until now reported asymmetric approach to this class of epoxides. In several cases, sulfides illustrated in Schemes 3.120 and 3.122 were recovered in good-to-high yields (70–98%) during the purification and recycled to make the process more convenient. With chiral sulfides being synthetic compounds, efforts were devoted to make the

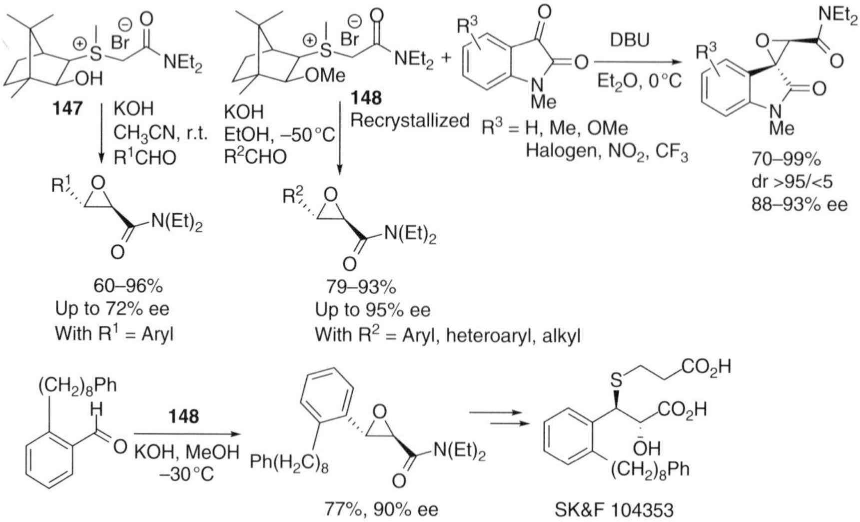

Scheme 3.122 Asymmetric synthesis of *trans*-glycidic amides via camphor-derived ylide-mediated epoxidation.

epoxidation cycle catalytic in sulfide loading. The first reports were illustrated by Metzner and coworkers with sulfide of type **145** [386h] or selenium analogs [397]. However, under basic conditions, the most problematic issue to solve was the alkylation step, which proved to be very slow at 10 mol% loading of sulfide to prolong the reaction time for days. The best results, comparable to that of the stoichiometric procedure, were obtained for the synthesis of *trans*-diaryl epoxides [386k].

Aggarwal and coworkers exploited an alternative method for the generation of ylides via metal-catalyzed decomposition of a diazo compound to give a metallocarbene, which in turn reacts with the sulfide to provide the required ylide (Scheme 3.123). To avoid the manipulation of potentially explosive diazo compounds, their *in situ* generation has been developed starting from tosyl hydrazones, which are decomposed by a phase-transfer catalyst. The same group demonstrated that a complete sequence could directly start with the condensation of tosyl hydrazine with an aldehyde as the first step. The most effective sulfide **140** was successfully used as low as 5 mol% loading with 1 mol% of $Rh_2(OAc)_4$ to produce *trans*-diaryl, aryl–alkyl, aryl–heteroaryl, and a limited number of vinyl–aryl epoxides in good-to-high yield, diastereo-, and enantioselectivity [386c]. This methodology has some advantages when compared to those previously reported: (i) neutral reaction conditions and wider aldehyde scope are applicable, (ii) sulfide **140** can be eventually recovered at the end of the process, and both enantiomeric epoxides are accessible from both enantiomers of the sulfide, which in turn available from the corresponding commercial camphorsulfonyl chlorides.

Most of the recent efforts in the asymmetric ylide-mediated epoxidation were focused on the synthesis of readily accessible optically pure sulfides able to improve and extend the scope of epoxidation in order to apply this methodology in multistep synthesis. An innovative scaffold for the starting sulfide has been reported by Sarabia and coworkers, based on the use of easily available amino acids L- and D-methionines (Scheme 3.124) [398].

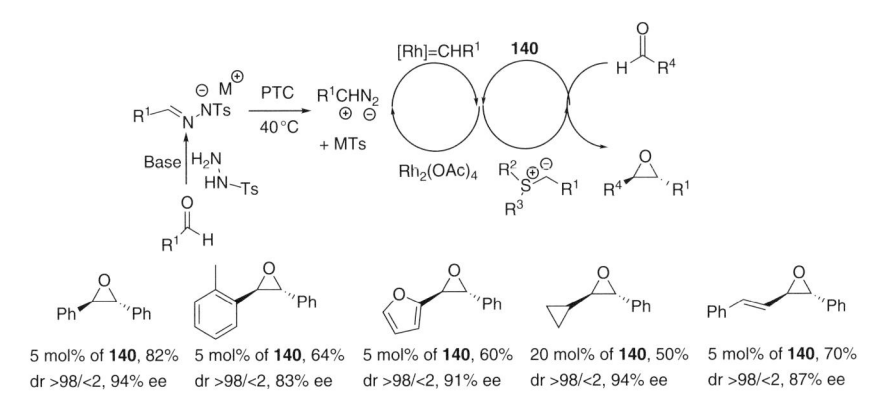

Scheme 3.123 Catalytic cycle for the sulfur-ylide-mediated epoxidation via metallocarbene.

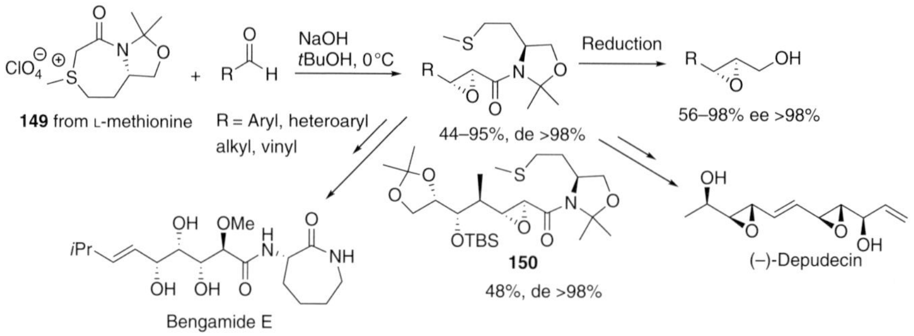

Scheme 3.124 Asymmetric synthesis of *trans*-epoxy amides via amino-acid-derived sulfur ylides and synthetic application.

Sulfonium salt **149** was obtained in four steps in 70% overall yield from L-methionine after crystallization. The reactions were performed with stoichiometric amount of the salt and aromatic or aliphatic aldehydes under basic conditions to exclusively give the *trans*-epoxides in almost diastereoisomerically pure form in satisfactory yield. Reduction of the epoxy amides enabled access to epoxy allylic alcohols with excellent values of enantioselectivity. Similar results have been illustrated when using the opposite enantiomer of salt **149**. The entire process is an alternative route to the asymmetric Sharpless epoxidation of *trans*-allylic alcohols to access epoxy alcohols of both absolute configurations. Employment of enantiopure aldehydes as reagents enabled the preparation of complex epoxy amides with excellent control of the diastereoselectivity, as illustrated for compound **150**, a useful intermediate for the synthesis of macrolide-type natural products. The epoxy amides were also susceptible to functionalizations via ring-opening reactions with nitrogen-, sulfur-, and carbon-centered nucleophiles with high and controllable regioselectivity for either the C-2 or C-3 position. The same group recently reported an in-depth study on the bicyclic core ring size of sulfonium salts of type **149**, demonstrating that it was possible to increase the scope of simple and chiral aldehydes employable in the epoxidation, for instance, heteroaromatic, vinyl, hemiacetal [399]. The versatility of this methodology has been demonstrated in the total syntheses of bengamide analogs [400], a family of marine natural products isolated from sponges, showing prominent antitumor, anthelmintic, antibiotic activities, and natural product (−)-depudecin [401], an antiangiogenic microbial polyketide. Further applications include the synthesis of cyclodepsipeptides globomycin and SF-1902 A5 [402] and sphingoid-type bases [403]. In the same context, good diastereoselectivity (dr >88/12) has been observed, when using stoichiometrically, aryl and alkyl oxazolidine-based sulfonium salts derived from (R)-2-phenylglycinol and aromatic and aliphatic aldehydes under basic conditions to access *trans*-epoxy amides [404].

Aggarwal's group introduced (+)- or (−)-isothiocineole as optically pure sulfides produced at multigram scale in one step from both readily available enantiomers of limonene (Scheme 3.125) [405]. Benzylic and allylic sulfonium salts of (−)-isothiocineole **151**, treated under basic conditions in the presence of aromatic, α,β-unsaturated and alkyl aldehydes, afforded 1,2-aryl-alkyl and α,β-unsaturated epoxides in good yield with high *trans*-diastereo- and enantioselectivity. The stereocontrol has been ascribed to the rigid bicyclic structure of isothiocineole that would dictate the following: (i) the exclusive attack of one sulfur lone pair by the alkylating agent to give the salt; (ii) the most favorable conformation of the ylides by nonbonded steric interactions; and (iii) facial selectivity of the ylide by face shielding provided by one of the two geminal dimethyl groups. The synthetic utility of this simple methodology has been demonstrated in a convergent and stereoselective synthesis of cinchona alkaloids, quinine and quinidine. Further investigations on this system enabled extension of the scope of the aldehydes, and by modulation of reaction parameters, the stereoselectivity was optimized to the highest level reported to date for 1,2-diaryl, aryl–heteroaryl, aryl–alkyl, and α,β-unsaturated epoxides [406]. Although being a stoichiometric method, practical and low-cost access of the sulfide makes the methodology competitive with the classical asymmetric epoxidation reactions.

In 2013, Chein and coauthors illustrated an expeditious preparation of the enantiopure sulfide (−)-(thiolan-2-yl)diphenylmethanol at multigram scale as a precursor of O-benzylated sulfide **152**. Efforts were directed at optimization of the reaction parameters to develop a catalytic process, with *in situ* generation of the sulfur ylide. Sulfide **152** was used at 20 mol% loading in the presence of tetrabutyl ammonium iodide (TBAI), LiOTf, and water as additives, NaOH as base, with benzyl bromide and aromatic aldehydes as the reagents in *tert*-butanol (Scheme 3.126) [407]. Under these conditions, *trans*-1,2-phenyl-aryl-substituted epoxides were obtained in high yield with variable diastereoselectivity and moderate to fairly good enantioselectivity, which, as previously observed, was significantly influenced by the substitution pattern of the aromatic aldehyde. Although the methodology appears specific in scope, it demonstrates that a simple sulfide of facile access can be used catalytically in this type of epoxidation.

Similarly, Aggarwal and coworkers synthesized a variety of chiral thiomorpholines, potentially applicable at catalytic scale, for the sulfur-ylide-mediated epoxidation of aromatic aldehydes. *trans*-1,2-Diarly epoxides were obtained with comparable results in terms of enantiocontrol, but better level of diastereoselectivity, working at lower temperatures [408]. Interestingly, the thiomorpholines were easily recovered in high yield (up to 97%) simply by acid/base extraction. A significant improvement in the area of asymmetric epoxidations mediated by sulfur ylides has been achieved by Connon and coworkers [409]. In 2012, they developed a sterically demanding C_2-symmetric methyl sulfonium salt **153** as an efficient reagent, whose ylide, obtained after deprotonation with a commercially available phosphazene base at −78 °C, attacked aromatic, heteroaromatic, and α,β-unsaturated aldehydes to produce challenging terminal epoxides in good-to-high yield with high ee values (Scheme 3.127). A rationale for the high enantioselectivity could be explained by the preferential approach of the ylide to the *Si*-face

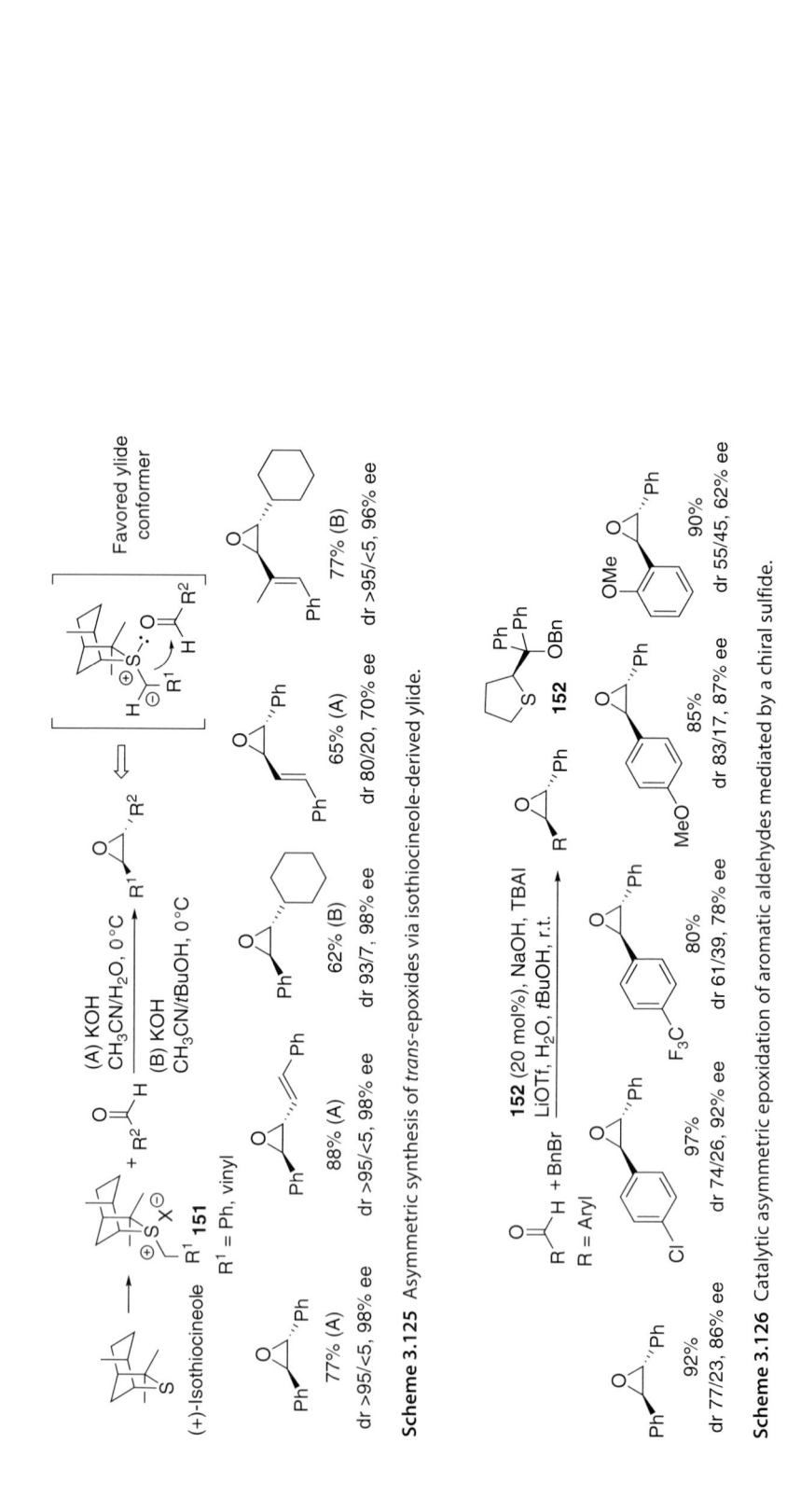

Scheme 3.125 Asymmetric synthesis of *trans*-epoxides via isothiocineole-derived ylide.

Scheme 3.126 Catalytic asymmetric epoxidation of aromatic aldehydes mediated by a chiral sulfide.

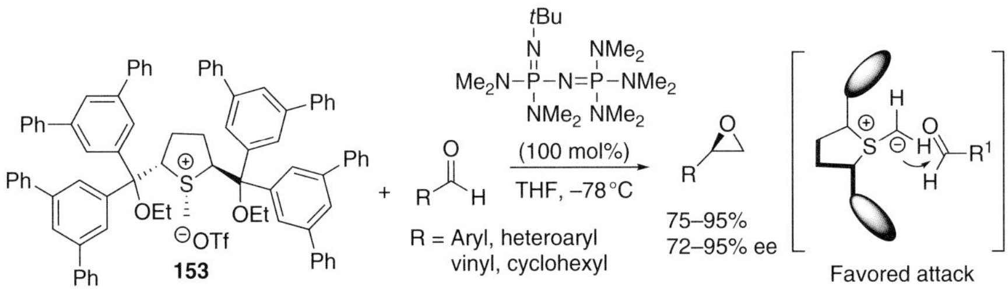

Scheme 3.127 Asymmetric synthesis of terminal epoxides via C_2-symmetric methyl sulfonium salts.

of the aldehyde, where all the interactions among the sterically demanding groups of the ylide and aldehyde are minimized. Terminal epoxides of opposite absolute configuration could also be prepared using salt **153** of opposite absolute configuration. The sulfide was conveniently recovered after chromatography in >70% yield.

3.7 Asymmetric Darzens-Type Epoxidations

3.7.1 Chiral Auxiliary- and Reagent-Mediated Darzens Reactions

An important transformation for synthesis of epoxides bearing electron-withdrawing groups, for instance, α,β-epoxycarbonyl, epoxysulfonyl, and epoxy amide products, is the Darzens reaction (Scheme 3.128) [410]. Together with the sulfur-ylide-mediated epoxidation of carbonyl compounds, the Darzens reaction represents a highly valuable process when oxidative conditions are not applicable to form the epoxide. On the other hand, in this process, a strong acid is formed as side product, which could catalyze other reactions comprising a racemic pathway and quenching of the basic catalyst when working under homogeneous catalytic conditions. Consequently, the control of different reaction parameters is necessary to develop a selective process.

The first step involves an aldol-type addition of an α-haloenolate to an aldehyde or a ketone, followed by S_N2-type ring closure of the diastereoisomeric alkoxy adducts to give the epoxide. The control of stereochemistry in the Darzens reaction is not a trivial task as being a two-step reaction involving a potentially

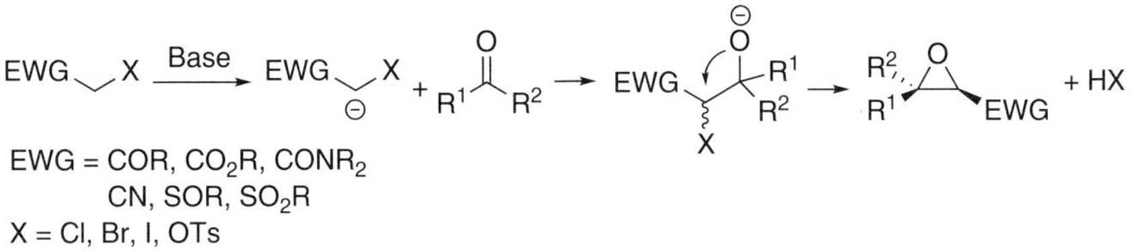

EWG = COR, CO_2R, $CONR_2$
CN, SOR, SO_2R
X = Cl, Br, I, OTs

Scheme 3.128 General pathway for the Darzens reaction.

reversible aldol reaction. It has been generally suggested that the relative and absolute configurations of the two stereocenters are established in the aldol step. Hence, typical models for the transition states in the aldol step of the Darzens reaction have been proposed [411]. The first approaches to control the diastereo- and enantioselectivity of the Darzens reaction benefited from the knowledge acquired on aldol reactions. Consequently, several methods based on chiral auxiliaries and chiral reagents were investigated in the 1980s and 1990s. Some of the most representative examples are illustrated in Scheme 3.129. Evans' oxazolidinones, reported by Prigden and coworkers [412], were treated under different conditions to generate metal (boron, tin(II), tin(IV), titanium, lithium, zinc) enolates with aromatic and aliphatic aldehydes. According to the reaction conditions

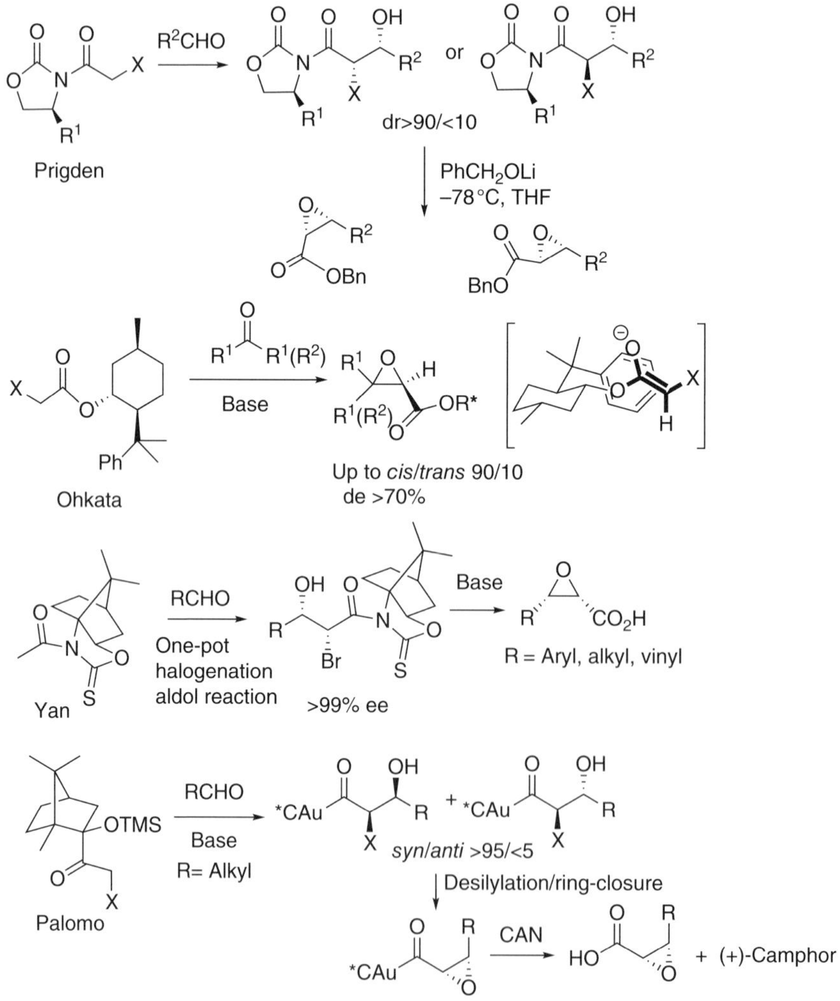

Scheme 3.129 Representative examples of Darzens reactions using chiral auxiliaries.

and nature of the aldehyde, *syn*-aldol (with boron enolates) or *anti*-aldol (with tin(IV)) enolates were preferentially obtained with good-to-high diastereoselectivity (Scheme 3.129). The following base-catalyzed ring closure and chiral auxiliary removal afforded the enantiomerically enriched *cis*- or *trans*-epoxides without epimerization. 8-Phenylmenthyl halogenoacetates were described by Ohkata and coauthors to react, under basic conditions, with ketones to preferentially give *cis*-epoxides, whereas when reacted with asymmetrically substituted ketones, moderate-to-high diastereocontrol was observed [413]. Modest results were achieved using aldehydes as the reagents. The face control in the enolate attack was explained by face shielding provided by the cumyl group via π–π interactions. Camphor-based thioamide has been applied by Yan's group in a one-pot halogenation/aldol reaction sequence with aromatic, alkyl, and vinyl aldehydes to enable highly diastereoselective formation of the *syn*-halo aldols [414]. Treatment under basic aqueous conditions afforded the *cis*-epoxy acids or eventually their esters. The camphor auxiliary has been applied to develop a diastereoselective aldol reaction with aliphatic aldehydes by Palomo and coworkers, giving the haloaldols with high *syn*-selectivity [415]. By following a silyl deprotection and base-catalyzed ring-closure procedure, *cis*-epoxides were finally obtained. In this case, camphor auxiliary was removed under oxidative conditions with cerium (IV) ammonium nitrate (CAN) to give the *cis*-glycidic acids and recoverable (+)-camphor. It has to be noted that the camphor-auxiliary-derived Darzens reactions illustrated in Scheme 3.129 are synthetically complementary to *trans*-selective sulfur-ylide-mediated epoxidation of carbonyl compounds.

In 2004, Ghosh and coworkers synthesized a *cis*-1-tosylamido-2-indanyl chloroacetate as the starting ester involved in Ti-catalyzed aldol reaction with aliphatic aldehydes and benzaldehyde (Scheme 3.130) [416]. The aldol reaction

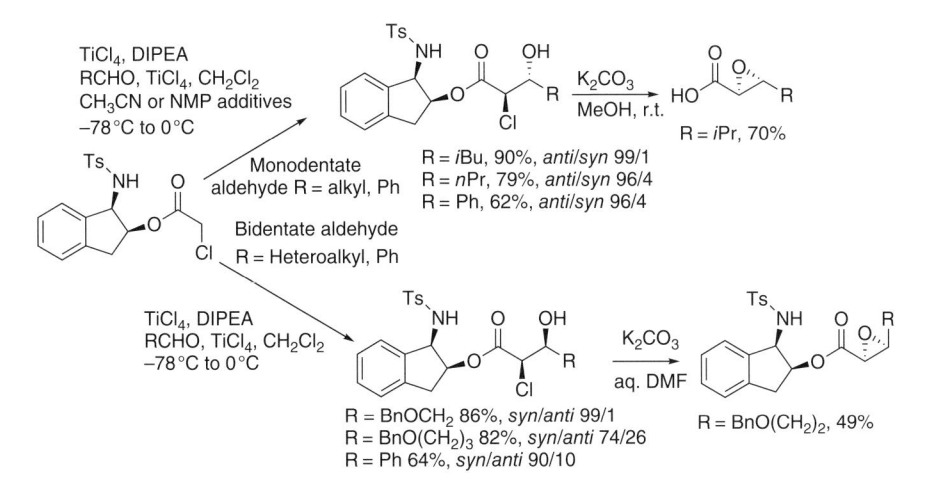

Scheme 3.130 Darzens reaction using a *cis*-1-tosylamido-2-indanyl-based chiral auxiliary.

was sensitive to the presence of additives and nature of the aldehyde. In the presence of acetonitrile or NMP as additive, simple aliphatic aldehydes and benzaldehydes afforded the haloaldols in satisfactory yield and with high *anti*-selectivity. Alternatively, in the absence of additives, heteroaliphatic aldehydes, able to achieve better complexation with the metal, achieved preferential *syn*-selectivity in the product. Matching and mismatching effects when using chiral racemic starting aldehydes were also observed and usefully applied for a kinetic resolution of racemic aldehydes. Epoxide formation and removal of the chiral auxiliary were performed by reacting the aldols with potassium carbonate in methanol to give the epoxy acids, whereas exclusive epoxy ester formation could be obtained using DMF as the solvent. The use of achiral haloesters and aldehydes in the presence of a stoichiometric loading of a chiral controller has been similarly investigated to set up an asymmetric Darzens reaction. This route has the advantage of reducing the synthetic steps avoiding the installation and removal of the chiral auxiliary on the halo derivative. The precursor of the chiral controller can be generally recovered, thus making the process more convenient.

C_2-Symmetric chiral borane **154** was employed as useful reagent to promote the aldol reaction of *tert*-butylbromoacetate with aliphatic, aromatic, and α,β-unsaturated aldehydes with high diastereo- and enantioselectivity (Scheme 3.131) [417]. The *anti*-aldols were converted into *trans*-glycidic esters under basic conditions. An interesting approach to the Darzens reaction has seen the application of the (S)-valine-derived oxazaborolidinone **155** to stoichiometrically promote the aldol reaction of β-bromo-β-methylketene silyl acetal with a few aldehydes, affording the haloaldols bearing a quaternary stereocenter in high *syn*-selectivity with excellent enantioselectivity (Scheme 3.131) [418]. The usual base-catalyzed epoxide formation enabled the access to highly enantioenriched trisubstituted *cis*-epoxy esters.

Recently, steroidal haloketones have been used as reagents in the Darzens reaction with aromatic aldehydes in water in the presence of Li_2CO_3 as the base, Aliquat 336 as phase-transfer catalyst, and granular polytetrafluoroethylene under mechanical stirring (Scheme 3.132) [419]. The reactions proceeded faster and more selectively in water compared to organic solvents. Different enantiopure steroidal haloketones reacted with aromatic aldehydes, affording the spiroepoxides with high level of diastereoselectivity. The protocol is very practical as the epoxides were filtered and crystallized from the reaction mixture.

3.7.2 Catalytic Asymmetric Darzens Reactions

The first methods illustrating a catalytic asymmetric Darzens reaction relied on the employment of PTCs [420]. In these reactions, α-halogenated esters were usually reacted with aldehydes with a chiral PTC under biphasic conditions with carbonates of hydroxides as the bases to give the final epoxides. PTCs derived from cinchona and ephedra alkaloids were initially pioneered for this reaction as attested by the work of Wynberg's and Colonna's groups, respectively, although with unsatisfactory results (ee <20%) [421]. A deep investigation on the ability of differently substituted ammonium salts derived from cinchonine and quinine has been

Scheme 3.131 Darzens reaction using boron-derived chiral promoters.

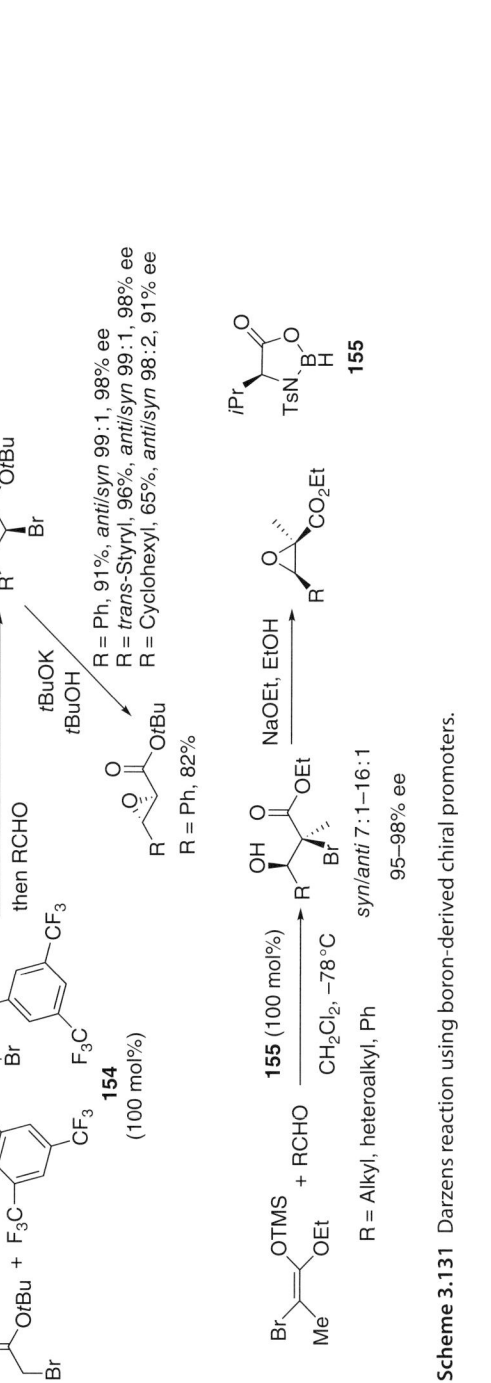

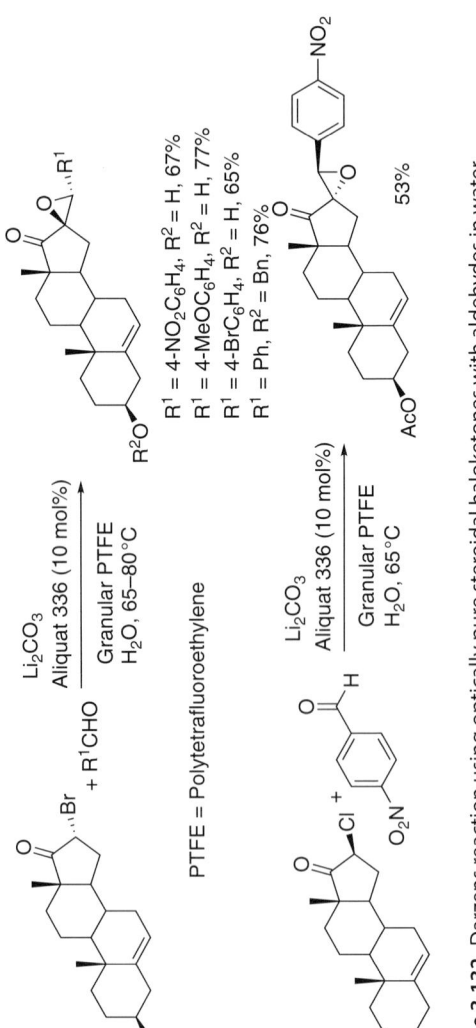

R¹ = 4-NO₂C₆H₄, R² = H, 67%
R¹ = 4-MeOC₆H₄, R² = H, 77%
R¹ = 4-BrC₆H₄, R² = H, 65%
R¹ = Ph, R² = Bn, 76%

PTFE = Polytetrafluoroethylene

Scheme 3.132 Darzens reaction using optically pure steroidal haloketones with aldehydes in water.

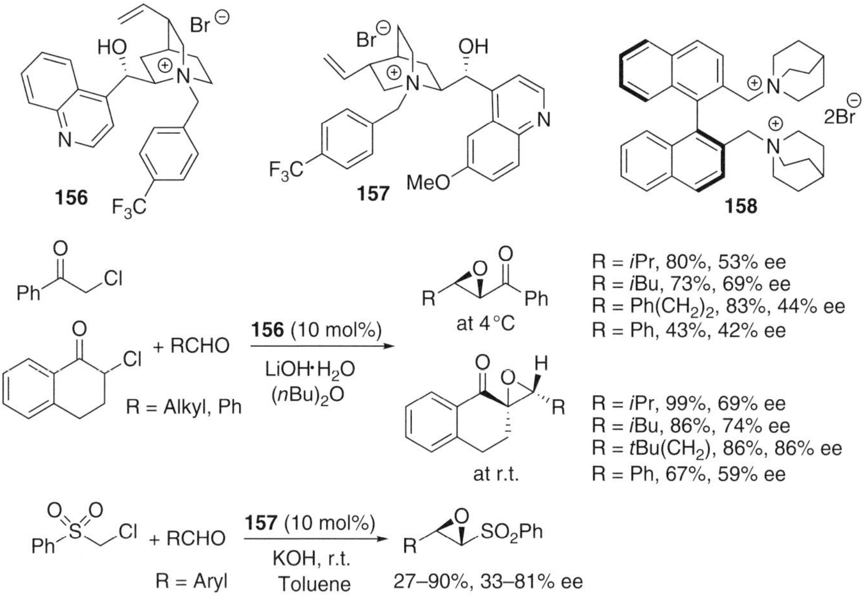

Scheme 3.133 Asymmetric catalytic Darzens reaction using cinchona-alkaloid-derived quaternary ammonium salts.

reported by Arai's group (Scheme 3.133). Ammonium salt **156**, used at 10 mol% loading, catalyzed the Darzens reaction of acyclic [422] and cyclic [423] chloroketones with aliphatic aldehydes and benzaldehydes in the presence of LiOH·H$_2$O as the base. The acyclic chloroketones gave the corresponding *trans*-epoxides in good yield and with modest enantioselectivity (up to 69% ee). Consistently, better yield and enantioselectivity were observed with the cyclic chloroketone as the reagent (up to 86% ee). The importance of the secondary OH group of the catalyst to achieve a satisfactory level of enantioselectivity was ascertained. Mechanistically, it has been demonstrated that a combination of the retro-aldol reaction and kinetic resolution of the intermediate haloaldols governed the stereochemical outcome of the reaction in the case of acyclic chloroketone. Chloromethyl phenyl sulfone was also tested with aromatic aldehydes in the presence of ammonium salt **157** with KOH as the base, leading to *trans*-epoxysulfones in moderate-to-good yield and enantioselectivity [424]. More recently, N-(2,3,4-trifluorobenzyl)quinidinium bromide showed to significantly improve the level of enantioselectivity (71–97% ee) when used as catalyst in the Darzens reaction to access *trans*-epoxysulfones [425]. Arai and coworkers synthesized a BINOL-derived bis-ammonium salt **158**, which was used in the Darzens reaction of haloamides with aldehydes, leading the corresponding epoxides with modest control of the diastereo- (*cis/trans* up to 8:1) and enantioselectivity (up to 70% ee) [426].

In 2011, Deng and coworkers developed a highly enantioselective Darzens reaction of wide general scope by reacting α-chloroketones and aromatic aldehydes with catalytic loading of cupreinium salt **159** in dichloromethane at 0 °C (Scheme 3.134) [427]. This catalyst, bearing a phenolic OH group and the

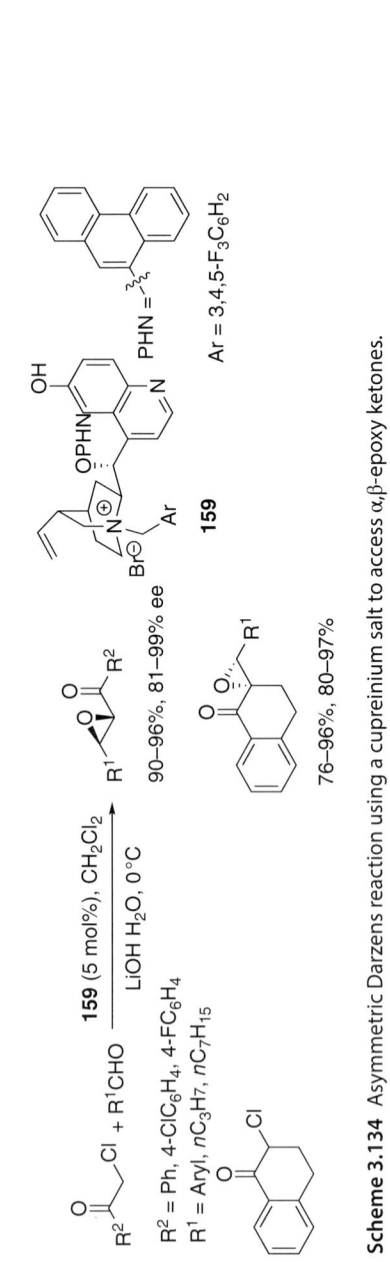

Scheme 3.134 Asymmetric Darzens reaction using a cupreinium salt to access α,β-epoxy ketones.

C-9 OH, protected with a sterically demanding 9-phenanthracenyl group, enabled the formation of *trans*-epoxy chalcones in excellent yield with high enantioselectivity (up to 99% ee). An aliphatic aldehyde afforded the epoxide with appreciable 81% ee value. Less-reactive cyclic ketone delivered the *trans*-epoxides with similar efficiency using 10 mol% of catalyst **159**. The crucial role played by the phenolic OH group in enhancing the activity and asymmetric induction, likely via H-bonding with the electrophile as an additional and directing effect besides ion-pairing interactions, has been demonstrated.

Chiral crown ethers represent another class of PTCs used in the Darzens reaction, firstly reported at the end of the 1990s by the group of Tőke and Bakó [428]. More recently, the same group illustrated an in-depth study on the scope of the Darzens reaction catalyzed by crown ethers bearing pendant sugars (Scheme 3.135). Catalyst **160**, which was used in PTC-catalyzed asymmetric epoxidation of α,β-unsaturated ketones (see Section 3.4.1), promoted the reaction of some chloroketones with aromatic and heteroaromatic aldehydes in the presence of 30% water solution of NaOH in toluene [429]. The epoxides were recovered with complete diastereoselectivity with moderate-to-good ee values (up to 96% ee), which were significantly dependent on the substitution pattern of the aromatic aldehyde.

Aminocatalysis can be exploited as an additional approach to the Darzens reaction, by taking advantage of L-proline-derived compounds to act as catalysts in aldol reactions [430]. Specifically, the Nájera's group developed a direct aldol reaction between α-chloroacetone and aromatic aldehydes by employing 10 mol% of recoverable (*S*)-BINAM-L-prolinamide **161**, in combination with benzoic acid as co-catalyst (Scheme 3.136) [431]. The reaction can be performed in different solvents, including water. After optimization of the aldol reaction to selectively afford the *anti*-chloro aldols with good regio-, diastero-, and enantioselectivity in DMF/H_2O as medium, the crude reaction mixture was directly treated with Et_3N in CH_2Cl_2 to give *trans*-epoxy ketones in satisfactory overall yield and with good-to-high enantioselectivity. In these examples, the feasibility of an asymmetric Darzens reaction even starting from acyclic aliphatic haloketones has been demonstrated.

Watanabe and coauthors employed a Lewis acid/Brønsted base catalyst formed by C_2-symmetric chiral selenide bearing the isoborneol skeleton, readily prepared from (1*S*)-10-camphor sulfonic acid to promote an asymmetric Darzens reaction [432]. Phenacyl bromide was reacted with a few aromatic aldehydes in the presence of 30 mol% of selenide **162** and LiOH as the base in $CHCl_3$ to furnish the *trans*-oxiranes in moderate yield and with up to 62% ee (Scheme 3.137). To figure out if the reaction proceeded via a Darzens type mechanism or a selenium-derived ylide, the corresponding selenonium salt, intermediate to ylide formation, was synthesized and treated under the same reaction conditions reported in Scheme 3.137 with benzaldehyde. However, only traces of racemic epoxide were detected at the end of the reaction. Hence, compound **162**, bearing Brønsted base and Lewis acid features, is likely to serve as a catalyst that is able to activate the reagents in a Darzens-type mechanism, via formation of a hypervalent organoselenium species.

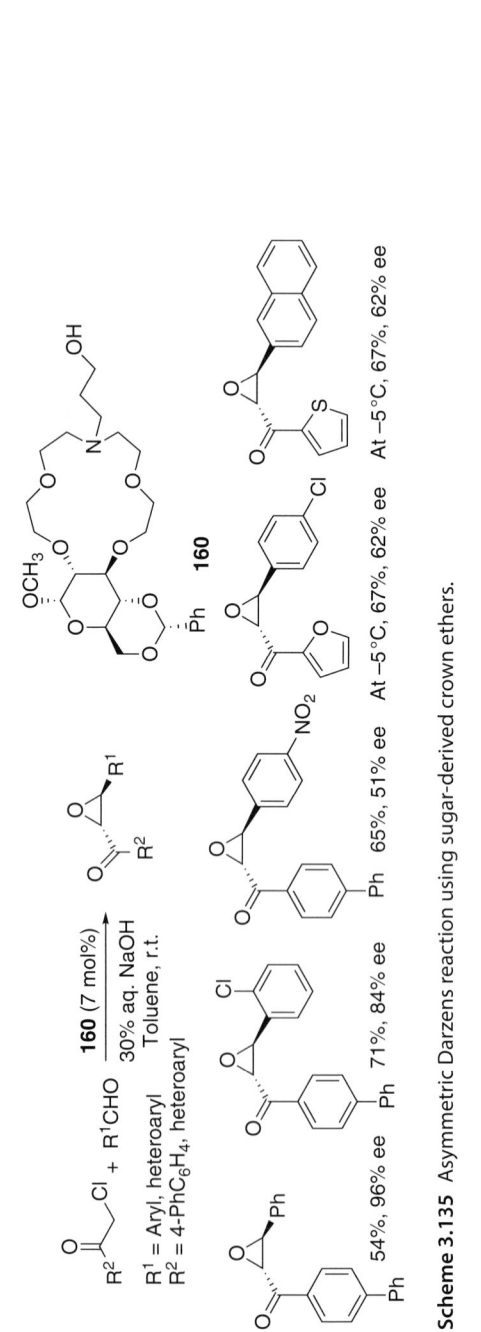

Scheme 3.135 Asymmetric Darzens reaction using sugar-derived crown ethers.

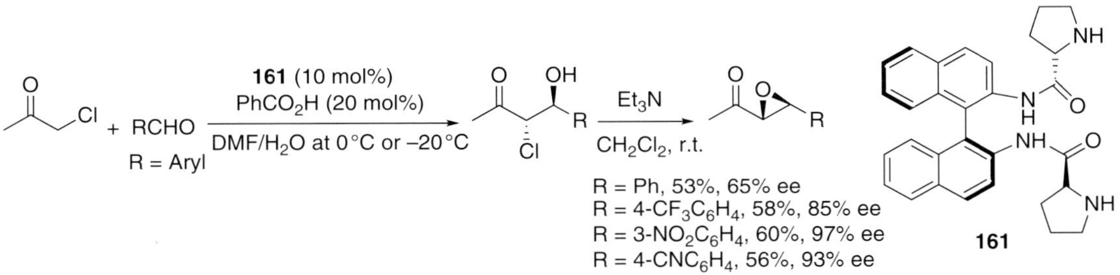

Scheme 3.136 Asymmetric Darzens reaction via amino-catalyzed aldol reaction/ring-closure sequence.

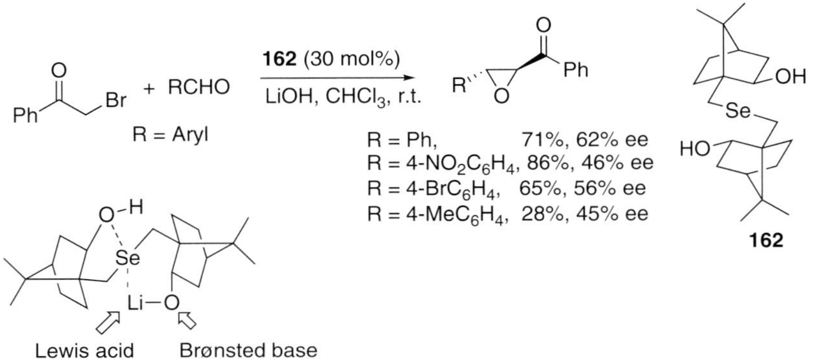

Scheme 3.137 Asymmetric Darzens reaction via C_2-symmetric chiral selenide.

North and coworkers were the first to report an asymmetric metal-catalyzed Darzens reaction. They used catalytic loadings of a Co(II)–salen complex with haloamides and aldehydes in the presence of hydroxides as the bases at room temperature. Although variable, moderate *trans/cis* ratios were observed for the epoxides, which were isolated in good yield but with low-to-modest enantioselectivity (up to 50% ee) [433]. A notable example of metal-catalyzed Darzens reaction has been described by Gong's group in 2009 [434]. An *in situ* formed complex by mixing commercially available Ti(O*i*Pr)$_4$ and (*R*)-BINOL catalyzed the Darzens reaction of diazoacetamides with a large variety of aromatic, heteroaromatic, aliphatic, and unsaturated aldehydes in dichloromethane as the solvent at 0 °C in the presence of molecular sieves (Scheme 3.138). A complete control of the diastereoselectivity was observed for a great variety of aldehydes, and the *cis*-glycidic amides were isolated in high yield with excellent enantioselectivity. This is a first general method reported to access challenging *cis*-glycidic amides with excellent stereocontrol. This class of epoxides are particularly useful for further elaboration as demonstrated in the formal synthesis of protease inhibitor (−)-bestatin. The Gong group envisaged that a more practical Darzens reaction to produce *cis*-glycidic amides could be developed by using an air-stable and storable chiral zirconium Lewis acid catalyst [435]. Hence, the catalyst, preformed according to a previous report by mixing Zr(O*n*Bu)$_4$ and (*R*)-3,3′-diiodo-BINOL **163**, was used at 10 mol% loading with a variety of aromatic, heteroaromatic, and aliphatic aldehydes and *N*-phenyl-diazoacetamide in acetonitrile at room temperature to exclusively give the *cis*-glycidic amides in high-to-excellent yield and enantioselectivity (Scheme 3.138). Interestingly, the product of opposite absolute configuration was obtained starting from the same *R*-configured BINOL. A gram-scale synthesis of a model glycidic amide demonstrated the robustness of the protocol. In 2013, Sun and coworkers investigated the same system by testing different easily accessible diols from the chiral pool as the ligands [436]. The (+)-pinane diol **164** has been found to be the most efficient, and working under the conditions previously reported by Gong, *cis*-glycidic amides were isolated with complete diastereoselectivity (Scheme 3.138). Concerning the enantioselectivity, slightly better ee values were detected when using aliphatic aldehydes, but lower values when reacting aromatic aldehydes.

In 2014, Feng and coworkers developed an asymmetric Darzens reaction between phenacyl bromides and N-protected isatins to synthesize spiroepoxyoxindoles [437]. The optically active *trans*-products were obtained in moderate-to-good yield and enantioselectivity when catalyzed by chiral *N,N*′-dioxide–Co(acac)$_2$ complex **165** in the presence of K$_3$PO$_4$/K$_2$HPO$_4$, molecular sieves as additives, working in THF/acetone mixture at −30 °C (Scheme 3.139). The substitution pattern on the aromatic group of the α-haloketone as well as the type of protecting group of isatins significantly influenced the enantiocontrol. Moreover, fine-tuning of the reaction conditions was necessary in order to suppress deleterious base-catalyzed background reaction. In order to have some clues on the stereochemical outcome of the reaction, a model intermediate bromoaldol was synthesized in racemic and enantioenriched forms and in relative (*S**,*S**)-configuration. By treatment under usual or achiral basic conditions, a retro-aldol process

Scheme 3.138 Asymmetric Darzens reaction to *cis*-glycidic amides catalyzed by Ti(O*i*Pr)$_4$/(*R*)-BINOL.

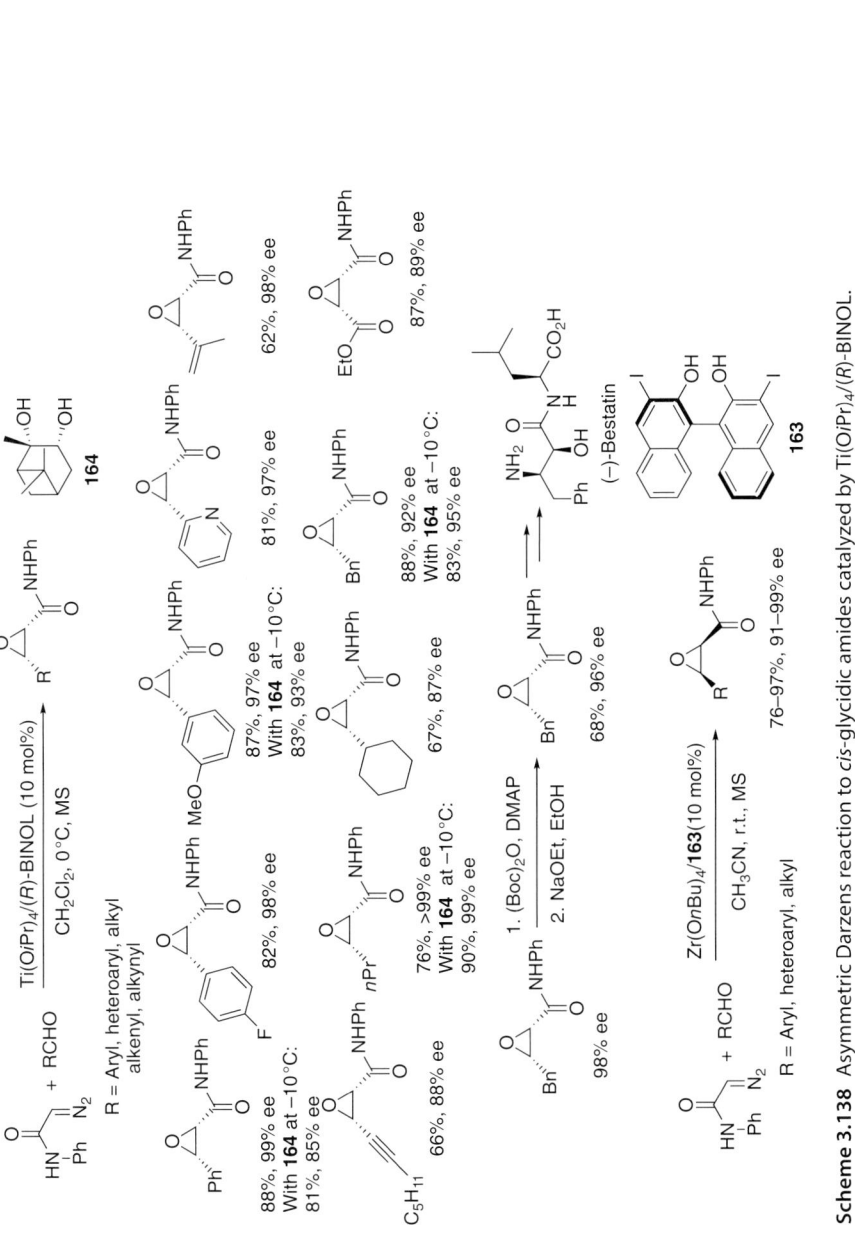

Scheme 3.139 Asymmetric Darzens reaction to *trans*-spiroepoxyindoles catalyzed by Co(acac)$_2$/ligand system.

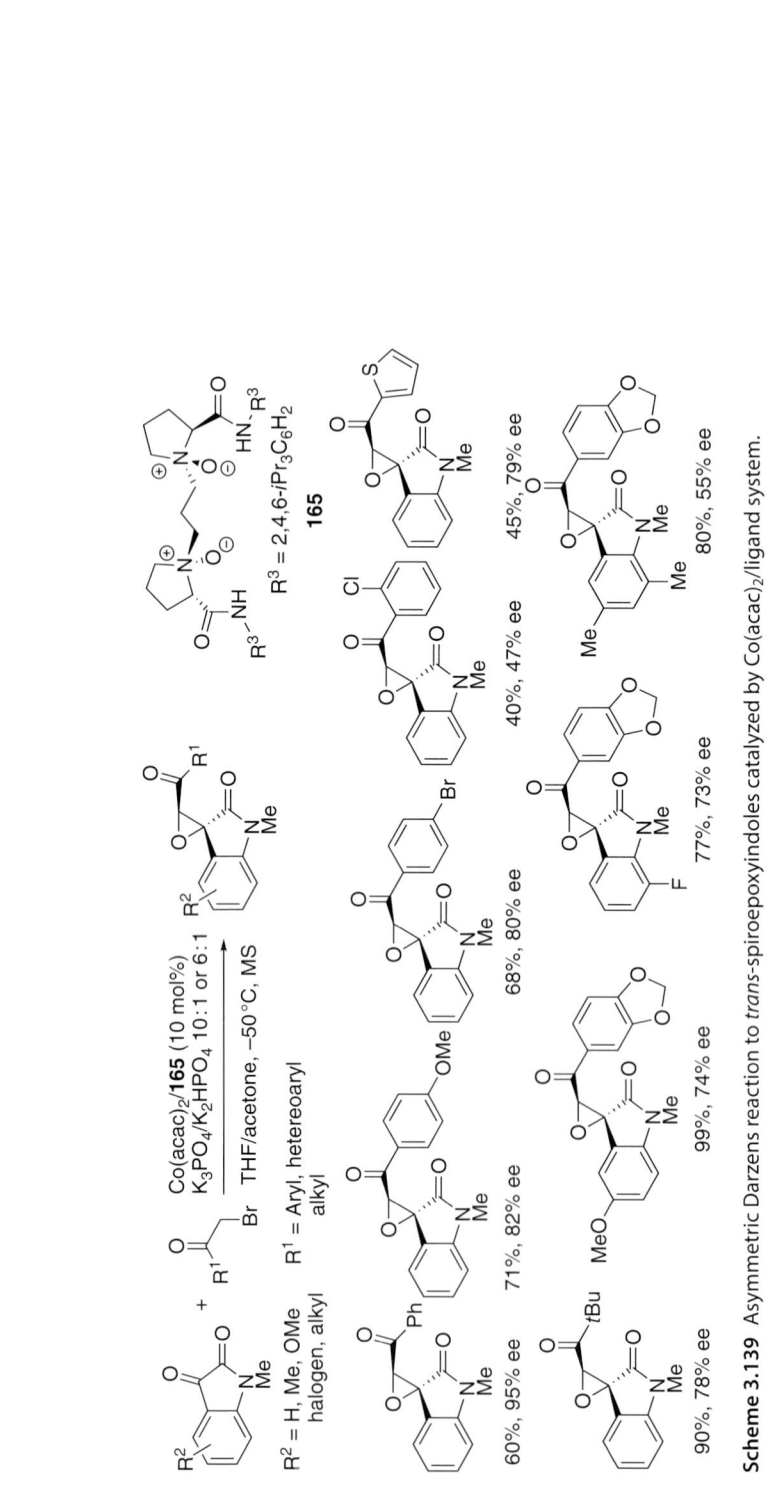

accompanying the ring-closure step was observed, and it was assessed that the initial aldol addition was the stereoselectivity-determining step. The Darzens reaction illustrated in Scheme 3.139 is an asymmetric catalytic process to access synthetically useful *trans*-spiroepoxindoles, which widens the structural diversity of the epoxides obtainable via the stoichiometric approach involving sulfur ylide **148** depicted in Scheme 3.122.

NHCs are efficiently used as ligands in metal catalysis [438] and as organocatalysts [439] to promote a variety of stereoselective carbon–carbon, carbon–oxygen, and carbon–nitrogen bond-forming reactions, involving umpolung of the carbonyl carbon atom functional group, then acting as acyl anion equivalents. In 2015, Sudalai and coworkers revealed an interesting novel example of Darzens reaction catalyzed by NHCs, based on an oxidative coupling of styrenes or alternatively α-bromoacetophenones with aldehydes to give α,β-epoxy ketones [440]. The reaction was initially optimized to set up a new highly diastereoselective methodology to access racemic *trans*-α,β-epoxy ketones in satisfactory yield, which required the employment of styrenes, aromatic or aliphatic aldehydes as reagents, stoichiometric amount of N-bromosuccinimide (NBS) and DBU in the presence of 10 mol% of NHC precatalyst **166** in dimethyl sulfoxide (DMSO) as the solvent (Scheme 3.140).

Mechanistic investigations enabled detection of α-bromoacetophenones as intermediates formed during the reaction, and consequently, a classical Darzens-type reaction starting from α-bromoacetophenones and aldehydes as reagents working under catalytic loading of **166** and DBU was efficiently developed. A first example of an asymmetric version of the epoxidation was performed with *cis*-indanol amine-derived NHC precatalyst **167** and DBU both at 10 mol% loadings, reacting phenacyl bromide and 4-nitrobenzaldehyde under usual conditions. The final epoxide was isolated in acceptable yield and with 56% ee. A ketodeoxy Breslow intermediate was also isolated during the investigations, and taking all the data together, a mechanistic picture of the catalytic cycle was formulated involving the oxidative generation of phenacyl bromides when starting with alkenes as the reagents. The NHC prefers to react with phenacyl bromide to give the salt then transformed into the ketodeoxy Breslow intermediate, which attacks the aldehyde, followed by regeneration of NHC and epoxide release. Further improvements on the scope and stereocontrol of an asymmetric version of this novel epoxidation are expected in the future, given the rich variety of chiral NHC precatalysts available for testing.

3.8 Other Ylide-Mediated Epoxidations

The synthetic usefulness of nitrogen-based ylides [441] has been successfully documented in the asymmetric preparation of cyclopropanes by Gaunt's work with cinchona-alkaloid-derived ammonium salts as the precursors [442]. However, nitrogen-based ylides are low-reactive species, which significantly discouraged studies for their application in epoxide and aziridine synthesis. Indeed, Aggarwal's group highlighted a critical issue in ylide reactivity in the leaving-group ability [443]. A scale of onium group decreasing leaving-group ability can

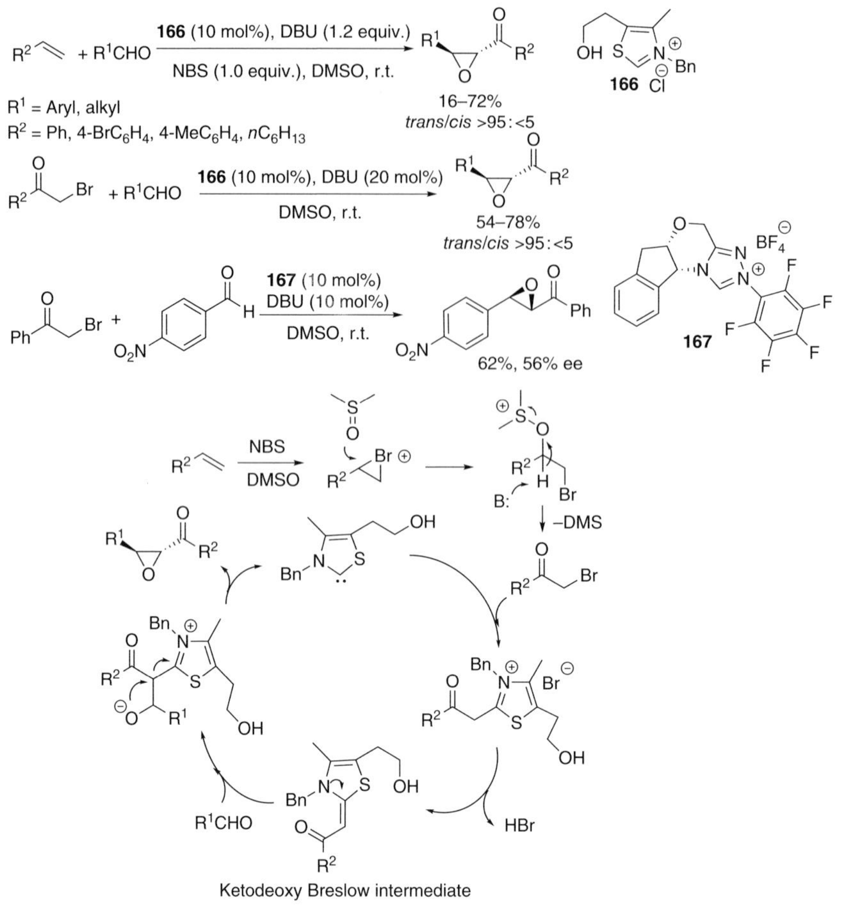

Scheme 3.140 Darzens epoxidation via *N*-heterocyclic carbenes.

be traced in the order of O > S > N > P, to end up, in the case of phosphonium ylides, with a mechanistic change taking place with the Wittig reaction. Consequently, a high energetic barrier is involved in the ring-closure step to epoxides when using nitrogen-based ylides (Scheme 3.121) [444]. A successful example of an asymmetric epoxidation of aromatic aldehydes, based on the generation of a nitrogen ylide from brucine benzyl ammonium chlorides, has been illustrated by Kimachi's group in 2010 (Scheme 3.141) [445]. A one-pot procedure was developed, which did not require the isolation of the ammonium salt, as its stoichiometric *in situ* generation was followed by basic treatment with *t*BuOK at low temperature, leading to a better control of the stereoselectivity. The reaction proved to be sensitive to the substitution pattern in the benzyl chloride used, as only electron-withdrawing groups were tolerated. Better yield and stereoselectivity were obtained when using aromatic aldehydes bearing electron-donating substituents. Epoxides were isolated in modest-to-satisfactory yield preferentially as *trans*-isomers, with moderate to fairly good level of enantioselectivity (up to 84% ee).

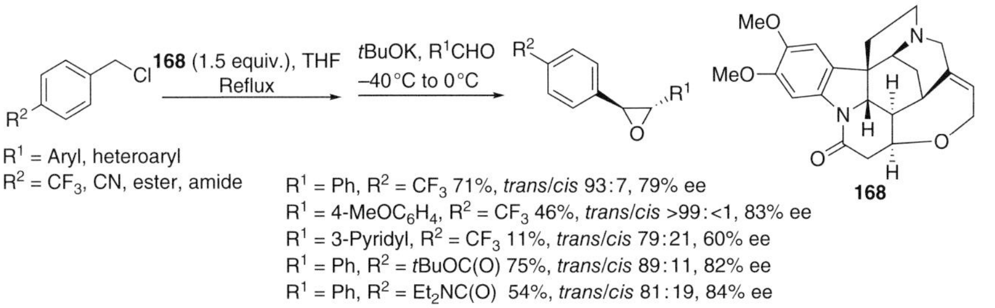

Scheme 3.141 Asymmetric epoxidation of aldehydes via ylide derived from a chiral ammonium salt of brucine.

Although this methodology is limited in scope, the results are encouraging, and they will likely inspire future investigations.

3.9 Asymmetric Biocatalyzed Synthesis of Epoxides

In Nature, several enzymes, distributed in plants and animals, catalyze the synthesis of optically active epoxide in highly stereoselective manner. This is certainly attested by a significant number of natural products, bearing chiral nonracemic oxirane units in their scaffold. The asymmetric epoxidation of alkenes is catalyzed by monooxygenases, namely chloroperoxidases (CPO), whereas epoxide hydrolases (EH) operate the HKR of racemic epoxides [446]. These biocatalysts can be extremely useful to access epoxides difficult to obtain by classical chemical methods with high enantioselectivity. The substrate scope of monooxygenases and hydrolases encompass several structurally different epoxides obtainable in virtually enantiomerically pure form. According to what is generally observed in biocatalysis, high substrate specificity and reagent solubility in the water medium are principal disadvantages associated with the choice of enzymes for the synthesis of a target epoxide. Monooxygenases, which are involved in many biological processes, comprising degradation of aromatic compounds and drug detoxification, work in the presence of molecular oxygen and NADPH (nicotinamide adenine dinucleotide phosphate) as the reducing agent. The heme-iron or FAD (flavin adenine dinucleotide) is the most common cofactor of monooxygenases, necessary for the activation of molecular oxygen. Among them, the most effective are styrene monooxygenase (SMO) and xylene monooxygenase (XMO). Styrene monooxygenase contains a two-component flavoenzyme composed of a FAD-specific styrene epoxidase and NADH-specific flavin reductase (dihydronicotinamide adenine dinucleotide). SMO catalyzes the epoxidation under mild conditions with high regio- and enantioselectivity, using molecular oxygen as the most environmentally friendly and cheap oxidant. SMO serves in epoxidation of styrene to produce (*S*)-styrene oxide in >99% ee [447], applied at pilot-scale production [448]. SMOs similarly efficiently epoxidize styrenes bearing different substitution patterns in the aromatic ring, although sterically demanding or electron-withdrawing groups are not tolerated [447a, 449]. In 2011, Wu's group demonstrated that a novel SMO enzyme, designated as StyAB2 contained in recombinant *Escherichia coli* BL21 cells, catalyzed the epoxidation other than styrenes and specifically the epoxidation of racemic α-substituted secondary allylic alcohols (Scheme 3.142) [450].

The epoxidation of racemic secondary alcohols proceeded via kinetic resolution in high yield, with good-to-high preference for the *erythro*-epoxide and generally excellent enantioselectivity for aromatic epoxy alcohols. The unreacted (*R*)-allylic alcohol was also recovered with high ee value. The methodology could not be applied to *ortho*-substituted aromatic allylic alcohols, which did not react, whereas somewhat inferior stereoselectivity was observed with aliphatic derivatives. This methodology is a valuable alternative to the Ti–tartrate-catalyzed or vanadium/hydroxamic acid-catalyzed epoxidations, although the access to the enantiomeric series of the epoxy alcohols showed in Scheme 3.142 is not feasible by using biocatalysis. The same group ingeniously developed a one-pot sequential reduction/epoxidation of starting α,β-unsaturated ketones in order to obtain the final epoxides in higher yields (>50% theoretical yield of a kinetic resolution) [451]. First, an (*S*)-specific ketoreductase *Ch*KRED03 was employed to produce (*S*)-allylic alcohols in nearly enantiopure form, followed by the addition of SMO, able to epoxidize them with excellent levels of diastereo- and enantioselectivity (Scheme 3.142). Very recently, they developed further this one-pot biocatalyzed process to set up a formal asymmetric epoxidation of α,β-unsaturated ketones with excellent stereocontrol via a reduction/epoxidation/dehydrogenation cascade (Scheme 3.143) [452]. A recombinant *E. coli* co-expressing carbonyl reductase (READH) and SMO (REAB) was developed as a highly effective biosystem, which can be chemically switched by modulation of

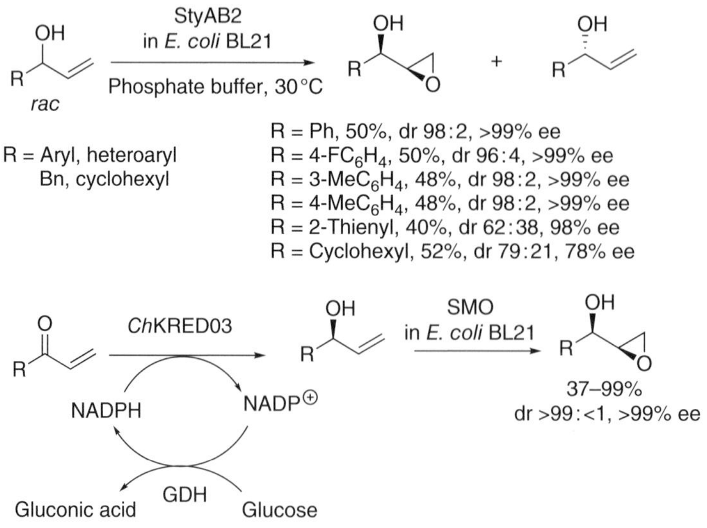

Scheme 3.142 Asymmetric epoxidation of racemic secondary allylic alcohols catalyzed by StyAB2.

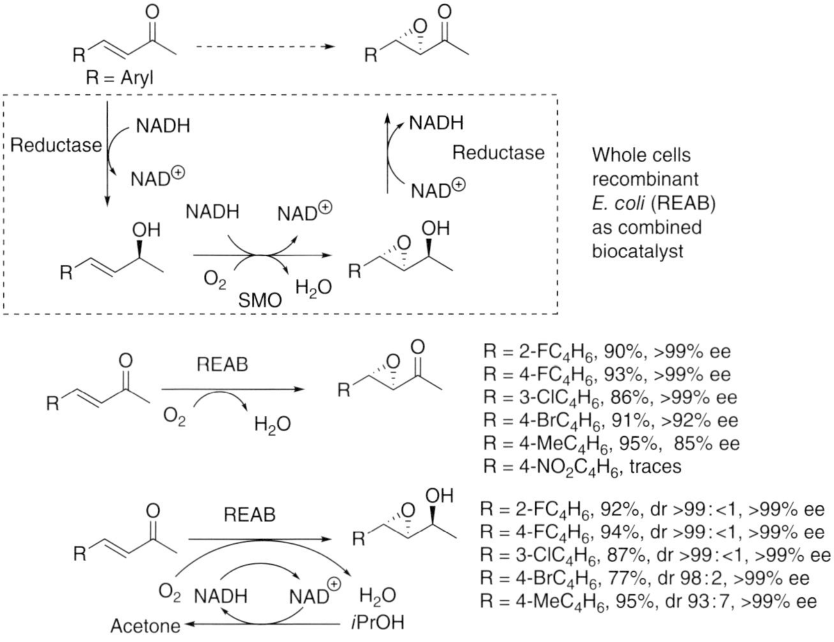

Scheme 3.143 Asymmetric switchable bioepoxidation of α,β-unsaturated ketones.

NADH and *i*PrOH to produce either epoxy ketones or allylic epoxy alcohols in excellent yield and with up to >99% ee values.

Electron-withdrawing groups such as halogen atoms are tolerated at different positions in the β-aromatic ring, with the exception of the nitro group, similarly heteroaromatic groups at the β-position of the enone. The same group applied the StyAB2 to synthesize a range of chiral epoxides from styrene derivatives (Scheme 3.144) [453]. Increasing the sterics of the substituents placed at the α- or β-position of the carbon–carbon double bond played a major role in decreasing the conversion to the epoxide product. Interestingly, *E*-isomer was obtained with a remarkable level of enantioselectivity when compared to chemical methodologies.

In 2013, Li and coworkers developed a novel enzyme, triple mutant of P450pyr monooxygenase (P450pyrTM), particularly effective for the epoxidation of *para*-substituted styrenes with electron-withdrawing groups, thus greatly improving the previous performance of SMO (Scheme 3.145) [454]. The (*R*)-epoxides were isolated in high yield with excellent enantioselectivity. Interestingly, when starting from *ortho*-substituted styrenes, although the conversions to the product decreased significantly, the (*S*)-epoxides were also recovered with excellent ee values. These results are notable as these substituted styrene epoxides are important pharmaceutical intermediates that are not accessible using other asymmetric epoxidation reactions.

Another useful class of enzymes are CPOs, and among them, the most studied is CPO EC 1.11.1.10 from the fungus *Caldariomyces fumago*. Besides halogenation of

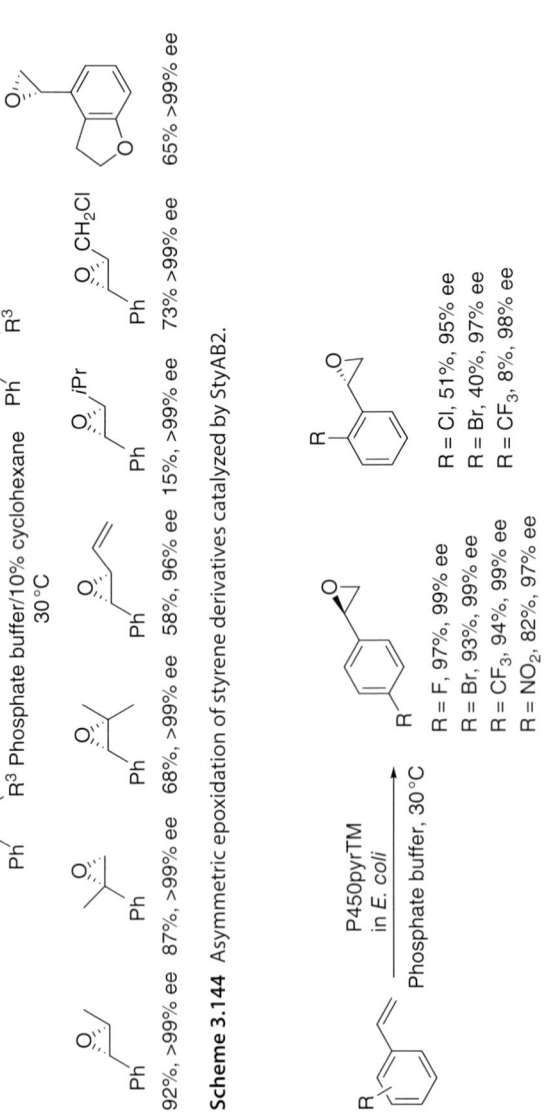

Scheme 3.144 Asymmetric epoxidation of styrene derivatives catalyzed by StyAB2.

Scheme 3.145 Asymmetric epoxidation of styrene derivatives catalyzed by P450pyrTM.

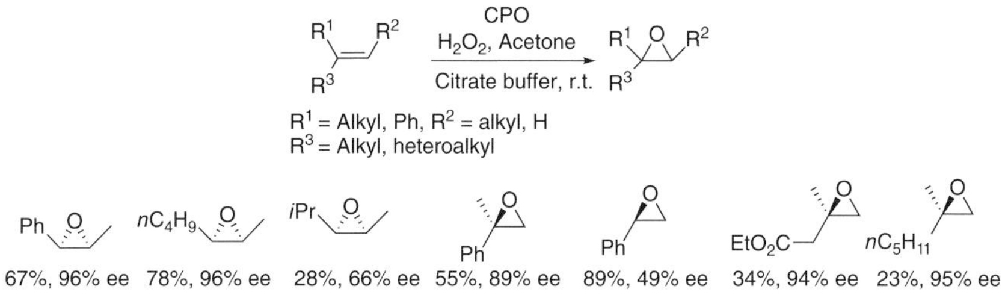

Scheme 3.146 Asymmetric epoxidation of alkenes catalyzed by CPO.

Reaction: R¹R²C=CR³ with CPO, H_2O_2, Acetone, Citrate buffer, r.t. → epoxide

R¹ = Alkyl, Ph, R² = alkyl, H
R³ = Alkyl, heteroalkyl

Products:
- Ph–CH(O)CH–: 67%, 96% ee
- nC_4H_9–CH(O)CH–: 78%, 96% ee
- iPr–CH(O)CH–: 28%, 66% ee
- Ph-substituted epoxide: 55%, 89% ee
- Ph-substituted epoxide: 89%, 49% ee
- EtO_2C–: 34%, 94% ee
- nC_5H_{11}–: 23%, 95% ee

compounds, CPO mediates hydroxylation, epoxidation, sulfoxidation reactions when H_2O_2 and halide ions are present in solution [455]. In the 1990s, the groups of Jacobsen and Hager showed the synthetic potential of CPO to catalyze the asymmetric epoxidation of Z-alkenes and terminal disubstituted olefins [456]. The substrate scope of CPO has demonstrated to be restricted and susceptible to fine steric effects (Scheme 3.146). High enantioselectivity was observed with a methyl substituent on the starting Z-alkene, and the other group has to be not sterically demanding in order to observe a satisfactory conversion and ee value. Terminal monosubstituted alkenes are epoxidized with low efficiency by CPO, whereas when disubstituted with a methyl group, better yield and high enantioselectivity were attained.

EHs are cofactor-independent ubiquitous enzymes, involved in the kinetic resolution of racemic epoxides. Moreover, optically active 1,2-diols can also be obtained by exploiting enantioconvergent hydrolysis of racemic epoxides [457]. In the 1990s, studies were focused on the characterization of these microbial enzymes to elucidate the mechanism of hydrolysis, and from the synthetic point of view, they were applied as catalysts in the kinetic resolution of terminal epoxides, for instance, styrene oxides, 2,2-disubstituted epoxides, and aryl glycidyl ethers [458]. A two-step mechanism has been suggested for the catalysis, where the epoxide would be activated via H-bonding by two tyrosine residues of the reactive site, whereas a nucleophilic aspartate group would attack the epoxide at the least-hindered carbon atom. Then, the ester intermediate complex is hydrolyzed. In most of the examples, low activities of the whole cell prevented a facile preparative-scale application of EHs for industrial production. In the epoxide hydrolysis of aryl glycidyl ethers, protein engineering has been applied to enhance the activity and enantioselectivity [459]. Recently, a new epoxide hydrolase (BMEH), cloned from *Bacillus megaterium* ECU1001, showed high activity and uncommon selectivity for (R)-enantiomer of the starting epoxide, with respect to most of the known EHs (Scheme 3.147) [460]. The reaction time ranges from 2 min to 2 h for

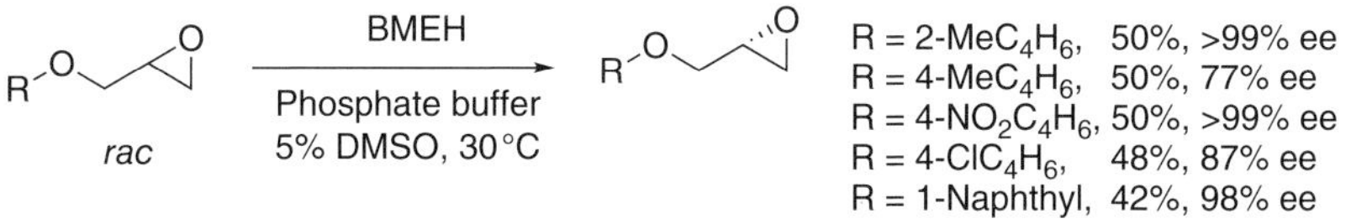

Reaction: rac R–O–CH₂–epoxide → BMEH, Phosphate buffer, 5% DMSO, 30°C → R–O–CH₂–epoxide

- R = 2-MeC_4H_6, 50%, >99% ee
- R = 4-MeC_4H_6, 50%, 77% ee
- R = 4-$NO_2C_4H_6$, 50%, >99% ee
- R = 4-ClC_4H_6, 48%, 87% ee
- R = 1-Naphthyl, 42%, 98% ee

Scheme 3.147 Epoxide-hydrolase-catalyzed kinetic resolution of aryl glycidyl ethers.

optimum conversion around 50% of a variety of aryl glycidyl ethers, generally isolated with very high ee values. Semipreparative 200 mg scale resolutions of racemic epoxides using the crude enzyme were successfully performed affording (S)-epoxides as useful starting material for the synthesis of β-blockers.

An interesting example of EH in substrate-specific application has been reported by Furstoss and coworkers in the kinetic resolution of 1-chloro-2-(2,4-difluorophenyl)-2,3-epoxypropane by using commercially available *Aspergillus niger* EH (Scheme 3.148) [461].

The (S)-epoxide and the (R)-diol are building blocks for the synthesis of new triazole drug derivative, namely D0870, known to display activity against candidiasis-type infections. The hydrolytic kinetic resolution of 1-chloro-2-(2,4-difluorophenyl)-2,3-epoxypropane proceeded in plain water, at room temperature, using a two-phase reactor at preparative 2 g scale, affording the epoxide and the diol in fairly good yield and with excellent enantioselectivity. The process was also demonstrated to be enantioconvergent as the (R)-diol was chemically transformed into the (S)-epoxide, thus obtained in nearly quantitative yield.

The same group developed the kinetic resolution of *trans*-spiroepoxide **169**, a product of interest as building block for the synthesis of 11-heterosteroids. The kinetic resolution was performed at 1g scale using two enantiocomplementary enzymes, *A. niger* EH and limonene 1,2-epoxide hydrolase (LEH) [462]. *A. niger* EH led to the recovery of (R,R)-epoxide **169**, whereas LEH afforded (S,S)-epoxide **169** both in comparable yield and with excellent ee values (>98% ee). It is interesting to note that LEH behaved with opposite regiocontrol when compared to common EHs as the attack of water occurred at the most sterically demanding carbon atom, giving diol (R,R)-**170** (Scheme 3.149).

Recently, LEHs showing complementary stereoselectivity were recombinantly expressed in *E. coli* and used in a highly convenient procedure for the kinetic resolution of (+)-*cis/trans* limonene oxide and (−)-*cis/trans*-limonene oxide to prepare all limonene oxide enantiomers [463]. 1,2-Disubstituted epoxides have been occasionally investigated as substrates in the EH-catalyzed kinetic resolution in place of common monosubstituted epoxides. Kotik and coworkers analyzed two hydrolases from a metagenomic DNA library of biofilter-derived biomass (Scheme 3.150) [464].

A Kau2 EH overexpressed in *E. coli* cells was found to be highly selective in a preparative-scale hydrolysis of racemic *trans*-1-phenyl-1,2-epoxypropane **171**,

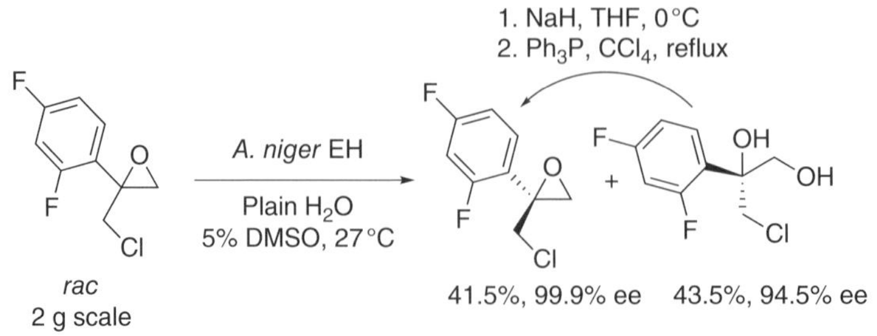

Scheme 3.148 Epoxide-hydrolase-catalyzed kinetic resolution of a 2,2-disubstituted epoxide.

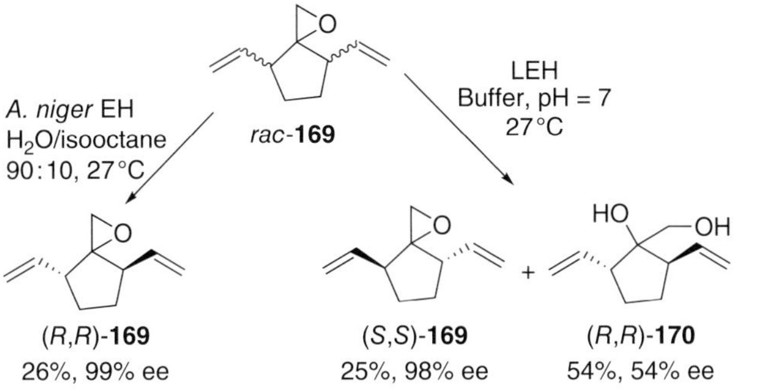

Scheme 3.149 Epoxide hydrolase catalyzed kinetic resolution of *trans*-spiroepoxide catalyzed by A. niger EH and LEH.

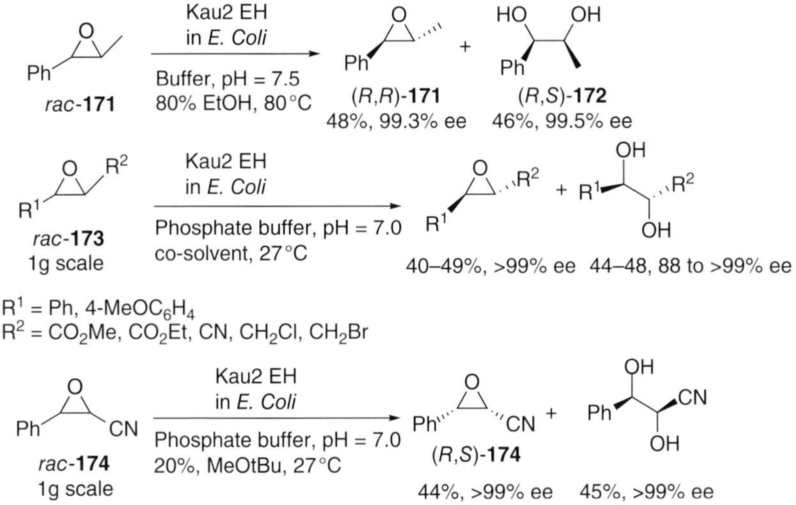

Scheme 3.150 Epoxide-hydrolase-catalyzed kinetic resolution of simple and functionalized *trans*-epoxides.

where (R,R)-epoxide and (R,S)-diol **172** were both produced in excellent yield and with >99% ee value. The same group recently employed Kau2 EH in the hydrolysis of racemic *trans*-disubstituted electron-poor alkenes **173**, namely *trans*-3-phenyl glycidates, *trans*-3-phenyl oxirane-2-carbonitrile, and *trans*-2-(halomethyl)-3-phenyloxiranes (Scheme 3.150) [465]. Highly valuable epoxides and 1,2-diols were isolated in excellent yield and enantiomerically pure form, working at 1g preparative scale of the starting racemic epoxide. It is interesting to note that while *rac*-methyl *cis*-3-phenyl glycidate unexpectedly proved to be unreactive when treated under standard conditions, the corresponding *rac-cis*-3-phenyloxirane-2-carbonitrile **174** was converted into diol in 45% yield and with >99% ee, and (R,S)-epoxide **174** recovered in 44% yield and with >99% ee. This astonishing change in reactivity has been tentatively explained considering the polar nature of the ester group, as not being compatible with the active site of the Kau2 EH, with respect to less polar cyano, methyl, and phenyl groups.

Finally, another class of enzymes, which has been less explored in the asymmetric synthesis of epoxides, are halohydrin dehalogenases (HDHHs), bacterial enzymes (with *Agrobacterium radiobacter AD1* (HheC) being the most studied) able to catalyze the conversion of 1,2-halohydrins into epoxides releasing HCl by-product and the reverse reaction. They can also catalyze the kinetic resolution of racemic epoxides by enantioselective ring opening with different nucleophiles, such as cyanide, azide, or nitrite, providing important targets, such as optically active α-substituted alcohols [466]. In general, the HDHHs have been prevalently used to synthesize the α-substituted alcohols as the compounds of interest, also exploiting multistep biocatalyzed reactions [467]. The ability of HDHHs in the kinetic resolution of racemic epoxides is exemplified in Scheme 3.151, starting from terminal epoxide **175** treated with HheC in water with NaN_3 to yield unreacted (S)-epoxide **175** and the azido alcohol **176** with >99% ee and 96% ee, respectively [468]. The same process was recently exploited in a one-pot sequence, via click reaction of the enantioenriched azido alcohol with alkynes to prepare optically active triazoles as the final products [469].

It is interesting to point out that most of the processes illustrated in the previous Schemes operate with whole cells, without the need of expensive purification of the enzymes and under environmentally friendly reaction conditions. All these features make biocatalysis a highly appealing tool for the industrial production of optically active epoxides.

3.10 Conclusions

In the past decade, the area of asymmetric synthesis of epoxides has witnessed many developments. First of all, new chiral metal-based systems suitable to catalyze the epoxidation of functionalized alkenes such as allylic, homoallylic, and *bis*-homoallylic alcohols had grown on the grounds of the knowledge acquired on titanium and vanadium well-established protocols. In this respect, a great variety of functionalized epoxy alcohols, frequently used as building blocks in total synthesis of biologically active compounds, can be prepared with high level of stereocontrol. Significant results were also achieved in the employment of environmentally friendly and more abundant manganese and iron complexes with tetradentate nitrogen ligands in the epoxidation of unfunctionalized alk-

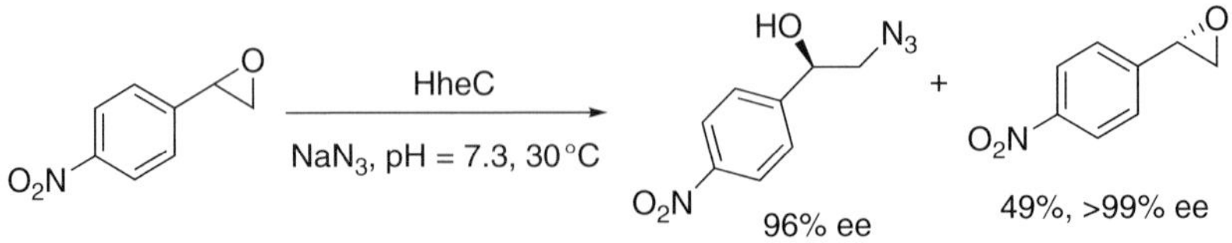

Scheme 3.151 Halohydrin-dehalogenase-catalyzed kinetic resolution of a terminal epoxide.

enes using hydrogen peroxide as a convenient oxidant. However, it appears still challenging to devise protocols of general scope and satisfactory level of enantioselectivity. Efforts in the area of manganese–salen-catalyzed epoxidation reactions were mainly focused on the immobilization of the catalysts for recycling, whereas important achievements have been observed in the Ti–salen, -salalen, and -salan complexes for the epoxidation of unfunctionalized and terminal alkenes. Novel metal–chiral ligand systems have expanded the arsenal of methods to perform the asymmetric epoxidation of an increasing number of olefins. However, the most notable advances have been observed in the organocatalyzed epoxidation reactions catalyzed by new effective scaffolds as PTCs and most of all secondary and primary amines. Indeed, for the first time, they can serve as catalyst to effect highly stereoselective epoxidation reactions of challenging α,β-unsaturated aldehydes, aliphatic and cyclic α,β-unsaturated ketones. Bifunctional chiral organic molecules are also employable in the asymmetric epoxidation of a good variety of electron-poor alkenes, including *trans*-di- and trisubstituted α,β-unsaturated ketones. The epoxidation of alkenes by chiral dioxiranes system has seen further improvements of the substrate scope, and it can be likely considered the most versatile and general method to apply in the asymmetric epoxidation of alkenes, clearly attested by numerous applications reporting this key step in total synthesis. Over the past years, the HKR performed on immobilized Co–salen catalysts has demonstrated to be the most useful and practical approach to obtain highly enantioenriched terminal epoxides.

Concerning nonoxidative methodologies, chiral sulfur-ylide-mediated epoxidations have reached a certain degree of maturity in terms of mechanistic knowledge of factors regulating the stereochemical outcome and structural requirements of the chiral sulfide. Efforts to make the catalytic version more effective and extension of substrate scope have improved the synthetic potential of this approach over the years, especially for the synthesis of *trans*-glycidic amides and 1,2-disubstituted aromatic epoxides. The Darzens reaction has been implemented, especially for the preparation of difficult-to-access α,β-epoxyesters and amides. However, synthetic application of this reaction still relies on chiral auxiliary- or chiral reagent-mediated approach, with only a handful of effective catalytic protocols. Consequently, there is plenty of space to await further developments. The biocatalyzed reactions showed their great utility as an additional opportunity to produce specific oxiranes. Terminal epoxides can be prevalently obtained with excellent ee values, working under simple and environmentally friendly conditions, by using whole cells as "catalysts" with the advantage of reducing costs in preparative synthesis. As a general remark, although nowadays chemists have different efficient tools to synthesize a large variety of chiral epoxides, they should design novel asymmetric catalytic systems suitable to produce *cis*-epoxides, tri- and tetrasubstituted epoxides, which still remain the most challenging classes of oxiranes to be obtained in stereocontrolled fashion. Finally, considering the increasing environmental concerns facing the productions of fine chemicals, forthcoming efforts in this area are expected to focus on the development of evermore practical (supported catalysts) and sustainable systems based on either low-cost, less toxic metals or readily accessible and renewable organocatalysts with hydrogen peroxide as the terminal oxidant.

References

1 Morgan, K.M., Ellis, J.A., Lee, J., Fulton, A., Wilson, S.L., Dupart, P.S., and Dastoori, R. (2013) *J. Org. Chem.*, 78, 4303–4311.

2 (a) Yudin, A. (ed.) (2006) *Aziridines and Epoxides in Organic Synthesis*, Chapters 7–9, Wiley-VCH Verlag GmbH, Weinheim; (b) Pineschi, M. (2006) *Eur. J. Org. Chem.*, 4979–4988; (c) Johnson, J.B. (2011) in *Science of Synthesis, Stereoselective Synthesis*, vol. 3 (eds J.G. de Vries, G.A. Molander, and P.A. Evans), Georg Thieme, Stuttgart, pp. 759–827.

3 (a) Heravi, M.M., Lashaki, T.B., and Poorahmad, N. (2015) *Tetrahedron: Asymmetry*, 26, 405–495; (b) Vilotijevic, I. and Jamison, T.F. (2011) in *Biomimetic Organic Synthesis*, Chapter 2, 1st edn (eds E. Poupon and B. Nay), Wiley-VCH Verlag GmbH, Weinheim, pp. 537–590; (c) Das, B. and Damodar, K. (2011) in *Heterocycles in Natural Product Synthesis*, 1st edn (eds K.C. Majumdar and S.K. Chattopadhyay), Wiley-VCH Verlag GmbH, Weinheim, pp. 63–95; (d) Vilotijevic, I. and Jamison, T.F. (2009) *Angew. Chem. Int. Ed.*, 48, 5250–5281; (e) Martín, T., Padrón, J.I., and Martín, V.S. (2014) *Synlett*, 25, 12–32.

4 (a) Jacobsen, E.N. and Wu, M.N. (1999) in *Comprehensive Asymmetric Catalysis I–III*, vol. 3 (eds E.N. Jacobsen, A. Pfaltz, and H. Yamamoto), Springer, Berlin, pp. 1309–1326; (b) Pastor, I.M. and Yus, M. (2005) *Curr. Org. Chem.*, 9, 1–29; (c) Schneider, C. (2006) *Synthesis*, 23, 3919–3944; (d) Matsunaga, S. (2012) in *Comprehensive Chirality*, vol. 5 (eds E. Carreira and H. Yamamoto), Elsevier, Oxford, pp. 534–579; (e) Wang, P.A. (2013) *Beilstein J. Org. Chem.*, 9, 1677–1695; For reviews on chiral ligands, see: (f) Mellah, M., Voituriez, A., and Schulz, E. (2007) *Chem. Rev.*, 107, 5133–5209; (g) Pellissier, H. (2007) *Tetrahedron*, 63, 1297–1330; (h) Zappia, G. (2012) in *Comprehensive Chirality*, vol. 3 (eds E. Carreira and H. Yamamoto), Elsevier, Oxford, pp. 408–485; (i) Kouklovsky, C. (2012) in *Comprehensive Chirality*, vol. 3 (eds E. Carreira and H. Yamamoto), Elsevier, Oxford, pp. 487–525; (j) Meninno, S. and Lattanzi, A. (2016) *Chem. Eur. J.*, 22, 3632–3642.

5 For reviews, see: (a) Sharpless, K.B., Behrens, C.H., Katsuki, T.L., Albert, W.M., Martin, V.S., Takatani, M., Viti, S.M., Walker, F.J., and Woodard, S.S. (1983) *Pure Appl. Chem.*, 55, 589–604; (b) Kumar, P., Naidu, V., and Gupta, P. (2007) *Tetrahedron*, 63, 2745–2785; (c) Keith, J.M., Larrow, J.F., and Jacobsen, E.N. (2001) *Adv. Synth. Catal.*, 343, 5–26; (d) Tokunaga, M., Larrow, J.F., Kakiuchi, F., and Jacobsen, E.N. (1997) *Science*, 277, 936–938.

6 Sonawane, S.P., Patil, G.D., and Gurjar, M.K. (2011) *Org. Process Res. Dev.*, 15, 1365–1370.

7 Wei, S., Messerer, R., and Tsogoeva, S.B. (2011) *Chem. Eur. J.*, 17, 14380–14384.

8 (a) Ryu, S.-E., Choi, H.-J., and Kim, D.H. (1997) *J. Am. Chem. Soc.*, 119, 38–41; (b) Lee, M. and Kim, D.H. (2002) *Bioorg. Med. Chem.*, 10, 913–922.

9 (a) Chen, W.L., Fu, X., Lin, L.L., Yuan, X., Luo, W.W., Feng, J.H., Liu, X.H., and Feng, X.M. (2014) *Chem. Commun.*, 50, 11480–11483; (b) Chen, W.L., Xia, Y., Lin, L.L., Yuan, X., Guo, S.S., Liu, X.H., and Feng, X.M. (2015) *Chem. Eur. J.*, 21, 15104–15107; (c) D'Elia, V., Pelletier, J.D.A., and Basset, J.-M. (2015) *ChemCatChem*, 7, 1906–1917; (d) Yuan, X., Lin, L., Chen, W., Wu, W., Liu, X., and Feng, X. (2016) *J. Org. Chem.*, 81, 1237–1243.

10 (a) Bierl, B.A., Beroza, M., and Collier, C.W. (1970) *Science*, 170, 87–89; (b) Prasad, K.R. and Anbarasan, P. (2007) *J. Org. Chem.*, 72, 3155–3157.
11 (a) Ayer, W.A., Lee, S.P., Tsuneda, A., and Hiratsuka, Y. (1980) *Can. J. Microbiol.*, 26, 766–773; (b) Garbaccio, R.M., Stachel, S.J., Baeschlin, D.K., and Danishefsky, S.J. (2001) *J. Am. Chem. Soc.*, 123, 10903–10908.
12 (a) García Ruano, J.L., Martin Castro, A.M., and Rodríguez, J.H. (1994) *J. Org. Chem.*, 59, 533–536; (b) Jiménez, O., Bosch, M.P., and Guerrero, A. (1997) *J. Org. Chem.*, 62, 3496–3499; (c) Bondzic, B.P., Urushima, T., Ishikawa, H., and Hayashi, Y. (2010) *Org. Lett.*, 12, 5434–5437.
13 (a) Yamaguchi, J. and Hayashi, Y. (2011) *Chem. Eur. J.*, 16, 3884–3901; (b) Nicolaou, K.C., Roschangar, F., and Vourloumis, D. (1998) *Angew. Chem. Int. Ed. Engl.*, 37, 2014–2045.
14 (a) Bäckvall, J.-E. (ed.) (2010) *Modern Oxidation Methods*, Chapters 2–4, 6, 7, 9, 11, Wiley-VCH Verlag GmbH, Weinheim; (b) Müller, C., Grover, N., Cokoja, M., and Kühn, F.E. (2013) in *Advances in Inorganic Chemistry*, vol. 65, Chapter 2 (eds R. van Eldik and C.D. Hubbard), Elsevier, Oxford, pp. 33–83; (c) Amini, M., Haghdoost, M.M., and Bagherzadeh, M. (2014) *Coord. Chem. Rev.*, 268, 83–100; (d) Huber, S., Cokoja, M., and Kühn, F.E. (2014) *J. Organomet. Chem.*, 751, 25–32; (e) Kück, J.W., Reich, R.M., and Kühn, F.E. (2016) *Chem. Rec.*, 16, 349–364; (f) Saisaha, P., de Boer, J.W., and Browne, W.R. (2013) *Chem. Soc. Rev.*, 42, 2059–2074.
15 Davies, S.G., Fletcher, A.M., and Thomson, J.E. (2014) *Org. Biomol. Chem.*, 12, 4544–4549.
16 Gnas, Y. and Glorius, F. (2006) *Synthesis*, 12, 1899–1930.
17 For a review on chiral auxiliary controlled epoxidations, see: Adam, W. and Zhang, A. (2005) *Synlett*, 1047–1072.
18 (a) Pastor, A., Adam, W., Wirth, T., and Tóth, G. (2005) *Eur. J. Org. Chem.*, 3075–3084; (b) Adam, W., Pastor, A., Peters, K., and Peters, E.-M. (2000) *Org. Lett.*, 2, 1019–1022; For diastereoselective epoxidation of chiral enecarbamates, see: (c) Adam, W., Bosio, S.G., Turro, N.J., and Wolff, B.T. (2004) *J. Org. Chem.*, 69, 1704–1715.
19 Zhang, S.-J., Chen, Y.-K., Li, H.-M., Huang, W.-Y., Rogatchov, V., and Metz, P. (2006) *Chin. J. Chem.*, 24, 681–688.
20 For reviews, see: (a) Jacobsen, E.N. and Wu, M.N. (1999) in *Comprehensive Asymmetric Catalysis I–III*, vol. 2 (eds E.N. Jacobsen, A. Pfaltz, and H. Yamamoto), Springer, Berlin, pp. 649–677; (b) Bäckvall, E.J. (ed.) (2004) *Modern Oxidation Methods*, Chapters 2, 3, 10, Wiley-VCH Verlag GmbH, Weinheim; (c) Matsumoto, K. and Katsuki, T. (2009) in *Asymmetric Synthesis*, 2nd edn (eds M. Christmann and S. Brase), Wiley-VCH Verlag GmbH, Weinheim, pp. 123–127; (d) Burke, A.J. and Carreiro, E.P. (2013) in *Comprehensive Inorganic Chemistry II* (eds J. Reedijk and K. Poeppelmeier), Elsevier, Amsterdam, pp. 309–382.
21 Katsuki, T. and Sharpless, K.B. (1980) *J. Am. Chem. Soc.*, 102, 5974–5976.
22 (a) Hanson, R.M. and Sharpless, K.B. (1986) *J. Org. Chem.*, 51, 1922–1925; (b) Gao, Y., Klunder, J.M., Hanson, R.M., Ko, S.Y., Masamune, H., and Sharpless, K.B. (1987) *J. Am. Chem. Soc.*, 109, 5765–5780; (c) Gawley, R.E. and Aubé, J. (1996) *Principles of Asymmetric Synthesis, Tetrahedron Organic Chemistry Series*, vol. 14, Chapter 8, Pergamon, Oxford, pp. 325–363; (d) Johnson, R.A.

and Sharpless, K.B. (2000) in *Catalytic Asymmetric Synthesis*, Chapter 6A, 2nd edn (ed. I. Ojima), Wiley-VCH Verlag GmbH, Weinheim, pp. 231–280; (e) Johnson, R.A. and Sharpless, K.B. (1991) in *Comprehensive Organic Synthesis*, vol. 7 (eds B.M. Trost and I. Fleming), Pergamon Press, Oxford, pp. 389–436.

23 (a) Martin, V.S., Woodard, S.S., Katsuki, T., Yamada, Y., Ikeda, M., and Sharpless, K.B. (1981) *J. Am. Chem. Soc.*, 103, 6237–6240; (b) Lumbroso, A., Cooke, M.L., and Breit, B. (2013) *Angew. Chem. Int. Ed.*, 52, 1890–1932.

24 (a) Carlier, P.R., Mungail, W.S., Schroder, G., and Sharpless, K.B. (1988) *J. Am. Chem. Soc.*, 110, 2978–2979; (b) McKee, B.H., Kalantar, T.H., and Sharpless, K.B. (1991) *J. Org. Chem.*, 56, 6966–6968; For examples of kinetic resolution, see: (c) Kagan, H.B. and Fiaud, J.C. (1988) *Top. Stereochem.*, 18, 249–330.

25 Schreiber, S.L., Schreiber, T.S., and Smith, D.B. (1987) *J. Am. Chem. Soc.*, 109, 1525–1529.

26 (a) Woodard, S.S., Finn, M.G., and Sharpless, K.B. (1991) *J. Am. Chem. Soc.*, 113, 106–113; (b) Finn, M.G. and Sharpless, K.B. (1991) *J. Am. Chem. Soc.*, 113, 113–126; (c) Wu, Y.-D. and Lai, D.K.W. (1995) *J. Am. Chem. Soc.*, 117, 11327–11336; (d) Bilenko, V., Jiao, H., Spannenberg, A., Fischer, C., Reinke, H., Kösters, J., Komarov, I., and Börner, A. (2007) *Eur. J. Org. Chem.*, 758–767.

27 Rossiter, B.E. and Sharpless, K.B. (1984) *J. Org. Chem.*, 49, 3707–3711.

28 For recent reviews, see: (a) Riera, A. and Moreno, M. (2010) *Molecules*, 15, 1041–1073; (b) Farina, V., Reeves, J.T., Senanayake, C.H., and Song, J.J. (2006) *Chem. Rev.*, 106, 2734–2793.

29 Carreras, J., Livendahl, M., McGonigal, P.R., and Echavarren, A.M. (2013) *Angew. Chem. Int. Ed.*, 53, 4896–4899.

30 (a) Moreira, I.C., Lago, J.H.G., Young, N.C.M., and Roque, N.F. (2003) *J. Braz. Chem. Soc.*, 14, 828–831; (b) Gaspar-Marques, C., Simões, M.F., and Rodríguez, B. (2004) *J. Nat. Prod.*, 67, 614–621; (c) De Tommasi, N., Pizza, C., Conti, C., Orsi, N., and Stein, M.L. (1990) *J. Nat. Prod.*, 53, 830–835.

31 Viswanadh, N., Mujumdar, P., Sasikumar, M., Kunte, S.S., and Muthukrishnan, M. (2016) *Tetrahedron Lett.*, 57, 861–863.

32 Voight, E.A., Yin, H., Downing, S.V., Calad, S.A., Matsuhashi, H., Giordano, I., Hennessy, A.J., Goodman, R.M., and Wood, J.L. (2010) *Org. Lett.*, 12, 3422–3425.

33 Bhadra, S., Akakura, M., and Yamamoto, H. (2015) *J. Am. Chem. Soc.*, 137, 15612–15615.

34 (a) Arends, I.W.C.E. and Sheldon, R.A. (2002) *Top. Catal.*, 19, 133–141; (b) Grigoropoulou, G., Clark, J.H., and Elings, J.A. (2003) *Green Chem.*, 5, 1–7; For reviews on asymmetric epoxidation and oxidation using this oxidant, see: (c) De Faveri, G., Ilyashenko, G., and Watkinson, M. (2011) *Chem. Soc. Rev.*, 40, 1722–1760; (d) Russo, A., De Fusco, C., and Lattanzi, A. (2012) *ChemCatChem*, 4, 901–916; (e) Wang, C. and Yamamoto, H. (2015) *Chem. Asian J.*, 10, 2056–2068.

35 (a) Massumoto, K., Sawada, Y., Saito, K., and Katsuki, T. (2005) *Angew. Chem. Int. Ed.*, 44, 4935–4939; (b) Sawada, Y., Matsumoto, K., and Katsuki, T. (2007) *Angew. Chem. Int. Ed.*, 46, 4559–4561.

36 Saito, B. and Katsuki, T. (2001) *Tetrahedron Lett.*, 42, 3873–3876.

37 Matsumoto, K., Kubo, T., and Katsuki, T. (2009) *Chem. Eur. J.*, 15, 6573–6575.

38 (a) Sawada, Y., Matsumoto, K., Kondo, S., Watanabe, H., Ozawa, T., Suzuki, K., Saito, B., and Katsuki, T. (2006) *Angew. Chem. Int. Ed.*, 45, 3478–3480; For

reviews, see: (b) Matsumoto, K., Sawada, Y., and Katsuki, T. (2008) *Pure Appl. Chem.*, 80, 1071–1077; (c) Matsumoto, K. and Katsuki, T. (2010) in *Catalytic Asymmetric Synthesis*, 3rd edn (ed. I. Ojima), John Wiley & Sons, Inc., Hoboken, NJ, pp. 839–890.

39 (a) Matsumoto, K., Sawada, Y., and Katsuki, T. (2006) *Synlett*, 3545–3547; (b) Shimada, Y., Kondo, S., Ohara, Y., Matsumoto, K., and Katsuki, T. (2007) *Synlett*, 2445–2447.

40 Berkessel, A., Bradenburg, M., Leitterstorf, E., Frey, J., and Schäfer, M. (2007) *Adv. Synth. Catal.*, 349, 2385–2391.

41 Berkessel, A., Bradenburg, M., and Schäfer, M. (2008) *Adv. Synth. Catal.*, 350, 1287–1294.

42 Kondo, S., Saruhasi, K., Kenichi, S., Matsubara, K., Miyaji, K., Kubo, T., Matsumoto, K., and Katsuki, T. (2008) *Angew. Chem. Int. Ed.*, 47, 10195–10198.

43 Matsumoto, K., Oguma, T., and Katsuki, T. (2009) *Angew. Chem. Int. Ed.*, 48, 7432–7435.

44 Ikegami, S., Katsuki, T., and Yamaguchi, M. (1987) *Chem. Lett.*, 16, 83–84.

45 For application in oxidative desymmetrization, see: Kramer, R., Berkenbusch, T., and Brückner, R. (2008) *Adv. Synth. Catal.*, 350, 1131–1148.

46 Okachi, T., Murai, N., and Onaka, M. (2003) *Org. Lett.*, 5, 85–87.

47 Li, Z. and Yamamoto, H. (2010) *J. Am. Chem. Soc.*, 132, 7878–7880.

48 Makita, N., Hoshino, Y., and Yamamoto, H. (2003) *Angew. Chem. Int. Ed.*, 42, 941–943.

49 Jang, J.-H., Asami, Y., Jang, J.-P., Kim, S.-O., Moon, D.O., Shin, K.-S., Hashizume, D., Muroi, M., Saito, T., Oh, H., Kim, B.Y., Osada, H., and Ahn, J.S. (2011) *J. Am. Chem. Soc.*, 133, 6865–6867.

50 Kohyama, A., Kanoh, N., Kwon, E., and Iwabuchi, Y. (2016) *Tetrahedron Lett.*, 57, 517–519.

51 Olivares-Romero, J.L., Li, Z., and Yamamoto, H. (2012) *J. Am. Chem. Soc.*, 134, 5440–5443.

52 Olivares-Romero, J.L., Li, Z., and Yamamoto, H. (2013) *J. Am. Chem. Soc.*, 135, 3411–3413.

53 Luo, L. and Yamamoto, H. (2015) *Org. Biomol. Chem.*, 13, 10466–10470.

54 Michaelson, R.C., Palermo, R.E., and Sharpless, K.B. (1977) *J. Am. Chem. Soc.*, 99, 1990–1992.

55 Bryliakov, K.P., Talsi, E.P., Kühn, T., and Bolm, C. (2003) *New J. Chem.*, 27, 609–614.

56 (a) Bolm, C. and Kühn, T. (2000) *Synlett*, 899–901; (b) Bolm, C. and Kühn, T. (2001) *Isr. J. Chem.*, 41, 263–269.

57 Murase, N., Hoshino, Y., Oishi, M., and Yamamoto, H. (1999) *J. Org. Chem.*, 64, 338–339.

58 Hoshino, Y. and Yamamoto, H. (2000) *J. Am. Chem. Soc.*, 122, 10452–10453.

59 Wu, H.-L. and Uang, B.-J. (2002) *Tetrahedron: Asymmetry*, 13, 2625–2628.

60 (a) Adam, W., Beck, A.K., Pichota, A., Chantu, R., Saha-Möller, C.R., Seebach, D., Vogl, N., and Zhang, R. (2003) *Tetrahedron: Asymmetry*, 14, 1355–1361; (b) Lattanzi, A., Piccirillo, S., and Scettri, A. (2005) *Eur. J. Org. Chem.*, 1669–1674; (c) Lattanzi, A. and Scettri, A. (2006) *J. Organomet. Chem.*, 691, 2072–2082.

61 Zhang, W., Basak, A., Kosugi, Y., Hoshino, Y., and Yamamoto, H. (2005) *Angew. Chem. Int. Ed.*, 44, 4389–4391.
62 Li, Z., Zhang, W., and Yamamoto, H. (2008) *Angew. Chem. Int. Ed.*, 47, 7520–7522.
63 Zhang, W. and Yamamoto, H. (2007) *J. Am. Chem. Soc.*, 129, 286–287.
64 Noji, M., Kobayashi, T., Uechi, Y., Kikuchi, A., Kondo, H., Sugiyama, S., and Ishii, K. (2015) *J. Org. Chem.*, 80, 3203–3210.
65 Tietze, L.F., Brasche, G., and Gericke, K. (2006) *Domino Reactions in Organic Synthesis*, Wiley-VCH Verlag GmbH, Weinheim.
66 Han, L., Liu, C., Zhang, W., Shi, X.-X., and You, S.-L. (2014) *Chem. Commun.*, 50, 1231–1233.
67 Han, L., Zhang, W., Shi, X.-X., and You, S.-L. (2015) *Adv. Synth. Catal.*, 357, 3064–3068.
68 (a) Bouranhi, Z. and Malkov, A.V. (2005) *Chem. Commun.*, 4592–4594; (b) Bouranhi, Z. and Malkov, A.V. (2006) *Synlett*, 3525–3528.
69 Malkov, A.V., Czemerys, L., and Malyshev, D.A. (2009) *J. Org. Chem.*, 74, 3350–3355.
70 Wang, J., Zhao, L., Shi, H., and He, J. (2011) *Angew. Chem. Int. Ed.*, 50, 9171–9176.
71 (a) Zhao, L.-W., Shi, H.-M., Wang, J.-Z., and He, J. (2012) *Chem. Eur. J.*, 18, 9911–9918; (b) Zhao, L.-W., Shi, H.-M., An, Z., Wang, J.-Z., and He, J. (2013) *Chem. Eur. J.*, 19, 12350–12355.
72 Li, F., Wang, Z.-H., Zhao, L., Xiong, F.-J., He, Q.-Q., and Chen, F.-E. (2011) *Tetrahedron: Asymmetry*, 22, 1337–1341.
73 Egami, H. and Katsuki, T. (2008) *Angew. Chem. Int. Ed.*, 47, 5171–5174.
74 Egami, H., Ogumu, T., and Katsuki, T. (2010) *J. Am. Chem. Soc.*, 132, 5886–5895.
75 Meunier, D., Piechaczyk, A., de Mallmann, A., and Basset, J.-M. (1999) *Angew. Chem. Int. Ed.*, 38, 3540–3542.
76 For a review, see: McGarrigle, E.M. and Gilheany, D.G. (2005) *Chem. Rev.*, 105, 1563–1602.
77 Daly, A.M., Renehan, M.F., and Gilheany, D.G. (2001) *Org. Lett.*, 3, 663–666.
78 Imanishi, H. and Katsuki, T. (1997) *Tetrahedron Lett.*, 38, 251–254.
79 Yamada, S.-i., Mashiko, T., and Terashima, S. (1977) *J. Am. Chem. Soc.*, 99, 1988–1990.
80 Kagan, H.B., Mimoun, H., Mark, C., and Schurig, V. (1979) *Angew. Chem. Int. Ed. Engl.*, 18, 485–486.
81 For reviews, see: (a) Burke, A.J. (2008) *Coord. Chem. Rev.*, 252, 170–175; (b) Jain, K.R., Herrmann, W.A., and Kühn, F.E. (2008) *Coord. Chem. Rev.*, 252, 556–568; (c) Sanz, R. and Pedrosa, M.R. (2009) *Curr. Org. Synth.*, 6, 239–263.
82 (a) Barlan, A.U., Basak, A., and Yamamoto, H. (2006) *Angew. Chem. Int. Ed.*, 45, 5849–5852; (b) Barlan, A.U., Zhang, W., and Yamamoto, H. (2007) *Tetrahedron*, 63, 6075–6087.
83 (a) Sharpless, K.B., Townsend, J.M., and Williams, D.R. (1972) *J. Am. Chem. Soc.*, 94, 295–296; (b) Chong, A.O. and Sharpless, K.B. (1977) *J. Org. Chem.*, 42, 1587–1590.
84 Wang, Y., Wu, Z., Li, Z., and Zhou, X.-G. (2009) *Tetrahedron Lett.*, 50, 2509–2511.

85 (a) Costa, A.P., Reis, P.M., Gamelas, C., Romão, C.C., and Royo, B. (2008) *Inorg. Chim. Acta*, 361, 1915–1921; (b) Neves, P., Gago, S., Pereira, C.C.L., Figueiredo, S., Lemos, A., Lopes, A.D., Gonçalves, I.S., Pillinger, M., Silva, C.M., and Valent, A.A. (2009) *Catal. Lett.*, 132, 94–103; (c) Reis, P.M., Gamelas, C.A., Brito, J.A., Saffon, N., Gómez, M., and Royo, B. (2011) *Eur. J. Inorg. Chem.*, 2011, 666–673.

86 Herrmann, W.A., Haider, J.J., Fridgen, J., Lobmaier, G.M., and Spiegler, M. (2000) *J. Organomet. Chem.*, 603, 69–79.

87 Wang, X.-Y., Shi, H.-C., Sun, C., and Zhang, Z.-G. (2004) *Tetrahedron*, 60, 10993–10998.

88 Wang, C. and Yamamoto, H. (2014) *J. Am. Chem. Soc.*, 136, 1222–1225.

89 Hachiya, H., Kon, Y., Ono, Y., Takumi, K., Sasagawa, N., Ezaki, Y., and Sato, K. (2012) *Synthesis*, 44, 1672–1678.

90 Zhang, W., Loebach, J.L., Wilson, S.R., and Jacobsen, E.N. (1990) *J. Am. Chem. Soc.*, 112, 2801–2803.

91 Irie, R., Noda, K., Ito, Y., Matsumoto, N., and Katsuki, T. (1990) *Tetrahedron Lett.*, 31, 7345–7348.

92 (a) Katsuki, T. (2002) *Adv. Synth. Catal.*, 344, 131–147; (b) Larrow, J.F. and Jacobsen, E.N. (2004) *Top. Organomet. Chem.*, 6, 123–152; (c) Xia, Q.-H., Ge, H.-Q., Ye, C.-P., Liu, Z.-M., and Su, K.-X. (2005) *Chem. Rev.*, 105, 1603–1662.

93 Srinivasan, K., Michaud, P., and Kochi, J.K. (1986) *J. Am. Chem. Soc.*, 108, 2309–2320.

94 (a) Jacobsen, E.N., Zhang, W., Muci, A.R., Ecker, J.R., and Deng, L. (1991) *J. Am. Chem. Soc.*, 113, 7063–7064; (b) Brandes, B.D. and Jacobsen, E.N. (1994) *J. Org. Chem.*, 59, 4378–4380; (c) Brandes, B.D. and Jacobsen, E.N. (1995) *Tetrahedron Lett.*, 36, 5123–5126; (d) Palucki, M., McCormick, G.J., and Jacobsen, E.N. (1995) *Tetrahedron Lett.*, 36, 5457–5460; (e) Sasaki, H., Irie, R., and Katsuki, T. (1994) *Synlett*, 356–358; (f) Palucki, M., Pospisil, P.J., Zhang, W., and Jacobsen, E.N. (1994) *J. Am. Chem. Soc.*, 116, 9333–9334; (g) Fukuda, T., Irie, R., and Katsuki, T. (1995) *Synlett*, 197–198.

95 For a recent review, see: Krishnan, K.K., Thomas, A.M., Sindhu, K.S., and Anilkumar, G. (2016) *Tetrahedron*, 72, 1–16.

96 (a) Martinez, A., Hemmert, C., and Meunier, B. (2005) *J. Catal.*, 234, 250–255; (b) Maity, N.C., Abdi, S.H.R., Kureshy, R.I., Khan, N.H., Suresh, E., and Dangi, G.P. (2011) *J. Catal.*, 277, 123–127.

97 (a) Borriello, C., Litto, R.D., Panunzi, A., and Ruffo, F. (2004) *Tetrahedron: Asymmetry*, 15, 681–686; (b) Borriello, C., Litto, R.D., Panunzi, A., and Ruffo, F. (2005) *Inorg. Chem. Commun.*, 8, 717–721; (c) Ruffo, F., Bismuto, A., Carpentieri, A., Cucciolito, M.E., and Lega, M. (2013) *Inorg. Chim. Acta*, 405, 288–294; (d) Shanshan, Z., Jiquan, Z., and Dongmin, Z. (2007) *Carbohydr. Res.*, 342, 254–258; (e) Zhao, J., Zhang, Y., Han, F., and Zhao, S. (2009) *Carbohydr. Res.*, 344, 61–66.

98 Ballistreri, F.P., Brinchi, L., Germani, R., Savelli, G., Tomaselli, G.A., and Toscano, R.M. (2008) *Tetrahedron*, 64, 10239–10243.

99 Havare, N. and Plattner, D.A. (2009) *Helv. Chim. Acta*, 92, 623–628.

100 Miura, K. and Katsuki, T. (1999) *Synlett*, 783–785.

101 Liao, S. and List, B. (2010) *Angew. Chem. Int. Ed.*, 49, 628–631.

102 (a) Huang, J., Fu, X., and Miao, Q. (2011) *Appl. Catal., A*, 407, 163–172; (b) Meng, X., Qin, C., Wang, X.L., Su, Z.M., Li, B., and Yang, Q.H. (2011) *Dalton Trans.*, 40, 9964–9966; (c) Lou, L.-L., Jiang, S., Yu, K., Gu, Z., Ji, R., Dong, Y., and Liu, S. (2011) *Microporous Mesoporous Mater.*, 142, 214–220; (d) Zou, X., Fu, X., Li, Y., Tu, X., Fu, S., Luo, Y., and Wu, X. (2010) *Adv. Synth. Catal.*, 352, 163–170.

103 Zhang, H., Zhang, Y., and Li, C. (2006) *J. Catal.*, 238, 369–381.

104 Kureshy, R.I., Ahmad, I., Khan, N.H., Abdi, S.H.R., Singh, S., Pandia, P.H., and Jasra, R.V. (2005) *J. Catal.*, 235, 28–34.

105 (a) Kureshy, R.I., Ahmad, I., Khan, N.H., Abdi, S.H.R., Pathak, K., and Jasra, R.V. (2006) *J. Catal.*, 238, 134–141; (b) Trusso Sfrazzetto, G., Millesi, S., Pappalardo, A., Toscano, R.M., Ballistreri, F.P., Tomaselli, G.A., and Gulino, A. (2015) *Catal. Sci. Technol.*, 5, 673–679.

106 Smith, K., Liu, C.H., and El-Hiti, G.A. (2006) *Org. Biomol. Chem.*, 4, 917–927.

107 Song, C.E., Roh, E.J., Yu, B.M., Chi, D.Y., Kim, S.C., and Lee, K.J. (2000) *Chem. Commun.*, 837–838.

108 (a) Lou, L.L., Ding, Y.K.F., Peng, X.J., Dong, M.M., Zhang, C., and Liu, S.X. (2007) *J. Catal.*, 249, 102–110; (b) Yu, K., Gu, Z., Ji, R., Lou, L.L., Ding, F., Zhang, C., and Liu, S.X. (2007) *J. Catal.*, 252, 312–320; (c) Zhang, H.D., Wang, Y.-M., Zhang, L., Gerritsen, G., Abbenhuis, H.C.L., van Santen, R.A., and Li, C. (2008) *J. Catal.*, 256, 226–236; (d) Roy, T., Kureshy, R.I., Khan, N.H., Abdi, S.H.R., Sadhukhan, A., and Bajaj, H.C. (2012) *Tetrahedron*, 68, 6314–6322.

109 (a) Tu, X.B., Fu, X.K., Hu, X.Y., and Li, Y.D. (2010) *Inorg. Chem. Commun.*, 13, 404–407; (b) Zhang, H., Zou, Y., Wang, Y.-M., Shen, Y., and Zheng, X. (2014) *Chem. Eur. J.*, 20, 7830–7841.

110 Federsel, H.-J. (2006) in *Stereoselective Synthesis of Drugs – An Industrial Perspective*, vol. 33 (eds E. Francotte and W. Lindner), Wiley-VCH Verlag GmbH, Weinheim, pp. 29–65.

111 Gerlach, U., Brendel, J., Lang, H.-J., Paulus, E.F., Weidmann, K., Brüggemann, A., Busch, A., Suessbrich, H., Bleich, M., and Greger, R. (2001) *J. Med. Chem.*, 44, 3831–3837.

112 Lynch, J.E., Choi, W.-B., Churchill, H.R.O., Volante, R.P., Reamer, R.A., and Ball, R.G. (1997) *J. Org. Chem.*, 62, 9223–9228.

113 (a) Chen, Y., Evarts, J.B., Torres, E., and Fuchs, P.L. (2002) *Org. Lett.*, 4, 3571–3574; (b) Torres, E., Chen, Y., Kim, I.C., and Fuchs, P.L. (2003) *Angew. Chem. Int. Ed.*, 42, 3124–3131; (c) El-Awa, A. and Fuchs, P.L. (2006) *Org. Lett.*, 8, 2905–2908; (d) El-Awa, A., du Jourdin, X.M., and Fuchs, P.L. (2007) *J. Am. Chem. Soc.*, 129, 9086–9093; (e) Hong, W.P., Noshi, M.N., El-Awa, A., and Fuchs, P.L. (2011) *Org. Lett.*, 13, 6342–6345; (f) Ebrahimian, G.R., du Jourdin, X.M., and Fuchs, P.L. (2012) *Org. Lett.*, 14, 2630–2633; (g) Sikervar, V., Fleet, J.C., and Fuchs, P.L. (2012) *Chem. Commun.*, 48, 9077–9079.

114 (a) Murphy, A., Dubois, G., and Stack, T.D.P. (2003) *J. Am. Chem. Soc.*, 125, 5250–5251; (b) Murphy, A., Pace, A., and Stack, T.D.P. (2004) *Org. Lett.*, 6, 3119–3122.

115 Gómez, L., Garcia-Bosch, I., Company, A., Sala, X., Fontrodona, X., Ribas, X., and Costas, M. (2007) *Dalton Trans.*, 5539–5545.

116 Garcia-Bosch, I., Ribas, X., and Costas, M. (2009) *Adv. Synth. Catal.*, 351, 348–352.
117 Wu, M., Wang, B., Wang, S., Xia, C., and Sun, W. (2009) *Org. Lett.*, 11, 3622–3625.
118 Ottenbacher, R.V., Bryliakov, K.P., and Talsi, E.P. (2011) *Adv. Synth. Catal.*, 353, 885–889.
119 (a) Garcia-Bosch, I., Gómez, L., Polo, A., Ribas, X., and Costas, M. (2012) *Adv. Synth. Catal.*, 354, 65–70; (b) Cussó, O., Garcia-Bosch, I., Font, D., Ribas, X., Lloret-Fillol, J., and Costas, M. (2013) *Org. Lett.*, 15, 6158–6161.
120 (a) Wang, B., Wang, S.F., Xia, C.G., and Sun, W. (2012) *Chem. Eur. J.*, 18, 7332–7335; (b) Wang, B., Miao, C.X., Wang, S.F., Xia, C.G., and Sun, W. (2012) *Chem. Eur. J.*, 18, 6750–6753; (c) Wang, X., Miao, C., Wang, S., Xia, C., and Sun, W. (2013) *ChemCatChem*, 5, 2489–2494.
121 (a) Lyakin, O.Y., Ottenbacher, R.V., Bryliakov, K.P., and Talsi, E.P. (2012) *ACS Catal.*, 2, 1196–1202; (b) Bryliakov, K.P. and Talsi, E.P. (2012) *Coord. Chem. Rev.*, 256, 1418–1434.
122 For first studies on Mn-chiral diamino-bisoxazoline ligands complexes mediated epoxidation, see: Guillemot, G., Neuburger, M., and Pfaltz, A. (2007) *Chem. Eur. J.*, 13, 8960–8970.
123 (a) Dai, W., Li, J., Li, G., Yang, H., Wang, L., and Gao, S. (2013) *Org. Lett.*, 15, 4138–4141; (b) Dai, W., Shang, S., Chen, B., Li, G., Wang, L., Ren, L., and Gao, S. (2014) *J. Org. Chem.*, 79, 6688–6694.
124 Ottenbacher, R.V., Samsonenko, D.G., Talsi, E.P., and Bryliakov, K.P. (2014) *ACS Catal.*, 4, 1599–1606.
125 Shen, D., Qiu, B., Xu, D., Miao, C., Xia, C., and Sun, W. (2016) *Org. Lett.*, 18, 372–375.
126 For a review and selected papers, see: (a) Herrmann, W., Fischer, R.W., and Marz, D.W. (1991) *Angew. Chem. Int. Ed. Engl.*, 30, 1638–1641; (b) Saladino, R., Neri, V., Pelliccia, A.R., Caminiti, R., and Sadun, C. (2002) *J. Org. Chem.*, 67, 1323–1332; (c) Neumann, R. and Wang, T.-J. (1997) *Chem. Commun.*, 1915–1916.
127 Haider, J.J., Kratzer, R.M., Herrmann, W.A., Zhao, J., and Kühn, F.E. (2004) *J. Organomet. Chem.*, 689, 3735–3740.
128 For reviews on asymmetric chiral Fe-catalyzed reactions, see: (a) Ollevier, T. (2016) *Catal. Sci. Technol.*, 6, 41–48; (b) Enthaler, S., Junge, K., and Beller, M. (2008) *Angew. Chem. Int. Ed.*, 47, 3317–3321.
129 (a) Cussó, O., Ribas, X., and Costas, M. (2015) *Chem. Commun.*, 51, 14285–14298; (b) Gelalcha, F.G. (2014) *Adv. Synth. Catal.*, 356, 261–299.
130 Francis, M.B. and Jacobsen, E.N. (1999) *Angew. Chem. Int. Ed.*, 38, 937–941.
131 Gelalcha, F.G., Bitterlich, B., Anilkumar, G., Tse, M.K., and Beller, M. (2007) *Angew. Chem. Int. Ed.*, 46, 7293–7296.
132 (a) White, M.C., Doyle, A.G., and Jacobsen, E.N. (2001) *J. Am. Chem. Soc.*, 123, 7194–7195; (b) Mas-Ballesté, R., Costas, M., van der Berg, T., and Que, L. Jr. (2006) *Chem. Eur. J.*, 12, 7489–7500; (c) Mas-Ballést, R. and Que, L. Jr. (2007) *J. Am. Chem. Soc.*, 129, 15964–15972.
133 Wu, M., Miao, C.-X., Wang, S., Hu, X., Xia, C., Kühn, F.E., and Sun, W. (2011) *Adv. Synth. Catal.*, 353, 3014–3022.

134 Cussó, O., Garcia-Bosch, I., Ribas, X., Lloret-Fillol, J., and Costas, M. (2013) *J. Am. Chem. Soc.*, 135, 14871–14878.
135 Cussó, O., Ribas, X., Lloret-Fillol, J., and Costas, M. (2015) *Angew. Chem. Int. Ed.*, 54, 2729–2733.
136 Nishikawa, Y. and Yamamoto, H. (2011) *J. Am. Chem. Soc.*, 133, 8432–8435.
137 Luo, L. and Yamamoto, H. (2014) *Eur. J. Org. Chem.*, 7803–7805.
138 Dai, W., Li, G., Chen, B., Wang, L., and Gao, S. (2015) *Org. Lett.*, 17, 904–907.
139 Niwa, T. and Nakada, M. (2012) *J. Am. Chem. Soc.*, 134, 13538–13541.
140 (a) Monnier, J.R. (2001) *Appl. Catal., A*, 221, 73–91; (b) Cheng, Q.F., Xu, X.Y., Ma, W.X., Yang, S.J., and You, T.P. (2005) *Chin. Chem. Lett.*, 16, 1467–1470.
141 Mägerlein, W., Dreisbach, C., Hugl, H., Kin Tse, M., Klawonn, M., Bhor, S., and Beller, M. (2007) *Catal. Today*, 121, 140–150.
142 (a) Murahashi, S.-I. and Komiya, N. (2004) in *Ruthenium in Organic Synthesis* (ed. S.-I. Murahashi), Wiley-VCH Verlag GmbH, Weinheim, pp. 53–87; (b) Barf, G.A. and Sheldon, R.A. (1995) *J. Mol. Catal. A: Chem.*, 102, 23–39.
143 Balvione, G., Eskenazi, F., Meunier, F., and Rivière, H. (1984) *Tetrahedron Lett.*, 25, 3187–3190.
144 (a) Kureshy, R.I., Khan, N.H., Abdi, S.H.R., and Bhatt, K.N. (1993) *Tetrahedron: Asymmetry*, 4, 1693–1701; (b) Kureshy, R.I., Khan, N.H., and Abdi, S.H.R. (1995) *J. Mol. Catal. A: Chem.*, 96, 117–122.
145 (a) Kureshy, R.I., Khan, N.H., Abdi, S.H.R., and Iyer, A.P. (1997) *J. Mol. Catal. A: Chem.*, 124, 91–97; (b) Takeda, T., Irie, R., Shinoda, Y., and Katsuki, T. (1999) *Synlett*, 1157–1159.
146 Chatterjee, D., Basak, S., Mitra, A., Sengupta, A., Le Bras, J., and Muzart, J. (2006) *Inorg. Chim. Acta*, 359, 1325–1328.
147 (a) End, N. and Pfaltz, A. (1998) *Chem. Commun.*, 589–590; (b) Nishiyama, H., Shimada, T., Itoh, H., Sugiyama, H., and Motoyama, Y. (1997) *Chem. Commun.*, 1863–1864.
148 (a) Bhor, S., Klawonn, M., Anilkumar, M.K., Tse, M.K., Klawonn, M., Döbler, C., Bitterlich, B., Grotevendt, A., and Beller, M. (2005) *Org. Lett.*, 7, 3393–3396; (b) Tse, M.K., Bhor, S., Klawonn, M., Anilkumar, G., Jiao, H., Döbler, C., Spannenberg, A., Mägerlein, W., Hugl, H., and Beller, M. (2006) *Chem. Eur. J.*, 12, 1875–1888; (c) Tse, M.K., Jiao, H., Anilkumar, G., Bitterlich, B., Gelalcha, F.G., and Beller, M. (2006) *J. Organomet. Chem.*, 691, 4419–4433.
149 Tanaka, H., Nishikawa, H., Uchida, T., and Katsuki, T. (2010) *J. Am. Chem. Soc.*, 132, 12034–12041.
150 Koya, S., Nishioka, Y., Mizoguchi, H., Uchida, T., and Katsuki, T. (2012) *Angew. Chem. Int. Ed.*, 51, 8243–8246.
151 (a) Gross, Z. and Ini, S. (1997) *J. Org. Chem.*, 62, 5514–5521; (b) Berkessel, A. and Frauenkron, M. (1997) *J. Chem. Soc., Perkin Trans. 1*, 2265–2266.
152 Berkessel, A., Kaiser, P., and Lex, J. (2003) *Chem. Eur. J.*, 9, 4746–4756.
153 (a) Fackler, P., Berthold, C., Voss, F., and Bach, T. (2010) *J. Am. Chem. Soc.*, 132, 15911–15913; (b) Voss, F., Vogt, F., Herdtweck, E., and Bach, T. (2011) *Synthesis*, 2011, 961–971.

154 Wang, C., Shalyaev, K.V., Bonchio, M., Carofiglio, T., and Groves, J.T. (2006) *Inorg. Chem.*, 45, 4769–4782.

155 (a) Pizzo, E., Sgarbossa, P., Scarso, A., Michelin, R.A., and Strukul, G. (2006) *Organometallics*, 25, 3056–3062; (b) Colladon, M., Scarso, A., Sgarbossa, P., Michelin, R.A., and Strukul, G. (2006) *J. Am. Chem. Soc.*, 128, 14006–14007.

156 (a) Baccin, C., Gusso, A., Pinna, F., and Strukul, G. (1995) *Organometallics*, 14, 1161–1167; (b) Zanardo, A., Michelin, R.A., Pinna, F., and Strukul, G. (1989) *Inorg. Chem.*, 28, 1648–1653.

157 Colladon, M., Scarso, A., Sgarbossa, P., Michelin, R.A., and Strukul, G. (2007) *J. Am. Chem. Soc.*, 129, 7680–7689.

158 (a) Enders, D., Zhu, J., and Raabe, G. (1996) *Angew. Chem. Int. Ed. Engl.*, 35, 1725–1728; (b) Enders, D., Kramps, L., and Zhu, J. (1998) *Tetrahedron: Asymmetry*, 9, 3959–3962.

159 Bian, Z., Marvin, C.C., Pettersson, M., and Martin, S.F. (2014) *J. Am. Chem. Soc.*, 136, 14184–14192.

160 Tsuda, M., Kasai, Y., Komatsu, K., Sone, T., Tanaka, M., Mikami, Y., and Kobayashi, J. (2004) *Org. Lett.*, 6, 3087–3089.

161 (a) Weitz, E. and Scheffer, A. (1921) *Chem. Ber.*, 54, 2327–2344; For a review on the synthetic importance of epoxides derived from electron-poor alkenes, see: (b) Lauret, C. (2001) *Tetrahedron: Asymmetry*, 12, 2359–2383.

162 Yu, H.-B., Zheng, X.-F., Lin, Z.-M., Hu, Q.-S., Huang, W.-S., and Pu, L. (1999) *J. Org. Chem.*, 64, 8149–8155.

163 Minatti, A. and Dötz, K.H. (2006) *Eur. J. Org. Chem.*, 268–276.

164 Wang, H., Wang, Z., and Ding, K. (2009) *Tetrahedron Lett.*, 50, 2200–2203.

165 (a) Rowley Kelly, A., Lurain, A.E., and Walsh, P.J. (2005) *J. Am. Chem. Soc.*, 127, 14668–14674; (b) Hussain, M.M. and Walsh, P.J. (2008) *Acc. Chem. Res.*, 41, 883–893.

166 (a) Bougauchi, M., Watanabe, S., Arai, T., Sasai, H., and Shibasaki, M. (1997) *J. Am. Chem. Soc.*, 119, 2329–2330; (b) Daikai, K., Kamaura, M., and Inanaga, J. (1998) *Tetrahedron Lett.*, 39, 7321–7322; (c) Nemoto, T., Kakei, H., Ganadesikan, V., Tosaki, S.-Y., Ohshima, T., and Shibasaki, M. (2001) *J. Am. Chem. Soc.*, 123, 2725–2732.

167 Watanabe, S., Arai, T., Sasai, H., Bougauchi, M., and Shibasaki, M. (1998) *J. Org. Chem.*, 63, 8090–8091.

168 (a) Nemoto, T., Kakei, H., Gnanadesikan, V., Tosaki, S., Ohshima, T., and Shibasaki, M. (2002) *J. Am. Chem. Soc.*, 124, 14544–14545; (b) Nemoto, T., Ohshima, T., and Shibasaki, M. (2001) *J. Am. Chem. Soc.*, 123, 9474–9475; (c) Kinoshita, T., Okada, S., Park, S.-R., Matsunaga, S., and Shibasaki, M. (2003) *Angew. Chem. Int. Ed.*, 42, 4680–4684; (d) Kakei, H., Nemoto, T., Ohshima, T., and Shibasaki, M. (2004) *Angew. Chem. Int. Ed.*, 43, 317–320.

169 Wang, X., Shi, L., Li, M., and Ding, K. (2005) *Angew. Chem. Int. Ed.*, 44, 6362–6366.

170 Kakei, H., Tsuji, R., Ohshima, T., and Shibasaki, M. (2005) *J. Am. Chem. Soc.*, 127, 8962–8963.

171 Hara, K., Park, S.-Y., Yamagiwa, N., Matsunaga, S., and Shibasaki, M. (2008) *Chem. Asian J.*, 3, 1500–1504.

172 Kanai, M., Kato, N., Ichikawa, E., and Shibasaki, M. (2005) *Synlett*, 1491–1508.

173 Nemoto, T., Ohshima, T., and Shibasaki, M. (2000) *Tetrahedron Lett.*, 41, 9569–9574.

174 Tosaki, S., Nemoto, T., Ohshima, T., and Shibasaki, M. (2003) *Org. Lett.*, 5, 495–498.

175 Qian, Q., Tan, Y., Zhao, B., Feng, T., Shen, Q., and Yao, Y. (2014) *Org. Lett.*, 16, 4516–4519.

176 Zeng, C., Yuan, D., Zhao, B., and Yao, Y. (2015) *Org. Lett.*, 17, 2242–2245.

177 Chu, Y., Liu, X., Li, W., Hu, X., Lin, L., and Feng, X. (2012) *Chem. Sci.*, 3, 1996–2000.

178 Chu, Y., Hao, X., Lin, L., Chen, W., Li, W., Tan, F., Liu, X., and Feng, X. (2014) *Adv. Synth. Catal.*, 356, 2214–2218.

179 (a) Dalko, P.I. (ed.) (2013) *Comprehensive Enantioselective Organocatalysis: Catalysts, Reactions and Applications*, vols. 1–3, Wiley-VCH Verlag GmbH, Weinheim; (b) Atodiresei, I., Vila, C., and Rueping, M. (2015) *ACS Catal.*, 5, 1972–1985; (c) Pihko, P.M. (ed.) (2009) *Hydrogen Bonding in Organic Synthesis*, Wiley-VCH Verlag GmbH, Weinheim.

180 Giacalone, F., Gruttadauria, M., Agrigento, P., and Noto, R. (2012) *Chem. Soc. Rev.*, 41, 2406–2447.

181 For reviews, see: (a) Davis, R., Stiller, J., Naicker, T., Jiang, H., and Jørgensen, K.A. (2014) *Angew. Chem. Int. Ed.*, 53, 7406–7426; (b) Colonna, S. and Perdicchia, D. (2011) in *Science of Synthesis, Stereoselective Synthesis*, vol. 1 (eds J.G. De Vries, G.A. Molander, and P. Evans), Thieme, pp. 123–153; (c) Weiss, K. and Tsogoeva, S.B. (2011) *Chem. Rec.*, 11, 18–39; (d) Lattanzi, A. (2008) *Curr. Org. Synth.*, 5, 117–133; (e) Kelly, D.R. and Roberts, S.M. (2006) *Biopolymers*, 84, 74–89.

182 For selected recent reviews on asymmetric phase transfer catalysis, see: (a) Hashimoto, T. and Maruoka, K. (2007) *Chem. Rev.*, 107, 5656–5682; (b) Jew, S.-S. and Park, H.-G. (2009) *Chem. Commun.*, 7090–7103; (c) Shirakawa, S. and Maruoka, K. (2013) *Angew. Chem. Int. Ed.*, 52, 4312–4348.

183 (a) Tan, J. and Yasuda, N. (2015) *Org. Process Res. Dev.*, 19, 1731–1746; (b) Albanese, D.C.M., Foschi, F., and Penso, M. (2016) *Org. Process Res. Dev.*, 20, 129–139.

184 Helder, R., Hummelen, J.C., Laane, R.W.P.M., Wiering, J.S., and Wynberg, H. (1976) *Tetrahedron Lett.*, 17, 1831–1834.

185 Arai, S., Tsuge, H., Oku, M., Miura, M., and Shioiri, T. (2002) *Tetrahedron*, 58, 1623–1630.

186 Corey, E.J. and Zhang, F.-Y. (1999) *Org. Lett.*, 1, 1287–1290.

187 (a) Lygo, B. and Wainwright, P.G. (1998) *Tetrahedron Lett.*, 39, 1599–1602; (b) Lygo, B. and Wainwright, P.G. (1999) *Tetrahedron*, 55, 6289–6300.

188 (a) Ye, J., Wang, Y., Liu, R., Zhang, G., Zhang, Q., Chen, J., and Liang, X. (2003) *Chem. Commun.*, 2714–2715; (b) Yoo, M.-S., Kim, D.-G., Ha, M.W., Jew, S.-s., Park, H.-g., and Jeong, B.-S. (2010) *Tetrahedron Lett.*, 51, 5601–5603.

189 Dorow, R.L. and Tymonko, S.A. (2006) *Tetrahedron Lett.*, 47, 2493–2495.

190 Vidal-Albalat, A., Świderek, K., Izquierdo, J., Rodríguez, S., Moliner, V., and González, F.V. (2016) *Chem. Commun.*, 52, 10060–10063.

191 Berkessel, A., Guixà, M., Schmidt, F., Neudörfl, J.M., and Lex, J. (2007) *Chem. Eur. J.*, 13, 4483–4498.
192 Jew, S.-s., Lee, J.-H., Jeong, B.-S., Yoo, M.-S., Kim, M.-J., Lee, Y.-J., Lee, J., Choi, S.-h., Lee, K., Lah, M.-S., and Park, H.-g. (2005) *Angew. Chem. Int. Ed.*, 44, 1383–1385.
193 Lv, J., Wang, X., Liu, J., Zhang, L., and Wang, Y. (2006) *Tetrahedron: Asymmetry*, 17, 330–335.
194 Ooi, T., Ohara, D., Tamura, M., and Maruoka, K. (2004) *J. Am. Chem. Soc.*, 126, 6844–6845.
195 Hori, K., Tamura, M., Tani, K., Nishiwaki, N., Ariga, M., and Tohda, Y. (2006) *Tetrahedron Lett.*, 47, 3115–3118.
196 Bakó, P., Bakó, T., Mészáros, A., Keglevich, G., Szöllösy, À., Bodor, S., Makó, A., and Töke, L. (2004) *Synlett*, 643–646.
197 Maruoka, K. (ed.) (2008) *Asymmetric Phase Transfer Catalysis*, Wiley-VCH Verlag GmbH, Weinheim.
198 (a) Kawai, H., Okusu, S., Yuan, Z., Tokunaga, E., Yamano, A., Shiro, M., and Shibata, N. (2013) *Angew. Chem. Int. Ed.*, 52, 2221–2225; (b) Wu, S., Pan, D.; Cao, C., Wang, Q., and Chen, F.-X. (2013) *Adv. Synth. Catal.*, 355, 1917–1923.
199 Ashokkumar, V., Balasaravanan, R., Sadhasivam, V., Jenofar, S.M., and Siva, A. (2015) *J. Mol. Catal. A: Chem.*, 409, 127–136.
200 Allingham, M.T., Howard-Jones, A., Murphy, P.J., Thomas, D.A., and Caulkett, P.W.R. (2003) *Tetrahedron Lett.*, 44, 8677–8680.
201 (a) Kita, T., Shin, B., Hashimoto, Y., and Nagasawa, K. (2007) *Heterocycles*, 73, 241–247; (b) Shin, B., Tanaka, S., Kita, T., Hashimoto, Y., and Nagasawa, K. (2008) *Heterocycles*, 74, 801–810.
202 Tanaka, S. and Nagasawa, K. (2009) *Synlett*, 667–670.
203 Sohtome, Y. and Nagasawa, K. (2012) *Chem. Commun.*, 48, 7777–7789.
204 Lygo, B., Gardiner, S.D., and To, D.C.M. (2006) *Synlett*, 2063–2066.
205 For recent reviews, see: (a) Colby Davie, E.A., Mennen, S.M., Xu, Y., and Miller, S.J. (2007) *Chem. Rev.*, 107, 5759–5812; (b) Wennemers, H. (2011) *Chem. Commun.*, 47, 12036–12041; (c) Megens, R.P. and Roelfes, G. (2011) *Chem. Eur. J.*, 17, 8514–8523.
206 (a) Krattiger, P., McCarthy, C., Pfaltz, A., and Wennemers, H. (2003) *Angew. Chem. Int. Ed.*, 42, 1722–1724; (b) Wiesner, M., Revell, J.D., and Wennemers, H. (2008) *Angew. Chem. Int. Ed.*, 47, 1871–1874.
207 (a) Juliá, S., Masana, J., and Vega, J.C. (1980) *Angew. Chem. Int. Ed. Engl.*, 19, 929–931; (b) Juliá, S., Guixer, J., Masana, J., Rocas, J., Colonna, S., Annunziata, R., and Molinari, H. (1982) *J. Chem. Soc., Perkin Trans. 1*, 1317–1324; (c) Banfi, S., Colonna, S., Molinari, H., Juliá, S., and Guixer, J. (1984) *Tetrahedron*, 40, 5207–5211.
208 (a) Geller, T., Gerlach, A., Krüger, C.M., and Militzer, H.-C. (2004) *Tetrahedron Lett.*, 45, 5065–5067; (b) Geller, T., Krüger, C.M., and Militzer, H.-C. (2004) *Tetrahedron Lett.*, 45, 5069–5071; (c) Gerlach, A. and Geller, T. (2004) *Adv. Synth. Catal.*, 346, 1247–1249; (d) Lopez-Pedrosa, J.-M., Pitts, M.R., Roberts, S.M., Saminathan, S., and Whittall, J. (2004) *Tetrahedron Lett.*, 45, 5073–5075.
209 (a) Bentley, P.A., Bergeron, S., Cappi, M.W., Hibbs, D.E., Hurst-House, M.B., Nugent, T.C., Pulido, R., Roberts, S.M., and Wu, L.E. (1997) *J. Chem. Soc.,*

Chem. Commun., 739–740; (b) Allen, J.V., Drauz, K.-H., Flood, R.W., Roberts, S.M., and Skidmore, J. (1999) *Tetrahedron Lett.*, 40, 5417–5420.

210 (a) Geller, T.P. and Roberts, S.M. (1999) *J. Chem. Soc., Perkin Trans. 1*, 1397–1398; (b) Yi, H., Zou, G., Li, Q., Chen, Q., Tang, J., and He, M.-y. (2005) *Tetrahedron Lett.*, 46, 5665–5668.

211 Qiu, W., He, L., Chen, Q., Luo, W., Yu, Z., Yang, F., and Tang, J. (2009) *Tetrahedron Lett.*, 50, 5225–5227.

212 Miranda, R.-A., Llorca, J., Medina, F., Sueiras, J.E., and Segarra, A.M. (2011) *J. Catal.*, 282, 65–73.

213 Licini, G., Bonchio, M., Broxterman, Q.B., Kaptein, B., Moretto, A., Toniolo, C., and Scrimin, P. (2006) *Biopolymers*, 84, 97–104.

214 Flood, R.W., Geller, T.P., Petty, S.A., Roberts, S.M., Skidmore, J., and Volk, M. (2001) *Org. Lett.*, 3, 683–686.

215 (a) Bentley, P.A., Flood, R.W., Roberts, S.M., Skidmore, J., Smith, C.B., and Smith, J.A. (2001) *Chem. Commun.*, 1616–1617; (b) Berkessel, A., Gasch, N., Glaubitz, K., and Koch, C. (2001) *Org. Lett.*, 3, 3839–3842; (c) Weyer, A., Díaz, D., Nierth, A., Schlörer, N.E., and Berkessel, A. (2012) *ChemCatChem*, 4, 337–340; (d) Yamagata, N., Demizu, Y., Sato, Y., Doi, M., Tanaka, M., Nagasawa, K., Okuda, H., and Kurihara, M. (2011) *Tetrahedron Lett.*, 52, 798–801.

216 Nagano, M., Doi, M., Kurihara, M., Suemune, H., and Tanaka, M. (2010) *Org. Lett.*, 12, 3564–3566.

217 (a) Carrea, G., Colonna, S., Meek, A.D., Ottolina, G., and Roberts, S.M. (2004) *Tetrahedron: Asymmetry*, 15, 2945–2949; (b) Mathew, S.P., Gunathilagan, S., Roberts, S.M., and Blackmond, D.G. (2005) *Org. Lett.*, 7, 4847–4850; (c) Carrea, G., Colonna, S., Kelly, D.R., Lazcano, A., Ottolina, G., and Roberts, S.M. (2005) *Trends Biotechnol.*, 23, 507–513.

218 Akagawa, K. and Kudo, K. (2011) *Adv. Synth. Catal.*, 353, 843–847.

219 Adger, B.M., Barkley, J.V., Bergeron, S., Cappi, M.W., Flowerdew, B.E., Jackson, M.P., MacCague, R., Nugent, T.C., and Roberts, S.M. (1997) *J. Chem. Soc., Perkin Trans. 1*, 3501–3508.

220 Peris, G., Jakobsche, C.E., and Miller, S.J. (2007) *J. Am. Chem. Soc.*, 129, 8710–8711.

221 (a) Pirkle, W.H. and Rinaldi, P.L. (1977) *J. Org. Chem.*, 42, 3217–3219; (b) Barbieri, G., Davoli, V., Moretti, I., Montanari, F., and Torre, G. (1969) *J. Chem. Soc. C*, 1969, 731–735.

222 Jakobsche, C.E., Peris, G., and Miller, S.J. (2008) *Angew. Chem. Int. Ed.*, 47, 6707–6711.

223 Kolundzic, F., Noshi, M.N., Tjandra, M., Movassaghi, M., and Miller, S.J. (2011) *J. Am. Chem. Soc.*, 133, 9104–9111.

224 Lichtor, P.A. and Miller, S.J. (2012) *Nat. Chem.*, 4, 990–995.

225 Lichtor, P.A. and Miller, S.J. (2014) *J. Am. Chem. Soc.*, 136, 5301–5308.

226 Furka, A., Sebestyen, F., Asgedom, M., and Dibo, G. (1991) *Int. J. Pept. Protein Res.*, 37, 487–493.

227 Curci, R., D'Accolti, L., and Fusco, C. (2006) *Acc. Chem. Res.*, 39, 1–9.

228 For recent reviews, see: (a) Wong, O.A. and Shi, Y. (2008) *Chem. Rev.*, 108, 3958–3987; (b) Wong, O.A., Shi, Y., Ramirez, T., and Shi, Y. (2012) in

Comprehensive Chirality (eds E.M. Carreira and H. Yamamoto), Elsevier, Amsterdam, pp. 528–553; (c) Zhu, Y., Wang, Q., Cornwall, R.G., and Shi, Y. (2014) *Chem. Rev.*, 114, 8199–8256.

229 Curci, R., Fiorentino, M., and Serio, M.R. (1984) *J. Chem. Soc., Chem. Commun.*, 155–156.

230 (a) Yang, D., Yip, Y.-C., Tang, M.-W., Wong, M.-K., Zheng, J.-H., and Cheung, K.-K. (1996) *J. Am. Chem. Soc.*, 118, 491–492; (b) Yang, D., Wong, M.-K., Yip, Y.-C., Wang, X.-C., Tang, M.-W., Zheng, J.-H., and Cheung, K.-K. (1998) *J. Am. Chem. Soc.*, 120, 5943–5952.

231 Imashiro, R. and Seki, M. (2004) *J. Org. Chem.*, 69, 4216–4226.

232 Denmark, S.E. and Matsuhashi, H. (2002) *J. Org. Chem.*, 67, 3479–3486.

233 Denmark, S.E. and Wu, Z. (1997) *J. Org. Chem.*, 62, 8288–8289.

234 Armstrong, A. and Hayter, B.R. (1998) *Chem. Commun.*, 621–622.

235 (a) Yang, D., Yip, Y.-C., Chen, J., and Cheung, K.-K. (1998) *J. Am. Chem. Soc.*, 120, 7659–7660; (b) Solladié-Cavallo, A. and Bouérat, L. (2000) *Org. Lett.*, 2, 3531–3534.

236 Armstrong, A. and Hayter, B.R. (1999) *Tetrahedron*, 55, 11119–11126.

237 Romney, D.K. and Miller, S.J. (2012) *Org. Lett.*, 14, 1138–1141.

238 Chan, W.-K., Yu, W.-Y., Che, C.-M., and Wong, M.-K. (2003) *J. Org. Chem.*, 68, 6576–6582.

239 (a) Wang, Z.-X. and Shi, Y. (1997) *J. Org. Chem.*, 62, 8622–8623; (b) Wang, Z.-X., Miller, S.M., Anderson, O.P., and Shi, Y. (1999) *J. Org. Chem.*, 64, 6443–6458.

240 Tu, Y., Wang, Z.-X., and Shi, Y. (1996) *J. Am. Chem. Soc.*, 118, 9806–9807.

241 Wang, Z.-X., Tu, Y., Frohn, M., Zhang, J.-R., and Shi, Y. (1997) *J. Am. Chem. Soc.*, 119, 11224–11235.

242 Wang, Z.-X. and Shi, Y. (1998) *J. Org. Chem.*, 63, 3099–3104.

243 (a) Frohn, M., Dalkiewicz, M., Tu, Y., Wang, Z.-X., and Shi, Y. (1998) *J. Org. Chem.*, 63, 2948–2953; (b) Wang, Z.-X., Cao, G.-A., and Shi, Y. (1999) *J. Org. Chem.*, 64, 7646–7650.

244 (a) Zhu, Y., Manske, K.J., and Shi, Y. (1999) *J. Am. Chem. Soc.*, 121, 4080–4081; (b) Feng, X., Shu, L., and Shi, Y. (1999) *J. Am. Chem. Soc.*, 121, 11002–11003.

245 Warren, J.D. and Shi, Y. (1999) *J. Org. Chem.*, 64, 7675–7677.

246 Wong, O.A. and Shi, Y. (2009) *J. Org. Chem.*, 74, 8377–8380.

247 Lorenz, J.C., Frohn, M., Zhou, X., Zhang, J.-R., Tang, Y., Burke, C., and Shi, Y. (2005) *J. Org. Chem.*, 70, 2904–2911.

248 Shu, L. and Shi, Y. (1999) *Tetrahedron Lett.*, 40, 8721–8724.

249 Tian, H., She, X., and Shi, Y. (2001) *Org. Lett.*, 3, 715–718.

250 (a) Wu, X.-Y., She, X., and Shi, Y. (2002) *J. Am. Chem. Soc.*, 124, 8792–8793; (b) Wang, B., Wu, X.-Y., Wong, O.A., Nettles, B., Zhao, M.-X., Chen, D., and Shi, Y. (2009) *J. Org. Chem.*, 74, 3986–3989; (c) Peng, X., Li, P., and Shi, Y. (2012) *J. Org. Chem.*, 77, 701–703. For an example of chiral auxiliary approach using achiral dioxirane as the oxygen source, see: (d) Hajra, S. and Bhowmick, M. (2010) *Tetrahedron: Asymmetry*, 21, 2223–2229.

251 (a) Tian, H., She, X., Shu, L., Yu, H., and Shi, Y. (2000) *J. Am. Chem. Soc.*, 122, 11551–11552; (b) Tian, H., She, X., Yu, H., Shu, L., and Shi, Y. (2002) *J. Org. Chem.*, 67, 2435–2446.

252 (a) Shu, L., Wang, P., Gan, Y., and Shi, Y. (2003) *Org. Lett.*, 5, 293–296; (b) Zhao, M.-X., Goeddel, D., Li, K., and Shi, Y. (2006) *Tetrahedron*, 62, 8064–8068; (c) Wong, O.A. and Shi, Y. (2006) *J. Org. Chem.*, 71, 3973–3976.

253 Burke, C.P. and Shi, Y. (2006) *Angew. Chem. Int. Ed.*, 45, 4475–4478.

254 Burke, C.P. and Shi, Y. (2007) *J. Org. Chem.*, 72, 4093–4097.

255 Burke, C.P. and Shi, Y. (2009) *Org. Lett.*, 11, 5150–5153.

256 (a) Goeddel, D., Shu, L., Yuan, Y., Wong, O.A., Wang, B., and Shi, Y. (2006) *J. Org. Chem.*, 71, 1715–1717; (b) Hickey, M., Goeddel, D., Crane, Z., and Shi, Y. (2004) *Proc. Natl. Acad. Sci. U.S.A.*, 101, 5794–5798.

257 (a) Shen, Y.-M., Wang, B., and Shi, Y. (2006) *Angew. Chem. Int. Ed.*, 45, 1429–1432; (b) Shen, Y.-M., Wang, B., and Shi, Y. (2006) *Tetrahedron Lett.*, 47, 5455–5458.

258 Wang, B., Shen, Y.-M., and Shi, Y. (2006) *J. Org. Chem.*, 71, 9519–9521.

259 (a) Wang, B., Wong, O.A., Zhao, M.-X., and Shi, Y. (2008) *J. Org. Chem.*, 73, 9539–9543; (b) Wong, O.A., Wang, B., Zhao, M.-X., and Shi, Y. (2009) *J. Org. Chem.*, 74, 6335–6338.

260 Boutureira, O., McGouran, J.F., Stafford, R.L., Emmerson, D.P.G., and Davis, B.G. (2009) *Org. Biomol. Chem.*, 7, 4285–4288.

261 Schöberl, C. and Jäger, V. (2012) *Adv. Synth. Catal.*, 354, 790–796.

262 Vega-Pérez, J.M., Vega-Holm, M., Martínez, M.L., Blanco, E., and Iglesias-Guerra, F. (2009) *Eur. J. Org. Chem.*, 6009–6018.

263 Vega-Pérez, J.M., Periñán, I., Vega-Holm, M., Palo-Nieto, C., and Iglesias-Guerra, F. (2011) *Tetrahedron*, 67, 7057–7065.

264 Shi, Y. (2010) in *Modern Oxidation Methods*, Chapter 3 (ed. J.-E. Bäckvall), Wiley-VCH Verlag GmbH, Weinheim, pp. 85–115.

265 Cachoux, F., Isarno, T., Wartmann, M., and Altmann, K.-H. (2005) *Angew. Chem. Int. Ed.*, 44, 7469–7473.

266 Kumar, V.P. and Chandrasekhar, S. (2013) *Org. Lett.*, 15, 3610–3613.

267 Ager, D.J., Anderson, K., Oblinger, E., Shi, Y., and VanderRoest, J. (2007) *Org. Process Res. Dev.*, 11, 44–51.

268 Tong, R., Valentine, J.C., McDonald, F.E., Cao, R., Fang, X., and Hardcastle, K.I. (2007) *J. Am. Chem. Soc.*, 129, 1050–1051.

269 For a review, see: Page, P.C.B. and Buckley, B.R. (2008) in *Mechanisms in Homogeneous and Heterogeneous Epoxidation Catalysis*, Chapter 5 (ed. S.T. Oyama), Elsevier, Oxford, pp. 177–216.

270 (a) Hanquet, G., Lusinchi, X., and Milliet, P. (1987) *Tetrahedron Lett.*, 28, 6061–6064; (b) Hanquet, G., Lusinchi, X., and Milliet, P. (1988) *Tetrahedron Lett.*, 29, 3941–3944.

271 Bohe, L., Hanquet, G., Lusinchi, M., and Lusinchi, X. (1993) *Tetrahedron Lett.*, 34, 7271–7274.

272 Aggarwal, V.K. and Wang, M.F. (1996) *Chem. Commun.*, 191–192.

273 (a) Armstrong, A., Ahmed, G., Garnett, I., Gioacolou, K., and Wailes, J.S. (1999) *Tetrahedron*, 55, 2341–2352; (b) Wong, M.-K., Ho, L.-M., Zheng, Y.-S., Ho, C.-Y., and Yang, D. (2001) *Org. Lett.*, 16, 2587–2590.

274 Lacour, J., Monchaud, D., and Marsol, C. (2002) *Tetrahedron Lett.*, 43, 8257–8260.

275 (a) Bulman Page, P.C., Buckley, B.R., and Blacker, A.J. (2004) *Org. Lett.*, 6, 1543–1546; (b) Page, P.C.B., Buckley, B.R., Farah, M.M., and Blacker, A.J. (2009) *Eur. J. Org. Chem.*, 3413–3426.

276 Page, P.C.B., Barros, D., Buckley, B.R., Ardakani, A., and Marples, B.A. (2004) *J. Org. Chem.*, 69, 3595–3597.

277 Page, P.C.B., Buckley, B.R., Heaney, H., and Blacker, A.J. (2005) *Org. Lett.*, 7, 375–377.

278 Page, P.C.B., Parker, P., Rassias, G.A., Buckley, B.R., and Bethell, D. (2008) *Adv. Synth. Catal.*, 350, 1867–1874.

279 Page, P.C.B., Parker, P., Buckley, B.R., Rassias, G.A., and Bethell, D. (2009) *Tetrahedron*, 65, 2910–2915.

280 Page, P.C.B., Marken, F., Williamson, C., Chan, Y., Buckley, B.R., and Bethell, D. (2008) *Adv. Synth. Catal.*, 350, 1149–1154.

281 Page, P.C.B., Appleby, L.F., Day, D.P., Chan, Y., Buckley, B.R., Allin, S.M., and McKenzie, M.J. (2009) *Org. Lett.*, 11, 1991–1993.

282 Page, P.C.B., Appleby, L.F., Chan, Y., Day, D., Buckley, B.R., Slawin, A.M.Z., Allin, S.M., and McKenzie, M.J. (2013) *J. Org. Chem.*, 78, 8074–8082.

283 Bartlett, C.J., Day, D.P., Chan, Y., Allin, S.M., McKenzie, M.J., Slawin, A.M.Z., and Page, P.C.B. (2012) *J. Org. Chem.*, 77, 772–774.

284 Page, P.C.B., Rassias, G.A., Barros, D., Ardakani, A., Bethell, D., and Merifield, E. (2002) *Synlett*, 580–582.

285 Vachon, J., Rentsch, S., Martinez, A., Marsol, C., and Lacour, J. (2007) *Org. Biomol. Chem.*, 5, 501–506.

286 Page, P.C.B., Bartlett, C.J., Chan, Y., Day, D., Parker, P., Buckley, B.R., Rassias, G.A., Slawin, A.M.Z., Allin, S.M., Lacour, J., and Pinto, A. (2012) *J. Org. Chem.*, 77, 6128–6138.

287 Page, P.C.B., Pearce, C.A., Chan, Y., Parker, P., Buckley, B.R., Rassias, G.A., and Elsegood, M.R.J. (2015) *J. Org. Chem.*, 80, 8036–8045.

288 Adam, W., Rao, P.B., Degen, H.-G., and Saha-Möller, C.R. (2000) *J. Am. Chem. Soc.*, 122, 5654–5655.

289 (a) Dwyer, C.L., Gill, C.D., Ichihara, O., and Taylor, R.J.K. (2000) *Synlett*, 704–706; (b) Bundu, A., Berry, N.G., Gill, C.D., Dwyer, C.L., Stachulski, A.V., Taylor, R.J.K., and Whittall, J. (2005) *Tetrahedron: Asymmetry*, 16, 283–293.

290 Aoki, M. and Seebach, D. (2001) *Helv. Chim. Acta*, 84, 187–207.

291 Lattanzi, A., Iannece, P., and Scettri, A. (2004) *Tetrahedron: Asymmetry*, 15, 1779–1785.

292 Lattanzi, A., Cocilova, M., Iannece, P., and Scettri, A. (2004) *Tetrahedron: Asymmetry*, 15, 3751–3755.

293 Kośnik, W., Bocian, W., Kozerski, L., Tvaroška, I., and Chmielewski, M. (2008) *Chem. Eur. J.*, 14, 6087–6097.

294 Tanaka, Y., Nishimura, K., and Tomioka, K. (2003) *Tetrahedron*, 59, 4549–4556.

295 For recent reviews, see: (a) Selig, P. (2013) *Synthesis*, 45, 703–718; (b) Leow, D. and Tan, C.-H. (2009) *Chem. Asian J.*, 4, 488–507.

296 For a recent review, see: Taylor, J.E., Bull, S.D., and Williams, J.M.J. (2012) *Chem. Soc. Rev.*, 41, 2109–2121.

297 (a) Schlessinger, R.H., Bebernitz, G.R., Lin, P., and Poss, A.J. (1985) *J. Am. Chem. Soc.*, 107, 1777–1778; (b) Yadav, V.K. and Kapoor, K.K. (1995) *Tetrahedron*, 51, 8573–8584.

298 (a) MacManus, J.C., Carey, J.S., and Taylor, R.J.K. (2003) *Synlett*, 365–368; (b) MacManus, J.C., Genski, T., Carey, J.S., and Taylor, R.J.K. (2003) *Synlett*, 369–371.

299 Kumamoto, T., Ebine, K., Endo, M., Araki, Y., Fushimi, Y., Miyamoto, I., Ishikawa, T., Isobe, T., and Fukuda, K. (2005) *Heterocycles*, 66, 347–359.
300 Terada, M. and Nakano, M. (2008) *Heterocycles*, 76, 1049–1055.
301 Lattanzi, A. (2005) *Org. Lett.*, 7, 2579–2582.
302 Lattanzi, A. (2006) *Adv. Synth. Catal.*, 348, 339–346.
303 (a) Lattanzi, A. and Russo, A. (2006) *Tetrahedron*, 62, 12264–12269; (b) Russo, A. and Lattanzi, A. (2008) *Eur. J. Org. Chem.*, 2767–2773.
304 Capobianco, A., Russo, A., Lattanzi, A., and Peluso, A. (2012) *Adv. Synth. Catal.*, 354, 2789–2796.
305 Li, Y., Liu, X., Yang, Y., and Zhao, G. (2007) *J. Org. Chem.*, 72, 288–291.
306 Liu, X., Li, Y., Wang, G., Chai, Z., Wu, Y., and Zhao, G. (2006) *Tetrahedron: Asymmetry*, 17, 750–755.
307 Cui, H., Li, Y., Zheng, C., Zhao, G., and Zhu, S. (2008) *J. Fluorine Chem.*, 129, 45–50.
308 Lu, J., Xu, Y.-H., Liu, F., and Loh, T.-P. (2008) *Tetrahedron Lett.*, 49, 6007–6008.
309 Zheng, C., Li, Y., Yang, Y., Wang, H., Cui, H., Zhang, J., and Zhao, G. (2009) *Adv. Synth. Catal.*, 351, 1685–1691.
310 (a) Russo, A. and Lattanzi, A. (2010) *Org. Biomol. Chem.*, 8, 2633–2638; (b) De Fusco, C., Tedesco, C., and Lattanzi, A. (2011) *J. Org. Chem.*, 76, 676–679.
311 Yoakim, C., Ogilvie, W.W., Goudreau, N., Naud, J., Haché, B., O'Meara, J.A., Cordingley, M.G., Archambault, J., and White, P.W. (2003) *Bioorg. Med. Chem. Lett.*, 13, 2539–2541.
312 Palumbo, C., Mazzeo, G., Mazziotta, A., Gambacorta, A., Loreto, M.A., Migliorini, A., Superchi, S., Tofani, D., and Gasperi, T. (2011) *Org. Lett.*, 13, 6248–6251.
313 Wang, X., Reisinger, C.M., and List, B. (2008) *J. Am. Chem. Soc.*, 130, 6070–6071.
314 Lu, X., Liu, Y., Sun, B., Cindric, B., and Deng, L. (2008) *J. Am. Chem. Soc.*, 130, 8134–8135.
315 Reisinger, C.M., Wang, X., and List, B. (2008) *Angew. Chem. Int. Ed.*, 47, 8112–8115.
316 Marigo, M., Franzén, G., Poulsen, T.B., Zhuang, W., and Jørgensen, K.A. (2005) *J. Am. Chem. Soc.*, 127, 6964–6965.
317 Zhuang, W., Marigo, M., and Jørgensen, K.A. (2005) *Org. Biomol. Chem.*, 3, 3883–3885.
318 Davis, R.L., Jensen, K.L., Gschwend, B., and Jørgensen, K.A. (2014) *Chem. Eur. J.*, 20, 64–67.
319 Duarte, F.J.S. and Santos, A.G. (2013) *Org. Biomol. Chem.*, 11, 7179–7191.
320 Sundén, H., Ibrahem, I., and Córdova, A. (2006) *Tetrahedron Lett.*, 47, 99–103.
321 Zhao, G.-L., Dziedzic, P., Ibrahem, I., and Córdova, A. (2006) *Synlett*, 3521–3524.
322 Zhao, G.-L., Ibrahem, I., Sundén, H., and Córdova, A. (2007) *Adv. Synth. Catal.*, 349, 1210–1224.
323 Varela, M.C., Dixon, S.M., Lam, K.S., and Schore, N.E. (2008) *Tetrahedron*, 64, 10087–10090.
324 Albrecht, Ł., Jiang, H., Dickmeiss, G., Gschwend, B., Hansen, S.G., and Jørgensen, K.A. (2010) *J. Am. Chem. Soc.*, 132, 9188–9196.
325 Jiang, H., Gschwend, B., Albrecht, Ł., and Jørgensen, K.A. (2010) *Org. Lett.*, 12, 5052–5055.

326 Albrecht, Ł., Ransborg, L.K., Gschwend, B., and Jørgensen, K.A. (2010) *J. Am. Chem. Soc.*, 132, 17886–17893.
327 Bénary, E. (1911) *Chem. Ber.*, 44, 489–493.
328 Sparr, C., Schweizer, W.B., Senn, H.M., and Gilmour, R. (2009) *Angew. Chem. Int. Ed.*, 48, 3065–3068.
329 Sparr, C., Tanzer, E.-M., Bachmann, J., and Gilmour, R. (2010) *Synthesis*, 2010, 1394–1397.
330 Tanzer, E.M., Zimmer, L.E., Schweizer, W.B., and Gilmour, R. (2012) *Chem. Eur. J.*, 18, 11334–11342.
331 Xu, M.-H., Tu, Y.-Q., Tian, J.-M., Zhang, F.-M., Wang, S.-H., Zhang, S.-H., and Zhang, X.-M. (2016) *Tetrahedron: Asymmertry*, 27, 294–300.
332 Lee, S. and MacMillan, D.W.C. (2006) *Tetrahedron*, 62, 11413–11424.
333 (a) Mayer, S. and List, B. (2006) *Angew. Chem. Int. Ed.*, 45, 4193–4195; (b) Hamilton, G.L., Kang, E.J., Mba, M., and Toste, F.D. (2007) *Science*, 317, 496–499.
334 Wang, X. and List, B. (2008) *Angew. Chem. Int. Ed.*, 47, 1119–1122.
335 (a) Berkessel, A. and Gröger, H. (eds) (2005) *Asymmetric Organocatalysis: From Biomimetic Concepts to Applications in Asymmetric Synthesis*, Wiley-VCH Verlag GmbH, Weinheim; (b) Dalko, P.I. (ed.) (2007) *Enantioselective Organocatalysis: Reactions and Experimental Procedures*, Wiley-VCH Verlag GmbH, Weinheim.
336 Lifchits, O., Reisinger, C. M., and List, B. (2010) *J. Am. Chem. Soc.*, 132, 10227–10229.
337 For recent reviews, see: (a) Inokuma, T. and Takemoto, Y. (2012) in *Science of Synthesis, Asymmetric Organocatalysis*, vol. 2 (eds B. List and K. Maruoka), Thieme, pp. 437–497; (b) Connon, S.J. (2008) *Chem. Commun.*, 2499–2510; (c) Marcelli, T., van Maarseveen, J.H., and Hiemstra, H. (2006) *Angew. Chem. Int. Ed.*, 45, 7496–7504.
338 Okino, T., Hoashi, Y., and Takemoto, Y. (2003) *J. Am. Chem. Soc.*, 128, 12672–12673.
339 Russo, A., Galdi, G., Croce, G., and Lattanzi, A. (2012) *Chem. Eur. J.*, 18, 6152–6157.
340 Meninno, S., Vidal-Albalat, A., and Lattanzi, A. (2015) *Org. Lett.*, 17, 4348–4351.
341 Kobayashi, Y., Li, S., and Takemoto, Y. (2014) *Asian J. Org. Chem.*, 3, 403–407.
342 For a recent review, see: Pellissier, H. (2011) *Adv. Synth. Catal.*, 353, 1613–1666.
343 (a) Larrow, J.F., Schaus, S.E., and Jacobsen, E.N. (1996) *J. Am. Chem. Soc.*, 118, 7420–7421; (b) Lebel, H. and Jacobsen, E.N. (1999) *Tetrahedron Lett.*, 40, 7303–7306.
344 For reviews, see: (a) Tokunaga, M. and Hamasaki, A. (2012) in *Comprehensive Chirality*, vol. 5 (eds E. Carreira and H. Yamamoto), Elsevier, Oxford, pp. 421–435; (b) Larrow, J.F. and Quigley, P.F. (2012) in *Comprehensive Chirality*, vol. 9 (eds E. Carreira and H. Yamamoto), Elsevier, Oxford, pp. 129–146.
345 (a) Ready, J.M. and Jacobsen, E.N. (1999) *J. Am. Chem. Soc.*, 121, 6086–6087; (b) Wu, M.H., Hansen, K.B., and Jacobsen, E.N. (1999) *Angew. Chem. Int. Ed.*, 38, 2012–2014.
346 (a) Hansen, K.B., Leighton, J.L., and Jacobsen, E.N. (1996) *J. Am. Chem. Soc.*, 118, 10924–10925; (b) Konsler, R.G., Karl, J., and Jacobsen, E.N. (1998) *J. Am.*

Chem. Soc., 120, 10780–10781; (c) Nielsen, L.P.C., Stevenson, C.P., Blackmond, D.G., and Jacobsen, E.N. (2004) *J. Am. Chem. Soc.*, 126, 1360–1362; (d) Nielsen, L.P.C., Zuend, S.J., Ford, D.D., and Jacobsen, E.N. (2012) *J. Org. Chem.*, 77, 2486–2495; For the importance of counteranion in HKR, see: (e) Shepperson, I., Cavazzini, M., Pozzi, G., and Quici, S. (2004) *J. Fluorine Chem.*, 125, 175–180.

347 (a) Breinbauer, R. and Jacobsen, E.N. (2000) *Angew. Chem. Int. Ed.*, 39, 3604–3607; (b) Beigi, M., Roller, S., Haag, R., and Liese, A. (2008) *Eur. J. Org. Chem.*, 2135–2141.

348 (a) Ready, J.M. and Jacobsen, E.N. (2001) *J. Am. Chem. Soc.*, 123, 2687–2688; (b) Ready, J.M. and Jacobsen, E.N. (2002) *Angew. Chem. Int. Ed.*, 41, 1374–1377.

349 White, D.E. and Jacobsen, E.N. (2003) *Tetrahedron: Asymmetry*, 14, 3633–3638.

350 White, D.E., Tadross, P.M., Lu, Z., and Jacobsen, E.N. (2014) *Tetrahedron*, 70, 4165–4180.

351 Rajapaksa, N.S., McGowan, M.A., Rienzo, M., and Jacobsen, E.N. (2013) *Org. Lett.*, 15, 706–709.

352 (a) Zheng, X., Jones, C.W., and Weck, M. (2006) *Chem. Eur. J.*, 12, 576–583; (b) Holbach, M. and Weck, M. (2006) *J. Org. Chem.*, 71, 1825–1836.

353 (a) Zheng, X., Jones, C.W., and Weck, M. (2008) *Adv. Synth. Catal.*, 350, 255–261; (b) Goyal, P., Zheng, X., and Weck, M. (2008) *Adv. Synth. Catal.*, 350, 1816–1822; (c) Gill, C.S., Venkatasubbaiah, K., Phan, N.T.S., Weck, M., and Jones, C.W. (2008) *Chem. Eur. J.*, 14, 7306–7313; (d) Venkatasubbaiah, K., Gill, C.S., Takatani, T., Sherrill, C.D., and Jones, C.W. (2009) *Chem. Eur. J.*, 15, 3951–3955; For other examples of macrocyclic oligomeric catalysts, see: (e) Zheng, X., Jones, C.W., and Weck, M. (2007) *J. Am. Chem. Soc.*, 129, 1105–1112.

354 Rossbach, B.M., Leopold, K., and Weberskirch, R. (2006) *Angew. Chem. Int. Ed.*, 45, 1309–1312.

355 Kim, Y.-S., Guo, X.-F., and Kim, G.-J. (2010) *Catal. Today*, 92, 91–99.

356 Kim, Y.-S., Guo, X.-F., and Kim, G.-J. (2009) *Chem. Commun.*, 4296–4298.

357 Solodenko, W., Jas, G., Kunz, U., and Kirschning, A. (2007) *Synthesis*, 2007, 583–589.

358 Belser, T. and Jacobsen, E.N. (2008) *Adv. Synth. Catal.*, 350, 967–971.

359 (a) Haak, R.M., Belmonte, M.M., Escudero-Adan, E.C., Benet-Buchholz, J., and Kleij, A.W. (2010) *Dalton Trans.*, 39, 593–602; (b) Wezenberg, S.J. and Kleij, A.W. (2010) *Adv. Synth. Catal.*, 352, 85–91.

360 (a) Thakur, S.S., Li, W.-J., Shin, C.-K., and Kim, G.-J. (2006) *Chirality*, 18, 37–43; (b) Thakur, S.S., Chen, S.-W., Li, W., Shin, C.-K., Kim, S.-J., Koo, Y.-M., and Kim, G.-J. (2006) *J. Organomet. Chem.*, 691, 1862–1872; (c) Patel, D., Kurrey, G.R., Shinde, S.S., Kumar, P., Kim, G.-J., and Thakura, S.S. (2015) *RSC Adv.*, 5, 82699–82703.

361 Kawthekar, R.B. and Kim, G.-J. (2008) *Helv. Chim. Acta*, 91, 317–332.

362 (a) Kim, Y.J. and Tae, J. (2006) *Synlett*, 61–64; (b) Santhosh, R., Chouthaiwale, P.V., Suryavanshi, G., Chavan, V.B., and Sudalai, A. (2010) *Chem. Commun.*, 46, 5012–5014.

363 Devalankar, D.A., Karabal, P.U., and Sudalai, A. (2013) *Org. Biomol. Chem.*, 11, 1280–1285.

364 Reddy, A., Mujahid, M., Sasikumar, M., and Muthukrishnan, M. (2014) *Synthesis*, 46, 1751–1756.

365 Narina, S.V. and Sudalai, A. (2007) *Tetrahedron*, 63, 3026–3030.

366 Gupta, P. and Kumar, P. (2008) *Eur. J. Org. Chem.*, 1195–1202.

367 Tripathi, D., Pandey, S.K., and Kumar, P. (2009) *Tetrahedron*, 65, 2226–2231.

368 Dyer, B.S., Jones, J.D., Ainge, G.D., Denis, M., Larsen, D.S., and Painter, G.F. (2007) *J. Org. Chem.*, 72, 3282–3288.

369 Sasikumar, M., Nikalje, M.D., and Muthukrisnan, M. (2009) *Tetrahedron: Asymmetry*, 20, 2814–2817.

370 Czerwonka, R., Reddy, K.R., Baum, E., and Knölker, H.-J. (2006) *Chem. Commun.*, 711–713.

371 Gadakh, S.K., Santhosh Reddy, R., and Sudalai, A. (2012) *Tetrahedron: Asymmetry*, 23, 898–903.

372 Bartoli, G., Bosco, M., Carlone, A., Locatelli, M., Massaccesi, M., Melchiorre, P., and Sambri, L. (2004) *Org. Lett.*, 6, 2173–2176.

373 Kureshy, R.I., Kumar, M., Agrawal, S., Khan, N.H., Abdi, S.H.R., and Bajaj, H.C. (2010) *Tetrahedron: Asymmetry*, 21, 451–456.

374 Kureshy, R.I., Singh, S., Khan, N.H., Abdi, S.H.R., Agrawal, S., and Jasra, R.V. (2006) *Tetrahedron: Asymmetry*, 17, 1638–1643.

375 Kureshy, R.I., Prathap, K.J., Singh, S., Agrawal, S., Khan, N.H., Abdi, S.H.R., and Jasra, R.V. (2007) *Chirality*, 19, 809–815.

376 Bartoli, G., Bosco, M., Carlone, A., Locatelli, M., Melchiorre, P., and Sambri, L. (2004) *Org. Lett.*, 6, 3973–3975.

377 Bartoli, G., Bosco, M., Carlone, A., Locatelli, M., Melchiorre, P., and Sambri, L. (2005) *Org. Lett.*, 7, 1983–1985.

378 Kureshy, R.I., Prathap, K.J., Agrawal, S., Kumar, M., Khan, N.H., Abdi, S.H.R., and Bajaj, H.C. (2009) *Eur. J. Org. Chem.*, 2863–2871.

379 For the kinetic resolution of terminal epoxides with CO_2 catalyzed by Jacobsen's catalyst, see: (a) Yamada, W., Kitaichi, Y., Tanaka, H., Kojima, T., Sato, M., Ikeno, T., and Yamada, T. (2007) *Bull. Chem. Soc. Jpn.*, 80, 1391–1401; (b) Chen, S.-W., Kawthekar, R.B., and Kim, G.-J. (2007) *Tetrahedron Lett.*, 48, 297–300.

380 Bandini, M., Cozzi, P.G., Melchiorre, P., and Umani-Ronchi, A. (2004) *Angew. Chem. Int. Ed.*, 43, 84–87.

381 Gayet, A. and Andersson, P.G. (2005) *Tetrahedron Lett.*, 46, 4805–4807.

382 Dickmeiss, G., De Sio, V., Udmark, J., Poulsen, T.B., Marcos, V., and Jørgensen, K.A. (2009) *Angew. Chem. Int. Ed.*, 48, 6650–6653.

383 (a) Newman, H. and Angier, R.B. (1970) *Tetrahedron*, 26, 825–836; (b) Vankar, Y.D., Shah, K., Bawa, A., and Singh, S.P. (1991) *Tetrahedron*, 47, 8883–8906.

384 Meninno, S., Napolitano, L., and Lattanzi, A. (2015) *Catal. Sci. Technol.*, 5, 124–128.

385 For reviews, see: (a) Li, A.-H., Dai, L.-X., and Aggarwal, V.K. (1997) *Chem. Rev.*, 97, 2341–2372; (b) McGarrigle, E.M. and Aggarwal, V.K. (2007) in *Enantioselective Organocatalysis: Reactions and Experimental Procedures*, Chapter 10 (ed. P.I. Dalko), Wiley-VCH Verlag GmbH, Weinheim, pp. 357–389; (c) McGarrigle, E.M., Myers, E.L., Illa, O., Shaw, M.A., Riches, S.L., and Aggarwal, V.K. (2007) *Chem. Rev.*, 107, 5841–5883.

386 (a) Breau, L. and Durst, T. (1991) *Tetrahedron: Asymmetry*, 2, 367–370; (b) Aggarwal, V.K., Ford, J.G., Thompson, A., Jones, R.V.H., and Standen, M.C.H. (1996) *J. Am. Chem. Soc.*, 118, 7004–7005; (c) Aggarwal, V.K., Alonso, E., Bae, I., Hynd, G., Lydon, K.M., Palmer, M.J., Patel, M., Porcelloni, M., Richardson, J., Stenson, R.A., Studley, J.R., Vasse, J.-L., and Winn, C.L. (2003) *J. Am. Chem. Soc.*, 125, 10926–10940; (d) Hayakawa, R. and Shimizu, M. (1999) *Synlett*, 1328–1330; (e) Solladié-Cavallo, A., Roje, M., Isarno, T., Sunjic, V., and Vinkovic, V. (2000) *Eur. J. Org. Chem.*, 1077–1080; (f) Saito, T., Akiba, D., Sakairi, M., and Kanazawa, S. (2001) *Tetrahedron Lett.*, 42, 57–59; (g) Myllymäki, V.T., Lindvall, M.K., and Koskinen, A.M.P. (2001) *Tetrahedron*, 57, 4629–4635; (h) Zanardi, J., Leriverend, C., Aubert, D., Julienne, K., and Metzner, P. (2001) *J. Org. Chem.*, 66, 5620–5623; (i) Winn, C.L., Bellenie, B.R., and Goodman, J.M. (2002) *Tetrahedron Lett.*, 43, 5427–5430; (j) Davoust, M., Brière, J.-F., and Metzner, P. (2006) *Org. Biomol. Chem.*, 4, 3048–3051; (k) Davoust, M., Brière, J.-F., Jaffrès, P.-A., and Metzner, P. (2005) *J. Org. Chem.*, 70, 4166–4169.

387 (a) Solladié-Cavallo, A., Diep-Vohuule, A., Sunjic, V., and Vinkovic, V. (1996) *Tetrahedron: Asymmetry*, 7, 1783–1788; (b) Solladié-Cavallo, A., Bouérat, L., and Roje, M. (2000) *Tetrahedron Lett.*, 41, 7309–7312.

388 Aggarwal, V.K., Bae, I., Lee, H.-Y., Richardson, J., and Williams, D.T. (2003) *Angew. Chem. Int. Ed.*, 42, 3274–3278.

389 Morales-Serna, J.A., Llaveria, J., Díaz, Y., Matheu, M.I., and Castillón, S. (2008) *Org. Biomol. Chem.*, 6, 4502–4506.

390 Kotoku, N., Narumi, F., Kato, T., Yamaguchi, M., and Kobayashi, M. (2007) *Tetrahedron Lett.*, 48, 7147–7150.

391 (a) Aggarwal, V.K. and Richardson, V.K. (2003) *Chem. Commun.*, 2644–2651; (b) Aggarwal, V.K., Harvey, J.N., and Richardson, J. (2002) *J. Am. Chem. Soc.*, 124, 5747–5756.

392 Solladié-Cavallo, A. and Diep-Vohuule, A. (1995) *J. Org. Chem.*, 60, 3494–3498.

393 (a) Bellenie, B.R. and Goodman, J.M. (2004) *Chem. Commun.*, 1076–1077; (b) Aggarwal, V.K., Coogan, M.P., Stenson, R.A., Jones, R.V.H., Fieldhouse, R., and Blacker, J. (2002) *Eur. J. Org. Chem.*, 319–326.

394 Zhou, Y.-G., Hou, X.-L., Dai, L.-X., Xia, L.-J., and Tang, M.-H. (1999) *J. Chem. Soc., Perkin Trans. 1*, 77–80.

395 (a) Aggarwal, V.K., Hynd, G., Picoul, W., and Vasse, J.-L. (2002) *J. Am. Chem. Soc.*, 124, 9964–9965; (b) Aggarwal, V.K., Charmant, J.P.H., Fuentes, D., Harvey, J.N., Hynd, G., Ohara, D., Picoul, W., Robiette, R., Smith, C., Vasse, J.-L., and Winn, C.L. (2006) *J. Am. Chem. Soc.*, 124, 2105–2114.

396 Boucherif, A., Yang, Q.-Q., Wang, Q., Chen, J.-R., Lu, L.-Q., and Xiao, W.-J. (2014) *J. Org. Chem.*, 79, 3924–3929.

397 Takada, H., Metzner, P., and Philouze, C. (2001) *Chem. Commun.*, 2350–2351.

398 Sarabia, F., Chammaa, S., García-Castro, M., and Martín-Gálvez, F. (2009) *Chem. Commun.*, 5763–5765.

399 Sarabia, F., Vivar-García, C., García-Castro, M., García-Ruiz, C., Martín-Gálvez, F., Sánchez-Ruiz, A., and Chammaa, S. (2012) *Chem. Eur. J.*, 18, 15190–15201.

400 Sarabia, F., Martín-Gálvez, F., García-Ruiz, C., Sánchez-Ruiz, A., and Vivar-García, C. (2013) *J. Org. Chem.*, 78, 5239–5253.
401 García-Ruiz, C., Cheng-Sánchez, I., and Sarabia, F. (2015) *Org. Lett.*, 17, 5558–5561.
402 Sarabia, F., Chammaa, S., and García-Ruiz, C. (2011) *J. Org. Chem.*, 76, 2132–2144.
403 Sarabia, F., Vivar-García, C., García-Ruiz, C., Sánchez-Ruiz, A., Pino-González, M.S., García-Castro, M., and Chammaa, S. (2014) *Eur. J. Org. Chem.*, 3847–3867.
404 Gordillo, P.G., Aparicio, D.M., Flores, M., Mendoza, A., Orea, L., Juárez, J.R., Huelgas, G., Gnecco, D., and Terán, J.L. (2013) *Eur. J. Org. Chem.*, 5561–5565.
405 Illa, O., Arshad, M., Ros, A., McGarrigle, E.M., and Aggarwal, V.K. (2010) *J. Am. Chem. Soc.*, 132, 1828–1830.
406 Illa, O., Namutebi, M., Saha, C., Ostovar, M., Chen, C.C., Haddow, M.F., Nocquet-Thibault, S., Luisi, M., McGarrigle, E.M., and Aggarwal, V.K. (2013) *J. Am. Chem. Soc.*, 135, 11951–11966.
407 Wu, H.-Y., Chang, C.-W., and Chein, R.-J. (2013) *J. Org. Chem.*, 78, 5788–5793.
408 Hansch, M., Illa, O., McGarrigle, E.M., and Aggarwal, V.K. (2008) *Chem. Asian J.*, 3, 1657–1663.
409 Piccinini, A., Kavanagh, S.A., and Connon, S.J. (2012) *Chem. Commun.*, 48, 7814–7816.
410 (a) Newmann, M.S. and Magerlein, B.J. (1949) *Org. React.*, 5, 413–441; (b) Ballester, M. (1955) *Chem. Rev.*, 55, 283–300; (c) Rosen, T. (1991) in *Comprehensive Organic Synthesis*, vol. 2 (eds B.M. Trost, I. Fleming, and C.H. Heathcock), Pergamon, Oxford, pp. 409–439.
411 Yliniemelä, A., Brunov, G., Flügge, J., and Teleman, O. (1996) *J. Org. Chem.*, 61, 6723–6726.
412 (a) Abdel-Magid, A., Prigden, L.N., Eggleston, D.S., and Lantos, I. (1986) *J. Am. Chem. Soc.*, 108, 4595–4602; (b) Prigden, L.N., Abdel-Magid, A., Lantos, I., Shilcrat, S., and Eggleston, D.S. (1993) *J. Org. Chem.*, 58, 5107–5117.
413 (a) Takagi, R., Kimura, J., Ohba, Y., Takezono, K., Hiraga, Y., Kojima, S., and Ohkata, K. (1998) *J. Chem. Soc., Perkin Trans. 1*, 689–698; (b) Ohkata, K., Kimura, J., Shinohara, Y., Takagi, R., and Yoshkazu, H. (1996) *Chem. Commun.*, 2411–2412.
414 Wang, Y.-C., Li, C.-L., Tseng, H.-L., Chuang, S.-C., and Yan, T.-H. (1999) *Tetrahedron: Asymmetry*, 10, 3249–3251.
415 Palomo, C., Oiarbide, M., Sharma, A.K., González-Rego, M.C., Linden, A., García, J.M., and González, A. (2000) *J. Org. Chem.*, 65, 9007–9012.
416 Ghosh, A.K. and Kim, J.-H. (2004) *Org. Lett.*, 6, 2725–2728.
417 Corey, E.J. and Choi, S. (1991) *Tetrahedron Lett.*, 32, 2857–2860.
418 Kiyooka, S.-i. and Shahid, K.A. (2000) *Tetrahedron: Asymmetry*, 11, 1537–1542.
419 Li, B. and Li, C. (2014) *J. Org. Chem.*, 79, 8271–8277.
420 For a recent review, see: Bakó, P., Rapi, Z., and Keglevich, G. (2014) *Curr. Org. Synth.*, 11, 361–376.

421 (a) Humelen, J.C. and Wynberg, H. (1978) *Tetrahedron Lett.*, 12, 1089–1092; (b) Colonna, S., Fornasier, R., and Pfeiffer, U. (1978) *J. Chem. Soc., Perkin Trans. 1*, 8–12.

422 (a) Arai, S. and Shioiri, T. (1998) *Tetrahedron Lett.*, 39, 2145–2148; (b) Arai, S., Shirai, Y., Ishida, T., and Shioiri, T. (1999) *Tetrahedron*, 55, 6375–6386.

423 Arai, S., Shirai, Y., Ishida, T., and Shioiri, T. (1999) *Chem. Commun.*, 49–50.

424 Arai, S. and Shioiri, T. (2002) *Tetrahedron*, 58, 1407–1413.

425 Ku, J.-M., Yoo, M.-S., Park, H.-G., Jew, S.-S., and Jeong, B.-S. (2007) *Tetrahedron*, 63, 8099–8103.

426 Arai, S., Tokumaru, K., and Aoyama, T. (2004) *Tetrahedron Lett.*, 45, 1845–1848.

427 Liu, Y., Provencher, B.A., Bartelson, K.J., and Deng, L. (2011) *Chem. Sci.*, 2, 1301–1304.

428 (a) Bakó, P., Szöllõsy, Á., Bombicz, P., and Tõke, L. (1997) *Synlett*, 291–292; (b) Bakó, P., Vizvárdi, K., Bajor, Z., and Tõke, L. (1998) *Chem. Commun.*, 1193–1194; (c) Bakó, P., Czinege, E., Bakó, T., Czugler, M., and Tõke, L. (1999) *Tetrahedron: Asymmetry*, 10, 4539–4551.

429 (a) Bakó, P., Rapi, Z., Keglevich, G., Szabó, T., Sóti, P.L., Vígh, T., Grűn, A., and Holczbauer, T. (2011) *Tetrahedron Lett.*, 52, 1473–1476; (b) Rapi, Z., Szabó, T., Keglevich, G., Szöllősky, Á., Drahos, L., and Bakó, P. (2011) *Tetrahedron: Asymmetry*, 22, 1189–1196; (c) Rapi, Z., Bakó, P., Keglevich, G., Szöllősky, Á., Drahos, L., Botyánszki, A., and Holczbauer, T. (2012) *Tetrahedron: Asymmetry*, 23, 489–496.

430 (a) List, B. (2004) in *Amine-Catalyzed Aldol Reactions in Modern Aldol Reactions*, vol. 1 (ed. R. Mahrwald), Wiley-VCH Verlag GmbH, Berlin, pp. 161–200; (b) Guillena, G. (2013) in *Modern Methods in Stereoselective Aldol Reactions* (ed. R. Mahrwald), Wiley-VCH Verlag GmbH, Weinheim, pp. 155–268; (c) Mase, N. and Hayashi, Y. (2014) in *Comprehensive Organic Synthesis*, vol. 2, 2nd edn (eds P. Knochel, G.A. Molander, and K. Mikami), Elsevier, Amsterdam, pp. 273–339.

431 Guillena, G., Hita, M.C., and Nájera, C. (2007) *Tetrahedron: Asymmetry*, 18, 1272–1277.

432 Watanabe, S.-I., Hasebe, R., Ouchi, J., Nagasawa, H., and Kataoka, T. (2010) *Tetrahedron Lett.*, 51, 5778–5780.

433 Achard, T.J.R., Belokon', Y.N., Ilyin, M., Moskalenko, M., North, M., and Pizzato, F. (2007) *Tetrahedron Lett.*, 48, 2965–2969.

434 Liu, W.-J., Lv, B.-D., and Gong, L.-Z. (2009) *Angew. Chem. Int. Ed.*, 48, 6503–6506.

435 He, L., Liu, W.J., Ren, L., Lei, T., and Gong, L.Z. (2010) *Adv. Synth. Catal.*, 352, 1123–1127.

436 Liu, G., Zhang, D., Li, J., Xu, G., and Sun, J. (2013) *Org. Biomol. Chem.*, 11, 900–904.

437 Kuang, Y., Lu, Y., Tang, Y., Liu, X., Lin, L., and Feng, X. (2014) *Org. Lett.*, 16, 4244–4247.

438 (a) Perry, M.C. and Burgess, K. (2003) *Tetrahedron: Asymmetry*, 14, 951–961; (b) César, V., Bellemin-Laponnaz, S., and Gade, L.H. (2004) *Chem. Soc. Rev.*, 33, 619–636.

439 (a) Niemeier, O., Henseler, A., and Enders, D. (2007) *Chem. Rev.*, 107, 5606–5655; (b) Bugaut, X. and Glorius, F. (2012) *Chem. Soc. Rev.*, 41, 3511–3522.

440 Reddi, R.N., Prasad, P.K., and Sudalai, A. (2015) *Angew. Chem. Int. Ed.*, 54, 14150–14153.

441 For a recent review, see: Jiang, K. and Chen, Y.-C. (2014) *Tetrahedron Lett.*, 55, 2049–2055.

442 (a) Gaunt, M.J., Papageorgiou, C.D., and Ley, S.V. (2003) *Angew. Chem. Int. Ed.*, 42, 828–831; (b) Bremeyer, N., Smith, S.C., Ley, S.V., and Gaunt, M.J. (2004) *Angew. Chem. Int. Ed.*, 43, 2681–2684.

443 Aggarwal, V.K., Harvey, J.N., and Robiette, R. (2005) *Angew. Chem. Int. Ed.*, 44, 5468–5471.

444 Rosier, L., Robiette, R., and Waser, M. (2016) *Synlett*, 27, 1963–1968.

445 Kinoshita, H., Ihoriya, A., Ju-ichi, M., and Kimachi, T. (2010) *Synlett*, 2330–2334.

446 For recent reviews, see: (a) Lin, H., Liu, J.-Y., Wang, H.-B., Ahmed, A.A.Q., and Wu, Z.-L. (2011) *J. Mol. Catal. B: Enzym.*, 72, 77–89; (b) Yudin, A. (ed.) (2006) *Aziridines and Epoxides in Organic Synthesis*, Chapter 10, Wiley-VCH Verlag GmbH, Weinheim.

447 (a) Bernasconi, S., Orsini, F., Sello, G., Colmegna, A., Galli, E., and Bestetti, G. (2000) *Tetrahedron Lett.*, 41, 9157–9161; (b) Park, S.M., Bae, J.W., Han, J.H., Lee, E.Y., Lee, S.G., and Park, S. (2006) *J. Microbiol. Biotechnol.*, 16, 1041–1046.

448 Panke, S., Held, M., Wubbolts, M.G., Witholt, B., and Schmid, A. (2002) *Biotechnol. Bioeng.*, 80, 33–41.

449 (a) Schmid, A., Hofstetter, K., Feiten, H.J., Hollmann, F., and Witholt, B. (2001) *Adv. Synth. Catal.*, 343, 732–737; (b) Bernasconi, S., Orsini, F., Sello, G., and Di Gennaro, P. (2004) *Tetrahedron: Asymmetry*, 15, 1603–1606.

450 Lin, H., Liu, Y., and Wu, Z.L. (2011) *Chem. Commun.*, 47, 2610–2612.

451 Liu, Y.-C., Liu, Y., and Wu, Z.-L. (2015) *Org. Biomol. Chem.*, 13, 2146–2152.

452 Liu, Y.-C. and Wu, Z.-L. (2016) *Chem. Commun.*, 52, 1158–1161.

453 Lin, H., Liu, Y., and Wu, Z.-L. (2011) *Tetrahedron: Asymmetry*, 22, 134–137.

454 Li, A., Liu, J., Pham, S.Q., and Li, Z. (2013) *Chem. Commun.*, 49, 11572–11574.

455 Dembitsky, V.M. (2003) *Tetrahedron*, 59, 4701–4720.

456 (a) Allain, E.J., Hager, L.P., Deng, L., and Jacobsen, E.N. (1993) *J. Am. Chem. Soc.*, 115, 4415–4416; (b) Dexter, A.F., Lakner, F.J., Campbell, R.A., and Hager, L.P. (1995) *J. Am. Chem. Soc.*, 117, 6412–6413.

457 For reviews, see: (a) Weijers, C. and de Bont, J.A.M. (1999) *J. Mol. Catal. B: Enzym.*, 6, 199–214; (b) Lee, E.Y. (2008) *Biotechnol. Lett*, 30, 1509–1514; (c) Nestl, B.M., Hammer, S.C., Nebel, B.A., and Hauer, B. (2014) *Angew. Chem. Int. Ed.*, 53, 3070–3095.

458 (a) Osprian, I., Kroutil, W., Mischitz, M., and Faber, K. (1997) *Tetrahedron: Asymmetry*, 8, 65–71; (b) Krenn, W., Osprian, I., Kroutil, W., Braunegg, G., and Faber, K. (1999) *Biotechnol. Lett*, 21, 687–690; (c) Deregnaucourt, J., Archelas, A., Barbirato, F., Paris, J.-M., and Furstoss, R. (2006) *Adv. Synth. Catal.*, 348, 1165–1169; (d) Bala, N. and Chimni, S.S. (2010) *Tetrahedron: Asymmetry*, 21, 2879–2898.

459 (a) Reetz, M.T. (2009) *J. Org. Chem.*, 74, 5767–5778; (b) Jochens, H., Stiba, K., Savile, C., Fujii, R., Yu, J.-G., Gerassenkov, T., Kazlauskas, R.J., and Bornscheuer, U.T. (2009) *Angew. Chem. Int. Ed.*, 48, 3532–3535.

460 Zhao, J., Chu, Y.-Y., Li, A.-T., Ju, X., Kong, X.-D., Pan, J., Tang, Y., and Xu, J.-H. (2011) *Adv. Synth. Catal.*, 353, 1510–1518.

461 Monfort, N., Archelas, A., and Furstoss, R. (2004) *Tetrahedron*, 60, 601–605.

462 Bottalla, A.-L., Ibrahim-Ouali, M., Santelli, M., Furstoss, R., and Archelas, A. (2007) *Adv. Synth. Catal.*, 349, 1102–1110.

463 Ferrandi, E.E., Marchesi, C., Annovazzi, C., Riva, S., Monti, D., and Wohlgemuth, R. (2015) *ChemCatChem*, 7, 3171–3178.

464 Kotik, M., Stepanek, V., Grulich, M., Kyslik, P., and Archelas, A. (2010) *J. Mol. Catal. B: Enzym.*, 65, 41–48.

465 Zhao, W., Kotik, M., Iacazio, G., and Archelas, A. (2015) *Adv. Synth. Catal.*, 357, 1895–1908.

466 Schallmey, A. and Schallmey, M. (2016) *Appl. Microbiol. Biotechnol.*, 100, 7827–7839.

467 (a) Majerić Elenkov, M., Tang, L., Hauer, B., and Janssen, D.B. (2006) *Org. Lett.*, 8, 4227–4229; (b) Szymanski, W., Postema, C.P., Tarabiono, C., Berthiol, F., Campbell-Verduyn, L., de Wildeman, S., de Vries, J.G., Feringa, B.L., and Janssen, D.B. (2010) *Adv. Synth. Catal.*, 352, 2111–2115.

468 Lutje Spelberg, J.H., van Hylckama Vlieg, J.E.T., Tang, L., Janssen, D.B., and Kellogg, R.M. (2001) *Org. Lett.*, 3, 41–43.

469 Campbell-Verduyn, L.S., Szymanski, W., Postema, C.P., Dierckx, R.A., Elsinga, P.H., Janssen, D.B., and Feringa, B.L. (2010) *Chem. Commun.*, 46, 898–900.

4

Asymmetric Oxaziridination

4.1 Introduction

Since their discovery in 1956 by Emmons [1], oxaziridines have become a class of versatile oxidants widely employed in organic synthesis, whose characteristic feature is the presence of two electronegative heteroatoms within a strained three-membered ring [2]. These compounds behave as oxygen or nitrogen donors, depending on their structural electronic features and the nature of the nucleophile employed [3]. For example, oxidations and aminations of alkanes, alkenes, arenes, amines, sulfides, phosphines, and alkoxides are typical reactions using oxaziridines as oxidants. Despite the diverse range of substrates oxidized in these reactions, the mechanisms involved are close. These transformations generally involve a substantially concerted atom transfer from the oxidant to the substrate, driven by the release of ring strain and by the formation of a strong carbonyl, imine, or oxometal π-bond [4]. In particular, chiral oxaziridines serve as stereoselective oxidants [3e, 5], but also as substrates in cycloadditions [6] and rearrangements [7]. Oxaziridines can be easily prepared by a variety of procedures on a multigram scale. Additionally, they present the advantage to be manipulated without precautions against air or moisture and can be stored indefinitely at reduced temperatures without noticeable decomposition. Research involving oxaziridines over the past five decades has been motivated by the unusual physical properties of these compounds as well as their distinctive reactivity as electrophilic oxygen atom transfer reagents. Even if several procedures for the synthesis of chiral oxaziridines are known, they are still limited. Most of those reported until recently rely upon diastereoselective oxidations of commercially available enantiopure reagents, and it must be noted that direct catalytic enantioselective oxidations of imines, which constitute a straightforward process to reach optically active oxaziridines, are still rare. The goal of this chapter is to collect the advances in asymmetric oxaziridination using chiral substrates (auxiliaries) as well as chiral catalysts, focusing on those published in the last 12 years. This chapter is subdivided into three parts, dealing successively with asymmetric oxaziridination based on the use of chiral substrates (auxiliaries), asymmetric oxaziridination based on the use of chiral catalysts, and kinetic resolutions.

Asymmetric Synthesis of Three-Membered Rings, First Edition.
Hélène Pellissier, Alessandra Lattanzi and Renato Dalpozzo.
© 2017 Wiley-VCH Verlag GmbH & Co. KGaA. Published 2017 by Wiley-VCH Verlag GmbH & Co. KGaA

4.2 Oxaziridination Using Chiral Substrates

The most used strategy to reach chiral oxaziridines consists in the diastereoselective oxidation of commercially available enantiopure reagents, such as chiral imines, using *meta*-chloroperoxybenzoic acid (MCPBA) or oxone as oxidant. One limitation of this approach is the restricted structural diversity of the products formed. An early and beautiful illustration of this methodology was reported by Davis *et al.*, dealing with the diastereoselective oxidation of camphor-derived *N*-sulfonylimines using oxone [5b, 8]. The oxidation took place from the endo-face of the C=N double bond due to the steric blocking of the exo-face, which resulted in the formation of a single oxaziridine diastereomer obtained in 90% yield, as shown in Scheme 4.1. Comparable results have been obtained with related camphor-derived cyclic sulfonylimines [8, 9]. In the same context, an optically active *N*-sulfamyloxaziridine was also reported by these authors [10]. A range of chiral *N*-sulfonyl oxaziridines have been successfully employed in various enantioselective oxidations, such as α-hydroxylations of enolates [11], oxidations of sulfides [9b], disulfides [12], selenides [13], alkenes [11, 14], and sulfenimines [5, 15]. Unlike *N*-alkyl and *N*-aryloxaziridines, which are only weak oxidants, these oxaziridines bearing electron-withdrawing groups react with a broad range of nucleophiles, such as enolates, silyl enol ethers, organometallic compounds, alkenes, arenes, tertiary amines, thiols, thioethers, and selenides [3].

In the same area, Lete *et al.* later developed the oxidation of chiral 3-oxocamphor-derived *N*-sulfonylimine by using MCPBA as oxidant, which afforded the corresponding enantiopure oxaziridine in almost quantitative yield (98%), as shown in Scheme 4.2 [16]. This compound was further successfully employed as oxidant in the sulfur oxidation of 1,3-dithiane and 2-acyl-1,3-dithiane derivatives.

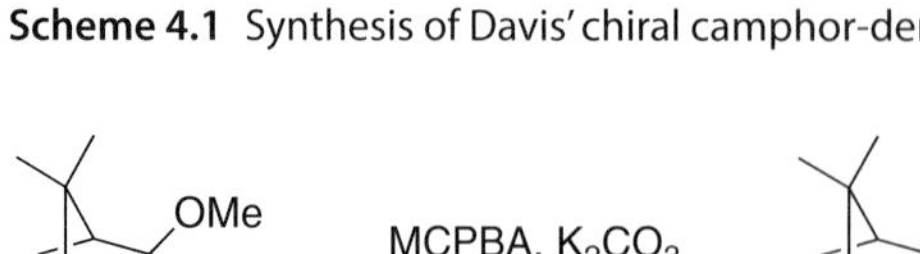

90%, de = 100%

Scheme 4.1 Synthesis of Davis' chiral camphor-derived oxaziridine.

98%, de = 100%

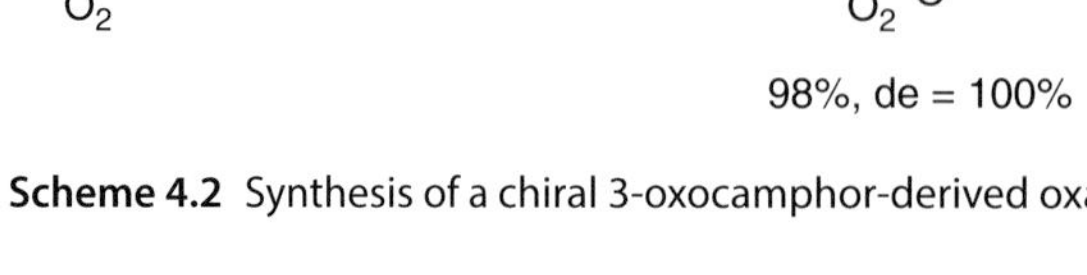

Scheme 4.2 Synthesis of a chiral 3-oxocamphor-derived oxaziridine.

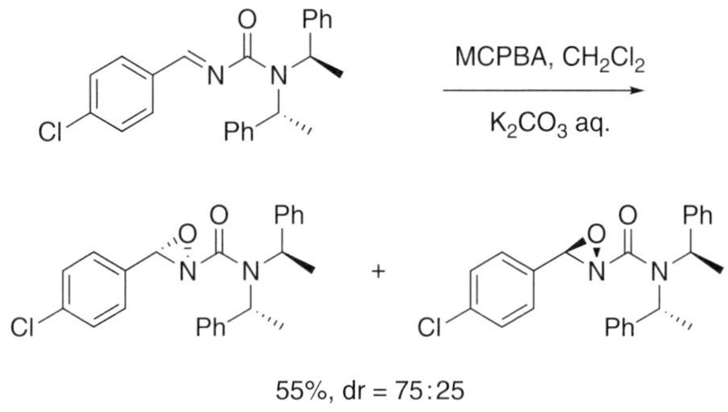

Scheme 4.3 Oxidation of a chiral acyclic imine.

While the best diastereoselectivities were generally achieved with cyclic substrates [15c, d, 17], acyclic imines usually afford mixtures of diastereomers [8, 9, 15e, 18]. As an example, the chiral acyclic imine depicted in Scheme 4.3 submitted to oxidation with MCPBA led to a 75:25 mixture of diastereomers in 55% yield [15a].

In another context, oxaziridines bearing *N*-phosphinoyl groups are typically accessed via the reaction of aryl oximes with dichlorodiphenylphosphine, followed by oxidation of the rearranged *N*-phosphinoyl imines with MCPBA. In 1994, Jennings *et al.* reported the synthesis of chiral *N*-phosphinoyloxaziridines with a stereogenic phosphorus center, which is depicted in Scheme 4.4 [19]. Due to the potential of chlorophosphorus reagents to racemize by chloride exchange, the corresponding chiral *N*-phosphinoyl imines were *in situ* prepared from the optically active phosphinic amides. For example, condensation of the amide depicted in Scheme 4.4 to an aryl aldehyde proceeded cleanly in the presence of titanium(IV) chloride and triethylamine, followed by oxidation of the resulting *N*-phosphinoyl imine with MCPBA/KF that provided the corresponding oxaziridine as a 72:28 mixture of diastereomers in 51% yield.

In 2011, Pellacani *et al.* reported the synthesis of chiral trifluoromethyl oxaziridines based on the asymmetric oxidation of the corresponding chiral

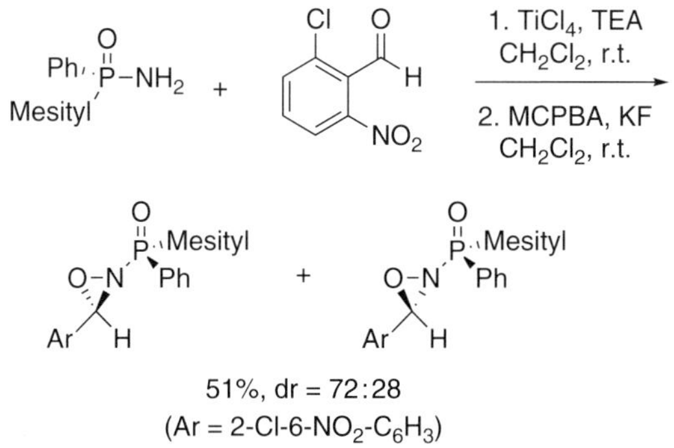

Scheme 4.4 Synthesis of a chiral *N*-phosphinoyloxaziridine.

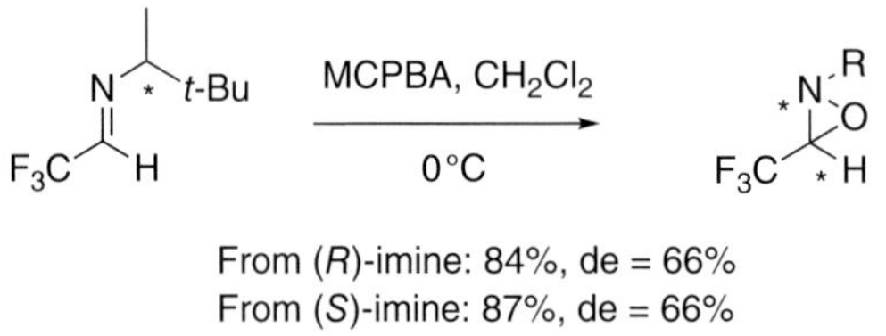

Scheme 4.5 Synthesis of chiral trifluoromethyl oxaziridines.

From (R)-imine: 84%, de = 66%
From (S)-imine: 87%, de = 66%

trifluoromethyl imines with MCPBA as oxidant [20]. As shown in Scheme 4.5, the oxaziridines were obtained in good yields (84–87%) with good diastereoselectivity of 66% de. The diastereoselectivity observed was found to be related to the steric hindrance of the N-substituent of the imine. These chiral products can be regarded as efficient oxygenating agents since fluorinated oxaziridines are considered effective as much as dioxiranes and certainly more effective than unfluorinated analogs, as a consequence of the strong electron-withdrawing effect of fluoroalkyl groups. Consequently, these compounds will be used in various reactions.

Oxaziridines bearing unsubstituted nitrogen atoms are generally more difficult to prepare due to the instability of the corresponding N–H imine precursors. In this context, Page et al. have described the synthesis of chiral N–H oxaziridines derived from camphor [17c, 21]. Due to the steric hindrance around the ketone moiety of camphor, the typical routes to access the N–H oxaziridine were unsuccessful. An alternative sequence involving nitrosation and rearrangement of the oxime depicted in Scheme 4.6, followed by ammonolysis of the resulting nitromine, provided access to the corresponding expected primary imine. Oxidation of the latter from the endo-face using MCPBA as oxidant then afforded the corresponding chiral N–H oxaziridine as a 60:40 mixture of diastereomers in nitrogen in 97% yield. It is noteworthy that unlike many N–H oxaziridines, this product could be isolated in pure form and stored up to 6 months at 5 °C without noticeable decomposition.

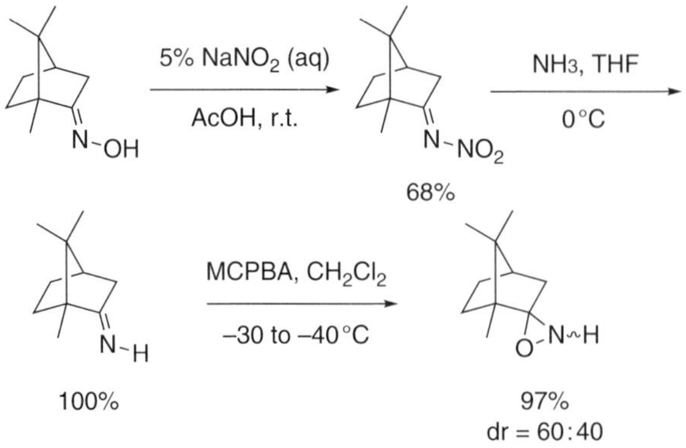

Scheme 4.6 Synthesis of a chiral camphor-derived N-H oxaziridine.

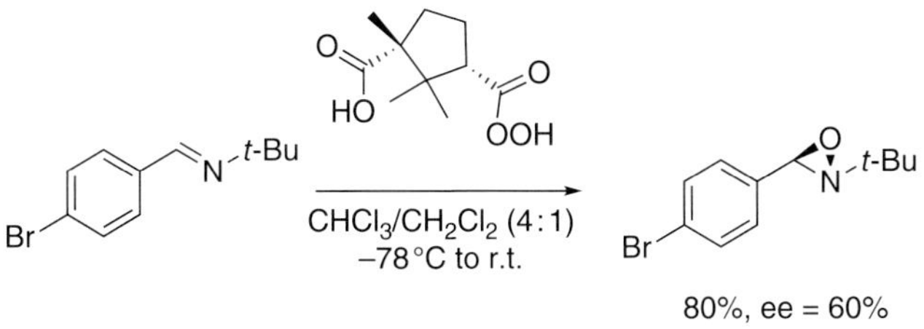

4 Asymmetric Oxaziridination

Scheme 4.7 Synthesis of a chiral N-alkyloxaziridine using a chiral oxidant.

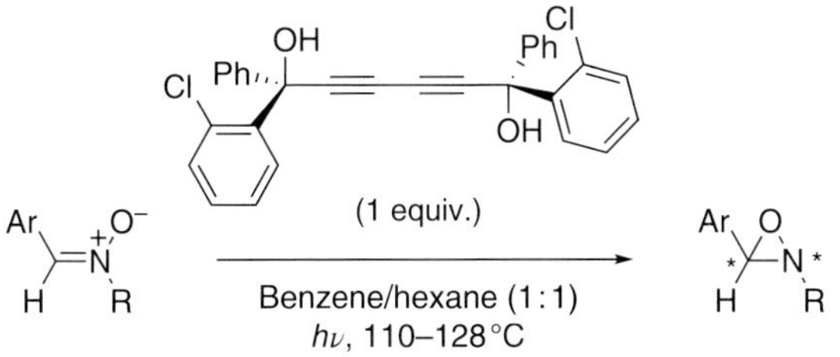

Scheme 4.8 Synthesis of chiral N-alkyloxaziridines through asymmetric photocyclization of nitrones.

Ar = o-ClC$_6$H$_4$, R = t-Bu: 51%, ee = 100%
Ar = Ph, R = CHMe(i-Pr): 40%, ee = 100%
Ar = p-ClC$_6$H$_4$, R = t-Bu: 74%, ee = 30%
Ar = 3,4-(OCH$_2$O)C$_6$H$_3$, R = t-Bu: 52%, ee = 94%
Ar = 3,4-(OCH$_2$O)C$_6$H$_3$, R = i-Pr: 63%, ee = 28%
Ar = Ph, R = t-Bu: 41%, ee = 10%

On the other hand, the asymmetric synthesis of oxaziridines based on the use of a chiral oxidizing agent proved to be an unsuccessful approach. For example, the few oxidations performed with monoperoxycamphoric acid generally yielded the products in low-to-moderate enantioselectivities [22], as reported in the example depicted in Scheme 4.7 for which an enantioselectivity of 60% ee was achieved [23].

In another context, an original asymmetric synthesis of oxaziridines was reported by Toda and Tanaka on the basis of the enantioselective photocyclization of nitrones in a crystalline inclusion complex with a chiral 2,4-diyne-2,6-diol [24]. In some cases of substrates, enantioselectivities of up to 100% ee were achieved in combination with moderate yields of up to 74%, as shown in Scheme 4.8. However, the performances were found strongly substrate-sensitive, spreading out in the 10–100% ee range.

In 2012, Hassine et al. reported the use of a combination of a chiral α-bromonitrile and hydrogen peroxide as oxidizing system in the asymmetric oxidation of (S)-1-phenylethylarylimines into the corresponding optically active oxaziridines [25]. As shown in Scheme 4.9, high diastereoselectivities of up to 94% de were achieved under this double induction. Under similar conditions, only moderate diastereoselectivities were, however, observed in the oxidation of N-t-butylarylimines.

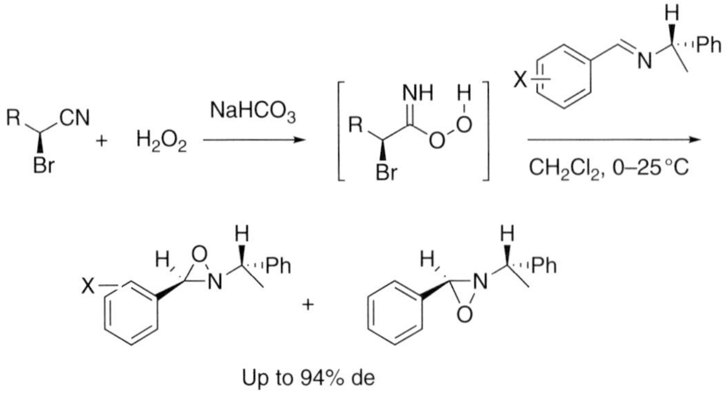

Scheme 4.9 Synthesis of chiral N-alkyloxaziridines using a combination of H_2O_2 and chiral α-bromonitrile.

4.3 Oxaziridination Using Chiral Catalysts

As demonstrated in Section 4.2, the asymmetric synthesis of oxaziridines can be performed starting from some specific scaffold of the chiral pool. Therefore, general methods applicable to a large array of substrates are still highly desirable. Until recently, no method based on the asymmetric catalysis starting from achiral imines was available. In the past few years, several efficient catalytic methods have been successfully developed, but all of them are limited to the synthesis of chiral N-sulfonyl oxaziridines. Actually, the first catalytic enantioselective synthesis of oxaziridines was reported by Jørgensen et al. in 2011, describing the oxidation of acyclic N-tosyl aldimines using a cinchona-alkaloid-based bifunctional phase-transfer catalyst [26]. This chiral base catalyst was employed to activate the nucleophilic attack of MCPBA. In this study, various dihydro derivatives of cinchona alkaloids were evaluated for this purpose. Moreover, alternative oxidants were found to be not suitable for the reaction. The presence of a hydrogen-bonding donor group in the catalyst structure proved to be an essential requisite to achieve a high level of enantioselectivity. This result was explained by assuming that the reaction proceeded through a highly organized transition state, wherein the organocatalyst was also able to activate the imine through coordination with the sulfonyl group. The anthracenylmethyl-modified catalyst depicted in Scheme 4.10 allowed the best enantioselectivities of up to 94% ee to be achieved combined with good yields (59–96%). A wide range of aromatic trans-N-tosyl oxaziridines were prepared through oxidation of the corresponding N-tosyl aldimines with MCPBA. This method was also successfully applied to aliphatic N-tosyl aldimines, provided that a bulky tertiary alkyl group was not present. Mechanistic study on concentration dependence effects and on the kinetics of differently substituted imines were not supportive of a concerted nucleophilic attack on the peroxidic oxygen, typical of metal-catalyzed epoxidations. Indeed, a stepwise mechanism with the former nucleophilic attack of the oxygen atom of the peracid to the imine, followed by ring closure favored by the heterolytic cleavage of the O—O bond, was suggested (Scheme 4.10). In the rate-determining

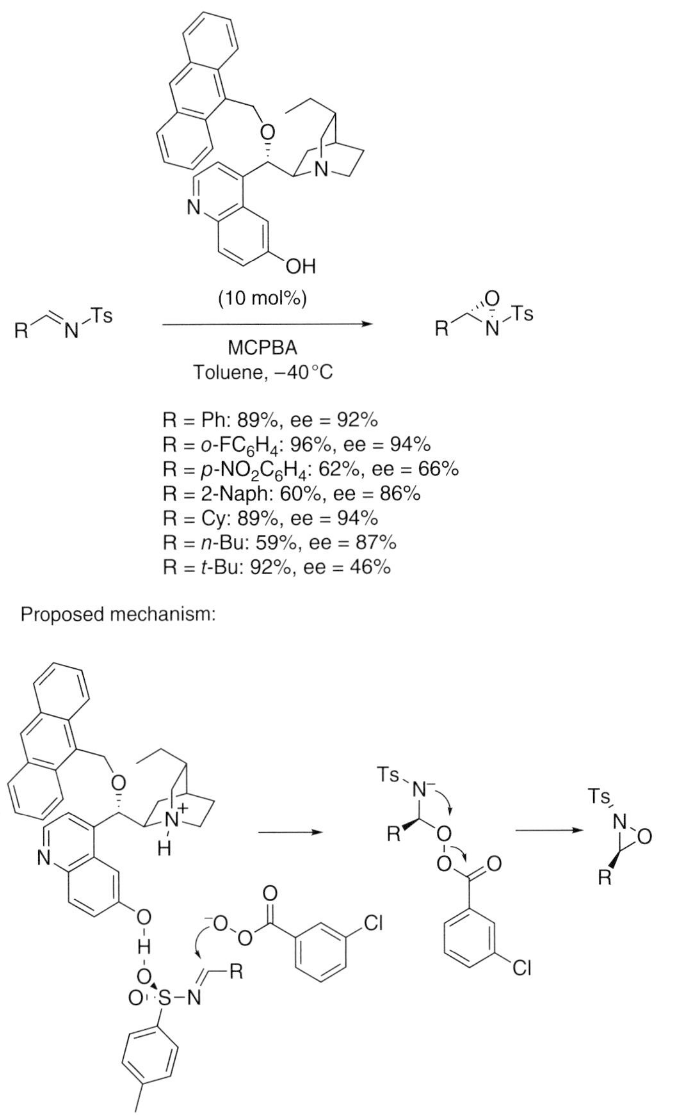
OH
(10 mol%)
MCPBA
Toluene, −40 °C
R = Ph: 89%, ee = 92%
R = o-FC6H4: 96%, ee = 94%
R = p-NO2C6H4: 62%, ee = 66%
R = 2-Naph: 60%, ee = 86%
R = Cy: 89%, ee = 94%
R = n-Bu: 59%, ee = 87%
R = t-Bu: 92%, ee = 46%
Proposed mechanism:

Scheme 4.10 First (organo)catalytic enantioselective oxaziridination.

step, both the imine and the peracid would be activated by the organocatalyst through hydrogen bonding and ion pairing, respectively. Undoubtedly, this methodology was attractive in terms of broad substrate scope and excellent level of enantioselectivity, although a modest performance remained with hindered *N*-sulfonyl imines.

Later, Jin *et al.* reinvestigated this process screening a variety of organocatalysts elaborated at 9-OH, 6′-OH, and terminal vinyl groups of quinidine and quinine compounds [27]. The quinidine derivative depicted in Scheme 4.11, bearing a free 9-OH group and a sulfur-containing chain linked to the quinuclidine core, catalyzed the oxidation, providing a range of chiral aromatic N-tosyl oxaziridines in high yields and moderate-to-excellent enantioselectivities of 53–95% ee. It must be noted that these enantioselectivities were generally lower than those obtained in the precedent study reported by Jørgensen (Scheme 4.10). The 9-OH group apparently played a crucial role in the bifunctional mechanism of the reaction, in place of the phenolic 6′-OH group, but they proved that the results were also positively affected by the presence of the sulfur-containing chain. A drawback of this methodology was the inability to reach both enantiomers of oxaziridines. Although the *trans*-(R,R)-oxaziridine was efficiently produced under these conditions, the use of the enantiomeric catalyst derived from quinine provided a racemic mixture.

With the aim of extending the substrate scope of asymmetric oxaziridination, the same authors later investigated a series of cinchona-alkaloid-ester derivatives as organocatalysts under similar reaction conditions [28]. It was shown that different esters exhibited some differences in enantioselectivity. Therefore, the electronic effect of the acyl group of ester in catalysts showed some degree of influence on the stereoinduction of the oxaziridination of N-tosyl benzaldimine. When a methyl group in the ester catalyst was changed into an ethyl group or an isopropyl group, the enantioselectivities decreased from 86% ee to 81% ee, and 79% ee, respectively, while when electron-withdrawing 4-nitrophenyl ester group

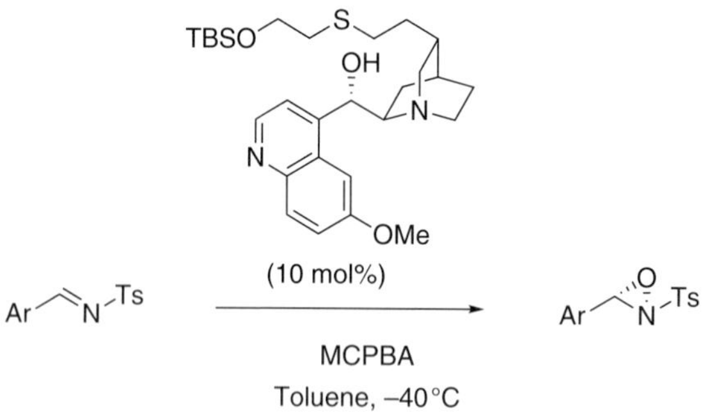

Ar = Ph: 93%, ee = 88%
Ar = m-Tol: 87%, ee = 95%
Ar = m-MeOC$_6$H$_4$: 83%, ee = 92%
Ar = p-(t-Bu)C$_6$H$_4$: 80%, ee = 91%
Ar = m-BrC$_6$H$_4$: 85%, ee = 90%
Ar = p-(i-Pr)C$_6$H$_4$: 78%, ee = 88%
Ar = m-FC$_6$H$_4$: 85%, ee = 86%
Ar = m-ClC$_6$H$_4$: 83%, ee = 86%
Ar = p-FC$_6$H$_4$: 78%, ee = 80%
Ar = o-FC$_6$H$_4$: 90%, ee = 75%
Ar = p-Tol: 72%, ee = 74%
Ar = o-BrC$_6$H$_4$: 92%, ee = 67%
Ar = o-ClC$_6$H$_4$: 91%, ee = 61%
Ar = o-Tol: 89%, ee = 60%
Ar = p-ClC$_6$H$_4$: 76%, ee = 57%
Ar = 1-Naph: 82%, ee = 53%

Scheme 4.11 Cinchona-alkaloid-catalyzed oxidation of aromatic N-tosyl aldimines with MCPBA.

was introduced in the catalyst, the highest enantioselectivity of 89% ee was reached. Having selected the cinchona alkaloid derivative bearing a 4-nitrophenyl ester group as optimal organocatalyst, it was further applied to the asymmetric oxaziridination of a range of aromatic *N*-tosyl aldimines with MCPBA as oxidant to provide the corresponding three-membered rings in high yields (86–94%) and good-to-excellent enantioselectivities of 70–98% ee, as shown in Scheme 4.12. The enantioselectivity values were significantly affected by the nature and the position of the substituents on the aromatic ring. The reaction of *meta*-substituted substrates furnished the corresponding products with a high level of enantioselectivities (90–97% ee). Furthermore, electron-donating groups at the *para*-position appeared to have a favorable effect on stereoinduction. It is worth noting that the use of this ester catalyst showed increased enantioselectivities (10–11% ee) on the *ortho*-substituted substrates in comparison with the previous study (Scheme 4.11). Additionally, 1-naphthyl-based imine gave the corresponding oxaziridine with 87% ee, which was much higher than the value of 53% ee in the previous work (Scheme 4.11).

In addition, remarkable results were reported by Ooi *et al.* in an enantioselective organocatalytic oxidation of *N*-sulfonyl aldimines using a mixture of hydrogen peroxide and trichloroacetonitrile as the oxidizing system [29]. The process

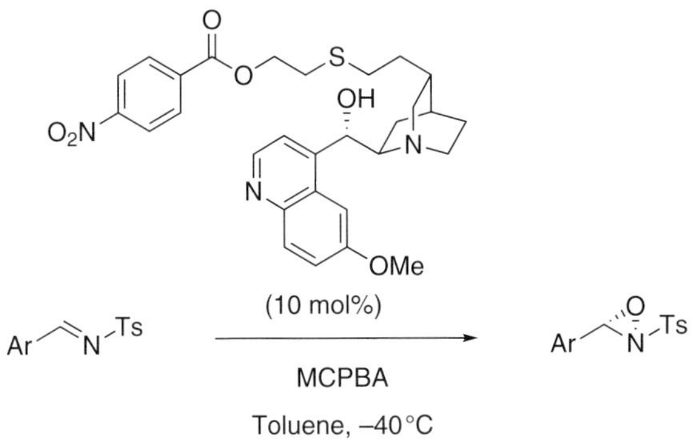

Ar = Ph: 94%, ee = 89%
Ar = *o*-FC$_6$H$_4$: 92%, ee = 86%
Ar = *o*-BrC$_6$H$_4$: 92%, ee = 78%
Ar = *o*-Tol: 89%, ee = 70%
Ar = *m*-ClC$_6$H$_4$: 88%, ee = 91%
Ar = *m*-BrC$_6$H$_4$: 86%, ee = 90%
Ar = *m*-Tol: 90%, ee = 97%
Ar = *m*-MeOC$_6$H$_4$: 94%, ee = 93%
Ar = 3,5-(MeO)$_2$C$_6$H$_3$: 92%, ee = 95%
Ar = *p*-FC$_6$H$_4$: 89%, ee = 82%
Ar = *p*-Tol: 90%, ee = 74%
Ar = *p*-(*i*-Pr)C$_6$H$_4$: 88%, ee = 92%
Ar = *p*-(*t*-Bu)C$_6$H$_4$: 90%, ee = 98%
Ar = 1-Naph: 91%, ee = 87%

Scheme 4.12 Cinchona-alkaloid-ester-catalyzed oxidation of aromatic *N*-tosyl aldimines with MCPBA.

was catalyzed by a strong basic organocatalyst, such as a chiral P-spiro triaminoiminophosphorane. Under such conditions, known as Payne modification [30], the base-activated hydrogen peroxide reacted with trichloroacetonitrile, generating a chiral aminophosphonium peroxyimidate as the active oxidant (Scheme 4.13). This species could be considered as an analog of the ammonium peroxycarboxylate nucleophilic oxidant operating in the mechanism proposed by Jørgensen (Scheme 4.10). Its attack on the Re face of the imine would selectively occur, favored by the high stereocontrol exerted by the chiral aminophosphonium counterion. The chiral peroxyhemiaminal intermediate formed would undergo smooth cyclization to the chiral oxaziridine, releasing the good leaving-group amidate ion. Finally, proton shift, with formation of the neutral trichloroacetamide side product, and regeneration of the catalyst, would complete the catalytic cycle. It must be highlighted that this nice method proved to be outstandingly high-performing, providing equally excellent yield and enantioselectivity in the oxidation of a broad range of aromatic, heteroaromatic, and aliphatic N-sulfonyl aldimines, regardless of the substituent bulkiness. In particular, primary but also α-branched aliphatic aldimines were oxidized in high yields if N-mesitylenesulfonyl was installed as the protecting group instead of an N-tosyl moiety (Scheme 4.13). Great advantages of the Payne-type conditions are the highest convenience and atom economy of H_2O_2 as the oxygen source over the use of peracids and the chemoselectivity of the reaction when potentially oxidizable groups are present.

The usefulness of this methodology was demonstrated in the synthesis of a prolinol derivative depicted in Scheme 4.14 [29]. Indeed, the application of closely related reaction conditions to the chemo- and enantioselective oxidation of an unsaturated N-mesitylene aldimine led to the corresponding oxaziridine in 83% yield with 96% ee when the reaction was performed in dichloromethane instead of toluene as solvent. This oxaziridine was subsequently subjected to intramolecular epoxidation under heating, yielding a chiral intermediate, which provided the final expected prolinol derivative in 61% overall yield. The overall synthesis occurred with complete preservation of the optical purity.

Later in 2014, the synthetic utility of the oxaziridination methodology was also highlighted by the same authors through the fruitful derivatization of a chiral unsaturated N-mesitylene oxaziridine into a pyrrolidine methanol derivative, as shown in Scheme 4.15 [31]. Indeed, this chiral secondary alcohol was prepared in 80% yield from the corresponding chiral oxaziridine (95% ee) via N-sulfonyl imine with an internal cis-epoxide moiety. Furthermore, a chiral aromatic N-tosyl oxaziridine (95% ee) was subjected to copper-catalyzed oxyamination of N-acetyl indole, providing the corresponding tricyclic indoline derivative in 76% yield as a 1.2 : 1 mixture of two diastereomers obtained without loss of enantiomeric excess (95% and 94% ee), as shown in Scheme 4.15.

In 2012, Yamamoto et al. reported the only method so far available for enantioselective metal-catalyzed oxaziridination of N-tosyl imines, which is depicted in Scheme 4.16 [32]. The oxidation of aromatic as well as aliphatic N-tosyl aldimines with cumene hydroperoxide was catalyzed by an in situ generated hafnium complex from a chiral C_2-symmetric bishydroxamic acid (Scheme 4.16) and $Hf(Ot\text{-}Bu)_4$, providing the corresponding trans-oxaziridines in good yields

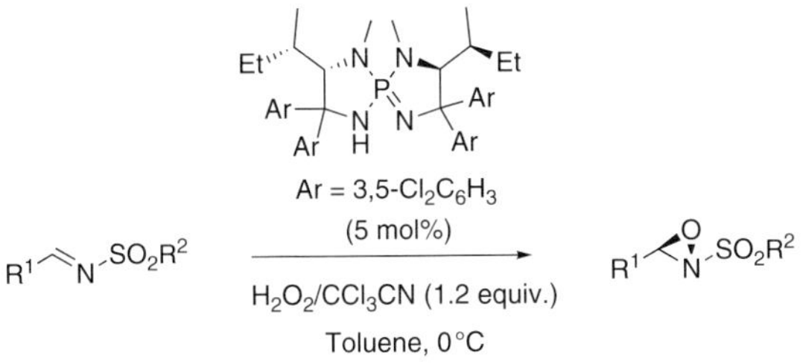

4 Asymmetric Oxaziridination

R¹ = Ph, R² = p-Tol: 90%, ee = 95%
R¹ = o-FC₆H₄, R² = p-Tol: 83%, ee = 95%
R¹ = m-BrC₆H₄, R² = p-Tol: 84%, ee = 97%
R¹ = p-ClC₆H₄, R² = p-Tol: 89%, ee = 96%
R¹ = o-BrC₆H₄, R² = p-Tol: 78%, ee = 95%
R¹ = o-Tol, R² = p-Tol: 86%, ee = 96%
R¹ = m-MeOC₆H₄, R² = p-Tol: 99%, ee = 95%
R¹ = p-NO₂C₆H₄, R² = p-Tol: 89%, ee = 98%
R¹ = p-ClC₆H₄, R² = p-Tol: 89%, ee = 96%
R¹ = p-Tol, R² = p-Tol: 83%, ee = 95%
R¹ = 2-Naph, R² = p-Tol: 99%, ee = 90%
R¹ = 2-furyl, R² = p-Tol: 91%, ee = 95%
R¹ = 3-furyl, R² = p-Tol: 96%, ee = 93%
R¹ = 3-thienyl, R² = p-Tol: 91%, ee = 94%
R¹ = Cy, R² = p-Tol: 91%, ee = 98%
R¹ = t-Bu, R² = p-Tol: 85%, ee = 98%
R¹ = (E)-PhCH=CH, R² = p-Tol: 85%, ee = 96%
R¹ = CH₂Bn, R² = Mesityl: 87%, ee = 96%
R¹ = i-Pr, R² = Mesityl: 78%, ee = 95%
R¹ = i-Bu, R² = Mesityl: 92%, ee = 95%
R¹ = n-Pent, R² = Mesityl: 87%, ee = 94%
R¹ = CH₂=CH(CH₂)₂, R² = Mesityl: 83%, ee = 96%

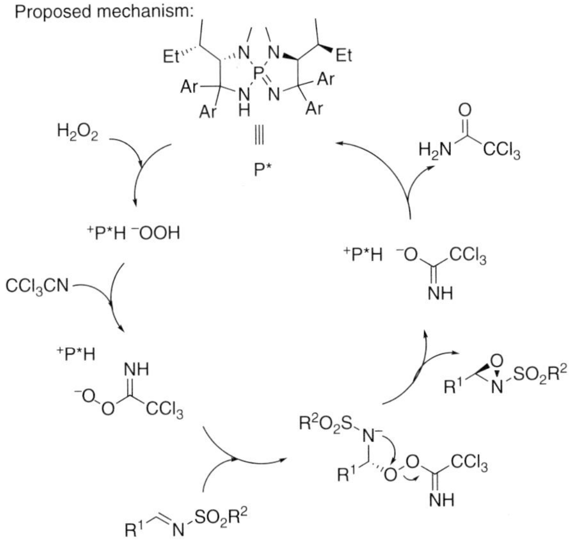

Scheme 4.13 Triaminoiminophosphorane-catalyzed oxidation of N-sulfonyl aldimines with H₂O₂/CCl₃CN.

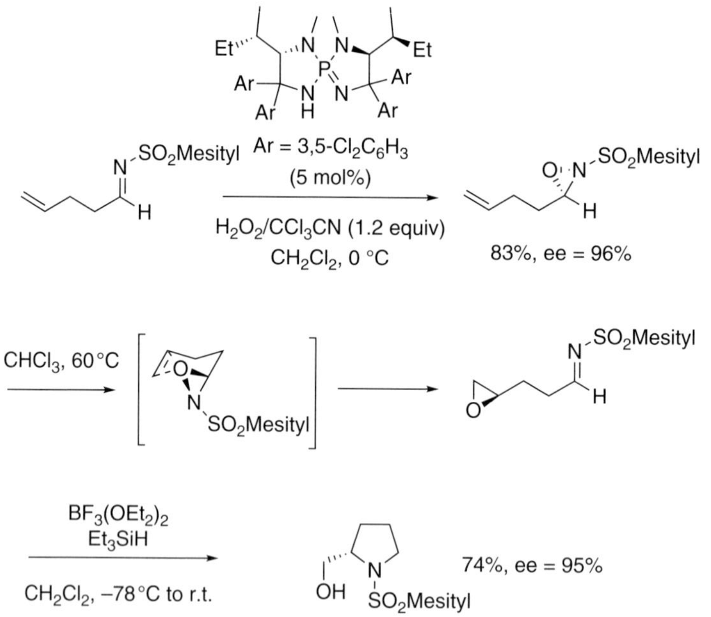

Scheme 4.14 Triaminoiminophosphorane-catalyzed oxidation of an unsaturated N-mesyl aldimine with H_2O_2/CCl_3CN and synthesis of a prolinol derivative.

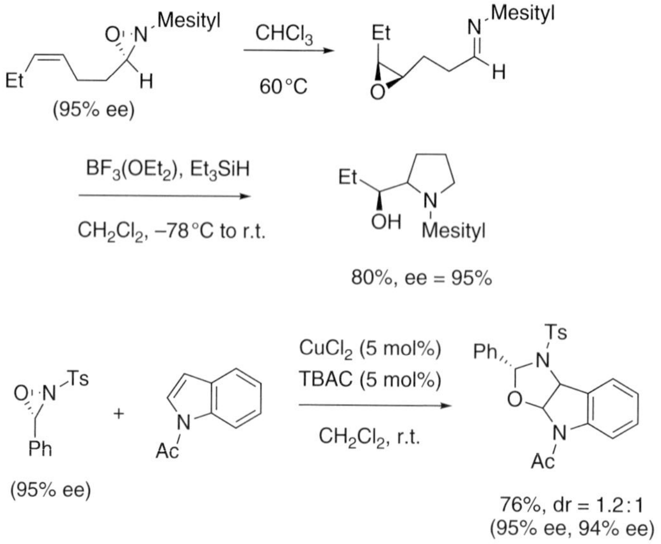

Scheme 4.15 Other derivatizations of chiral oxaziridines prepared from Ooi's methodology.

with very high enantioselectivities of 91–98% ee. One drawback of this study is that only a few substrates were investigated. Remarkably, a less reactive N-tosyl methylketimine could be oxidized into the corresponding oxaziridine with excellent enantioselectivity of 98% ee albeit in moderate yield of 38%. This feature is

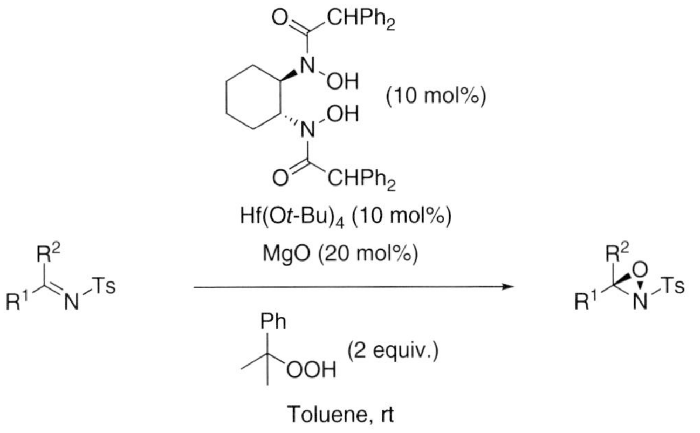

Scheme 4.16 Hafnium-catalyzed oxaziridination of N-tosyl imines including a ketoimine.

R¹ = Ph, R² = H: 78%, ee = 98%
R¹ = p-BrC₆H₄, R² = H: 84%, ee = 95%
R¹ = Cy, R² = H: 82%, ee = 91%
R¹ = Ph, R² = Me: 38%, ee = 98%

unique related to the fact that synthesizing oxaziridines from ketoimines constitutes a challenging task, highlighting the great potential of metal catalysis over organocatalysis.

4.4 Kinetic Resolutions

Despite the increased industrial demand for enantiomerically pure compounds, to date, only a few catalytic asymmetric processes have found commercial application [33], among them, rare exceptions are catalytic kinetic resolutions [34]. Indeed, kinetic resolution as one of the most powerful tools in asymmetric catalysis has found wide applications in both academics and industry, complementing approaches such as asymmetric synthesis and classical resolution. When an asymmetric route to an enantiomerically pure oxaziridine cannot be achieved, the possibility of performing a kinetic resolution of the racemic oxaziridine mixture still exists. In recent years, several examples of nonoxidative catalytic approach to chiral oxaziridines based on kinetic resolution of racemic oxaziridines have been successfully developed. As an example, Ye et al. reported in 2010 an asymmetric formal Lewis-base-catalyzed [3+2] cycloaddition of racemic N-tosyloxaziridines to give a variety of disubstituted ketenes, as shown in Scheme 4.17 [35]. In this study, the use of a chiral N-heterocyclic carbene derived from imidazolium salt chiral catalyst allowed the formation of chiral cis-oxazolin-4-ones to be achieved with high enantioselectivities of 85–95% ee along with enantioenriched oxaziridines recovered with good-to-excellent enantioselectivities of 76–99% ee, highly depending on the ketene employed. It must be noted that the reaction conditions were preferentially optimized to achieve the highest yield in cycloadducts, affording the chiral oxaziridines in low yields (15–26%). While the scope of ketenes was widely explored, however, only one oxaziridine was investigated in the process. The use of an O-silylated chiral N-heterocyclic carbene derived from L-pyroglutamic acid was found to provide decreased enantioselectivities (10–67% ee), as shown in Scheme 4.17.

In 2012, a related formal [3+2] cycloaddition of racemic N-nosyl oxaziridines with styrenes was developed by Williamson and Yoon [36]. In this case, the reaction was catalyzed by a novel iron(II) bisoxazoline complex depicted in Scheme 4.18. For example, the reaction of styrene bearing a methyl substituent at the *para*-

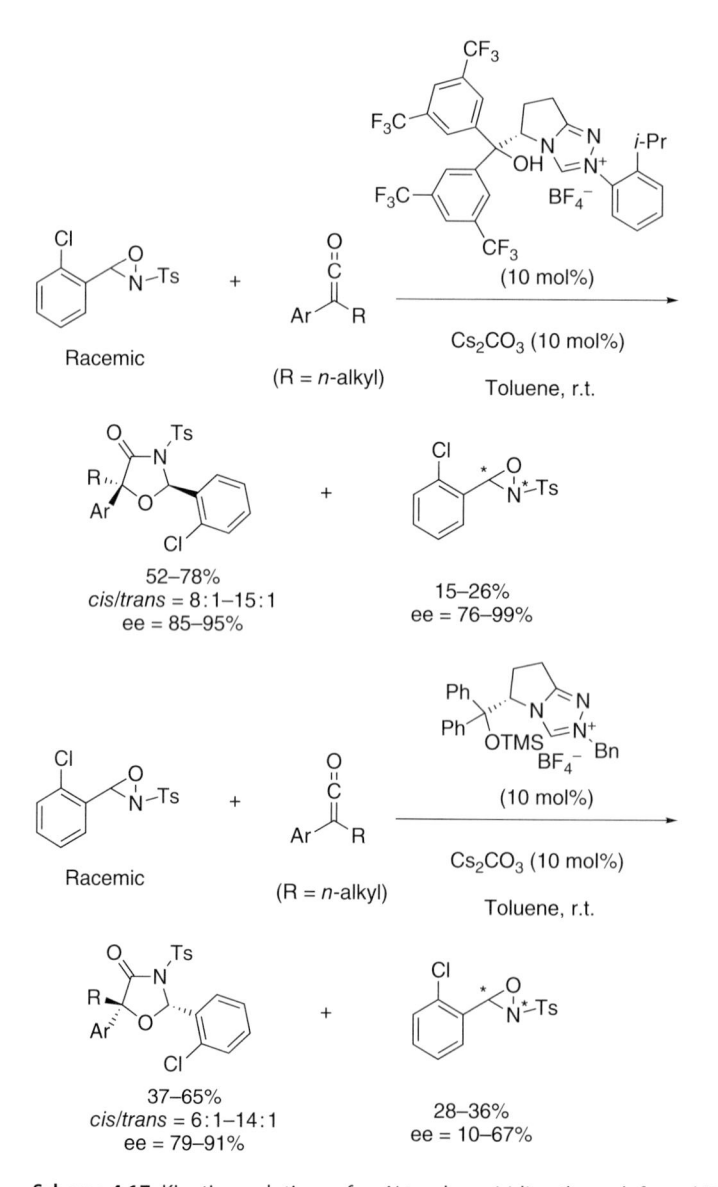

Scheme 4.17 Kinetic resolutions of an N-tosyl oxaziridine through formal [3+2] cycloaddition to ketenes catalyzed by N-heterocyclic carbenes.

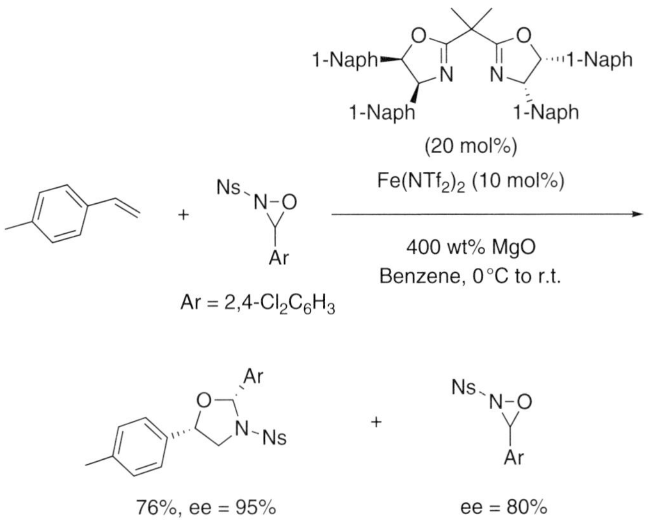

Scheme 4.18 Kinetic resolution of an *N*-nosyl oxaziridine through formal [3+2] cycloaddition to a styrene catalyzed by a chiral bisoxazoline iron complex.

position with a racemic aromatic *N*-nosyl oxaziridine led to the corresponding oxazolidine cycloadduct in 76% yield with 95% ee along with recovered significantly enantioenriched oxaziridine (80% ee) arisen from kinetic resolution. A drawback of this study was its limitation to only one example since the aim of the work was the synthesis of chiral oxazolidines.

The principle of kinetic resolution was fully exploited for preparative purposes by Feng et al. in 2013 in another formal [3+2] cycloaddition occurring between azlactones and racemic *N*-sulfonyl oxaziridines under catalysis with a chiral bisguanidinium salt [37]. In this process, the authors took advantage of the ambiphilic nature of azlactones, which, after enolization, could act as a nucleophile at the α-carbon and as an electrophile at the carbonyl group. The reaction with oxaziridines led to the corresponding formal [3+2] cycloadducts as intermediates (Scheme 4.19). The latter were subsequently subjected to a ring-opening reaction to afford final chiral oxazolidin-4-ones. During the reaction, one enantiomeric oxaziridine was preferentially consumed, allowing an efficient kinetic resolution to be achieved. Therefore, chiral aryl, heteroaryl, and alkyl oxaziridines were recovered in high enantioselectivities of 82–99% ee. Despite the high stereoselectivity of the reaction, it was found that both enantiomers of starting azlactones were consumed at an equal rate because of the concomitant fast racemization under the reaction conditions. The scope of both azlactones and oxaziridines was found to be broad, giving generally excellent results. Oxazolin-4-ones were obtained with good-to-high diastereo- and enantioselectivities (70–94% de, 55–99% ee), except when an *N*-nosyl oxaziridine was employed. To explain the results, the authors have proposed the involvement of a bifunctional mechanism in which the guanidinium moiety would serve to recognize and activate the (*S,S*)-oxaziridine, whereas the guanidine might function as a Brønsted base in the activation of the azlactone.

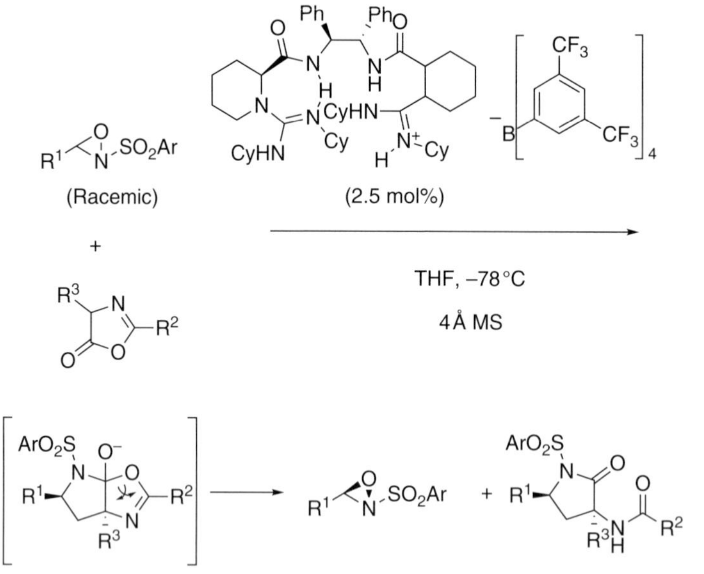

R¹ = Ph, Ar = p-Tol: 36%, ee = 99%
R¹ = Ph, Ar = p-NO₂C₆H₄: 29%, ee = 98%
R¹ = Ph, Ar = p-(i-Pr)C₆H₄: 44%, ee = 93%
R¹ = m-FC₆H₄, Ar = p-Tol: 39%, ee = 98%
R¹ = o-Tol, Ar = p-Tol: 46%, ee = 87%
R¹ = 2-furyl, Ar = p-Tol: 36%, ee = 85%
R¹ = 2-thienyl, Ar = p-Tol: 45%, ee = 98%
R¹ = 1-Naph, Ar = p-Tol: 33%, ee = 85%
R¹ = Bn, Ar = p-Tol: 40%, ee = 82%
R¹ = Et, Ar = p-Tol: 39%, ee = 82%

41–67%
de = 70–94%
ee = 55–99%

Scheme 4.19 Kinetic resolution of N-sulfonyl oxaziridines through formal [3+2] cycloaddition to azlactones catalyzed by a chiral bisguanidinium salt.

4.5 Conclusions

This chapter collects the major progress achieved in the synthesis of chiral oxaziridines based on the use of chiral reagents, chiral catalysts, and kinetic resolutions. Initial interest in the unique structural and physical properties of these strained heterocyclic ring compounds became overshadowed in the 1980s by the recognition that electron-deficient oxaziridines could be employed as convenient, stable, and neutral sources of electrophilic oxygen. The use of oxaziridines in oxygen atom transfer reactions continues to be the most broadly appreciated application of oxaziridine chemistry. However, many of the most recent advances in their reactivity concern nitrogen atom transfer, oxyamination, cycloadditions, and skeletal rearrangement reactions, and consequently, novel methodologies for their asymmetric synthesis are needed. The most significant progress achieved in the past decade in the asymmetric synthesis of oxaziridines is the development of catalytic methodologies while literature precedents almost entirely relied on classical stoichiometric chiral substrate/auxiliary-controlled approaches. In particular, the availability of various types of organocatalysts has opened a way to explore well-established or novel routes to access chiral oxaziridines by taking advantage of covalent or noncovalent activation strategies of the reagents. Further developments of the field of asymmetric synthesis of oxaziridines, which is in its infancy, are expected in the future, such as improvements in limited substrate scopes and design of more powerful catalytic systems to reduce the catalyst loadings.

References

1 Emmons, W.D. (1956) *J. Am. Chem. Soc.*, 78, 6208–6209.
2 (a) Williamson, K.S., Michaelis, D.J., and Yoon, T.P. (2014) *Chem. Rev.*, 114, 8016–8036; (b) Della Sala, G. and Lattanzi, A. (2014) *ACS Catal.*, 4, 1234–1245.
3 (a) Davis, F.A. and Sheppard, A.C. (1989) *Tetrahedron*, 45, 5703–5742; (b) Andreae, S. and Schmitz, E. (1991) *Synthesis*, 5, 327–341; (c) Davis, F.A. and Thimma Reddy, R. (1996) in *Comprehensive Heterocyclic Chemistry II*, vol. 1 (eds A.R. Katritsky, C.W. Rees, and E.F.V. Scriven), Pergamon Press, Oxford, pp. 365–413; (d) Petrov, V.A. and Resnati, G. (1996) *Chem. Rev.*, 96, 1809–1823; (e) Mishra, J.K. (2005) *Synlett*, 3, 543–544; (f) Davis, F.A., Chen, B.-C., and Zhou, P. (2008) in *Comprehensive Heterocyclic Chemistry III*, vol. 1 (eds A.R. Katritsky, C.A. Ramsden, E.F.V. Scriven, and R.J.K. Taylor), Elsevier, Oxford, pp. 559–622; (g) Kumar, K.M. (2012) *Synlett*, 17, 2572–2573; (h) Buglioni, L. (2013) *Synlett*, 20, 2773–2774.
4 Davis, F.A., Billmers, J.M., Gosciniak, D.J., Towson, J.C., and Bach, R.D. (1986) *J. Org. Chem.*, 51, 4240–4245.
5 (a) Davis, F.A. (2006) *J. Org. Chem.*, 71, 8993–9003; (b) Davis, F.A. and Chen, B.-C. (1992) *Chem. Rev.*, 92, 919–934; (c) Davis, F.A., Reddy, R.T., Han, W., and Reddy, R.E. (1993) *Pure Appl. Chem.*, 65, 633–640.
6 (a) Troisi, L., De Lorenzis, S., Fabio, M., Rosato, F., and Granito, C. (2008) *Tetrahedron: Asymmetry*, 19, 2246–2251; (b) Troisi, L., Caccamese, S., Pilati, T., Videtta, V., and Rosato, F. (2010) *Synthesis*, 2010, 3211–3216.
7 Aubé, J. (1997) *Chem. Soc. Rev.*, 26, 269–278.
8 (a) Davis, F.A., Haque, M.S., Ulatowski, T.G., and Towson, J.C. (1986) *J. Org. Chem.*, 51, 2402–2404; (b) Davis, F.A., Towson, J.C., Weismiller, M.C., Lal, S., and Carroll, P.J. (1988) *J. Am. Chem. Soc.*, 110, 8477–8482; (c) Davis, F.A. and Weismiller, M.C. (1990) *J. Org. Chem.*, 55, 3715–3717.
9 (a) Davis, F.A., ThimmaReddy, R., and Weismiller, M.C. (1989) *J. Am. Chem. Soc.*, 111, 5964–5965; (b) Davis, F.A., ThimmaReddy, R., Han, W., and Carroll, P.J. (1992) *J. Am. Chem. Soc.*, 114, 1428–1437; (c) Davis, F.A., Weismiller, M.C., Lal, G.S., Chen, B.-C., and Przeslawski, R.M. (1989) *Tetrahedron Lett.*, 30, 1613–1616; (d) Davis, F.A., Kumar, A., and Chen, B.-C. (1991) *J. Org. Chem.*, 56, 1143–1145; (e) Verfürth, U. and Ugi, I. (1991) *Chem. Ber.*, 124, 1627–1634.
10 Davis, F.A., McCauley, J.P., Chattopadhyay, S., Harakal, M.E., Towson, J.C., Watson, W.H., and Tavanaiepour, I. (1987) *J. Am. Chem. Soc.*, 109, 3370–3371.

11 Davis, F.A., Sheppard, A.C., Chen, B.-C., and Haque, M.S. (1990) *J. Am. Chem. Soc.*, 112, 6679–6690.
12 Davis, F.A., Jenkins, R.H., Awad, S.B., Watson, O.D., and Galloy, W.H. (1982) *J. Am. Chem. Soc.*, 104, 5412–5418.
13 Davis, F.A. and Thimma Reddy, R. (1992) *J. Org. Chem.*, 57, 2599–2606.
14 Aggarwal, V.K. and Wang, M.F. (1996) *Chem. Commun.*, 2, 191–192.
15 (a) Amstrong, A., Edmonds, I.D., and Swarbrick, M.E. (2003) *Tetrahedron*, 44, 5335–5338; (b) Amstrong, A., Edmonds, I.D., and Swarbrick, M.E. (2005) *Tetrahedron Lett.*, 46, 2207–2210; (c) Bohé, L., Lusinchi, M., and Lusinchi, X. (1999) *Tetrahedron*, 55, 141–154; (d) Bohé, L., Lusinchi, M., and Lusinchi, X. (1999) *Tetrahedron*, 55, 155–166; (e) Akhatou, A., Rahimi, M., Cheboub, K., Ghosez, L., and Hanquet, G. (2007) *Tetrahedron*, 63, 6232–6240; (f) Davis, F.A., Thimma Reddy, R., and Reddy, R.E. (1992) *J. Org. Chem.*, 57, 6387–6389.
16 Camarero, C., Arrasate, S., Sotomayor, N., and Lete, E. (2010) *Arkivoc*, 2010 (iii), 45–55.
17 (a) Davis, F.A., ThimmaReddy, R., McCauley, J.P., Przeslawski, R.M., and Harakal, M.E. (1991) *J. Org. Chem.*, 56, 809–815; (b) Lattes, A., Oliveros, E., Riviére, M., Belzecki, C., Mostowicz, D., Abramskj, W., Piccini-Leopardi, C., Germain, G., and Van Meerssche, M. (1982) *J. Org. Chem.*, 104, 3929–3934; (c) Page, P.C.B., Limousin, C., and Murrell, V.L. (2002) *J. Org. Chem.*, 67, 7787–7796; (d) Aubé, J. and Wang, Y. (1988) *Tetrahedron Lett.*, 29, 6407–6408.
18 (a) Bjørgo, J. and Boyd, D.R. (1973) *J. Chem. Soc., Perkin Trans.*, 11, 1575–1577; (b) Montanari, F., Moretti, I., and Torre, G. (1969) *J. Chem. Soc. D, Chem. Commun.*, 19, 1086b–1087; (c) Jennings, W.B., Watson, S.P., and Tolley, M.S. (1987) *J. Am. Chem. Soc.*, 109, 8099–8100.
19 Jennings, W.B., Kochanewycz, M.J., Lovely, C.J., and Boyd, D.R. (1994) *J. Chem. Soc., Chem. Commun.*, 22, 2569–2570.
20 Carrocia, L., Fioravanti, S., Pellacani, L., Sadun, C., and Tardella, P.A. (2011) *Tetrahedron*, 67, 5375–5381.
21 (a) Page, P.C.B., Murrell, V.L., Limousin, C., Laffan, D.D.P., Bethell, D., Slawin, A.M.Z., and Smith, T.A.D. (2010) *J. Org. Chem.*, 66, 4204–4207; (b) Blanc, S., Bordogna, C.A.C., Buckley, B.R., Elsegood, M.R.J., and Page, P.C.B. (2010) *Eur. J. Org. Chem.*, 2010, 882–889.
22 (a) Forni, A., Moretti, I., and Torre, G. (1987) *J. Chem. Soc., Perkin Trans.*, 6, 699–704; (b) Bucciarelli, M., Forni, A., Marcaccioli, S., Moretti, I., and Torre, G. (1983) *Tetrahedron*, 39, 187–192.
23 Pirkle, W.H. and Rinaldi, P.L. (1977) *J. Org. Chem.*, 42, 2080–2082.
24 Toda, F. and Tanaka, K. (1987) *Chem. Lett.*, 11, 2283–2284.
25 Tka, N., Kraïem, J., and Hassine, B.B. (2012) *Synth. Commun.*, 42, 2994–3003.
26 Lykke, L., Rodriguez-Escrich, C., and Jørgensen, K.A. (2011) *J. Am. Chem. Soc.*, 133, 14932–14935.
27 Zhang, T., He, W., Zhao, X., and Jin, Y. (2013) *Tetrahedron*, 69, 7416–7422.
28 Jin, Y., Zhang, T., Zhang, W., Chang, S., and Feng, B. (2014) *Chirality*, 26, 150–154.
29 (a) Uraguchi, D., Tsutsumi, R., and Ooi, T. (2013) *J. Am. Chem. Soc.*, 135, 8161–8164; (b) Tsutsumi, R., Kim, S., Uraguchi, D., and Ooi, T. (2014) *Synthesis*, 46, 871–878.

30 (a) Payne, G.B., Deming, P.H., and Williams, P.H. (1961) *J. Org. Chem.*, 26, 659–663; (b) Payne, G.B. (1962) *Tetrahedron*, 18, 763–765.
31 Uraguchi, D., Tsutsumi, R., and Ooi, T. (2014) *Tetrahedron*, 70, 1691–1701.
32 Olivares-Romero, J.L., Li, Z., and Yamamoto, H. (2012) *J. Am. Chem. Soc.*, 134, 5440–5443.
33 Blaser, H.U. and Schmidt, E. (2004) *Asymmetric Catalysis on Industrial Scale*, Wiley-VCH Verlag GmbH, Weinheim.
34 (a) Kagan, H.B. and Fiaud, J.C. (1988) *Top. Stereochem.*, 18, 249–330; (b) Hoveyda, A.H. and Didiuk, M.T. (1998) *Curr. Org. Chem.*, 2, 489–526; (c) Cook, G.R. (2000) *Curr. Org. Chem.*, 4, 869–885; (d) Keith, M., Larrow, J.F., and Jacobsen, E.N. (2001) *Adv. Synth. Catal.*, 343, 5–26; (e) Robinson, D.E.J.E. and Bull, S.D. (2003) *Tetrahedron: Asymmetry*, 14, 1407–1446; (f) Vedejs, E. and Jure, M. (2005) *Angew. Chem. Int. Ed.*, 44, 3974–4001; (g) Jarvo, E.R. and Miller, S.J. (2004) in *Comprehensive Asymmetric Catalysis: Supplement 1* (eds E.N. Jacobsen, A. Pfaltz, and H. Yamamoto), Springer, Berlin, pp. 189–206.
35 Shao, P.-L., Chen, X.-Y., and Ye, S. (2010) *Angew. Chem. Int. Ed.*, 49, 8412–8416.
36 Williamson, K.S. and Yoon, T.P. (2012) *J. Am. Chem. Soc.*, 134, 12370–12373.
37 Dong, S., Liu, X., Zhu, Y., He, P., Lin, L., and Feng, X. (2013) *J. Am. Chem. Soc.*, 135, 10026–10029.

5
Asymmetric Azirination and Thiirination

5.1 Introduction

The goal of this chapter is to collect the advances in asymmetric azirination and thiirination using chiral substrates (auxiliaries) as well as chiral catalysts, focusing on those published in the last 12 years. This chapter is subdivided into two parts, dealing successively with asymmetric azirination and asymmetric thiirination. Even if this chapter is entitled "Asymmetric Azirination and Thiirination," it must be noted that it includes some examples of synthesis of chiral 2H-azirines and thiiranes in which there is no creation of stereogenic centers. Indeed, in these works, the chirality of the substrates is just maintained during the course of the reactions without racemization. These studies could also interest the readers looking for all types of methods to reach these chiral products.

5.2 Asymmetric Azirination

2H-Azirines constitute the smallest nitrogen unsaturated heterocyclic system, with two carbon atoms and one double bond in a three-membered ring. Their stability can be attributed not only to the combined effects of bond shortening and angle compression but also to the presence of the electron-rich nitrogen atom. The strain energy associated with these heterocycles is principally due to the deformation of the normal bond angles between the atoms of the ring. The total ring-strain energy of azirine has been estimated at 48 kcal mol^{-1} [1]. The theoretical and biological applications and the synthetic chemistry of these heterocycles have been extensively explored [2]. Substituted 2H-azirines are versatile compounds and have been used for the preparation of functionalized aziridines. The chemistry of 2H-azirines is related to their ring strain, reactive π-bond, and ability to undergo regioselective ring cleavage. Azirines are capable of acting not only as nucleophiles and electrophiles in organic reactions but also as dienophiles and dipolarophiles in cycloaddition reactions. For example, they readily participate in cycloaddition reactions as 2π-components and undergo ring cleavage upon photochemical excitation to give nitrile ylides. These dipoles then undergo a subsequent 1,3-dipolar cycloaddition reaction with a variety of π-bonds. Thermal ring cleavage produces vinyl nitrenes by cleavage of the N—C$_2$

Asymmetric Synthesis of Three-Membered Rings, First Edition.
Hélène Pellissier, Alessandra Lattanzi and Renato Dalpozzo.
© 2017 Wiley-VCH Verlag GmbH & Co. KGaA. Published 2017 by Wiley-VCH Verlag GmbH & Co. KGaA

bond, which then underwent ring-expansion reactions. Azirines constitute useful precursors for the synthesis of a variety of nitrogen-containing heterocyclic systems. Furthermore, 2H-azirines containing a carboxylic ester group are constituents of naturally occurring antibiotics.

A number of synthetic methods are available for forming 2H-azirines, including photo- and thermal extrusion of nitrogen from vinyl azides, intramolecular rearrangements of N-functionalized imines, vinyl azides, isoxazoles, and oxazaphospholes, and bimolecular reactions between nitriles and carbenes or nitrenes and acetylenes. 2H-Azirines can also be prepared by elimination reaction of N-substituted aziridines, namely, dehydrochlorination of N-chloroaziridines and elimination from N-sulfinylaziridines. Moreover, the Swern oxidation of NH aziridine esters constitutes a powerful synthetic methodology to achieve 2H-azirines. Another significant preparative pathway to the 2H-azirines is the Neber rearrangement of oxime sulfonates.

5.2.1 Neber Approaches

An important route to 2H-azirines is the Neber reaction of oxime sulfonates discovered in 1926 [3]. The presence of strong electron-withdrawing groups in the α-position to the oxime increases the acidity of those protons and thus favors the cycloelimination reaction under mild conditions. The Neber reaction occurs either through an internal concerted nucleophilic displacement or via a vinyl nitrene (Scheme 5.1).

In 1993, Vosekalna et al. reported the first asymmetric version of this reaction, which involved a chiral O-mesyl derivative of amidoxime derived from ethyl ester of phenylglycine as chiral substrate [4]. By treatment with NaOMe, this amidoxime led to the corresponding optically active 3-amino-2H-azirine in 74% yield as a 96:4 mixture of two diastereomers, as shown in Scheme 5.2.

Later in 1996, Zwanenburg et al. demonstrated that optically enriched 2H-azirines could be prepared through a catalytic asymmetric version of the Neber

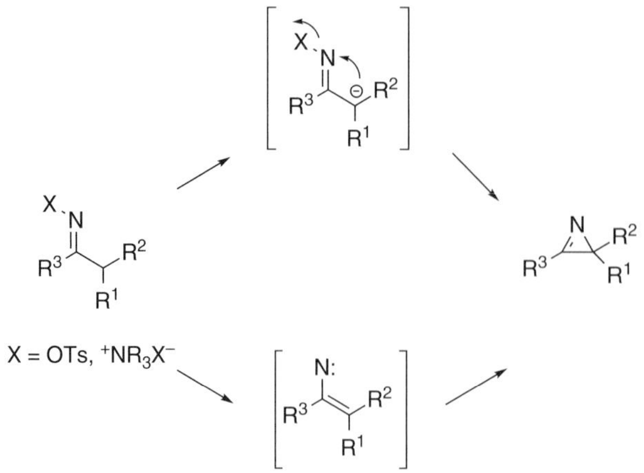

Scheme 5.1 Synthesis of 2H-azirines through Neber reaction.

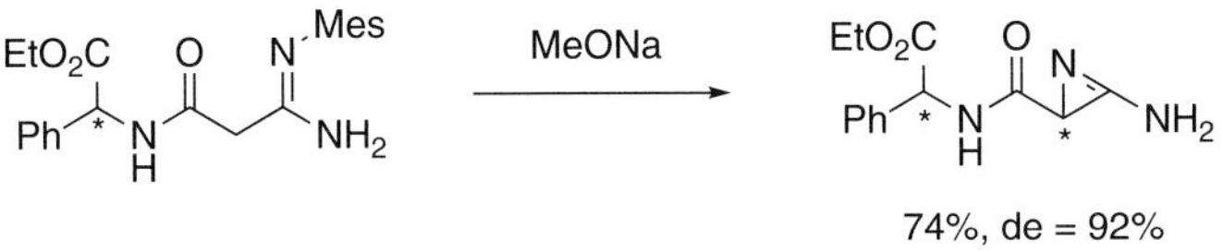
EtO2C
O
N
Mes
Ph
*
N
H
NH2
MeONa
EtO2C
O
N
Ph
*
N
H
*
NH2
74%, de = 92%

Scheme 5.2 First asymmetric version of Neber reaction.

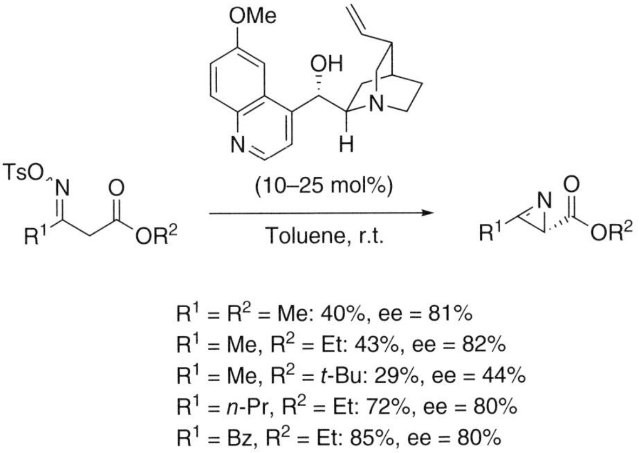
OMe
OH
N
N
H
TsO
N
O
R1
OR2
(10–25 mol%)
Toluene, r.t.
N
O
R1
OR2
R1 = R2 = Me: 40%, ee = 81%
R1 = Me, R2 = Et: 43%, ee = 82%
R1 = Me, R2 = t-Bu: 29%, ee = 44%
R1 = n-Pr, R2 = Et: 72%, ee = 80%
R1 = Bz, R2 = Et: 85%, ee = 80%

Scheme 5.3 First catalytic asymmetric version of Neber reaction.

reaction using cinchona alkaloids [5]. As shown in Scheme 5.3, ketoxime tosylates derived from 3-oxocarboxylic esters submitted to a large excess of potassium carbonate, and a catalytic amount of quinidine afforded the corresponding Neber products in moderate-to-good yields (29–85%) with enantioselectivities of 44–82% ee. Quinidine was selected as optimal organocatalyst among various chiral tertiary bases including sparteine, brucine, strychnine, cinchonine, cinchonidine, quinine, and dihydroquinidine. To explain the results, the authors suggested that the alkaloid base formed a tightly bound complex with the ketoxime tosylate. Since the presence of an alcohol function in the catalyst seemed to be a prerequisite, it was proposed that hydrogen bonding of this base and the substrate through this hydroxyl group was a governing factor in the enantiodifferentiation during the abstraction of the methylene protons.

In 2000, a similar strategy was developed by Palacios and coworkers for the preparation of alkyl- and aryl-substituted 2*H*-azirines bearing a phosphonate group in the 2-position of the ring [6]. As shown in Scheme 5.4, the Neber reaction of β-phosphorylated tosyloximes provided the corresponding 2*H*-azirines in excellent yields along with moderate-to-good enantioselectivities of up to 82% ee when employing quinidine as catalyst. The scope of this methodology was extended to the synthesis of enantioenriched 2*H*-azirines derived from phosphonates albeit with lower enantioselectivities ranging from 20% to 52% ee along with high yields (85–95%) [7]. Later, better enantioselectivities of up to 72% ee were reported by the same authors in the synthesis of chiral 2*H*-azirine phosphonates performed with the same catalyst in the presence of K_2CO_3

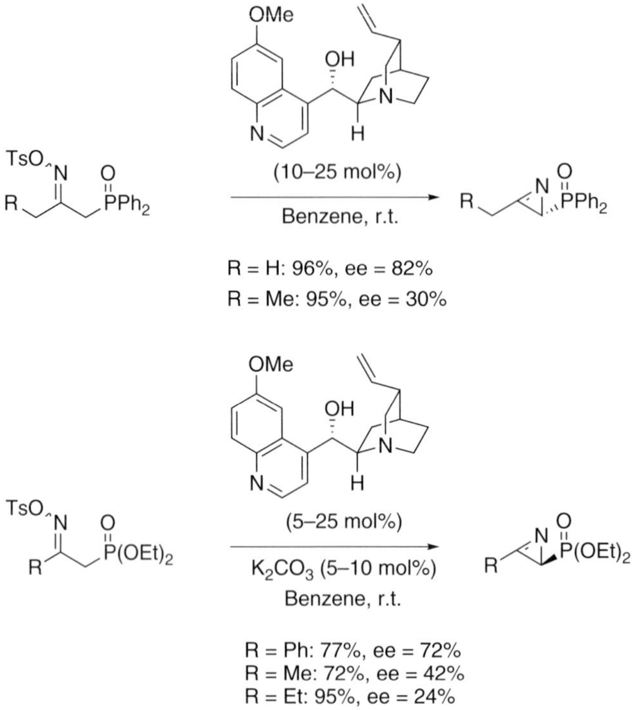

Scheme 5.4 Quinidine-catalyzed Neber reactions of β-phosphorylated tosylimines.

(Scheme 5.4) [8]. These three-membered heterocycles constitute important building blocks in the preparation of biologically active compounds of interest in medicinal chemistry.

In 2002, Maruoka et al. reported asymmetric induction in the Neber rearrangement of simple ketoxime sulfonates under chiral phase-transfer catalysis (Scheme 5.5) [9]. In the presence of a chiral quaternary ammonium bromide, the reaction was presumed to evolve through removal of an α-proton from the ketoxime sulfonate to give an anionic intermediate, followed by loss of the tosylate moiety to give the corresponding intermediate 2H-azirine. The latter was directly converted into the corresponding amino ketone in good yield (81%) with moderate enantioselectivity (51–70% ee).

In 2008, Molinski et al. reported a total synthesis of marine natural and antifungal product (−)-Z-dysidazirine based on a quinidine-catalyzed Neber reaction [10]. As shown in Scheme 5.6, the reaction of a tosyloxime derived from pentadecyne submitted to quinidine at 0 °C in toluene afforded the corresponding enantioenriched azirine in 84% yield with moderate enantioselectivity of 59% ee. This product was further converted into expected (−)-Z-dysidazirine through partial hydrogenation using Lindlar's catalyst.

Later, the same authors applied these conditions to the synthesis of shorter chain analogs **1** and **2** of (−)-Z-dysidazirine to be evaluated as antifungal agents [11]. As depicted in Scheme 5.7, these products were achieved in 61% and

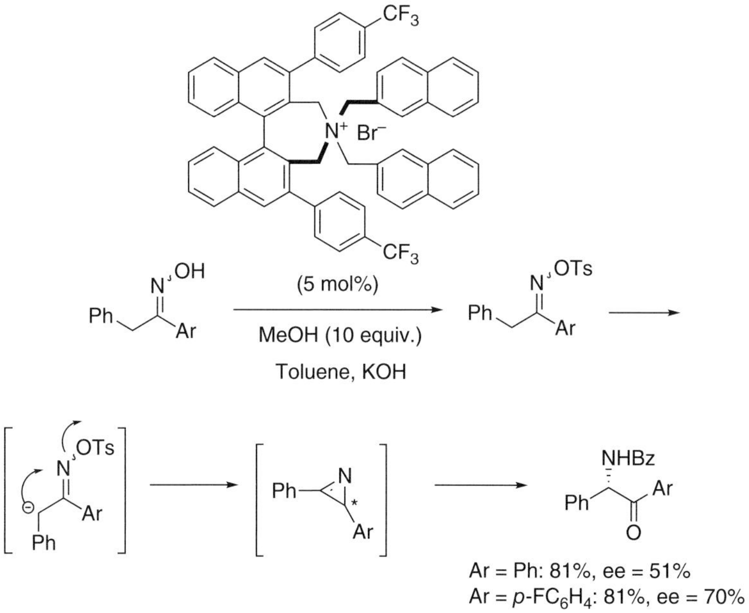

Scheme 5.5 Neber reaction under chiral phase-transfer catalysis.

86% ee, respectively. Moreover, their antifungal activity was found comparable to that of (−)-Z-dysidazirine.

In 2011, Takemoto *et al.* reported the first enantioselective Neber reaction of β-ketoxime sulfonates catalyzed by a bifunctional thiourea [12]. The process required only 5 mol% of catalyst loading and 10 equiv. of Na_2CO_3 to provide the

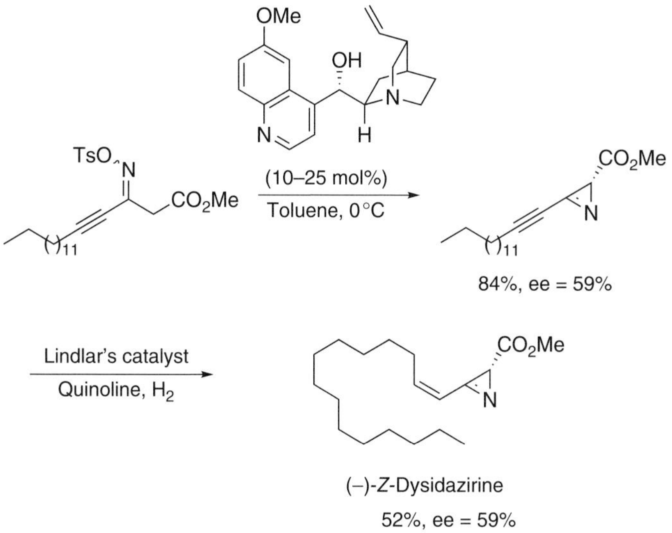

Scheme 5.6 Synthesis of (−)-Z-dysidazirine based on a Neber reaction.

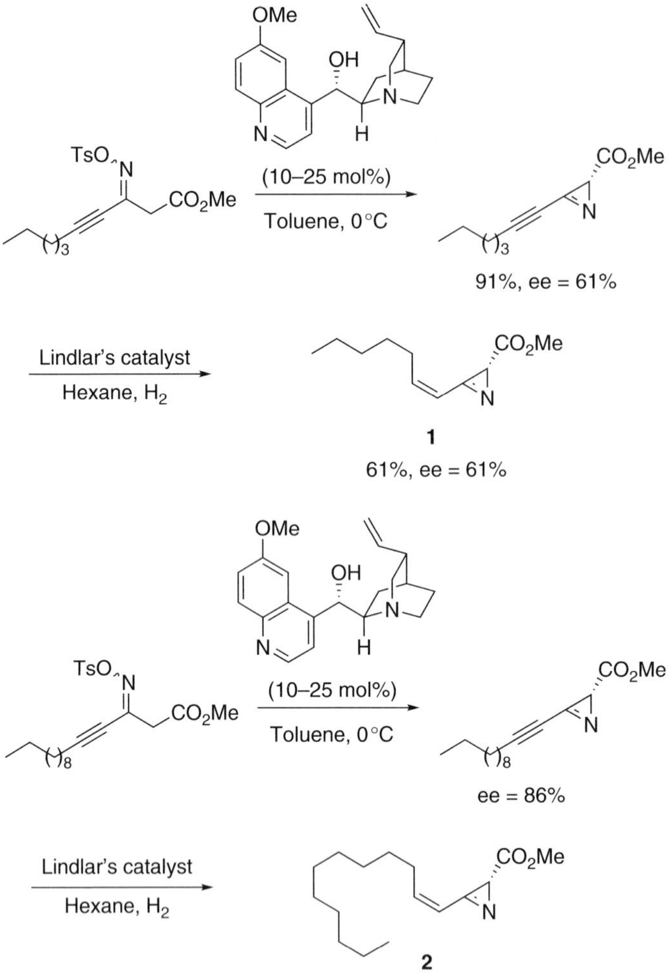

Scheme 5.7 Synthesis of shorter chain analogs of (−)-Z-dysidazirine based on Neber reactions.

corresponding 2H-azirine carboxylic esters (amide) in good yields of up to 82% with moderate-to-high enantioselectivities of up to 93% ee (Scheme 5.8). The thiourea catalyst was selected as optimal among various types of bifunctional organocatalysts, showing that amine and sulfonamide catalysts possessing no thiourea resulted in a dramatic decrease in enantioselectivity. A range of β-ketoxime sulfonate esters were found to be compatible to the process as well as a β-ketoxime sulfonate tertiary amide, which provided the corresponding 2H-azirine in 64% ee. The best enantioselectivities for the synthesis of 2H-azirine carboxylic esters were generally found for substrates with long alkyl chains. The authors have demonstrated that it was effective to use the 3,5-bis(trifluoromethyl)-benzenesulfonyl group as a leaving group in the substrate for achieving high enantioselectivity. Moreover, the utility of this novel nice methodology was illustrated by an asymmetric synthesis of (+)-dysidazarine.

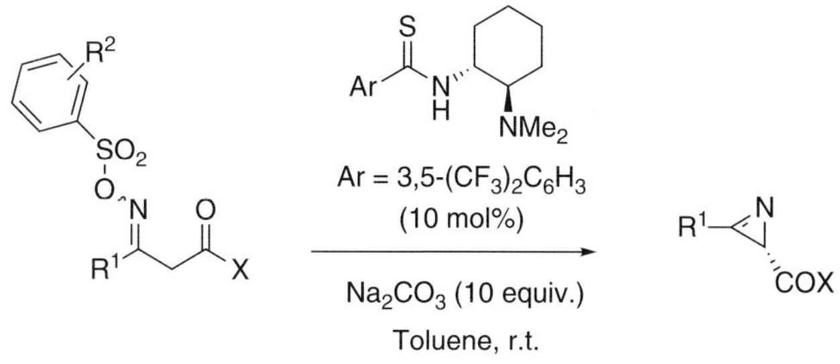

R¹ = Et, R² = 4-Me, X = Ot-Bu: 79%, ee = 64%
R¹ = Et, R² = 4-NO₂, X = Ot-Bu: 78%, ee = 68%
R¹ = Et, R² = 2-NO₂, X = Ot-Bu: 78%, ee = 62%
R¹ = Et, R² = 2-Me, X = Ot-Bu: 78%, ee = 49%
R¹ = Et, R² = 3,5-Me₂, X = Ot-Bu: 81%, ee = 54%
R¹ = Et, R² = 3,5-(CF₃)₂, X = Ot-Bu: 80%, ee = 72%
R¹ = Et, R² = 4-Me, X = NPh₂: 80%, ee = 64%
R¹ = Et, R² = 4-Me, X = OBn: 81%, ee = 67%
R¹ = Et, R² = 4-Me, X = OEt: 75%, ee = 82%
R¹ = Et, R² = 3,5-(CF₃)₂, X = OEt: 72%, ee = 86%
R¹ = n-Pr, R² = 3,5-(CF₃)₂, X = OEt: 71%, ee = 86%
R¹ = Bn, R² = 3,5-(CF₃)₂, X = OEt: 78%, ee = 93%
R¹ = BnCH₂, R² = 3,5-(CF₃)₂, X = OEt: 82%, ee = 88%
R¹ = i-Pr, R² = 4-Me, X = OEt: 82%, ee = 85%
R¹ = (E)-PhCH=CH, R² = 3,5-(CF₃)₂, X = OEt: 96%, ee = 90%
R¹ = (E)-4-MeOC₆H₄CH=CH, R² = 3,5-(CF₃)₂, X = OEt: 73%, ee = 93%
R¹ = (E)-4-FC₆H₄CH=CH, R² = 3,5-(CF₃)₂, X = OEt: 77%, ee = 90%
R¹ = Me, R² = 3,5-(CF₃)₂, X = OCHPh₂: 64%, ee = 80%

Scheme 5.8 Synthesis of shorter chain analogs of (−)-Z-dysidazirine based on Neber reactions.

The tetrazole ring has been successfully incorporated into pharmacological drug formulations, namely, cardiovascular or hypertension drugs. Furthermore, 1,5-disubstituted tetrazoles are conformational mimics of *cis* blocked peptide bonds, similar to those found in a wide variety of biologically important peptides [13]. In this context, Pinho e Melo *et al.* have recently developed a Neber approach to 2-(tetrazol-5-yl)-2*H*-azirines bearing various aryl and heteroaryl substituents at C3 [14]. The achiral reaction of the corresponding tosylates of β-ketoxime-1*H*-tetrazoles was performed in the presence of triethylamine as base, providing moderate-to-good yields (59–77%). An asymmetric version of this process was developed using 1 equiv. of quinidine, which provided a moderate enantioselectivity of 54% ee, as illustrated in Scheme 5.9.

5.2.2 Elimination Approaches

2*H*-Azirines can be prepared by elimination reaction of N-substituted aziridines, namely, dehydrochlorination of *N*-chloroaziridines and elimination from *N*-sulfinylaziridines. In 1995, Davis *et al.* described the first synthesis of chiral 2*H*-azirines based on a base-induced elimination of sulfenic acid from chiral

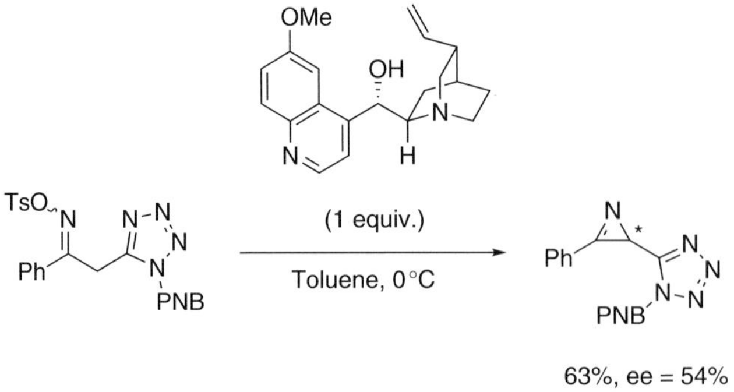

Scheme 5.9 Neber approach to a chiral 2-(tetrazol-5-yl)-2H-azirine.

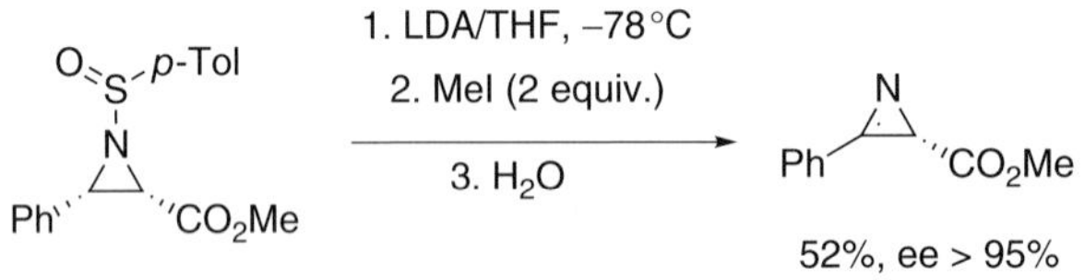

Scheme 5.10 First asymmetric synthesis of 2H-azirine based on elimination from an N-sulfinylaziridine.

N-sulfinylaziridine 2-carboxylate esters [15]. As shown in Scheme 5.10, treatment of chiral methyl 3-phenyl-N-p-toluenesulfinylaziridine-2-carboxylate by lithium diisopropylamide (LDA) at −78 °C in tetrahydrofuran (THF) in the presence of iodomethane followed by quenching with H_2O afforded the corresponding chiral 2H-azirine in 52% yield with enantiomeric purity of >95% ee. Even if there is no creation of asymmetric center in the processes described in Schemes 5.10–5.15, it was decided to include them in this chapter related to the fact that they allow access to chiral 2H-azirines without racemization and could interest readers looking for general methods of synthesis of these chiral products.

Later, these authors investigated a closely related methodology to achieve chiral trisubstituted 2H-azirines [16]. However, treatment of chiral methyl 2-methyl-3-phenyl-N-p-toluenesulfinylaziridine-2-carboxylate esters by LDA in THF at −78 °C led to a complex mixture of products. On the other hand, treatment of the corresponding N-tosylaziridine (Scheme 5.11) with 1.25 equiv. of LDA afforded the corresponding chiral trisubstituted 2H-azirine in 87% yield and 96% ee. In

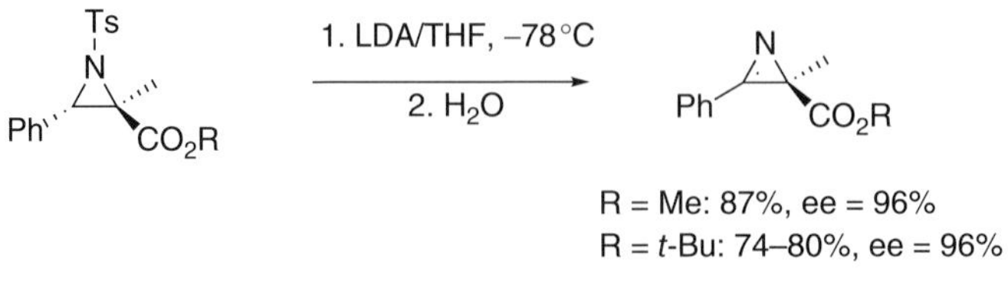

Scheme 5.11 Asymmetric synthesis of chiral trisubstituted 2H-azirines based on elimination from N-tosylaziridines.

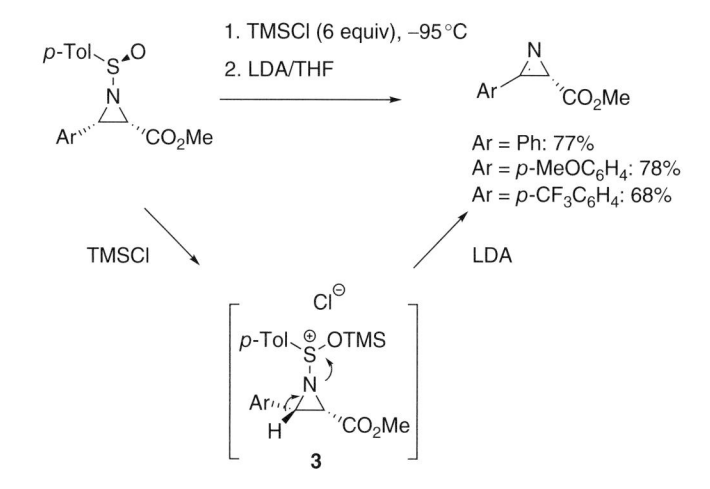

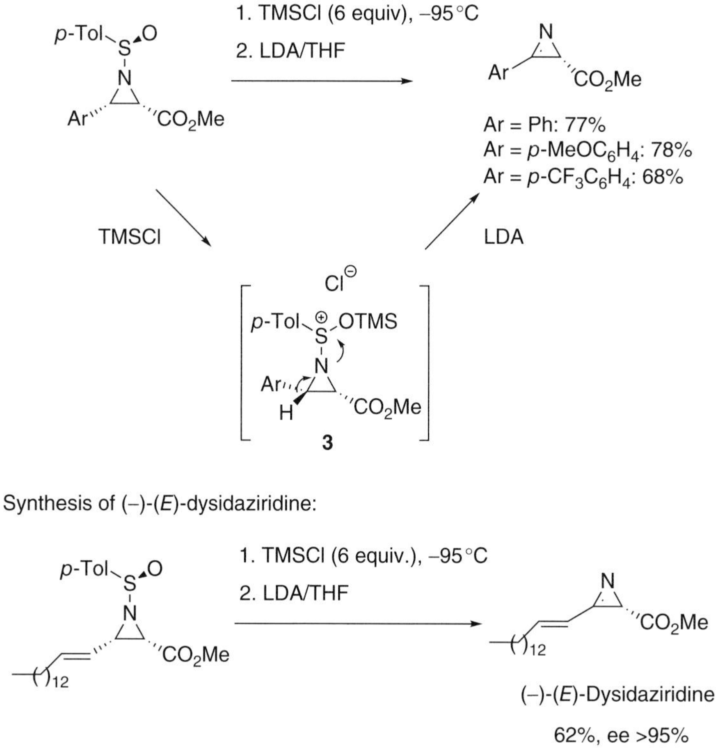

Scheme 5.12 Asymmetric synthesis of 2H-azirines based on elimination from N-sulfinylaziridine 2-carboxylate esters using TMSCl and LDA.

addition, the scope of this methodology was extended to the analogous *t*-butyl ester, which provided the corresponding 2H-azirine in 74–80% yield and same enantiomeric excess (Scheme 5.11).

In 1999, the authors reinvestigated their early asymmetric synthesis of 2H-azirine from chiral methyl 3-phenyl-N-p-toluenesulfinylaziridine-2-carboxylate by using another methodology consisting in the successive treatment of the substrate by trimethylsilyl chloride (TMSCl) at −95 °C followed by LDA instead of only LDA in the early work [17]. Under these novel reaction conditions, the authors achieved several methyl 3-aryl-N-p-toluenesulfinylaziridine-2-carboxylates in much better yields (68–78%) and as enantiopure products, as shown in Scheme 5.12. The increased yields were attributed to the improved leaving-group ability of intermediate silicon-oxonium species **3** arisen from the treatment of the N-sulfinylaziridine with TMSCl. The utility of this novel methodology was illustrated in the first total synthesis of the marine cytotoxic antibiotic (−)-dysidazirine, as shown in Scheme 5.12.

In 1995, Zwanenburg *et al.* demonstrated that chiral 2H-azirine-2-carboxylic esters could be synthesized by Swern oxidation of the corresponding optically active aziridine-2-carboxylic esters [18]. As shown in Scheme 5.13, the process consisted of the treatment of the starting aziridine with dimethyl sulfoxide

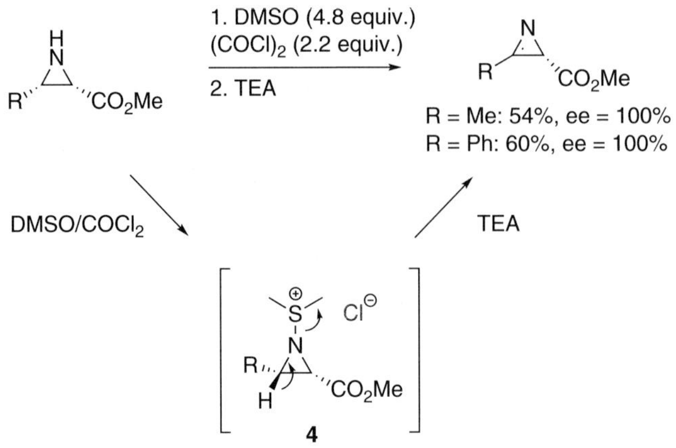

Scheme 5.13 Swern oxidation of NH aziridine 2-carboxylate esters.

(DMSO) (4.8 equiv.) and oxalyl chloride (2.2 equiv.) at −70 °C to generate a sulfonium intermediate salt **4**. The latter was subsequently submitted to a base, such as TEA, which underwent elimination of SMe_2 to give the final 2H-azirine. In this study, the authors found that oxidation of either (Z)- or (E)-aziridine-2-carboxilic esters provided the same 2H-azirine-2-carboxylates, where the integrity of the stereogenic center at C2 was retained. This regioselectivity resulted from the unexpected removal of the apparently less acidic C3 proton during the base-induced *syn*-elimination of the N-dimethylsulfonium intermediate. As shown in Scheme 5.13, the products were obtained in moderate yields (54–60%) with enantioselectivities of up to 100% ee.

In 2002, Davis developed asymmetric synthesis of 2H-azirine-3-phosphonates through Swern oxidation of the corresponding chiral aziridines [19]. These products were generated as major products in moderate-to-good yields (68–72%) along with the corresponding 2H-azirine-2-phosphonates as minor products (Scheme 5.14). The major products were further used as chiral iminodienophiles in Diels–Alder reactions to provide optically pure quaternary piperidine phosphonates after subsequent hydrogenation.

On the other hand, Zwanenburg *et al.* demonstrated in 1992 that 2H-azirines could be generated through base-induced dehydrochlorination of the corresponding N-chloroaziridines [20]. In 2007, Davis and Deng reported the first asymmetric dehydrochlorination of methyl 2-chloroaziridine 2-carboxylates to

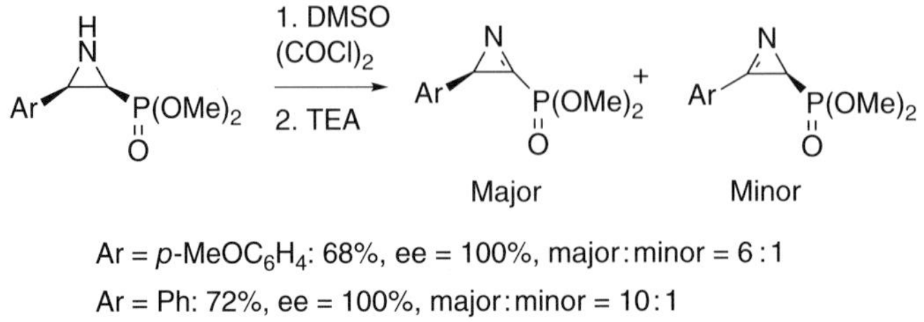

Scheme 5.14 Swern oxidation of NH aziridine 2-phosphonates.

give the corresponding enantiopure 2-substituted 2H-azirine 3-carboxylates [21]. Actually, the process involved methyl 2-chloro-N-p-toluenesulfinylaziridine 2-carboxylate **5** as starting material, the N-sulfinyl group of which was firstly removed by treatment with a Grignard reagent to give the corresponding anionic intermediate **6**. The nitrogen anion of the latter lost the chloride anion to provide the corresponding azirine **7**, which actually was not isolated because it was not stable enough under the reaction conditions but was directly added to the Grignard reagent from the least hindered direction to give the corresponding final trisubstituted aziridines in moderate yields (55–67%), as shown in Scheme 5.15. On the other hand, photocleavage of the N-sulfinyl group of **5a** allowed the corresponding methyl NH aziridine 2-chloro-2-carboxylate **8** to be isolated, which was subsequently submitted to dehydrochlorination by treatment with Hunig's base. This elimination generated intermediate 2H-azirine 3-carboxylate, which directly reacted with a 1,3-diene to give the final Diels–Alder cycloadduct. As shown in Scheme 5.15, the reaction of this intermediate azirine with 2,3-dimethyl-1,3-butadiene, *trans*-piperylene, or cyclopentadiene afforded the corresponding chiral cycloadducts in good yields (46–74%).

5.2.3 Other Approaches

Despite the increased industrial demand for enantiomerically pure compounds, to date, only a few catalytic asymmetric processes have found commercial

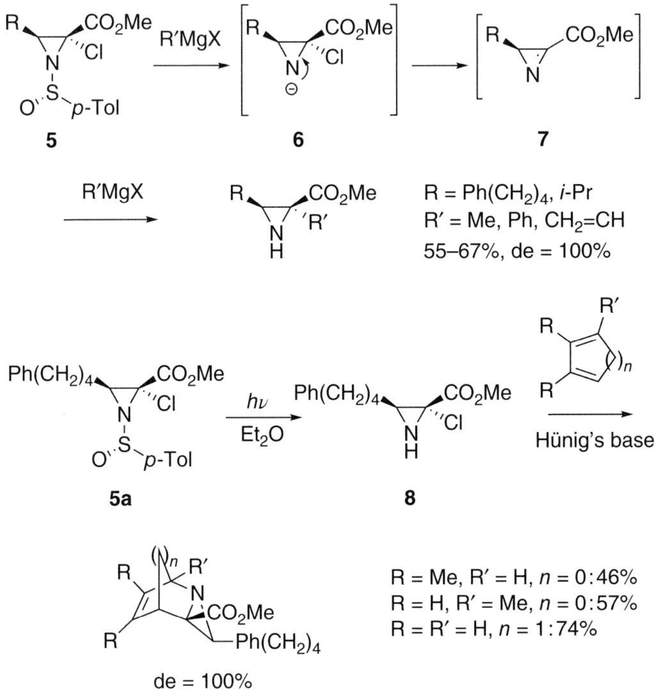

Scheme 5.15 Asymmetric synthesis of 2H-azirine 3-carboxylates through dehydrochlorination of aziridines.

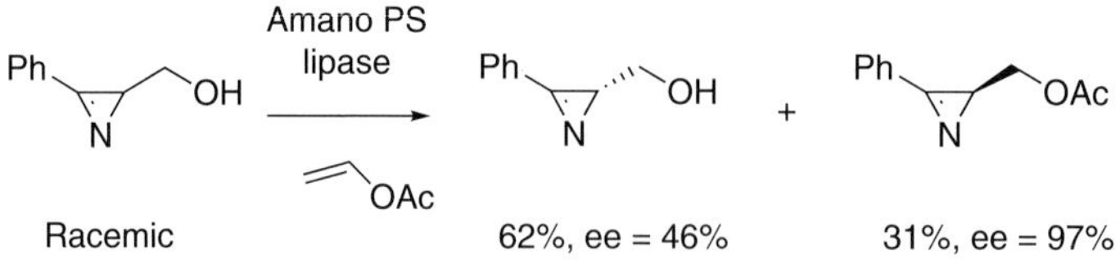

Scheme 5.16 Lipase-catalyzed kinetic resolution of 3-phenyl-2H-azirine-2-methanol.

application [22], among them, rare exceptions are catalytic kinetic resolutions [23]. Indeed, kinetic resolution as one of the most powerful tools in asymmetric catalysis has found wide applications in both academics and industry, complementing approaches such as asymmetric synthesis and classical resolution. When an asymmetric route to an enantiomerically pure 2H-azirine cannot be achieved, the possibility to perform a kinetic resolution of the racemic 2H-azirine mixture still exists. As an early example, Sakai and Utaka reported the lipase-catalyzed kinetic resolution of 3-phenyl-2H-azirine-2-methanol [24]. As shown in Scheme 5.16, using lipase from *Pseudomonas cepacia* (Amano PS) at −40 °C allowed the best enantioselectivities of up to 97% ee to be reached. Indeed, at this temperature, the reaction afforded (S)-3-phenyl-2H-azirine-2-methanol in 62% yield and 46% ee along with its (R)-acetate in 31% yield and 97% ee.

Finally, Suarez et al. developed the synthesis of chiral monosubstituted 3-alkyl-2H-azirines on the basis of intramolecular thermal rearrangement of the corresponding chiral vinyl azides [25] derived from carbohydrates [26]. As shown in Scheme 5.17, heating a series of chiral vinyl azides at reflux of toluene allowed the corresponding enantiopure 2H-azirines to be achieved in good yields (58–87%). Even if there is no creation of asymmetric center in the processes described in Scheme 5.17, it was decided to include them in this chapter related to the fact that they allow access to chiral 2H-azirines without racemization and could interest readers looking for general methods of synthesis of these chiral products.

5.3 Asymmetric Thiirination

Sulfur-containing compounds are widespread among natural products and bioactive substances and also useful ligands in asymmetric catalysis [27]. Therefore, considerable efforts have been devoted to develop stereocontrolled C—S-bond-forming procedures [28]. In particular, thiiranes are suitable precursors of numerous products including biologically active substances and polymers. Several methods for thiirane preparation [2i, 29] have been reported, such as conversion of epoxides to thiiranes, condensation of carbonyl compounds with sulfur-stabilized carbanions, intramolecular cyclization by nucleophilic displacement of β-substituted alkanethiols, among others. However, it must be recognized that a general methodology to prepare enantiomerically pure thiiranes, which could open the way to the synthesis of a number of important chiral products, is still challenging, and more especially a catalytic approach.

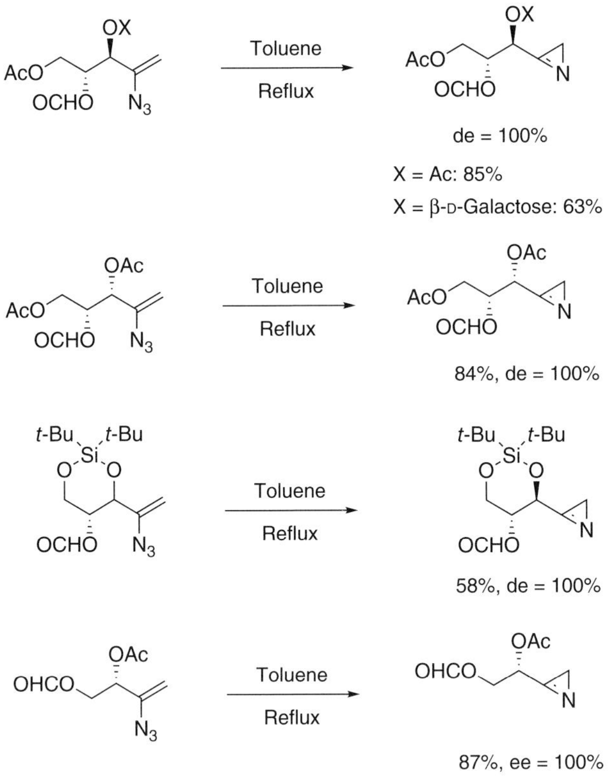

Scheme 5.17 Thermal rearrangements of vinyl azides derived from carbohydrates.

5.3.1 Conversion of Epoxides

The most convenient route to achieve thiiranes consists of the conversion of oxiranes into the corresponding thiiranes by an oxygen–sulfur exchange reaction. Various sulfur reagents, such as inorganic thiocyanates [30], thiourea [30, 31], phosphine sulfide [32], potassium thiocyanate [33] supported on silica gel, or dimethylthioformamide [34], among others, have been used to this end. However, most of these methods led to racemic products. However, optically active thiiranes were obtained almost quantitatively in a similar reaction of chiral oxiranes with thiourea in methanol solution at room temperature by Wong et al., in 1990 [35]. This procedure did not affect the chirality of the stereogenic center since no epimerization was observed. Another asymmetric procedure reported by Iranpoor and Kazemi in 1997 was based on Ru(III)-catalyzed conversion of optically active oxiranes into the corresponding thiiranes in the presence of ammonium thiocyanate [36]. In this case, (S)-styrene sulfide was obtained from (R)-styrene oxide in enantiomeric excess of 78% ee along with excellent yield (93%), as shown in Scheme 5.18. In this study, the authors have investigated other Lewis acids to induce the process and found that only AlCl$_3$ was able to convert (R)-styrene oxide into (S)-styrene sulfide albeit in only 65% yield and 51% ee.

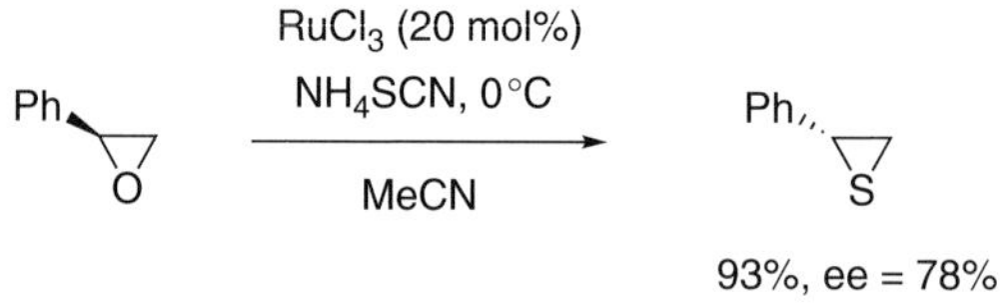

Scheme 5.18 Ru-catalyzed conversion of (R)-styrene oxide into (S)-styrene sulfide by reaction with NH$_4$SCN.

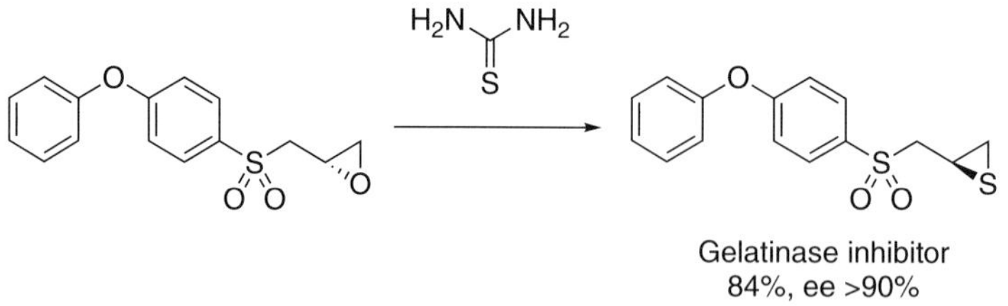

Scheme 5.19 Synthesis of a gelatinase inhibitor by reaction of an epoxide with thiourea.

Later in 2005, Mobashery et al. described an asymmetric synthesis of 1,2-(4-phenoxyphenylsulfonylmethyl)thiirane, which is a selective gelatinase inhibitor active for cancer metastasis [37]. The key step of the synthesis consisted of the reaction of (S)-epoxide depicted in Scheme 5.19 with thiourea to give the corresponding (R)-1,2-(4-phenoxyphenylsulfonylmethyl)thiirane in 84% yield with an enantiomeric excess of >90% ee.

5.3.2 Condensation of Sulfur-Stabilized Carbanions to Carbonyl Compounds

Thiiranes can also be prepared by condensation of carbonyl compounds with sulfur-stabilized carbanions [38]. For example, Meyers and Ford described in 1975 a one-carbon homologation of carbonyl compounds to thiiranes, which was achieved using 2-(methylthio)oxazolines as precursors of sulfur-stabilized carbanions [39]. Application of this thiirane synthesis to chiral derivatives was successfully implemented. Indeed, a chiral 2-(methylthio)oxazoline could be reacted with various aldehydes to give the corresponding chiral thiiranes in good yields (73–78%) albeit with low enantioselectivities of 19–32% ee, as shown in Scheme 5.20.

Later in 1979, Soai and Mukaiyama reported the addition of lithium salt of 2-(methylthio)thiazoline to benzaldehyde carried out at −100 °C in the presence of 4 equiv. of lithium salt of (2S,2′S)-2-hydroxymethyl-1-[1-methyl-2-pyrrolidinyl)methyl]pyrrolidine as chiral ligand [40]. With this method, chiral styrene sulfide was obtained in 49% yield with only 20% optical purity (Scheme 5.21).

In 1993, Jiménez et al. investigated the reaction between a [2,1-b]thiazolium-4-olate system and benzaldehydes in refluxing toluene [41]. Surprisingly, the process afforded only one diastereomer of the corresponding thiiranes in low yields (16–31%), as depicted in Scheme 5.22. On the other hand, the reaction of 4-(N,N-dimethylamino)benzaldehyde with the same chiral imidazo

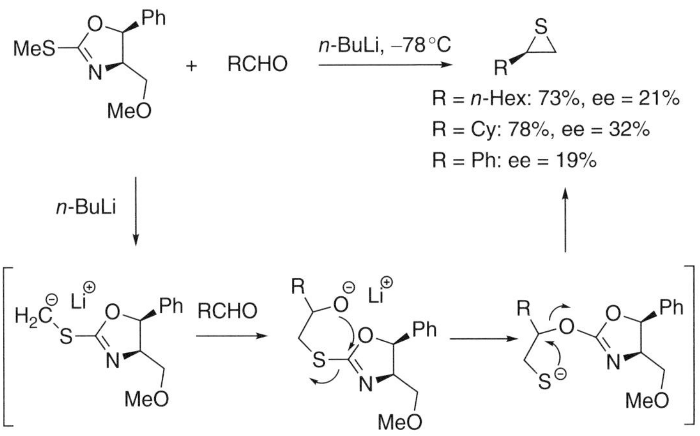

Scheme 5.20 Reaction of a chiral 2-(methylthio)oxazoline with aldehydes.

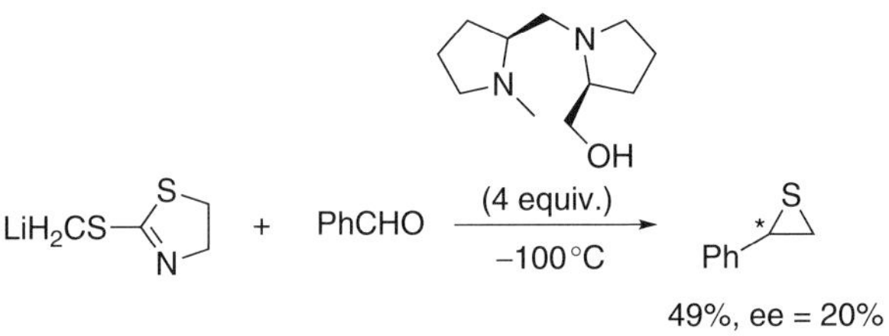

Scheme 5.21 Reaction of 2-(methylthio)thiazoline with benzaldehyde in the presence of 4 equiv. of a chiral ligand.

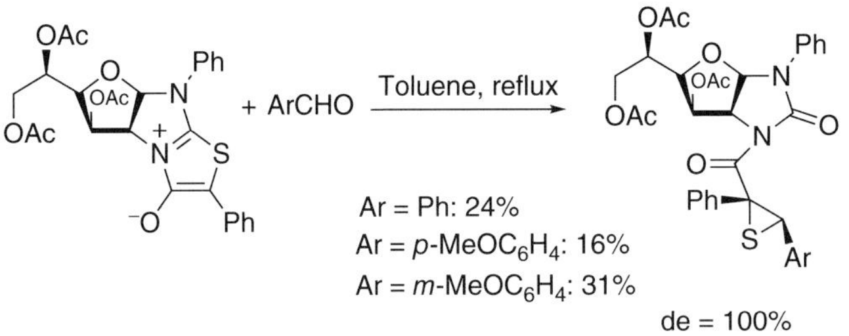

Scheme 5.22 Reaction of a chiral imidazo[2,1-b]thiazolium-4-olate system with benzaldehydes.

[2,1-b]thiazolium-4-olate led to the corresponding thiirane in 15% yield along with 11% of another diastereomer.

Recently, Palomo et al. reported the use of chiral N-(diazoacetyl)oxazolidin-2-thiones as new sulfur-donor reagents that in combination with aldehydes and a rhodium(II) catalyst were capable of producing the corresponding chiral α,β-thioepoxy carbonyls in a highly stereoselective manner [42]. Screening of various metal catalysts demonstrated that $Rh_2(OAc)_4(H_2O)_2$ provided the best results employed at 2 mol% of catalyst loading at −20 °C. As shown in Scheme 5.23,

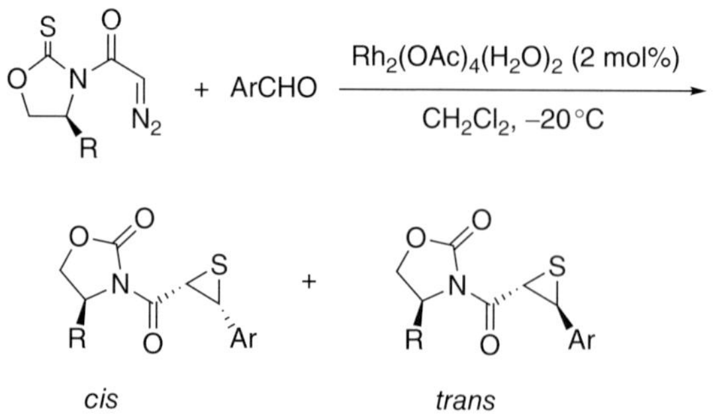

Scheme 5.23

Rh$_2$(OAc)$_4$(H$_2$O)$_2$ (2 mol%), CH$_2$Cl$_2$, −20 °C

R = i-Pr, Ar = Ph: 62% (cis), cis/trans = 97:3
R = t-Bu, Ar = Ph: 60% (cis), cis/trans = 94:6
R = Ar = 45% (cis), cis/trans = 92:8
R = i-Pr, Ar = p-BrC$_6$H$_4$: 56% (cis), cis/trans = 91:9
R = i-Pr, Ar = p-NO$_2$C$_6$H$_4$: 61% (cis), cis/trans = 92:8
R = i-Pr, Ar = p-CNC$_6$H$_4$: 56% (cis), cis/trans = 91:9
R = i-Pr, Ar = p-MeOC$_6$H$_4$: 61% (trans), cis/trans = 1:99
R = i-Pr, Ar = p-TBSC$_6$H$_4$: 31% (trans), cis/trans = 1:99

Scheme 5.23 Rhodium-catalyzed reaction of chiral N-(diazoacetyl)oxazolidin-2-thiones with benzaldehydes.

a range of aldehydes reacted with variously substituted N-(diazoacetyl)-2-oxazolidinethiones to give a mixture of the corresponding *cis*- and *trans*-thiiranes in moderate-to-good yields. Interestingly, in most cases, *cis*-thiiranes were obtained as the major isomers with *cis/trans* ratios ranging from 82:18 to 97:3, whereas in the case of benzaldehydes bearing electron-donating substituents, such as *p*-anisidine, and *p-tert*-butyldimethylsilyloxybenzaldehyde, the *trans*-configured thiiranes were formed almost exclusively (*cis/trans* = 1:99). This work constituted the first rhodium-catalyzed reaction of a diazoacetyl compound with aldehydes that afforded thiiranes instead of oxiranes.

5.3.3 Intramolecular Nucleophilic Substitution

Another general method for the preparation of thiiranes is based on intramolecular cyclization by nucleophilic displacement of β-substituted alkanethiols. For example, treatment of optically active 1-(benzothiazol-2-ylsulfanyl)-2-alkanols, prepared by reduction of the corresponding ketones with baker's yeast, with sodium hydride afforded the corresponding chiral thiiranes in good yields with high enantioselectivities (91–99% ee), as shown in Scheme 5.24 [43]. The mechanism of the process is depicted in this scheme, showing that the reaction occurred purely by inversion of the (R)-configuration at the stereocenter of the alkanols, providing the corresponding (S)-thiiranes.

In 2007, Plenkiewicz and Lukowska studied the asymmetric synthesis of chiral thiiranes based on intramolecular nucleophilic substitution of chiral 1-aryloxy-3-thiocyanatopropan-2-ols and 1-arylsulfanyl-3-thiocyanatopropan-2-yl acetates

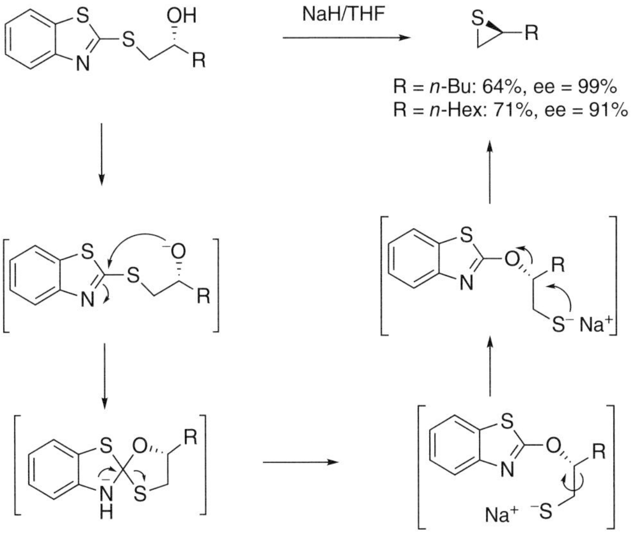

Scheme 5.24 Intramolecular nucleophilic substitution of 1-(benzothiazol-2-ylsulfanyl)-2-alkanols.

R = n-Bu: 64%, ee = 99%
R = n-Hex: 71%, ee = 91%

in the presence of a 60% aqueous solution of potassium hydroxide at room temperature [44]. Unfortunately, only polymers were formed under these conditions. In order to avoid the thiirane polymerization, the authors changed the procedure by making the base reacting not with the alcohols but with their more stable acetates. Moreover, a solution of lithium hydroxide was used as a base instead of potassium hydroxide. The intramolecular nucleophilic substitution of chiral 1-aryloxy-3-thiocyanatopropan-2-yl acetates and 1-arylsulfanyl-3-thiocyanatopropan-2-yl acetates was performed in a two-phase aqueous LiOH/THF system, allowing the contact time of the formed thiiranes and the base to be considerably reduced. Under these optimal conditions, a range of chiral thiiranes were synthesized in good yields (67–87%) with moderate-to-excellent enantioselectivities of up to 97% ee, as shown in Scheme 5.25. It must be noted that the best enantioselectivities were reached in the case of 1-aryloxy-3-thiocyanatopropan-2-yl acetates as substrates (86–97% ee) while lower enantioselectivities (53–86% ee) were achieved in the reaction of 1-arylsulfanyl-3-thiocyanatopropan-2-yl acetates.

Later, these authors could apply the same reaction conditions to the intramolecular nucleophilic substitution of chiral 1-aryloxy-3-thiocyanatopropan-2-ols, which provided the corresponding thiiranes in good-to-high yields (76–91%) with very good enantioselectivities (92–99% ee), as shown in Scheme 5.26 [45].

In 2009, Denmark and Vogler reported the synthesis of enantiomerically enriched thiiranium ions on the basis of silver-promoted ionization of the corresponding chloro sulfides [46]. As shown in Scheme 5.27, these products were obtained in quantitative yields in the presence of silver hexafluoroantimonate. They were found configurationally stable in solution up to room temperature

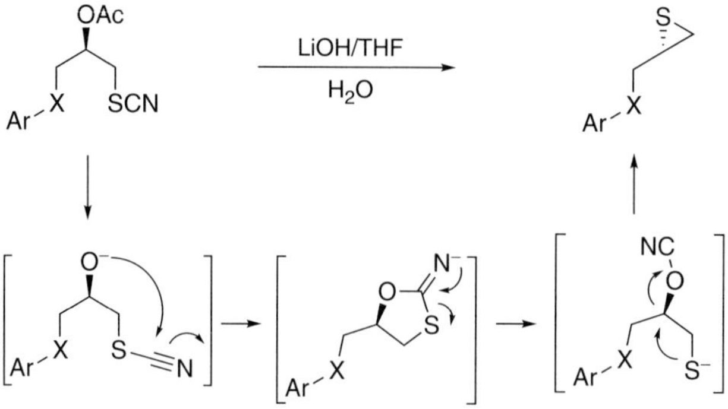

5.3.4 Miscellaneous Reactions

Scheme 5.25 Intramolecular nucleophilic substitution of 1-aryloxy-3-thiocyanatopropan-2-yl acetates and 1-arylsulfanyl-3-thiocyanatopropan-2-yl acetates.

Ar = Ph, X = S: 79%, ee (substrate) = 90%, ee (product) = 86%
Ar = p-ClC$_6$H$_4$, X = S: 67%, ee (substrate) = 58%, ee (product) = 53%
Ar = p-BrC$_6$H$_4$, X = S: 84%, ee (substrate) = 61%, ee (product) = 59%
Ar = Ph, X = O: 87%, ee (substrate) = 97%, ee (product) = 97%
Ar = p-ClC$_6$H$_4$, X = O: 83%, ee (substrate) = 90%, ee (product) = 90%
Ar = p-Tol, X = O: 76%, ee (substrate) = 86%, ee (product) = 86%

Scheme 5.26 Intramolecular nucleophilic substitution of 1-aryloxy-3-thiocyanatopropan-2-ols.

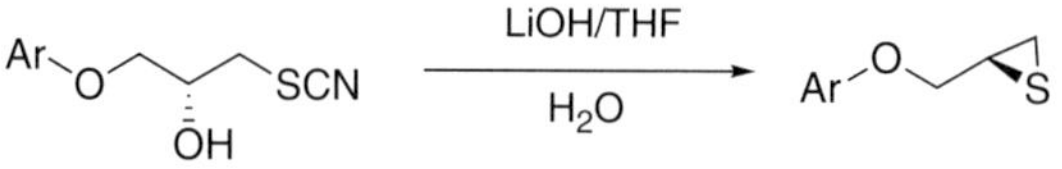

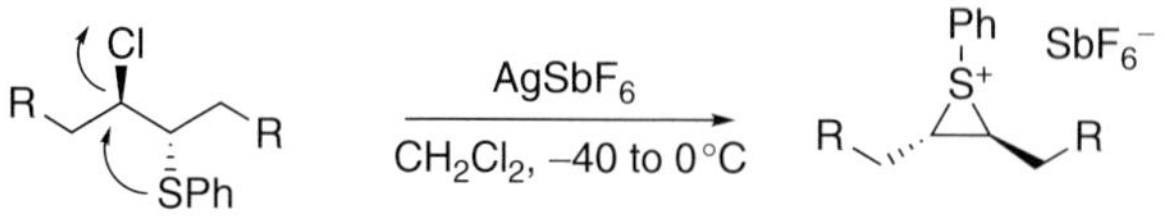

Ar = Ph: 77%, ee = 93%
Ar = p-ClC$_6$H$_4$: 86%, ee = 99%
Ar = p-Tol: 79%, ee = 92%
Ar = o-ClC$_6$H$_4$: 76%, ee = 95%
Ar = o-Tol: 91%, ee = 96%

Scheme 5.27 Silver-promoted intramolecular nucleophilic substitution of chiral chloro sulfides.

R = n-Pr: >99%, ee (substrate) = 94%, ee (product) = 94%
R = Ph: >99%, ee (substrate) = 92%, ee (product) = 92%

as demonstrated by their subsequently stereospecific capture by various oxygen- and nitrogen-based nucleophiles.

5.3.4 Miscellaneous Reactions

Fu and Ma have studied the electrophilic halocyclization of chiral allenes bearing a nucleophilic functionality. In this context, they developed highly regio- and stereoselective bromohydroxylation reactions of chiral 1,2-allenyl sulfones [47].

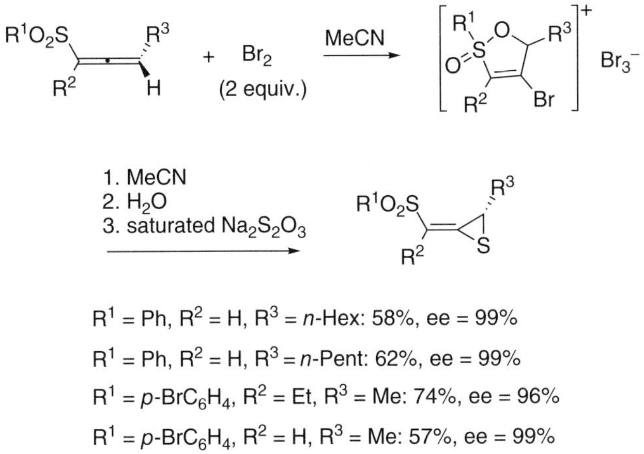
R1O2S
R3
R2
H
+ Br2
(2 equiv.)
MeCN
R1
O
R3
O=S
R2
Br
+
Br3−
1. MeCN
2. H2O
3. saturated Na2S2O3
R1O2S
R3
R2
S
R1 = Ph, R2 = H, R3 = n-Hex: 58%, ee = 99%
R1 = Ph, R2 = H, R3 = n-Pent: 62%, ee = 99%
R1 = p-BrC6H4, R2 = Et, R3 = Me: 74%, ee = 96%
R1 = p-BrC6H4, R2 = H, R3 = Me: 57%, ee = 99%

Scheme 5.28 Thiirination of chiral 1,2-allenyl sulfones with Br_2.

The process employed Br_2 as electrophilic agent followed by sequential treatment with water and an aqueous solution of $Na_2S_2O_3$, which afforded the corresponding almost enantiopure sulfonyl alkylidenethiiranes with moderate-to-good yields (57–74%). The axial chirality in the allenes was transferred with high efficiency to the thiiranes in the form of central chirality, as shown in Scheme 5.28. The authors have proposed that the process evolved through the five-membered intermediate depicted in Scheme 5.28 having Br_3^- as the counter ion, which could be isolated and its structure established by X-ray diffraction analysis. This intermediate was further attacked by $S_2O_3^{2-}$ with ring opening to afford the final thiirane. This novel methodology should be useful in organic synthesis because optically active starting allenes, which are readily available from propargylic alcohols, can be transformed into densely functionalized products.

Finally, Kawashima *et al.* reported the synthesis of a tetracoordinated sulfurane bearing a $1\lambda^4$,2-dithietane moiety, whose 3- and 4-positions were chiral carbon centers [48]. They demonstrated that its thermolysis provided the corresponding thiirane as a single diastereomer in 94% yield, as shown in Scheme 5.29.

5.4 Conclusions

Because of their ring strain, reactive π-bond, and ability to undergo regioselective ring-cleavage reactions, 2*H*-azirines constitute versatile chemical compounds. Indeed, they can act as nucleophiles as well as electrophiles in organic

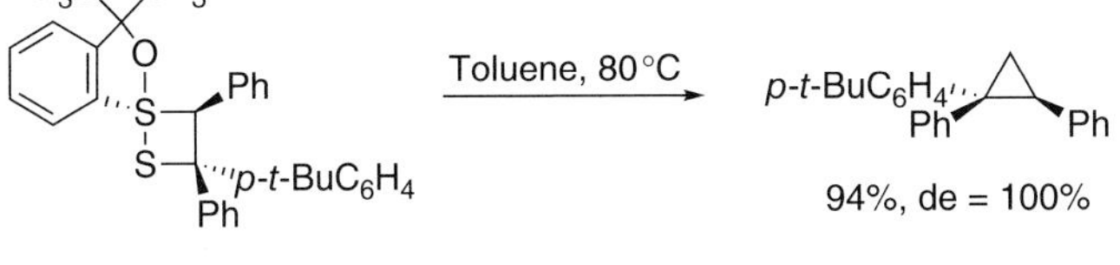
F3C
CF3
O
S
Ph
S
p-t-BuC6H4
Ph
Toluene, 80 °C
p-t-BuC6H4
Ph
Ph
94%, de = 100%

Scheme 5.29 Thermolysis of a chiral $1\lambda^4$,2-dithietane.

reactions, but can also acta as dienophiles and dipolarophiles in cycloaddition chemistry. Consequently, chiral 2*H*-azirines represent key precursors for the synthesis of a variety of important chiral nitrogen-containing ring systems, including complex natural products. The facility with which these small-ring nitrogen-containing compounds can be converted to important pharmaceutical products under mild conditions with wide functional group compatibility makes these molecules quite useful for heterocyclic chemistry. The synthesis of chiral 2*H*-azirines has been the topic of many researches in the past decade. Among significant preparative pathways to chiral 2*H*-azirines is the organocatalyzed Neber rearrangement of oxime sulfonates, which has recently allowed high enantioselectivities of up to 93% ee to be achieved using a chiral bifunctional thiourea. Other good results were previously reported using chiral cinchona alkaloids. A less-interesting approach to chiral 2*H*-azirines is based on elimination reactions of chiral N-substituted aziridines, namely, dehydrochlorination of *N*-chloroaziridines and elimination from *N*-sulfinylaziridines.

On the other hand, because sulfur-containing compounds are widespread among natural products and bioactive molecules and also useful ligands in asymmetric catalysis, several methods for chiral thiirane preparation have been reported. Among them are the conversion of chiral epoxides to chiral thiiranes, the condensation of chiral sulfur-stabilized carbanions to carbonyl compounds, the intramolecular cyclization by nucleophilic displacement of chiral β-substituted alkanethiols. In the past decade, even if a few conversions of chiral oxiranes into the corresponding chiral thiiranes by an oxygen–sulfur exchange reaction have been successfully described, the most developed routes to chiral thiiranes have been based on condensation of chiral sulfur-stabilized carbanions to carbonyl compounds. For example, the reaction between a chiral imidazo[2,1-*b*]thiazolium-4-olate system and benzaldehydes was found to afford only one diastereomer of the corresponding thiiranes. Another nice result obtained from the use of chiral *N*-(diazoacetyl)oxazolidin-2-thiones as new sulfur-donor reagents that in combination with aldehydes and a rhodium(II) catalyst were capable of producing the corresponding chiral α,β-thioepoxy carbonyls in a highly stereoselective manner. Moreover, another general method for the preparation of thiiranes is based on intramolecular cyclization by nucleophilic displacement of β-substituted alkanethiols. In this context, several excellent results were recently reported, such as the treatment of optically active 1-(benzothiazol-2-ylsulfanyl)-2-alkanols with sodium hydride, the intramolecular nucleophilic substitution of chiral 1-aryloxy-3-thiocyanatopropan-2-ols and 1-arylsulfanyl-3-thiocyanatopropan-2-yl acetates in the presence of a 60% aqueous solution of potassium hydroxide, and the silver-promoted ionization of chiral chloro sulfides. Among other successful methodologies of synthesis of chiral thiiranes recently described is a regio- and stereoselective bromohydroxylation reaction of chiral 1,2-allenyl sulfones to provide the corresponding enantiopure sulfonyl alkylidenethiiranes.

However, in spite of the results discussed in the chapter, it must be recognized that a general methodology to prepare enantiomerically pure 2*H*-azirines and thiiranes, which could open the way to the synthesis of a number of important products, is still challenging, and more especially catalytic approaches.

References

1 (a) Heimgartner, H. (1991) *Angew. Chem. Int. Ed. Engl.*, 30, 238–264;(b) Würthwein, E.U., Hergenröther, T., and Quast, H. (2002) *Eur. J. Org. Chem.*, 2002, 1750–1755.
2 (a) Padwa, A. (2010) *Adv. Heterocycl. Chem.*, 99, 1–31;(b) Lemos, A. (2009) *Molecules*, 14, 4098–4119;(c) Padwa, A. (2008) in *Comprehensive Heterocyclic Chemistry III*, Chapter 1 (eds A. Ramsden, E.F.V. Scriven, and R.J.K. Taylor), Elsevier, Oxford, pp. 1–104;(d) Pinho e Melo, T.M.V.D. and Rocha Gonsalves, A.M.d'A. (2004) *Curr. Org. Synth.*, 1, 275–292;(e) Palacios, F., Ochoa de Retana, A.M., de Marigorta, E.M., and de los Santo, J.M. (2002) *Org. Prep. Proced. Int.*, 34, 219–269;(f) Palacios, F., Ochoa de Retana, A.M., de Marigorta, E.M., and de los Santos, J.M. (2001) *Eur. J. Org. Chem.*, 2001, 2401–2414;(g) Zwanenburg, B. and ten Holte, P. (2001) *Top. Curr. Chem.*, 216, 93–124;(h) Pearson, W.H., Lian, B.W., and Bergmeier, S.C. (1996) in *Comprehensive Heterocyclic Chemistry II*, vol. 1A (eds A.R. Katritzky, C.W. Rees, and E.F.V. Scriven), Elsevier, Oxford, pp. 1–60;(i) Nair, V. (1983) in *Small Ring Heterocycles, Part 1: Aziridines, Azirines, Thiiranes, Thiirenes* (ed. A. Hassner), John Wiley & Sons, Inc., New York, pp. 215–332.
3 Neber, P.W. and Friedsheim, A.V. (1926) *Liebigs Ann. Chem.*, 449, 109–134.
4 Piskunova, I.P., Eremeev, A.V., Mishnev, A.F., and Vosekalna, I.A. (1993) *Tetrahedron*, 49, 4671–4676.
5 Verstappen, M.M.H., Ariaans, G.J.A., and Zwanenburg, B. (1996) *J. Am. Chem. Soc.*, 118, 8491–8492.
6 Palacios, F., Ochoa de Retana, A.M., Gil, J.I., and Ezpeleta, J.M. (2000) *J. Org. Chem.*, 65, 3213–3217.
7 Palacios, F., Ochoa de Renata, A.M., and Gil, J.I. (2000) *Tetrahedron Lett.*, 41, 5363–5366.
8 Palacios, F., Aparicio, D., Ochoa de Renata, A.M., de los Santos, J.M., Gil, J.I., and Lopez de Munain, R. (2003) *Tetrahedron: Asymmetry*, 14, 689–700.
9 Ooi, T., Takahashi, M., Doda, K., and Maruoka, K. (2002) *J. Am. Chem. Soc.*, 124, 7640–7641.
10 Skepper, C.K., Dalisay, D.S., and Molinski, T.F. (2008) *Org. Lett.*, 10, 5269–5271.
11 Skepper, C.K., Dalisay, D.S., and Molinski, T.F. (2010) *Bioorg. Med. Chem. Lett.*, 20, 2029–2032.
12 Sakamoto, S., Inokuma, T., and Takemoto, Y. (2011) *Org. Lett.*, 13, 6374–6377.
13 Logyda-Chruscinska, E. (2011) *Chem. Rev.*, 111, 1824–1833.
14 Cardoso, A.L., Gimeno, L., Lemos, A., Palacios, F., and Pinho e Melo, T.M.V.D. (2013) *J. Org. Chem.*, 78, 6983–6991.
15 Davis, F.A., Reddy, G.V., and Liu, H. (1995) *J. Am. Chem. Soc.*, 117, 3651–3652.
16 Davis, F.A., Liang, C.-H., and Liu, H. (1997) *J. Org. Chem.*, 62, 3796–3797.
17 Davis, F.A., Liu, H., Liang, C.-H., Reddy, G.V., Zhang, Y., Fang, T., and Titus, D.D. (1999) *J. Org. Chem.*, 64, 8929–8935.
18 Gentilucci, L., Grijzen, Y., Thijs, L., and Zwanenburg, B. (1995) *Tetrahedron Lett.*, 36, 4665–4668.
19 Davis, F.A., Wu, Y., Yan, H., Prasad, K.R., and McCoull, W. (2002) *Org. Lett.*, 4, 655–658.

20 Legters, J., Thijs, L., and Zwanenburg, B. (1992) *Recl. Trav. Chim. Pays-Bas*, 111, 75–78.
21 Davis, F.A. and Deng, J. (2007) *Org. Lett.*, 9, 1707–1710.
22 Blaser, H.U. and Schmidt, E. (2004) *Asymmetric Catalysis on Industrial Scale*, Wiley-VCH Verlag GmbH, Weinheim.
23 (a) Kagan, H.B. and Fiaud, J.C. (1988) *Top. Stereochem.*, 18, 249–330;(b) Hoveyda, A.H. and Didiuk, M.T. (1998) *Curr. Org. Chem.*, 2, 489–526;(c) Cook, G.R. (2000) *Curr. Org. Chem.*, 4, 869–885;(d) Keith, M., Larrow, J.F., and Jacobsen, E.N. (2001) *Adv. Synth. Catal.*, 343, 5–26;(e) Robinson, D.E.J.E. and Bull, S.D. (2003) *Tetrahedron: Asymmetry*, 14, 1407–1446;(f) Vedejs, E. and Jure, M. (2005) *Angew. Chem. Int. Ed.*, 44, 3974–4001;(g) Jarvo, E.R. and Miller, S.J. (2004) in *Comprehensive Asymmetric Catalysis: Supplement 1* (eds E.N. Jacobsen, A. Pfaltz, and H. Yamamoto), Springer, Berlin, pp. 189–206.
24 Sakai, T., Kawabata, I., Kishimoto, T., Ema, T., and Utaka, M. (1997) *J. Org. Chem.*, 62, 4906–4907.
25 Smolinsky, G. (1962) *J. Org. Chem.*, 27, 3557–3559.
26 Alonso-Cruz, C.R., Kennedy, A.R., Rodriguez, M.S., and Suarez, E. (2003) *Org. Lett.*, 5, 3729–3732.
27 (a) Pellissier, H. (2009) *Chiral Sulfur Ligands, Asymmetric Catalysis*, Royal Society of Chemistry, Cambridge;(b) Toru, T. and Bolm, C. (eds) (2008) *Organosulfur Chemistry in Asymmetric Synthesis*, Wiley-VCH Verlag GmbH, Weinheim;(c) Mellah, M., Voituriez, A., and Schulz, E. (2007) *Chem. Rev.*, 107, 5133–5209.
28 (a) Clayden, J. and MacLellan, P. (2011) *Beilstein J. Org. Chem.*, 7, 582–595;(b) Zhang, Y. and Wang, J. (2010) *Coord. Chem. Rev.*, 254, 941–953.
29 (a) Sander, M. (1996) *Chem. Rev.*, 66, 297–339;(b) Vedejs, E. and Krafft, G.A. (1982) *Tetrahedron*, 38, 2857–2881.
30 Bouda, H., Borredon, M.E., Delman, M., and Gaset, A. (1989) *Chem. Commun.*, 19, 491–500.
31 Ketcham, R. and Shah, V.P. (1963) *J. Org. Chem.*, 28, 229–230.
32 Korn, A.C. and Morder, L. (1995) *J. Chem. Res., Synop.*, 3, 102–103.
33 Brimeyer, M.O., Mehrota, A., Quici, S., Nigman, A., and Regen, S.L. (1980) *J. Org. Chem.*, 45, 4254–4255.
34 Takido, T., Kobayashi, Y., and Itabashi, K. (1986) *Synthesis*, 1986, 779–780.
35 Pederson, R.L., Liu, K.K.C., Rutan, J.F., Chen, L., and Wong, C.H. (1990) *J. Org. Chem.*, 55, 4897–4901.
36 Iranpoor, N. and Kazemi, F. (1997) *Tetrahedron*, 53, 11377–11382.
37 Lee, M., Bernardo, M.M., Meroueh, S.O., Brown, S., Fridman, R., and Mobashery, S. (2005) *Org. Lett.*, 7, 4463–4465.
38 (a) Johnson, R., Nakanishi, A., Nakanishi, N., and Tanaka, K. (1975) *Tetrahedron Lett.*, 16, 2865–2868;(b) Meyers, A. and Mihelich, E.D. (1976) *Angew. Chem. Int. Ed. Engl.*, 15, 270–281;(c) Jonson, C.R. and Tanaka, K. (1976) *Synthesis*, 1976, 413–414;(d) Hoppe, D. and Follmann, R. (1977) *Angew. Chem.*, 89, 478–479.
39 (a) Meyers, A.I. and Ford, M.E. (1975) *Tetrahedron Lett.*, 16, 2861–2864;(b) Meyers, A.I. and Ford, M.E. (1976) *J. Org. Chem.*, 41, 1735–1742.
40 Soai, K. and Mukaiyama, T. (1979) *Bull. Chem. Soc. Jpn.*, 52, 3371–3376.

41 Aceres, P., Avalos, M., Babiano, R., Gonzalez, L., Jiménez, J.L., Mendez, M.M., and Palacios, J.C. (1993) *Tetrahedron Lett.*, 34, 2999–3002.
42 Cano, I., Gomez-Bengoa, E., Landa, A., Maestro, M., Mielgo, A., Olaizola, I., Oiarbide, M., and Palomo, C. (2012) *Angew. Chem. Int. Ed.*, 51, 10856–10860.
43 Di Nunno, L., Franchini, C., Nacci, A., Scilimati, A., and Sinicropi, M.S. (1999) *Tetrahedron: Asymmetry*, 10, 1913–1926.
44 Lukowska, E. and Plenkiewicz, J. (2007) *Tetrahedron: Asymmetry*, 18, 493–499.
45 Lukowska, E. and Plenkiewocz, J. (2007) *Tetrahedron: Asymmetry*, 18, 1202–1209.
46 Denmark, S.E. and Vogler, T. (2009) *Chem. Eur. J.*, 15, 11737–11745.
47 Zhou, C., Fu, C., and Ma, S. (2007) *Angew. Chem. Int. Ed.*, 46, 4379–4381.
48 Kusaka, S., Kano, N., and Kawashima, T. (2010) *Heteroat. Chem*, 21, 412–417.

Index

a

1,2-amino halides   278–282
aryl aldimines   333
aryldiazo reagents   19
asymmetric counteranion-directed catalysis (ACDC)   463, 465
asymmetric epoxidation
  chiral auxiliary approach   380
  metal-catalyzed procedures
    Cr-, Mo-, W-catalyzed epoxidations   397–398
    Mn-, Re-, Fe-, Ru-catalyzed epoxidations   400, 404–406, 408–410, 412
    Pt-, Zn-, lanthanoid-catalyzed epoxidations   412–413, 416, 418
    Ti-, Zr-, Hf-catalyzed epoxidations   381–383, 385, 387–388, 391
    V-, Nb-, Ta-catalyzed epoxidations   391, 393, 395–396
asymmetric organocatalysis. *see also* chiral amines
  chiral dioxiranes, iminium salts and alkyl hydroperoxides   431, 433, 436, 438–442, 444, 446
  polyamino acids and aspartate-derived peracids   423, 425, 427, 429–430
  PTC   419–420, 422
aza-Michael-initiated ring closure (aza-MIRC) reaction   222

aziridination
  addition/elimination processes
    allenylzinc reagent   259–260
    chiral α-amino ketimine   258
    chiral α-chloro-*N*-*p*-toluenesulfinylaldimines   264
    chiral α-chloro *N*-sulfinyl aldimines   256
    chiral α-chloro *N*-sulfinylimines   258
    chiral α-chloro *N*-sulfinylketimines   257
    chiral α-chlorovinylsulfoxides   221
    chiral α-chloro-*N*-*p*-toluenesulfinyliso-butyraldimine   263
    chiral-chloro *N*-sulfinyl aldimines reaction   256
    chiral methyl *N*-*tert*-butanesulfinyl-2-chloroethanimidate   261
    chiral (*E*)-nitroalkenes   223–224
    chiral *N*-sulfinylaldimines   261
    chiral *N*-*tert*-butanesulfinylaldimines   266
    chiral protected amino ester   265
    cyclic α-bromo enones   220
    Gabriel–Cromwell aziridine synthesis   219–220
    chiral tricyclic 1,2-oxazine   257
    organocerium reagent   259
    α,β-unsaturated olefins   223

Asymmetric Synthesis of Three-Membered Rings, First Edition.
Hélène Pellissier, Alessandra Lattanzi and Renato Dalpozzo.
© 2017 Wiley-VCH Verlag GmbH & Co. KGaA. Published 2017 by Wiley-VCH Verlag GmbH & Co. KGaA

aziridination (*Contd.*)
  Aza–Darzens reaction
    camphorsultam-derived
      α-bromoenolate 254–255
    chiral aldimines 252
    chiral
      N,N-dibenzyl-*N*-
      phosphonylaldimines 251
    chiral
      N,N-diisopropyl-*N*-
      phosphonylaldimines 251
    chiral
      N-tert-
      butylsulfinylaldimine 248
    chiral *N*-(2,4,6-
      trimethylphenylsulfinyl)
      aldimines 249
    chiral phenylalaninal-derived
      aldimine 254
    chiral 2-pyridinealdimines 253
    chiral *tert*-butanesulfinylimines 250
    2-quinolinealdimine 253
  carbene transfer to Imines
    carbene methodology 332–350
    miscellaneous
      reaction 353–357
    sulfur-ylide-mediated
      aziridination 350–353
  chiral aziridines 356–363
  chiral substrates, alkenes
    addition 206–219
  Cu-catalyzed aziridination 317–318
    alkenes 299
    allylglycine derivatives 298
    chalcones 302
    1,3-dienes 304
    heterogeneous styrene
      aziridination 298
    (R,R)-β-methoxytyrosine 297
    methyl cinnamate 305
    olefins 307
    styrenes 309
    2,2,2-trichloroethyl
      tosyloxycarbamate 304
    unsaturated sulfamates 301, 308
  Fe-catalyzed aziridination 319
  four-step synthesis 355
  intramolecular substitution
    1,2-amino alcohols 270
    1,2-amino halides 278
    1,2-amino sulfides and 2-amino
      selenides 1, 285
    1,2-azido alcohols 282
    epoxides 286
  kinetic resolutions 357
  Lewis-acid-catalyzed aza-Diels–Alder
    reactions 354
  Lewis acid-mediated aza-Diels–Alder
    reactions 271
  miscellaneous reactions 266, 287
  nitrene-transfer reagent
    (+)-agelastatin A synthesis 207
    Cu-catalyzed aziridination 208,
      210
    L-daunosamine and L-ristosamine
      glycosides synthesis 217–218
    Pb-catalyzed aziridination 214
    Rh-catalyzed intramolecular
      aziridinations 211–212, 214
  nitrogen heterocycles reaction, chiral
    2*H*-azirine-2-carboxylic
      ester 268
  nitrogen heterocycles reaction,
    chiral 2*H*-azirine-3-
      carboxylates 269
  organocatalyzed aziridination
    α-acyl acrylates 327
    chalcone 322
    cyclic α,β-unsaturated
      aldehydes 329
    cyclic enones 324, 327
    heteroaromatic-substituted
      enones 322
    olefins 321, 325
    2-(phenylsulfanyl)-2-
      cycloalkenones 320
    trans-chalcone 327
    α,β-unsaturated aldehydes 328
    α,β-unsaturated ketones 323
  Rh-catalyzed aziridination 310
  Ru-catalyzed aziridination 312
  sulfur ylide-mediated
    aziridination 350–353
aziridinomitosene 277

b

Baeyer–Villiger oxidation 436
   of ketone 433
Bamford–Stevens reaction 240
BINOL 336
BINOL-derived ligand 339
biocatalyzed synthesis, epoxides
   monooxygenases 505
   natural products 505
   styrene derivatives, P450pyrTM 509
BIRT-377 336
bis(2-pyridylimino)isoindoles complex 38
bis(oxazolinyl)phenyl ruthenium complex 73
brevipolide H 5
Buchner cyclization 81
butane-2,3-diacetal of (L)-threitol 82

c

carbohydrate-based organocatalysts 436
carzinophilin 277
Charette's ligand 15
Charette's method 17–18
C–H insertion reaction 162
chiral acetal-directed cyclopropanations 7–9
chiral 3-acetoxyaminoquinazolinones 214
chiral aldimines 342
chiral allyl acetoacetate 82
chiral allylic alcohols 3
chiral allylic amines 7
chiral amines
   cinchona-derived amine salt/ROOH systems 457
   cinchona thiourea/TBHP system 468
   diaryl prolinols/TBHP system 453
   electron-poor 2-hydroxyalkyl furanes 461
   TMS-prolinol/H2O2 system 457
   with trans-chalcone 449
   α,β-unsaturated ketones 453
chiral aminoalcohols 55
chiral auxiliaries 11, 28
chiral azidoformate 217
chiral aziridinecarboxamides 292
chiral aziridine 2-phosphonates 274
chiral 3-aziridinofarnesol 2, 284
chiral aziridinomitosane 283
chiral aziridino-γ-lactones 294
chiral bicyclic aziridines 293
chiral β-lactams 347
chiral catalysts
   Charette's ligand 15–20
   copper, enantioselective cyclopropanation reactions 39
   iridium-salen catalysts 78
   palladium complexes 77
   rhodium 56
   ruthenium 69
   stoichiometric ligands 20
   Walsh' procedure 22–24
chiral cis-aziridines 282
chiral 2-diaryl aziridines 1, 280
chiral diazophosphonates 82
chiral dirhodium catalysts 57
chiral γ-oxazolidine derivatives 102
chiral guanidines
   trans-chalcone 449
chiral imidazolidine ligand 55
chiral Mn-salen catalysts 402
chiral N-allyl hydroxylamine-O-sulfamates 212
chiral N-benzyl-2-hydroxymethylaziridine 229, 276
chiral N-ferrocenyl aziridino alcohols 275
chiral N-sulfinylaziridine 2-phosphonates 279
chiral N-sulfinylaziridines 280
chiral N-tert-butylsulfinyl aziridines 281
chiral N-tosyl-2-trifluoromethyl aziridines 275
chiral 2-oxazines 1, 293
chiral oxazoline ligands 55
chiral oxiranes
   asymmetric epoxy nitrile coupling 149

chiral oxiranes (*Contd.*)
  Et$_3$Al-mediated intramolecular
    opening   150
  LIDAKOR-promoted
    rearrangements   151
  mechanisms and formation   148
  methylene-transfer   152
  α,β-unsaturated Fischer carbene
    complexes   150
  terminal enantioenriched
    epoxides   152
chiral silyloxycycloheptene   144
chiral spiroaziridines   282
chiral 4-substituted
    tetrahydroisoquinolines   281
chiral substrates
  chiral acetal-directed
    cyclopropanations   7
  chiral allylic alcohols   3
  chiral allylic amines   7
  simple chiral alkenes   9
chiral *tert*-butyl-pybox   333
chiral *tert*-butylsulfinylketimines   231
chiral 2-(trifluoromethyl)acrylates   292
chiral unsaturated
    aminoquinazolinones   215
chiral unsaturated lactams   102
cobalt
  cobalt-porphyrin catalysis   32
  cobalt-porphyrin complex   33
  cobalt-salen complexes   36
copper
  α-amino acid-bound borate
    anions   56
  bidentate ligands   50
  chiral imidazolidine ligand   55
  chiral oxazoline ligands   55
  copper-bisoxazoline-catalyzed
    cyclopropanation   44, 47
  Cu-catalyzed aziridination   296
  Cu-salen catalysts   54
  enantioselective cyclopropanation
    reactions   39
  nitrogen ligands   53
  copper-bisoxazoline-catalyst   47
Cr/salen-catalyzed asymmetric
    epoxidation

trans-β-methyl styrene   397
cyclopropanated ketene
    hemithioacetal   171
cyclopropenes   164–167

d

Darzens-type epoxidations
  catalytic asymmetric
    amino-catalyzed aldol reaction/
      ring-closure sequence   497
    chiral PTC   492
    α,β-epoxy ketones   497
    haloamides   495
    cis-glycidic amides, Ti(OiPr)$_4$/
      (R)-BINOL   503
    C$_2$-symmetric chiral selenide   500
  chiral auxiliary-and
    reagent-mediated
    aldol-type addition,
      α-haloenolate   489
    camphor-based thioamide   491
    mechanical stirring   492
dehydroalanine acceptor   105
density functional theory (DFT)   18
D-glucose   294
diazocarbonyl compounds   82
divinyldiazoacetates   88
domino Knoevenagel/aza-MIRC
    reactions   225

e

electrophilic halocyclization, chiral
    allenes   576
Eliel's oxathiane   240
enantioselective cyclopropanation   20
  chiral perhydrobenzoxazine   25
  chiral zinc BINOL reagent   21
  chiral zinc dipeptide catalyst   25
  chiral zinc dipeptide reagent   22
  chiral zinc phosphate reagent   20
  chiral zinc TADDOL reagents   21
  sugar zinc reagent   22
1,6-enynes   155, 158
1,5-enynes cycloisomerization   152
epoxides   286
Evans' bisoxazoline ligand   301
Evans's chiral auxiliary   114

f

fat-derived chiral aziridine   284
Fe/catalyzed asymmetric epoxidation
  trans-stilbenes and cinnamic alcohols   411
  trisubstituted α,β-unsaturated enones and esters   410
Fe/chiral tetradentate nitrogen ligand complex
  α-alkyl substituted styrenes   409
Fe/simple nitrogen-based ligand complexes   408
fluorinated cyclopropanes   63

g

Gabriel–Cromwell aziridine synthesis   219–220
glycals   170
Grignard reagents   256
guanidine   287

h

halicholactone   4
halodicarbonyl compounds   134
Hashimoto's model   58, 66
homoallylic alcohols   145
HOMO-raising strategy   329
hydroxylated enynes   153

i

in situ generated Ti-salalen complexes   388
intermolecular cyclopropanation
  chiral auxiliaries   28
  chiral catalysts
    cobalt complexes   32
    copper   38
    rhodium   56
    ruthenium   69
intramolecular cyclopropanation
  chiral auxiliaries and chiral compounds   80
  chiral catalysts
    allenic diazoacetates   92
    allyldiazoacetates   89
    Co-salen complexes   83
    bis-diazoacetate   91

dirhodium catalysts   88
dirhodium(II) carboxamidate catalysts   82
divinyldiazoacetates   88
epi-tremulane skeleton   90
α-diazo-β-keto sulfones   85
gold(I)-catalyzed cyclopropanations   159
iodocyclopropanation   18–19
iridium-salen catalysts   78
isopropylidenediphenylsulfonium ylide   96

j

Jacobsen and Katsuki Mn-salen catalysts   401
Jacobsen's ligand   308
Jørgensen–Haiashi catalyst   134
Juliá–Colonna reaction   425

k

Katsuki–Sharpless asymmetric epoxidation (KSAE)
  of allylic alcohols   381–382
  2-allylic phenols   385
  of homoallylic alcohols   385
Kulinkovich method   168

l

levomilnacipran   152
limonene aziridines   283
LUMO-lowering strategy   329

m

MEDAM aldimines   337
menthyl esters   101
metal catalysis   139–140
methylenedimethylsulfoxonium ylide   96
methyl *N*-arylaziridine-2-carboxylates   359
Michael-initiated ring-closing (MIRC) reactions
  allylsilane homoallylic alcohols   144
  allyl sulfones   142
  boron–silicon bifunctional cyclopropanes   148

Michael-initiated ring-closing (MIRC) reactions (*Contd.*)
  chiral auxiliaries, Michael acceptors 100
  chiral benzyl alcohols 147
  chiral dicarboxylates 141
  chiral homoallylic alcohols 144
  chiral nucleophiles
    camphor-derived sulfonium ylides 109
    chiral oxathiane 108
    chiral phosphonamide 111
    chiral selenonium ylides 110
    Evans's chiral auxiliary 114
    lithiated sulfoximines 108
    pyridinium ylide-mediated cyclopropanation 113
  chiral substrates
    chiral cyclopropyl amino acids 100
    chiral (*Z*)-oxazolone and nitroalkenes 99
    cyclopropyl nucleoside derivatives 98
    D-glucose-derived enone 100
    diastereoselective cyclopropanation, of bis(unsaturated esters) 99
    enone 98
    methylenedimethylsulfoxonium ylide 96
    sulfur ylides 97
  metal catalysis 139–140
  organocatalysis
    halocarbonyl compounds 128–139
    nitrocyclopropanes 118–128
    ylides 116–118
  phosphorus-mediated asymmetric cyclopropanations 143
  tributylstannane homoallylic alcohols 145
  vicinal diol cyclic sulfur derivatives 141
Mn-salen-catalyzed epoxidation
  biologically active compounds 403
monofluorocyclopropanes 19

n

nitrocyclopropanes 118
nitrogen-based ylides 503
nitrogen ligands 53
nitroxyl-introduced olefinic aldehyde 290

o

Oppolzer's chiral sultam 30
organocatalysis
  halocarbonyl compounds 128–139
  nitrocyclopropanes 118–128
  ylides 116–118
organocatalyzed aziridination
  α-acyl acrylates 327
  chalcone 322
  cyclic enones 324, 327
  cyclic α,β-unsaturated aldehydes 329
  heteroaromatic-substituted enones 322
  olefins 321, 325
  2-(phenylsulfanyl)-2-cycloalkenones 320
  trans-chalcone 327
  α,β-unsaturated aldehydes 328
  α,β-unsaturated ketones 323
oseltamivir phosphate 278
Oxa–Michael addition, of nucleophilic peroxide 451
oxaziridination
  chiral *N*-alkyloxaziridines 544
  chiral substrates 540
  discovery 539
oxidative cyclization 128

p

palladium complexes 77
phase transfer catalysis (PTC) 419, 422
phenyl imines 333
platinum 153
2-(2′-phosphonocyclopropyl) glycines 105
PNNP chiral ligand 333
polyamino acids and aspartate-derived peracids 423, 425–427, 429–430

porphyrin-inspired Mn-complexe 407
pyrazoline
  photolysis 160
  sulfonylpyrazolines, thermolysis of 161

q

quantitative structure-selectivity relationship (QSSR) modeling 38

r

racemic epoxides, kinetic resolution
  Jacobsen's Co-salen complex 471, 473
  β-elimination to allylic alcohols 480
rhodium 31, 69
  alkenes 67
  chiral dirhodium catalysts 57
  cis-β-azidocyclopropane esters 62
  cis-cyclopropane α-aminoacid precursors 64
  α-cyanodiazophosphonate and α-cyanodiazoacetate 60
  diastereoselective rhodium-catalyzed cyclopropanation 58
  α-diazopropionate 65
  fluorinated cyclopropanes 63
  N-sulfonyl-1,2,3-triazoles 67
  pyrroles/furans 61
  Rh-catalyzed aziridination 310
  $Rh_2(S\text{-biTISP})_2$ 65
  $Rh_2(S\text{-NTTL})_4$ 66
  styryldiazoacetate cyclopropanation 63
Ru/oxo complex and precatalysts 411
ruthenium
  chiral catalyst 69–77
  Ru-catalyzed aziridination 312

s

Sharpless-type epoxidizing system 430
Shibasaki's epoxidizing systems 418
Shi's epoxidation
  pure (+)-ambrisentan 440
  ent-nakorone synthesis 440
Simmons–Smith cyclopropanation
  chiral auxiliaries 11

chiral catalysts
  catalytic procedures 24–27
  Charette's ligand 15–20
  stoichiometric ligands 20–22
  Walsh' procedure 22–24
chiral substrates
  chiral acetal-directed cyclopropanations 7–9
  chiral allylic alcohols 3–7
  chiral allylic amines 7
  simple chiral alkenes 9–10
  mechanisms 3
solandelactones E 5
solandelactones F 5
spirocyclopropyloxindoles 65
styrene derivatives 300
styryldiazoacetates 63
sulfinyl auxiliaries 106
sulfinyl compounds 104
1,1'-sulfonylbisaziridines 280
sulfonylpyrazolines 161
sulfur ylide-mediated epoxidations
  diastereo-and enantioselectivities 482
  metal-catalyzed decomposition 485
  oxathiane 482
  reactive pathway 482
  spiroepoxyindoles 484
  stoichiometric process 482
sulfur ylides 119
Suzuki–Miyaura coupling 147

t

telluronium ylide 100
thiirination
  epoxides conversion 571–572
  intramolecular nucleophilic substitution 574–576
  sulfur-stabilized carbanions 572–574

v

V/bishydroxamic acid-catalyzed asymmetric epoxidation
  homoallylic alcohols and synthetic application 393
  indole derivatives 394

w

Wadsworth–Emmons cyclopropanation   149
Walsh' procedure   22–24
Weitz–Scheffer epoxidation   409, 447

y

yildes   116
- allylic telluronium ylides reaction   237
- (−)-balanol synthesis   236
- chiral allyl aminosulfoxonium ylide reaction   244–245
- chiral CF_3-substituted N-tert-butanesulfinylketimines reaction   239
- chiral Eliel's oxathiane-derived ylide reaction   242–243
- chiral isothiocineole-derived allylic sulfonium salt reaction   246
- chiral N-tert-butanesulfinylaldimines reaction   238
- dimethylsulfonium/dimethyloxosulfonium ylide reacation   234
- dimethylsulfonium methylide reaction   231
- S-allyl tetrahydrothiophenium bromide reaction   232
- (S)-dimethylsulfonium-(p-tolylsulfinyl)methylide reaction   244
- sulfur allyl ylides reaction   233
- sulfur ylides reaction   235

z

$ZnEt_2$/(R)-BINOL/CHP system   415